Modeling and Simulation of Energy Systems

Modeling and Simulation of Energy Systems

Special Issue Editor

Thomas A. Adams II

MDPI • Basel • Beijing • Wuhan • Barcelona • Belgrade

MDPI

Special Issue Editor
Thomas A. Adams II
McMaster University
Canada

Editorial Office
MDPI
St. Alban-Anlage 66
4052 Basel, Switzerland

This is a reprint of articles from the Special Issue published online in the open access journal *Processes* (ISSN 2227-9717) from 2018 to 2019 (available at: https://www.mdpi.com/journal/processes/special_issues/simulation_energy)

For citation purposes, cite each article independently as indicated on the article page online and as indicated below:

LastName, A.A.; LastName, B.B.; LastName, C.C. Article Title. *Journal Name* **Year**, *Article Number, Page Range.*

ISBN 978-3-03921-518-8 (Pbk)
ISBN 978-3-03921-519-5 (PDF)

Cover image courtesy of Jaffer Ghouse.

Contents

About the Special Issue Editor

Thomas A. Adams II, P.Eng, is Associate Professor and Associate Chair in Chemical Engineering at McMaster University. His research interests are in Process Systems Engineering, particularly with regards to the modeling, simulation, and the design of sustainable chemical and energy processes. This research encompasses different areas of application, including power generation, carbon capture, synthetic fuels, alternative fuels, advanced distillation technology, mobile and modular chemical plants, energy conversion, and eco-techno-economic analyses. Adams is Section Editor-in-Chief of the journal *Processes*, the chair of the *Systems & Control Division of the Canadian Society for Chemical Engineering (CSChE)*, and author of the popular textbook Learn Aspen Plus® in 24 Hours. He has recently been recognized with awards such as the *Canadian Journal of Chemical Engineering* Lectureship Award, the *CSChE* Emerging Leader Award, and the Ontario Early Researcher Award, and received the title of University Scholar at McMaster in addition to being named by *Industrial & Engineering Chemistry Research's 2018 Class of Most Influential Researchers*.

processes

MDPI

Editorial

Special Issue: Modeling and Simulation of Energy Systems

Thomas A. Adams II

Department of Chemical Engineering, McMaster University, 1280 Main St W, Hamilton, ON L8S4L7, Canada; tadams@mcmaster.ca; Tel.: +1-905-525-9140

Received: 1 August 2019; Accepted: 2 August 2019; Published: 8 August 2019

Abstract: This editorial provides a brief overview of the Special Issue "Modeling and Simulation of Energy Systems." This Special Issue contains 21 research articles describing some of the latest advances in energy systems engineering that use modeling and simulation as a key part of the problem-solving methodology. Although the specific computer tools and software chosen for the job are quite variable, the overall objectives are the same—mathematical models of energy systems are used to describe real phenomena and answer important questions that, due to the hugeness or complexity of the systems of interest, cannot be answered experimentally on the lab bench. The topics explored relate to the conceptual process design of new energy systems and energy networks, the design and operation of controllers for improved energy systems performance or safety, and finding optimal operating strategies for complex systems given highly variable and dynamic environments. Application areas include electric power generation, natural gas liquefaction or transportation, energy conversion and management, energy storage, refinery applications, heat and refrigeration cycles, carbon dioxide capture, and many others. The case studies discussed within this issue mostly range from the large industrial (chemical plant) scale to the regional/global supply chain scale.

Keywords: modeling; simulation; energy; energy systems; process systems engineering; optimization; process design; operations

1. Introduction

Energy systems are currently a subject of rapidly growing interest within the engineering research community. Energy conversion and consumption impacts nearly all aspects of our lives, including the food we eat, the water we drink, the products we buy, how we battle the elements, how we communicate, how we move people and goods from place to place, how we work, and even how we are entertained. Although this has always been true throughout human history, the scale at which energy is consumed today is larger and expanding more quickly than ever before. The associated impacts of our energy consumption on our planet are now becoming so significant that the makeup of the atmosphere itself, particularly with regard to atmospheric CO_2 concentration, is being impacted.

Since the possible consequences are so alarming, energy systems engineering has become an extremely important area of research since one key aspect of solving this problem relates to the development of energy systems with far lower environmental impacts. Although energy is used in very diverse ways at scales from large to very small, large-scale systems, such as electric power plants, chemical plants, refineries, and oil and gas supply chains, are the easiest targets for improvement and the likeliest places where meaningful environmental impact reductions can be achieved. This is why almost all of the systems discussed in this Special Issue are in these application areas and, at large scales, range from 100 MW to 1000 MW class plants to massive international supply chains. Moreover, about half of the studies in this issue concern electric power generation, in a large part because fossil-based combustion systems tend to be the largest single-point sources of CO_2 emissions in the world. To address these concerns, the articles in this Special Issue took a variety of approaches, including the design of new energy systems and networks, improved control strategies for existing systems, and improved daily or hourly operational strategies for very complex systems. As a consequence of the large scales involved, even relatively small percentage improvements to efficiency or emissions can result in meaningful large-scale impacts.

2. Modeling Types

This issue focuses on the modeling and simulation of energy systems, or more precisely, research which relies heavily on mathematical models in order to address critical issues within energy systems. The issue begins with an extensive review of how modeling and simulation is used in energy systems research by Subramanian et al. [1], which examined and categorized over 300 papers on the subject. They proposed the modeling taxonomy shown in Figure 1 and noted that the "Process Systems Engineering Approach" to modeling energy systems focuses on mathematical modeling using the bottom-up approach. This means that mathematical models of individual process units, pieces of equipment, or process sections are written in the form of equations that describe the thermo-physical phenomena associated with it.

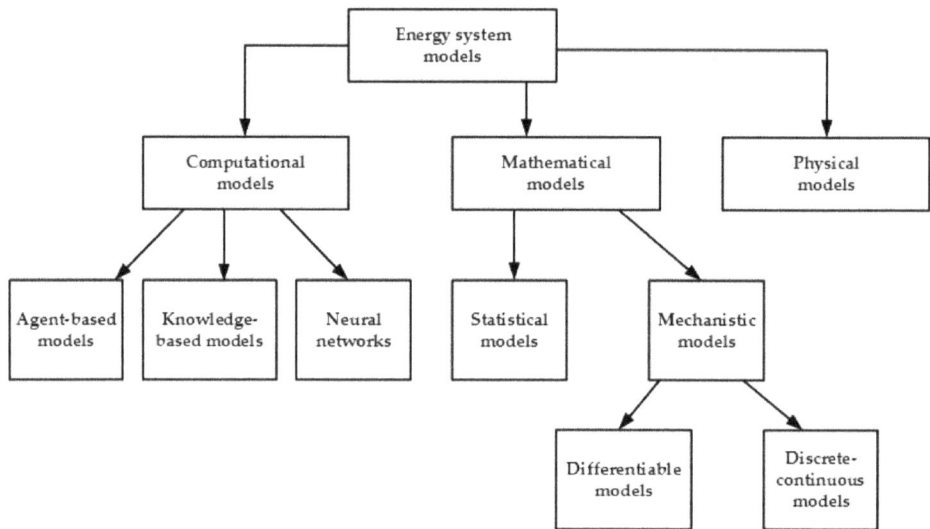

Figure 1. Taxonomy of energy systems modeling proposed by Subramanian et al. [1]. Reproduced with permission from MDPI.

Most of the articles in this issue use mechanistic models via a "first principles" approach, in which the equations and constraints derive from fundamental theory related to the first and second laws of thermodynamics, such as mass, energy, and momentum balances. These are usually coupled with equations that represent the physical properties of various chemicals or mixtures under different conditions, as well as equations describing physical or mechanical behaviour of the process equipment. The model parameters for physical property and equipment models are usually empirically determined in prior studies and are readily available through physical property databases or other sources.

As noted in the review by Subramanian et al. [1], statistical models are becoming increasingly more important in energy systems due to the increasing availability of data and computational capabilities in data analytics. Statistical models attempt to capture important characteristics of processes or process units without the use of fundamental first principles models. The benefits are usually improved computational speed at the risk of losing model rigor, extrapolative power, or certain nuances. For example, in this issue, Riboldi and Nord [2] use Kriging-type statistical models to create a surrogate of a much larger and more complex first principles model. The surrogate model is used for optimization purposes in place of the more rigorous one to help significantly reduce the computation time of optimization, which would be mostly intractable when using the fully-rigorous model. Similarly, Zimmerman et al. [3] create a statistical model from a more rigorous one, which is used for model predictive control (MPC). MPC requires very fast model solution times since it must re-solve the model frequently and repeatedly in order to determine ongoing control actions.

3. Implementation and Solution Frameworks

Interestingly, the software and implementation frameworks on which the models were built and simulated in this Special Issue varied widely from article to article. The list of software and packages includes, but is not limited to, the following: Aspen Custom Modeler, Aspen Exchanger and Design Rating, Aspen HYSYS, Aspen Plus, Aspen Plus Dynamics, Aspen Properties, casADi, Dymola, EcoInvent, GAMS, JuliaPro, JuMP, LINGO, MATLAB, Minitab, Modellica, Plant Engineering And Construction Estimator, PVWatts, Thermoflex, and other software developed in-house specifically for the articles in this issue, such as SoLCAT and EVA. There were generally two approaches for construction of the models. Most rigorous models of chemical processes were constructed with flowsheeting software (most commonly with the Aspen suite), in which the software builds the overall flowsheet model from a convenient model library containing models of the individual unit operations and connections. Models with a lower resolution (often because the boundaries of the model are at a much larger scale, such as a supply chain), models not based on mass and energy balances, and models with less rigour intended for use in optimization, tended to be implemented in general equation solving software such as GAMS or MATLAB, in which all of the equations needed to be strictly written out by the user. However, the diversity of software packages and implementation methods indicates the wide variety of problem types that were considered throughout this Special Issue.

4. Issue Summary

A summary of the articles in this Special Issue is provided in Table 1. It is an interesting snapshot of important research in energy systems and demonstrates both the breadth of problems considered and the depth of detail and understanding involved. Almost all articles use mathematical optimization to some degree, whether to find optimal designs, optimal controllers, or optimal operational strategies.

Table 1. Summary of articles in this Special Issue, categorized by problem type.

Authors/Ref	Application	Models and Software	Comments
Reviews			
Subramanian, Gundersen, and Adams [1]	Field-wide survey of models in energy systems.	Modelling taxonomy proposed	Proposed connecting the PSE-style bottom-up approach with top-down approach used in energy economics.
Energy System Design			
Riboldi and Nord [2]	Offshore power plants, integrated with renewables.	1st Principles + Kriging. Thermoflex, Plant Engineering, and Construction Estimator, MATLAB.	Dynamic considerations with regard to wind and electricity demand. Surrogate models used for optimization purposes.
Surindra, Caesarendra, Prasetyo, Mahlia, and Taufik [4]	Organic Rankine cycles in geothermal energy systems.	1st Principles of thermodynamic cycles.	Blends physical models (experimental apparatus) with mathematical ones.
Mussati, Mansouri, Gernaey, Morosuk, and Mussati [5]	Adsorption refrigeration cycles.	1st Principles. GAMS.	Optimal design with a superstructure approach.

Table 1. *Cont.*

Authors/Ref	Application	Models and Software	Comments
Yadav, Fabiano, Soh, Zimmerman, Sen, and Seider [6]	Transesterification of triolein to methyl-oleate (biofuels).	1st Principles. Aspen Plus with custom models.	Experimental validation of models in some conditions. Models used to predict performance in other conditions.
Vikse, Watson, Gundersen, and Barton [7]	Multi-stream heat exchanger (MHEX) design for natural gas liquefaction.	1st Principles. Julia. Aspen Plus for comparison.	Presents nonsmooth framework and algorithm for designing optimal MHEXs when standard methods fail.
Ridha, Li, Gençer, Siirola, Miller, Ribeiro, and Agrawal [8]	Shale gas condensate to oligomers and alkanes at the wellhead.	1st Principles. Aspen Plus, Aspen Economic Analyzer.	Techno-economic analysis. Premise: Cheaper to transport oligomers than Natural Gas Liquids.
Stuber [9]	Concentrated solar power with thermal energy storage.	1st Principles with empirical elements. JuliaPro/JuMP.	Equation oriented, differentiable model for determination of optimal design params.
Al-Aboosi and El-Halwagi [10]	Integrated water and energy between systems.	Mostly empirical models. LINGO.	Optimal design of integrated multi-product, multi-source systems considering time-varying solar.
Li, Demirel, and Hasan [11]	Automatically generate work-heat exchanger networks (WHEN).	1st Principles. GAMS. Phenomena level models.	Algorithm to create optimal WHENs from sources and sinks using building block superstructures.
Control Systems			
Sarda, Hedrick, Reynolds, Bhattacharyya, Zitney, and Omell [12]	Load-following Supercritical pulverized coal (SCPC).	1st Principles with reduced models. Aspen Plus Dynamics, Aspen Custom Modeler, Aspen Exchanger, and Design Rating.	Plant-wide dynamic model for designing and simulating plant-wide control system.
Zimmerman, Kyprianidis, and Lindberg [3]	Combustion of fuel derived from waste (refuse).	1st Principles. Modellica.	MPC with feedforward system developed. Soft sensors. Experimental validation.
Rahman, Zaccaria, Zhao, and Kyprianidis [13]	Micro gas turbine systems.	1st Principles with data-driven model tuning. EVA (in-house).	Dynamic models. Fault detection and diagnostics.
Pravin, Guidi, and Bhartiya [14]	Integrated reformer-membrane fuel cell systems.	1st Principles ODEs with some empirical characteristics. MATLAB.	Controllability analysis. Certain design considerations must be made for controllability purposes.
Decardi-Nelson, Liu, and Liu [15]	Flexible post-combustion CO_2 capture systems.	1st Principles. casADI, Python, Aspen Properties.	Economic MPC for disturbances. Look-up table made from Aspen Properties for fast use.
Flexible Operations and Operational Strategies			
Chen and Bollas [16]	Flexible, load-following subcritical coal power plant.	1st Principles. Dymola. Modelon Thermal-Power Library, MATLAB.	Dynamic optimization of transitions during load changes.
Corengia and Torres [17]	Optimal operating schedule of grid-scale battery energy storage.	1st Principles. GAMS.	Considers degradation of the batteries, demand cycles, and local tariff policies.
Kazda and Li [18]	Optimal operations of natural gas transport networks.	1st Principles. GAMS.	Created piecewise linear models to capture nonlinearities with optimization problem tractability.
Du and Cluett [19]	Operational improvements to existing Naphtha recovery units.	1st Principles and statistical models (Principle Component Analysis). Aspen Plus, Minitab.	Aspen Models released. Statistical models suggest unintuitive options, explained by Aspen model.

Table 1. *Cont.*

Authors/Ref	Application	Models and Software	Comments
	Systems Analysis		
Miller, Gençer, and O'Sullivan [20]	Life cycle analysis (LCA) of integrated solar PV, wind, and batteries.	Empirical/data driven models. SoLCAT (in-house). Ecoinvent. PVWatts.	LCA focused on emissions from use/manufacture of various power sources in several case studies.
Siddiqui, Taimoor, and Almitan. [21]	Supercritical CO_2 Brayton cycles coupled with bottoming cycles.	1st Principles. Aspen HYSYS.	Energy and exergy cycle analysis for working fluid screening.

Funding: This research received no external funding.

Conflicts of Interest: The author declares no conflict of interest.

References

1. Subramanian, A.S.R.; Gundersen, T.; Adams, T.A., II. Modeling and simulation of energy systems: A review. *Processes* **2018**, *6*, 238. [CrossRef]
2. Riboldi, L.; Nord, L.O. Offshore power plants integrating a wind farm: Design optimization and techno-economic assessment based on surrogate modeling. *Processes* **2018**, *6*, 249. [CrossRef]
3. Zimmerman, N.; Kyprianidis, L.; Lindberg, C.-F. Waste fuel combustion: Dynamic modeling and control. *Processes* **2018**, *6*, 222. [CrossRef]
4. Surindra, M.D.; Caesarendra, W.; Prasetyo, T.; Mahlia, T.M.I. Comparison of the utilization of 110 °C and 120 °C heat sources in a geothermal energy system using organic Rankine cycle (ORC) with R235fa, R123, and mixed-ratio fluids as working fluids. *Processes* **2019**, *7*, 113. [CrossRef]
5. Mussati, S.F.; Mansouri, S.S.; Gernaey, K.V.; Morosuk, T.; Mussati, M.C. Model-based cost optimization of double-effect water-lithium bromide absorption refrigeration systems. *Processes* **2019**, *7*, 50. [CrossRef]
6. Yadav, G.; Fabiano, L.A.; Soh, L.; Zimmerman, J.; Sen, R.; Seider, W.D. Supercritical CO_2 transesterification of triolein to methyl-oleate in a batch reactor: Experimental and simulation results. *Processes* **2019**, *7*, 16. [CrossRef]
7. Vikse, M.; Watson, H.A.J.; Gundersen, T.; Barton, P.I. Simulation of dual mixed refrigerant natural gas liquefaction processes using a nonsmooth framework. *Processes* **2018**, *6*, 193. [CrossRef]
8. Ridha, T.; Li, Y.; Gençer, E.; Siirola, J.J.; Miller, J.T.; Ribeiro, F.H.; Agrawal, R. Valorization of shale gas condensate to liquid hydrocarbons through catalytic dehydrogenation and oligomerization. *Processes* **2018**, *6*, 139. [CrossRef]
9. Stuber, M.D. A differentiable model for optimizing hybridization of industrial process heat systems with concentrating solar thermal power. *Processes* **2018**, *6*, 76. [CrossRef]
10. Al-Aboosi, F.Y.; El-Halwagi, M.M. An integrated approach to water-energy nexus in shale-gas production. *Processes* **2018**, *6*, 52. [CrossRef]
11. Li, J.; Demirel, S.E.; Hasan, M.M.F. Building block-based synthesis and intensification of work-heat exchanger networks (WHENS). *Processes* **2019**, *7*, 23. [CrossRef]
12. Sarda, P.; Hedrick, E.; Reynolds, K.; Bhattacharyya, D.; Zitney, S.E.; Omell, B. Development of a dynamic model and control system for load-following studies of supercritical pulverized coal power plants. *Processes* **2018**, *6*, 226. [CrossRef]
13. Rahman, M.; Zaccaria, V.; Zhao, X.; Kyprianidis, K. Diagnostics-oriented modelling of micro gas turbines for fleet monitoring and maintenance operation. *Processes* **2018**, *6*, 216. [CrossRef]
14. Pravin, P.S.; Gudi, R.D.; Bhartiya, S. Dynamic modeling and control of an integrated reformer-membrane-fuel cell system. *Processes* **2018**, *6*, 169.
15. Decardi-Nelson, B.; Liu, S.; Liu, J. Improving flexibility and energy efficiency of post-combustion CO_2 capture plants using economic model predictive control. *Processes* **2018**, *6*, 135. [CrossRef]
16. Chen, C.; Bollas, G.M. Dynamic optimization of a subcritical steam power plant under time-varying power load. *Processes* **2018**, *6*, 114. [CrossRef]

17. Corengia, M.; Torres, A.I. Effect of tariff policy and battery degradation on optimal energy storage. *Processes* **2018**, *6*, 204. [CrossRef]
18. Kazda, K.; Li, X. Approximating nonlinear relationships for optimal operation of natural gas transport networks. *Processes* **2018**, *6*, 198. [CrossRef]
19. Du, J.; Cluett, W.R. Modelling of a naphtha recovery unit (NRU) with implications for process optimization. *Processes* **2018**, *6*, 74. [CrossRef]
20. Miller, I.; Gençer, E.; O'Sullivan, F.M. A general model for estimating emissions from integrated power generation and energy storage. Case study: Integration of solar photovoltaic power and wind power with batteries. *Processes* **2018**, *6*, 267. [CrossRef]
21. Siddiqui, M.E.; Taimoor, A.A.; Almitani, K.H. Energy and exergy analysis of the S-CO_2 Brayton cycle coupled with bottoming cycles. *Processes* **2018**, *6*, 153. [CrossRef]

processes

MDPI

Review

Modeling and Simulation of Energy Systems: A Review

Avinash Shankar Rammohan Subramanian [1], Truls Gundersen [1] and Thomas Alan Adams II [2,*]

[1] Department of Energy and Process Engineering, Norwegian University of Science and Technology (NTNU), Kolbjørn Hejes vei 1B, NO-7491 Trondheim, Norway; avinash.subramanian@ntnu.no (A.S.R.S.); truls.gundersen@ntnu.no (T.G.)
[2] Department of Chemical Engineering, McMaster University, 1280 Main St. W, Hamilton, ON L8S 4L7, Canada
*   Correspondence: tadams@mcmaster.ca; Tel.: +1-905-525-9140 (ext. 24782)

Received: 12 October 2018; Accepted: 18 November 2018; Published: 23 November 2018

Abstract: Energy is a key driver of the modern economy, therefore modeling and simulation of energy systems has received significant research attention. We review the major developments in this area and propose two ways to categorize the diverse contributions. The first categorization is according to the modeling approach, namely into computational, mathematical, and physical models. With this categorization, we highlight certain novel hybrid approaches that combine aspects of the different groups proposed. The second categorization is according to field namely Process Systems Engineering (PSE) and Energy Economics (EE). We use the following criteria to illustrate the differences: the nature of variables, theoretical underpinnings, level of technological aggregation, spatial and temporal scales, and model purposes. Traditionally, the Process Systems Engineering approach models the technological characteristics of the energy system endogenously. However, the energy system is situated in a broader economic context that includes several stakeholders both within the energy sector and in other economic sectors. Complex relationships and feedback effects exist between these stakeholders, which may have a significant impact on strategic, tactical, and operational decision-making. Leveraging the expertise built in the Energy Economics field on modeling these complexities may be valuable to process systems engineers. With this categorization, we present the interactions between the two fields, and make the case for combining the two approaches. We point out three application areas: (1) optimal design and operation of flexible processes using demand and price forecasts, (2) sustainability analysis and process design using hybrid methods, and (3) accounting for the feedback effects of breakthrough technologies. These three examples highlight the value of combining Process Systems Engineering and Energy Economics models to get a holistic picture of the energy system in a wider economic and policy context.

Keywords: energy systems; modeling and simulation; multi-scale systems engineering; sustainable process design; energy economics; top-down models; hybrid Life Cycle Assessment

1. Introduction

Energy is one primary driver of the modern economy that involves several stakeholders such as energy production and distribution firms, energy investors, end users, as well as government regulators. Population growth and improving standards of living, especially in developing countries, are expected to significantly increase energy consumption. The IEA predicts an increase in total primary energy demand (TPED) from 13.8 billion tonnes of oil equivalent (toe) in 2016 to 19.3 billion toe under its "current policies" scenario in 2040 [1]. Alternatively, in the "sustainable development" scenario where policies are enacted in order to achieve the objectives of the COP 21 Paris agreement (2015) together with universal access to energy services and a large reduction in energy-related pollution, TPED grows to 14.1 billion toe in 2040. In this period, CO_2 emissions would increase from 32.1 billion tonnes to 42.7 billion tonnes under the current policies scenario or would have to decrease to 18.3 billion tonnes in the sustainable development scenario. Transitioning to a sustainable energy future through the accelerated adoption of clean energy technologies and energy efficiency practices requires the engagement of various decision makers from the scientific, financial, industrial, and public-policy communities with an interdisciplinary approach that combines engineering, economics, and environmental perspectives [2].

An energy system is defined by the Intergovernmental Panel on Climate Change (IPCC) in its fifth Assessment Report as a "system [that] comprises all components related to the production, conversion, delivery, and use of energy" [3]. Figure 1 shows the different components of an energy system. First, primary energy stored in natural resources (such as fossil fuels, uranium, renewable resources) is harvested and transported to the conversion site(s) in which a wide range of processes (such as combustion, refining, bioconversion, etc.) may take place to transform energy to more usable forms such as electricity and liquid fuels. This conversion process may integrate with local utilities such as the water distribution network. The usable energy is then transported and distributed through a potentially large number of infrastructure components to the final user. Final energy demand can be disaggregated into homogeneous categories of users such as transportation, residential, industrial, and commercial users. However, the energy production, conversion, transportation, and distribution steps combined typically also consume the largest amount of energy as a result of generally low efficiencies [4].

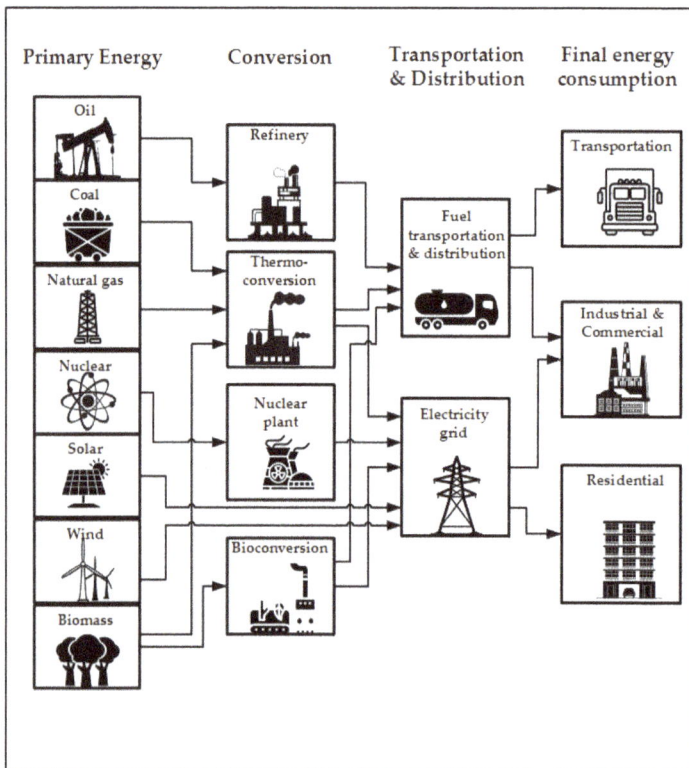

Figure 1. Energy system showing the flow of energy from primary energy supply to final energy consumption.

Trade between suppliers and consumers occurs in energy markets with the primary energy price depending on a large number of factors such as supply and demand quantities, geopolitics and international trade policies, interaction with other economic sectors, technological changes, and even natural disasters. Figure 2 presents a comparison of the normalized prices (in \$/GJ) of natural gas, oil, and coal fuels in the United States in order to illustrate the complex interdependencies between the energy sector and other sectors. For example, historically speaking, gas and oil have followed approximately the same price trends when expressed on a normalized per-energy basis up until the shale gas boom, which caused an unprecedented and sustained decoupling of oil and gas prices that has persisted for the past decade. Other recognizable events within the past generation include the gulf war of 1990, and the energy crisis and subsequent Great Recession during 2008–2009 that had major impacts on prices, although often temporary. Coal, on the other hand, has followed a relatively stable and consistent trend independent of world events, and has consistently been the lowest cost form of energy. However, it is possible that if gas continues its steady decline further below what are now the lowest prices in a generation, gas could even overtake coal as the cheapest form of energy within about 6–10 years at current rates.

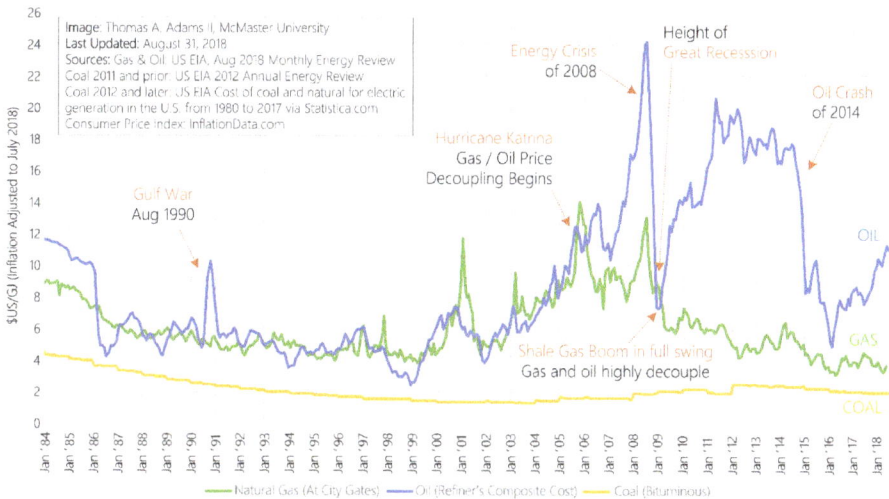

Figure 2. The variation of the normalized prices of natural gas, oil, and coal fuels with a variety of factors such as supply and demand quantities, geopolitics and international trade policies, interaction with other economic sectors, technological changes, and even natural disasters. Coal, gas, and oil prices were collected from various publications from the US Energy Information Administration (see inset) depending on fuel type and year (See inset). Note that a small change in the standard indexing procedure for coal explains a slight jump in coal price at the beginning of 2012. Gas prices are for natural gas located at the city gates (i.e., prior to "last mile" transportation). Oil prices are the refiner's composite cost of oil, which includes transportation and storage of oil, factoring in both domestic and imported crudes. Coal prices are free-on-board prices and do not include shipping or insurance. Prices are normalized by the consumer price index and converted to an energy basis using the following assumed energy densities: 32 GJ per tonne of coal (using medium-volatility bituminous), 6.118 GJ per barrel of oil, and 1037 BTU per standard ft^3 of natural gas.

The need to understand and predict the functioning and performance of individual components of the energy system or the overall system behavior motivates the development of models. Several modeling approaches have been proposed in the literature for different purposes. In Section 2, we propose a classification of these approaches into computational, mathematical, and physical models, and outline the capabilities of the different formalisms in describing different phenomena.

With this categorization, we highlight novel hybrid approaches that combine aspects of the different groups proposed. Simulation, on the other hand, involves solving the set of equations of mathematical models in order to determine the unknown variables so as to obtain key insights into the system's behavior. The value of simulation lies in exploring the system's behavior under a range of operational domains which may be expensive or even infeasible to do with the real-world system. Commonly, a further optimization step may be added in order to determine the conditions of optimal system performance with respect to a particular objective which may be economical, environmental, social, or a combination of the three [5].

Energy systems are studied by researchers in two different fields: Process Systems Engineering (PSE) and Energy Economics (EE). Figure 3 and Table 1 illustrate the division using different criteria such as the nature of variables, theoretical underpinnings, level of technological aggregation, spatial and temporal scales, and model purpose. Energy system models in PSE are typically at the unit operation, processing plant, or supply chain scale. Each of these scales represents a level of aggregation of technologies: Different unit operations are aggregated to give an overall conversion process at the plant scale and the conversion process together with the feedstock supply and product distribution network are aggregated at the supply chain scale. The purpose of modeling energy systems in PSE is to obtain insight into their technological performance for optimal decision making at the design, operations, and control level. Thus, the technological characteristics of the system components are modeled endogenously (i.e., are dependent on other variables or parameters in the model). The economic, environmental, or social parameters may be modeled exogenously (i.e., are independent of other variables or parameters in the model) depending on the optimization objective. Classically, modeling and simulation in PSE has been used for strategic, tactical, and operational decision-making at low levels of technological aggregation as illustrated in Figure 3 and detailed in Table 1. Energy economics approaches, on the other hand, use models with a high level of aggregation of technologies: All technologies comprising the entire energy sector at a regional, national or global scale may be studied (e.g., in bottom-up models) or even other sectors of the economy such as manufacturing, mining, construction, etc. (e.g., in top-down models). EE models are based on economic theory such as the laws of supply, demand and market equilibrium. Thus, the economic characteristics of the system components are modeled endogenously while technological, environmental, or social parameters may be modeled exogenously. The classic purpose of modeling energy systems in EE is to aid in making strategic decisions at regional, national, or global scales. EE models typically work with long-term time scales for a couple of reasons: first, the capital intensiveness, long gestation periods, and long payback periods motivate long term thinking; second, addressing sustainability issues requires planning of energy transformation pathways that may take decades to mature [6]. However, the advent of intermittent-supply renewable technologies and deregulation of the electricity market have led to the recent use of EE models for operational decision making as well. Application areas include determining the optimal operation of energy management systems and modeling electricity markets. To improve firm competitiveness, these EE models include several features such as seasonal demand variability, weather prediction, price spikes, etc. Reviews of electricity market modeling are given in [7,8].

To further explain the differences between PSE and EE models, we distinguish between a national energy supply chain PSE model (e.g., the biomass-to-bioenergy supply chain published by Elia et al. [9]) and an EE model for a national energy system (e.g., reviewed in [10]): While these two models are at the same scale, the PSE model contains information about the spatial (geographical) distribution of the different supply chain components while the EE model only abstracts their economic characteristics. In addition, while the PSE model only includes the biomass-to-bioenergy supply chain, the EE model would usually include all possible energy sources of the nation. Finally, the purpose of the two models may be different: EE models are typically used for strategic decision making by national planners and public policy officials, PSE supply chain models may be used for strategic, tactical, and operational decision-making typically by energy supply firms or enterprises.

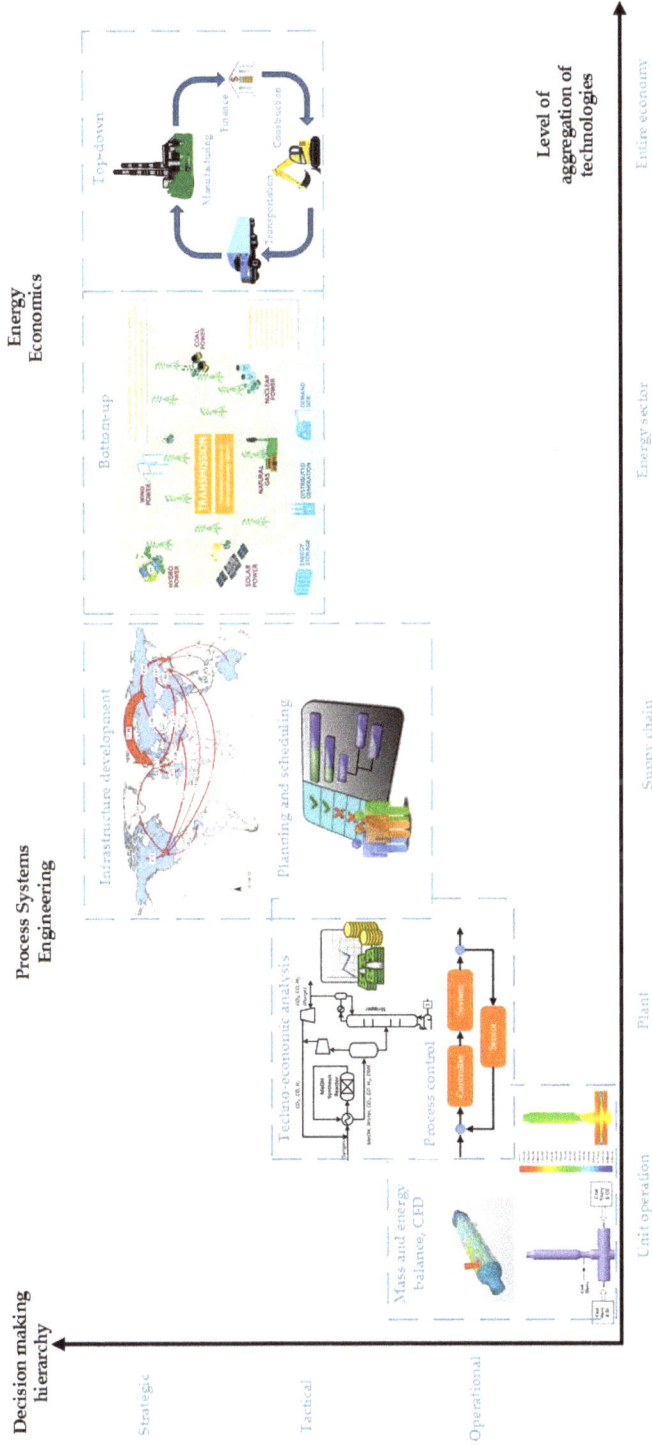

Figure 3. Classification of energy system models according to discipline and level of technological aggregation. Computational Fluid Dynamics (CFD) Gasifier model reprinted from [11] with permission from Elsevier, Process flow diagram reprinted from [12] with permission of Adams, Thato, Le Feuvre and Swartz Copyright (2018), global supply chain reprinted from [13] with permission from Elsevier.

Processes **2018**, 6, 238

Table 1. Classification of energy system models according to field.

Field	Process Systems Engineering (PSE)			Energy Economics (EE)
Nature of variables				
Endogenous	Technological (e.g., temperature, pressure, enthalpy, Gibbs free energy, process size)			Economic
Exogenous	Economic (e.g., raw material prices, equipment prices, product demand, interest rates) Environmental (e.g., Global Warming Potential, Ecotoxity, resource depletion, terrestrial acidification)			Technological, Environmental
Theoretical underpinnings	Thermodynamics, Fluid mechanics, Kinetics			Economics (producer theory, consumer theory, and market equilibrium)
Level of aggregation of technologies	Unit operation, Processing plant, Supply chain			Entire energy sector, All economic sectors
Spatial scale	Local, Regional, National, Global			Regional, National, Global
Decision making hierarchy	Strategic	Tactical	Operational	Strategic [1]
Temporal scale	Several years	Days-Weeks	Seconds-Hours	Several years
Classic purposes	*Process design & integration:* Reviews: [14–21] High Impact/Seminal: [22–28] Open Models: [29] *, [30] *, [31–33] *Supply chain design/infrastructure:* Reviews: [34–40] High Impact/Seminal: [9,41–45]	*Production Planning and Scheduling:* Reviews: [46–52], High Impact/Seminal: [53–58]	*Process control:* Reviews: [59–63] High Impact: [64–66] *Flexible operation:* Reviews: [67–70] High Impact: [71–76]	*Sustainable energy policy planning:* Reviews: [10], [77–81] High Impact: [82–89], [90–94], [95] [2] *Long term energy forecasting:* Reviews: [96–102] High Impact/Seminal: [103–108]
Recent trends	*Multi-scale systems engineering:* Reviews: [38,39,109–112], High Impact: [9,11,41,113–120] *Sustainable process analysis and design:* Reviews: [5], [121–132], High Impact: [133–142], Open Models: [143–145]			• Outside the scope of this paper

* indicates that the open access model is available in the LAPSE archive at http://psecommunity.org/lapse. [1] Energy economics approaches have been used to aid in tactical and operation decision making as well, for instance in electricity generation planning and electricity market modeling. However, a discussion of these applications is outside the scope of this paper. [2] A comprehensive list of open energy modeling software for sustainable energy policy planning is provided in [90].

Modeling and simulation of energy systems has received significant research interest from the PSE and EE communities. The methodology for this review paper began with a search using Engineering Village for journal papers published between 2015 and 2019 that contained the terms "energy system(s)" and "model" or "simulation" in the subject, title, or abstract which resulted in 5071 records, with the majority belonging to the EE field. In addition, we noticed that a large number of relevant papers from the PSE community were not included in this search because these used more plant-specific terms such as coal fired power plant, LNG processes, polygeneration system, etc. Considering this enormous body of literature, the scope of this paper is necessarily broad; we scrutinized for relevance according to the following criteria:

- Emphasis on critical review papers of the different sub-fields within EE and PSE. These are denoted using blue in Table 1.
- Emphasis on the most impactful papers (measured by number of citations) or seminal works that presented novel approaches. These are denoted using red in Table 1.
- Emphasis on the modeling approach used rather than the application area.
- Emphasis on open access models or modeling tools to reflect the open source ethos of the Processes journal. These are denoted using black in Table 1.

Considering the scope and audience of the *Processes* journal, we provide a detailed treatment of the PSE approach to energy system modeling and simulation, while we only provide a fundamental treatment of EE approaches. We discuss emerging trends in the PSE field, such as multi-scale systems engineering and sustainable process engineering, in detail. On the other hand, a discussion of the state-of-the-art in EE is outside the scope of this paper.

Previous reviews have primarily been from the EE field: Ringkjøb et al. presented a recent review of currently available modeling tools for energy systems containing significant renewable energy sources [146]. They provide an overview of the features and properties of the different tools, such as model purpose, approach, methodology, temporal resolution, modeling horizon, and geographical coverage, with the purpose of aiding modelers to choose the right tool. Lopion et al. presented a recent review of current challenges and trends in modeling national energy systems for the purpose of supporting governmental decision-making [10]. They provide a review of the different modeling tools used, the methodologies used, analytic approach, and the temporal and spatial resolutions. Hall and Buckley conducted a systematic review of energy systems modeling tools in the UK and found that certain modeling tools were preferred by academic and policy users [147]. They provide a review of the different purposes, model structure, geographical coverage, sectoral coverage, and temporal scale of prevalent UK energy systems models. Van Beeck proposed several ways of categorizing EE models of energy systems e.g., according to purpose, model structure and assumptions, analytical approach, underlying methodology, geographical coverage, and time horizon [148]. Jebaraj and Iniyan provided a review of different EE models used for a variety of purposes e.g., planning and emissions reductions, supply-demand forecasting, incorporating renewable energy sources, etc. [149]. Connolly et al. presented a review of computer tools used for integrating renewable energy sources into EE models, with the aim of helping future decision makers pick the right modeling tool [81]. They provide an analysis of the properties of 37 modeling tools with regards to criteria such as availability, tool capabilities, geographical coverage, and future time frame of uncertainty scenarios. Pfenninger et al. reviewed models used for national and international policy making and point out existing challenges in developing high spatial and temporal resolution models, handling uncertainty and complexity issues and incorporating human behavior and social risks [94]. Bhattacharyya and Timilsina provided a review of energy system models and illustrate the challenges for their use by energy, environment, and climate policy decision makers in developing countries [100]. Herbst et al. provided an introduction and overview of EE models and point out the strengths of bottom-up and top-down models as well as their weaknesses to motivate the development of linked models [150]. Nakata reviewed various EE models applied at local, national, and global scales and points out the issues arising from their incorrect application [151]. In another work, Nakata et al. reviewed EE models used for transforming to a low-carbon society and illustrate the need to consider a trans-disciplinary approach incorporating social, economic, and environmental considerations [79]. Weijermars et al. provided a review of models used to determine the optimal energy mix in order to mitigate climate change [77]. They provide a description of several modeling approaches such as forecasting and back-casting, and scenario and systems analysis, and point out the value of using integrated approaches.

In the PSE community, review papers have focused on individual feedstocks and their value chains such as: biomass [152,153], renewable sources [154,155], or hybrid feedstocks [156]. Other reviews have focussed on analysis along different scales such as district energy systems [157] and urban energy systems [158]. Zeng et al. provide a review of research done on modeling and optimization of energy systems for planning and emissions reduction purposes. Some reviewed papers use a PSE approach while the majority use an EE approach. Emphasis is on different optimization problem formulations that take parametric uncertainty into account: fuzzy, stochastic, and interval programming [159]. Liu et al. present an overview of different formulations of the optimal design and operation problem for a number of applications such as polygeneration energy systems, hydrogen infrastructure planning, commercial building systems as well as biofuels and biorefineries [160]. Adams provides a perspective on the opportunities for the design of new efficient energy systems. In particular, opportunities to integrate unrelated processes in order to exploit certain synergies, as well as the potentials to exploit waste feedstocks such as petroleum coke and flare gas, are highlighted [161]. Recently, Martin and Adams presented a perspective on near-term opportunities in PSE including the use of big data approaches, integrating process design, control and scheduling, and supply chain management [162]. In addition, they highlighted the potential for exploitation of new energy sources, production of new products for energy storage as well as the need for sustainable process design. Finally, Gani et al. provide an encyclopedic account of modeling and simulation approaches in PSE [19]. They provide a detailed description of the approach to modeling, the different model types and their properties and utility in representing various physical phenomena, different numerical methods and process simulation techniques. In addition, they present a review of general-purpose process simulation software with a discussion of their specific capabilities, strengths and weaknesses. To the authors' knowledge, this review paper is the first to present the key contributions from both the PSE and EE fields to energy system modeling and simulation. The original contributions of this paper are as follows:

- First, we propose a categorization according to modeling approach namely into computational, mathematical, and physical approaches. With this categorization, we highlight certain novel hybrid approaches that combine aspects of the different groups proposed.
- Second, we propose a categorization according to field namely Process Systems Engineering (PSE) and Energy Economics (EE). We use the following criteria to illustrate the difference: the nature of variables, theoretical underpinnings, level of technological aggregation, spatial and temporal scales, and model purpose. With this categorization, we present the interaction between the PSE and EE fields and make the case for combining these two complementary approaches to get a more holistic picture of energy systems.

The remaining sections are organized as follows: Section 2 presents the categorization according to the modeling approach, Section 3 provides an overview of the PSE approach including a discussion of emerging trends, while Section 4 provides a basic overview of the EE approach, Section 5 makes the case for combining the PSE and EE approaches, and Section 6 presents our conclusions.

2. Categorization According to Modeling Approach

In this section, we present an overview of the different classes of energy system models. The categorization is made according to the modeling approach rather than the system boundary.

Figure 4 illustrates the division into three classes: computational, mathematical, and physical models. The dichotomy between computational and mathematical models is explained by Fisher and Henzinger: computational models are a sequence of instructions that can be executed by a computer, yet mathematical models are a series of equations that denote relationships between different meaningful variables and parameters [163]. The development of mathematical models preceded the use of computers. However, approximations to mathematical models are usually developed using numerical techniques, which can then be solved (or simulated) using a computer. A variety of specialized numerical algorithms (e.g., Newton's method, Secant method, etc. for root-finding) are available to solve the numerical approximation with different levels of accuracy and speed of convergence. In contrast, computational models are in the form of a single algorithm, which is immediately implementable by the computer. Thus, Fisher and Henzinger propose using the term "execution" for computational models in contrast to "simulation" for mathematical models implemented numerically on a computer. In physical models, on the other hand, the phenomena of the real-world system actually occur albeit at a smaller scale or with less complexity. For example, a physical model of a solid oxide fuel cell system has been built at the US Department of Energy National Energy Technology Laboratory for hardware simulation purposes as described by Tucker et al. [164,165].

A large class of computational models, termed expert or intelligent systems, are programmed to imitate intelligent behavior. A general review of expert systems is provided by Liao [166], while a review of specific applications in process engineering is provided by Stephanopoulos and Han [167]. Expert systems include agent-based, neural networks, knowledge-based, fuzzy models, etc. Agent-based models are typically used to mimic the human element of energy systems. They consist of a number of autonomous, self-interested entities represented in computer code, which act according to certain rules. Thus, agent-based models are used to model supply chain entities [56,168], electricity markets [169], or technological change [170–172]. The interaction between multiple agents is simulated in order to determine the overall system behavior, which may highlight certain non-intuitive aspects not predictable by first-principles modeling [173]. Knowledge-based systems consist of a knowledge base of domain-specific expertise, an inference engine that deduces new knowledge based on certain rules, and a user interface [174]. PROSYN is an example of a commercially available knowledge-based system that contains a database of specialist information and heuristics to aid in conceptual process design [175]. Unlike knowledge-based systems, neural networks are generic and do not have explicit rules but instead consist of a collection of nodes that processes input and output information. Neural networks are trained by adjusting the weights of the connections between nodes. Neural networks have had wide application in energy systems, for instance, in modeling biomass gasifiers [176] and control systems [177], as described in the book by Baughman and Liu [178].

Mathematical models can be classified into statistical (empirical or black-box) and mechanistic (theoretical, first-principles or white box) models. Statistical methods use techniques such as regression and optimization [179], kriging [180], self-optimizing control [181], and neural networks to derive a set of simple mathematical relations from input and output data [182]. Mechanistic models, on the other hand, employ fundamental discipline-specific theories such as fluid mechanics, thermodynamics, economics, mass and energy balances, etc. that provide the model structure and generate equations to describe the phenomena of the real-world system. There are several ways of categorizing mathematical models, for instance, into discrete and continuous models according to the nature of the variables, into steady state and dynamic models according to whether the variables vary with time, or into deterministic or stochastic models according to the uncertainty of parameters. The division of mechanistic models into differentiable and discrete-continuous models is proposed by Watson [183]. Modeling different phenomena with differentiable equations requires different mathematical formalisms: steady state phenomena with no spatial variation are modeled by algebraic equations, dynamic phenomena, or steady state phenomena with spatial variation in one dimension are modeled by ordinary differential equations, and finally dynamic phenomena with spatially distributed phenomena are modeled by partial differential equations. A detailed description of the different formalisms for representing various physical phenomena is provided by Gani et al. [19]. While differentiable models are sufficient for modeling continuous phenomena, several energy system components also exhibit discrete phenomena such as thermodynamic phase changes, flow reversals, and changes in flow regimes such as from laminar to turbulent flow, etc. [22]. In addition, external actions on the system such as through the use of digital controllers, plant start-up and shut-down in batch processes, or mode changes in semicontinuous processes result in discrete behavior. Mathematical formalisms for modeling discrete-continuous phenomena include hybrid automata [184], disjunctive models [185], hybrid Petri nets [186], etc. as detailed in a review by Barton and Lee [23].

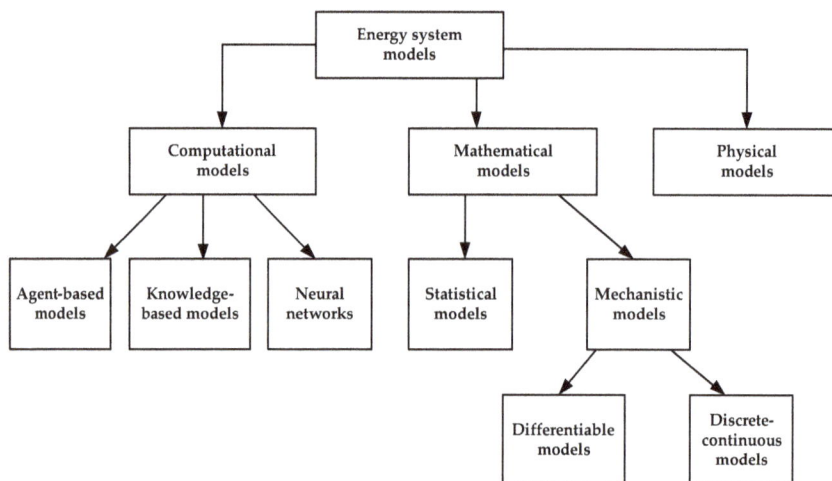

Figure 4. Classification of energy system models according to modeling approach.

With this categorization, we note that several emerging approaches in energy systems modeling combine elements from two or more of the categories proposed in Figure 4. For instance, recent developments in non-smooth modeling by Barton and coworkers bridge the gap between differentiable and discrete-continuous models. Local sensitivity information for non-smooth models is obtained by calculating generalized derivatives automatically as explained in [187]. This approach has been applied to develop compact models for multi-stream heat exchangers [188], which have been incorporated into flowsheets of natural gas liquefaction [26] such as the simple mixed refrigerant [189,190] and dual mixed refrigerant processes [191,192].

Another emerging approach is hybrid modeling that combines aspects of mechanistic and statistical models, i.e., physical insights and constraints obtained from first principles are incorporated synergistically to adjust the statistical model. Significant advances have been made by Cozad et al. who developed the ALAMO framework to develop surrogate models by constrained regression such that first principles limitations such as. mass and energy balances, physical limitations, variable bounds, etc. are obeyed [179]. Similarly, Straus and Skogestad propose an approach that introduces auxiliary variables, such as the extent of reaction or the separation coefficient, in order to reduces the dimension of the surrogate model and satisfy mass balance constraints [193,194]. In addition, a hybrid approach of first principles and neural networks for modeling chemical processes has been developed by Psichogios and Ungar [27], and Thompson and Kramer [28], with Guo et al. [195] applying this approach to model biomass gasification processes. Finally, certain aspects of the model, such as demand and supply forecasting in EE models, may rely on statistical data with the other part based on economic theory.

3. The PSE Approach to Energy System Modeling and Simulation

Traditionally, PSE has focused on modeling chemical conversion systems at the processing plant scale. Any system can be modeled as a collection of sub-systems with interconnections depicting different types of flows to give a complex network [25]. The input to the system denotes the influence of factors outside the system (i.e., in the surroundings) and the output denotes the influence of the system on the surroundings. In a processing plant, the sub-systems are unit operations and the interconnections are mass, energy, or information flows between processing units. Modeling unit operations requires knowledge of the fundamentally occurring phenomena taking places such as thermodynamic processes, transport phenomena (mass transfer, fluid dynamics and heat transfer), and chemical reaction kinetics [196]. The mathematical formulation that incorporates all these phenomena and connects the molecular and continuum descriptions of matter is the differential balance equation (consisting of hold-up, transport and source terms) as developed by Bird, Stewart, and Lightfoot [24] that gave rise to the dominant BSL modeling paradigm of chemical engineering science [197]. Unit operation models are usually used together with simulation and optimization for a variety of purposes such as reaction path synthesis, design of the entire process flowsheet or subsections such as the reactor networks, heat exchanger networks, recycle systems, or separation networks [123]. If the purpose is operation or control modeling, transient effects and dynamic simulation is necessary. An alternative is to use empirical models, which rely on input and output experimental data for a wide range of the process operational domain, as discussed in Section 2. As a result of the chemical engineering science base, PSE models of energy systems typically focus on describing the technological characteristics of the components—economic, environmental, and social characteristics are passed on to the model as exogenous variables. For instance, at the process scale, the operating conditions (e.g., stream flow rates, temperatures, pressures) are modeled explicitly but parameters such as fuel prices and availability, product prices, and environmental impacts are considered exogenously. An encyclopedic account of the steps involved in model development in PSE is given by Gani et al. [19], while Marquardt provides a survey and tutorial of modeling procedures [20].

The scope of PSE has been expanded by Grossmann and Westerberg to encompass all approaches for improved decision making in the creation and operation of the chemical supply chain [198]. This necessitates analysis of energy systems over a range of scales from the unit operation to the supply chain scale (as shown in Figure 3), thus motivating the trend towards multi-scale systems engineering, which we describe in Section 3.1. The scope of PSE has been expanded in both directions to finer and coarser scales. Traditionally, in order to reduce the computational cost of model solution, lumped parameter models are used together with a number of assumptions (such as equilibrium, plug flow, perfect or ideal mixing, ignoring dispersion and other transport phenomena) in order to reduce complexity. The need for higher fidelity models that incorporate these effects has motivated modeling at finer scales [111]. On the other hand, there is significant economic potential in coordinating logistics with production planning thus it is necessary to include the supply chain level in decision-making [198,199]. Another emerging trend is the inclusion of sustainability factors quantified by the triple bottom line of economics, environment, and social criteria as described in Section 3.2 [121].

3.1. Multi-Scale Systems Engineering

Multi-scale systems engineering is a new paradigm in engineering science with the key idea of linking a network of models across different spatial and temporal scales such that information computed in one model can be used in another. The most common goal is to develop a high-fidelity model that accurately captures the overall system behavior at a large (coarse) scale by utilizing information provided by higher resolution models at smaller (finer) scales [197]. This approach is termed 'upscaling' because of the flow of information from smaller to larger scales with a corresponding reduction in degrees of freedom. However, the relatively under-examined 'down-scaling' approach, in which a desired outcome is determined at a larger scale and passed down to smaller scale models in order to determine a feasible technological path, may also be relevant. Multi-scale systems engineering has been aided by recent advances in computational capability, including progress in parallel computing, as well as by improvements in experimental techniques particularly at finer scales [200].

Floudas et al. present an outlook of the opportunities to apply multi-scale systems engineering at various time and length scales in order to address pressing energy and environmental challenges, such as generation of affordable energy in an environmentally sustainable way and ensuring future energy security [109]. A number of important design and operational problems in the energy systems engineering necessitate multi-scale modeling with upscaling or downscaling approaches. We present two applications next.

3.1.1. Designing Novel Conversion Processes for Heterogeneous Feedstocks

Unlocking the potential of the large amounts of lignocellulosic biomass available in agricultural, industrial, and forestry residues has enormous economic and environmental value [201]. However, biomass feedstocks have a variable and complex structure, thus accurate modeling of novel biomass conversion processes is challenging. Mettler et al. note that this problem is particularly severe for lignocellulosic biomass unlike other feedstocks like petroleum, natural gas, or coal: lignocellulosic biomass possesses a multiscale structure spanning eleven orders of magnitude over which different degradation phenomena can occur [202]. Thus, developing high fidelity thermoconversion process models (e.g., pyrolysis, combustion, gasification, liquefaction, and hydrogenation [152]) requires a deeper understanding of the underlying reaction mechanisms and transport phenomena. Similarly, multi-scale analysis, beginning at the molecular scale, has the potential to develop novel biochemical conversion pathways through systems and synthetic biology approaches. For example, Kumar et al. suggest that recombinant DNA technology, metabolic engineering, and genomics approaches have great potential to advance processes for bioconversion of cellulose to useful products [203]. Lee et al. suggest that such approaches of manipulating microbial activity may be used to efficiently produce drop in biofuels that are similar to existing petroleum-based fuels—an advance that would save significant capital by reusing existing transportation infrastructure [204].

Accurate modeling of the multi-scale phenomena occurring within the biomass conversion unit operation is particularly relevant because all further downstream cleaning and processing units utilize product stream information such as composition and flow rate. In addition, the conversion unit is typically a significant heat source or sink, thus optimal design of the heat exchanger network and utility system requires an accurate conversion model. Thus, the overall process economics may strongly depend on using a rigorous and more realistic model for the conversion process. An example of applying the multi-scale systems engineering approach to design novel biomass conversion processes is given in a series of papers by Baliban et al. In [205], Baliban et al. developed a novel stoichiometry-based mathematical model for the biomass gasifier in order to evaluate syngas compositions. The model consisted of several unknown parameters that were tuned to experimental data by non-linear parameter estimation. This reactor level model was dynamically linked to lumped parameter Aspen Plus models of the rest of the biomass-based conversion process. At the process plant level, Baliban et al. then used the gasifier model developed in [205], as part of the process flowsheet [113]. Several alternative biomass conversion flowsheets were modeled to generate a superstructure which could be used to determine the optimal process design. Finally, the heat, power, and water networks were integrated with the rest of the process resulting in an economical and environmentally optimal plant design [114].

Similarly for municipal solid waste (MSW), Onel et al. developed a generic model for a gasifier with an optimization based monomer model for the pyrolysis zone, and a detailed thermodynamic model for the oxidation and reduction zones [206]. Unknown parameters were fixed by non-linear parameter estimation using experimental data. In the following paper, Niziolek et al. used the MSW gasifier model as part of a superstructure for optimal synthesis of a MSW to liquid fuel conversion process [115].

Thermo-conversion of coal has received considerable attention from PSE researchers with models made with the aid of commercial software such as Aspen Plus [207–209] and multi-scale models. Coal gasification and combustion are complex processes: Singh et al. note that several spatially distributed phenomena occur between the gases and solid coal particles such as multiphase fluid flow, heterogeneous and homogeneous reactions, as well as heat and mass transfer [210]. In addition, key characteristics such as the flame shape, flow recirculation, and thus the flow field variables (gas composition, temperature, pressure and velocity) are dependent on the reactor geometry [211]. Computational fluid dynamics (CFD) tools have the potential to be used to develop more accurate models that offer insight into the inner workings of the combustor or gasifier unit operation and to predict the syngas product composition [210–213]. These models can be integrated with the rest of the process flowsheet for more rigorous simulation and optimization. For example, Shi et al. developed a computational fluid dynamics model of a two-stage, oxygen blown, entrained flow, coal slurry gasifier that predicted syngas compositions similar to restricted equilibrium reactor models tuned with experimental data [214]. Lang et al. developed a reduced order model of this detailed CFD model and converted it to an Aspen Plus module which was simulated as part of an integrated gasification combined cycle (IGCC) process [11]. Optimization of the process showed an increase in power output of 5–7% compared with conventional simplified unit operation models—a result that illustrates the value of the multi-scale systems engineering approach. Considering that gasification and combustion unit operations centerpieces of several processes, Zitney [118] and Biegler and Lang [215] note the potential of application of high-fidelity models to other coal based energy systems such as the oxy-fuel combustion of pulverized coal for carbon capture [216], integrated gasification fuel cell systems [217], and polygeneration plants [218,219].

The approach discussed so far is termed the "upscaling" approach in that there is an upward flow of information from detailed finer resolution models to coarser models. Multi-scale system engineering using a "downscaling" approach to design novel energy systems may also have significant value, but has received less research attention. A series of papers by Adams and colleagues illustrates the potential of the downscaling approach. First, Adams and Barton propose a novel scheme in which a coal gasifier is heat integrated with a natural gas reformer to produce syngas streams with different H_2/CO ratios [220]. Blending these streams gives the correct ratios for downstream polygeneration of electricity, methanol, and liquid fuels, thereby eliminating the inefficient water gas shift reaction. However, the actual technology required to achieve the heat integration was purely theoretical, and a plant-level techno-economic analysis was performed on the assumption that such a technology could be created. The results of a techno-economic analysis performed at the plant scale suggested that the proposed polygeneration process is economically viable and more robust to market uncertainties, thus motivating further study of the design details of the heat integration scheme. Next, Ghouse and Adams developed a multi-scale two-dimensional dynamic heterogeneous model of the natural gas steam reforming reactor that accounted for both intra-particle and inter-phase mass transfer limitations [116]. This model illustrated the feasibility of the heat integration concept proposed in [220] where the endothermic natural gas reforming reaction takes place in the tube side of the radiant syngas cooling section of the entrained-flow coal gasifier. The model was used to aid in the design of the device [221] such that it would be able to meet the process requirements of the previous work [220]. The model was then used to design control schemes to respond to syngas composition requirements for downstream synthesis processes [117,222].

3.1.2. Modeling and Optimal Design of Supply Chains for Distributed Energy Sources

A typical energy supply chain consists of three sections: feedstock harvesting and transportation, conversion plant, and product distribution to end user. Analysis at the supply chain scale is essential for distributed energy sources because logistical costs are significant and thus a key determining factor of the economic viability of the overall process. There are several papers that provide an overview of the challenges and opportunities in supply chain modeling and optimization including the works of Garcia and You [34], Papageorgiou [36], Shah [35], Barbosa-Povoa [37], Nikolopoulou and Ierapetritou [38], and Lainez and Puigjaner [39]. More focused review on biomass-to-bioenergy supply chains are given by Sharma et al. [223], Yue et al. [224] and Hosseini and Shah [112], on waste biomass-to-energy by Iakovou et al. [225] and on shale gas supply chains by Cafaro and Grossmann [226]. These works highlight the complexity involved in modeling and optimal design of supply chains for distributed energy sources: Several constraints, such as the variable availability, composition, and geographical distribution of feedstock sources, feedstock degradation, the need for storage, logistical costs, conversion plant capacities, product demand locations and specifications need to be taken into account. Furthermore, supply chain components may be spread across large distances and thus require analysis at several different spatial scales: community energy systems and district heating, cooling, and power systems [227] require detailed modeling at the regional level, while various fuels and chemicals are produced at production plants worldwide and exchanged in international markets thus requiring modeling at the global level. Thus, moving up on the level of technological aggregation from unit operations to supply chains necessitates accounting for a large number of externalities.

Supply chain related decision-making spans all three temporal scales: strategic (long-term or over multiple years), tactical (medium term or over multiple months), and operational (short term or day-to-day) [225]. Strategic decisions include: determining how to make the optimal investments in supply and product distribution infrastructure, choosing the right conversion technology, choosing the right site location and capacity, choosing the types of transportation, and addressing sustainability issues over the plant lifetime. Tactical decisions deal with optimization of production planning and scheduling [46,47], e.g., determining the right time to run batch processes, managing the transportation fleet and inventory levels, etc. Operational decisions at the supply chain scale are made several times each day, such as determining production operating conditions, detailed logistical issues, and response to weather disruptions. In addition to this complexity, supply chains also face significant uncertainties such as supply disruption, transportation failures, feedstock and product price changes, emissions policies etc. [34]. Thus, supply chain optimization involves determining the feedstock types and quantities, transportation types and routes, conversion plant types and capacities, and final product distribution options that minimize the overall system cost under these uncertainties.

Considering the high logistical costs associated with distributed energy sources, supply chain modeling and optimization may be an important determining factor in overall project viability. For example, Iakovou et al. suggest that logistical costs are a crucial bottleneck limiting biomass utilization [225]. Papageorgiou highlights the opportunity supply chain optimization offers for huge economic savings in the process industry [36]—an observation corroborated by Min and Zhou who suggest that many firms have realized the value of planning, controlling and designing the entire supply chain rather than focusing on separate functions [40]. Example of multi-scale modeling that includes analysis at the supply chain level is the work of Elia et al. [9] who used the conversion process models of [113] as technology options to optimize the nationwide energy supply chain network and Niziolek et al. who developed a supply chain model for a municipal solid waste (MSW) to liquid fuel process [41]. The newly emerging field of enterprise-wide optimization is a further step that expands the scope of PSE past supply chain management to aid in making decisions in firm R&D efforts, corporate finance and management, demand modeling, etc. [228,229].

Several enabling technologies have facilitated multi-scale systems engineering. Work done by Zitney and co-workers at the US Department of Energy National Energy Technology Laboratories led to the development of the Advanced Process Engineering Co-Simulator (APECS) software framework that co-simulates unit operations modeled using CFD together with lumped parameter models for the rest of the process [118]. In this way, one can integrate CFD unit operation models developed with software such as FLUENT into a process flowsheet developed with software such as Aspen Plus or HYSYS. In subsequent work, Lang et al. developed reduced order models from the CFD models that were then used for simulation at the plant process scale, with the aim of reducing computational costs [119,120]. Finally, Biegler and Lang propose a framework to develop reduced order models that when used in a multi-scale model for optimization, guarantee convergence to the same solution as using the full model [111]. At the supply chain level, Lam et al. propose model reduction techniques such as eliminating unnecessary variables and constraints, and merging certain nodes, that lowered computational time by several orders of magnitude [230].

3.2. Modeling Sustainability Criteria

Sustainability is the concept of "meeting the needs of the present without compromising the ability of future generations to meet their own needs" [231]. Sustainability criteria can be divided into three dimensions: Economic, Environmental, and Social, together termed the "triple bottom line" of sustainability, as discussed in the next sub-sections [121,124,232]. The modeling of sustainability criteria for decision-making in PSE is a relatively recent trend—the vast majority of previous research efforts have focused on modeling the techno-economic performance of energy systems [5,133]. Bakshi and Fiksel suggest that the inclusion of sustainability objectives in a firm's decision-making strategy represents a shift in thinking: environmental and social aspects may not necessarily be conflicting with economic objectives but instead represent a business opportunity to increase shareholder profits [121,124]. Environmentally-friendly energy projects that also present economic opportunities include: exploitation of waste feedstocks [233,234] such as petcoke [30] or coke oven and blast furnace gas from steel making [235,236], improving energy efficiency through process integration [237], for instance by work and heat integration [238], freshwater use minimization [239,240], etc. In addition, adopting a proactive approach to including sustainability considerations voluntarily may have pragmatic value in preventing crippling regulations and political backlash [121].

Incorporating sustainability considerations is important in every stage of the energy system project including through further retrofit projects. However, Cano-Ruiz and McRae point out that it is particularly valuable in the early design stage [123]. This is because regarding sustainability concerns as a design objective rather than a constraint on operations can motivate the search for novel processes with improved economic and environmental performance [134]. Several metrics have been proposed to incorporate the three sustainability criteria into process design decision-making, for instance, GREENSCOPE [241], and the indicators of Azapagic and Perdan [242]. An important challenge in sustainable process design is to quantify an objective function that includes metrics measuring all three of the competing economic, environmental, and social criteria. It may not be possible or even desirable to reduce all the metrics that measure different factors with different units into a single score. Several approaches have been proposed including assigning a dollar value to environmental and social criteria, instituting a carbon tax that in order to link economic and environmental criteria for a social good, and multi-objective optimization [243]. Bakshi presents a review of current approaches to sustainable process design as well as future research challenges [122]. Requirements for a process to claim to be sustainable are proposed: the overall demand for raw materials from the ecosystem and the release of emissions should not exceed nature's regenerative capacity. In other work, Bakshi and Fiksel provide a perspective on why expertise built in the PSE community, for instance in mathematical programming techniques, is uniquely useful to addressing challenges in developing sustainable energy systems [121]. Grossmann and Guillén-Gosálbez present a review of the application of mathematical programming techniques to aid in sustainable process decision-making [5]. Hugo and Pistikopoulos present an optimization framework for inclusion of sustainability criteria in supply chain network design and planning [134].

3.2.1. Economic Criteria

Economic criteria are used to ensure the long-term profitability of firms taking part in the energy system project. Techno-economic modeling involves determining the economic performance of the designed process. In PSE, the technological characteristics of the process (e.g., pressure, temperature, flowrate, equipment size, etc.) are modeled endogenously while the economic characteristics (e.g., raw material and product prices, equipment transportation costs, etc.) are specified exogenously.

Modeling the economic characteristics can be divided into two activities: determining the capital and operational costs of the project (economic input) and determining the revenue from sales (economic output). Methods for economic analysis are detailed in widely used PSE textbooks such as [14,244–246]. Project capital costs includes several components such as land, offsite infrastructure, equipment costs, etc. For a given process flowsheet, accurate equipment costs can be determined by obtaining quotes from vendors. However, vendor quotes may not be available at the preliminary process design stage, thus estimates based on historical (base) costs are used together with certain empirically evaluated indices (e.g., CEPCI, Marshall and Swift equipment cost index) in order to account for the change in costs with time. The variation of capital cost with process capacity is determined either from actual vendor quotes or with smooth-curve generalizations based on historical data together with economy of scale indices. Cost parameters are available in literature references such as [247,248] for several pieces of equipment. Alternatively, commercial software such as Aspen Capital Cost Estimator, that contains a large database of cost quotes and scaling factor indices, may be incorporated directly into process flowsheet simulations as explained in [249]. Project operational costs include several expenses such as utilities, raw materials, labor, maintenance, etc. Raw material prices can be determined from published estimates such as [250]. Utility costs depend on the process flowsheet e.g., the heat exchanger network design. Utility costs are commonly estimated by using assumed constant cost rates per unit of energy delivered or unit of service. These constant rates can be estimated based on current market energy prices, as described in [244]. At the supply chain scale, costing can be done by determining the costs of the transportation and distribution system and adding these to the process costs.

Revenue from sales is determined from product flow rates and published product price data. Both project expenditures and revenues are accounted for in cash flow analysis, which involves determining the project's overall economic performance. Methods used are categorized into non-discounted and discounted methods: Non-discounted methods do not consider the time value of money and are used for short-cut analysis of a project's feasibility compared to competing projects by calculating metrics such as the return on investment or the payback time, while discounted methods are typically used for more detailed analysis and consider the time value of money with metrics such as net present value (NPV), breakeven time of a project, or annualized rate of return [14]. Pintarič and Kravanja provide a review of the different economic metrics such as total annual cost, profit, payback time, equivalent annual cost, net present value, and the internal rate of return that could be used as objective functions for optimal process design and conclude that net present value (or net present worth) with a discount rate equal to the minimum acceptable rate of return is the most suitable [125]. They mention that qualitative profitability measures, such as the internal rate of return and the payback time, favor cheaper projects with small cash flows and high profitability, while other quantitative measures, such as total annual cost or profit, favor solutions with higher cash flows but low profitability. They conclude that compromise measures, such as the NPV, the equivalent annual cost, and the modified profit account for both criteria resulting in a solution with relatively large cash flows and a promising internal rate of return.

One pertinent issue in techno-economic modeling is accounting for uncertainty. This uncertainty can arise either from components within the process plant boundary (e.g., uncertainties in kinetics or transfer coefficients of equipment, product yields, reservoir sizes, etc.) or from externalities (e.g., in product demands, market prices, emissions policies, etc.). It is essential to account for uncertainty because the optimal solution found using the nominal model may no longer be optimal or even feasible when applied in practice. One approach to account for uncertainty is to perform a sensitivity analysis to determine the impact on the process performance as a result of changes in the uncertain parameters. Another widely used approach is design under uncertainty that incorporates uncertain parameters directly into the design formulation. The main mathematical programming techniques for optimization of problems with parametric uncertainty include: stochastic programming [251], robust optimization [252], chance-constrained programming [253], and dynamic programming [254]. To address uncertainty, processes are designed to be flexible in that they are able to maintain feasible operation over a wide range of realizations of uncertainty. Flexible design strategies have been applied to several processes including distillation columns [12], air separation units [255], polygeneration systems [73], and supply chain networks [256], with details given in a review by Grossmann et al. [257]. In other work, Grossmann et al. also provide a review on the use of mathematical programming techniques for optimal process design under uncertainty [21].

3.2.2. Environmental Criteria

Modeling the effect of energy systems on the surrounding ecosystem requires accounting for two factors: The environmental impacts as a result of emission of pollutants and greenhouse gases, and the primary resources used by the system [121]. All the different components of the energy system from primary energy harvesting to final energy consumption contribute to these two environmental effects. Thus, analysis with a system boundary that is broader than the traditional PSE processing plant boundary is necessary to include all such systemic effects.

Life Cycle Assessment (LCA) is a commonly used framework used to systematically account for the environmental impacts and resource use of the different stages of the energy system [126–129,132,258]. The purpose of modeling with a comprehensive life cycle perspective is to maintain accounting consistency in the decision-making process by preventing burden shifting between different life cycle stages. In other words, one should ensure that efforts to lower environmental impacts or resource use in one lifecycle stage are not overwhelmed by unaccounted increased effects at another stage. Thus, it is essential to account for both the direct effects from within the processing plant and the indirect effects from relevant activities outside the processing plant.

Figure 5 illustrates the complexity involved in performing LCA to account for all systemic effects. For a given energy product (such as electricity or fuel), there are direct emissions associated with the processing plant as well as inflows of primary resources. However, each component of this processing plant requires certain inputs, usually from other economic sectors such as mining, manufacturing, logistics, etc. For instance, the processing equipment is supplied by manufacturers, the feedstock is supplied by primary energy harvesting firms (such as coal mining) and requires transportation to the processing plant, etc. Furthermore, each component of the upstream supply chain (Tier 1) requires its own supply chain (Tier 2), and so on to higher tiers. Figure 5 also illustrates the circular interactions between components of different tiers of the supply chain, which results in additional complexity. Each of these supply chain tiers is associated with environmental impacts that need to be proportionally included to the plant level impacts. Similarly, resource utilization in higher order supply chains need to be accounted for. Adding effects from higher supply chain tiers to the LCA model corresponds to expanding the system boundary until eventually all the inputs and outputs correspond to elementary flows, which are defined as material or energy flows drawn from the environment without previous human transformation, or released into the environment without subsequent human transformation [135]. Thus, LCA requires analysis at a high level of technological aggregation in order to account for the complex interdependencies between multiple supply chain tiers. These interdependencies may give rise to non-intuitive effects as illustrated by Nease and Adams, who performed a detailed cradle-to-grave life cycle analysis of a natural gas combined cycle plant with carbon capture [137]. They found that capturing 90% of the CO_2 only reduced lifecycle GHG emissions by approximately 65%, but simultaneously increases the environmental damage from nearly every other category by 10–25%. This result showed the importance of accounting for all global effects using cradle-to-grave LCA with a comprehensive system boundary.

In order to provide a common basis for LCA modeling, the ISO framework was developed [259]. The motivation was to ensure comparability and consistency in assumptions, data, and methodologies as well as to provide transparency and thus more credibility to resulting decisions. The LCA framework consists of four steps [130]. The first is goal and scope definition which involves specifying the system boundary, the system's inputs and outputs, and the purpose of the analysis. The second step is inventory analysis, which involves obtaining data on the mass and energy flows across the system boundary. For a given energy system, it is desirable to use specific local data corresponding to the modeled system components. However, data is often unavailable, especially for higher order supply chains. In these cases, the inventory can still be collected from simplified aggregate data obtained from the weighted-average of the different supply chain alternatives in the corresponding sector. The third step is impact assessment which involves determining the potential detrimental environmental effects such as global warming potential, terrestrial acidification, freshwater eutrophication, ozone depletion, nitrification, resource depletion, etc. [138]. Each assessment requires an impact assessment method, e.g., ReCiPe [260], TRACI [261], eco99 [262], etc. The final step is interpretation which involves using the results of impact assessment to answer the questions set out in the LCA scope including where to focus redesign efforts, or which new design or policy decisions to make [130].

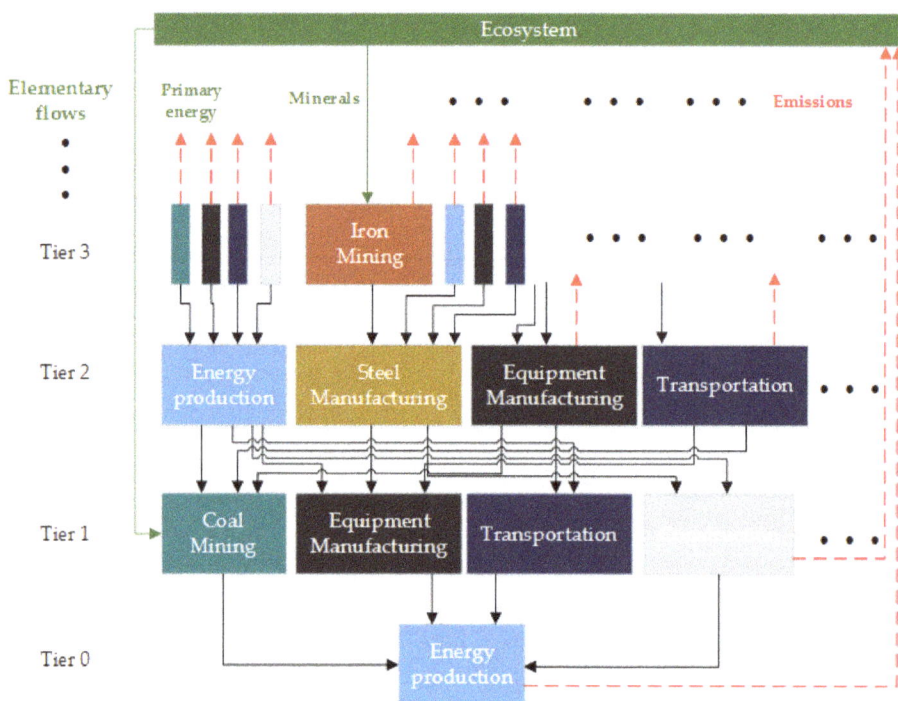

Figure 5. The multi-tiered supply chain associated with energy production.

Software such as SimaPro and OpenLCA [143] can help determine inventories and impacts. Inventory data for generic process steps common in many supply chains (such as transportation, energy distribution, bulk chemicals and materials production) are often available through databases such as The LCA Commons hosted by the National Agricultural Library at the US Department of Agriculture [263] and Ecoinvent [264]. Further details of the LCA methodology are presented in a review by Finnveden et al. [129]. In addition, a collection of guidelines, inventory and impact assessment data, and a registry of LCA software tools is available at the European LCA knowledge base [139].

In PSE, the purpose of LCA modeling is to make nuanced decisions that compare different process and technological options. Thus, the process LCA approach is used in contrast with input-output LCA of EE, which is explained in Section 4. Process LCA maps onto the process flow diagram model and requires specific data on the inputs, outputs and environmental impacts of each unit operation or process section. As far as possible, localized (often proprietary) data is used. Averaged data obtained at the sector level of aggregation is only used where localized data is unavailable or if the impact on the analysis is negligible. For this reason, process LCA models typically have a high level of granularity and are suited for detailed decision-making. The environmental impacts of components of progressively higher tiers of the supply chain are manually added until particular cut-off criteria are reached. The cut-off criteria quantify a threshold (e.g., <1–5% of total impacts) beyond which deletion of certain supply chain components is allowed. However, Suh et al. suggest that it may be difficult to choose the appropriate system boundary. They mention that deciding which components to exclude from the analysis may be challenging to do for the general case because the negligibility of their impacts cannot be guaranteed [135]. Furthermore, it may be time consuming and expensive to include all the complex upstream impacts suggested in Figure 5, especially with regards to higher tier supply chain components. This motivates the use of hybrid LCA methods as discussed in Section 5.2.

3.2.3. Social Criteria

Social criteria involve the cultural and personal connection to the technology and the way it impacts society (such as convenience, politics, personal values and beliefs, human behavior, and emotional factors). Thus, social criteria metrics measure the well-being and quality of life of the local communities served by the energy system. Several factors are included such as health and safety, access to education, access to energy services, employment situation, social equity, tolerance and diversity, etc. A list of social metrics for sustainability is provided by the Institute of Chemical Engineers [265]. However, the social dimension of the triple bottom line has been studied the least in the context of PSE as a result of being harder to quantify and outside the training and traditions of most systems engineers. As a result, the intersection between PSE and the social sciences in this way is not well studied today. A notable exception to this trend is the work of Othman et al. who include models of social criteria, such as safety considerations (operational safety, safe start-up and shut-down) and societal impacts of project (technology transfer, employment, effect on other industries, and regulations), in the process design stage [133]. A review of methodologies for social life cycle assessment is provided by Jørgensen et al. [266].

4. The EE Approach to Energy System Modeling and Simulation

The field of energy economics deals with the optimal allocation of scarce energy resources to satisfy consumer demand. It received increasing recognition as a distinct branch of economics during the energy crisis of the 1970's which highlighted the central role of energy to economic development [6,99]. The scope has since widened to range from studies on the economic performance of a regional or national energy sector through to modeling of trade in international energy markets and finally to analysis on a global scale of the long-term implications of public policies, such as the emissions reduction goals agreed to at the Conference of Parties (COP) 21 in Paris [267]. The distinguishing feature of EE models is that the principles of economics are used as the underlying theoretical basis: consumer theory, producer theory and market equilibrium [268]. Consumer theory explains that rational consumers spend their income to choose certain goods that will maximize their "utility" i.e., provide the most satisfaction. Thus, the quantity of certain goods demanded (including electricity and fuels) can be determined by drawing relations to factors such as income levels, prices, emissions-related tax credit policies, etc. Producer theory, on the other hand, explains that supply firms will produce the optimal quantity of products that maximizes profitability. Market equilibrium resolves these two conflicting optimization problems by introducing the notion of "equilibrium price" at which the quantity of goods supplied equals the quantity demanded, i.e., the market clears.

Thus, in EE, economic characteristics (such as energy prices, demand quantities, supply capacities for each technological option, etc.) of the energy system are modeled endogenously together with their relationship to non-energy related variables such as income, population growth, growth of other industries, etc. A comprehensive treatment of the EE approach to energy systems modeling is presented in the book by Bhattacharyya [6]. To explain the EE approach to energy systems, we first present an overview of energy demand and supply forecasting models, followed by the bottom-up and finally top-down EE energy system models.

4.1. Demand and Supply Forecasting Models

Demand forecasting models use consumer theory and statistical data to predict the quantity of energy products demanded. Bhattacharyya and Timilsina [99] as well as Suganthi and Samuel [97] provide comparative reviews of various approaches to energy demand forecasting. Commonly, two approaches may be used: econometric approaches and end-use accounting [99]. The econometric approach uses statistical analysis of historical data to derive the relationship between the energy quantity demanded and several other driving variables such as GDP, average income, energy price, technology characteristics, macroeconomic influences, etc. Using estimates of the change in the driving variables (once again attained from historical data), the energy demand can be forecasted using the derived relationship [102]. The end-use accounting approach, on the other hand, divides the energy demand into a number of homogeneous categories such as transportation, residential, industrial, or commercial, after which historical data on energy demands and primary drivers in each of these categories is used to predict category-wise energy demand [99]. The difference between these two approaches is that the end-use approach forecasts demand with a higher level of granularity. Thus, the end-use approach has more explaining power than econometric approaches but is more data intensive and may not include all inter-sector interactions. In practice, a hybrid of both these approaches that includes complicated relationships between various factors is used, for example by the US Department of Energy to prepare the Annual Energy Outlook [269].

Supply forecasting models vary depending on the primary energy source under consideration. For renewable energy sources like wind and solar energy, meteorological models (e.g., [104,270]) may be required to model the day-to-day variability. Detailed sourcing models may be used to predict biomass availability [271,272]. Supply from non-renewable reserves can be forecasted using exponential decay production models together with fossil-fuel exploration data or by analyzing firm investment trends [6]. The IEA uses a data intensive forecasting approach that includes information about fossil fuel field developments worldwide to prepare the World Energy Outlook [6]. Energy supply forecasting can be also be done by using an econometric approach in a similar way to demand forecasting.

4.2. Bottom-Up Models

Bottom-up models are characterized by their high level of technological detail and are used to study the entire energy sector on the regional, national, or global scale. We explain the bottom-up approach with a schematic of a model used at the national scale shown in Figure 6 [273]. The inputs to the model are shown with inward pointing arrows: the bottom-up model requires modules that forecast the national energy demand (for instance, by using the end-use accounting approach) and primary energy supply both from imports and national energy resources. In addition, the bottom-up model connects to an extensive database containing the economic characteristics (such as the conversion process investment and operating costs, maintenance costs, logistical costs, processing plant capacities, etc.) of a wide range of technological components that may be involved in the national energy system. For the national energy system shown in Figure 6, the technological components considered in the bottom-up model comprising both the conversion processes (coal processing, refineries, power plants, CHP plants) and corresponding distribution processes (gas networks, transportation, district heat networks) are shown in red. Primary energy both from domestic sources and imports are shown in blue. The final user categories are shown in orange (industry, commercial and tertiary, households, and transportation) and the services provided by energy are shown in green (process energy, heating area, etc.).

Bottom-up models are typically used together with multi-period optimization schemes to aid in strategic planning and policymaking. The optimization problem is defined as follows:

- Given a set of end users and forecasts for their demand over a certain (usually long term) time horizon, a set of candidate primary energy sources, and a set of corresponding conversion and distribution technologies; determine the optimal energy system configuration that minimizes overall costs (or maximizes overall efficiency) such that energy demand is satisfied by supply in each time period.

Typically, time horizons in the order of a few decades are used. In this period, both existing technological components as well as future technologies available through investments are included. The temporal resolution of the model should be sufficient to capture daily energy demand variations as well as seasonal variations. Thus, bottom-up models divide the time horizon into a number of time slices (such as summer day, summer night, winter day, winter night) and at each time slice the quantity of energy demanded should be equal to the quantity supplied. Thus, the key principle of bottom-up models is the market clearing condition together with energy balance satisfied by each technological component. The solution to the multi-period optimization problem highlights the national energy system configuration with the optimal primal energy mix, and optimal choice of present and future conversion and distribution technologies. In addition, the model can be augmented to include sustainability criteria, for instance by quantifying the emissions associated with each technological option [91]. Including an emissions penalty to the objective function (as shown in [91]) gives a solution representing the long-term energy system configuration that optimizes the trade-offs between economics and environmental factors.

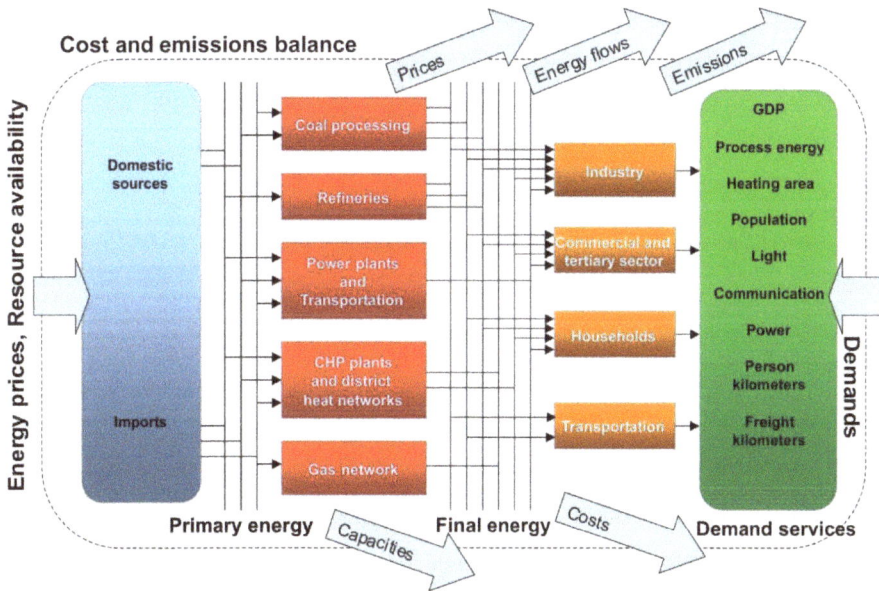

Figure 6. A schematic of a bottom-up model used at the national scale. The inward pointing arrows indicate the inputs to the model while the outward pointing arrows show the outputs. Reprinted by permission from Springer: Operations Research Proceedings [273] Copyright (2001).

The advantages of bottom-up models arise as a result of the high level of technological detail considered. First, new breakthrough technologies can be included into existing models and their impact and market penetration studied [274]. Thus, bottom-up models provide a plausible roadmap for adoption of new technologies, which can be used to inform policymaking decisions. In addition, bottom-up models have a lot of explaining power and can be used to understand why a certain outcome arises. However, the large amount of data required together with the high computational costs are drawbacks to bottom-up energy models. In addition, the mechanistic nature of bottom-up approaches suggests a disadvantage that factors which cannot be easily described or predicted (such as human agency) are not included.

The most widely used bottom-up models are the MARKAL (MARket ALlocation) family of models [274] developed by the International Energy Agency's Energy Technology Systems Analysis Program (IEA-ETSAP), and the MESSAGE (Model for Energy Supply Strategy Alternatives and their General Environmental Impacts) family of models [275,276] developed by the International Atomic Energy Agency (IAEA) and the International Institute for Applied Systems Analysis (IIASA) [147]. MARKAL type models have been used in more than 250 institutions in 70 countries for various purposes including economic analysis of climate policies, studies on the potential of hydrogen fuel cells, nuclear power, etc. [277]. The MESSAGE family of models has been augmented with several other modules (such as a macro-economic module that predicts the interaction with other economic sectors, a climate module, an air pollution module, an agriculture and forestry module to predict land-use changes, etc.) to give the IIASA Integrated Assessment framework [278,279]. Such an Integrated Assessment Model (IAM) is a suite of tools that bring together knowledge and data from a variety of disciplines such as climate change modeling, energy economics, social studies, forestry and agriculture, etc. to aid in decision support in public policy making [280–283]. In order to keep the IAM computationally tractable, a very high level of technological aggregation is used typically at the national or global scale. Considering that energy models have profound public policy implications, several open source models have been developed. Efforts have included the launch of the open energy modeling initiative [90], and the recent OSeMOSYS framework [91,92] which has been extended for a global system boundary [93]. The open energy modeling initiative has a list of the different bottom-up modeling tools available. Open energy system models have the additional advantage of providing transparency when used to make public policies [95,284].

4.3. Top-Down Models

The top-down approach, on the other hand, models all the sectors within the entire economy. Energy is an input to virtually every other sector of the economy, thus top-down energy system models aim to capture the intersectoral connections and feedback that would not be predicted from a bottom-up analysis. For example, growth in industrial sectors (e.g., in the steel industry) may correspond to increased energy demand and higher energy prices. However, any productivity increases in the industrial sector may also spill over to reduce energy generation costs and energy prices. Thus, modeling the interaction between the energy sector and other economic sectors has value in highlighting the often counter-intuitive impact on the overall aggregated economy of specific decisions made in the energy sector and vice versa. Modeling such interconnections is particularly relevant for energy systems because of the strong impact of energy on other sectors. Furthermore, energy system decision-making is closely connected with sustainability considerations which necessitates analysis with broader system boundaries and a higher level of technological aggregation as discussed in Section 3.2. These factors make top-down models suitable tools to aid in general policy making such as tax, energy subsidy, or climate change policy at regional, national, or global scales. Advantages of top-down models include the ability to consistently take into account the entire economy, while their disadvantages include generalized results without much explaining power, inadequate accounting of technological progress, and insufficient detail at smaller scales as a result of using highly aggregated empirical data [285]. Details on the top-down modeling approach are presented in a review by Dannenberg et al. [286]. Top-down models can be divided into input-output models and equilibrium models as discussed next.

4.3.1. Input-Output (IO) Models

Input-output models were introduced by the Nobel laureate Vasily Leontief, and IO analysis is one of the most widely used methods in economics [287]. For instance, IO methods are used by the US department of commerce for planning at local, regional, and national scale as well as by the UN for planning on a global scale [287]. IO models are used to represent the fact that for an economic sector to make a unit of output, certain inputs from other economic sectors are required, in addition to inputs from non-sectoral actors such as labor. By analyzing these input-output connections between different economic sectors, the complex relationships between sectors can be studied. We illustrate how to build an input-output model using a toy example of an economy consisting of only sectors aggregated from the multi-tiered supply chain presented in Figure 5: energy production, mining (coal, iron, etc.), manufacturing, transportation, and construction. The input-output model consists of a system of linear equations describing the flows of products from each economic sector (acting as a producer) to itself and other economic sectors (acting as consumers), as well as to final non-sectoral demand. These flows can be represented in input-output tables as shown in Table 2. The intersectoral flows are shown in the shaded grey section of the table, in either monetary or physical units. Let the indices i and j denote row (producer) and column (consumer) sectors respectively, then the intersectoral flow of an output from producer sector i to consumer sector j is z_{ij}. Thus, the rows of the table show the distribution of a sector's output to inputs of other sectors or to satisfy the final demand, while the columns describe the various inputs required by a particular sector to produce its outputs. Therefore, for a given sector the corresponding column represents all its expenses while the row represents all its revenues assuming the table is constructed in monetary units. The final demand for the output of sector i, f_i, includes demand from households, government, and foreign trade, i.e., from non-sectoral actors. This final demand is not modeled endogenously in IO analysis; instead, final demand data from other sources is used to specify this column. The value added accounts for inputs to the different sectors that do not arise from other economic sectors, such as labor, government services, etc. A country's gross domestic product (GDP) can be calculated as a sum of the value-added row or equivalently as the sum of the final demand column. The total output from a given sector is given by the row sum, i.e., a sector's total output consists of the portion that is an input to other sectors and the portion that is used to satisfy final demand. For a single sector i, the input-output model equation can thus be written:

$$x_i = \sum_j z_{ij} + f_i \tag{1}$$

For all sectors, the input-output model can be written in matrix form. Let vector $\mathbf{X}$ denote the total output from the different economic sectors, i.e., the components of $\mathbf{X}$ are equal to the row sums, and let the vector $\mathbf{F}$ denote final demands. For each column sector, the intersectoral flows are normalized by that sector's total output to give the "direct requirements matrix" $\mathbf{A}$ with elements a_{ij}, as shown in Table 2. Thus, the linear equations representing the input-output model can be rewritten as:

$$\mathbf{X} = \mathbf{AX} + \mathbf{F} \tag{2}$$

The key value of IO models is that they automatically account for all the complex interdependencies between various sectors suggested in Figure 5. Thus, the direct requirements matrix, $\mathbf{A}$, can be used to explicitly derive the inputs and outputs of all the components of the multi-tiered supply chain [127]. The inputs and outputs of the first-tier supply chain are related by the equation $\mathbf{X} = \mathbf{AF}$, and of the second-tier supply chain by the equation $\mathbf{X} = \mathbf{A}^2\mathbf{F}$ and so on to higher tiers. Thus, Equation (1) can be written as [127]:

$$\mathbf{X} = \left(\mathbf{I} + \mathbf{A} + \mathbf{A}^2 + \ldots\right)\mathbf{F} \tag{3}$$

The infinite geometric series can be replaced by the matrix inverse thus the IO model is re-written:

$$\mathbf{X} = (\mathbf{I} - \mathbf{A})^{-1}\mathbf{F} \tag{4}$$

The $(\mathbf{I} - \mathbf{A})^{-1}$ matrix is called the total requirements matrix or the Leontief inverse matrix.

Table 2. Input-output table showing the intersectoral transactions, value-added and the final demand.

		ECONOMIC SECTORS AS CONSUMERS					Row Sums	
		Energy Production	Mining	Manufacturing	Transportation	Construction	Final Demand	Total Output
ECONOMIC SECTORS AS PRODUCERS	Energy Production							
	Mining							
	Manufacturing			$z_{ij} = a_{ij} \cdot x_j$			f_i	x_i
	Transportation							
	Construction							
VALUE ADDED	Labor						GDP	
	Government services							

Data on intersectoral interactions, the value added, and final demands is routinely collected in several countries; in the US data on 79 sectors is made available by the Bureau of economic analysis [288]. A comprehensive treatment of IO modeling is presented in the book by Miller and Blair [287]. Miller and Blair also outline application areas of IO in aiding energy related strategic decision-making. For instance, they highlight the work of Bullard and Herendeen who studied the potential impact of a tax on energy use in the system and showed that the tax would be distributed such that the prices of energy-intensive process products are substantially increased [289]. In addition, Miller and Blair point out the work of Just in 1974 who studied the potential impacts of the then new energy technologies of coal gasification or combined gas-and-steam-cycle electric power generation on the US economy [290]. A final more recent application of IO modeling is to aid in sustainability analysis using the IO LCA approach discussed next.

4.3.2. Input-Output LCA Models

IO LCA models extend economic IO models to account for direct, indirect, and total environmental effects of the various sectors. Thus, IO LCA models exploit the key advantage of IO models in that they inherently account for all environmental impacts associated with the multi-tiered supply chain. Thus, IO LCA models work with a system boundary that includes the entire economy of interconnected industrial sectors to give a comprehensive top-down view of environmental impacts [291]. Herein lies the key advantage of IO LCA models: the challenge faced by the process LCA approach in determining the appropriate system boundary is removed. Equation (4) can be modified to give:

$$E = RX = R(I - A)^{-1}F \tag{5}$$

E is the vector of economy-wide environmental impacts and R is the environmental burden matrix, which gives the impacts of a sector per unit output of the corresponding sector. Thus, E gives the total impacts over all economic sectors associated with satisfying the overall demand F [140].

IO LCA models operate at a high level of technological aggregation. Averaged data obtained at the sectoral level of aggregation is typically used to determine the environmental impacts. While this fact enables relatively quick and cheap assessment with publicly available data, it also implies that IO LCA models are usually not nuanced enough for detailed decision-making on choosing the optimal process or technological options. This is the major disadvantage of IO LCA models that motivated the development of hybrid LCA models as explained in Section 5.2. Further details on IO LCA models are provided in a review paper and book by Hendrickson et al. [292,293]. Software for IO LCA analysis is available from the Green design institute of Carnegie Mellon University [144] (free for non-commercial use) and the EPA (open source) [145].

4.3.3. Equilibrium Modeling

Equilibrium models use the notion of an equilibrium price to resolve the two conflicting optimization problems of consumers (who maximize utility) and producers (who maximize profitability). At this equilibrium price, the quantity of goods supplied by producers equals the quantity demanded, and it is not possible to change this quantity without a decrease in producer profitability or consumer utility. Thus, equilibrium models explicitly model product prices and final demand quantities unlike IO models. The purpose of equilibrium modeling is to determine the influence of changes such as new government policies, trade agreements, technological breakthroughs, or shocks such as natural disasters, which would result in a new equilibrium [269]. Equilibrium models are classified into either general equilibrium or partial equilibrium models: General equilibrium models include all sectors of the economy while partial equilibrium models only include the subset of sectors with significant interactions. A prominent example of an equilibrium model is the National Energy Modeling System (NEMS) developed by the U.S. Energy Information Administration to generate the Annual Energy Outlook [294]. The general structure of NEMS is documented in [294] while the mathematical structure and numerical convergence issues are described by Gabriel et al. [89].

5. Combining PSE and EE Approaches

In this section, we make the case for leveraging the features of both PSE and EE models to aid in optimal decision-making. Thus, we present the value of analysis of energy systems using models with different system boundaries, levels of technological aggregation, underlying theoretical frameworks, and purposes. Three opportunities are identified as explained next.

5.1. Optimal Design and Operation of Flexible Processes Using Demand and Price Forecasts

The idea behind this application is to use demand forecasts for a certain future time period obtained through techniques such as end-use accounting and econometrics, as discussed in Section 4.1, to determine the optimal design and operation of processing plants that maximizes profit over that time period. Thus, a demand forecasting model operating at a high level of technological aggregation in order to account for a number of external factors is used to influence plant level design and operation decisions. Such an approach is valuable for decision-making in processes that involve inputs and products with highly variable demand in the short-term (e.g., electricity) or medium to long-term (e.g., chemicals and fuels).

This approach is followed by Mitra et al. who propose a multi-period optimization framework to determine investment decisions (both for the initial design and for yearly retrofits) as well as operational decisions suitable for power-intensive processes such as air separation units, cement production, and steel and aluminum production [295]. For the case study of air separation units, a demand forecasting model over a 10-year horizon is developed for both electricity and the liquid oxygen and nitrogen products. The electricity demand forecasting model is developed using historical data from PJM Interconnection [296] together with EIA long-term price forecasts [297] with an hourly temporal resolution, while the product demand forecasting model is based on historical data assuming an annual growth rate. In order to account for uncertainty, stochastic demand models were used in addition to deterministic demand models, and the corresponding investment and operational decisions were determined. The results illustrated the value of initial design flexibility as well as the value of flexibility due to retrofitting (e.g., adding a new liquefaction train or a new storage tank) with both stochastic and deterministic demand forecasting models. Since the optimal investment and operational decisions as well as the net profitability depend strongly on the actual demand, there is significant value in developing accurate forecasting models that account for a wide range of external factors. Rolfsman also utilizes this idea of using forecasts to determine the optimal investment and operational decisions. In [298], a regression model that includes factors such as the outdoor temperature is used to forecast the electricity spot price with a temporal resolution of 3 h in the city of Linkoping, Sweden. This price information is used to make investment decisions on the optimal capacity of energy storage over a period of 1 year. In other work, Rolfsman also builds a model with regression analysis to determine Linkoping's district heating demand profile with a temporal resolution of 3 h and uses it to determine the optimal investment over a 20-year period both in combined heat and power plants as well as demand-side measures such as wall insulation or triple glazing windows [298]. It is also suggested that detailed MARKAL-based models could be used to forecast the district heating demand. The results suggested that the quality of the forecasting model had a significant impact on economic performance thus motivating research to develop accurate models with higher temporal resolutions.

Accurate demand forecasting models may be relevant in plant scale operational decision-making as well. For instance, Nease and Adams used demand predictions from the Independent Electricity System Operator of Ontario (IESO) [299] to schedule the electricity supply of a solid-oxide fuel cell plant with compressed air energy storage using a rolling horizon optimization (RHO) framework with a forecast period of 24 h and resolution of 1 h [300]. The results showed that, over a time period of 1 week, the mismatch between electricity supply and demand (measured by the sum-of-squared errors) decreases by 68% when RHO is implemented. Uncertainties in the forecasted demand, however, reduced this improvement.

5.2. Sustainability Analysis and Process Design Using Hybrid Methods

Hybrid LCA models enable more accurate determination of environmental impacts by leveraging the key advantages of process and IO LCA models: they combine the specificity and process-level detail of process LCA models with the comprehensive assessment capability of IO LCA models. Hendrickson et al. provide a comparison of the IO LCA method and the process LCA method [301]. An overview of challenges that may arise in selecting the system boundary of LCA models is provided by Suh et al. [135], while Lenzen presents a more detailed analysis with several case studies of environmental impacts of a range of sectors such as agriculture, energy, transportation, finance, etc. [136]. The percentage errors associated with performing process LCA (with a system boundary containing the 1st, 2nd, and 3rd tiers of the supply chain) and IO LCA analysis are compared in an Australian context for a wide range of products. The results showed the errors of process LCA were smallest for energy-intensive processes, especially with a 3rd tier supply chain boundary, but were still significant. For instance, the errors of process LCA with a 1st, 2nd, and 3rd tier system boundary respectively compared to IO LCA for electricity production were 6%, 4%, and 2% vs. 7%. Similarly, the errors for basic chemical production were 29%, 16%, and 9% vs. 32% [136]. However, the errors for gas production and distribution were 27%, 20%, and 14% vs. 7%. These results illustrate the strengths and weaknesses of the two LCA approaches depending on the product type and motivate hybrid approaches.

Commonly used hybrid LCA models include tiered and IO-based models. Tiered models use the process LCA approach to determine the direct impacts of the main process facility and certain key upstream components, and use the IO approach for the indirect impacts of the remaining upstream processes. Thus, localized data is used for specific process equipment or process sections while aggregated data is used for upstream impacts, with special attention given to avoid double counting. IO-based models disaggregate the sectors of IO models into more homogenous categories that can be mapped onto particular processes. For example, the electricity production sector can be disaggregated into generation, transmission, and distribution categories [302]. In this way, emissions data of a representative plant can be used for the disaggregated categories. Further details on hybrid LCA methods to account for environmental impacts are presented in the book by Hendrickson et al. [293].

In addition to accounting for the broader environmental impacts of the system, the depletion of natural resources can also be accounted for by hybrid methods. Cornelissen and Hirs point out the weakness of conventional LCA analysis in quantifying the depletion of natural resources and propose combining LCA with thermodynamics-based methods [142]. Ukidwe et al. [131] present an overview and Rocco [132] provides a taxonomy of thermodynamics-based resource accounting methods, including those that quantify mass flow (such as Material Flow Analysis [303]), energy flow (such as Net Energy Analysis [304]), and exergy flows. In particular, exergy is valuable in providing a fair basis to measure the quality and value of an array of different material and energy resource streams [305]. Traditionally, exergy-based methods were used primarily in PSE for energy system analysis and optimization at plant scale as detailed in the books by Kotas [306] and Szargut et al. [307]. A historical account of the use of the concept of exergy is provided in the critical review (with more than 2600 references) by Sciubba and Wall [308].

However, exergy analysis at the plant level of technological aggregation only accounts for the direct resource utilization; several approaches have been proposed to expand the spatial and temporal boundaries of the model to account for the indirect resource depletion in the multi-tiered supply chain. These approaches fall within the broad field termed thermo-economics (in the US) or exergo-economics (in Europe), as reviewed in [132,308]. Two notable early (and ultimately equivalent) approaches were calculating the Cumulative Exergy Consumption (CEC) and the Exergy Cost. The CEC approach, developed by Szargut and Morris, is based on dividing the production chain from natural resources to final products into a series of industrial processes each of which contributes a certain proportion of its exergy input to the final product with the rest being destroyed [309]. Exergy Cost Theory (ECT), developed by Valero and coworkers, generates a function that relates the exergy in the final products to input exergy in resources, depending on the production chain configuration [310–312]. However, these two approaches may not account for all the complex effects of the different components in the multi-tiered supply chain: for instance, they may not account for goods and services that are not amenable to an exergy-based characterization, or for the different lifecycle stages of the energy system [132].

In order to address these challenges, the combination of exergy-based methods with LCA techniques was proposed to give the Exergy Life Cycle Assessment (ELCA) framework [132,142,313]. In this framework, first the Exergy Cost approach is applied at the level of the energy conversion system in order to generate a set of proposals to improve the design by minimizing the exergy cost of the products. Keshavarzian et al. provide a review of practical approaches to apply ECT to energy conversion systems [314]. Second, the impacts of these proposals on the broader system are verified by using ELCA to ensure a reduction in primary resource utilization in the different lifecycle stages. Rocco illustrates this framework with an example of a Waste-to-Energy plant situated in the Italian economic context [132]. First, ECT is applied to the plant and several process improvement proposals were highlighted focusing on redesign of the grate furnace, steam turbine, super heater, and preheater components. Next, the ELCA method is applied using the IO-based hybrid LCA approach, considering both the construction and operation phases, in order to account for the complex interaction of the energy system and other sectors of the Italian economy. The process improvement proposals made in the first phase were then independently verified based on the ELCA results. This two-step methodology was applied iteratively. The final result suggested a 4.5% reduction in both direct and primary exergy costs of the electricity generation implying that reducing conversion process inefficiencies had a ripple effect in lowering primary resource consumption. Other approaches to combine exergy-based methods with IO-based hybrid LCA methods were proposed by Hau and Bakshi [315] and Ukidwe et al. [131]. It is also worth noting other thermo-economic approaches that use the mathematical structure of IO analysis for design and analysis at the conversion plant level [314,316,317]

Hanes and Bakshi point out that although most research efforts have used hybrid approaches for process analysis, there may also be significant value in their use for sustainable process design [140,141]. With this motivation, they developed the "process to planet" (P2P) framework which includes sub-models that operate at different levels of technological aggregation: overall economy scale sub-models are combined with value chain scale and unit operation scale sub-models. As a result of its multi-scale nature, the P2P framework can be used for a variety of purposes including process design and supply chain network planning. Furthermore, the framework offers a way of evaluating the impact of decisions made at the overall economy scale of aggregation (such as the institution of a carbon tax) on detailed decisions made on the processing plant scale. The influence of public policies on process design decisions may be significant as illustrated by Adams and Barton, who analyzed the potential impacts of the American Clean Energy and Security (ACES) Act of 2009 (which was not passed) on the choice of power production technologies [318]. The ACES Act proposed using a cap-and-trade scheme to provide economic incentives for lowering emissions, and had a complex system of quotas, credits, and taxes. Adams and Barton calculated the change in the levelized cost of electricity (LCOE) of different power plant configurations with different generation technologies (such as gas turbines, solid oxide fuel cells, and combined cycles), different feedstocks (natural gas or coal), and different implementations of carbon capture and sequestration (CCS) technologies, with the lifetime average price of carbon credits. They found that pulverized coal without CCS was no longer the cheapest above a carbon credit price of $15/tonne, after which coal or natural gas-based processes with solid oxide fuel cells would become the most profitable. Hanes and Bakshi applied the P2P framework to aid in the sustainable design of a corn to ethanol plant. They compared the potential lifetime emissions with a plant designed with the state-of-the-art approach that only includes process and value chain emissions but not overall economy scale emissions [141]. They found that the P2P approach lowered overall emissions by 17%; the state-of-the-art approach had lower process and value chain emissions but neglected significant overall economy scale emissions. Other work on using hybrid approaches for sustainable process design was done by Yue et al. who combined a hybrid LCA approach with a multi-objective optimization framework [319]. They applied this framework to design a biomass to ethanol supply chain in the UK and found that indirect emissions were significant (up to 58.4%).

5.3. Accounting for Feedback Effects of Breakthrough Technologies

Innovation that leads to possible breakthrough energy technologies (for instance, in rechargeable batteries for energy storage, fuel or solar cells, carbon capture and sequestration or utilization (CCSU), novel catalysts for conversion processes, production of novel energy carriers such as hydrogen or ammonia, etc.) typically occurs at low levels of technological aggregation and relies on progress in basic science research [320]. Thereafter, the technological and economic feasibility of the innovation is analyzed usually at the unit operation or plant scale. Thus, a techno-economic model is built that studies the innovation in the context of several externalities such as market conditions, emissions policies, etc. resulting in information on its investment costs, operating costs, capacities, availability factors, fixed maintenance costs, etc. Such techno-economic analyses are typically used to influence decisions made at plant scale, such as whether to invest in the innovation. However, breakthrough technologies can have significant effects at higher levels of technological aggregation, for instance if public policy decisions are made to promote the use of the innovation. For this reason, developing approaches to transfer information about breakthrough technologies from smaller scales to the overall economy scale for strategic planning purposes is valuable. Similarly, it is valuable to analyze the feedback effects of these breakthroughs. In this way, the long-term trajectory of the use of the innovation can be mapped. Approaches suggested so far have linked PSE models to both bottom-up and top-down EE models as we explain next.

Bolat and Thiel present an approach to build a new technological module (for a hydrogen supply chain, in this case) for use in bottom-up models [321,322]. They surveyed different hydrogen production technologies and abstracted information from techno-economic models to develop a hydrogen production module compatible with the MARKAL family of bottom-up models. Such a model can be used to evaluate the feasibility of the hydrogen supply chain in the context of other technologies in the energy system. For instance, Sgobbi et al. used a MARKAL-type model with this hydrogen module to evaluate the potential for hydrogen in a future European energy system [323]. They concluded that hydrogen could become a viable option in 2030, and supply 5–6% of the final energy consumption in the transport and industrial sector in 2050 if certain emissions policies are enacted.

In addition to influencing favorable policies, a second key feedback effect is technological learning: this is the fact that the cost of a technological component falls with increasing total installed capacity. This cost reduction can be due to increased expertise, lower risk and uncertainty, improved manufacturing capacities, economies of scale, etc. Based on historical data, a technological learning curve can be derived which predicts the reduction in the cost of that technological component with increasing installed capacity. Quantifying the technological learning curve is done at a high level of aggregation since the overall global installed capacity is relevant. Details on estimating the learning curve and its incorporation into energy system models are presented in a review by Kahouli-Brahmi [324]. Models incorporating technological learning can be used to analyze the long-term prospects of certain technological options. Rubin et al. provide a comprehensive list of learning rates of a range of electricity supply technologies, including nuclear, fossil fuel, and renewable options [325]. M. van den Broek et al. used energy system models with technological learning to project the future (2040) LCOE of natural gas combined cycle power plants with CCS (71 €$_{2012}$/MWh), concentrated solar power (68 €$_{2012}$/MWh), offshore wind (82 €$_{2012}$/MWh), and photovoltaic systems (104 €$_{2012}$/MWh) [326]. In other work, M. van den Broek et al. evaluated the potential for technological learning and suggested that IGCC plants with CCS have the largest potential for reduced costs by 2050 because an expected 3100 GW of combined cycle capacity would have been installed [327]. They also highlight the role emissions policies that motivate the wider installation of clean technologies could play in lowering mitigation costs due to technological learning. The value of modeling technological learning is also highlighted by Creutzig et al. who suggest that ignoring learning effects has led to underestimating the potential of solar PV [328]. They calculate a range of LCOE values in 2050 from 0.02 to 0.06 $/kWh. Huang et al. use a MARKAL type model to study the influence of technological learning on the future use of low carbon power generation technologies [329]. They conclude that the introduction of endogenous technological learning could increase global installed capacity of wind power, solar PV, coal CCS, and natural gas CCS by 970, 312, 221, and 150 GW, respectively, by 2050.

Despite the potentially significant effect of the cost reductions on plant scale decisions, the inclusion of technological learning in optimization frameworks for process design remains an underexplored research area. In the vast majority of PSE studies covered in this review, techno-economic analyses of energy systems typically use an "Nth plant" approach to estimating technology costs. This means that the cost estimates assume that the technology is sufficiently "mature" such that enough prior attempts at constructing the technology have been made to get over the learning curve. A good example of this in energy systems is the IGCC process, in which Nth plant studies typically predict that capital costs are on the order of $1 bln [318] for the >500 MW scale. However, the two most recent (and only) IGCC plants constructed at the >500 MW scale in the US were made in Edwardsport, Indiana and in Kemper County, Mississippi, which had capital costs of about $3.4 bln [330] and $7.1 bln [331], respectively. In the latter case, operators gave up on the gasification portion and operated the downstream portion of the process solely on natural gas. Thus, the learning curve is a key barrier to adoption that should be considered in energy systems design directly.

To maximize the learning effects, Reiner highlighted the value of coordinated decision-making between energy firms on the design and development of new industrial projects [332]. For the particular case of lowering mitigation costs associated with CCSU technologies, Reiner proposed coordination to enable the global deployment of a portfolio of CCSU projects to facilitate learning by replication and learning by project diversity. Firms participating in this coordinated effort would make certain design decisions at the plant scale (for instance, choosing certain CCSU options from those reviewed by Adams et al. [333]) that maximize overall technological learning at the global scale over a future time period.

Feedback effects of breakthrough technologies can also involve other sectors of the economy, which motivates analysis using top-down models. For example, developments in biomass-to-biofuel processes and their wider adoption could have significant impacts on the wood market, which may need to be accounted for in making plant scale design decisions [334]. Voll et al. combined a preliminary model of a wood to 2-methyltetrahydrofuran process with a simple partial equilibrium model of the German wood market used to predict future wood prices [334]. The results showed that depending on the size of the biofuel plant, the price of wood could increase by up to a factor of 30, presenting significant changes to the plant's NPV. Such results present the value of analysis over different levels of technological aggregation in order to account for the influence of externalities on plant scale decisions. In other words, it may be important to account for the impact of the implementation of the technology on the market price, rather than just the impact of the market price on the choice or design of the technology.

6. Concluding Remarks

The sustainable generation of energy is a key challenge involving several stakeholders including government regulators, energy production and distribution firms, intermediate suppliers, and final consumers. In view of the primary importance of energy to the modern economy, energy systems have been studied by researchers in both the Process Systems Engineering (PSE) and Energy Economics (EE) fields using modeling and simulation approaches with different nature of variables, theoretical

Processes **2018**, *6*, 238

underpinnings, level of technological aggregation, spatial and temporal scales, and model purpose. In addition, several modeling approaches have been proposed which can be categorized into computational, mathematical, and physical models. Although computational and hybrid modeling approaches are increasingly relevant, the majority of PSE models have a mechanistic mathematical structure and draw from theories in chemical engineering science. Thus, the PSE approach models the technological characteristics of energy systems endogenously. Traditionally, PSE tools have been used to aid in the design, operation, and control at the processing plant level. However, the processing plant is situated in a broader economic context that includes several actors, such as competing energy firms, research and technology developers, final consumers with evolving needs, and regulatory agencies. These externalities manifest as energy price and demand uncertainties, changes in emissions policies, and breakthroughs in competing technologies, which may have significant impacts on plant level decisions and profitability. For this reason, leveraging the expertise developed in the EE field on modeling these complexities that arise at higher levels of technological aggregation may be valuable to PSE engineers. Thus, this paper aims to build a bridge between these two communities in order to get a holistic picture of the long-term performance of the energy system in a wider economic and policy context. We point out three specific application areas in which combining the PSE and EE approaches is valuable: (1) optimal design and operation of flexible processes using demand and price forecasts, (2) sustainability analysis and process design using hybrid methods, and (3) accounting for the feedback effects of breakthrough technologies. With these examples, we submit that approaches linking the PSE and EE fields warrant more research attention.

Author Contributions: Conceptualization, A.S.R.S. and T.A.A.II; writing—original draft preparation, A.S.R.S.; writing—review and editing, T.A.A.II and T.G.; supervision, T.A.A.II and T.G.; project administration, T.G.; funding acquisition, T.G.

Funding: The first author gratefully acknowledges the financial support of the Ph.D. scholarship from NTNU's Department of Energy and Process Engineering.

Acknowledgments: We thank Rahul Anantharaman of SINTEF Energy Research for his helpful insights.

Conflicts of Interest: The authors declare no conflict of interest.

References

1. International Energy Agency. *World Energy Outlook*; International Energy Agency: Paris, France, 2017.
2. Chu, S.; Majumdar, A. Opportunities and challenges for a sustainable energy future. *Nature* **2012**, *488*, 294–303. [CrossRef] [PubMed]
3. Allwood, J.M.; Bosetti, V.; Dubash, N.K.; Gómez-Echeverri, L.; von Stechow, C. Glossary. In *Climate Change 2014: Mitigation of Climate Change*; Contribution of Working Group III to the Fifth Assessment Report of the Intergovernmental Panel on Climate Change; Edenhofer, O., Pichs-Madruga, R., Sokona, Y., Farahani, E., Kadner, S., Seyboth, K., Adler, A., Baum, I., Brunner, S., Eickemeier, P., et al., Eds.; Cambridge University Press: Cambridge, UK; New York, NY, USA, 2014.
4. Bruckner, T.; Bashmakov, I.A.; Mulugetta, Y.; Chum, H.; Navarro, A.D.; Edmonds, J.; Faaij, A.; Fungtammasan, B.; Garg, A.; Hertwich, E.; et al. Energy Systems. In *Climate Change 2014: Mitigation of Climate Change*; Contribution of Working Group III to the Fifth Assessment Report of the Intergovernmental Panel on Climate Change; Edenhofer, O., Pichs-Madruga, R., Sokona, Y., Farahani, E., Kadner, S., Seyboth, K., Adler, A., Baum, I., Brunner, S., Eickemeier, P., et al., Eds.; Cambridge University Press: Cambridge, UK; New York, NY, USA, 2014.
5. Grossmann, I.E.; Guillén-Gosálbez, G. Scope for the application of mathematical programming techniques in the synthesis and planning of sustainable processes. *Comput. Chem. Eng.* **2010**, *34*, 1365–1376. [CrossRef]
6. Bhattacharyya, S.C. *Energy Economics: Concepts, Issues, Markets and Governance*; Springer: London, UK, 2011; ISBN 978-0-85729-267-4.
7. Foley, A.M.; Gallachóir, B.Ó.; Hur, J.; Baldick, R.; McKeogh, E.J. A strategic review of electricity systems models. *Energy* **2010**, *35*, 4522–4530. [CrossRef]

8.   Ventosa, M.; Baíllo, Á.; Ramos, A.; Rivier, M. Electricity market modeling trends. *Energy Policy* **2005**, *33*, 897–913. [CrossRef]

9.   Elia, J.A.; Baliban, R.C.; Floudas, C.A. Nationwide energy supply chain analysis for hybrid feedstock processes with significant CO_2 emissions reduction. *AIChE J.* **2012**, *58*, 2142–2154. [CrossRef]

10. Lopion, P.; Markewitz, P.; Robinius, M.; Stolten, D. A review of current challenges and trends in energy systems modeling. *Renew. Sustain. Energy Rev.* **2018**, *96*, 156–166. [CrossRef]

11. Lang, Y.; Zitney, S.E.; Biegler, L.T. Optimization of IGCC processes with reduced order CFD models. *Comput. Chem. Eng.* **2011**, *35*, 1705–1717. [CrossRef]

12. Adams, T.A., II; Thatho, T.; Feuvre Le, M.C.; Swartz, C.L.E. The Optimal Design of a Distillation System for the Flexible Polygeneration of Dimethyl Ether and Methanol under Uncertainty. *Front. Energy Res.* **2018**, *6*, 41. [CrossRef]

13. Chen, G.Q.; Wu, X.F. Energy overview for globalized world economy: Source, supply chain and sink. *Renew. Sustain. Energy Rev.* **2017**, *69*, 735–749. [CrossRef]

14. Biegler, L.T.; Grossmann, I.E.; Westerberg, A.W. *Systematic Methods for Chemical Process Design*; Prentice Hall PTR: Upper Saddle River, NJ, USA, 1997; ISBN 0-13-492422-3. Available online: https://www.osti.gov/biblio/293030-systematic-methods-chemical-process-design (accessed on 12 October 2018).

15. Smith, R. *Chemical Process: Design and Integration*; John Wiley & Sons: Hoboken, NJ, USA, 2016; ISBN 9781118699096.

16. Cameron, I.T.; Gani, R. *Product and Process Modelling: A Case Study Approach*; Elsevier: Amsterdam, The Netherlands, 2011.

17. Nishida, N.; Stephanopoulos, G.; Westerberg, A.W. A review of process synthesis. *AIChE J.* **1981**, *27*, 321–351. [CrossRef]

18. Grossmann, I.E. Mixed-integer programming approach for the synthesis of integrated process flowsheets. *Comput. Chem. Eng.* **1985**, *9*, 463–482. [CrossRef]

19. Gani, R.; Cameron, I.; Lucia, A.; Sin, G.; Georgiadis, M. Process Systems Engineering, 2. Modeling and Simulation. In *Ullmann's Encyclopedia of Industrial Chemistry*; American Cancer Society: Atlanta, GA, USA, 2012; ISBN 978-3-527-30673-2.

20. Marquardt, W. Trends in computer-aided process modeling. *Comput. Chem. Eng.* **1996**, *20*, 591–609. [CrossRef]

21. Grossmann, I.E.; Apap, R.M.; Calfa, B.A.; García-Herreros, P.; Zhang, Q. Recent advances in mathematical programming techniques for the optimization of process systems under uncertainty. *Comput. Chem. Eng.* **2016**, *91*, 3–14. [CrossRef]

22. Barton, P.I.; Pantelides, C.C. Modeling of combined discrete/continuous processes. *AIChE J.* **1994**, *40*, 966–979. [CrossRef]

23. Barton, P.I.; Lee, C.K. Modeling, simulation, sensitivity analysis, and optimization of hybrid systems. *ACM Trans. Model. Comput. Simul.* **2002**, *12*, 256–289. [CrossRef]

24. Bird, R.B.; Stewart, W.E.; Lightfoot, E.N. Transport phenomena. *Appl. Mech. Rev.* **2002**, *55*, R1–R4. [CrossRef]

25. Marquardt, W. Towards a Process Modeling Methodolgy. In *Methods of Model Based Process Control*; Springer: Dordrecht, The Netherlands, 1995; pp. 3–40.

26. Watson, H.A.; Vikse, M.; Gundersen, T.; Barton, P.I. Reliable Flash Calculations: Part 2. Process flowsheeting with nonsmooth models and generalized derivatives. *Ind. Eng. Chem. Res.* **2017**, *56*, 14848–14864. [CrossRef]

27. Psichogios, D.C.; Ungar, L.H. A hybrid neural network-first principles approach to process modeling. *AIChE J.* **1992**, *38*, 1499–1511. [CrossRef]

28. Thompson, M.L.; Kramer, M.A. Modeling chemical processes using prior knowledge and neural networks. *AIChE J.* **1994**, *40*, 1328–1340. [CrossRef]

29. Hoseinzade, L.; Adams, T.A., II. Modeling and simulation of an integrated steam reforming and nuclear heat system. *Int. J. Hydrogen Energy* **2017**, *42*, 25048–25062. [CrossRef]

30. Okeke, I.J.; Adams, T.A., II. Combining petroleum coke and natural gas for efficient liquid fuels production. *Energy* **2018**, *163*, 426–442. [CrossRef]

31. Miller, D.C.; Agarwal, D.; Bhattacharyya, D.; Boverhof, J.; Chen, Y.; Eslick, J.; Leek, J.; Ma, J.; Mahapatra, P.; Ng, B.; et al. Innovative computational tools and models for the design, optimization and control of carbon capture processes. In *Process Systems and Materials for CO₂ Capture: Modelling, Design, Control and Integration*; Papadopoulos, A.I., Seferlis, P., Eds.; John Wiley & Sons Ltd.: Chichester, UK, 2017; pp. 311–342. Available online: https://github.com/CCSI-Toolset/ (accessed on 12 October 2018).
32. Yu, M.; Miller, D.C.; Biegler, L.T. Dynamic Reduced Order Models for Simulating Bubbling Fluidized Bed Adsorbers. *Ind. Eng. Chem. Res.* **2015**, *54*, 6959–6974. [CrossRef]
33. Modekurti, S.; Bhattacharyya, D.; Zitney, S.E. Dynamic modeling and control studies of a two-stage bubbling fluidized bed adsorber-reactor for solid–sorbent CO₂ capture. *Ind. Eng. Chem. Res.* **2013**, *52*, 10250–10260. [CrossRef]
34. Garcia, D.J.; You, F. Supply chain design and optimization: Challenges and opportunities. *Comput. Chem. Eng.* **2015**, *81*, 153–170. [CrossRef]
35. Shah, N. Process industry supply chains: Advances and challenges. *Comput. Chem. Eng.* **2005**, *29*, 1225–1235. [CrossRef]
36. Papageorgiou, L.G. Supply chain optimisation for the process industries: Advances and opportunities. *Comput. Chem. Eng.* **2009**, *33*, 1931–1938. [CrossRef]
37. Barbosa-Póvoa, A.P. Progresses and challenges in process industry supply chains optimization. *Curr. Opin. Chem. Eng.* **2012**, *1*, 446–452. [CrossRef]
38. Nikolopoulou, A.; Ierapetritou, M.G. Optimal design of sustainable chemical processes and supply chains: A review. *Comput. Chem. Eng.* **2012**, *44*, 94–103. [CrossRef]
39. Laínez, J.M.; Puigjaner, L. Prospective and perspective review in integrated supply chain modelling for the chemical process industry. *Curr. Opin. Chem. Eng.* **2012**, *1*, 430–445. [CrossRef]
40. Min, H.; Zhou, G. Supply chain modeling: Past, present and future. *Comput. Ind. Eng.* **2002**, *43*, 231–249. [CrossRef]
41. Niziolek, A.M.; Onel, O.; Tian, Y.; Floudas, C.A.; Pistikopoulos, E.N. Municipal solid waste to liquid transportation fuels—Part III: An optimization-based nationwide supply chain management framework. *Comput. Chem. Eng.* **2017**, in press. [CrossRef]
42. You, F.; Grossmann, I.E. Design of responsive supply chains under demand uncertainty. *Comput. Chem. Eng.* **2008**, *32*, 3090–3111. [CrossRef]
43. You, F.; Grossmann, I.E. Mixed-integer nonlinear programming models and algorithms for large-scale supply chain design with stochastic inventory management. *Ind. Eng. Chem. Res.* **2008**, *47*, 7802–7817. [CrossRef]
44. You, F.; Tao, L.; Graziano, D.J.; Snyder, S.W. Optimal design of sustainable cellulosic biofuel supply chains: Multiobjective optimization coupled with life cycle assessment and input–output analysis. *AIChE J.* **2012**, *58*, 1157–1180. [CrossRef]
45. Liu, S.; Papageorgiou, L.G. Multiobjective optimisation of production, distribution and capacity planning of global supply chains in the process industry. *Omega* **2013**, *41*, 369–382. [CrossRef]
46. Verderame, P.M.; Elia, J.A.; Li, J.; Floudas, C.A. Planning and Scheduling under Uncertainty: A Review Across Multiple Sectors. *Ind. Eng. Chem. Res.* **2010**, *49*, 3993–4017. [CrossRef]
47. Maravelias, C.T.; Sung, C. Integration of production planning and scheduling: Overview, challenges and opportunities. *Comput. Chem. Eng.* **2009**, *33*, 1919–1930. [CrossRef]
48. Floudas, C.A.; Lin, X. Continuous-time versus discrete-time approaches for scheduling of chemical processes: A review. *Comput. Chem. Eng.* **2004**, *28*, 2109–2129. [CrossRef]
49. Li, Z.; Ierapetritou, M. Process scheduling under uncertainty: Review and challenges. *Comput. Chem. Eng.* **2008**, *32*, 715–727. [CrossRef]
50. Méndez, C.A.; Cerdá, J.; Grossmann, I.E.; Harjunkoski, I.; Fahl, M. State-of-the-art review of optimization methods for short-term scheduling of batch processes. *Comput. Chem. Eng.* **2006**, *30*, 913–946. [CrossRef]
51. Kallrath, J. Planning and scheduling in the process industry. *OR Spectr.* **2002**, *24*, 219–250. [CrossRef]
52. Reklaitis, G.V. Overview of scheduling and planning of batch process operations. In *Batch Processing Systems Engineering*; Springer: Berlin/Heidelberg, Germany, 1996; pp. 660–705.
53. Mitra, S.; Grossmann, I.E.; Pinto, J.M.; Arora, N. Optimal production planning under time-sensitive electricity prices for continuous power-intensive processes. *Comput. Chem. Eng.* **2012**, *38*, 171–184. [CrossRef]
54. Mitra, S.; Sun, L.; Grossmann, I.E. Optimal scheduling of industrial combined heat and power plants under time-sensitive electricity prices. *Energy* **2013**, *54*, 194–211. [CrossRef]

55. Birewar, D.B.; Grossmann, I.E. Simultaneous production planning and scheduling in multiproduct batch plants. *Ind. Eng. Chem. Res.* **1990**, *29*, 570–580. [CrossRef]

56. Julka, N.; Srinivasan, R.; Karimi, I. Agent-based supply chain management—1: Framework. *Comput. Chem. Eng.* **2002**, *26*, 1755–1769. [CrossRef]

57. Karimi, I.A.; McDonald, C.M. Planning and Scheduling of Parallel Semicontinuous Processes. 2. Short-Term Scheduling. *Ind. Eng. Chem. Res.* **1997**, *36*, 2701–2714. [CrossRef]

58. McDonald, C.M.; Karimi, I.A. Planning and Scheduling of Parallel Semicontinuous Processes. 1. Production Planning. *Ind. Eng. Chem. Res.* **1997**, *36*, 2691–2700. [CrossRef]

59. Morari, M.; Zafiriou, E. Robust process control. *Chem. Eng. Res. Des.* **1987**, *65*, 462–479.

60. Luyben, W.L. *Process Modeling, Simulation and Control for Chemical Engineers*, 2nd ed.; McGraw-Hill: New York, NY, USA, 1989; ISBN 978-0-07-039159-8.

61. Skogestad, S.; Postlethwaite, I. *Multivariable Feedback Control: Analysis and Design*; Wiley: New York, NY, USA, 2007; Volume 2, pp. 359–368.

62. Hussain, M.A. Review of the applications of neural networks in chemical process control—Simulation and online implementation. *Artif. Intell. Eng.* **1999**, *13*, 55–68. [CrossRef]

63. Bequette, B.W. Nonlinear control of chemical processes: A review. *Ind. Eng. Chem. Res.* **1991**, *30*, 1391–1413. [CrossRef]

64. Downs, J.J.; Vogel, E.F. A plant-wide industrial process control problem. *Comput. Chem. Eng.* **1993**, *17*, 245–255. [CrossRef]

65. Skogestad, S. Plantwide control: The search for the self-optimizing control structure. *J. Process Control* **2000**, *10*, 487–507. [CrossRef]

66. Skogestad, S. Control structure design for complete chemical plants. *Comput. Chem. Eng.* **2004**, *28*, 219–234. [CrossRef]

67. Grossmann, I.E.; Halemane, K.P.; Swaney, R.E. Optimization strategies for flexible chemical processes. *Comput. Chem. Eng.* **1983**, *7*, 439–462. [CrossRef]

68. Grossmann, I.E.; Morari, M. Operability, Resiliency, and Flexibility: Process Design Objectives for a Changing World. 1983. Available online: https://pdfs.semanticscholar.org/2b7c/85a9ff57ba9322910fc00128bca66ba0b544.pdf (accessed on 13 November 2018).

69. Halemane, K.P.; Grossmann, I.E. Optimal process design under uncertainty. *AIChE J.* **1983**, *29*, 425–433. [CrossRef]

70. Gonzalez-Salazar, M.A.; Kirsten, T.; Prchlik, L. Review of the operational flexibility and emissions of gas-and coal-fired power plants in a future with growing renewables. *Renew. Sustain. Energy Rev.* **2018**, *82*, 1497–1513. [CrossRef]

71. Meerman, J.C.; Ramírez, A.; Turkenburg, W.C.; Faaij, A.P.C. Performance of simulated flexible integrated gasification polygeneration facilities. Part A: A technical-energetic assessment. *Renew. Sustain. Energy Rev.* **2011**, *15*, 2563–2587. [CrossRef]

72. Liu, P.; Pistikopoulos, E.N.; Li, Z. A multi-objective optimization approach to polygeneration energy systems design. *AIChE J.* **2010**, *56*, 1218–1234. [CrossRef]

73. Chen, Y.; Adams, T.A., II; Barton, P.I. Optimal Design and Operation of Flexible Energy Polygeneration Systems. *Ind. Eng. Chem. Res.* **2011**, *50*, 4553–4566. [CrossRef]

74. Swaney, R.E.; Grossmann, I.E. An index for operational flexibility in chemical process design. Part I: Formulation and theory. *AIChE J.* **1985**, *31*, 621–630. [CrossRef]

75. Grossmann, I.E.; Floudas, C.A. Active constraint strategy for flexibility analysis in chemical processes. *Comput. Chem. Eng.* **1987**, *11*, 675–693. [CrossRef]

76. Yunt, M.; Chachuat, B.; Mitsos, A.; Barton, P.I. Designing man-portable power generation systems for varying power demand. *AIChE J.* **2008**, *54*, 1254–1269. [CrossRef]

77. Weijermars, R.; Taylor, P.; Bahn, O.; Das, S.R.; Wei, Y.-M. Review of models and actors in energy mix optimization-can leader visions and decisions align with optimum model strategies for our future energy systems? *Energy Strateg. Rev.* **2012**, *1*, 5–18. [CrossRef]

78. Strantzali, E.; Aravossis, K. Decision making in renewable energy investments: A review. *Renew. Sustain. Energy Rev.* **2016**, *55*, 885–898. [CrossRef]

79. Nakata, T.; Silva, D.; Rodionov, M. Application of energy system models for designing a low-carbon society. *Prog. Energy Combust. Sci.* **2011**, *37*, 462–502. [CrossRef]

80.   Pohekar, S.D.; Ramachandran, M. Application of multi-criteria decision making to sustainable energy planning—A review. *Renew. Sustain. Energy Rev.* **2004**, *8*, 365–381. [CrossRef]
81.   Connolly, D.; Lund, H.; Mathiesen, B.V.; Leahy, M. A review of computer tools for analysing the integration of renewable energy into various energy systems. *Appl. Energy* **2010**, *87*, 1059–1082. [CrossRef]
82.   Hafez, O.; Bhattacharya, K. Optimal planning and design of a renewable energy based supply system for microgrids. *Renew. Energy* **2012**, *45*, 7–15. [CrossRef]
83.   Manne, A.; Mendelsohn, R.; Richels, R. MERGE: A model for evaluating regional and global effects of GHG reduction policies. *Energy Policy* **1995**, *23*, 17–34. [CrossRef]
84.   Hennicke, P. Scenarios for a robust policy mix: The final report of the German study commission on sustainable energy supply. *Energy Policy* **2004**, *32*, 1673–1678. [CrossRef]
85.   Kydes, A.S. The Brookhaven Energy System Optimization Model: Its Variants and Uses. In *Energy Policy Modeling: United States and Canadian Experiences*; Springer: Dordrecht, The Netherlands, 1980; pp. 110–136.
86.   Naill, R.F. A system dynamics model for national energy policy planning. *Syst. Dyn. Rev.* **1992**, *8*, 1–19. [CrossRef]
87.   Jacobsson, S.; Lauber, V. The politics and policy of energy system transformation—Explaining the German diffusion of renewable energy technology. *Energy Policy* **2006**, *34*, 256–276. [CrossRef]
88.   Lund, H.; Mathiesen, B.V. Energy system analysis of 100% renewable energy systems—The case of Denmark in years 2030 and 2050. *Energy* **2009**, *34*, 524–531. [CrossRef]
89.   Gabriel, S.A.; Kydes, A.S.; Whitman, P. The National Energy Modeling System: A Large-Scale Energy-Economic Equilibrium Model. *Oper. Res.* **2001**, *49*, 14–25. [CrossRef]
90.   Available online: https://openmod-initiative.org/ (accessed on 12 October 2018).
91.   Howells, M.; Rogner, H.; Strachan, N.; Heaps, C.; Huntington, H.; Kypreos, S.; Hughes, A.; Silveira, S.; DeCarolis, J.; Bazillian, M. OSeMOSYS: The open source energy modeling system: An introduction to its ethos, structure and development. *Energy Policy* **2011**, *39*, 5850–5870. [CrossRef]
92.   Howells, M.; Welsch, M. *OSeMOSYS-The Open Source Energy Modelling System*; International Energy Workshop: Stockholm, Sweden, 2010.
93.   Löffler, K.; Hainsch, K.; Burandt, T.; Oei, P.-Y.; Kemfert, C.; von Hirschhausen, C. Designing a Model for the Global Energy System—GENeSYS-MOD: An Application of the Open-Source Energy Modeling System (OSeMOSYS). *Energies* **2017**, *10*, 1468. [CrossRef]
94.   Pfenninger, S.; Hawkes, A.; Keirstead, J. Energy systems modeling for twenty-first century energy challenges. *Renew. Sustain. Energy Rev.* **2014**, *33*, 74–86. [CrossRef]
95.   Pfenninger, S.; Hirth, L.; Schlecht, I.; Schmid, E.; Wiese, F.; Brown, T.; Davis, C.; Gidden, M.; Heinrichs, H.; Heuberger, C.; et al. Opening the black box of energy modelling: Strategies and lessons learned. *Energy Strateg. Rev.* **2018**, *19*, 63–71. [CrossRef]
96.   Kannan, R.; Strachan, N. Modelling the UK residential energy sector under long-term decarbonisation scenarios: Comparison between energy systems and sectoral modelling approaches. *Appl. Energy* **2009**, *86*, 416–428. [CrossRef]
97.   Suganthi, L.; Samuel, A.A. Energy models for demand forecasting—A review. *Renew. Sustain. Energy Rev.* **2012**, *16*, 1223–1240. [CrossRef]
98.   Torriti, J. A review of time use models of residential electricity demand. *Renew. Sustain. Energy Rev.* **2014**, *37*, 265–272. [CrossRef]
99.   Bhattacharyya, S.C.; Timilsina, G.R. *Energy Demand Models for Policy Formulation: A Comparative Study of Energy Demand Models*; Policy Research Working Papers; The World Bank: Washington, DC, USA, 2009.
100. Bhattacharyya, S.C.; Timilsina, G.R. A review of energy system models. *Int. J. Energy Sect. Manag.* **2010**, *4*, 494–518. [CrossRef]
101. Craig, P.P.; Gadgil, A.; Koomey, J.G. What Can History Teach us? A Retrospective Examination of Long-Term Energy Forecasts for the United States. *Annu. Rev. Energy Environ.* **2002**, *27*, 83–118. [CrossRef]
102. Werbos, P.J. 2.1. Econometric techniques: Theory versus practice. *Energy* **1990**, *15*, 213–236. [CrossRef]
103. Huang, Y.; Bor, Y.J.; Peng, C.Y. The long-term forecast of Taiwan's energy supply and demand: LEAP model application. *Energy Policy* **2011**, *39*, 6790–6803. [CrossRef]

104. Ulbricht, R.; Fischer, U.; Lehner, W.; Donker, H. First steps towards a systematical optimized strategy for solar energy supply forecasting. In Proceedings of the European Conference on Machine Learning and Principles and Practice of Knowledge Discovery in Databases (ECMLPKDD 2013), Prague, Czech Republic, 23–27 September 2013; Volume 2327.

105. Ekonomou, L. Greek long-term energy consumption prediction using artificial neural networks. *Energy* **2010**, *35*, 512–517. [CrossRef]

106. Kankal, M.; Akpınar, A.; Kömürcü, M.İ.; Özşahin, T.Ş. Modeling and forecasting of Turkey's energy consumption using socio-economic and demographic variables. *Appl. Energy* **2011**, *88*, 1927–1939. [CrossRef]

107. Kandananond, K. Forecasting electricity demand in Thailand with an artificial neural network approach. *Energies* **2011**, *4*, 1246–1257. [CrossRef]

108. Hamzaçebi, C. Forecasting of Turkey's net electricity energy consumption on sectoral bases. *Energy Policy* **2007**, *35*, 2009–2016. [CrossRef]

109. Floudas, C.A.; Niziolek, A.M.; Onel, O.; Matthews, L.R. Multi-scale systems engineering for energy and the environment: Challenges and opportunities. *AIChE J.* **2016**, *62*, 602–623. [CrossRef]

110. Hasan, M.F. Multi-scale Process Systems Engineering for Carbon Capture, Utilization, and Storage: A Review. In *Process Systems and Materials for CO$_2$ Capture: Modelling, Design, Control and Integration*; John Wiley & Sons Ltd.: Hoboken, NJ, USA, 2017.

111. Biegler, L.T.; Lang, Y.; Lin, W. Multi-scale optimization for process systems engineering. *Comput. Chem. Eng.* **2014**, *60*, 17–30. [CrossRef]

112. Hosseini, S.A.; Shah, N. Multi-scale process and supply chain modelling: From lignocellulosic feedstock to process and products. *Interface Focus* **2011**. [CrossRef] [PubMed]

113. Baliban, R.C.; Elia, J.A.; Floudas, C.A. Optimization framework for the simultaneous process synthesis, heat and power integration of a thermochemical hybrid biomass, coal, and natural gas facility. *Comput. Chem. Eng.* **2011**, *35*, 1647–1690. [CrossRef]

114. Baliban, R.C.; Elia, J.A.; Floudas, C.A. Simultaneous process synthesis, heat, power, and water integration of thermochemical hybrid biomass, coal, and natural gas facilities. *Comput. Chem. Eng.* **2012**, *37*, 297–327. [CrossRef]

115. Niziolek, A.M.; Onel, O.; Hasan, M.M.F.; Floudas, C.A. Municipal solid waste to liquid transportation fuels—Part II: Process synthesis and global optimization strategies. *Comput. Chem. Eng.* **2015**, *74*, 184–203. [CrossRef]

116. Ghouse, J.H.; Adams, T.A., II. A multi-scale dynamic two-dimensional heterogeneous model for catalytic steam methane reforming reactors. *Int. J. Hydrogen Energy* **2013**, *38*, 9984–9999. [CrossRef]

117. Seepersad, D.; Ghouse, J.H.; Adams, T.A., II. Dynamic simulation and control of an integrated gasifier/reformer system. Part I: Agile case design and control. *Chem. Eng. Res. Des.* **2015**, *100*, 481–496. [CrossRef]

118. Zitney, S.E. Process/equipment co-simulation for design and analysis of advanced energy systems. *Comput. Chem. Eng.* **2010**, *34*, 1532–1542. [CrossRef]

119. Lang, Y.; Malacina, A.; Biegler, L.T.; Munteanu, S.; Madsen, J.I.; Zitney, S.E. Reduced order model based on principal component analysis for process simulation and optimization. *Energy Fuels* **2009**, *23*, 1695–1706. [CrossRef]

120. Lang, Y.-D.; Biegler, L.T.; Munteanu, S.; Madsen, J.I.; Zitney, S.E. *Advanced Process Engineering Co-Simulation Using CFD-Based Reduced Order Models*; National Energy Technology Laboratory (NETL): Pittsburgh, PA, USA; Morgantown, WV, USA; Albany, OR, USA, 2007.

121. Bakshi, B.R.; Fiksel, J. The quest for sustainability: Challenges for process systems engineering. *AIChE J.* **2003**, *49*, 1350–1358. [CrossRef]

122. Bakshi, B.R. Methods and tools for sustainable process design. *Curr. Opin. Chem. Eng.* **2014**, *6*, 69–74. [CrossRef]

123. Cano-Ruiz, J.A.; McRae, G.J. Environmentally Conscious Chemical Process Design. *Annu. Rev. Energy Environ.* **1998**, *23*, 499–536. [CrossRef]

124. Fiksel, J. Designing Resilient, Sustainable Systems. *Environ. Sci. Technol.* **2003**, *37*, 5330–5339. [CrossRef] [PubMed]

125. Pintarič, Z.N.; Kravanja, Z. Selection of the Economic Objective Function for the Optimization of Process Flow Sheets. *Ind. Eng. Chem. Res.* **2006**, *45*, 4222–4232. [CrossRef]

126. Guinée, J.B. Handbook on life cycle assessment operational guide to the ISO standards. *Int. J. Life Cycle Assess.* **2002**, *7*, 311. [CrossRef]

127. Matthews, H.S.; Hendrickson, C.T.; Matthews, D.H. Life Cycle Assessment: Quantitative Approaches for Decisions that Matter. 2015. Available online: http://www.lcatextbook.com (accessed on 12 October 2018).

128. Curran, M.A. Environmental life-cycle assessment. *Int. J. Life Cycle Assess.* **1996**, *1*, 179. [CrossRef]

129. Finnveden, G.; Hauschild, M.Z.; Ekvall, T.; Guinée, J.; Heijungs, R.; Hellweg, S.; Koehler, A.; Pennington, D.; Suh, S. Recent developments in Life Cycle Assessment. *J. Environ. Manag.* **2009**, *91*, 1–21. [CrossRef] [PubMed]

130. Klöpffer, W.; Grahl, B. *Life Cycle Assessment (LCA): A Guide to Best Practice*; John Wiley & Sons: Hoboken, NJ, USA, 2014.

131. Ukidwe, N.U.; Hau, J.L.; Bakshi, B.R. Thermodynamic Input-Output Analysis of Economic and Ecological Systems. In *Handbook of Input-Output Economics in Industrial Ecology*; Suh, S., Ed.; Eco-Efficiency in Industry and Science; Springer: Dordrecht, The Netherlands, 2009; pp. 459–490, ISBN 978-1-4020-5737-3.

132. Rocco, M.V. *Primary Exergy Cost of Goods and Services*; Springer Briefs in Applied Sciences and Technology; Springer International Publishing: Cham, Switzerland, 2016; ISBN 978-3-319-43655-5.

133. Othman, M.R.; Repke, J.-U.; Wozny, G.; Huang, Y. A Modular Approach to Sustainability Assessment and Decision Support in Chemical Process Design. *Ind. Eng. Chem. Res.* **2010**, *49*, 7870–7881. [CrossRef]

134. Hugo, A.; Pistikopoulos, E.N. Environmentally conscious long-range planning and design of supply chain networks. *J. Clean. Prod.* **2005**, *13*, 1471–1491. [CrossRef]

135. Suh, S.; Lenzen, M.; Treloar, G.J.; Hondo, H.; Horvath, A.; Huppes, G.; Jolliet, O.; Klann, U.; Krewitt, W.; Moriguchi, Y.; et al. System Boundary Selection in Life-Cycle Inventories Using Hybrid Approaches. *Environ. Sci. Technol.* **2004**, *38*, 657–664. [CrossRef] [PubMed]

136. Lenzen, M. Errors in Conventional and Input-Output—Based Life—Cycle Inventories. *J. Ind. Ecol.* **2000**, *4*, 127–148. [CrossRef]

137. Nease, J.; Adams, T.A., II. Life cycle analyses of bulk-scale solid oxide fuel cell power plants and comparisons to the natural gas combined cycle. *Can. J. Chem. Eng.* **2015**, *93*, 1349–1363. [CrossRef]

138. Singh, B.; Strømman, A.H.; Hertwich, E.G. Comparative life cycle environmental assessment of CCS technologies. *Int. J. Greenh. Gas Control* **2011**, *5*, 911–921. [CrossRef]

139. Available online: http://eplca.jrc.ec.europa.eu/ (accessed on 12 October 2018).

140. Hanes, R.J.; Bakshi, B.R. Process to planet: A multiscale modeling framework toward sustainable engineering. *AIChE J.* **2015**, *61*, 3332–3352. [CrossRef]

141. Hanes, R.J.; Bakshi, B.R. Sustainable process design by the process to planet framework. *AIChE J.* **2015**, *61*, 3320–3331. [CrossRef]

142. Cornelissen, R.L.; Hirs, G.G. The value of the exergetic life cycle assessment besides the LCA. *Energy Convers. Manag.* **2002**, *43*, 1417–1424. [CrossRef]

143. Available online: http://www.openlca.org/ (accessed on 13 November 2018).

144. Carnegie Mellon University, Green Design Institute. *Economic Input-Output Life Cycle Assessment (EIO-LCA) Model*; Carnegie Mellon University: Pittsburgh, PA, USA, 2003.

145. Yang, Y.; Ingwersen, W.W.; Hawkins, T.R.; Srocka, M.; Meyer, D.E. USEEIO: A new and transparent United States environmentally-extended input-output model. *J. Clean. Prod.* **2017**, *158*, 308–318. [CrossRef] [PubMed]

146. Ringkjøb, H.-K.; Haugan, P.M.; Solbrekke, I.M. A review of modelling tools for energy and electricity systems with large shares of variable renewables. *Renew. Sustain. Energy Rev.* **2018**, *96*, 440–459. [CrossRef]

147. Hall, L.M.H.; Buckley, A.R. A review of energy systems models in the UK: Prevalent usage and categorisation. *Appl. Energy* **2016**, *169*, 607–628. [CrossRef]

148. Van Beeck, N. *Classification of Energy Models*; Tilburg University: Tilburg, The Netherlands, 2000.

149. Jebaraj, S.; Iniyan, S. A review of energy models. *Renew. Sustain. Energy Rev.* **2006**, *10*, 281–311. [CrossRef]

150. Herbst, A.; Toro, F.; Reitze, F.; Jochem, E. Introduction to Energy Systems Modelling. *Swiss J. Econ. Stat.* **2012**, *148*, 111–135. [CrossRef]

151. Nakata, T. Energy-economic models and the environment. *Prog. Energy Combust. Sci.* **2004**, *30*, 417–475. [CrossRef]

152. Goyal, H.B.; Seal, D.; Saxena, R.C. Bio-fuels from thermochemical conversion of renewable resources: A review. *Renew. Sustain. Energy Rev.* **2008**, *12*, 504–517. [CrossRef]

153. Zaimes, G.G.; Vora, N.; Chopra, S.S.; Landis, A.E.; Khanna, V. Design of Sustainable Biofuel Processes and Supply Chains: Challenges and Opportunities. *Processes* **2015**, *3*, 634–663. [CrossRef]

154. Deshmukh, M.K.; Deshmukh, S.S. Modeling of hybrid renewable energy systems. *Renew. Sustain. Energy Rev.* **2008**, *12*, 235–249. [CrossRef]

155. Shivarama Krishna, K.; Sathish Kumar, K. A review on hybrid renewable energy systems. *Renew. Sustain. Energy Rev.* **2015**, *52*, 907–916. [CrossRef]

156. Floudas, C.A.; Elia, J.A.; Baliban, R.C. Hybrid and single feedstock energy processes for liquid transportation fuels: A critical review. *Comput. Chem. Eng.* **2012**, *41*, 24–51. [CrossRef]

157. Allegrini, J.; Orehounig, K.; Mavromatidis, G.; Ruesch, F.; Dorer, V.; Evins, R. A review of modelling approaches and tools for the simulation of district-scale energy systems. *Renew. Sustain. Energy Rev.* **2015**, *52*, 1391–1404. [CrossRef]

158. Keirstead, J.; Jennings, M.; Sivakumar, A. A review of urban energy system models: Approaches, challenges and opportunities. *Renew. Sustain. Energy Rev.* **2012**, *16*, 3847–3866. [CrossRef]

159. Zeng, Y.; Cai, Y.; Huang, G.; Dai, J. A Review on Optimization Modeling of Energy Systems Planning and GHG Emission Mitigation under Uncertainty. *Energies* **2011**, *4*, 1624–1656. [CrossRef]

160. Liu, P.; Georgiadis, M.C.; Pistikopoulos, E.N. Advances in Energy Systems Engineering. *Ind. Eng. Chem. Res.* **2011**, *50*, 4915–4926. [CrossRef]

161. Adams, T.A., II. Future opportunities and challenges in the design of new energy conversion systems. *Comput. Chem. Eng.* **2015**, *81*, 94–103. [CrossRef]

162. Martín, M.; Adams, T.A., II. Future directions in process and product synthesis and design. *Comput. Aided Chem. Eng.* **2018**, *44*, 1–10.

163. Fisher, J.; Henzinger, T.A. Executable cell biology. *Nat. Biotechnol.* **2007**, *25*, 1239–1249. [CrossRef] [PubMed]

164. Tucker, D.; Shelton, M.; Manivannan, A. The Role of Solid Oxide Fuel Cells in Advanced Hybrid Power Systems of the Future. *Electrochem. Soc. Interface* **2009**, *18*, 45.

165. Tucker, D.; Liese, E.; VanOsdol, J.; Lawson, L.; Gemmen, R.S. Fuel Cell Gas Turbine Hybrid Simulation Facility Design. In *ASME 2002 International Mechanical Engineering Congress and Exposition*; American Society of Mechanical Engineers: New York, NY, USA, 2002; pp. 183–190. [CrossRef]

166. Liao, S.H. Expert system methodologies and applications—A decade review from 1995 to 2004. *Expert Syst. Appl.* **2005**, *28*, 93–103. [CrossRef]

167. Stephanopoulos, G.; Han, C. Intelligent systems in process engineering: A review. *Comput. Chem. Eng.* **1996**, *20*, 743–791. [CrossRef]

168. Mele, F.D.; Guillén, G.; Espuña, A.; Puigjaner, L. An agent-based approach for supply chain retrofitting under uncertainty. *Comput. Chem. Eng.* **2007**, *31*, 722–735. [CrossRef]

169. Weidlich, A.; Veit, D. A critical survey of agent-based wholesale electricity market models. *Energy Econ.* **2008**, *30*, 1728–1759. [CrossRef]

170. Ma, T.; Nakamori, Y. Agent-based modeling on technological innovation as an evolutionary process. *Eur. J. Oper. Res.* **2005**, *166*, 741–755. [CrossRef]

171. Ma, T.; Nakamori, Y. Modeling technological change in energy systems–from optimization to agent-based modeling. *Energy* **2009**, *34*, 873–879. [CrossRef]

172. Ma, T.; Grubler, A.; Nakamori, Y. Modeling technology adoptions for sustainable development under increasing returns, uncertainty, and heterogeneous agents. *Eur. J. Oper. Res.* **2009**, *195*, 296–306. [CrossRef]

173. Bonabeau, E. Agent-based modeling: Methods and techniques for simulating human systems. *Proc. Natl. Acad. Sci. USA* **2002**, *99*, 7280–7287. [CrossRef] [PubMed]

174. Dhaliwal, J.S.; Benbasat, I. The use and effects of knowledge-based system explanations: Theoretical foundations and a framework for empirical evaluation. *Inf. Syst. Res.* **1996**, *7*, 342–362. [CrossRef]

175. Available online: https://www.process-design-center.com/prosyn.html (accessed on 12 October 2018).

176. Puig-Arnavat, M.; Hernández, J.A.; Bruno, J.C.; Coronas, A. Artificial neural network models for biomass gasification in fluidized bed gasifiers. *Biomass Bioenergy* **2013**, *49*, 279–289. [CrossRef]

177. Hunt, K.J.; Sbarbaro, D.; Żbikowski, R.; Gawthrop, P.J. Neural networks for control systems—A survey. *Automatica* **1992**, *28*, 1083–1112. [CrossRef]

178. Baughman, D.R.; Liu, Y.A. *Neural Networks in Bioprocessing and Chemical Engineering*; Academic Press: Cambridge, MA, USA, 2014.

179. Cozad, A.; Sahinidis, N.V.; Miller, D.C. A combined first-principles and data-driven approach to model building. *Comput. Chem. Eng.* **2015**, *73*, 116–127. [CrossRef]

180. Lophaven, S.N.; Nielsen, H.B.; Søndergaard, J. *DACE: A Matlab Kriging Toolbox*; Technical Report; Technical University of Denmark: Lyngby, Denmark, 2002.

181. Straus, J.; Skogestad, S. Surrogate model generation using self-optimizing variables. *Comput. Chem. Eng.* **2018**, in press. [CrossRef]

182. Forrester, A.; Keane, A. *Engineering Design via Surrogate Modelling: A Practical Guide*; John Wiley & Sons: Hoboken, NJ, USA, 2008.

183. Watson, H.A.J. Robust Simulation and Optimization Methods for Natural Gas Liquefaction Processes. Ph.D. Thesis, Massachusetts Institute of Technology, Cambridge, MA, USA, 2018.

184. Henzinger, T.A. The theory of hybrid automata. In *Verification of Digital and Hybrid Systems*; Springer: Berlin, Germany, 2000; pp. 265–292.

185. Oldenburg, J.; Marquardt, W. Disjunctive modeling for optimal control of hybrid systems. *Comput. Chem. Eng.* **2008**, *32*, 2346–2364. [CrossRef]

186. David, R.; Alla, H. On hybrid Petri nets. *Discret. Event Dyn. Syst.* **2001**, *11*, 9–40. [CrossRef]

187. Barton, P.I.; Khan, K.A.; Stechlinski, P.; Watson, H.A. Computationally relevant generalized derivatives: Theory, evaluation and applications. *Optim. Methods Softw.* **2017**, 1–43. [CrossRef]

188. Watson, H.A.; Khan, K.A.; Barton, P.I. Multistream heat exchanger modeling and design. *AIChE J.* **2015**, *61*, 3390–3403. [CrossRef]

189. Watson, H.A.; Vikse, M.; Gundersen, T.; Barton, P.I. Optimization of single mixed-refrigerant natural gas liquefaction processes described by nondifferentiable models. *Energy* **2018**, *150*, 860–876. [CrossRef]

190. Vikse, M.; Watson, H.A.; Gundersen, T.; Barton, P.I. Versatile Simulation Method for Complex Single Mixed Refrigerant Natural Gas Liquefaction Processes. *Ind. Eng. Chem. Res.* **2017**, *57*, 5881–5894. [CrossRef]

191. Vikse, M.; Watson, H.A.; Barton, P.I.; Gundersen, T. Simulation of a Dual Mixed Refrigerant LNG Process using a Nonsmooth Framework. *Comput. Aided Chem. Eng.* **2018**, *44*, 391–396.

192. Vikse, M.; Watson, H.A.; Gundersen, T.; Barton, P.I. Simulation of Dual Mixed Refrigerant Natural Gas Liquefaction Processes using a Nonsmooth Framework. *Processes* **2018**, *6*, 193. [CrossRef]

193. Straus, J.; Skogestad, S. Variable reduction for surrogate modelling. In Proceedings of the Foundations of Computer-Aided Process Operations, Tucson, AZ, USA, 8–12 January 2017.

194. Straus, J.; Skogestad, S. Use of Latent Variables to Reduce the Dimension of Surrogate Models. *Comput. Aided Chem. Eng.* **2017**, *40*, 445–450.

195. Guo, B.; Li, D.; Cheng, C.; Lü, Z.; Shen, Y. Simulation of biomass gasification with a hybrid neural network model. *Bioresour. Technol.* **2001**, *76*, 77–83. [CrossRef]

196. Takamatsu, T. The nature and role of process systems engineering. *Comput. Chem. Eng.* **1983**, *7*, 203–218. [CrossRef]

197. Vlachos, D.G. A Review of Multiscale Analysis: Examples from Systems Biology, Materials Engineering, and Other Fluid–Surface Interacting Systems. In *Advances in Chemical Engineering*; Elsevier: Amsterdam, The Netherlands, 2005; Volume 30, pp. 1–61, ISBN 978-0-12-008530-9.

198. Grossmann, I.E.; Westerberg, A.W. Research challenges in process systems engineering. *AIChE J.* **2000**, *46*, 1700–1703. [CrossRef]

199. Grossmann, I.E. Challenges in the new millennium: Product discovery and design, enterprise and supply chain optimization, global life cycle assessment. *Comput. Chem. Eng.* **2004**, *29*, 29–39. [CrossRef]

200. Horstemeyer, M.F. Multiscale Modeling: A Review. In *Practical Aspects of Computational Chemistry*; Springer: Dordrecht, The Netherlands, 2009; pp. 87–135, ISBN 978-90-481-2686-6.

201. French, C.E. Synthetic biology and biomass conversion: A match made in heaven? *J. R. Soc. Interface* **2009**, *6*, S547–S558. [CrossRef] [PubMed]

202. Mettler, M.S.; Vlachos, D.G.; Dauenhauer, P.J. Top ten fundamental challenges of biomass pyrolysis for biofuels. *Energy Environ. Sci.* **2012**, *5*, 7797–7809. [CrossRef]

203. Kumar, R.; Singh, S.; Singh, O.V. Bioconversion of lignocellulosic biomass: Biochemical and molecular perspectives. *J. Ind. Microbiol. Biotechnol.* **2008**, *35*, 377–391. [CrossRef] [PubMed]

204. Lee, S.K.; Chou, H.; Ham, T.S.; Lee, T.S.; Keasling, J.D. Metabolic engineering of microorganisms for biofuels production: From bugs to synthetic biology to fuels. *Curr. Opin. Biotechnol.* **2008**, *19*, 556–563. [CrossRef] [PubMed]

205. Baliban, R.C.; Elia, J.A.; Floudas, C.A. Toward Novel Hybrid Biomass, Coal, and Natural Gas Processes for Satisfying Current Transportation Fuel Demands, 1: Process Alternatives, Gasification Modeling, Process Simulation, and Economic Analysis. *Ind. Eng. Chem. Res.* **2010**, *49*, 7343–7370. [CrossRef]
206. Onel, O.; Niziolek, A.M.; Hasan, M.M.F.; Floudas, C.A. Municipal solid waste to liquid transportation fuels—Part I: Mathematical modeling of a municipal solid waste gasifier. *Comput. Chem. Eng.* **2014**, *71*, 636–647. [CrossRef]
207. Field, R.P.; Brasington, R. Baseline flowsheet model for IGCC with carbon capture. *Ind. Eng. Chem. Res.* **2011**, *50*, 11306–11312. [CrossRef]
208. Adams, T.A., II; Barton, P.I. High-efficiency power production from coal with carbon capture. *AIChE J.* **2010**, *56*, 3120–3136. [CrossRef]
209. Kunze, C.; Spliethoff, H. Modelling, comparison and operation experiences of entrained flow gasifier. *Energy Convers. Manag.* **2011**, *52*, 2135–2141. [CrossRef]
210. Singh, R.I.; Brink, A.; Hupa, M. CFD modeling to study fluidized bed combustion and gasification. *Appl. Therm. Eng.* **2013**, *52*, 585–614. [CrossRef]
211. Zitney, S.E.; Syamlal, M. Integrated Process Simulation and CFD for Improved Process Engineering. *Comput. Aided Chem. Eng.* **2002**, *10*, 397–402.
212. Baruah, D.; Baruah, D.C. Modeling of biomass gasification: A review. *Renew. Sustain. Energy Rev.* **2014**, *39*, 806–815. [CrossRef]
213. Sloan, D.A.P.; Fiveland, W.A.P.; Zitney, S.E.; Osawe, M. Plant design: Integrating Plant and Equipment Models. *Power Mag.* **2007**, *151*, 8.
214. Shi, S.-P.; Zitney, S.E.; Shahnam, M.; Syamlal, M.; Rogers, W.A. Modelling coal gasification with CFD and discrete phase method. *J. Energy Inst.* **2006**, *79*, 217–221. [CrossRef]
215. Biegler, L.T.; Lang, Y. Multi-scale optimization for advanced energy processes. In *Computer Aided Chemical Engineering*; Elsevier: Amsterdam, The Netherlands, 2012; Volume 31, pp. 51–60.
216. Chen, L.; Yong, S.Z.; Ghoniem, A.F. Oxy-fuel combustion of pulverized coal: Characterization, fundamentals, stabilization and CFD modeling. *Prog. Energy Combust. Sci.* **2012**, *38*, 156–214. [CrossRef]
217. Adams, T.A., II; Barton, P.I. Combining coal gasification, natural gas reforming, and solid oxide fuel cells for efficient polygeneration with CO_2 capture and sequestration. *Fuel Process. Technol.* **2011**, *92*, 2105–2115. [CrossRef]
218. Adams, T.A., II; Ghouse, J.H. Polygeneration of fuels and chemicals. *Curr. Opin. Chem. Eng.* **2015**, *10*, 87–93. [CrossRef]
219. Jana, K.; Ray, A.; Majoumerd, M.M.; Assadi, M.; De, S. Polygeneration as a future sustainable energy solution—A comprehensive review. *Appl. Energy* **2017**, *202*, 88–111. [CrossRef]
220. Adams, T.A., II; Barton, P.I. Combining coal gasification and natural gas reforming for efficient polygeneration. *Fuel Process. Technol.* **2011**, *92*, 639–655. [CrossRef]
221. Ghouse, J.H.; Seepersad, D.; Adams, T.A., II. Modelling, simulation and design of an integrated radiant syngas cooler and steam methane reformer for use with coal gasification. *Fuel Process. Technol.* **2015**, *138*, 378–389. [CrossRef]
222. Seepersad, D.; Ghouse, J.H.; Adams, T.A., II. Dynamic simulation and control of an integrated gasifier/reformer system. Part II: Discrete and model predictive control. *Chem. Eng. Res. Des.* **2015**, *100*, 497–508. [CrossRef]
223. Sharma, B.; Ingalls, R.G.; Jones, C.L.; Khanchi, A. Biomass supply chain design and analysis: Basis, overview, modeling, challenges, and future. *Renew. Sustain. Energy Rev.* **2013**, *24*, 608–627. [CrossRef]
224. Yue, D.; You, F.; Snyder, S.W. Biomass-to-bioenergy and biofuel supply chain optimization: Overview, key issues and challenges. *Comput. Chem. Eng.* **2014**, *66*, 36–56. [CrossRef]
225. Iakovou, E.; Karagiannidis, A.; Vlachos, D.; Toka, A.; Malamakis, A. Waste biomass-to-energy supply chain management: A critical synthesis. *Waste Manag.* **2010**, *30*, 1860–1870. [CrossRef] [PubMed]
226. Cafaro, D.C.; Grossmann, I.E. Strategic planning, design, and development of the shale gas supply chain network. *AIChE J.* **2014**, *60*, 2122–2142. [CrossRef]
227. Rezaie, B.; Rosen, M.A. District heating and cooling: Review of technology and potential enhancements. *Appl. Energy* **2012**, *93*, 2–10. [CrossRef]
228. Varma, V.A.; Reklaitis, G.V.; Blau, G.E.; Pekny, J.F. Enterprise-wide modeling & optimization—An overview of emerging research challenges and opportunities. *Comput. Chem. Eng.* **2007**, *31*, 692–711. [CrossRef]

229. Zhang, Q.; Grossmann, I.E. Enterprise-wide optimization for industrial demand side management: Fundamentals, advances, and perspectives. *Chem. Eng. Res. Des.* **2016**, *116*, 114–131. [CrossRef]

230. Lam, H.L.; Klemeš, J.J.; Kravanja, Z. Model-size reduction techniques for large-scale biomass production and supply networks. *Energy* **2011**, *36*, 4599–4608. [CrossRef]

231. Our Common Future, Chapter 2: Towards Sustainable Development-A/42/427 Annex, Chapter 2-UN Documents: Gathering a Body of Global Agreements. Available online: http://www.un-documents.net/ocf-02.htm (accessed on 9 September 2018).

232. Elkington, J. *Cannibals with Forks: The Triple Bottom Line of 21st Century Business*; John Wiley and Sons: Hoboken, NJ, USA, 1999; ISBN 1841120847.

233. Belgiorno, V.; De Feo, G.; Della Rocca, C.; Napoli, R.M.A. Energy from gasification of solid wastes. *Waste Manag.* **2003**, *23*, 1–15. [CrossRef]

234. Psomopoulos, C.S.; Bourka, A.; Themelis, N.J. Waste-to-energy: A review of the status and benefits in USA. *Waste Manag.* **2009**, *29*, 1718–1724. [CrossRef] [PubMed]

235. Ghanbari, H.; Pettersson, F.; Saxén, H. Sustainable development of primary steelmaking under novel blast furnace operation and injection of different reducing agents. *Chem. Eng. Sci.* **2015**, *129*, 208–222. [CrossRef]

236. Deng, L.; Adams, T.A., II. Optimization of coke oven gas desulfurization and combined cycle power plant electricity generation. *Ind. Eng. Chem. Res.* **2018**, in press. [CrossRef]

237. Klemes, J.J. *Handbook of Process Integration (PI): Minimisation of Energy and Water Use, Waste and Emissions*; Elsevier: Amsterdam, The Netherlands, 2013.

238. Fu, C.; Vikse, M.; Gundersen, T. Work and Heat Integration: An emerging research area. *Energy* **2018**, *158*, 796–806. [CrossRef]

239. Karuppiah, R.; Grossmann, I.E. Global optimization for the synthesis of integrated water systems in chemical processes. *Comput. Chem. Eng.* **2006**, *30*, 650–673. [CrossRef]

240. Wang, Y.P.; Smith, R. Wastewater minimisation. *Chem. Eng. Sci.* **1994**, *49*, 981–1006. [CrossRef]

241. Smith, R.L.; Ruiz-Mercado, G.J.; Gonzalez, M.A. Using GREENSCOPE indicators for sustainable computer-aided process evaluation and design. *Comput. Chem. Eng.* **2015**, *81*, 272–277. [CrossRef]

242. Azapagic, A.; Perdan, S. Indicators of Sustainable Development for Industry. *Process Saf. Environ. Prot.* **2000**, *78*, 243–261. [CrossRef]

243. Bamufleh, H.S.; Ponce-Ortega, J.M.; El-Halwagi, M.M. Multi-objective optimization of process cogeneration systems with economic, environmental, and social tradeoffs. *Clean Technol. Environ. Policy* **2013**, *15*, 185–197. [CrossRef]

244. Sinnott, R.K.; Towler, G. *Chemical Engineering Design: SI Edition*; Elsevier: Amsterdam, The Netherlands, 2009.

245. Seider, W.D.; Seader, J.D.; Lewin, D.R. *Product and Process Design Principles: Synthesis, Analysis and Evaluation*; John Wiley & Sons: Hoboken, NJ, USA, 2009.

246. Turton, R.; Bailie, R.C.; Whiting, W.B.; Shaeiwitz, J.A. *Analysis, Synthesis and Design of Chemical Processes*; Pearson Education: London, UK, 2008.

247. Larson, E.D.; Jin, H.; Celik, F.E. Large-scale gasification-based coproduction of fuels and electricity from switchgrass. *Biofuels Bioprod. Biorefin.* **2009**, *3*, 174–194. [CrossRef]

248. Hamelinck, C.; Faaij, A.; Denuil, H.; Boerrigter, H. Production of FT transportation fuels from biomass; technical options, process analysis and optimisation, and development potential. *Energy* **2004**, *29*, 1743–1771. [CrossRef]

249. Adams, T.A., II. *Learn Aspen Plus in 24 Hours*; McGraw-Hill Education: New York, NY, USA, 2018; ISBN 978-1-260-11645-8.

250. ICIS Pricing. Chemical Industry Trends. Available online: https://www.icis.com (accessed on 10 September 2018).

251. Birge, J.R.; Louveaux, F. *Introduction to Stochastic Programming*; Springer Science & Business Media: Berlin, Germany, 2011.

252. Ben-Tal, A.; Nemirovski, A. Robust optimization–methodology and applications. *Math. Program.* **2002**, *92*, 453–480. [CrossRef]

253. Charnes, A.; Cooper, W.W. Chance-constrained programming. *Manag. Sci.* **1959**, *6*, 73–79. [CrossRef]

254. Bellman, R. *Dynamic Programming*; Courier Corporation: North Chelmsford, MA, USA, 2013.

255. Sirdeshpande, A.R.; Ierapetritou, M.G.; Andrecovich, M.J.; Naumovitz, J.P. Process synthesis optimization and flexibility evaluation of air separation cycles. *AIChE J.* **2005**, *51*, 1190–1200. [CrossRef]

256. Wang, H.; Mastragostino, R.; Swartz, C.L. Flexibility analysis of process supply chain networks. *Comput. Chem. Eng.* **2016**, *84*, 409–421. [CrossRef]

257. Grossmann, I.E.; Calfa, B.A.; Garcia-Herreros, P. Evolution of concepts and models for quantifying resiliency and flexibility of chemical processes. *Comput. Chem. Eng.* **2014**, *70*, 22–34. [CrossRef]

258. Horne, R.; Grant, T.; Verghese, K. *Life Cycle Assessment: Principles, Practice and Prospects*; Csiro Publishing: Clayton, Australia, 2009.

259. International Organization for Standardization. *Environmental Management: Life Cycle Assessment; Principles and Framework*; ISO: Geneva, Switzerland, 2006.

260. Huijbregts, M.A.J.; Steinmann, Z.J.N.; Elshout, P.M.F.; Stam, G.; Verones, F.; Vieira, M.D.M.; Hollander, A.; Zijp, M.; van Zelm, R. *ReCiPe 2016: A Harmonized Life Cycle Impact Assessment Method at Midpoint and Endpoint Level. Report I: Characterization*; RIVM Report 2016-0104; National Institute for Public Health and the Environment: Bilthofen, The Netherlands, 2016.

261. Bare, J.C. TRACI: The tool for the reduction and assessment of chemical and other environmental impacts. *J. Ind. Ecol.* **2002**, *6*, 49–78. [CrossRef]

262. *Eco-Indicator 99 Methodology Report*; Pre Consultants B.V.: Amersfoort, The Netherlands, 22 June 2011.

263. Available online: https://uslci.lcacommons.gov/ (accessed on 12 October 2018).

264. Available online: https://www.ecoinvent.org/ (accessed on 12 October 2018).

265. Azapagic, A.; Howard, A.; Parfitt, A.; Tallis, B.; Duff, C.; Hadfield, C.; Pritchard, C.; Gillett, J.; Hackitt, J.; Seaman, M.; et al. The Sustainability Metrics. Available online: http://nbis.org/nbisresources/metrics/triple_bottom_line_indicators_process_industries.pdf (accessed on 12 October 2018).

266. Jørgensen, A.; Le Bocq, A.; Nazarkina, L.; Hauschild, M. Methodologies for social life cycle assessment. *Int. J. Life Cycle Assess.* **2008**, *13*, 96. [CrossRef]

267. Kober, T.; Panos, E.; Volkart, K. Energy system challenges of deep global CO_2 emissions reduction under the World Energy Council's scenario framework. In *Limiting Global Warming to Well Below 2 °C: Energy System Modelling and Policy Development*; Springer: Berlin, Germany, 2018; pp. 17–31.

268. Perloff, J.M. *Microeconomics*; Pearson Addison Wesley: Boston, MA, USA, 2004.

269. EIA-The National Energy Modeling System: An Overview. 2009. Available online: https://www.eia.gov/outlooks/aeo/nems/overview/index.html (accessed on 12 October 2018).

270. Monteiro, C.; Bessa, R.; Miranda, V.; Botterud, A.; Wang, J.; Conzelmann, G. *INESC Porto Wind Power Forecasting: State-of-the-Art 2009*; Argonne National Lab.: Lemont, IL, USA, 2009.

271. E3Mlab of ICCS/NTUA. The PRIMES Model. Available online: https://ec.europa.eu/energy/sites/ener/files/documents/sec_2011_1569_2_prime_model_0.pdf (accessed on 12 October 2018).

272. Berndes, G.; Hoogwijk, M.; Van den Broek, R. The contribution of biomass in the future global energy supply: A review of 17 studies. *Biomass Bioenergy* **2003**, *25*, 1–28. [CrossRef]

273. Remme, U.; Goldstein, G.A.; Schellmann, U.; Schlenzig, C. MESAP/TIMES—Advanced Decision Support for Energy and Environmental Planning. In *Operations Research Proceedings 2001*; Chamoni, P., Leisten, R., Martin, A., Minnemann, J., Stadtler, H., Eds.; Springer: Berlin/Heidelberg, Germany, 2002.

274. Loulou, R.; Goldstein, G.; Noble, K. Documentation for the MARKAL Family of Models. *Energy Technol. Syst. Anal. Program.* **2004**, 65–73. Available online: http://iea-etsap.org/MrklDoc-I_StdMARKAL.pdf (accessed on 12 October 2018).

275. Schrattenholzer, L. *The Energy Supply Model MESSAGE*; International Institute for Applied Systems Analysis: Laxenburg, Austria, 1981; Available online: http://pure.iiasa.ac.at/id/eprint/1542/ (accessed on 12 October 2018).

276. Krey, V.; Havlik, P.; Fricko, O.; Zilliacus, J.; Gidden, M.; Strubegger, M.; Kartasasmita, I.; Ermolieva, T.; Forsell, N.; Gusti, M. *Message-Globiom 1.0 Documentation*; International Institute for Applied Systems Analysis: Laxenburg, Austria, 2016; Available online: http://data.ene.iiasa.ac.at/message-globiom (accessed on 12 October 2018).

277. Energy PLAN. MARKAL/TIMES. Available online: https://www.energyplan.eu/othertools/national/markaltimes/ (accessed on 12 October 2018).

278. Available online: http://www.iiasa.ac.at/web/home/research/researchPrograms/Energy/IAMF.en.html (accessed on 13 November 2018).

279. Huppmann, D.; Gidden, M.; Fricko, O.; Kolp, P.; Orthofer, C.; Pimmer, M.; Vinca, A.; Mastrucci, A.; Riahi, K.; Krey, V. The MESSAGEix Integrated Assessment Model and the ix modeling platform (ixmp). *Environ. Model. Softw.* **2018**. Available online: http://pure.iiasa.ac.at/15157 (accessed on 12 October 2018).

280. Dowlatabadi, H. Integrated assessment models of climate change: An incomplete overview. *Energy Policy* **1995**, *23*, 289–296. [CrossRef]
281. Parker, P.; Letcher, R.; Jakeman, A.; Beck, M.B.; Harris, G.; Argent, R.M.; Hare, M.; Pahl-Wostl, C.; Voinov, A.; Janssen, M. Progress in integrated assessment and modelling1. *Environ. Model. Softw.* **2002**, *17*, 209–217. [CrossRef]
282. Stanton, E.A.; Ackerman, F.; Kartha, S. Inside the integrated assessment models: Four issues in climate economics. *Clim. Dev.* **2009**, *1*, 166–184. [CrossRef]
283. Rotmans, J. Methods for IA: The challenges and opportunities ahead. *Environ. Model. Assess.* **1998**, *3*, 155–179. [CrossRef]
284. Pfenninger, S. Energy scientists must show their workings. *Nat. News* **2017**, *542*, 393. [CrossRef] [PubMed]
285. Herbst, M.A.; Toro, F.A.; Reitze, F.; Eberhard, J. Bridging Macroeconomic and Bottom up Energy Models-the Case of Efficiency in Industry. Available online: https://www.eceee.org (accessed on 12 October 2018).
286. Dannenberg, A.; Mennel, T.; Moslener, U. What does Europe pay for clean energy?—Review of macroeconomic simulation studies. *Energy Policy* **2008**, *36*, 1318–1330. [CrossRef]
287. Miller, R.E.; Blair, P.D. *Input-Output Analysis: Foundations and Extensions*, 2nd ed.; Cambridge University Press: Cambridge, UK, 2009.
288. US Bureau of Economic Analysis. Input-Output Accounts Data. Available online: https://www.bea.gov/industry/input-output-accounts-data (accessed on 12 October 2018).
289. Bullard, C.W.; Herendeen, R.A. Energy impact of consumption decisions. *Proc. IEEE* **1975**, *63*, 484–493. [CrossRef]
290. Just, J.E. Impacts of new energy technology using generalized input-output analysis. *Comput. Oper. Res.* **1974**, *1*, 97–109. [CrossRef]
291. Matthews, H.S.; Small, M.J. Extending the boundaries of life-cycle assessment through environmental economic input-output models. *J. Ind. Ecol.* **2000**, *4*, 7–10. [CrossRef]
292. Hendrickson, C.; Horvath, A.; Joshi, S.; Lave, L. Peer reviewed: Economic input–output models for environmental life-cycle assessment. *Environ. Sci. Technol.* **1998**, *32*, 184A–191A. [CrossRef]
293. Hendrickson, C.T.; Lave, L.B.; Matthews, H.S. *Environmental Life Cycle Assessment of Goods and Services: An Input-Output Approach*; Resources for the Future: Washington, DC, USA, 2006.
294. U.S. Energy Information Administration (EIA). Annual Energy Outlook 2016. Available online: https://www.eia.gov/outlooks/archive/aeo16/appendixe.php (accessed on 8 October 2018).
295. Mitra, S.; Pinto, J.M.; Grossmann, I.E. Optimal multi-scale capacity planning for power-intensive continuous processes under time-sensitive electricity prices and demand uncertainty. Part I: Modeling. *Comput. Chem. Eng.* **2014**, *65*, 89–101. [CrossRef]
296. PJM Interconnection, 2011. Daily Day-Ahead Locational Marginal Pricing. Available online: http://www.pjm.com (accessed on 12 October 2018).
297. U.S. Energy Information Administration (EIA). Electric Power Annual 2016. Available online: https://www.eia.gov/electricity/annual/ (accessed on 12 October 2018).
298. Rolfsman, B. Combined heat-and-power plants and district heating in a deregulated electricity market. *Appl. Energy* **2004**, *78*, 37–52. [CrossRef]
299. Independent Electricity System Operator. Ontario Demand and Market Prices. 2012. Available online: http://www.ieso.ca/imoweb/siteShared/demandprice.asp?sid=ic (accessed on 12 October 2018).
300. Nease, J.; Adams, T.A., II. Application of rolling horizon optimization to an integrated solid-oxide fuel cell and compressed air energy storage plant for zero-emissions peaking power under uncertainty. *Comput. Chem. Eng.* **2014**, *68*, 203–219. [CrossRef]
301. Hendrickson, C.T.; Horvath, A.; Joshi, S.; Klausner, M.; Lave, L.B.; McMichael, F.C. Comparing two life cycle assessment approaches: A process model vs. economic input-output-based assessment. In Proceedings of the 1997 IEEE International Symposium on Electronics and the Environment. ISEE-1997, San Francisco, CA, USA, 5–7 May 1997; pp. 176–181.
302. Joshi, S. Product environmental life-cycle assessment using input-output techniques. *J. Ind. Ecol.* **1999**, *3*, 95–120. [CrossRef]
303. Adriaanse, A.; Bringezu, S.; Hammond, A.; Moriguchi, Y.; Rodenburg, E.; Rogich, D.; Schütz, H. *Resource Flows: The Material Basis of Industrial Economies*; World Resources Institute: Washington, DC, USA, 1997.

304. Spreng, D.T. *Net-Energy Analysis and the Energy Requirements of Energy Systems*; Praeger: Westport, CT, USA, 1988.
305. Ukidwe, N.U.; Bakshi, B.R. Thermodynamic Accounting of Ecosystem Contribution to Economic Sectors with Application to 1992 U.S. Economy. *Environ. Sci. Technol.* **2004**, *38*, 4810–4827. [CrossRef] [PubMed]
306. Kotas, T.J. *The Exergy Method of Thermal Plant Analysis*; Elsevier: Amsterdam, The Netherlands, 2013.
307. Szargut, J.; Morris, D.R.; Steward, F.R. *Exergy Analysis of Thermal, Chemical, and Metallurgical Processes*; Springer: Berlin, Germany, 1987.
308. Sciubba, E.; Wall, G. A brief commented history of exergy from the beginnings to 2004. *Int. J. Thermodyn.* **2007**, *10*, 1–26.
309. Szargut, J.; Morris, D.R. Cumulative exergy consumption and cumulative degree of perfection of chemical processes. *Int. J. Energy Res.* **1987**, *11*, 245–261. [CrossRef]
310. Lozano, M.A.; Valero, A. Theory of the exergetic cost. *Energy* **1993**, *18*, 939–960. [CrossRef]
311. Valero, A.; Lozano, M.A.; Serra, L.; Torres, C. Application of the exergetic cost theory to the CGAM problem. *Energy* **1994**, *19*, 365–381. [CrossRef]
312. Valero, A.; Serra, L.; Uche, J. Fundamentals of exergy cost accounting and thermoeconomics. Part I: Theory. *Energy Resour. Technol.* **2006**, *128*, 1–8. [CrossRef]
313. Ayres, R.U.; Ayres, L.W.; Martinás, K. Exergy, waste accounting, and life-cycle analysis. *Energy* **1998**, *23*, 355–363. [CrossRef]
314. Keshavarzian, S.; Rocco, M.V.; Gardumi, F.; Colombo, E. Practical approaches for applying thermoeconomic analysis to energy conversion systems: Benchmarking and comparative application. *Energy Convers. Manag.* **2017**, *150*, 532–544. [CrossRef]
315. Hau, J.L.; Bakshi, B.R. Expanding Exergy Analysis to Account for Ecosystem Products and Services. *Environ. Sci. Technol.* **2004**, *38*, 3768–3777. [CrossRef] [PubMed]
316. Keshavarzian, S.; Gardumi, F.; Rocco, M.V.; Colombo, E. Off-Design Modeling of Natural Gas Combined Cycle Power Plants: An Order Reduction by Means of Thermoeconomic Input–Output Analysis. *Entropy* **2016**, *18*, 71. [CrossRef]
317. Keshavarzian, S.; Rocco, M.V.; Colombo, E. Thermoeconomic diagnosis and malfunction decomposition: Methodology improvement of the Thermoeconomic Input-Output Analysis (TIOA). *Energy Convers. Manag.* **2018**, *157*, 644–655. [CrossRef]
318. Adams, T.A., II; Barton, P.I. High-efficiency power production from natural gas with carbon capture. *J. Power Sources* **2010**, *195*, 1971–1983. [CrossRef]
319. Yue, D.; Pandya, S.; You, F. Integrating Hybrid Life Cycle Assessment with Multiobjective Optimization: A Modeling Framework. *Environ. Sci. Technol.* **2016**, *50*, 1501–1509. [CrossRef] [PubMed]
320. Dresselhaus, M.S.; Thomas, I.L. Alternative energy technologies. *Nature* **2001**, *414*, 332–337. [CrossRef] [PubMed]
321. Bolat, P.; Thiel, C. Hydrogen supply chain architecture for bottom-up energy systems models. Part 1: Developing pathways. *Int. J. Hydrogen Energy* **2014**, *39*, 8881–8897. [CrossRef]
322. Bolat, P.; Thiel, C. Hydrogen supply chain architecture for bottom-up energy systems models. Part 2: Techno-economic inputs for hydrogen production pathways. *Int. J. Hydrogen Energy* **2014**, *39*, 8898–8925. [CrossRef]
323. Sgobbi, A.; Nijs, W.; De Miglio, R.; Chiodi, A.; Gargiulo, M.; Thiel, C. How far away is hydrogen? Its role in the medium and long-term decarbonisation of the European energy system. *Int. J. Hydrogen Energy* **2016**, *41*, 19–35. [CrossRef]
324. Kahouli-Brahmi, S. Technological learning in energy–environment–economy modelling: A survey. *Energy Policy* **2008**, *36*, 138–162. [CrossRef]
325. Rubin, E.S.; Azevedo, I.M.L.; Jaramillo, P.; Yeh, S. A review of learning rates for electricity supply technologies. *Energy Policy* **2015**, *86*, 198–218. [CrossRef]
326. Van den Broek, M.; Berghout, N.; Rubin, E.S. The potential of renewables versus natural gas with CO_2 capture and storage for power generation under CO_2 constraints. *Renew. Sustain. Energy Rev.* **2015**, *49*, 1296–1322. [CrossRef]
327. Van den Broek, M.; Hoefnagels, R.; Rubin, E.; Turkenburg, W.; Faaij, A. Effects of technological learning on future cost and performance of power plants with CO_2 capture. *Prog. Energy Combust. Sci.* **2009**, *35*, 457–480. [CrossRef]

328. Creutzig, F.; Agoston, P.; Goldschmidt, J.C.; Luderer, G.; Nemet, G.; Pietzcker, R.C. The underestimated potential of solar energy to mitigate climate change. *Nat. Energy* **2017**, *2*, 17140. [CrossRef]

329. Huang, W.; Chen, W.; Anandarajah, G. The role of technology diffusion in a decarbonizing world to limit global warming to well below 2 °C: An assessment with application of Global TIMES model. *Appl. Energy* **2017**, *208*, 291–301. [CrossRef]

330. Patel, S. Duke Hit Hard by Exorbitant O&M Costs at Edwardsport IGCC Facility. *Power Magzine*. 27 September 2018. Available online: https://www.powermag.com/duke-hit-hard-by-exorbitant-om-costs-at-edwardsport-igcc-facility/?pagenum=3 (accessed on 12 October 2018).

331. Wagman, D. Three Factors that Doomed kemper County IGCC. *IEEE Spectrum*. 30 June 2017. Available online: https://spectrum.ieee.org/energywise/energy/fossil-fuels/the-three-factors-that-doomed-kemper-county-igcc (accessed on 12 October 2018).

332. Reiner, D.M. Learning through a portfolio of carbon capture and storage demonstration projects. *Nat. Energy* **2016**, *1*, 15011. [CrossRef]

333. Adams, T.A., II; Hoseinzade, L.; Madabhushi, P.B.; Okeke, I.J. Comparison of CO_2 Capture Approaches for Fossil-Based Power Generation: Review and Meta-Study. *Processes* **2017**, *5*, 44. [CrossRef]

334. Voll, A.; Sorda, G.; Optehostert, F.; Madlener, R.; Marquardt, W. Integration of market dynamics into the design of biofuel processes. *Comput. Aided Chem. Eng.* **2012**, *31*, 850–854.

processes

MDPI

Article

Offshore Power Plants Integrating a Wind Farm: Design Optimisation and Techno-Economic Assessment Based on Surrogate Modelling

Luca Riboldi * and Lars O. Nord

Department of Energy and Process Engineering, Norwegian University of Science and Technology—NTNU, NO-7491 Trondheim, Norway; lars.nord@ntnu.no
* Correspondence: luca.riboldi@ntnu.no; Tel.: +47-450-714-63

Received: 15 October 2018; Accepted: 30 November 2018; Published: 4 December 2018

Abstract: The attempt to reduce the environmental impact of the petroleum sector has been the driver for researching energy efficient solutions to supply energy offshore. An attractive option is to develop innovative energy systems including renewable and conventional sources. The paper investigates the possibility to integrate a wind farm into an offshore combined cycle power plant. The design of such an energy system is a complex task as many, possibly conflicting, requirements have to be satisfied. The large variability of operating conditions due to the intermittent nature of wind and to the different stages of exploitation of an oil field makes it challenging to identify the optimal parameters of the combined cycle and the optimal size of the wind farm. To deal with the issue, an optimisation procedure was developed that was able to consider the performance of the system at a number of relevant off-design conditions in the definition of the optimal design. A surrogate modelling technique was applied in order to reduce the computational effort that would otherwise make the optimisation process unfeasible. The developed method was applied to a case study and the resulting optimal designs were assessed and compared to other concepts, with or without wind power integration. The proposed offshore power plant returned the best environmental performance, as it was able to significantly cut the total carbon dioxide (CO_2) emissions in comparison to all the other concepts evaluated. The economic analysis showed the difficulty to repay the additional investment for a wind farm and the necessity of favourable conditions, in terms of gas and carbon dioxide (CO_2) prices.

Keywords: oil and gas; offshore wind; combined cycle; hybrid system; kriging; multi-objective optimisation

1. Introduction

Offshore wind energy is an attractive technology to reduce the local emissions of offshore oil and gas extraction. Environmental impact of offshore installations is an issue which is drawing an increased attention [1], especially for a country like Norway where a large share of the total greenhouse gas emissions is ascribable to the petroleum sector [2]. Several concepts have been assessed to ensure an efficient offshore energy supply, considering various options for the power plant [3] but also measures on the processing plant [4]. The utilisation of a combined cycle has been comprehensively investigated in the literature in terms of working fluids (steam, hydrocarbons and air evaluated in Reference [5], carbon dioxide in Reference [6]), design and off-design performance [7] and optimisation of the design (steam Rankine cycle in Reference [8], organic Rankine cycle in Reference [9]). The potential of combined cycles in a cogeneration mode has also been studied [10]. More recently, electrification of the offshore facilities raised interest. The analyses carried out so far showed that larger cuts in the lifetime CO_2 emissions could potentially be achieved in comparison to local power generation solutions [11].

The extent of these cuts was strongly dependent on the method to account for the emissions associated with power from shore [3]. Further, it was shown that the economic competitiveness of electrification could be disputable and would need strong support in terms of energy policies [12]. The uncertainties with offshore electrification were the drivers for investigating alternatives. The utilisation of offshore wind power does not require the laying of long subsea cable to ensure the connection to the onshore grid and the wind power can be accounted for as emission free (or close to emissions free from a life cycle assessment standpoint).

Norway displays large potentials related to offshore wind. Offshore applications can take advantage of excellent characteristics of the wind resource in comparison to onshore sites, for instance higher average wind speed, lower turbulence intensity and wind shear [13]. Offshore renewable resources guarantee an extremely low environmental impact during operation, with the main source of pollution being, in a life cycle assessment perspective, the manufacturing process [14]. The possibility of operating an offshore wind farm in parallel with gas turbines has been previously discussed [15], resulting in a 20 MW wind farm being integrated with a plant whose power consumption varied between 20 MW and 35 MW. A capacity utilisation factor of 43% for the wind farm was obtained, with an annual CO_2 emissions reduction of 53.8 kt (approximately 40%) compared to the reference case based on the utilisation of two gas turbines. This included an operating strategy where one of the gas turbines was allowed to shut-down according to specified criteria. Further, no considerations were made with regard to the process heat to be supplied to the plant. The dynamic simulations were used to establish the maximum amount of wind power integration, which resulted to be between 20 MW and 25 MW [13]. An additional step towards efficient offshore energy supply involved the combination of combined cycles and wind farms. This concept has been investigated in Reference [16], where a wind farm of 10 MW was integrated to three combined cycle units constituted of a gas turbine (rated for 16.5 MW) and a 4.5 MW organic Rankine cycle (ORC) module. The performance of the combined cycle units was compared to that of simple cycle gas turbine units. Even though a couple of co-generative solutions were discussed, the necessity to supply heat in parallel to power was not simulated in detail. In a follow-up paper [17], an economic analysis was proposed, comparing the economic performance of the wind farm coupled with three combined cycles to that of the wind farm coupled with three gas turbines. The results showed that the first concept (wind power and combined cycle) becomes more convenient when fuel cost increases or when the CO_2 tax increase. A comparison between the integration of wind power and an independent combined cycle was not provided. The papers referenced in the literature review investigated the coupling of arbitrary wind power capacities into local power generation units. No assessments have been performed to establish the optimal wind capacity to be installed. Small installed wind capacities could limit the environmental benefits associated with the exploitation of wind power. On the other hand, large installed wind capacities, apart endangering the grid stability and the economic feasibility, could result in dissipation of large fractions of wind power in periods of low power demand. To add up to the complexity, large wind farms could lead to operation of the combined cycle at very low part-loads with low efficiency. Moreover, the power plant needs to be able, at all operating conditions, to supply heat to the process, an issue which is often neglected in the literature.

The novel contribution of this paper is twofold. First, it presents an advanced procedure to identify the optimal design of an offshore power plant integrating a wind farm, taking into account constraints specific to the offshore environment. Second, it provides a comprehensive evaluation of its techno-economic feasibility by considering the expected working conditions characterising its lifetime operation.

The developed optimisation procedure identifies the optimal design of the offshore power plant, in terms of optimal integration of wind power and optimal characteristics of the combined cycle to work in parallel to the wind farm. A multi-objective approach is adopted to define the optima, where the three objective functions are: (i) the cumulative CO_2 emissions; (ii) the total cost to supply energy to the plant; and (iii) the weight of the onsite power cycle. Another key characteristic of the optimisation procedure is that it measures the performance in all the significant operating conditions at which the power plant is expected to operate. The importance of considering several relevant operating conditions in the definition of a design was demonstrated in a previous paper [18], in which the novel design succeeded in decreasing the lifetime CO_2 emissions by 17.4 kt with respect to a standard design. Because of the complexity of such an optimisation process, requiring a very large number of simulations of the system, the model needs to be simple enough for reasonable computational time. On the other hand, a good level of accuracy has to be guaranteed in order to obtain reliable results. The necessity of finding a balance between these contrasting requirements is typical for such optimisation problems [19]. Surrogate modelling techniques could serve the purpose to accomplish this [20] and were applied to the current analysis. The optimisation procedure described was applied to a case study and a Pareto front of optimal solutions was obtained. The results were analysed and a specific design pinpointed to be further investigated. A techno-economic assessment was performed with the objective to provide a comprehensive evaluation on the effectiveness of the wind power integration in comparison to more standard concepts. The assessment covered the long time span of expected lifetime of an offshore installation and thus future scenarios on the development of economic parameters and energy policies needed to be considered.

The paper is structured as follows: the methods developed are first presented in Section 2. The application of those methods to a case study is described in Section 3 and the related results are presented in Section 4. Section 5 analyses and discusses the results. Conclusive remarks are given in Section 6.

2. Methods

In this section, the methods developed are described so to provide the basis for understanding the results obtained.

2.1. Process Modelling of the Offshore Power Plant

The offshore power plant in the study consists of a combined cycle integrating a wind farm (see Figure 1). The wind power contribution is modelled by considering relevant measurements of wind speed, which are converted into power outputs through an ideal wind turbine power curve based on existing technologies. A process model of the combined cycle is developed in Thermoflex (Thermoflow Inc., version 26.0, Fayville, MA, USA, 2016) [21]. Thermoflex is a program specifically designed for design and off-design simulation of thermal systems. It is also able to provide estimations of weights and costs of the major equipment through the utility PEACE (Plant Engineering And Construction Estimator). The combined cycle supplies heat and power to the processing block of an offshore installation. The topping unit is an aero-derivative gas turbines (GT) typically used in offshore platforms according to their reliability, flexibility and high power-to-weight ratio. The operation of the GTs is simulated through data-defined models, based on the tabulated data from actual installations and manufacturers. The models cover the entire operating range of a GT (10–100%). The thermal energy of the GT exhaust gas is first exploited in a waste heat recovery unit (WHRU) to meet the heat demand of the plant. The WHRU is modelled as a counter-flow vertical finned tube heat exchanger. The gas stream leaving the WHRU is directed to a once-through heat recovery steam generator (OTSG), where the residual thermal energy is used to produce superheated steam. The once-through technology is selected in accordance with the indications provided by Nord and Bolland [22]. The superheated steam leaving the OTSG is expanded in a steam turbine. The steam is condensed in a deaerating condenser, modelled as a shell-and-tube heat exchanger. The cooling medium is sea water at constant temperature. The combined cycle is set to integrate the power contribution from the wind farm. Given a specific plant power demand to supply and a variable wind power output (depending on the wind speed), a simple operating strategy was assumed where the combined cycle always provides back-up power to deal with the irregular contribution of wind and its load is regulated accordingly. The power contribution of the combined cycle is modified through changes in the GT load. The control strategy of the GT is a combination of variable guide vanes (VGV) control and turbine inlet temperature control (TIT). A sliding pressure control mode applies to the steam cycle. A steam turbine bypass control valve ensures not to overpass the maximum established live-steam pressure. The live-steam temperature is controlled by the feedwater flow to the OTSG, as suggested for heat recovery steam generators of the once-through type [23]. Limitations on the minimum load at which the combined cycle can be operated are considered. The limitations include minimum combined cycle loads to ensure the ability to supply heat to the processing plant and minimum GT loads to meet environmental obligations (NO_x and CO emissions).

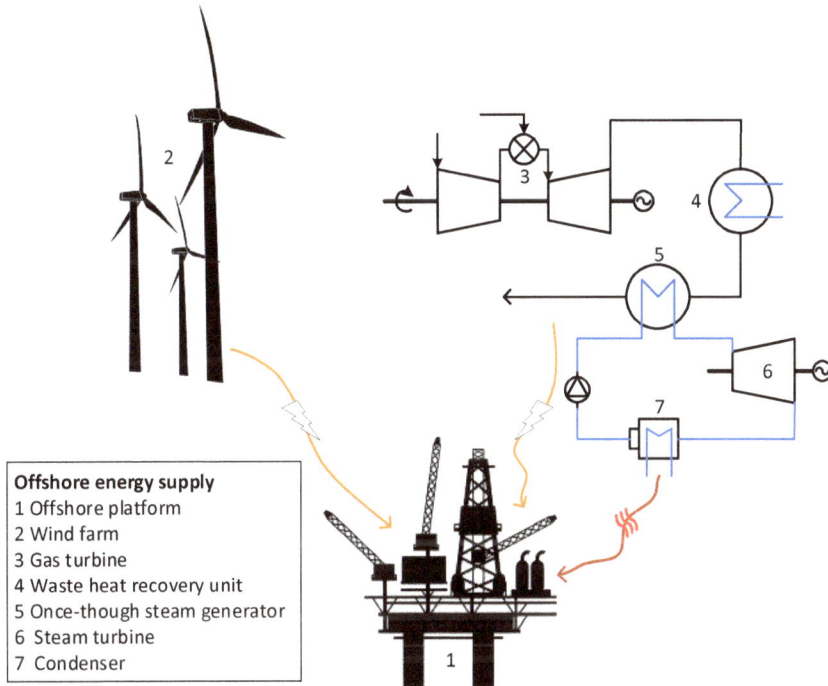

Figure 1. Layout of the offshore energy system integrating a wind farm and a combined cycle to supply energy to the processing block of a platform.

2.2. Surrogate Model Based on Kriging and Off-Design Correlation

The surrogate model has to predict the behaviour of the power cycle (defined as a set of dependent variables—the output parameters) at different operating conditions (defined as a set of independent variables—the input parameters). The high-fidelity model (that developed in Thermoflex) is simulated at a set of input conditions and its results recorded. The combination of inputs and outputs from the rigorous simulations were the sampling data set used to train the surrogate model. The input variables are those having the largest influence on the power cycle performance and their bounds are selected to represent its expected operation. For an effective mapping of the entire space where the model needs to operate, a combination of deterministic (Box-Behnken and central composite—113 points) and randomized (Latin hypercube—1000 points) sampling was performed. The output variables, whose variation at different input conditions is monitored, are those able to fully describe the operation of the power cycle, thus those the surrogated model has to be able to accurately estimate. The building of the surrogate model is performed through the Kriging technique, implemented in the Matlab ooDACE toolbox (Matlab R2015a, The MathWorks Inc., Natick, MA, USA, 2016) [24]. A short introduction on

Kriging as surrogate modelling technique is provided in Appendix A. The defined Kriging model is then validated at a set of independent testing conditions.

The validated Kriging model is used to characterize a design. Once the design is fixed, the operation at different off-design conditions has to be simulated. The performance of various components (e.g., heat exchangers, turbines, pumps) is affected by changes in operating conditions. The resulting deviation from the design performance is estimated through simplified off-design correlations. The complete set of equations used is reported in Appendix B. The off-design simulation of the GTs is based on the data-defined models of the engine selected. The heat exchanging components included in the cycle (i.e., the WHRU and the OTSG) are modelled through the relation from Incropera et al. [25] that evaluates the heat transfer coefficient at off-design conditions. The dominant heat transfer resistance is assumed to be that of the hot gas side for the WHRU and for the economizer and evaporator sections of the OTSG [26]. This simplification allows neglecting the conductive term and the heat transfer resistance of the cold water side. Conversely, in the superheater section of the OTSG the water side is assumed to dominate the heat transfer process [5], while the conductive term and the heat transfer resistance of the hot gas side are neglected. The heat transfer coefficient is not estimated for the condenser as a simplified representation is used. The condenser is modelled as a fixed pressure component to provide the cooling duty to condense the expanded steam for all operating conditions. This simplification is supported by the large availability of sea water as cooling medium. The nonlinear dependence between the inlet conditions and extraction pressure in the steam turbine is modelled in accordance with the Stodola's cone law. The performance changes of the steam turbine at off-design are evaluated through the relation proposed by Schobeiri [27]. A correction in the generator efficiency at off-design is considered, based on the formula proposed by Haglind and Elmegaard [28]. The pumps performance at off-design are computed according to the method described by Veres [29]. The pressure drops (Δp) are accounted for by correlations assuming a quadratic dependence of the mass flow rate.

2.3. Design Optimisation Procedure Considering Off-Design Performance

The reduced computational effort associated with the utilisation of a surrogate model allows to develop an advanced optimisation procedure. The flowchart in Figure 2 shows its simplified representation.

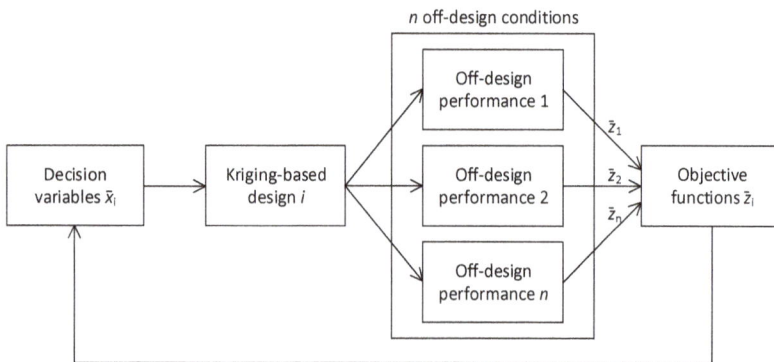

Figure 2. Flowchart of the optimisation procedure developed.

A multi-objective constrained optimisation problem has to be defined. An array of decision variables is first established:

$$\bar{x} = [GT\ load, p_{steam}, T_{steam}, \Delta T_{OTSG}, p_{cond}, \Delta T_{CW}, wind_{PW}] \tag{1}$$

The decision variables are the same input variables used in the definition of the surrogate model. The same bounds also apply, ensuring that the optimisation algorithm only search for an optimal solution in the space where the surrogate model is able to provide reliable outputs. The $wind_{PW}$ is the wind power capacity installed. In this study, the wind integration is assumed as always possible, regardless the size of the wind farm. A design i is defined by the Kriging model after assigning a value to each of the decision variables. The performance of the specific design is then evaluated at n off-design operating conditions. The set of operating conditions is selected to represent the relevant modes of plant operation during its lifetime. The off-design performances are obtained starting from the information provided by the Kriging model and applying the off-design correlations. The values of selected objective functions are calculated, so to define the array $\bar{z}$ of objective functions to minimize:

$$\bar{z}(\bar{x}) = [CO_2^*, W^*, cost^*] \tag{2}$$

$$CO_2^*(\bar{x}) = \sum_{y=1}^{years} m_{CO2,y} \tag{3}$$

$$cost^*(\bar{x}) = TCR + \sum_{y=1}^{years} DCF_y \tag{4}$$

$$W^*(\bar{x}) = \sum_{components} W_{component} \tag{5}$$

Three domains are considered when defining the objective functions: (i) the cumulative CO_2 emissions, (ii) the total cost to supply energy to the plant and (iii) the weight of the onsite power cycle. The first two objectives are typical indicators of the sustainability of a project. The third objective considered was believed of significance for offshore applications as the very limited deployment of offshore combined cycles is likely due to issues with their sizes and weights.

The environmental performance is measured as the total amount of CO_2 emissions (CO_2^*). It is calculated as the summation of the annual CO_2 emissions ($m_{CO2,y}$) over the plant's lifetime. Every year is described by a power demand that is to be covered by a combination of wind power and combined cycle power. Given the irregularity of the wind power contribution, the year will be further characterized by several off-design conditions at which the combined cycle has to be operated to meet the power demand. A specific design will perform differently at those various off-design conditions, resulting in a correspondent number of mass flow rates of emitted CO_2 ($m_{CO2,i}$). The annual CO_2 emission is then the summation of those emissions ($m_{CO2,i}$) weighed over the equivalent number of hours (h_{eq}) at which an off-design condition is expected to apply to one year:

$$m_{CO2,y} = \sum_{i=1}^{NOC} m_{CO2,i} h_{eq,i} \tag{6}$$

The economic performance is measured as the total cost to supply energy to the plant ($cost^*$). The calculations are based on the principles of the net present value (NPV) method and the economic metric can be seen as the NPV of the offshore energy system at the end of the lifetime. This term is thereby composed by two parts: the total capital requirement (TCR) and the summation of the annual discounted cash flows (DCF_y). The TCR is assumed to be made before the installation starts operation. The total investment for the power cycle (TCR_{cc}) is calculated in accordance with [30], as the summation of direct and indirect costs, estimated by a factor method. Table 1 shows a breakdown of the TCR_{cc} together with the factors used. The purchased-equipment cost (PEC) is an output of the surrogate model. The factors are selected based on the indications provided by Bejan et al. [30], applying a rather high contingency factor (25% of the total cost) and are in line with another paper that performed an estimation of the TCR to install an offshore combined cycle [9]. With regard to the wind farm, the estimation of the TCR_{wind} (4503 \$/kW), including direct and indirect costs, is based on the information retrieved from the European Commission's report ETRI 2014 [31].

Table 1. Breakdown of the costs included in the total capital requirement [9] (Reprinted with permission from Pierobon, L; et al., Multi-objective optimization of organic Rankine cycles for waste heat recovery: Application in an offshore platform, published by Elsevier, 2013).

Direct Cost (DC)	Range from [30]	Factor Selected
Onsite cost		
Purchased-equipment costs (PEC)		
Purchased-equipment installation	20–90% of PEC	45%
Piping	10–70% of PEC	35%
Instrumentation and controls	6–40% of PEC	20%
Electrical equipment and materials	10–15% of PEC	11%
Offsite cost		
Civil, structural and architectural work	15–90% of PEC	30%
Service facilities	30–100% of PEC	65%
Indirect Cost (IC)		
Engineering and supervision	6–15% of DC	8%
Construction costs and constructors profit	15% of DC	15%
Contingencies	8–25% of total cost	25%

The annual DCF_y is calculated as:

$$DCF_y = \frac{CF_y}{(1+r)^y} \tag{7}$$

where CF is the cash flow, y is the year when the cash flow occurs and r is the discount rate (set to 7%). In the analysis, only negative cash flows are considered, thus adding up to the TCR. The annual cash flows associated with the onsite gas consumption (CF^{gas}) and with the CO_2 taxation (CF^{CO2}) are calculated as weighed summation over the off-design conditions characterising a specific year:

$$CF_y^{gas} = \sum_{i=1}^{NOC} \dot{m}_{gas,i} LHV_{gas} c_{gas} h_{eq,i} \tag{8}$$

$$CF_y^{CO2} = \sum_{i=1}^{NOC} \dot{m}_{CO2,i} c_{CO2} h_{eq,i} \tag{9}$$

where $\dot{m}_{gas}$ is the mass flow rate of natural gas used as fuel, LHV_{gas} is the lower heating value of the natural gas, c_{gas} is the gas price, $\dot{m}_{CO2}$ is the mass flow rate of the emitted CO_2, c_{CO2} is the CO_2 price and h_{eq} are the equivalent operating hours per year. An estimation of the gas price and of the CO_2 price needed for each year of plant's operation. Hence, a scenario for the future developments of those economic parameters has to be used. For the gas price, the new policies scenario developed by the International Energy Agency (IEA) is considered and the related annual gas price used [32]. The new policies scenario reflects the way the governments see their energy sectors developing in the coming decades. For the CO_2 price, the Norwegian situation is evaluated. The petroleum sector in Norway is subjected to a rather high CO_2 tax (0.12 \$/Sm^3 in 2016 [33]), while contemporary takes part to the European Union Emissions Trading System (EU ETS). In the recent years, the trend had been to adjust the CO_2 taxation in order to make up for the increase in the costs associated with the ETS so to keep the overall CO_2 price approximately constant. Assuming that the same strategy will apply in the years to come, the level of CO_2 price is kept constant and equal to 46 \$/t.

The last objective function to be estimated is the total weight of the bottoming cycle (*W**). It is calculated as the summation of the weights of the cycle components (*W*component), provided by the Kriging model.

Once evaluated the three objective functions for a given design, a new iteration is commenced. A Pareto front of optimal solutions is ultimately obtained. A genetic algorithm (GA), from the MATLAB Global Optimisation Toolbox [34], is implemented to solve the optimisation problem.

3. Case Study

The characteristics of the case study are highlighted. The offshore installations and their related power demands are presented, followed by the offshore power plants, with a combined cycle and a wind farm.

3.1. Offshore Installations

The joint development of two offshore installations in the North Sea was considered. The two offshore installations are named after the related oil fields, Edvard Grieg and Ivar Aasen. Both fields already began production and have an expected 20 year lifetime. The Edvard Grieg platform is currently equipped with two gas turbines in order to meet heat and power requirements of both fields. A dedicated alternating current (AC) cable from the Edvard Grieg platform will cover Ivar Aasen power demand, while oil and gas from Ivar Aasen will be channelled to the Edvard Grieg platform for processing and export. A more detailed description of the development scheme of the two installations and of the topside processes, responsible of the power and heat demand, was provided in a previous publication [3]. Annual power requirements were considered, based on the information retrieved from the relevant field development reports for Edvard Grieg [35] and Ivar Aasen [36]. These official documents provide an estimation of power and heat requirements to operate the facility based on the estimated production profiles. High quality forecasting techniques are applied as the development plan is the foundation for decision in all phases of the petroleum activity. However, a certain degree of uncertainty is to be expected as oil & gas fields are extremely complex systems. Figure 3 gives an overview of the obtained power demand profile throughout the years. The variability of the power demand is common for this type of plants and is due to the changing oil production rates during the different stage of exploitation of an oil field. To simplify the implementation of the optimisation procedure, the power demand profile was divided into 4 groups with reference to different stages of the plant's lifetime:

1. Early life—29.7 MW (year 2016)
2. Middle life—35.5 MW (years 2017 to 2018 and years 2021 to 2023)
3. Peak—39.9 MW (years 2019 and 2020)
4. Tail years—33.0 MW (years 2024 to 2034)

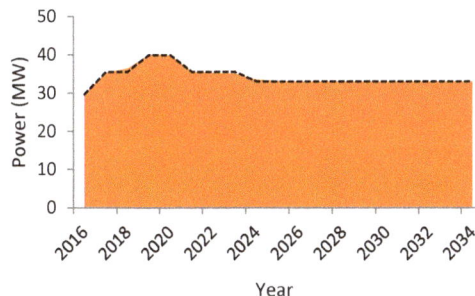

Figure 3. Lifetime power demand of the installation. The dashed line represents an approximation of the power demand profile.

The dashed line in Figure 3 shows how the power demand profile was approximated. The power demand values considered were the average of the power demands belonging to the group. Although the heat demand would be subjected to a similar trend during the plant's lifetime, in this paper such variability was not considered. The heat demand was set to be 11 MW and it was retained constant at different operating conditions. Including the variability of the heat demand would add a degree of complexity in the design optimisation and will be considered for further developments of the work.

3.2. Combined Cycle

The topping unit was set to be a General Electric (GE) LM2500+G4 or a GE LM6000 PF, two aero-derivative gas turbines (GT). The former is a smaller size machine (rated power 32.2 MW) in comparison to the latter (rated power 41.9 MW). In the remaining of the paper, the GE LM2500+G4 and the GE LM6000 PF will be referred as GT A and GT B, respectively. Thermoflow indicates that the maximum model errors for the two engines are lower than 0.5% for the exhaust mass flow rate, the power output and the heat rate and lower than 2.8 °C for the exhaust temperature (test range for ambient temperature: −18 to 49 °C). Such level of uncertainty could be also assumed to apply to this study. The performance of GT A was checked against real operational data, showing good agreement. The high-fidelity model of the combined cycle is based on that developed for a previous publication [3], to which reference should be made for a better insight. The model of the combined cycle was validated against the paper from Nord et al. [8], which in turn was validated against the 2012 Gas Turbine World Handbook. Table 2 shows the input parameters to the surrogate model. The lower and upper bounds were selected to result in feasible operation of the combined cycle by considering technical and operational limitations. The higher bound of the superheated steam temperature (T_{steam}) was constrained by the GT outlet temperature and, in fact, some differences can be noted when a different GT is used. The lower bound was set to ensure a reasonably high steam quality at the steam turbine outlet. The steam evaporation pressure (p_{steam}) and the condenser pressure (p_{cond}) were varied within a range which was sufficiently large to not exclude optimal solutions while guaranteeing feasible ones. The lower bound of p_{cond} was also selected in accordance with typical limitations of the vacuum and sealing systems. The upper load of the GTs was set at 0.95 in order to maintain a safety margin in case of sudden increase of plant load. The lower bound was limited to ensure the capability of the cycle to meet the process heat demand in any instance. The bounds to the pinch point differences (ΔT_{OTSG} and ΔT_{cw}) were defined in accordance with the practical limitation discussed by Nord et al. [8].

Table 2. List of the independent variables of the Kriging model, with the lower and upper bounds considered. The same bounds apply for the optimisation problem (OTSG: once-through heat recovery steam generator; GT: gas turbines).

Input Parameters		GT A		GT B	
Description	Symbol	Lower Bound	Upper Bound	Lower Bound	Upper Bound
Gas turbine load	GT load	0.80	0.95	0.60	0.95
Steam evaporation pressure (bar)	p_{steam}	15	40	15	40
Superheated steam temperature (°C)	T_{steam}	300	410	280	370
Pinch point temperature difference in the OTSG (°C)	ΔT_{OTSG}	10	30	10	30
Condenser pressure (bar)	p_{cond}	0.03	0.10	0.03	0.12
Condenser cooling water temperature difference (°C)	ΔT_{cw}	3	10	3	10

The list of output parameters that the surrogate model was trained to estimate can be found in Table A1 in Appendix C. In Appendices C and D, the validation of the Kriging model and of the off-design correlations are discussed, respectively. The decision variables, related to the combined cycle, used for the optimisation were the same as the input parameters to the surrogate model and were let range within the same bounds indicated in Table 2.

3.3. Wind Power

The data set of wind speeds considered refers to a location in the North Sea where the Norwegian Meteorological Institute performed measurements with a 20 min resolution [15]. The data set was further integrated by generating intermediate wind speeds instances using a distribution function for 10 min variations based on similar wind speed series from Norway. The resolution was further increased to 1 min by linear interpolation. The same data set was used in another publication to assess the integration of wind power to a generic offshore oil platform [13]. The wind speed data were considered of sufficient high quality for the purpose of this study. There is a degree of uncertainty related to the measurements on oil platforms. The platforms could produce disturbances of the wind field, potentially leading to an overestimation of the offshore wind energy potential [37]. However, undisturbed measurements are not available in the Norwegian sector for direct comparison. According to the data set, an average wind speed of 9.8 m/s at turbine hub height was calculated, in line with the values expected for offshore wind farms [13]. The wind speeds were converted to wind power outputs by means of an ideal wind turbine power curve, shown in Figure 4. The power curve was based on a three-bladed floating turbine concept [38]. The duration curve shown in Figure 5 displays the wind power made available throughout one year.

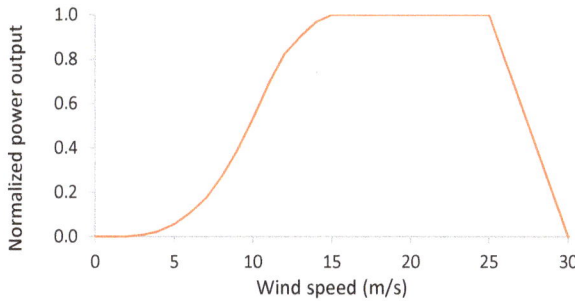

Figure 4. Ideal wind turbine power curve.

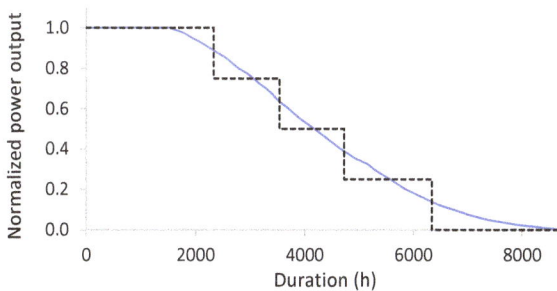

Figure 5. Duration curve for the wind power. The dashed line represents an approximation of the wind power available throughout one year.

The annual normalized power contribution from the wind power was then discretized. The instances of normalized wind power outputs were grouped into five intervals (0%, 25%, 50%, 75% and 100% of the rated capacity), to which a value of equivalent hours in a year was associated depending on the number of annual wind speed instances falling into the specific power output group. The dashed line in Figure 5 shows how the duration curve was approximated accordingly. The wind power capacity installed (*wind*$_{PW}$)—the remaining decision variable of the optimisation problem—was let range between 0 and 30 MW, with 5 MW step intervals. The annual contribution of wind power was then fully defined, further influencing the working conditions at which the combined cycle has to be operated to provide back-up power. Given the discretization of the lifetime power demand (4 instances as shown in Figure 3) and of the annual contribution of the wind power (5 instances as shown in Figure 5), for a selected wind farm size, a set of 20 off-design conditions at which the combined cycle has to operate was automatically defined. An example is given in Table 3, considering a wind power capacity installed of 10 MW.

Table 3. Off-design conditions to be tested by the optimisation procedure given a wind power capacity installed of 10 MW.

Power Demand Offshore	Wind Power	Combined Cycle Power
P_O	P_W	$P_{CC} = P_O - P_W$
MW	MW	MW
29.7 (year 2016)	10.0	19.7
	7.5	22.2
	5.0	24.7
	2.5	27.2
	0.0	29.7
35.5 (years 2017 to 2018 and years 2021 to 2023)	10.0	25.5
	7.5	28.0
	5.0	30.5
	2.5	33.0
	0.0	35.5
39.9 (years 2019 and 2020)	10.0	29.9
	7.5	32.4
	5.0	34.9
	2.5	37.4
	0.0	39.9
33.0 (years 2024 and 2034)	10.0	23.0
	7.5	25.5
	5.0	28.0
	2.5	30.5
	0.0	33.0

4. Results

Before reporting the simulation results, a brief premise on the advantage in terms of computational time for the use of Kriging: the computer used in this work has an Intel Core processor of 2.60 GHz and 16.0 GB of random-access memory (RAM). The Kriging model performed a simulation in, on average, 0.07 s, fully characterising a design. The simulation of the same design with the commercial software Thermoflex took on average 20.35 s. A significant computational saving could be realized with the use of the Kriging model, reducing the run time of a factor 285. This reduction in computational time was fundamental in order to be able to assess a very large number of designs at a number of operating conditions, like in the optimisation procedure implemented in the study.

A population size of 350 was established for the genetic algorithm (GA) and the maximum number of generations allowed was set to 25. The solver was stopped before if the spread of Pareto solutions, a measure of the movement of the Pareto frontier, was less than the function tolerance (10^{-3}) over a number of stall generations (5). The Pareto fronts obtained can be observed in Figure 6, where the decision map showing trade-offs between total cost and CO_2 emissions is represented. The third objective functions (i.e., the weight of the bottoming cycle) is shown through shades of colours: the darker the colour, the heavier the design. The two Pareto fronts refer to the cycles based on the GT A (blue) and on the GT B (green).

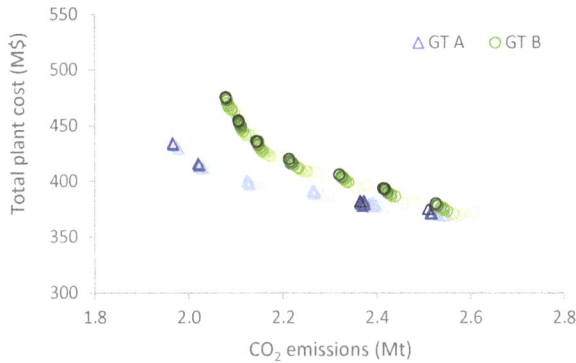

Figure 6. Decision map of the Pareto front showing trade-offs between total cost and CO_2 emissions. The shades of colour represent different levels of weights of the optimal designs: the darker the colour, the heavier the design.

The stepwise trend of the Pareto fronts is because of how the wind power capacity ($wind_{PW}$) was considered in the design optimisation. $wind_{PW}$ was allowed to take values multiple of 5 MW, within the bounds assigned (i.e., 0 and 30 MW). The steps in the Pareto fronts correspond to the various levels of $wind_{PW}$ and highlight the strong influence that the size of the wind farm had on the environmental and economic performance. Within each of these "Pareto steps" the heavier designs are generally those with the lower CO_2 emissions but higher costs. This makes sense as the heavier combined cycles are likely the most efficient ones but the related increased complexity translates in higher investment costs. By looking at the general trend, it can be noted that increasing $wind_{PW}$ meant worse economics compared to a lower value of $wind_{PW}$. Accordingly, the designs returning the best economic performance were those not integrating any wind capacity. In other words, the reduced operating expenses coming along with the exploitation of wind power were not sufficient to balance out the increased initial investment. On the other hand, increasing $wind_{PW}$ always led to a reduction of CO_2 emissions. Adding capacity to the wind farm increased the environmental performance of the plant more than what a refined—thus more expensive and bulkier—design of the combined cycle could possibly do. These considerations are confirmed by observing Figure 7 where the Pareto solutions with no wind integration and with maximum wind integration are highlighted. Those solutions showed to be those returning optimal economic and environmental performance, respectively.

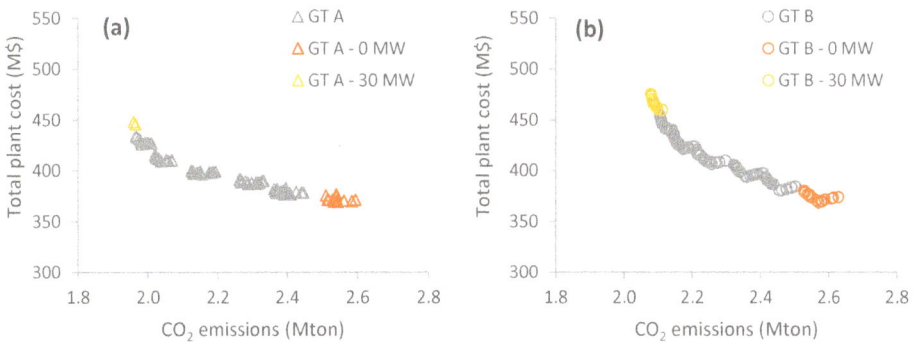

Figure 7. Decision map of the Pareto front showing trade-offs between total cost and CO_2 emissions for the designs based on GT A (**a**) and GT B (**b**). The results referring to no wind integration (0 MW) and maximum wind integration (30 MW) are highlighted.

The Pareto fronts reported refer to offshore power plants integrating a wind farm to a combined cycle. In a following section of the paper, the performance of these optimised systems are compared to more standard configurations, where the power generation unit installed on the platform is simply cycle gas turbines, either coupled with a wind farm or not. These solutions were not optimised as the performance of the gas turbines and the wind turbine are fixed by the commercial technologies considered (for which performance curves were available). The only parameter with an influence on the overall performance was the size of the wind farm, which defined the wind power contributions available for one year and, consequently, the operating conditions and, thus, the performance of the gas turbines. Such concepts were not the focus of the study, rather the basis for comparison for the more advanced power plants including the steam bottoming cycle.

Before analysing the results obtained, a few words on the limitations of the study. A comprehensive evaluation on the feasibility of the proposed integrated offshore power system should include two additional elements. First, an analysis on the offshore electric grid should be performed to evaluate the possibility to integrate the wind farm while guaranteeing frequency stability through a proper frequency control scheme. Second, the dynamic coupling between the intermittent wind resource and the offshore power cycle should be investigated. These issues were beyond the scope of the current study but will be considered in further work on the topic. The development of a dynamic model of offshore combined cycles [39] and the analyses of control strategies for fast load changes [40] were the first steps in this direction.

5. Discussion and Analysis of the Results

The multi-objective approach returned a number of Pareto optimal designs. A method to navigate through those various Pareto results was developed in order to be able to extract relevant information from them. The Pareto solutions were initially screened by setting constraints on the maximum CO_2 emissions and the maximum weight of the bottoming cycle. A weight threshold was set at 120 t, while the maximum allowable amount of CO_2 emissions ranged between 2.0 Mt and 2.6 Mt. Among the designs fulfilling the criteria indicated, the optimal one was then selected as that returning the best economic performance. Figure 8 gives a visual representation of the screening mechanism applied to the Pareto solutions with the GT A, for a maximum CO_2 emissions level of 2.3 Mt (and maximum weight of 120 t). The optimal design identified was termed Design A (CC+W). The same screening mechanism was applied to the Pareto solutions of GT B and the optimal design Design B (CC+W) was pinpointed. Tables 4 and 5 report the characteristics of these optimal designs that were further used for the following techno-economic analyses.

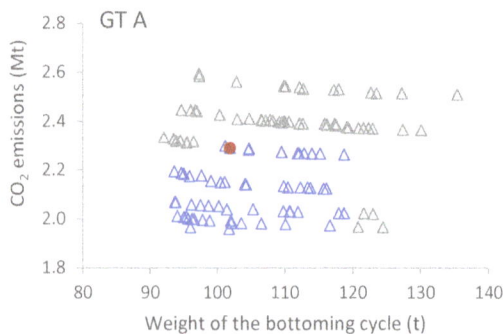

Figure 8. Visual representation of the screening mechanism of the Pareto designs. The grey empty diamonds are the designs screened out, the blue empty diamonds are the designs complying with the two criteria and the red full circle is the design selected.

Table 4. Characteristics of the optimal designs selected based on a GT A.

	GT A			
	Design A (CC+W)	Design A (CC)	Design A (GT+W)	Design A (GT)
Decision variables				
GT load	0.86	0.86	-	-
p_{steam} (bar)	17.7	17.7	-	-
T_{steam} (°C)	355.8	355.8	-	-
ΔT_{OTSG} (°C)	18.3	18.3	-	-
p_{cond} (bar)	0.09	0.09	-	-
ΔT_{cw} (°C)	6.1	6.1	-	-
$wind_{PW}$ (MW)	10	-	10	-
Objective functions				
CO_2^* (Mt)	2.3	2.6	2.8	3.3
W^* (t)	102	102	-	-
$cost^*$ (M$)	387	369	396	399

Table 5. Characteristics of the optimal designs selected based on a GT B.

	GT B			
	Design B (CC+W)	Design B (CC)	Design B (GT+W)	Design B (GT)
Decision variables				
GT load	0.62	0.62	-	-
p_{steam} (bar)	16.7	16.7	-	-
T_{steam} (°C)	323.3	323.3	-	-
ΔT_{OTSG} (°C)	24.7	24.7	-	-
p_{cond} (bar)	0.09	0.09	-	-
ΔT_{cw} (°C)	5.8	5.8	-	-
$wind_{PW}$ (MW)	15	-	10	-
Objective functions				
CO_2^* (Mt)	2.3	2.6	2.6	2.8
W^* (t)	104	104	-	-
$cost^*$ (M$)	407	369	377	356

The optimal designs obtained by ranging the CO_2 emissions constraint between 2.0 Mt and 2.6 Mt are shown in Figure 9. For each optimal design identified the lifetime economic performance and the wind farm size are reported. The set of results helped to make some considerations on the optimal wind power integration. A trade-off emerged between the extent of the environmental and economic aspects. The outcome confirmed what already hinted by the stepwise trend of the Pareto fronts. On one hand, the installation of offshore wind is economically challenging and, in fact, the total cost is consistently increasing with increasing wind power capacity installed. On the other hand, cutting the expected CO_2 emissions is challenging as well and the most effective way (also under an economical point of view) to meet more severe emissions limitations is to increase the size of the wind farm. Summing up, it can be argued that the optimal size of the wind farm should be selected by carefully defining and weighing the performance requirements (environmental and economic) that are to be achieved by the plant.

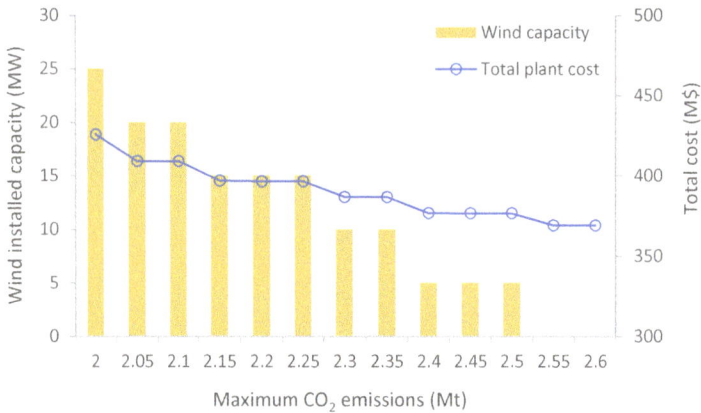

Figure 9. Characteristics of the designs identified through the selection process at the different maximum levels of CO_2 emissions. The bars represent the optimal wind capacity to be installed. The empty dots represent the total plant cost.

5.1. Comparison between the Cycles Based on the Two Gas Turbines

A qualitative comparison between the cycles based on the two different GTs demonstrated that better performance could be achieved by the GT A. The advantage of the cycle based on the GT A can be observed in Figure 6, where the GT A optimal designs outperform the GT B optimal designs under every metrics. In order to quantitatively assess the benefits associated with the utilisation of the smaller GT, the two designs previously pinpointed were considered, namely Design A (CC+W) and Design B (CC+W) (see Tables 4 and 5 for the related characteristics). The larger size of the GT B compared to the GT A was overall penalising. Most of the instances simulated required the larger GT B to operate at rather low part-load, either because of the availability of wind power or because of a reduced power demand, making the GT B particularly ineffective. Only at specific conditions the GT B entailed better performance, namely at peak power demand, when the wind gave a minor contribution due to low wind speeds. In those cases, the larger size of the GT allowed to meet the high power demand without starting a second backup GT, which was instead necessary with the GT A. The first effect described was predominant and the Design B (CC+W) was less efficient. That made necessary to have a larger wind farm to meet the same emission constraints, ultimately leading to a worse economic performance of Design B (CC+W) (about 19 M$ higher total cost to supply energy to the plant). The larger size of the GT B became more and more beneficial when increased shares of power had to be directly supplied by the GT. This can be noted by analysing the results of alternative concepts where the same two GTs are employed in different power plant configurations. The performance of the configuration involving a combined cycle without wind power (Design A (CC) vs. Design B (CC)) was practically identical with the two different GT sizes. The utilisation of simple GT cycles, both with (Design A (GT+W) vs. Design B (GT+W)) and without (*Design A (GT)* vs. *Design B (GT)*) a wind power farm integrated, favoured the utilisation of the larger GT B.

5.2. Performance Analysis of Offshore Power Plant

The performance of the Pareto optimal solution Design A (CC+W) was compared to alternative offshore power plant concepts. The following four power plants were defined and simulated throughout the entire expected installation's lifetime:

- Combined cycles with wind power—Design A (CC+W)
- Combined cycles—Design A (CC)
- Simple GT cycles with wind power—Design B (GT+W)

66

- Simple GT cycles—Design B (GT)

The related output results can be checked in Tables 4 and 5. Figure 10 shows the annual CO_2 emissions for each option simulated that ultimately add up to give the overall CO_2 footprint. Figure 11 shows the evolution of the total cost to supply energy to the plant (*cost**) during the years of plant operation that ultimately constitute the economic performance of the various concepts. The *cost** is always negative as only costs were considered in the analysis. The advanced offshore power plant proposed—Design A (CC+W)—reached the best environmental performance. The cuts of CO_2 emissions ranged between 272 kt (a 11.9% reduction) in comparison to Design A (CC) and 557 kt (a 24.4% reduction) in comparison to Design B (GT). Whilst advantageous in terms of environmental impact, the integration of wind power implied worse economics. Design B (GT) returned the lowest cost for the offshore energy supply, followed by Design A (CC). In comparison to their counterparts without wind power, Design A (CC+W) and Design B (GT+W) entailed an additional cost of 19 M$ and 21 M$, respectively. The operational costs were minimized with Design A (CC+W) but the savings achieved were not sufficient to repay the additional investment for the wind farm. Conversely, Design B (GT) showed the highest operational costs but the smaller initial investment guaranteed an overall better economic performance.

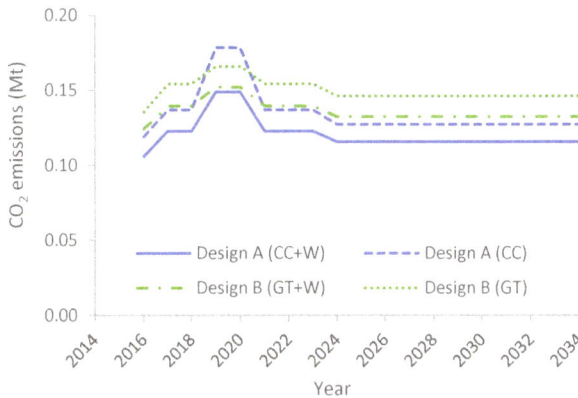

Figure 10. Annual CO_2 emissions for the concepts analysed: Design A (CC+W), Design A (CC), Design B (GT+W) and Design B (GT).

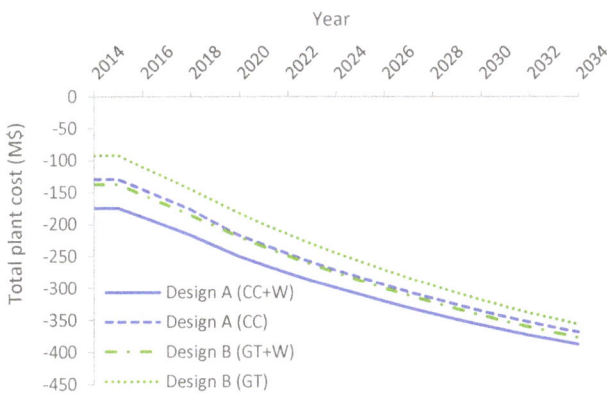

Figure 11. Evolution of the total plant cost throughout plant's lifetime for the concepts analysed: Design A (CC+W), Design A (CC), Design B (GT+W) and Design B (GT).

5.3. Sensitivity Analysis on Economic Parameters

The previous analyses showed the importance of the economic aspect for the studied offshore power plant in order to reach feasibility. To gain a better insight, a sensitivity analysis was carried out by acting on some main input parameters.

Wind farm total capital requirement (TCR_{wind}). The economic performances of Design A (CC+W) would have matched that of Design A (CC) if the specific value of the TCR_{wind} had dropped to 2611 \$/kW from the reference value of 4503 \$/kW. Albeit that number is in line with the most optimistic future scenarios, it is much lower than the current situation [31]. When the comparison was made with respect to Design B (GT+W) and Design B (GT), the *TCR* of the wind farm had to decrease down to, respectively, 3416 \$/kW and 1353 \$/kW before returning a better economic performance.

Combined cycle total capital requirement (TCR_{cc}). The calculation of TCR_{cc} was believed to be subjected to a large degree of uncertainty. The costs to install the necessary components offshore are significantly higher compared to typical onshore applications and difficult to estimate as very site specific. The numbers proposed, even though calculated taking into account a large contingency, could be under-estimated. In the comparative analysis between the various concepts, the impact of the high uncertainty level was limited by the fact that the options integrating wind power included the same power generation unit of the equivalent options without wind power. Larger differences could potentially arise between the concepts based on a combined cycle and those based on a simple GT cycle. In order to assess that, the TCR_{cc} was increased by a factor 2 and 5 (cases TCR2 and TCR5, respectively). Figure 12 shows the resulting *cost** trends. The concepts based on a GT simple cycle became more and more attractive since the gap of capital investment with respect to concepts based on a combined cycle increased. If in the base case Design A (CC) and Design B (GT+W) had similar performances, already at TCR2 Design B (GT+W) returned a better economic performance by about 29 M\$. Ultimately, the increase of the cost to install the power generation unit on the platform benefitted more conservative solutions (e.g., Design B (GT)) over more advanced ones (e.g., Design A (CC+W)).

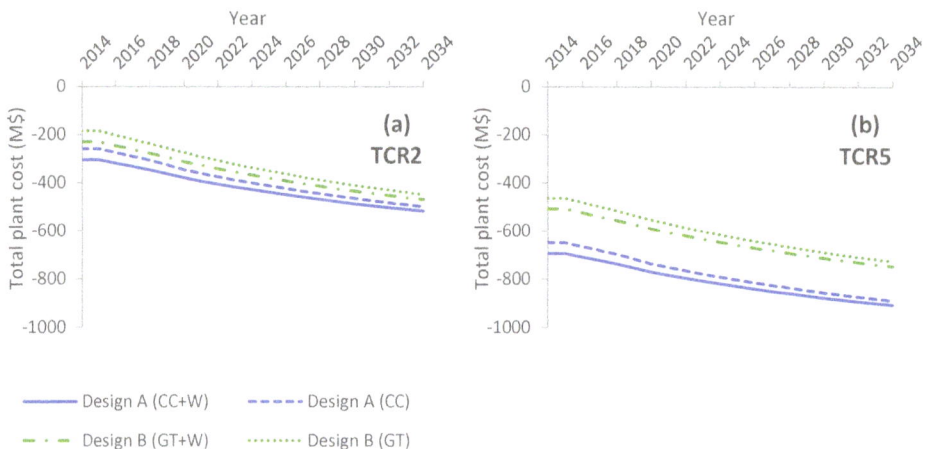

Figure 12. Evolution of the total plant cost throughout plant's lifetime when the total capital requirement of the onsite power generation unit is increased by (**a**) a factor 2 (TCR2) or (**b**) 5 (TCR5).

Discount rate (r). The effect of a lower (5%) and higher (9%) discount rate was evaluated. Figure 13 shows the related profile of the *cost** throughout the plant's lifetime. At lower discount rates, it becomes more important to minimise the operational costs as they will weigh more on the final economic performance. Accordingly, the concepts entailing lower operational costs—for instance, Design A (CC+W)—are favoured by lower values of the discount rate. Conversely, the concepts with lower investment costs but higher operational costs—for instance, Design B (GT)—are favoured by higher values of the discount rate. Ultimately, even though the economic gap between the various concepts changed with the different discount rates applied, the relative economic performance remained the same.

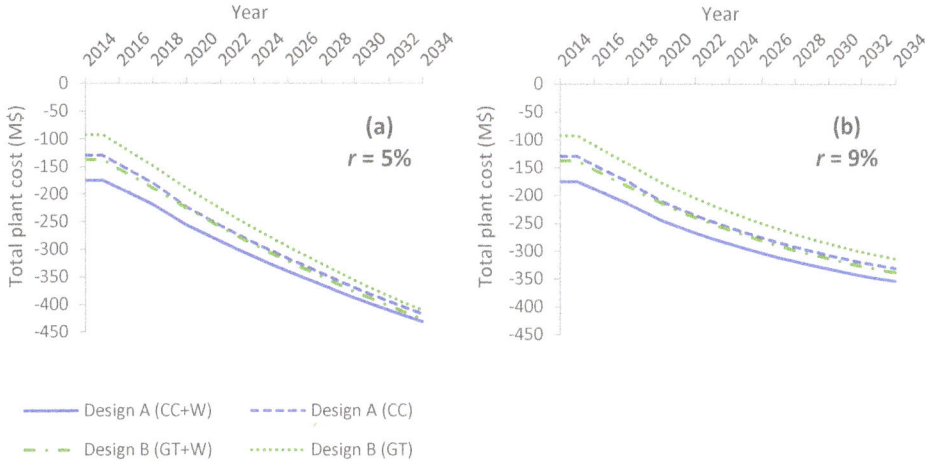

Figure 13. Evolution of the total plant cost throughout plant's lifetime when the discount rate is set to (**a**) 5% or (**b**) 9%.

CO$_2$ price (c_{CO2}). Figure 14 shows the relative effect of different CO$_2$ prices on the economic performance. Keeping Design A (CC+W) as the reference for comparison, the $\Delta cost^*$ obtained at the end of the lifetime is reported in the figure for the other concepts. A positive value indicates a better economic performance with respect to Design A (CC+W). Conversely, a negative value indicates a worse economic performance. The analysis showed that the CO$_2$ price had to exceed 174.1 $/t for Design A (CC+W) to entail an economic advantage over Design A (CC). Economic competitiveness could be achieved with relatively smaller levels of CO$_2$ price (i.e., 121.0 $/t and 158.8 $/t) with respect to Design B (GT+W) and Design B (GT). Such high levels of CO$_2$ price are foreseen in the future from some specific scenarios involving a strong international commitment on environmental issues (e.g., the 450 scenario by IEA displayed a CO$_2$ price of 140 $/t in 2040 [32]). However, they appear unlikely in the short term.

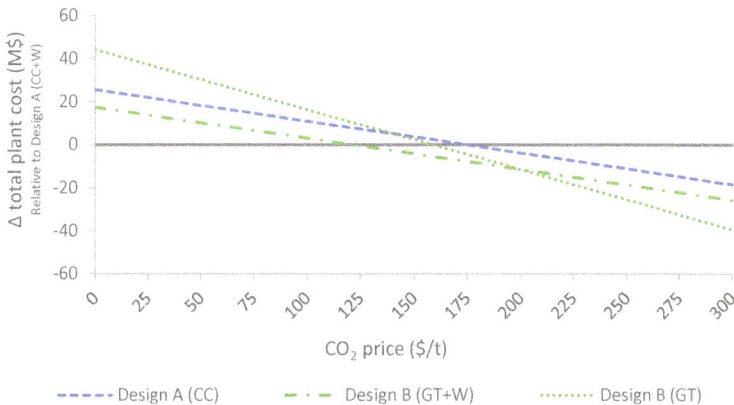

Figure 14. Variation of the total plant cost of Design A (CC), Design B (GT+W) and Design B (GT) relative to Design A (CC+W), as a function of the CO$_2$ price.

Gas price (c_{gas}). Figure 15 shows the effect of both higher (+25%) and lower (−25%) gas prices, alongside a variable c_{CO_2}. When the gas prices were increased, the situation became more favourable to the integration of wind power. Though, rather high levels of CO_2 prices (between 124.7 and 141.3 $/t) would still be needed to even the economic performance. The low levels of gas price seemed to rule out the possibility to achieve economic competitiveness for offshore wind power integration, as CO_2 prices around 200 $/t would be needed.

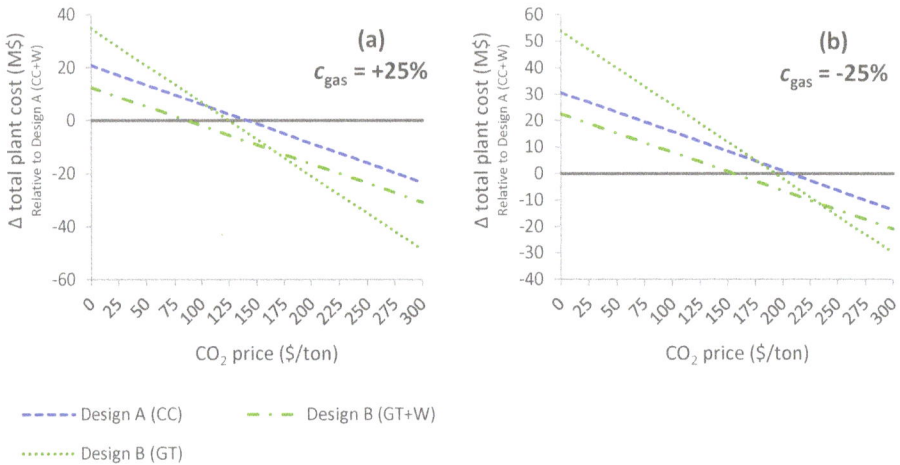

Figure 15. Variation of the total plant cost of Design A (CC), Design B (GT+W) and Design B (GT) relative to Design A (CC+W), as a function of the CO_2 price, when the gas price is increased by (**a**) +25% or decreased by (**b**) 25%.

6. Conclusions

A method is presented to define the optimal design of an offshore combined cycle power plant integrated with a wind farm. The defined design can be then used as a basis to evaluate the effectiveness of such concept to supply energy offshore. The optimisation procedure takes into account the performance at relevant off-design conditions in the definition of the optimal design. Given the complexity of the optimisation problem, a surrogate model is developed and validated. A multi-objective approach is applied including constraints specific to the offshore environment. The optimisation procedure presented was implemented on an actual installation in the North Sea for which a Pareto front of optimal solutions was obtained. An analysis of the results sparked some interesting considerations. A parameter to carefully select in the design phase was the size of the wind farm. As a trade-off between the environmental and economic performance emerged, the optimal level of wind integration should be based on the objectives to prioritise and on the constraints of that system. A comparison between the utilisation of two different size gas turbines (GT), either the smaller GT A or the larger GT B, showed that the GT A allowed to match better the power demand profile when additional power contributions were coming from the bottoming cycle and the wind farm. Conversely, the GT B appeared to be oversized for that application as it often operated at low part-load. Among the Pareto optimal designs obtained, one was selected for a techno-economic analysis. The selected design was compared to the same one without wind power integration and to offshore plants based on simple GT cycles, with and without wind power. The advanced offshore power plant proposed reduced the CO_2 emissions of 272 kt (−11.9%) and of 557 kt (−24.4%) with respect to the same combined cycle and to a simple GT cycle not integrating a wind farm. The economic performance was questionable. A wind farm meant an increased initial investment. Even though lower operational costs were obtained, paying back such additional investment proved to be challenging. With the current levels of gas and CO_2 prices, the final cost for the offshore energy supply was about 19 M\$ and 32 M\$ higher compared to the two concepts without wind power. The sensitivity analysis showed that very favourable price conditions would be needed to even out the difference. Conservative concepts, displaying a lower initial investment, demonstrated to be advantageous under an economic point of view. The results presented are case specific and cannot be generalized. For instance, in larger offshore projects with a longer lifetime, the reduced operational costs would result in a better economic outlook. In addition, it should be pointed out that the offshore power plant integrating a wind farm achieved a substantial cut in CO_2 emissions that affected the economic analysis only through a reduced cost for the CO_2 emitted. In an energy system including emission caps and penalties for the plants failing to fulfil such requirements, the better environmental performance could contribute to close the economic gap and, possibly, make offshore wind power integration economically feasible.

Author Contributions: L.R. developed the models, carried out the simulations and wrote the manuscript. L.O.N. contributed to the critical analysis of the results, reviewed the manuscript and supervised the work.

Funding: This research received no external funding.

Conflicts of Interest: The authors declare no conflicts of interest.

Nomenclature

A	Heat transfer area, m^2
c_{CO2}	CO_2 price, \$/t
c_{gas}	Gas price, \$/MWh
CF	Cash flow
CF^{CO2}	Cash flows associated with the CO_2 emissions, M\$
CF^{gas}	Cash flows associated with the onsite gas consumption, M\$
CO_2^*	Total CO_2 emissions, Mt
$cost^*$	Total cost to supply energy to the plant, M\$
C_S	Constant flow coefficient
DCF	Discounted cash flow, M\$
F_{CU}	Factor accounting for copper losses
$GT\ load$	Gas turbine load
h_{eq}	Equivalent hours per year, h
k_ε	Correction factor
LHV_{gas}	Lower heating value of the natural gas, kJ/kg
$load$	Mechanical load
$\dot{m}_{steam}$	Steam mass flow rate, kg/s
m_{CO2}	CO_2 emissions, Mt
$\dot{m}$	Mass flow rate, kg/s
$\dot{m}_{CO2}$	Mass flow rate of emitted CO_2, kg/s
$\dot{m}_{cw}$	Mass flow rate of cooling water, kg/s
$\dot{m}_{gas}$	Mass flow rate of natural gas, kg/s
$\dot{m}_{WHRU}$	Mass flow rate in the WHRU, kg/s
p_{cond}	Condenser pressure, bar
p_{in}	Turbine inlet pressure, bar
p_{out}	Turbine outlet pressure, bar
p_{steam}	Steam evaporation pressure, bar
P_{CC}	Combined cycle power requirement, MW
P_{net}	Net cycle power output, MW
P_{ST}	Steam power output, MW
P_O	Offshore power demand, MW
P_W	Wind power contribution, MW
PEC	Purchased-equipment cost, M\$

r	Discount rate
$T_{cond,in}$	Temperature at the condenser inlet, °C
T_{in}	Turbine inlet temperature, °C
T_{steam}	Superheated steam temperature, °C
TCR	Total capital requirement, M$
TCR_{CC}	Total capital requirement for the combined cycle, M$
TCR_{wind}	Total capital requirement for the wind farm, M$
U	Overall heat transfer coefficient, kW/K/m^2
UA_{ECO1}	UA coefficient of the 1st economizer, kW/K
UA_{ECO2}	UA coefficient of the 2nd economizer, kW/K
UA_{OTB}	UA coefficient of the evaporator, kW/K
UA_{SH}	UA coefficient of the superheater, kW/K
UA_{WHRU}	UA coefficient of the waste heat recovery unit, kW/K
$\dot{V}$	Volumetric flow rate, m^3/s
$wind_{PW}$	Wind power capacity installed, MW
$W_{component}$	Weight of the specific component of the power cycle, t
W^*	Total weight of the bottoming cycle, t
W_{OTSG}	Weight of the OTSG, t
W_{ST}	Weight of the steam turbine, t
W_{GEN}	Weight of the generator, t
W_{COND}	Weight of the condenser (wet), t
$\bar{x}$	Array of decision variables
$\bar{z}$	Array of objective functions
Greek Letters	
γ	Exponent of the Reynolds number in the heat transfer correlation
Γ	Marginal likelihood
$\Delta h_{T,is}$	Isentropic enthalpy difference, kJ/kg
Δp	Pressure drop, bar
Δp_{ECO1}	Pressure drop in the 1st economizer, bar
Δp_{ECO2}	Pressure drop in the 2nd economizer, bar
Δp_{OTB}	Pressure drop in the evaporator, bar
Δp_{OTSG}	Overall pressure drop in the OTSG, bar
Δp_{SH}	Pressure drop in the superheater, bar
ΔT_{cw}	Cooling water temperature difference, °C
ΔT_{OTSG}	Pinch point difference in the OTSG, °C
η_{cycle}	Net cycle efficiency
η_{gen}	Generator efficiency
η_{pump}	Pump isentropic efficiency
η_T	Isentropic steam turbine efficiency
ϑ_k	Hyperparameter
σ^2	Process variance
ψ	Correlation function
Ψ	Correlation matrix
Acronyms	
DC	Direct costs
GA	Genetic algorithm
GT	Gas turbine
IC	Indirect costs
MAE	Mean average error
NOC	Number of off-design conditions
NPV	Net present value
OTSG	Once-through steam generator
TIT	Turbine inlet temperature
WHRU	Waste heat recovery unit

Appendix A. Kriging Surrogate Modelling Technique

Kriging works as a locally weighted regression method based on a Gaussian process [41]. In its basic formulation, the function $y(x)$ is approximated as following [42]:

$$y(x) = f(x) + Z(x) \tag{A1}$$

$f(x)$ is a regression function that approximates globally the function. The regression function could be a known constant (simple Kriging), an unknown constant (ordinary Kriging) or a multivariate polynomial (universal Kriging). It is determined by the generalized least squares method. A constant term suffices in many applications [43] and ordinary Kriging was therefore used in the study. $Z(x)$ is a realization of a normally distributed Gaussian random process with zero mean, variance σ^2 and a correlation matrix Ψ. $Z(x)$ takes into account localized variations and ensures the interpolation of the training data. The correlation function ψ is parametrized by a set of hyperparameters ϑ_k, determined using the maximum likelihood estimation. The correlation function used is the following:

$$\psi(x_i, x_j) = \exp\left(-\sum_{k=1}^{d} \vartheta_k |x_i - x_j|^2\right) \tag{A2}$$

The natural logarithm of the marginal likelihood used to identify the hyperparameters is expressed in the ooDACE tool as [24]:

$$-\ln\left(\Gamma_{\text{marginal}}\right) = -\frac{n}{2}\ln\left(\sigma^2\right) - \frac{1}{2}\ln(|\Psi|) \tag{A3}$$

where Γ is the marginal likelihood, σ^2 is the process variance and Ψ is the n X n correlation matrix. While the first term represents the quality of the fit, the second term can be interpreted as a complexity penalty. The combination of the two allows balancing between flexibility and accuracy.

Kriging is a flexible surrogate modelling technique, well-suited for deterministic applications and has been successfully applied in several engineering design applications [44]. For these reasons, it was chosen for the study.

Appendix B. Correlations for the Off-Design Performance Predictions

The complete set of equations used to evaluate the off-design performance of the various components of the cycle is reported.

The heat transfer coefficient for the heat exchangers is calculated as [25]:

$$UA = k_\varepsilon UA_d \left(\frac{\dot{m}}{\dot{m}_d}\right)^{\gamma} \tag{A4}$$

where U is the overall heat transfer coefficient, A is the heat transfer area, $\dot{m}$ is a mass flow rate and γ is the exponent of the Reynolds number in the heat transfer correlation. γ was set equal to 0.6 in the WHRU and in the economizer and evaporator sections of the OTSG, equal to 0.8 in the superheater section of the OTSG. k_ε is a correction factor. It was noted that the error in the estimation of the heat transfer coefficients increased with the decrease of the plant load and became relevant at very low loads (of importance in the scenarios including wind power integration). The off-design model of the heat transfer coefficient was tuned in order to address this effect. The correction factor k_ε is defined as a function of the deviation of the flow rate from the design value. It is applied to all the heat exchange sections but the superheater of the OTSG, where it is not necessary.

The Stodola's cone law is applied for modelling the steam turbine behaviour [45]:

$$C_S = \frac{\dot{m}\sqrt{T_{in}}}{\sqrt{p_{in}^2 - p_{out}^2}} \tag{A5}$$

where C_S is the constant flow coefficient, $\dot{m}$ is the mass flow rate, T_{in} is the turbine inlet temperature, p_{in} is the turbine inlet pressure and p_{out} is the turbine outlet pressure.

The variation of the isentropic efficiency at part-load is evaluated as [27]:

$$\frac{\eta_T}{\eta_{T,d}} = 2\sqrt{\frac{\Delta h_{T,is,d}}{\Delta h_{T,is}} - \frac{\Delta h_{T,is,d}}{\Delta h_{T,is}}} \tag{A6}$$

where η_T is the isentropic efficiency of the turbine at off-design and $\Delta h_{T,is}$ is the isentropic enthalpy difference due to the expansion in the turbine.

The efficiency of the generator is calculated as [28]:

$$\eta_{gen} = \frac{load \cdot \eta_{gen,d}}{load \cdot \eta_{gen,d} + \left(1 - \eta_{gen,d}\right)\left[(1 - F_{CU}) + F_{CU}load^2\right]} \tag{A7}$$

where η_{gen} is the generator efficiency, *load* is the mechanical load and F_{CU} is a term representing the copper losses (produced in the winding of the stator). The term F_{CU} is set equal to 0.43 [28].

The off-design performance of the pumps is calculated as [29]:

$$\frac{\eta_{pump}}{\eta_{pump,d}} = -0.029265\left(\frac{\dot{V}}{\dot{V}_d}\right)^3 - 0.14086\left(\frac{\dot{V}}{\dot{V}_d}\right)^3 + 0.3096\left(\frac{\dot{V}}{\dot{V}_d}\right)^2 + 0.86387 \tag{A8}$$

where η_{pump} is the isentropic efficiency of the pump and $\dot{V}$ is the volumetric flow rate.

The pressure drops are modelled as [46]:

$$\Delta p = \Delta p_d \left(\frac{\dot{m}}{\dot{m}_d}\right)^2 \tag{A9}$$

where Δp is the pressure drop and $\dot{m}$ is the mass flow rate.

Appendix C. Validation of the Kriging Model

Once the Kriging model fitting was completed, the model had to be tested before being deemed reliable. A set of 30 testing conditions was randomly selected and a comparative analysis was performed between the outputs of the high-fidelity model and those of the Kriging model. The mean average error (MAE) obtained for each output variable is reported in Table A1. Figures A1–A4 show the parity plots of selected parameters in order to visualize the extent of the approximation introduced. The diamonds refer to the cycle based on the GT A, the circles on the cycle based on the GT B.

The Kriging model demonstrated to be able to capture the behaviour of the combined cycle with a reasonable accuracy. Most of the parameters were predicted with a MAE smaller than 1%. One parameter that demonstrated to be particularly difficult to predict was the weight of the condenser (W_{COND}). The related MAE were 3.18% and 2.30% and substantial prediction errors were highlighted by the parity plot in Figure A4. However, it was also noted that the W_{COND} contributed marginally (between 4% and 10%) to the total weight of the bottoming cycle, whose estimation resulted to be rather good with MAE of 0.37% and 0.34%. Therefore, a slightly worse accuracy in the prediction of W_{COND} was considered acceptable. Another term with MAE larger than 1% was the UA coefficient of the superheater (UA_{SH}). However, it was noted that UA_{SH} had a limited impact on the overall heat transfer process occurring in the OTSG, due to the limited degree of superheating implemented in the various designs. Overall, the accuracy demonstrated by the Kriging model was within reasonable levels and the model was, thus, considered validated.

Table A1. Mean average error (MAE) of the output parameters of the Kriging model.

Output Parameters		GT A	GT B
Description	Symbol	MAE	MAE
Net cycle efficiency	η_{cycle}	0.03%	0.07%
Net power output	P_{net}	0.05%	0.05%
Mass flow rate in the WHRU	$\dot{m}_{WHRU}$	0.01%	0.04%
UA coefficient of the WHRU	UA_{WHRU}	0.00%	0.02%
UA coefficient of the first economizer	UA_{ECO1}	0.58%	0.78%
UA coefficient of the second economizer	UA_{ECO2}	0.52%	0.68%
UA coefficient of the evaporator	UA_{OTB}	0.30%	0.44%
UA coefficient of the superheater	UA_{SH}	1.35%	1.28%
Pressure drop in the first economizer	Δp_{ECO1}	0.00%	0.00%
Pressure drop in the second economizer	Δp_{ECO2}	0.03%	0.04%
Pressure drop in the evaporator	Δp_{OTB}	1.02%	0.99%
Pressure drop in the superheater	Δp_{SH}	0.00%	0.00%
Steam mass flow rate	$\dot{m}_{steam}$	0.23%	0.29%
Isentropic steam turbine efficiency	η_T	0.39%	0.43%
Temperature at the condenser inlet	$T_{cond,in}$	0.00%	0.00%
Mass flow rate of cooling water	$\dot{m}_{cw}$	0.54%	0.56%
Weight of the OTSG	W_{OTSG}	0.42%	0.56%
Weight of steam turbine	W_{ST}	0.48%	0.41%
Weight of generator	W_{GEN}	0.19%	0.27%
Weight of the condenser (wet)	W_{COND}	3.18%	2.30%
Purchased-equipment cost	PEC	0.23%	0.27%

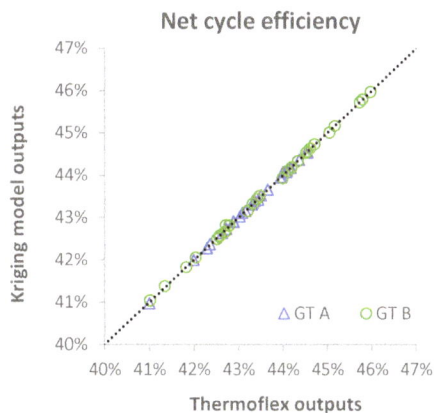

Figure A1. Parity plot of the net cycle efficiency for Kriging model validation.

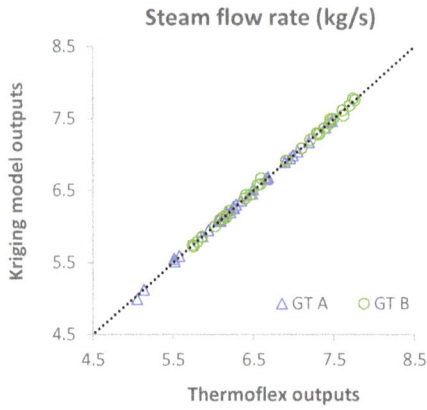

Figure A2. Parity plot of steam flow rate for Kriging model validation.

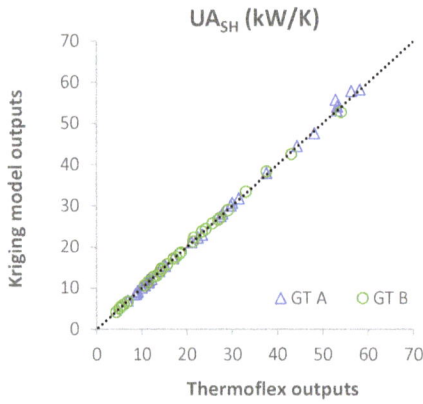

Figure A3. Parity plot UA coefficient of superheater for Kriging model validation.

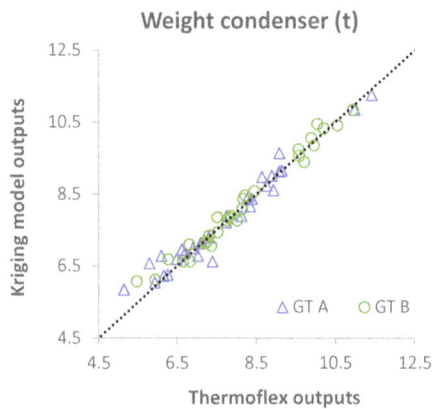

Figure A4. Parity plot of the weight of the condenser for Kriging model validation.

Appendix D. Validation of the Off-Design Correlations

A validation testing was carried out in order to verify the capability of the off-design correlations to predict the off-design performance. Three random designs were selected both for the cycle based on the GT A and for the cycle based on the GT B. The related values of the parameters are reported in Table A2 and were allowed to range within the bounds previously established.

Table A2. Designs selected to test the off-design correlations.

	GT A			GT B		
	Design #1	Design #2	Design #3	Design #1	Design #2	Design #3
GT load	0.90	0.82	0.93	0.90	0.69	0.82
p_{steam}	20	32	27	30	18	35
T_{steam}	328	390	360	350	320	290
ΔT_{OTSG}	25	20	15	25	12	18
p_{cond}	0.07	0.05	0.04	0.07	0.04	0.09
ΔT_{cw}	8	5	6	8	5	4

Each design proposed was simulated at 25 of off-design conditions, defined to cover the entire range of possible operating conditions the cycle could be subjected to. In particular, operation at GT loads as low as 40% and 30% were simulated, respectively for the GT A and the GT B. These values were of significance because they were evaluated as the minimum GT loads at which the exhaust gas were able to meet the heat duty (i.e., 11 MW) with a reasonable flexibility margin, given the specific design of the WHRU considered. Those low GT loads levels were expected to occur when the wind farm was integrated to the combined cycle. The same operating conditions was also simulated with the high-fidelity model. The outputs of the comparative analysis are shown in Table A3 as well as in Figures A5–A7 as parity plots of selected parameters.

Table A3. Mean average error (MAE) from the off-design simulations for some selected parameters.

	Design #1	Design #2	Design #3	Overall
	MAE	MAE	MAE	MAE
GT A				
η_{cycle}	0.20%	0.26%	0.27%	0.24%
P_{net}	0.20%	0.26%	0.27%	0.24%
$\dot{m}_{CO2}$	0.00%	0.00%	0.00%	0.00%
P_{ST}	0.87%	1.18%	1.21%	1.08%
p_{steam}	0.79%	1.13%	1.26%	1.06%
T_{steam}	0.45%	0.21%	0.29%	0.32%
$\dot{m}_{steam}$	0.66%	0.79%	1.06%	0.84%
GT B				
η_{cycle}	0.12%	0.17%	0.05%	0.11%
P_{net}	0.12%	0.17%	0.05%	0.12%
$\dot{m}_{CO2}$	0.00%	0.00%	0.01%	0.01%
P_{ST}	0.57%	1.00%	0.42%	0.66%
p_{steam}	0.80%	1.19%	0.50%	0.83%
T_{steam}	0.41%	0.12%	0.63%	0.39%
$\dot{m}_{steam}$	0.67%	0.69%	0.52%	0.63%

The off-design performance of the cycle was generally well captured by the surrogate model, relying on the Kriging model and on the off-design correlations. Important parameters for the optimisation processes, like the net cycle efficiency (η_{cycle}) and the CO_2 emissions ($\dot{m}_{CO2}$), were predicted with good accuracy even at low part-loads (see Figure A5 with the parity plot of η_{cycle}). The simulation of the steam cycle demonstrated to be somewhat challenging, especially the heat transfer process in the OTSG. At low part-loads the accuracy of the correlation for the heat transfer coefficients started to diminish, resulting in less precise values of the steam parameters (see for example Figure A7) and consequently in a larger error in the steam power output (P_{ST}) calculated. This was particularly evident by looking at the parity plot of P_{ST} in Figure A6, where the region of low part-loads (i.e., the bottom left corner of the parity plot) is characterised by a larger scattering of the results. However, a proper tuning of the correlations allowed to contain the maximum error within few percentage points and the MAE close to 1%. Considering that the contribution of the ST to the total power output is rather small, the performance predictions at off-designs were deemed as adequate to be used in an optimisation procedure.

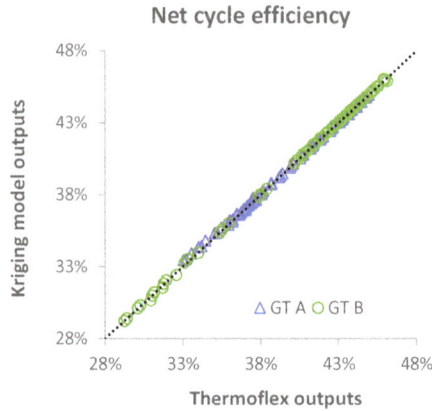

Figure A5. Parity plot of the net cycle efficiency for the off-design model validation.

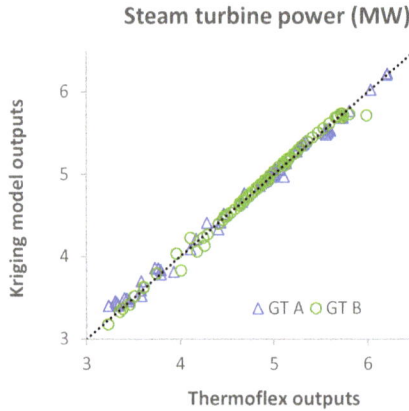

Figure A6. Parity plot of the steam turbine power output for the off-design model validation.

Steam flow rate (kg/s)

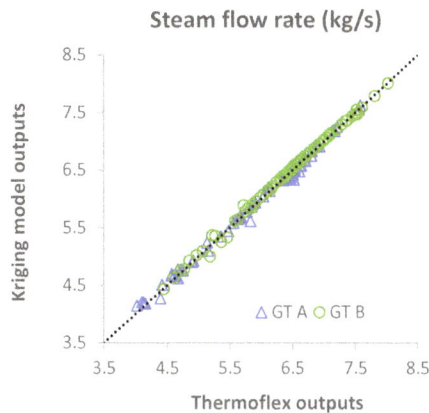

Figure A7. Parity plot of the steam flow rate for the off-design model validation.

References

1. Nguyen, T.-V.; Tock, L.; Breuhaus, P.; Maréchal, F.; Elmegaard, B. CO₂-mitigation options for the offshore oil and gas sector. *Appl. Energy* **2016**, *161*, 673–694. [CrossRef]
2. Statistisk Sentralbyrå. *Utslipp av klimagasser 2015*; Statistisk Sentralbyrå: Oslo, Norway, 2016.
3. Riboldi, L.; Nord, L.O. Concepts for lifetime efficient supply of power and heat to offshore installations in the North Sea. *Energy Convers. Manag.* **2017**, *148*, 860–875. [CrossRef]
4. Nguyen, T.-V.; Voldsund, M.; Breuhaus, P.; Elmegaard, B. Energy efficiency measures for offshore oil and gas platforms. *Energy* **2016**, *117*, 1–16. [CrossRef]
5. Pierobon, L.; Benato, A.; Scolari, E.; Haglind, F.; Stoppato, A. Waste heat recovery technologies for offshore platforms. *Appl. Energy* **2014**, *136*, 228–241. [CrossRef]
6. Mazzetti, J.M.; Nekså, P.; Walnum, H.T.; Hemmingsen, A.K.T. Energy-Efficient Technologies for Reduction of Offshore CO₂ Emmissions. In Proceedings of the Offshore Technology Conference, Houston, TX, USA, 6–9 May 2013.
7. Nord, L.O.; Bolland, O. Design and off-design simulations of combined cycles for offshore oil and gas installations. *Appl. Therm. Eng.* **2013**, *54*, 85–91. [CrossRef]
8. Nord, L.O.; Martelli, E.; Bolland, O. Weight and power optimization of steam bottoming cycle for offshore oil and gas installations. *Energy* **2014**, *76*, 891–898. [CrossRef]
9. Pierobon, L.; Van Nguyen, T.; Larsen, U.; Haglind, F.; Elmegaard, B. Multi-objective optimization of organic Rankine cycles for waste heat recovery: Application in an offshore platform. *Energy* **2013**, *58*, 538–549. [CrossRef]
10. Riboldi, L.; Nord, L.O. Lifetime assessment of combined cycles for cogeneration of power and heat in offshore oil and gas installations. *Energies* **2017**, *10*, 744. [CrossRef]
11. Econ Pöyry. *CO₂-Emissions Effect of Electrification*; Econ Pöyry: Helsinki, Finland, 2011.
12. Riboldi, L.; Cheng, X.; Farahmand, H.; Korpås, M.; Nord, L.O. Effective concepts for supplying energy to a large offshore oil and gas area under different future scenarios. *Chem. Eng. Trans.* **2017**, *61*, 1597–1602.
13. He, W.; Jacobsen, G.; Anderson, T.; Olsen, F.; Hanson, T.D.; Korpås, M.; Toftevaag, T.; Eek, J.; Uhlen, K.; Johansson, E. The potential of integrating wind power with offshore oil and gas platforms. *Wind Eng.* **2010**, *34*, 125–137. [CrossRef]
14. Elginoz, N.; Bas, B. Life Cycle Assessment of a multi-use offshore platform: Combining wind and wave energy production. *Ocean Eng.* **2017**, *145*, 430–443. [CrossRef]
15. Korpås, M.; Warland, L.; He, W.; Tande, J.O.G. A case-study on offshore wind power supply to oil and gas rigs. *Energy Procedia* **2012**, *24*, 18–26. [CrossRef]
16. Bianchi, M.; Branchini, L.; De Pascale, A.; Melino, F.; Orlandini, V.; Peretto, A.; Haglind, F.; Pierobon, L. Cogenerative performance of a wind-gas turbine-organic rankine cycle integrated system for offshore applications. In Proceedings of the ASME Turbo Expo 2016, Seoul, Korea, 13–17 June 2016; Volume 3.

17. Orlandini, V.; Pierobon, L.; Schløer, S.; De Pascale, A.; Haglind, F. Dynamic performance of a novel offshore power system integrated with a wind farm. *Energy* **2016**, *109*, 236–247. [CrossRef]
18. Riboldi, L.; Nord, L.O. Optimal design of flexible power cycles through Kriging-based surrogate models. In Proceedings of the ASME Turbo Expo 2018: Turbomachinery Technical Conference and Exposition, Oslo, Norway, 11–15 June 2018.
19. Kang, C.A.; Brandt, A.R.; Durlofsky, L.J. Optimal operation of an integrated energy system including fossil fuel power generation, CO_2 capture and wind. *Energy* **2011**, *36*, 6806–6820. [CrossRef]
20. Kang, C.A.; Brandt, A.R.; Durlofsky, L.J. A new carbon capture proxy model for optimizing the design and time-varying operation of a coal-natural gas power station. *Int. J. Greenh. Gas Control* **2016**, *48*, 234–252. [CrossRef]
21. *Thermoflex, Version 26.0*; Thermoflow Inc.: Fayville, MA, USA, 2016.
22. Nord, L.O.; Bolland, O. Steam bottoming cycles offshore—Challenges and possibilities. *J. Power Technol.* **2012**, *92*, 201–207.
23. Kehlhofer, R. *Combined-Cycle Gas & Steam Turbine Power Plants*; Pennwell Books: Houston, TX, USA, 1999.
24. Couckuyt, I.; Dhaene, T.; Demeester, P. ooDACE toolbox: A flexible object-oriented kriging implementation. *J. Mach. Learn. Res.* **2014**, *15*, 3183–3186.
25. Incropera, F.P.; DeWitt, D.P.; Bergman, T.L.; Lavine, A.S. *Fundamentals of Heat and Mass Transfer*; John Wiley & Sons: Hoboken, NJ, USA, 2007; Volume 6.
26. Haglind, F. Variable geometry gas turbines for improving the part-load performance of marine combined cycles - Combined cycle performance. *Appl. Therm. Eng.* **2011**, *31*, 467–476. [CrossRef]
27. Schobeiri, M. *Turbomachinery Flow Physics and Dynamic Performance*; Springer: Berlin, Germany, 2005.
28. Haglind, F.; Elmegaard, B. Methodologies for predicting the part-load performance of aero-derivative gas turbines. *Energy* **2009**, *34*, 1484–1492. [CrossRef]
29. Veres, J.P. *Centrifugal and Axial Pump Design and Off-Design Performance Prediction*; NASA Techincal Memo 106745; NASA Lewis Research Center: Cleveland, OH, USA, 1995; pp. 1–24.
30. Bejan, A.; Tsatsaronis, G.; Moran, M. *Thermal Design and Optimization*; John Wiley & Sons: Hoboken, NJ, USA, 1996.
31. Carlsson, J.; Fortes, M.D.; de Marco, G.; Giuntoli, J.; Jakubcionis, M.; Jäger-Waldau, A.; Lacal-Arantegui, R.; Lazarou, S.; Magagna, D.; Moles, C.; et al. *ETRI 2014—Energy Technology Reference Indicator projections for 2010–2050*; JRC Sci Policy Reports; European Commission: Brussels, Belgium, 2014; pp. 1–108.
32. International Energy Agency (IEA). *World Energy Outlook 2016*; International Energy Agency: Paris, France, 2016; pp. 1–684.
33. Directorate, Norwegian Petroleum. *Ministry of Petroleum and Energy*; Norwegian Petroleum: Stavanger, Norway, 2017.
34. MathWorks. *Global Optimization Toolbox, Version R2015a*; MathWorks: Natick, MA, USA, 2016.
35. Lundin; Wintershall; RWE. Plan for Utbygging, Anlegg og Drift av Luno—Del 2: Konsekvensutredning. 2011. Available online: https://www.lundin-petroleum.com/Documents/ot_no_Luno_EIA_2011.pdf (accessed on 15 October 2018).
36. Det Norske Oljeselskap ASA. Plan for Utbygging og Drift av Ivar Aasen—Del 2: Konsekvensutredning. 2012. Available online: https://docplayer.me/423297-Plan-for-utbygging-og-drift-av-ivar-aasen.html (accessed on 15 October 2018).
37. Berge, E.; Byrkjedal, Ø.; Ydersbond, Y.; Kindler, D. Modelling of offshore wind resources. Comparison of a meso-scale model and measurements from FINO 1 and North Sea oil rigs. *Eur. Wind Energy Conf. Exhib.* **2009**, *4*, 2327–2334.
38. StatoilHydro. *The World's Firs Full Scale Floating Wind Turbine*; Hywind by StatoilHydro: Stavanger, Norway, 2009.
39. Rúa, J.; Montañés, R.M.; Riboldi, L.; Nord, L.O. Dynamic Modeling and Simulation of an Offshore Combined Heat and Power (CHP) Plant. In Proceedings of the 58th Conference on Simulation and Modelling (SIMS 58), Reykjavik, Iceland, 25–27 September 2017; Linköping University Electronic Press: Linköping, Sweden, 2017; pp. 241–250.
40. Nord, L.O.; Montañés, R.M. Compact steam bottoming cycles: Model validation with plant data and evaluation of control strategies for fast load changes. *Appl. Therm. Eng.* **2018**, *142*, 334–345. [CrossRef]

41. Sacks, J.; Welch, W.J.; Mitchell, T.J.; Wynn, H.P. Design and analysis of computer experiments. *Stat. Sci.* **1989**, *4*, 409–423. [CrossRef]

42. Koziel, S.; Yang, X. *Computational Optimization, Methods and Algorithms*; Springer: Berlin, Germany, 2011.

43. Jones, D.R.; Schonlau, M.; Welch, W.J. Efficient global optimization of expensive black-box functions. *J. Glob. Optim.* **1998**, *13*, 455–492. [CrossRef]

44. Simpson, T.W.; Peplinski, J.D.; Koch, P.N.; Allen, J.K. Metamodels for computer-based engineering design: Survey and recommendations. *Eng. Comput.* **2001**, *17*, 129–150. [CrossRef]

45. Cooke, D.H. On prediction of off-design multistage turbine pressures by stodola's ellipse. *J. Eng. Gas Turbines Power* **1985**, *107*, 596–606. [CrossRef]

46. Lecompte, S.; Huisseune, H.; van den Broek, M.; De Schampheleire, S.; De Paepe, M. Part load based thermo-economic optimization of the Organic Rankine Cycle (ORC) applied to a combined heat and power (CHP) system. *Appl. Energy* **2013**, *111*, 871–881. [CrossRef]

![processes logo] *processes*

MDPI

Article

Comparison of the Utilization of 110 °C and 120 °C Heat Sources in a Geothermal Energy System Using Organic Rankine Cycle (ORC) with R245fa, R123, and Mixed-Ratio Fluids as Working Fluids

Mochamad Denny Surindra [1], Wahyu Caesarendra [2,3,*], Totok Prasetyo [1], Teuku Meurah Indra Mahlia [4] and Taufik [5]

[1] Mechanical Engineering, Politeknik Negeri Semarang, Semarang 50275, Indonesia; dennysurindra@polines.ac.id (M.D.S.); totok.prasetyo@polines.ac.id (T.P.)
[2] Faculty of Integrated Technologies, Universiti Brunei Darussalam, Jalan Tungku Link, Gadong BE1410, Brunei Darussalam
[3] Mechanical Engineering Department, Diponegoro University, Semarang 50275, Indonesia
[4] School of Information, Systems and Modelling, Faculty of Engineering and Information Technology, University of Technology Sydney, Sydney, NSW 2007, Australia; tmindra.mahlia@uts.edu.au
[5] Electrical Engineering Department, Cal Poly State University, San Luis Obispo, CA 93407, USA; taufik@calpoly.edu
* Correspondence: wahyu.caesarendra@ubd.edu.bn or w.caesarendra@gmail.com; Tel.: +673-7-345-623

Received: 12 January 2019; Accepted: 15 February 2019; Published: 21 February 2019

Abstract: Binary cycle experiment as one of the Organic Rankine Cycle (ORC) technologies has been known to provide an improved alternate scenario to utilize waste energy with low temperatures. As such, a binary geothermal power plant simulator was developed to demonstrate the geothermal energy potential in Dieng, Indonesia. To better understand the geothermal potential, the laboratory experiment to study the ORC heat source mechanism that can be set to operate at fixed temperatures of 110 °C and 120 °C is conducted. For further performance analysis, R245fa, R123, and mixed ratio working fluids with mass flow rate varied from 0.1 kg/s to 0.2 kg/s were introduced as key parameters in the study. Data from the simulator were measured and analyzed under steady-state condition with a 20 min interval per given mass flow rate. Results indicate that the ORC system has better thermodynamic performance when operating the heat source at 120 °C than those obtained from 110 °C. Moreover, the R123 fluid produces the highest ORC efficiency with values between 9.4% and 13.5%.

Keywords: Organic Rankine Cycle (ORC); geothermal energy; binary cycle; R245fa; R123; mixture ratio; Dieng; Indonesia

---

1. Introduction

Indonesia is the largest archipelago in the world with the number of islands exceeding 17,000, of which only 922 are inhabited. Indonesia is located in Southeast Asia, between the Pacific and the Indian oceans, and the Asian and Australian continents. The country lies on the equator, and thus falls in the tropical region. Geologically, Indonesia encompasses three active plates which are the Indo-Australian plate, the Euro-Asia plate, and the Pacific plate [1,2]. As shown in Figure 1, the volcanoes stretch from the Aceh province at the eastern most tip of the country, down through the Sumatra island, across to the Java island, Nusa Tenggara, Maluku, and end on the Sulawesi island [3]. The number of volcanoes has been recorded to total more than 200, 129 of which are considered active and have the potential to cause volcanic eruption and earthquake at any time. Many researchers such as Budi et al. (2014) [4],

Manfred et al. (2008) [5], Hall et al. (2002) [6], Simandjuntak et al. (1996) [7], and Hamilton (1979) [8] have studied Indonesia's geographical conditions and its volcanoes. One conclusion that their studies share is that Indonesia has the significant potential for earthquake disasters and volcanic eruptions. However, despite the disadvantage of being prone to natural disasters, the country's geological location does offer many benefits including one that pertains to today's increasing demand and interest on non-fossil based energy sources, namely geothermal energy.

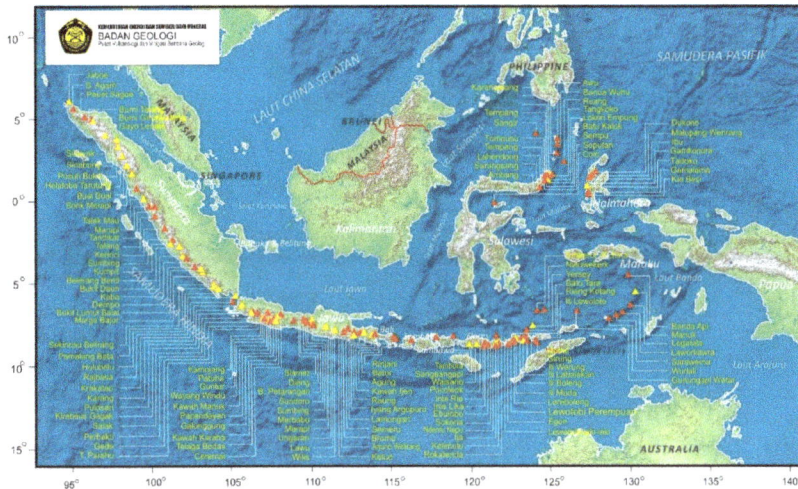

Figure 1. Geographical features of Indonesia archipelago and distribution map of volcanoes in Indonesia [9].

Geothermal energy is the thermal energy source from the earth. More specifically, the geothermal energy is generated from hot water and hot rock, which are stored a few miles beneath the earth's surface [10]. The water heats up and becomes pressurized steam underneath a permeable layer [11]. Prince Piero Ginori Conti was the first to utilize geothermal energy for conversion into electricity in July 1904. He was a pioneering scientist in Lardarello city Italy who created a mini geothermal power plant to power several incandescent lamps [12]. As an alternative energy source, geothermal energy has the important advantage of being one of the cleanest energy sources since the energy production process lacks CO_2 and/or greenhouse gas emissions, unlike its fossil-based energy source counterparts. Geothermal energy resources will never run out because if utilized in power plant, pressurized steam experiences a renewable and sustainable natural circulation process [13,14]. Geothermal energy is environmentally friendly, meaning it does not cause pollution (air pollution, noise pollution, gas pollution, liquids pollution, and other toxic materials) [15]. Compared with other alternative energy sources such as wind energy and solar energy, geothermal energy source is more stable even under weather and seasonal changes. In addition, the electrical energy generated from geothermal does not require the use of energy storage since geothermal energy source is dispatch able, and thus it operates according to the power plant capacity and load demands. Furthermore, geothermal power plants require a narrower physical area than the conventional power plant [16,17].

Dieng area has been identified, investigated, and explored for its geothermal potential since 1918. The Dieng geothermal field in the Dieng plateau sits at 2000 m above sea level in the Central Java province. Based on local meteorological data, the atmospheric pressure at Dieng is 78.06 kPa with an average annual ambient temperature of 18 °C [18]. The United States Agency for International Development (USAID) together with the United State Geological Survey (USGS), state owned utility (PLN) and Institute Teknologi Bandung (ITB) [5,8] teamed up from 1970 to 1972 to investigate the

Kawah Sikidang region of the Dieng area. According to Radja (1975) [19], in 1972 the team drilled several exploration holes with depths reaching 145 m and obtained temperatures reaching 175 °C, but the wells had been considered unproductive. Additionally, Pambudi [1] corroborated this finding by reporting that in 1973 the geothermal wells gained geothermal potential with temperatures ranging from 92 °C to 173 °C.

The binary cycle power plant has been used for geothermal reservoir with low operating temperature condition down to 100 °C. The fluid cannot be used to control the turbine directly; however, the geothermal reservoir can be applied as a heat source to vaporize working fluids [20,21]. The hot fluid from the geothermal reservoir flows through the pipe to a heat exchanger, which then combines with a working fluid such as butane or pentane hydrocarbon having lower boiling point temperature. The working fluid changes from liquid phase to vapor phase and then streamed through a pipe to drive a turbine. The turbine couples with a generator to produce electricity as the turbine turns. After driving the turbine, the vapor comes out and flows into the condenser. In the condenser, the vapor is cooled under liquid to yield waste water which is then injected back into the geothermal reservoir in the earth. Thus, two types of fluid, hot water and secondary fluid as working fluids, are required in this type of geothermal plant. When compared to other types of geothermal power plant, the binary cycle geothermal power plant works well in the lowest heat source condition and produces the best efficiency at the same temperature as the resources. If the secondary fluid is chosen to have a high-density vapor, then the dimensions of the turbine and heat exchanger could be made smaller. Another benefit of the binary cycle power plant is the absence of the flashing process with its associated issues in the condenser such as non-condensable gases resulting in decreased generated power and worsening of emissions. However, the use of the binary cycle power plant entails a technical challenge from the occurrence of scaling in the primary heat exchanger [22].

Geothermal resources with latent heat at 150 °C (423 °F) and medium temperatures have also utilized the binary cycle power plant. Such application of the binary cycle may make use of several thermodynamic systems including the Organic Rankine Cycles (ORC) and the Kalina cycles [23]. Some researchers such as Madhawa et al. (2007) [24], DiPippo (2012) [25], Bayer et al. (2013) [26], Guzovic' et al. (2014) [27], Liu et al. (2014) [28], and DiPippo (2015) [29] explained that the appropriate binary plant technology for low temperature should implement the ORC, especially to convert geothermal energy into electrical energy. Geothermal energy conversion process produces a hot fluid called brine which is streamed through a heat exchanger (evaporator) to transfer heat energy to a second working fluid and to further be re-injected to the earth. The working fluid changes to a superheated vapor when exiting the evaporator. The superheated vapor streams through a turbine and exits into a condenser. The working fluid comes out of the condenser as a feed liquid in the reservoir tank and is pumped back to the evaporator to complete the Rankine cycle [30]. Typically, the working fluid uses organic fluid which has a low boiling point temperature and high pressure vapor [31,32]. These conditions are needed to allow for size reduction of the turbine. Moreover, the binary system offers another benefit in terms of flexibility in the plant's power capacity which may vary from hundreds of megawatts to a few megawatts [33]. Bertanni [34] and Franco et. al. [35] revealed that about 70% of geothermal sources in the world have the potential for use with hot water with low enthalpy running at temperatures below 150 °C. This further demonstrates the importance of the binary cycle plant.

Implementation of the binary cycle technology for geothermal power plants will also require the use of the most suitable type of working fluid. To this extent, Franco et. al. [36] conducted further experiments by testing several working fluids in their ORC system. They concluded that working fluids isobutane, n-pentane, and R152a had better performance than others being used in their experiments. Liu et. al. [37] performed a similar study and were captivated by using isobutene and R245fa as the working fluids. Coskun et. al. [38] discovered that isobutene was the most appropriate as an ORC working fluid by utilizing geothermal heat sources. Along the same line, Budisulistyo et al. [39] recommended the use of n-pentane as a working fluid based on economic analysis reasons. Shengjun et al. [40] investigated the use of 16 different working fluids at 80 °C to 100 °C. Their study discovered that isobutene required the lowest cost to produce electricity and R152 required the smallest area of heat exchanger per unit of output power.

Based on the aforementioned studies, we conducted a research to investigate the potential of Dieng's geothermal source for electricity generation. The research utilized a laboratory setup to simulate the binary cycle plant in an ORC system. The potential heat source in the Dieng mountains was simulated using heated lubricant oil to obtain continuous controllable temperature. Refrigerants R245fa, R123, and mixtures of both refrigerants were selected as the working fluids in the ORC system. Comparison of thermodynamic performance with geothermal heat source operating at 110 °C and 120 °C as well as other pertinent results of the research are presented and discussed in this paper.

2. Thermodynamic Modeling

The principal operation and heat transfer of the ORC system are illustrated in Figure 2 which shows the T–S diagram of the thermodynamic cycle. The second law of thermodynamics, conservation of energy, conservation of mass, and thermodynamic parameters are parts of the thermodynamic analysis. Relevant to thermodynamics, major components of the ORC plant should consist of pumps, heat exchanger for the evaporator, scroll expander, and heat exchanger for the condenser. For testing purpose, thermodynamic properties of working fluids on each individual component in the ORC system can be evaluated by observing their pressure and temperature.

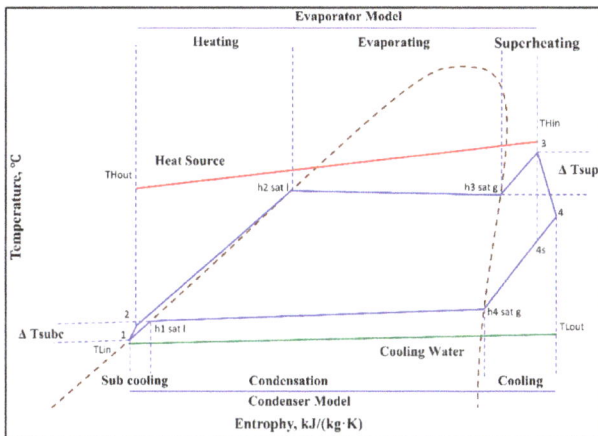

Figure 2. The principle operation and heat transfer of the Organic Rankine Cycle (ORC) system.

The mathematical model of each individual component is calculated as follows:

Input Power Pump

The input power pump increases working fluid pressure from state 1 to state 2 in order to match the operation rate of the evaporator. Volume control was built around the pump as a barrier for incoming and outgoing heat transfers with the surrounding. The pump power can be written as:

$$W_p = \dot{m}(h_2 - h_1),$$

(1)

where $\dot{m}$ represents mass flow rate for working fluid. Isentropic efficiency ($\eta_{is,\,pump}$) and mechanical efficiency ($\eta_{me,\,pump}$) can be expressed as:

$$\eta_{is,pump} = \frac{h_{2s} - h_1}{h_2 - h_1},$$

(2)

$$\eta_{me,pump} = \frac{\dot{m}(h_2 - h_1)}{W_{ele,pump}}, \tag{3}$$

where h_{2s} and h_2 are the specific enthalpies of the working fluid at the outlet of the pump under ideal and actual conditions, respectively.

Evaporation Process

The compression vapor flows throughout an evaporator at state 3. During heat exchange process (state 2 to state 3), the hot oil infiltrates heat into the working fluid. The total heat transfer rate (Q_{evap}) from the heat source to working fluid in the evaporator is represented in (4):

$$Q_{evap} = \dot{m}(h_3 - h_2), \tag{4}$$

where h_3 is the specific enthalpy of working fluid at the evaporator outlet.

Expansion Work of Scroll Expander

The superheated vapor at state 3 flows in scroll expander where it expands and produces output power by rotating the shaft. The pressure and temperature drop during this process then discharge to state 4. By neglecting heat transfer flow in and flow out to the surrounding, the scroll expander output power (W_t) can be written as:

$$W_t = \dot{m}(h_3 - h_4), \tag{5}$$

where h_4 is the specific enthalpy of working fluid at the expander outlet.

Condensation Process

A condenser facilitates heat transfer from steam to cooling water that flows in a separate channel. According to the principle of mass and energy balance for a controlled volume at a condenser, the output heat can be written as:

$$Q_{cond} = \dot{m}(h_1 - h_4). \tag{6}$$

where Q_{cond} is the condenser output heat and h_4 is the specific enthalpy of working fluid at the condenser inlet.

Thermal Efficiency

The thermal efficiency can be represented as the ratio between total output to input powers and heat transfer rate. The thermal efficiency can be described as:

$$\eta_{th} = \frac{W_t - W_p}{Q_{in}} = \frac{(h_3 - h_4) - (h_2 - h_1)}{h_3 - h_2}. \tag{7}$$

Since work input and output powers are equal to total operating heat, another expression of the thermal efficiency may be calculated as follows:

$$\eta_{th} = \frac{Q_{in} - Q_{out}}{Q_{in}} = 1 - \frac{Q_{out}}{Q_{in}} = 1 - \frac{(h_4 - h_1)}{(h_3 - h_2)}. \tag{8}$$

3. Experimental and Equipment Setup

The schematic flow diagram of an experimental ORC operation is displayed in Figure 3. The system consists of three sections: heating cycle, ORC cycle, and cooling cycle. Figure 4 depicts the laboratory implementation of the ORC experiment showing the various equipment used in the setup.

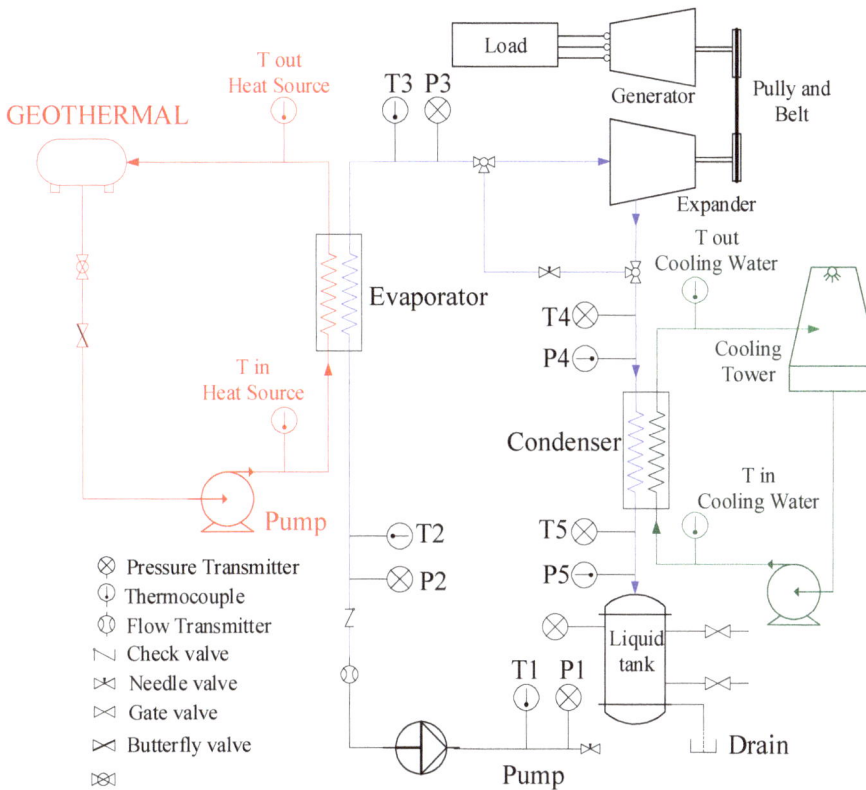

Figure 3. Flow diagram of an experimental ORC system.

3.1. Heat Cycle

An external heat source comprising four electric heating rods with capacity of 80 kW heats up lubricant oil (S-OIL Total Lubricants Co., LTD., Seoul, Korea) that serves as a heat source simulator to provide input heat to the ORC system. The selected lubricant oil has excellent thermal stability even at high temperature, providing a stable 120 °C operating temperature. The axial pump (SAER Elettropompe, Guastalla, Italia) adjusts the mass flow rate of lubricant oil. Moreover, the evaporator's heat transfer rate can be changed by adjusting the electric heating input power, controlled by the electric heating rods. The hot oil operation is controlled to yield a fixed 110 °C and 120 °C in the evaporator inlet.

3.2. ORC Cycle

The ORC system has four main components: pump (Wuli Agriculture Machine CO., LTD., Taichung, Taiwan), evaporator (Kaori Heat Treatment CO., LTD., Taoyuan, Taiwan), expander (Shenzhen Sino-Australia Refrigeration Equipment Co., Ltd., HuiZhou, Chinna), and condenser (Kaori Heat Treatment CO., LTD., Taoyuan, Taiwan). The main components are integrated into one closed cycle system to take advantage of low waste heat into electrical energy as seen in Figure 4a. The amount of electrical energy consumed by the piston pump could be identified by gauging current and voltage, whereas the pump shaft's power could be counted with thermodynamic indicators in the pump inlet and outlet. The piston pump passes the working fluid from the holding tank to the evaporator by escalating the working fluid pressure to match the operating pressure required by the evaporator. Pressurized working fluid enters the evaporator in which the working fluid is vaporized by hot oil as a heat resource. Vapor in high-pressure streamed into scroll expander to expand enthalpy to produce output power. After the expansion process, the vapor in low-pressure streams out from a scroll expander and leads to a condenser for releasing heat energy and then it turns into a liquid in the subcooled-phase position. The working fluid in subcooled-phase flows and it is collected into the reservoir tank to be re-pumped to start a new cycle.

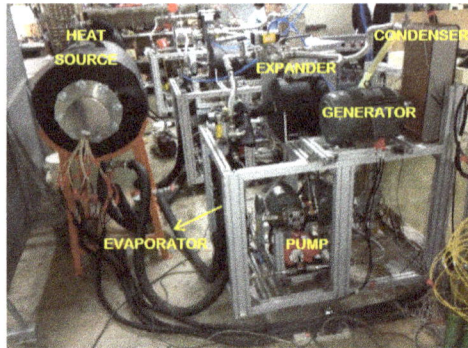
(a)

(b)

(c)

Figure 4. Experiment apparatus layout: (**a**) experimental set up; (**b**) electrical load; and (**c**) evaporator.

Evaporator and condenser use a heat exchanger with plate heat exchanger (PHE) type. Compared to other heat exchangers, PHE has several advantages such as flexible thermal size, easy cleaning to maintain extreme hygienic conditions, great approaching input heat temperature, and improved heat transfer performance [41]. For the interest of energy conservation and space saving, the PHE is used in the experimental ORC system. Since the refrigerant used in the ORC is highly corrosive and high-pressured substance, the brazing plate heat exchanger (BPHE) (Kaori Heat Treatment CO., LTD., Taoyuan, Taiwan) is particularly suitable to utilize where stainless steel vacuum brazing plates use copper as the brazing material, as shown in Figure 4c. The BPHE evaporator and condenser have heat transfer area of 4.157 m^2 which employs glass wool and barrier foam around the evaporator to prevent escaping heat.

The expansion process area utilizes a scroll expander which is taken from a cool storage compressor. The mechanical power of the expander rotates the shaft and the coupled generator (as displayed on Figure 4a) by adding pulleys and belts. An induction motor is used as the generator due to its low cost and availability. Re-magnetization in the rotor is generally sufficient to generate the initial voltage of the generator. To fulfill the reactive power requirement to generate a rotating magnetic field, an excitation capacitor is implemented, as depicted Figure 4b.

3.3. Cooling Cycle

The cooling cycle for the ORC system utilizes cooling towers with closed circuits operating in a counter-flow basis. The cooling tower system delivers cooling fluid to infiltrate heat energy from condenser and dissipates the heat energy into the ambient air through spray at the top of the cooling towers. There are two separate and distinct functions of fluid circuitry (as displayed in Figure 3). The external fluid circuit decreases the temperature of the ORC system and it is called the cooling cycle. The second circuit, the internal fluid circuit, is in the center of the ORC system and the fluid acts as a working fluid. The cooling tower is on the rooftop as part of the building cooling system. This causes the cooling water temperature to fluctuate due to the surrounding environment temperature. Cooling tower's water circulation is controlled by a needle valve to enable adjustment of speed and mass flow rate.

3.4. Measurement Equipment

The ORC laboratory experiments were conducted to simulate the operation of an actual ORC system. All parameters were considered to represent the whole ORC operation. The main indicators measured were temperature and pressure at the pump inlet and pump outlet, mass flow rate of working fluid, temperature and pressure on expander inlet and outlet, temperature inlet and outlet of heat resources, temperature inlet and outlet of cooling water, and swivel speed of the expander and power generator. The T-type thermocouple (Deange Industry Co., Ltd., New Taipei, Taiwan) was used to measure the temperature, and the piezo-resistive pressure transmitter (Jetec Electronics Co., Ltd., Taichung, Taiwan) was used to measure the pressure. The measurement results combined with NIST Refrigerant Properties (NIST REFPROP) can be used to determine the enthalpy value and entropy for each state. The NIST REFPROP database provides the most accurate thermo-physical property model for a range of industry-important fluids and fluid mixtures, including accepted standards. Based on the model, pump work input, pump isentropic efficiency, evaporator heat input, expander work output, condenser heat output, and thermal efficiency can then be calculated.

Measurement uncertainty is an expression of statistical dispersion of values associated with measured quantities. Error propagation theory revealed the measurement uncertainties is calculated using the root-sum-square method. Account results of the uncertainty U_y from variable Y are calculated as a function of the uncertainties U_{xi}, for each measured variable x_i, which are presented in Equation (9). Table 1 presents a list of the accuracy of the measuring instruments obtained from the manufacturer's data sheet.

$$U_y = \sqrt{\sum_i \left(\frac{\partial Y}{\partial x_i}\right)^2 U_{x_i}^2} \tag{9}$$

Table 1. Accuracy of measuring instrument.

No	Measuring Instrument	Type	Range	Accuracy
1	Pressure transmitter (Jetec Electronics Co., Ltd., Taichung, Taiwan)	JPT-131S	0–30 bar	±0.5% P.S
2	Temperature (Deange Industry Co., Ltd., New Taipei, Taiwan)	T-type	0–623.15 °K	±0.3 °C
3	Flowmeter (Great Plains Industries, Sydney, Australia)	GPI S050	1.9–37.9 L/min	±0.3% L/min
4	Rotation meter (Uni-Trend Technology (Dongguan) Limited, Dongguan, China)	UT-372	10–99,999 rpm	±0.3% rpm
5	Power meter (Arch Meter Corporation, Hsinchu, Taiwan)	PA310	V (0–300 VAC), 1 (0–400 A) Hz (50/60 Hz), PF(−1–1)	±0.5%

3.5. Working Fluids

The choice of a working fluid for the ORC operation is crucial since it affects the dimension of system components, design of the expansion machine, system efficiency and cost [42,43]. Safety of the working fluid is another main requirement, and so the important features of working fluid should be low toxicity, controlled explosion and flammable characteristics, chemical stability, thermal conductivity, boiling temperature, blow-off point, latent heat, and specific heat. Environmental hazards: GWP (Global Warming Potential) and ODP (Ozone Decrease Potential) are in fact the main issues for researchers in the ORC operation to determine the most suitable working fluid.

Working fluids R123 (Dupont Taiwan Ltd., Taipei, Taiwan) and R245fa (Hangzhou Xianglin Chemical Industry Co., Ltd., Hangzhou, China) are commonly applied for experimental ORC operation due to their preferred thermodynamic performance and environmental advantage (low GWP and ODP effects). Based on the slope of the T–S diagram, working fluids R245fa and R123 could be categorized into highly profitable dry fluids in the expander area because if applied in the wet working fluid, they naturally generate droplets which are affected in expander failure. Hence, the dry and isentropic working fluids will work well and should yield precise results [44,45].

As previously discussed, there are many working fluids used for low-temperature heat sources. However, this is not the case high-temperature heat sources. Chen et. al. [46] revealed that the temperature range contained in the heat source potential has a highly influential relationship. Heat source temperature provides an idea for researchers to determine the appropriate working fluid, one of which was done by Xu et. al. [47]. They recommended applying R245fa as a working fluid due to its capability to operate in a wide range of heat resource temperature. An application was directly carried out by Feng et. al. [48] using the working fluid R245fa by considering thermal efficiency and environmental performance. In addition, they revealed that the working fluid of R245fa was suitable for heat source with 125 °C temperature. Table 2 provides a list of other researchers, which in their study used working fluid R245fa and R123.

The idea of mixing working fluids to get different results and better impact to heat source potential in the ORC system was investigated further by Li et. al. [60]. Several studies have been conducted to compare pure and mixed working fluids such as Feng et. al. [48] and Pang et. al. [61] who used a mixed working fluid R123 with R245fa. Prior to conducting research experiments, Pang et. al. [61] conducted a safety test by heating the sealed container in two working fluid mixtures. In this study, we employed working fluids R245fa, R123, R245fa 1:1 R123 (admixture R245fa 50% and 50% R123), R245fa 2:1 R123 (admixture R245fa 66.6% and 33.3% R123), and R245fa 1:2 R123 (admixture R245fa 33.3% and 66.6% R123). We also determined the working fluid cycles in the system with mass flow rate set at 0.1 kg/s, 0.13 kg/s, 0.15 kg/s, 0.175 kg/s, and 0.2 kg/s. The thermos-physical characteristics of working fluid R245fa, R123 and mixture ratio are both listed in Table 3.

Table 2. Previous completed experiments using R245fa and R123 as working fluids.

No	Year	Researcher	Working Fluid	Expander Type
1	2018	Jiang et al. [49]	R123	-
2	2017	Feng et al. [50]	R123	Scroll expander
3	2017	Yang et al. [51]	R245fa	Scroll expander
4	2017	Shao et al. [52]	R123	Radial turbine
5	2017	Feng et al. [53]	R245fa	Scroll expander
6	2016	Eyerer [54]	R245fa	Scroll expander
7	2016	Shu et al. [55]	R123 & R245fa	Expansion valve
8	2015	Chang et al. [56]	R245fa	Scroll expander
9	2014	Chang et al. [57]	R245fa	Scroll expander
10	2013	Li et al. [58]	R123	Axial flow turbine
11	2012	Shu et al. [59]	R123	Turbine expander

Table 3. Thermo-physical properties of R245fa, R123, and mixture ratio.

Working Fluid	R245fa	R245fa 2:1 R123	R245fa 1:1 R123	R245fa 1:2 R123	R123
Type	Dry	Dry	Dry	Dry	Dry
Formula	$CHCl_2CF_3$	-	-	-	$CF_2CH_2CHF_2$
Molecular mass (g/mol)	134.03	139.8	142.87	146.07	152.93
Freezing point (°C)	<-107	-	-	-	-107
Critical Temperature	154	158,19	162.5	167.9	183.8
Critical pressure	36.504	36.435	36.638	36.59	36.6
Density (kg/m³)	537.03	568.11	550.07	510.09	550
Ozone Depletion Potential (ODP)	0	-	-	-	0.02
Global Warming Potential (GWP)	950–1030	-	-	-	77
Inflammability	nonflammable	-	-	-	nonflammable
Vapor Viscosity	10.3 cP	-	-	-	0.011 cP
Liquid Viscosity	402.7 cP	-	-	-	0.456 cP
Vapor Specific Heat	0.89 kJ/(kg·K)	-	-	-	0.72 kJ/(kg·K)
Liquid Specific Heat	1.36 kJ/kg	-	-	-	0.965 kJ/(kg·K)
Liquid Thermal Conductivity	0.081 W/(m·K)	-	-	-	0.096 W/(m·K)

The T–S diagram of working fluids could be explained as displayed in Figure 5 with data presented in Table 3 using the NIST REFPROP program to obtain the properties in saturated liquids, saturated gas, and entropy data.

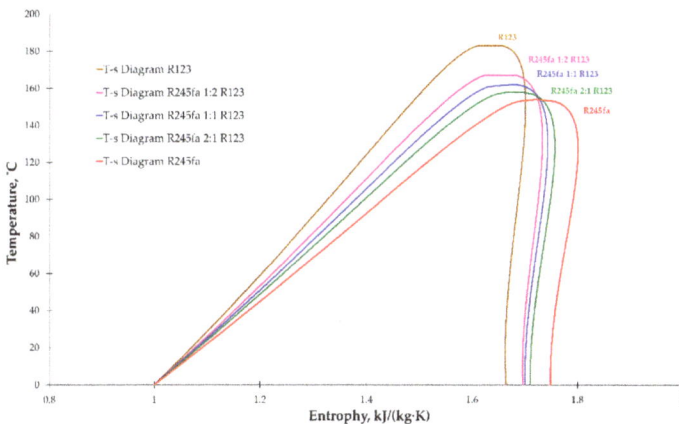

Figure 5. T–S diagram for R245fa, R123, and mixed working fluids.

4. Results and Discussion

Geothermal energy potential in the Dieng plateau in Indonesia was simulated using a laboratory setup with the heat source temperature set at 110 °C and 120 °C. The heat source is generated from several heaters to heat up lubricating oil to simulate heat energy from geothermal. The mass flow rate can also be changed by varying the pump frequency. Experimental data obtained from laboratory tests reveal the thermodynamic performance of the ORC operation.

4.1. General Experimental Conditions and Steady-State Measurements

The simulated heat source used 40 kW to heat the oil in the tube resources until the temperature reached 110 °C and 120 °C at the inlet evaporator. Hot oil circulation used an adjustable axial pump to convert the mass flow rate of hot oil to an evaporator. All experiments used an axial pump at speed of 55 Hz, and therefore, it was likely the heat flow conditions flowed into the evaporator at a constant speed without any change. The lubricating oil was from the TOTAL Company with SERIOLA K 3120 type with a specific heat (cp) of 0.535 kcal/(kg°C) at temperature 120 °C.

The mass flow rates of hot oil are presented in Table 4.

Table 4. The mass flow rates of hot oil.

R245fa kg/s	R123 kg/s	R245fa 2:1 R123 kg/s	R245fa 1:1 R123 kg/s	R245fa 1:2 R123 kg/s
3.51	7.55	3.52	3.23	3.66
3.45	8.27	3.72	3.68	3.74
3.51	9.39	3.65	3.48	3.97
3.60	10.05	3.89	4.05	4.37
3.95	10.30	4.28	4.55	4.81

Cooling utilized water from cooling towers with a mass flow rate ranging from 2.12 kg/s to 3 kg/s. According to data presented in Table 5, the mass flow rate in a condenser indicated a uniform value for all experiments. Temperature of air cooling tower fluctuated due to the influence from the surrounding environment. The cooling tower was located on the roof of a manufacturing factory building to serve the cooling system. Fluctuations in the operating condition of cooling water are important factors that affect for ORC system. To control the influence of the mass flow rate in the cooling cycle, a needle valve is installed and arranged the valve opening.

Table 5. Data condition of cooling cycle operation.

R245fa			R123			R245fa 1:1 R123			R245fa 2:1 R123			R245fa 1:2 R123		
Cold Water Inlet °C	Cold Water Outlet °C	Flow Rate Kg/s	Cold Water Inlet °C	Cold Water Outlet °C	Flow Rate Kg/s	Cold Water Inlet °C	Cold Water Outlet °C	Flow Rate Kg/s	Cold Water Inlet °C	Cold Water Outlet °C	Flow Rate Kg/s	Cold Water Inlet °C	Cold Water Outlet °C	Flow Rate Kg/s
27.67	30.55	2.14	25.65	27.96	2.12	30.21	30.25	2.48	19.28	21.61	2.43	25.04	27.04	2.61
28.46	32.04	2.12	16.27	18.27	3.17	28.50	31.20	2.44	19.96	22.82	2.46	25.35	27.89	2.55
28.85	33.01	2.13	16.25	18.54	3.12	28.29	31.02	2.74	21.50	24.78	2.47	25.63	28.45	2.61
29.06	33.52	2.15	16.33	18.70	3.04	28.31	31.15	2.83	21.57	25.03	2.58	25.66	28.60	2.74
29.11	33.70	2.20	18.42	20.97	3.00	28.37	31.35	2.99	21.09	24.66	2.71	25.66	28.63	3.03

The evaporator intake temperature was maintained at a constant value of 110 °C and 120 °C. Data were collected every 5 s under steady-state conditions for 20 min, resulting in a total of 240 data points for each experimental condition. Mass flow rate was varied in five conditions to produce 1200 data points. As previously stated, this study employed five working fluids: R245fa, R123, mixtures of R245fa and R123 with the three different compositions: (1) R245fa 2:1 R123, (2) R245fa 1:1 R123 and (3) R245fa 1:2 R123. Thus for the experiments, these five types of working fluids produced 6000 data points for further analysis.

Observing the working fluid behavior in the ORC cycle section is the focus of the experiment. To explore the experiments, it was carried out by varying the mass flow rate ranging from 0.1 kg/s, 0.125 kg/s, 0.15 kg/s, 0.175 kg/s, and 0.2 kg/s with temperature at the inlet evaporator at 110 °C and 120 °C. A number of datasets such as pump inlet temperature, pump outlet temperature, inlet expander temperature, outlet expander temperature, heat source inlet temperature, heat source outlet, cooling water channel, and cooling water outlet were recorded and presented in Figure 6. The error bar shows data variation due to error deviation or uncertainty in performing the measurements. The 10 data samples were in steady-state condition for 20 min operation, where the heat source inlet at evaporator was kept constant and recorded temperature (Thin) from 120.205 °C to 120.309 °C, then flowed out from the evaporator with heat source outlet temperature (Thout) recorded from 105.048 °C to 105.075 °C. The working fluid operated at the pump inlet temperature (T1) ranging from 28.637 °C to 28.666 °C, and the pump outlet temperature (T2) were recorded to range from 29.058 °C to 29.11 °C.

When the working fluid was in the expander the inlet temperature (T3) ranged from 117.231 °C to 117.482 °C, the working fluid underwent expansion and flowed out from the expander with the temperature (T4) ranging from 87.586 °C to 87.672 °C. In the cooling process, cooling water flowed in the condenser with temperatures (TLin) from 28.61 °C to 28.646 °C, then flowed out with temperatures (TLout) from 31.267 °C to 31.347 °C.

Figure 7 presents 10 experimental data for pressures and volume flow rates in steady state condition using R245fa 1:1 R123 working fluid, mass flow rate 0.125 kg/s and heat source 120° C with a 5% bar error. The volume flow rate of the working fluid was controlled by adjusting the frequency of the pump motor. The recorded data were 5.409 liters per minute (L/min) up to 5.430 L/min. The working fluid, before passing the pump, has an inlet pressure pump (P1) from 2.456 bar to 2.483 bar. Afterwards, the pump performed a compression process and made an outlet pressure pump (P2) from 10.513 bar to 10.589 bar. When the working fluid streamed into an expander, it should be noted that expander inlet pressure (P3) of 9.66 bar to over than 10 bar were measured, then the working fluid expanded processing onward and streamed out with an expander outlet pressure (P4) from 2.718 bar to 2.737 bar.

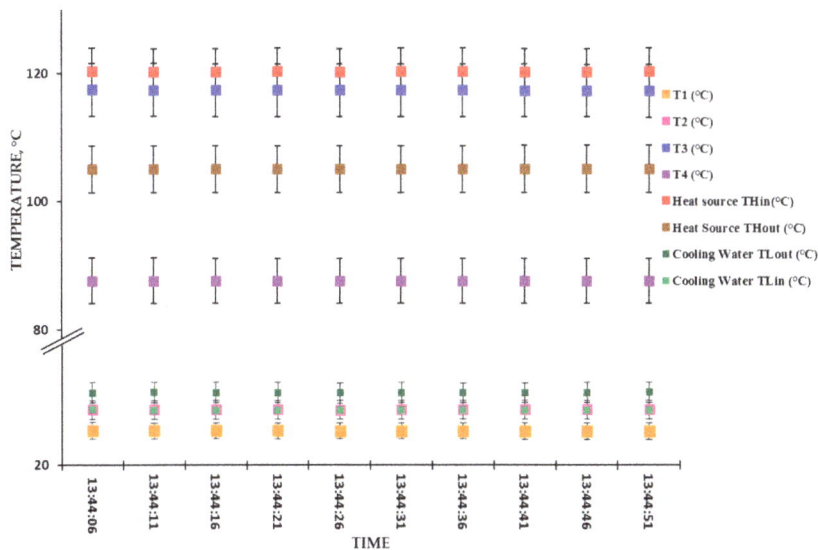

Figure 6. Temperature data in steady state condition for R245fa 1:1 R123 working fluid, mass flow rate of 0.125 kg/s and heat source 120 °C with error bar 5%.

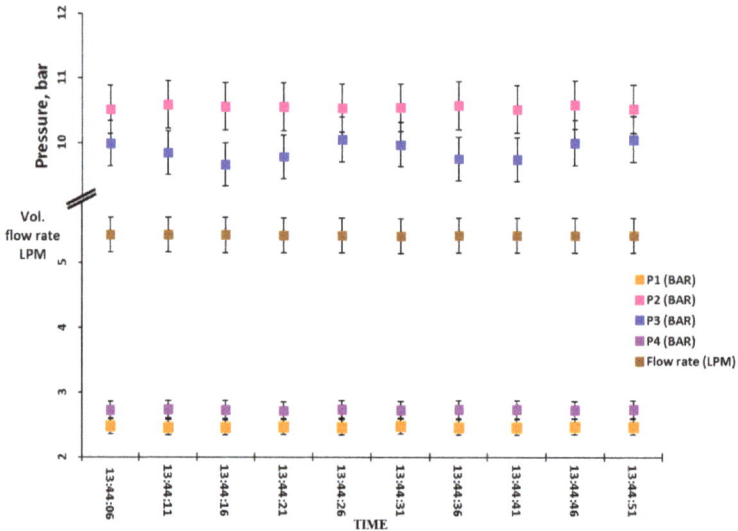

Figure 7. Pressure and volume flow rate data in steady state condition for R245fa 1:1 R123 working fluid, mass flow rate 0.125 kg/s and heat source 120 °C with error bar 5%.

The highest achievement of working fluid pressure occurred in state 2 because the subcooled working fluid was compressed by the pump to pressure P2 (indicated by the pink box in Figure 7) at constant temperature T2 (indicated by the pink box in Figure 6). Compressed working fluid enters the evaporator to be converted into superheated vapor at temperature T3 (indicated by the blue box in Figure 6) and constant pressure P3 (indicated by the blue box in Figure 7). If it works ideally, pressure P2 is equal to pressure P3, but in reality pressure P3 is lower than pressure P2. This is caused by the use of a plate heat exchanger (PHE) as an evaporator. When working fluid flows inside the PHE, pressure drops occur due to friction with a narrow corrugated wall.

The indicator chart in Figure 8 demonstrates the variation of pump inlet pressure (P1) and outlet (P2) as a function of mass flow rate for R245fa, R123, and mixed working fluids. As reviewed in Figure 8, the pump inlet pressure (P1) was not affected by changes in the mass flow rate due to almost remained unchanged. Individual working fluid was difficult to observe because the indicators showed coincide position with suppress each other. Meanwhile, the pink circle indicators are easier to see than others, representing R245fa 1:1 R123 with a slightly higher pump inlet pressure with solid circles for heat source 110 °C and hollow circles for heat source 120 °C. Adjustment of the heat source at 110 °C have produced pump inlet pressures ranging from 2.08 bar to 2.16 bar and when heat source changed to the temperature of 120 °C, pump inlet pressures ranged from 2.46 bar to 2.50 bar.

The main function of the ORC pump is to increase the working fluid pressure as can be observed from the pump outlet pressure (P2) in Figure 8. All working fluid presents P2 uniform movement and trends increase with the addition of mass flow rate. R123 generates the lowest P2 when the heat source temperature is at 110 °C, which are in the range from 7.73 bar to 8.00 bar. Furthermore, higher heat source witnessed higher pump outlet pressures, as noted R123 with heat source 120 °C produces P2 in the range of 8.88 bar to 9.89 bar. The highest of pump outlet pressures is obtained by R245fa, which recorded P2 with values ranging from 9.88 bar to 12.43 bar at heat source 110 °C. While using heat source 120 °C, R245fa yielded P2 values in the span of 8.89 bar to 13.01 bar.

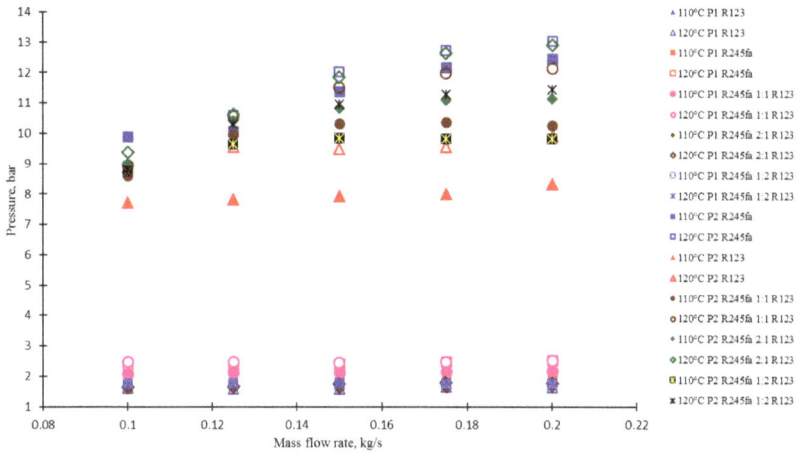

Figure 8. Pressure at pump inlet and outlet with mass flow rate for R245fa, R123, and mixed working fluids.

4.2. Thermodynamic Performance with Heat Source at 110 °C

Figure 9 demonstrates the transformation in mass flow rate with the pump work input for R245fa, R123, and mixed working fluid. The pump work input escalates with increasing mass flow rates, which is associated with large frequencies of the pump.

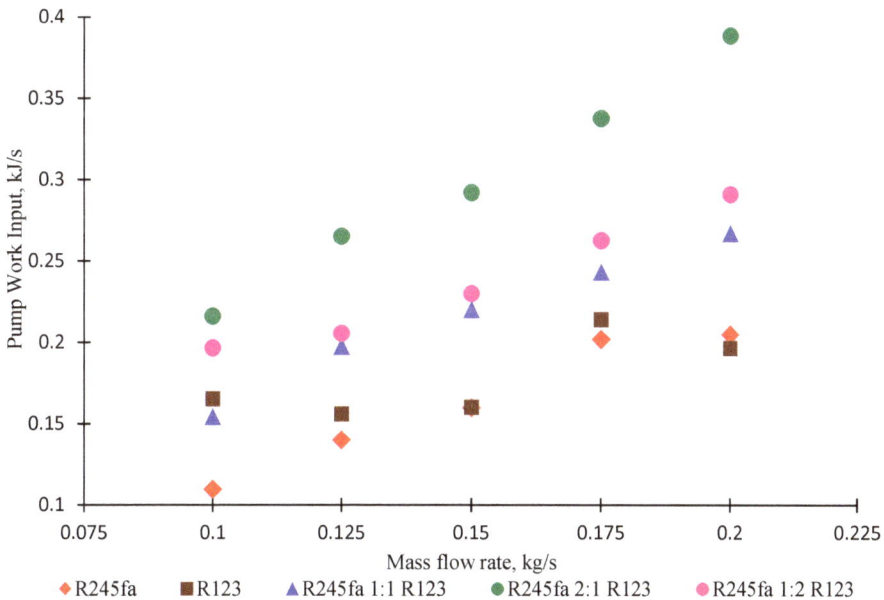

Figure 9. Pump work input as a function of mass flow rate for R245fa, R123, and mixed working fluids with heat source 110 °C.

The mixed working fluids produces pump work input ranging from 0.155 kJ/s to 0.388 kJ/s, while the pure working fluids give a lower value in the range 0.1095 kJ/s to 0.214 kJ/s. Notably, the three mixed working fluids (R245fa 2:1 R123, R245fa 1:1 R123, and R245fa 1:2 R123) have a relatively higher pump work input with spacing from 0.243 kJ/s to 0.388 kJ/s. When compared with working fluid R245fa, the pump work input for the mixed fluids are higher by 10%–70%, and if compared with R123 the values are higher by around 20%–80%. This phenomenon is due to mass flow rate and enthalpy changes in the pump. At a predetermined mass flow rate, R245fa 2:1 R123 has a specific heat of 1.2312 kJ/(kg·K) (18.041 °C and 8.34 bar), and therefore it is higher than R123 which has 1.0097 kJ/(kg·K) (18.041°C and 8.34 bar). Accordingly, R123 requires a pump work input smaller than R245fa 2:1 R123. Therefore, this reveals that the mixed working fluids require a pump work input greater than the pure working fluids. For a specific mass flow rate of 0.15 kg/s, the pump work input requirement for working fluids R245fa, R123, R245fa 2:1 R123, R245fa 1:1 R123, R245fa 1:2 R123, and R123 are at 0.1599 kJ/s, 0.1602 kJ/s, 0.2919 kJ/s, 0.2202 kJ/s, and 0.2299 kJ/s, respectively.

Dynamic changes of the pump isentropic efficiency are related to the mass flow rate for R245fa, R123, and mixed working fluids as displayed in Figure 10. By adjusting the mass flow rate from 0.1 kg/s to 0.2 kg/s, the pump output parameters and pump work input requirements changes and forces the pump's performance to approach isentropical operation. When using a pure working fluid, the pump isentropic efficiency has a similar behavior, where the trend increases from the beginning of a mass flow rate of 0.1 kg/s to the highest at a mass flow rate of 0.2 kg/s. The experimental data show that R245fa working fluid has a pump isentropic efficiency starting from 42.66% to 52.06%, while the lowest pump isentropic efficiency is obtained from R123 working fluid with magnitudes between 25.15% and 36.12%. Furthermore, the mixed working fluids produces the position of pump isentropic efficiency between pure working fluids (R245fa and R123) with values ranging from 25.15% to 45.88%. According to Feng et. al. [48], transformation in a fluid mass fraction will affect the density of a mixed working fluid, resulting in decreased pump isentropic efficiency. At specified mass flow rate of 0.15 kg/s, the pump isentropic efficiency for R245fa, R123, R245fa 2:1 R123, R245fa 1:1 R123, and R245fa 1:2 R123 are 47.68%, 35.27%, 45.25%, 40.66%, and 37.77%, respectively.

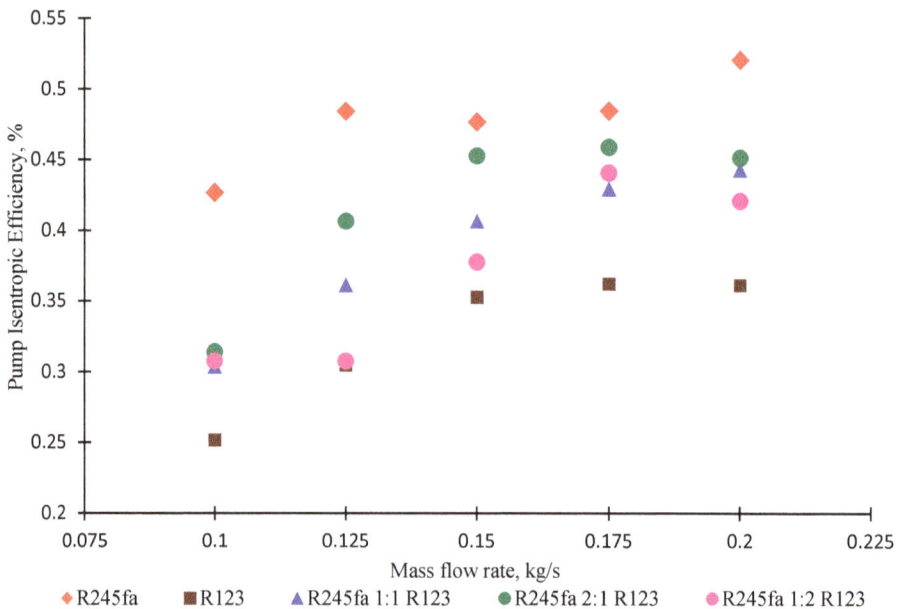

Figure 10. Pump isentropic efficiency as a function mass flow rate for R245fa, R123, and mixed working fluids with heat source at 110 °C.

The geothermal heat source provides some energy in the evaporator as illustrated in Figure 11. The figure also depicts the dynamic movement of evaporator heat input against changes in the mass flow rate. Increasing the mass flow rate for R245fa, R123, R245 2:1 R123, and R245 1:1 R123 produces a uniform heat input, except for the mixed R245 1:2 R123 working fluid which yields a constant increase of heat input with mass flow rate at 0.1 kg/s to 0.15 kg/s and at 0.175 kg/s. At the end of the experiments, an increase in the mass flow rate of 0.2 kg/s is observed. When a pure working fluid (R245fa or R123) is applied, the evaporator develops the highest heat input range from 27.79 kJ/s to 50.09 kJ/s, whereas the mixed working fluids (R245fa 2:1 R123, R245fa 1:1 R123, and R245fa 1:2 R123) produces the lowest evaporator heat input ranging from 27.79 kJ/s to 50.09 kJ/s. This is caused by R245fa, and R123 working fluids having the highest evaporator heat transfer coefficient and the largest increase in Logarithmic Mean Temperature Difference (LMTD) when compared with the mixed working fluids. For the specific mass flow rate of 0.15 kg/s, the evaporator heat input for R245fa, R123, R245fa 2:1 R123, R245fa 1:1 R123, and R245fa 1:2 R12 are 39.67 kJ/s, 38.82 kJ/s, 34.93 kJ/s, 32.59 kJ/s, and 31.44 kJ/s, respectively.

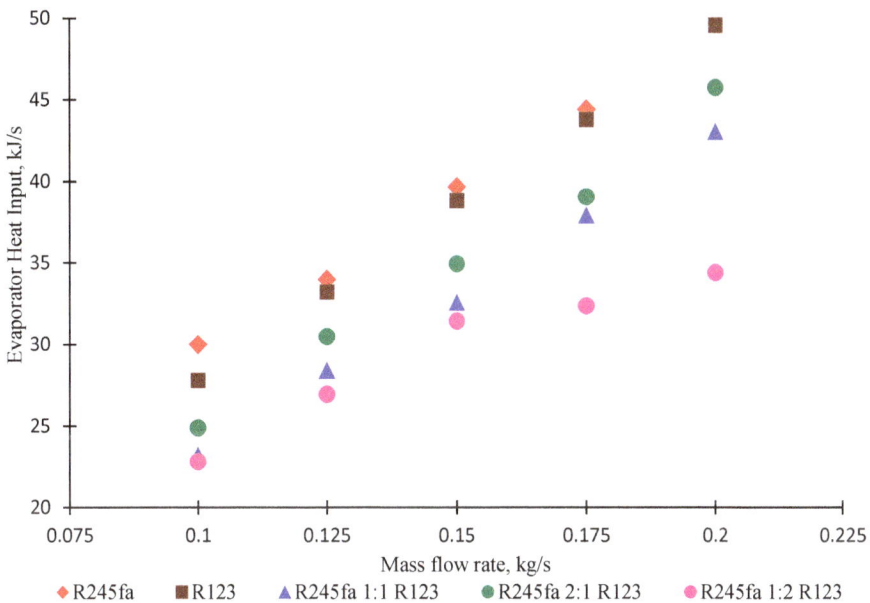

Figure 11. Evaporator heat input as a function mass flow rate for R245fa, R123, and mixed working fluids with heat source at 110 °C.

Figure 12 describes the variation in mass flow rate with the expander work output for R245fa, R123, and mixed working fluids. As seen in Figure 12, the trend for expander work output shows the same increasing behavior, specifically with mass flow rates of 0.1 kg/s to 0.2 kg/s for all working fluids. Increased mass flow rates provide more heat energy to create torque power in the expander. Furthermore, R123 causes an increase in one experimental parameters for expander work output with values ranging from 2.78 kJ/s to 6.5 kJ/s. In addition, R245fa generates the lowest expander work output, which is between 1.6 kJ/s and 5.02 kJ/s or a decrease of around 30%. Nevertheless, all working fluids produce work input values in the range of 1.84 kJ/s to 6.30 kJ/s or a decrease of about 20% in the overall mass flow rate. The work output of an expander is affected by the parameters from working fluid inlet and outlet, which represent how much energy can be expanded to the shaft. Experiments with specific mass flow rate of 0.15 kg/s produce an expander work output for R245fa, R123, R245fa 2:1 R123, R245fa 1:1 R123, and R245fa 1:2 R123 of 2.53 kJ/s, 4.14 kJ/s, 2.91 kJ/s, 3.66 kJ/s, and 3.604 kJ/s, respectively.

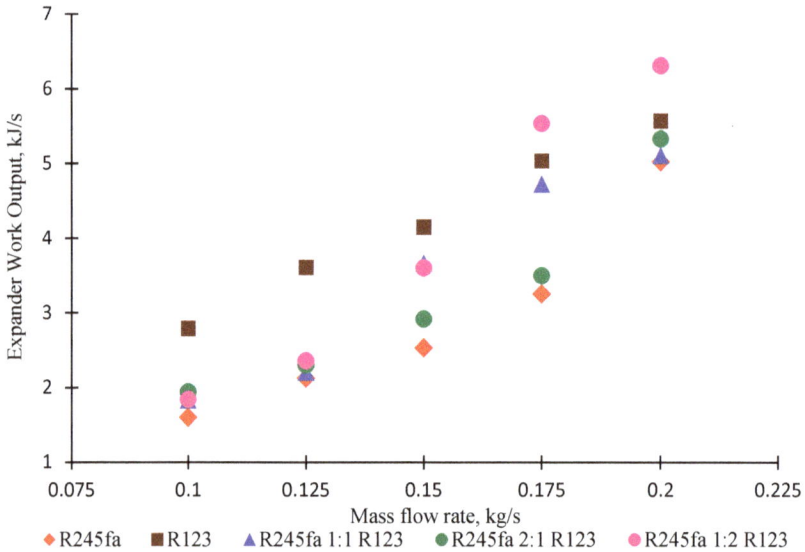

Figure 12. Expander work output as a function mass flow rate for R245fa, R123, and mixed working fluids with heat source at 110 °C.

Figure 13 discloses the effect of changes in mass flow rate on the condenser heat output when simulating a heat source at 110 °C. As seen in the graph, the trend of condenser heat output increases significantly and uniformly except for the R123 and the mixed R245fa 2:1 R123 working fluids. The line appears to fluctuate slightly which means that there is a change in the amount of energy of the condenser heat output. R245fa has the highest trend with results ranging from 26 kJ/s to 43.35 kJ/s and followed by R123 with 24 kJ/s to 40.35 kJ/s. However, the mixed R245fa 2:1 R123, R245fa 1:1 R123, and R245fa 1:2 R123 working fluids produce a condenser heat output with ranges of 22.85 kJ/s to 37.7 kJ/s, 21.36 kJ/s to 36.27 kJ/s, and 20.77 kJ/s to 35.72 kJ/s, respectively. The condenser heat output represents the amount of condenser energy transferred from the ORC system to the surrounding environment. The amount of energy released into the environment is influenced by the magnitude of coefficient heat transfer condenser and Logarithmic Mean Temperature Difference (LMTD) which are sensitive to increased mass flow rate. This phenomenon is caused by R245fa and R123 working fluids having the highest heat transfer coefficient and the largest increase in LMTD when compared to the mixed working fluids. Experiments with specific mass flow rate of 0.15 kg/s for R245fa, R123, R245fa 2:1 R123, R245fa 1:1 R123, and R245fa 1:2 R123 produce condenser heat outputs of 36.99 kJ/s, 31.38 kJ/s, 28.78 kJ/s, 27.65 kJ/s, and 30.99 kJ/s, respectively.

Figure 14 reveals ORC's efficiency which is produced by controlling mass flow rate of the working fluids. Each working fluid is found to have an effect on increasing and varying the ORC system efficiency. R123 working fluid produces the highest efficiency as compared to other working fluids with significant increase occurring from mass flow rates of 0.1 kg/s, 0.125 kg/s, and 0.15 kg/s with efficiency values of 7.38%, 8.59%, and 11.04%, respectively. Afterward, the efficiency decreases to 10.74% at mass flow rate of 0.175 kg/s and then sharply increases to 12.75% at the end of the experiment with mass flow rate of 0.2 kg/s. The lowest efficiency occurs from R245fa working fluid with mass flow rate of 0.1 kg/s, 0.125 kg/s, and 0.15 kg/s and efficiency of 6.21%, 6.34%, and 6.05%, respectively. Subsequently, the efficiency climbs from a mass flow rate of 0.175 kg/s with efficiency of 6.79% to the end of the experiments with a mass flow rate of 0.2 kg/s giving efficiency at 12.27%. On the other hand, the mixed working fluids are measured to have random efficiency with values between those of the two pure working fluids in the range of 6.32% to 12.33%.

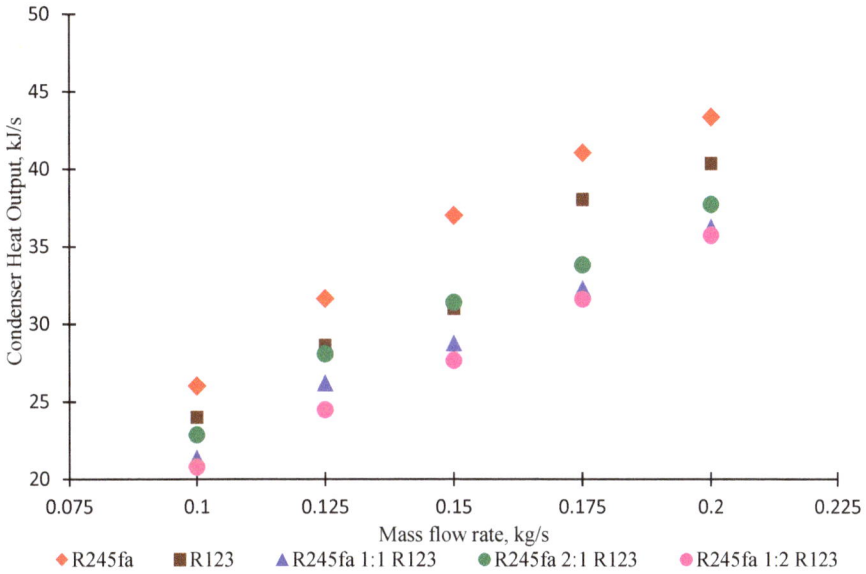

Figure 13. Condenser heat output as a function of mass flow rate for R245fa, R123, and mixed working fluids with heat source at 110 °C.

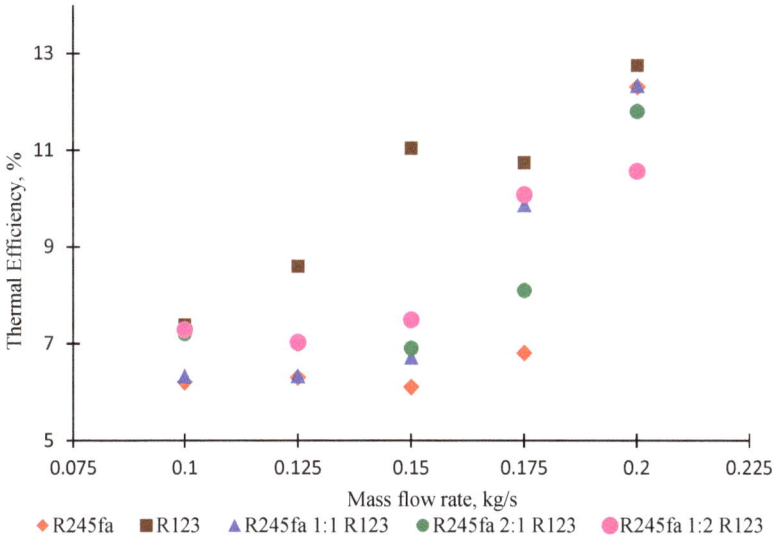

Figure 14. Efficiency ORC as a function mass flow rate for R245fa, R123, and mixed working fluids with heat source at 110 °C.

The changes in mass flow rate for all working fluids affect pump work input with heat source at 120 °C as presented in Figure 15. All working fluids also increase pump heat input, especially for mixed working fluids, which moves up dramatically with values ranging from 0.1571 kJ/s to 0.37 kJ/s. Pure working fluids have lower position compared to mixed working fluids. More specifically, R123 is followed by R245fa as producing the lowest pump work input. Starting with the highest pump work input at mass flow rate of 0.1kg/s, R123 generates an increased pump work input but located at relatively low position with the values ranging from 0.1765 kJ/s to 0.2176 kJ/s. Meanwhile, with R245fa, the ORC system generates the lowest position with pump work input ranging from 0.097 kJ/s to 0.216 kJ/s. At mass flow rate of 0.15 kg/s, the pump work input requirement for R245fa, R123, R245fa 2:1 R123, R245 1:1 R123, and R245 1:2 R123 are recorded at 0.162 kJ/s, 0.1937 kJ/s, 0.266 kJ/s, 0.246 kJ/s, and 0.248 kJ/s, respectively. Comparing the experimental results of heat source at 110 °C and 120 °C shows a similar trend. R245fa with mass flow rate of 0.1 kg/s up to 0.2 kg/s produces changes in efficiency ranging from 1.54% to 13.41%, while for R123 the changes are from 6.35% to 9.7%. In addition, when applying the mixed working fluids (R245fa 2:1 R123, R245fa 1:1 R123, and R245fa 1:2 R123) the efficiency increase varies from 1.69% to 27.67%.

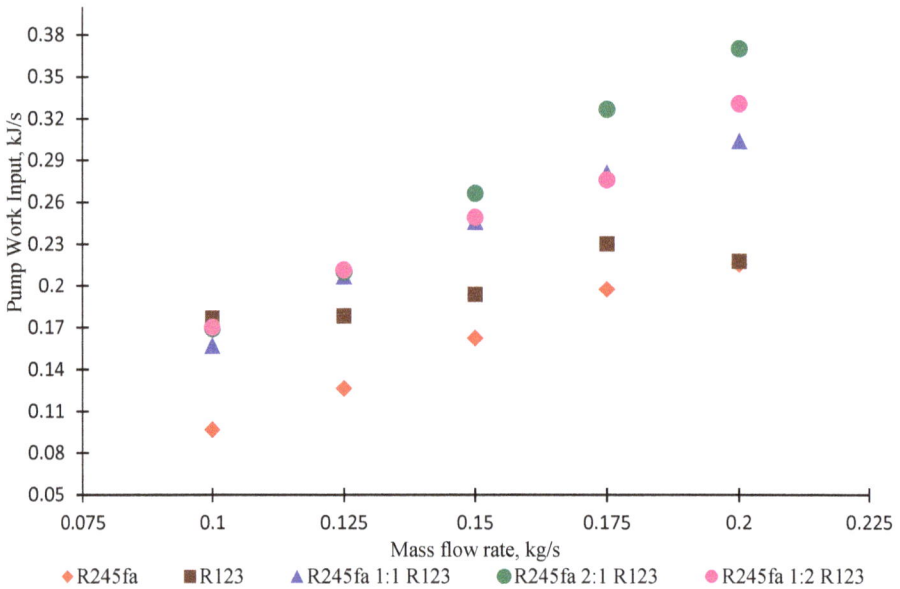

Figure 15. Pump work input as a function mass flow rate for R245fa, R123, and mixed working fluids with heat source at 120 °C.

The effect of changes in mass flow rate on the dynamic movement of pump isentropic efficiency with applied heat source at 120 °C for R245fa, R123, and mixed working fluids is shown in Figure 16. With mass flow rate of the working fluid varied from 0.1 kg/s to 0.2 kg/s, the line graph shows that the pump isentropic efficiency trend increases gradually. R245fa produces the highest pump isentropic efficiency, which starts with a mass flow rate of 0.1 kg/s resulting in an isentropic efficiency of 53.75%, then when the mass flow rate is increased to 0.15% the pump isentropic efficiency amounts to 69.79%. Moreover, experiments at a mass flow rate of 0.175 kg/s result in isentropic efficiency decreasing slightly to 69.28%; however, at a mass flow rate of 0.2 kg/s, the isentropic efficiency rises significantly to 78.2%. On the other hand, R123 gives pump isentropic efficiency in the lowest position ranging from 26.76% to 43.42%, while the mixed working fluid produces results ranging from 30.91% to 46.18%. For specific mass flow rate of 0.15 kg/s, R245fa, R123, R245 2:1 R123, R245 1:1 R123, and R245 1:2 R123 working fluids with heat source at 120 °C produce pump isentropic efficiency of 69.79%, 38.82%, 45.77%, 41.3%, and 42.11%, respectively. The addition of mass flow rate notably results in an energy increase. Changing the heat source from 110 °C to 120 °C changes the enthalpy to be higher. Consequently, magnitude of enthalpy at pump input and pump output are affected causing pump performance to be closer to its isentropic performance. If the experimental results of pump isentropic efficiency with heat source at 110 °C and 120 °C are compared, as an example for R245fa, both have the highest position, but the heat source at 120 °C yields higher pump isentropic efficiency as evidenced by Figures 10 and 16. If the differences are calculated at mass flow rates of 0.1 kg/s, 0.125 kg/s, 0.15 kg/s, 0.175 kg/s, and 0.2 kg/s, then difference values of 20.63%, 22.84%, 31.67%, 30.03%, and 33.42% are obtained, respectively.

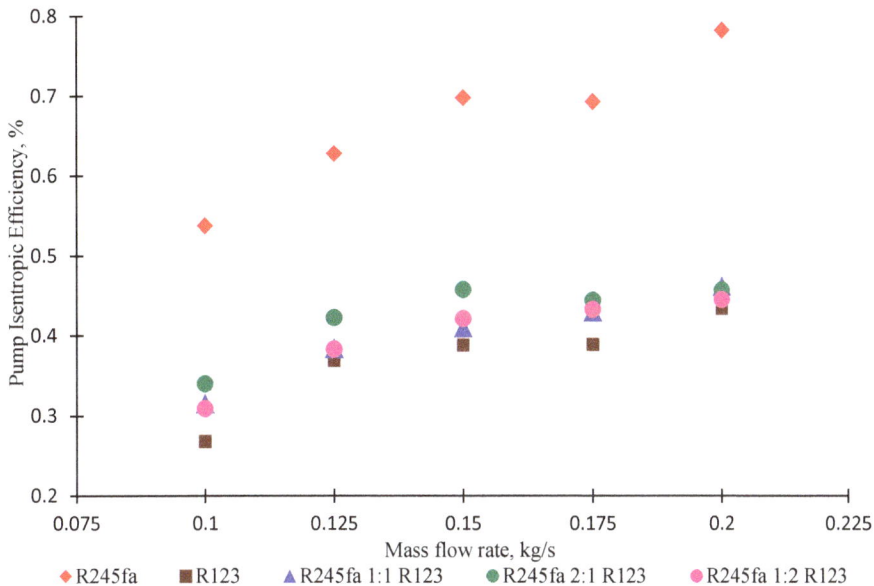

Figure 16. Pump isentropic efficiency as a function mass flow rate for R245fa, R123, and mixed working fluids with heat source at 120 °C.

Figure 17 demonstrates the effect of changes in mass flow rate against dynamic movement of heat input received by evaporator from heat source at 120 °C. Overall, the line graph increases sharply when the mass flow rate is varied from 0.1 kg/s to 0.2 kg/s. R245fa with R123 showing the highest value even though the line graph fluctuates and coincides between 27.85 kJ/s to 50.58 kJ/s. Whereas, for R245fa 2:1 R123, the values are observed to be from 25.81 kJ/s to 45.6 kJ/s. The lowest is obtained from R245fa 1:1 R123 and R245fa 1:2 R123 mixed working fluids. However, both fluids seem to increase with heat inputs of 23.86 kJ/s to 43.3 kJ/s. This is due to the heat source transferring some amount of energy to the evaporator. If mass flow rate is varied, then the heat transfer coefficient of the evaporator will change, and eventually will affect evaporator heat input whose value corresponds to enthalpy and LMTD. For specific mass flow rate of 0.15 kg/s, R245fa, R123, R245 2:1 R123, R245 1:1 R123, and R245 1:2 R123 working fluids with heat source at 120 °C produce evaporator heat inputs 39.86 kJ/s, 39.35 kJ/s, 36.86 kJ/s, 33.78 kJ/s, and 33.59 kJ/s, respectively. Comparing the results with heat source at 110 °C and 120 °C as presented in Figures 11 and 17, the evaporator heat input showed the same trend. The pure working fluids (R245fa and R123) produces the difference in heat input between 1% with 7%, while the mixed working fluids (R245fa 2:1 R123, R245fa 1:1 R123, and R245fa 1:2 R123) have the difference ranging from 1.5% to 8.7%.

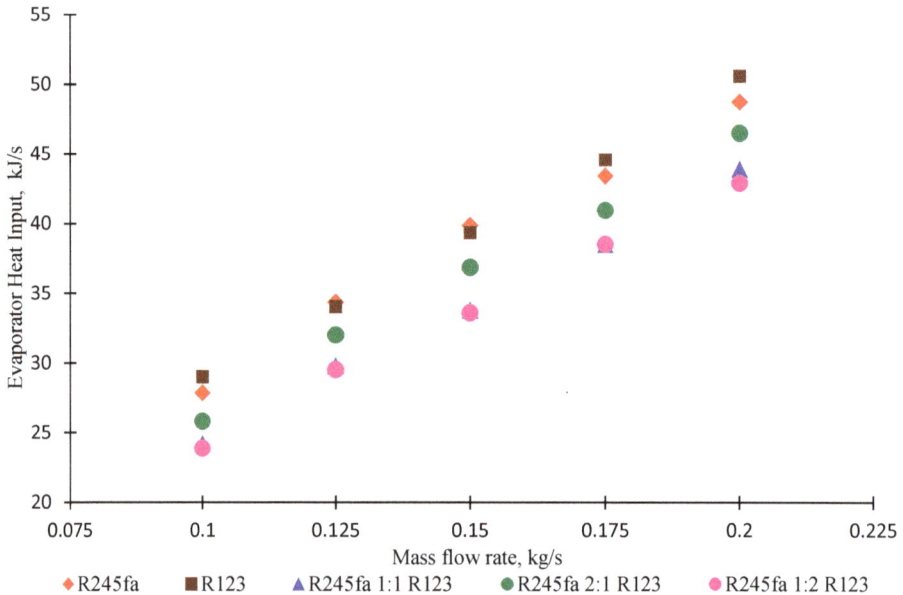

Figure 17. Evaporator heat input as a function mass flow rate for R245fa, R123, and mixed working fluids with heat source at 120 °C.

Figure 18 illustrates the effect of changing mass flow rate against expander work output. Overall, increasing the mass flow rate results in the rising movement of the expander working output. Expander power output for R123 moves with the highest, where at mass flow rate of 0.1 kg/s, 0.125 kg/s, 0.15 kg/s, 0.175 kg/s and 0.2 kg/s the move increases rapidly by 2.32 kJ/s, 3.11 kJ/s, 4.55 kJ/s, 5.7 kJ/s, and 6.5 kJ/s, respectively. Experimental result using R245fa with mass flow rate of 0.1 kg/s up to 0.15 kg/s yields work output position almost in the middle of the other working fluids with the amount recorded from 1.85 kJ/s to 2.57 kJ/s. Afterward mass flow rate 0.175 kg/s and 0.2 kg/s produce work output that slowly increases with the lowest values occur from 3.14 kJ/s and 5.19 kJ/s. For specific mass flow rate of 0.15 kg/s, R245fa, R123, R245 2:1 R123, R245 1:1 R123, and R245 1:2 R123 working fluids with heat source at 120 °C produce expander work output of 2.57 kJ/s, 4.55 kJ/s, 2.79 kJ/s, 2.51 kJ/s, and 2.76 kJ/s, respectively. Comparison of expander work output when the applied heat source are at 110 °C and 120 °C show a random movement as illustrated in the line graph of Figures 12 and 18. The R123 produces an increased on expander work output with values ranging from 8.85% to 19.68%, while R245fa generates an increase with 1.76% up to 13.79%. Lastly, R245fa 2:1 R123, R245fa 1:1 R123, and R245fa 1:2 R123 working fluids yield values ranging from 3.84% to 8.82%, 0.31% to 45.81%, and 3.18% to 17.91%, respectively.

Figure 19 reveals some important information which affects for the amount of expander work output. The box indicator shows a mass flow rate of 0.2 kg/s and a round indicator shows a mass flow rate of 0.125 kg/s. The picture graph clearly provides information that the mass flow rate of 0.2 kg/s has produced a greater output expander compared to the mass flow rate of 0.125 kg/s. This event is caused by the 0.2 kg mass of a substance which passes per unit of time are more weight than 0.125 kg mass, certainly when multiplied by enthalpy will result in a large expander work output. This means that the expander receives a larger mass to expand into mechanical power.

The comparison of the heat source 110 °C with 120 °C was also expressed in Figure 19. The blue indicator represented a heat source of 110 °C and the red color represented a heat source of 120 °C. Based on the picture showed a tiny difference of expander work output, even though in generally, the heat source 120 °C produced a slightly larger expander work output. This case is caused by temperature as a measure of the ability of a substance to transfer heat energy, consequently, the higher the temperature generated the more the expander power output.

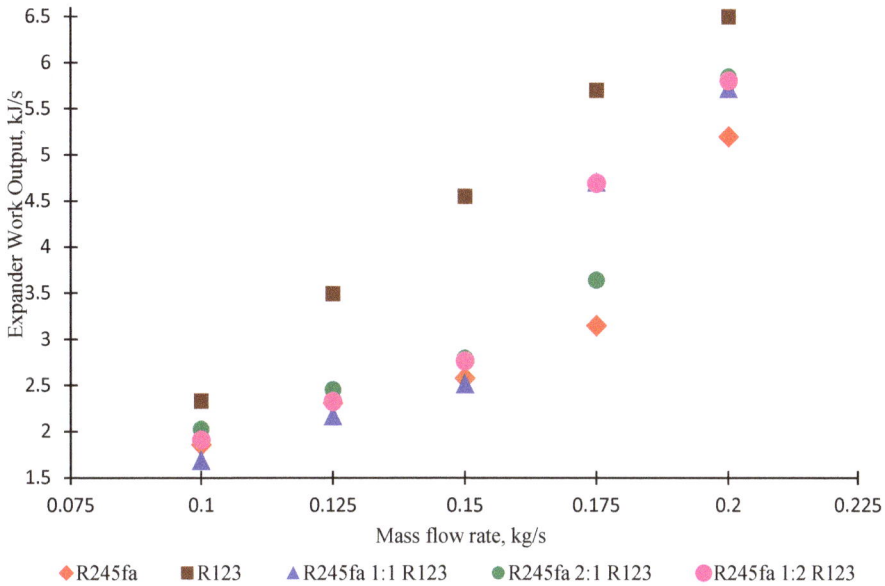

Figure 18. Expander work output as a function mass flow rate for R245fa, R123, and mixed working fluids with heat source at 120 °C.

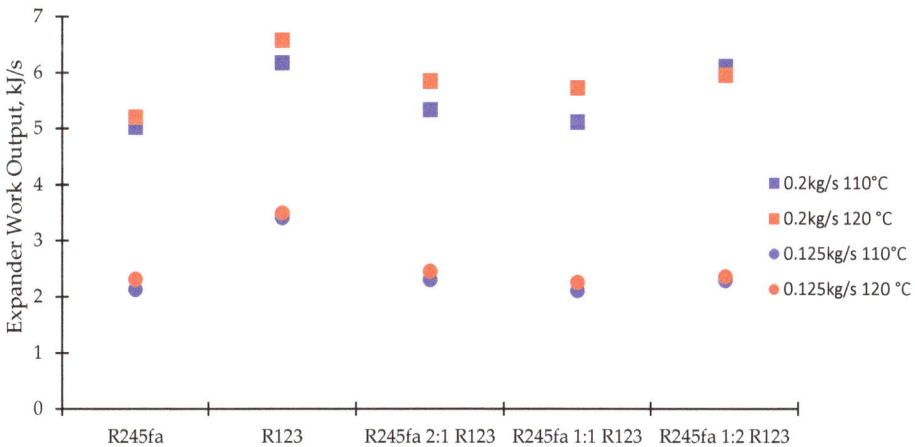

Figure 19. Expander work output with mass flow rate of 0.125 kg/s on a heat source 110 °C and 120 °C, compared to a mass flow rate of 0.2 kg/s on a heat source 110 °C and 120 °C.

Discussing working fluid, R123 has produced the highest of expander work output compared to other working fluids. Mass flow rate of 0.125 kg/s on the heat source 110 °C and 120 °C produced an expander work output of 3.41 kJ/s and 3.49 kJ/s, respectively, while for the mass flow rate 0.2 kg/s on the heat source 110 °C and 120 °C yielded of 6.17 kJ/s and 6.57 kJ/s, respectively.

Changes in the mass flow rate have an effect on the condenser heat output when heat source at 120 °C is applied as illustrated in Figure 20. The graphical trend illustrates that the greater mass flow rate contributes to the increase in energy heat output which transfers out of the system. Working fluid R245fa has the highest position with a range between 25.73 kJ/s and 42.3 kJ/s. The working fluid R123 produces heat output which coincides with those from R245fa at 42.2 kJ/s, with the mass flow rate of 0.2 kg/s. The results from mixed working fluids increase to a range from 21.76 kJ/s to 40.46 kJ/s. For a specific mass flow rate at 0.15 kg/s, the condenser heat output for R245fa, R123, R245fa 2:1 R123, R245fa 1:1 R123, and R245fa 1:2 R12 are at 37.08 kJ/s, 31.53 kJ/s, 33.88 kJ/s, 31.29 kJ/s, and 30.79 kJ/s, respectively. Comparison of condenser heat output when applying heat source at 110 °C and 120 °C shows overall increasing trend, but the value decreases for heat source at 120 °C as illustrated in Figures 13 and 20. The pure working fluids (R245fa and R123) have the difference ranging from 0.22% to 9.59%, while the mixed working fluids (R245fa 2:1 R123, R245fa 1:1 R123, and R245fa 1:2 R123) decrease by 3.3% to 11.36%.

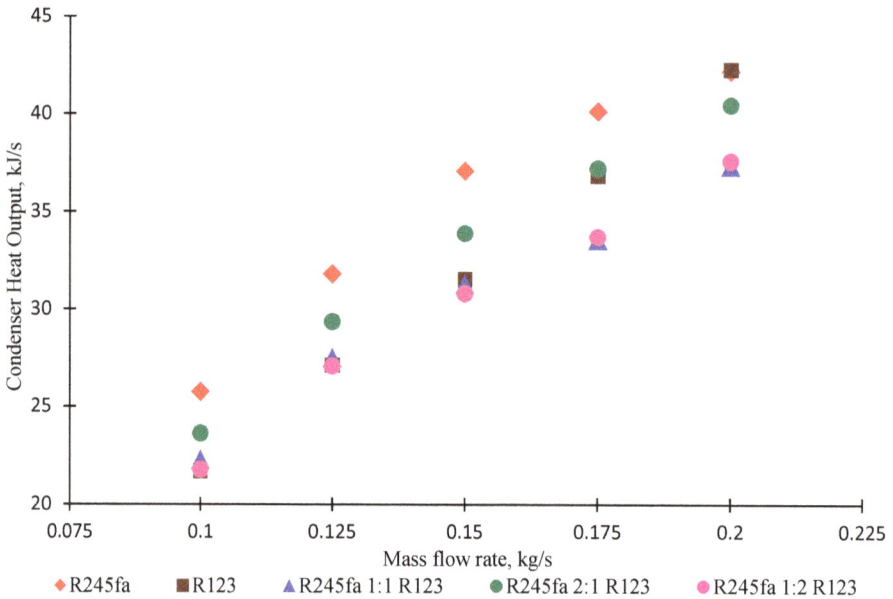

Figure 20. Condenser heat output as a function mass flow rate for R245fa, R123, and mixed working fluids with heat source at 120 °C.

Efficiency of the ORC exhibits the overall performance of the system, consisting of thermal efficiency and system-making efficiency for R245fa, R123, and mixed working fluids as presented in Figure 21. R123 shows the gradual increase in ORC efficiency with values from 9.4% to 13.5%. R245fa at mass flow rates of 0.1 kg/s, 0.125 kg/s, and 0.15 kg/s have relatively stable changes at 6.01%, 6.10%, and 5.8%, respectively. At a mass flow rate of 0.175 kg/s the ORC efficiency rises slowly by 6.73% and finally at a mass flow rate 0.2 kg/s the efficiency rises drastically by 12.03%. The ORC efficiency is influenced by heat transfer coefficient at evaporator and how much the expander produces work output as illustrated in Figure 12. This is because the addition of mass flow rate rises the heat transfer in evaporator and increases the amount of energy produced by the expander. For specific mass flow rate at 0.15 kg/s, the ORC efficiency for R245fa, R123, R245fa 2:1 R123, R245fa 1:1 R123, and R245fa 1:2 R123 working fluids are at 5.79%, 10.64%, 9.48%, 10.27%, and 8.59%, respectively. Comparing the ORC efficiency from heat source at 110 °C and 120 °C shows quite similar trend but with an increase for heat source at 120 °C as illustrated in Figures 14 and 21. R123 working fluid results in an increase in OCR efficiency with values ranging from 3.61% to 27.4%. R245fa has a lower trend with magnitudes ranging from 0.91% to 4.21%. Furthermore, when using mixed working fluids R245fa 2:1 R123, R245fa 1:1 R123, and R245fa 1:2 R123, the ORC efficiency increases with a wider range from 0.65% to 38.35%.

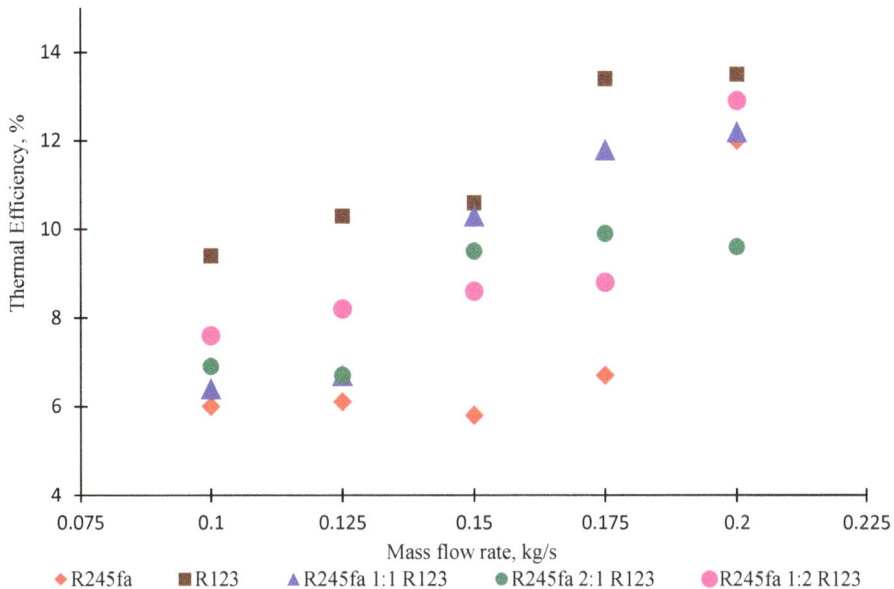

Figure 21. Efficiency ORC as a function mass flow rate for R245fa, R123, and mixed working fluids with heat source at 120 °C.

5. Conclusions

To study the potential of geothermal energy in Dieng Indonesia, an ORC system laboratory setup was constructed which enables us to simulate the geothermal heat source at 110 °C and 120 °C utilizing various working fluids. Several working fluids were used in the study including R245fa, R123, and three mixed ratios R245fa 2:1 R123, R245fa 1:1 R123, and R245fa 1:2 R123. Additionally, the frequency of the pump was also varied to obtain different mass flow rates; and thus, will change the main operating indicators of the Rankine cycle. Several main conclusions from the study are summarized as follows:

(1)   The heat source 110 °C and 120 °C does not influence significantly on behavior of expander work output and the other equipment, trend charts tend to be uniform when compared in one image. The expander work output presents an increase sharply with an escalation of mass flow rate.

(2)   The heat source at 120 °C has a higher pump isentropic efficiency from 0.5% to 33.5% when compared to the heat source at 110 °C, and the R245fa as a working fluid produces the highest efficiency ranging from 53.75% to 78.2%.

(3)   Expander work output increases when the applied heat source is at 120 °C, specifically with working fluid R123 which was observed to have the highest change with the range of 8.85% to 19.68%.

(4)   The highest evaporator heat input of 50.58 kJ/s is produced by R123 working fluid with heat source at 120 °C and mass flow rate of 0.2 kg/s, and when comparing the evaporator heat input between 110 °C and 120 °C, the difference ranges from 1% to 7%.

(5)   Condenser heat output changes uniformly for both heat source temperatures 110 °C and 120 °C, with the pure working fluid (R245fa and R123) having a difference ranging from 0.22% to 9.59% while mixed ratio fluids yield a decrease with the scale ranging from 3.3% to 11.36%.

(6) The highest thermal efficiency is generated when R123 working fluid is applied with heat source at 120 °C whose efficiency values range from 9.41% to 13.53%, but when comparing both heat source temperatures the highest change in thermal efficiency is produced by R245fa 1:1 R123 whose value is 53.01%.

(7) Working fluid R123 is feasible as working fluid recommendation due to produced significant expander work output with temperature 110 °C or 120 °C. When compared with other working fluid, R123 recorded a higher difference in the range of 10.39% to 61.89%.

Knowledge of the pump type and availability in the market is very important. When using PHE as an evaporator, a large pump power is required to reach the evaporator operating pressure. Furthermore, the challenge of getting the appropriate seal is very serious because working fluid can cause rubber seal in expand and be easily damaged.

Author Contributions: Conceptualization, M.D.S., W.C., and T.P.; methodology, M.D.S., W.C., and T.P.; software, W.C. and T.P.; validation, M.D.S., W.C., and T.M.I.M.; formal analysis, M.D.S., W.C., and T.P.; investigation, M.D.S., W.C., and T.M.I.M.; resources, M.D.S. and W.C.; data curation, M.D.S. and W.C.; writing—original draft preparation, M.D.S. and W.C.; writing—review and editing, M.D.S., W.C., T.P., T.M.I.M., and T.; visualization, M.D.S and W.C.; supervision, W.C., T.P., T.M.I.M., and T.; project administration, M.D.S.; funding acquisition, T.P. and W.C.

Funding: This research received no external funding.

Conflicts of Interest: The authors declare no conflicts of interest.

References

1. Pambudi, N.A. Geothermal power generation in Indonesia, a country within the ring of fire: Current status, future development, and policy. *Renew. Sustain. Energy Rev.* **2018**, *81*, 2893–2901. [CrossRef]
2. Nasruddin; Alhamid, M.I.; Yunus, D.; Arief, S.; Agus, S.; Aditya, H.B.; Mahlia, T.M.I. Potential of geothermal energy for electrical generation in Indonesia: A review. *Renew. Sustain. Energy Rev.* **2016**, *53*, 733–740. [CrossRef]
3. Manalu, P. Geothermal development in Indonesia. *Geothermics* **1988**, *17*, 415–420. [CrossRef]
4. Purnomo, B.J.; Pichler, T. Geothermal systems on the island of Java, Indonesia. *J. Volcanol. Geotherm. Res.* **2014**, *285*, 47–59. [CrossRef]
5. Hochstein, M.P.; Sudarman, S. History of geothermal exploration in Indonesia from 1970 to 2000. *Geothermics* **2008**, *37*, 220–266. [CrossRef]
6. Hall, R. Cenozoic geological and plate tectonic evolution of SE Asia and the SW pacific; computer-based reconstructions, model and animations. *J. Asia Earth Sci.* **2002**, *20*, 353–431. [CrossRef]
7. Simandjuntak, T.O.; Barber, A.J. Contrasting tectonic styles in the Neogene orogenic belts of Indonesia. *Geol. Soc. Spec. Publ.* **1996**, *106*, 185–201. [CrossRef]
8. Simkin, T.; Siebert, L. *Volcanoes of the World*; Geoscience Press: Tucson, AZ, USA, 1994; pp. 64–79. ISBN 0-945005-12-1.
9. Badan Geologi, Kementerian Energi dan Sumber Daya Mineral. Available online: www.esdm.go.id/id/media-center/arsip-berita/miliki-127-gunung-api-aktif-jadikan-indonesia-laboratorium-gunung-api-dunia (accessed on 31 January 2019). (In Bahasa)
10. Dickson, M.H.; Fanelli, M. *Geothermal Energy: Utilization and Technology*; Routledge: London, UK, 2003; ISBN 92-3-103915-6.
11. Muffler, P.; Cataldi, R. Methods for regional assessment of geothermal resources. *Geothermics* **1978**, *7*, 53–89. [CrossRef]
12. Bayu, R.; IbnuAtho, I.; Nugroho, A.P.; Cheng, C.C.; Reza, A.; Imran, M.; Saw, L.H.; Renanto, H. Preliminary analysis of dry-steam geothermal power plant by employing exergy assessment: Case study in Kamojang geothermal power plant, Indonesia. *Case Stud. Therm. Eng.* **2017**, *10*, 292–301. [CrossRef]
13. Aneke, M.; Agnew, B.; Underwood, C. Performance analysis of the Chena binary geothermal power plant. *Appl. Therm. Eng.* **2011**, *31*, 1825–1832. [CrossRef]

14. Alison Holm, D.J.; Blodgett, L. Geothermal Energy and Green-House Gas Emission Geothermal Energy Association. Available online: http://geo-energy.org/reports/ GeothermalGreenhouseEmissionsNov2012GEA_web.pdf (accessed on 31 January 2019).

15. Duffield, W.A.; Sass, J.H. *Geothermal Energy: Clean Power from the Earth's Heat*; DIANE Publishing: Darby, PA, USA, 2003.

16. Edrisi, B.H.; Michaelides, E.E. Effect of the working fluid on the optimum work of binary-flashing geothermal power plant. *Energy* **2013**, *50*, 389–394. [CrossRef]

17. Rosinski, S.; Coleman, T.; Cerezo, L. *Geothermal Power, Issue, Technologies, and Opportunities for Research, Development, Demonstration, and Deployment Electric Power Research Institute*; The Electric Power Research Institute, Inc. (EPRI): Palo Alto, CA, USA, 2010.

18. Pambudi, N.A.; Itoi, R.; Jalilinasrabady, S.; Jaelani, K. Exergy analysis and optimization of Dieng single-flash geothermal power plant. *Energy Convers. Manag.* **2014**, *78*, 405–411. [CrossRef]

19. Radja, V.T. Overview of geothermal energy studies in Indonesia. In Proceedings of the 2nd United Nations Symposium on Development and Use of Geothermal Resources, San Francisco, CA, USA, 20–29 May 1975; pp. 233–240.

20. DiPippo, R. Geothermal energy electricity generation and environmental impact. *Energy Policy* **1991**, *19*, 798–807. [CrossRef]

21. Gehringer, M.; Loksha, V. *Geothermal Handbook: Planning and Financing Power Generation*; Energy Sector Management Assistance Program (ESMAP) World Bank: Washington DC, WA, USA, 2012.

22. Efstathios, E.S.M. Future directions and cycles for electricity production from geothermal resources. *Energy Convers. Manag.* **2016**, *107*, 3–9. [CrossRef]

23. DiPippo, R. Second Law assessment of binary plants generating power from low-temperature geothermal fluids. *Geothermics* **2004**, *33*, 565–586. [CrossRef]

24. Madhawa Hettiarachchi, H.D.; Golubovica, M.; Worek, W.M.; Ikegami, Y. Optimum design criteria for an organic Rankine cycle using low-temperature geothermal heat sources. *Energy* **2007**, *32*, 1698–1706. [CrossRef]

25. DiPippo, R. *Geothermal Power Plants: Principles, Applications, Case Studies and Environmental Impact*, 3rd ed.; Elsevier: Amsterdam, The Netherlands, 2012; ISBN 978-008-0982-06-9.

26. Bayer, P.; Rybach, L.; Blum, P.; Brauchler, R. Review on life cycle environmental effects of geothermal power generation. *Renew. Sustain. Energy Rev.* **2013**, *26*, 446–463. [CrossRef]

27. Guzovic, Z.; Raškovic, P.; Blataric, Z. The comparison of a basic and a dual-pressure ORC (Organic Rankine Cycle): Geothermal power plant Velika Ciglena case study. *Energy* **2014**, *76*, 175–186. [CrossRef]

28. Liu, Q.; Duan, Y.Y.; Yang, Z. Effect of condensation temperature glide on the performance of organic Rankine cycles with zeotropic mixture working fluids. *Appl. Energy* **2014**, *115*, 394–404. [CrossRef]

29. DiPippo, R. Geothermal power plants: Evolution and performance assessments. *Geothermics* **2015**, *53*, 291–307. [CrossRef]

30. Basaran, A.; Ozgener, L. Investigation of the effect of different refrigerants on performances of binary geothermal power plants. *Energy Convers. Manag.* **2013**, *76*, 483–498. [CrossRef]

31. Yari, M. Exergetic analysis of various types of geothermal power plants. *Renew. Energy* **2010**, *35*, 112–121. [CrossRef]

32. Gu, Z.; Sato, H. Performance of supercritical cycles for geothermal binary design. *Energy Convers. Manag.* **2002**, *43*, 961–971. [CrossRef]

33. Guo, T.; Wang, H.X.; Zhang, S.J. Selection of working fluids for a novel low-temperature geothermally-powered ORC based cogeneration system. *Energy Convers. Manag.* **2011**, *52*, 2384–2391. [CrossRef]

34. Bertani, R. Geothermal power generation in the world 2010–2014 update report. *Geothermics* **2016**, *60*, 31–43. [CrossRef]

35. Franco, A.; Vaccaro, M. Numerical simulation of geothermal reservoirs for the sustainable design of energy plants: A review. *Renew. Sustain. Energy Rev.* **2014**, *30*, 987–1002. [CrossRef]

36. Franco, A.; Villani, M. Optimal design of binary cycle power plants for water-dominated, medium-temperature geothermal fields. *Geothermics* **2009**, *38*, 379–391. [CrossRef]

37. Liu, X.; Wang, X.; Zhang, C. Sensitivity analysis of system parameters on the performance of the Organic Rankine Cycle system for binary-cycle geothermal power plants. *Appl. Therm. Eng.* **2014**, *71*, 175–183. [CrossRef]

38. Coskun, A.; Bolatturk, A.; Kanoglu, M. Thermodynamic and economic analysis and optimization of power cycles for a medium temperature geothermal resource. *Energy Convers. Manag.* **2014**, *78*, 39–49. [CrossRef]

39. Budisulistyo, D.; Krumdieck, S. Thermodynamic and economic analysis for the prefeasibility study of a binary geothermal power plant. *Energy Convers. Manag.* **2015**, *103*, 639–649. [CrossRef]

40. Shengjun, Z.; Huaixin, W.; Tao, G. Performance comparison and parametric optimization of subcritical Organic Rankine Cycle (ORC) and transcritical power cycle system for low-temperature geothermal power generation. *Appl. Energy* **2011**, *88*, 2740–2754. [CrossRef]

41. Walraven, D.; Laenen, B.; D'haeseleer, W. Comparison of shell-and-tube with plate heat exchangers for the use in low-temperature organic Rankine cycles. *Energy Convers. Manag.* **2014**, *87*, 227–237. [CrossRef]

42. Feng, Y.Q.; Hung, T.C.; Greg, K.; Zhang, Y.N.; Li, B.X.; Yang, J.F. Thermoeconomic comparison between pure and mixture working fluids for low-grade organic Rankine cycles (ORCs). *Energy Convers. Manag.* **2015**, *106*, 859–872. [CrossRef]

43. Feng, Y.Q.; Hung, T.C.; Yaning, Z.; Li, B.X.; Yang, J.F.; Shi, Y. Performance comparison of low-grade organic Rankine cycles (ORCs) using R245fa, pentane and their mixtures based on the thermoeconomic multi-objective optimization and decision makings. *Energy* **2015**, *99*, 2018–2029. [CrossRef]

44. Hung, T.C.; Shai, T.Y.; Wang, S.K. A review of organic Rankine cycles (ORCs) for the recovery of low-grade waste heat. *Energy* **1997**, *22*, 661–667. [CrossRef]

45. Hung, T.C. Waste heat recovery of organic Rankine cycle using dry fluids. *Energy Convers. Manag.* **2001**, *42*, 539–553. [CrossRef]

46. Chen, Q.C.; Xu, J.L.; Chen, H.X. A new design method for Organic Rankine Cycles with constraint of inlet and outlet heat carrier fluid temperatures coupling with the heat source. *Appl. Energy* **2012**, *98*, 562–573. [CrossRef]

47. Xu, J.L.; Yu, C. Critical temperature criterion for selection of working fluids for subcritical pressure Organic Rankine cycles. *Energy* **2014**, *74*, 719–733. [CrossRef]

48. Feng, Y.Q.; Hung, T.C.; He, Y.L.; Wang, Q.; Wang, S.; Li, B.X.; Lin, J.R.; Zhang, W. Operation characteristic and performance comparison of organic Rankine cycle (ORC) for low-grade waste heat using R245fa, R123 and their mixtures. *Energy Convers. Manag.* **2017**, *144*, 2018–2029. [CrossRef]

49. Jiang, F.; Zhu, J.L.; Xin, G.L. Experimental investigation on Al_2O_3-R123 nanorefrigerant heat transfer performances in evaporator based on organic Rankine cycle. *Int. J. Heat Mass Transf.* **2018**, *127*, 145–153. [CrossRef]

50. Feng, Y.Q.; Hung, T.C.; Wu, S.L.; Lin, C.H.; Li, B.X.; Huang, K.C.; Qin, J. Operation characteristic of a R123-based organic Rankine cycle depending on working fluid mass flow rates and heat source temperatures. *Energy Convers. Manag.* **2017**, *131*, 55–68. [CrossRef]

51. Yang, S.C.; Hung, T.C.; Feng, Y.Q.; Wu, C.J.; Wong, K.W.; Huang, K.C. Experimental investigation on 3 kW organic Rankine cycle for low-grade waste heat under different operation parameters. *Appl. Therm. Eng.* **2017**, *113*, 756–764. [CrossRef]

52. Shao, L.; Zhu, J.; Meng, X.R.; Wei, X.L.; Ma, X.L. Experimental study of an organic Rankine cycle system with radial inflow turbine and R123. *Appl. Therm. Eng.* **2017**, *124*, 940–947. [CrossRef]

53. Feng, Y.Q.; Hung, T.C.; Su, T.Y.; Wang, S.; Wang, Q.; Yang, S.C.; Lin, J.R.; Lin, C.H. Experimental investigation of a R245fa-based organic Rankine cycle adapting two operation strategies: Stand alone and grid connect. *Energy* **2017**, *141*, 1239–1253. [CrossRef]

54. Eyerer, S.; Wieland, C.; Vandersickel, A.; Spliethoff, H. Experimental study of an ORC (Organic Rankine Cycle) and analysis of R1233zd-E as a drop-in replacement for 245fa for low temperature. *Energy* **2016**, *103*, 660–671. [CrossRef]

55. Shu, G.; Zhao, M.; Tian, H.; Huo, Y.Z.; Zhu, W. Experimental comparison of R123 and R245fa as working fluids for waste recovery from heavy-duty diesel engine. *Energy* **2016**, *115*, 756–769. [CrossRef]

56. Chang, J.C.; Hung, T.C.; He, Y.L.; Zhang, W. Experimental study on low-temperature organic Rankine cycle utilizing scroll type expander. *Appl. Energy* **2015**, *155*, 150–159. [CrossRef]

57. Chang, J.C.; Chang, C.W.; Hung, T.C.; Lin, J.R.; Huang, K.C. Experimental study and CFD approach for scroll type expander used in low-temperature organic Rankine cycle. *Appl. Therm. Eng.* **2014**, *155*, 1444–1452. [CrossRef]

58. Li, M.; Wang, J.F.; He, W.F.; Gao, L.; Wang, B.; Ma, S.; Dai, Y. Construction and preliminary test of a low-temperature regenerative Organic Rankine Cycle (ORC) using R123. *Renew. Energy* **2013**, *57*, 216–222. [CrossRef]

59. Shu, G.; Zhao, J.; Tian, H.; Liang, X.; Wei, H.Q. Parametric and exergetic analysis of waste heat recovery system based on thermoelectric generator and organic rankine cycle utilizing R123. *Energy* **2012**, *45*, 806–816. [CrossRef]

60. Li, T.; Zhu, J.; Fu, W.; Hu, K. Experimental comparison of R245fa and R245fa/R601a for organic Rankine cycle using scroll expander. *Int. J. Energy Res.* **2015**, *39*, 202–214. [CrossRef]

61. Pang, K.C.; Chen, S.C.; Hung, T.C.; Feng, Y.Q.; Yang, S.C.; Wong, K.W.; Lin, J.R. Experimental study on organic Rankine cycle utilizing R245fa, R123 and their mixtures to investigate the maximum power generation from low-grade heat. *Energy* **2017**, *133*, 636–651. [CrossRef]

processes

MDPI

Article

Model-Based Cost Optimization of Double-Effect Water-Lithium Bromide Absorption Refrigeration Systems

Sergio F. Mussati [1], Seyed Soheil Mansouri [2], Krist V. Gernaey [2], Tatiana Morosuk [3] and Miguel C. Mussati [1,*]

[1]   INGAR Instituto de Desarrollo y Diseño (CONICET-UTN), Avellaneda 3657, S3002GJC Santa Fe, Argentina; mussati@santafe-conicet.gov.ar
[2]   Process and System Engineering Center (PROSYS), Department of Chemical and Biochemical Engineering, Technical University of Denmark, Søltofts Plads, Building 229, 2800 Kgs. Lyngby, Denmark; seso@kt.dtu.dk (S.S.M.); kvg@kt.dtu.dk (K.V.G.)
[3]   Institute for Energy Engineering, Technische Universität Berlin, Marchstr. 18, 10587 Berlin, Germany; tetyana.morozyuk@tu-berlin.de
*   Correspondence: mmussati@santafe-conicet.gov.ar; Tel.: +54-342-453-4451

Received: 16 October 2018; Accepted: 17 January 2019; Published: 19 January 2019

Abstract: This work presents optimization results obtained for a double-effect H_2O-LiBr absorption refrigeration system considering the total cost as minimization criterion, for a wide range of cooling capacity values. As a model result, the sizes of the process units and the corresponding operating conditions are obtained simultaneously. In this paper, the effectiveness factor of each proposed heat exchanger is considered as a model optimization variable which allows (if beneficial, according to the objective function to be minimized) its deletion from the optimal solution, therefore, helping us to determine the optimal configuration. Several optimization cases considering different target levels of cooling capacity are solved. Among the major results, it was observed that the total cost is considerably reduced when the solution heat exchanger operating at low temperature is deleted compared to the configuration that includes it. Also, it was found that the effect of removing this heat exchanger is comparatively more significant with increasing cooling capacity levels. A reduction of 9.8% in the total cost was obtained for a cooling capacity of 16 kW (11,537.2 $\cdot$year^{-1} vs. 12,794.5 $\cdot$year^{-1}), while a reduction of 12% was obtained for a cooling capacity of 100 kW (31,338.1 $\cdot$year^{-1} vs. 35,613.9 $\cdot$year^{-1}). The optimization mathematical model presented in this work assists in selecting the optimal process configuration, as well as determining the optimal process unit sizes and operating conditions of refrigeration systems.

Keywords: absorption refrigeration; H_2O-LiBr working pair; double-effect system; cost optimization; nonlinear mathematical programming

1. Introduction

Compared to vapor compression cycles, the main advantage of absorption refrigeration systems (ARSs) such as water-lithium bromide (H_2O-LiBr) ARSs is that they are activated by low-level energy sources [1] (such as geothermal or solar energies) or low-grade waste heat rejected from various processes, as opposed to through the use of electric energy. On the other hand, compared to other working pairs, such as ammonia-water (NH_3-H_2O), a LiBr solution has no ozone-depleting potential or global warming effect reported in literature, in line with the Montreal, Kyoto, and Paris Accords.

The energy efficiency of a single-effect ARS is relatively low. To cope with this weakness, several papers have been published that aimed at improving the performance of single-effect

H_2O-LiBr ARSs based on energy [2–4], exergy [4–6], exergo-economic [7,8], or cost [5,9] studies. Other authors have addressed such limitations by investigating other process configurations instead, including advanced configurations of multi-effect systems [10]. Among them, the double-effect schemes have comparatively received more interest, and are, in fact, the most frequently applied in industry [11,12]. Many studies on the double-effect H_2O-LiBr ARS were conducted by performing energy analyses [13–15], exergy analyses [15,16], and exergo-economic analyses [1,17,18]. A special feature of the double-effect ARS is its capability of running in series, parallel, and reverse parallel flow schemes according to the working solution flow through the heat exchangers and generators [11–13,19].

Despite the fact that systematic computer-aided methods and mathematical programming techniques have been successfully employed to optimize energy processes [20–25], not that many publications can be found for ARS [5,9,26–30]. These methods and techniques make it possible to optimize large mathematical models considering at the same time all the continuous and discrete decisions, which is one of the major advantages over parametric optimization approaches.

Chahartaghi et al. [27] recently studied two novel arrangements of double-effect absorption chillers with series and parallel flow, which differ from earlier conventional absorption chillers by the fact that they have an additional solution heat exchanger. They investigated the effects on the coefficient of performance (COP) of the temperature and mass flow rate of the vapor entering the high-temperature generator (HTG) and water entering the absorber (ABS). One of the results indicated that for an inlet vapor temperature to the HTG lower than 150 °C, the series cycle has a higher COP than the parallel cycle.

Lee et al. [28] employed a multi-objective genetic algorithm (MOGA) and meta-models to optimize several generators for a H_2O-LiBr absorption chiller with multiple heat sources. The integrated generation system included a HTG, a low-temperature generator (LTG), and a waste heat recovery generator (WHRG). The optimization problem consisted of the minimization of the total generation volume and the maximization of the total generation rate. It was found that the WHRG is dominant for reducing the total volume, and the HTG is dominant for improving the total generation rate.

Sabbagh and Gomez [29] proposed an optimal control strategy to operate H_2O-LiBr absorption chillers. The aim of the control strategy was to keep the cold water flow at a desired temperature (11 °C). To this end, a dynamic model consisting of differential algebraic equations (DAE) was first developed and then reformulated into a set of algebraic equations by discretizing the state and control variables using orthogonal collocation on finite elements, by dividing the time horizon into finite elements. The resulting model was implemented in the General Algebraic Modeling System (GAMS) and solved with the Interior Point OPTimization (IPOPT) solver [31]. Both step and sinusoidal perturbations of the hot water inlet temperature were studied. The results obtained are promising because, through the implementation of the optimal control strategy, the COP was significantly improved, thus reducing the operational cost and maintaining the cold water outlet temperature at the desired level.

In this paper, a mathematical model of a double-effect system with series flow configuration presented by Mussati et al. [32] is modified to consider another double-effect configuration, where the stream leaving the absorber is now split into two streams: one is passed through a solution heat exchanger (the low-temperature heat exchanger LTSHE) that is placed before the LTG, and the other is passed through another solution heat exchanger (the high-temperature heat exchanger HTSHE) that is placed before the HTG. The effectiveness factor of each solution heat exchanger is a model variable, thus making it possible to remove the corresponding solution heat exchanger, if beneficial according to the objective function that is optimized. Therefore, improved cost-effective process configurations can be found. The cost model presented by Mussati et al. [32] is employed. To the best of our knowledge, few articles deal with the simultaneous optimization approach presented in this work in order to take into account all the trade-offs existing between the model variables, which include both operation conditions and process unit sizes. The application of the proposed optimization approach leads to the improved configuration, in terms of costs, of a double-effect H_2O-LiBr absorption refrigeration system, which is the main contribution of this paper.

2. Process Description

As shown in Figure 1, the stream #1 that leaves the ABS is split into two streams. A fraction (stream #1′) is directed to the LTSHE through the solution pump PUMP1; it is then fed to the LTG (stream #3). The other fraction (stream #1″) is conducted to the HTSHE through the solution pump PUMP2, and then fed to the HTG (stream #12). In both generators, a vapor stream of refrigerant and a stream of concentrated LiBr solution are obtained.

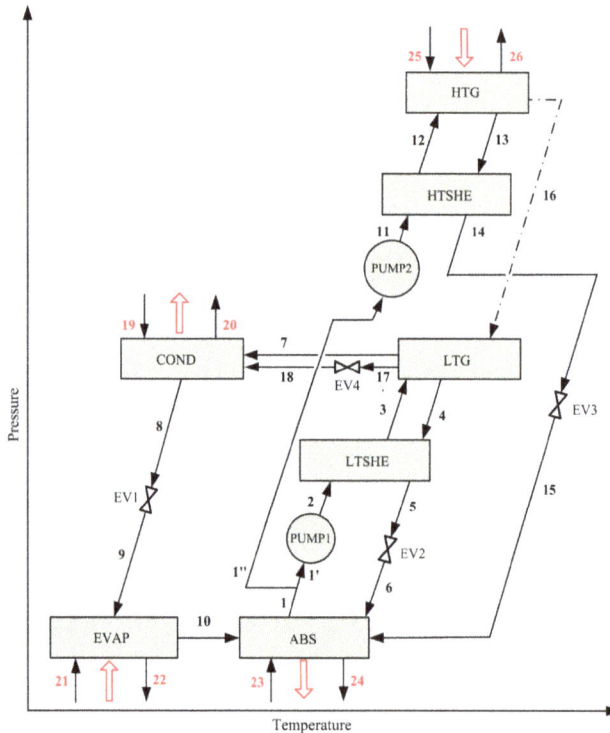

Figure 1. Schematic of the studied double-effect H_2O-LiBr ARS. EV1, EV2, EV3 and EV4 represent expansion valves; EVAP evaporator, ABS absorber; PUMP1 and PUMP2 solution pumps; LTSHE and HTSHE low and high temperature solution heat exchangers, respectively; LTG and HTG low- and high-temperature generators; COND condenser; dash-dotted line (stream #16) refers to an energy stream associated to the refrigerant formed in HTG.

The heat of the refrigerant generated in the HTG ('energy stream' #16)—represented by the dash-dotted line in Figure 1—is used in the LTG to produce refrigerant (stream #7) and the strong solution (stream #4). Also, low-grade waste heat rejected from other processes can be additionally used to increase the refrigerant production, which is, in fact, a remarkable feature of multi-stage configurations. This facilitates waste heat recovery as a means of implementing a circular economy strategy [33,34]. The streams #18 and #7 (refrigerant vapors) transfer their heat into the condenser COND. The condensed refrigerant (stream #8) is passed through the expansion valve EV1, and then fed to the evaporator EVAP that operates at the lowest pressure of the system. Finally, the stream #10 (vapor) is fed to the ABS and is absorbed in the resulting mixture of the strong solutions coming from LTSHE and HTSHE after passing through EV2 (stream #6) and EV3 (stream #15), respectively. The generated heat is rejected by using cooling water.

3. Mathematical Model

The mathematical model has been derived considering the following assumptions: (a) steady-state condition [12,19,35]; (b) no pressure drops and heat losses are taken into account [12,19,35]; (c) saturation condition for refrigerant streams that leave the condenser and evaporator [12,19]; (d) saturation condition for the diluted (weak) LiBr solution that leaves the absorber [12]; (e) the concentrated (strong) LiBr solutions leaving the generators are at equilibrium conditions [12]; and (f) isenthalpic process in expansion valves [19,35].

Each process unit is described by using a similar mathematical model presented by Mussati et al. [32]. The list of assumptions and the complete mathematical model (mass and energy balances) here employed are provided as Supplementary Materials related to this article. The correlations used to estimate the physicochemical properties of the LiBr solution (stream enthalpy) reported by ASHRAE [36] and the correlations used to describe the LiBr solution crystallization region given by Gilani and Ahmed [37] are also included as Supplementary Materials.

Optimization Problem: Total Annual Cost (TAC) Minimization

The optimal design consists of minimizing the TAC (Equation (1)), which accounts for the annualized capital expenditure (annCAPEX) and he operating expenditure (OPEX), while meeting the process design specifications and operation constraints for a wide range of cooling capacity levels.

$$TAC = annCAPEX + OPEX \tag{1}$$

The annCAPEX is given by Equation (2). The capital recovery factor (CRF) is given by Equation (3), which is computed for a lifetime (n) of 25 years and an interest rate (i) of 10.33% [5]. The investment (Z_k) of a process unit k is given by Equation (4).

$$annCAPEX = CRF \cdot \sum_k Z_k \tag{2}$$

$$CRF = \frac{i \cdot (1+i)^n}{(1+i)^n - 1} \tag{3}$$

$$Z_k = A_k \cdot (f \cdot HTA_k)^{B_k} + C_k \tag{4}$$

The OPEX is estimated by Equation (5), which includes costs associated with the heating (HU) and cooling (CU) utilities, consisting of steam (in $t \cdot year^{-1}$) and cooling water (in $t \cdot year^{-1}$), respectively. The unitary cost of vapor (C_{HU}) is 2.0 $\$ \cdot t^{-1}$ and for cooling water (C_{CU}) it is 0.0195 $\$ \cdot t^{-1}$ [5].

$$OPEX = C_{HU} \cdot HU + C_{CU} \cdot CU \tag{5}$$

The cooling capacity in EVAP (Q_{EVAP}) is the target design specification; it is a model parameter i.e., a known and fixed value in each optimization run. In this optimization study, Q_{EVAP} is parametrically varied from 16 kW to 100 kW. The optimization result provides the optimal distribution of annCAPEX and OPEX, the optimal sizes of the process units, and optimal operating conditions (stream pressure, temperature, concentration, and flow rate).

The computational tools to implement and solve the model equations were GAMS®v. 23.6.5 [38] and CONOPT 3 v. 3.14W [39], respectively. Since several nonlinear and non-convex constraints are present in the model and a local solver is used, it cannot be guaranteed that the obtained solutions correspond to the global optimum. However, based on the insights gathered from literature sources [2,5,32], the model was solved using different initial values obtaining the same solutions in all the cases. The latter forms a strong indication that the obtained solution is likely to correspond to the global optimum.

4. Results and Discussion

The optimization results obtained for a wide range of cooling capacity values and two (original and improved) process configurations are discussed. The main model parameter values are related with the cooling capacity, which is varied from 16 kW to 100 kW, and the global heat transfer coefficients, which are: $1.50 \text{ kW·m}^{-2}.°\text{C}^{-1}$ for the evaporator, $1.0 \text{ kW·m}^{-2}.°\text{C}^{-1}$ for the absorber, $2.50 \text{ kW·m}^{-2}.°\text{C}^{-1}$ for the condenser, $1.50 \text{ kW·m}^{-2}.°\text{C}^{-1}$ for the generators, and $1.0 \text{ kW·m}^{-2}.°\text{C}^{-1}$ for the solution heat exchangers.

The external design conditions are:

- High temperature generator (HTG): saturated steam at 160 °C.
- Absorber (ABS) and condenser (COND): cooling water at 20 °C.
- Evaporator (EVAP): Inlet and outlet chilled water temperatures: 13.0 °C and 10.0 °C, respectively; evaporator working temperature: 4.0 °C.

In addition, the following lower and upper bounds were imposed, respectively: 40% and 70% for LiBr concentrations, 0.1 kPa and 100 kPa for operating pressures, 0 kg·s^{-1} and 100 kg·s^{-1} for flow rates, and 75% and 100% for the effectiveness factors of the solution heat exchangers.

The optimization runs were performed by varying the cooling capacity from 16 kW to 100 kW. As shown in Figure 2, the minimum TAC value and the associated annCAPEX and OPEX values increase almost linearly with increasing cooling capacity levels. Also, it can be observed that the annCAPEX contribution to the TAC is significantly higher than the OPEX contribution, and that the difference between annCAPEX and OPEX increases as the cooling capacity increases. When the cooling capacity increases from 16 kW to 100 kW, the minimum TAC value and the optimal annCAPEX and OPEX values increase, respectively, 2.8, 2.5, and 6.4 times (from 12,794.5 $\text{\$·year}^{-1}$ to 35,613.9 $\text{\$·year}^{-1}$, from 12,013.6 $\text{\$·year}^{-1}$ to 30,644.4 $\text{\$·year}^{-1}$, and from 780.8 $\text{\$·year}^{-1}$ to 4969.5 $\text{\$·year}^{-1}$).

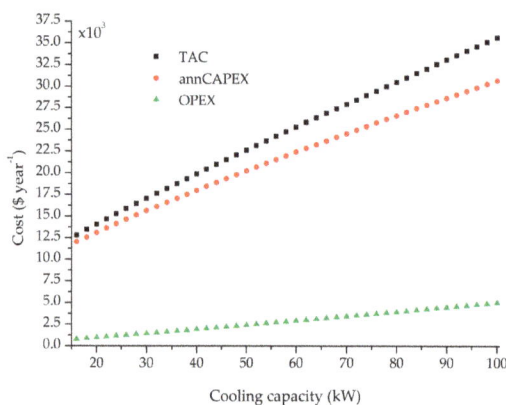

Figure 2. Optimal values of the TAC, annCAPEX, and OPEX versus cooling capacity.

Figure 3 illustrates the individual contributions of the process units to annCAPEX with increasing cooling capacity levels. It can be seen that the HTG and LTG have virtually the same annCAPEX values throughout the examined range, and that they are in the same order of magnitude as the EVAP for the lowest cooling capacity levels. These values are comparatively higher than the values obtained for the other process units. For cooling capacity values between 16 and 30 kW, the contributions of the ABS and COND to the annCAPEX are similar to each other, as is the case for the HTSHE and LTSHE. Also, Figure 3 shows that the contribution of EVAP is nonlinear while the contributions of the remaining process units are practically linear. For cooling capacities higher than 18 kW, EVAP is the largest contributor to annCAPEX. When the cooling capacity increases, EVAP and ABS are the process units that increase the most rapidly in annCAPEX compared to the other process units. Indeed, ABS and EVAP increase by around 11 and 3 times, respectively, when the cooling capacity increases from 19 to 100 kW.

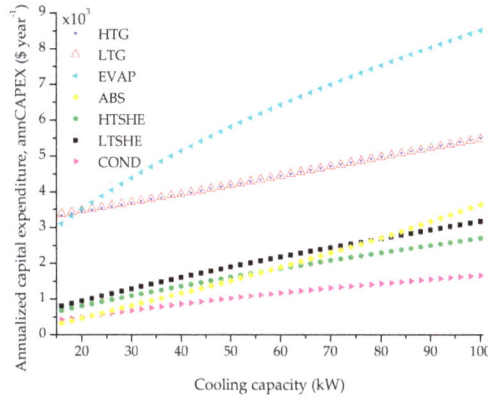

Figure 3. Optimal process-unit annCAPEX versus the cooling capacity.

The optimal values of the annualized investment cost for each process unit shown in Figure 3 correspond to the optimal values of the heat transfer areas, heat loads, and driving forces shown in Figure 4a–c, respectively.

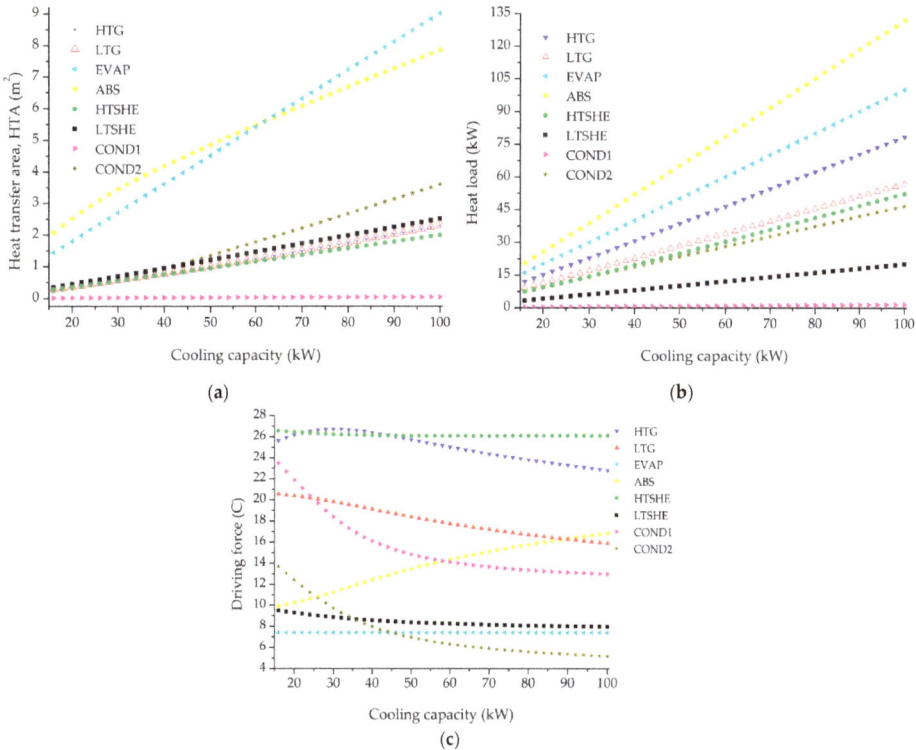

(a)

(b)

(c)

Figure 4. Optimal values for each process unit of (**a**) heat transfer area; (**b**) heat loads; (**c**) driving force, versus the cooling capacity level.

Regarding the OPEX distribution, Figure 5 shows that the contribution of the cost for steam required in the HGT as a heating source is slightly lower than the contribution of the cost for cooling water required in the COND and ABS, but the differences in cost increase with increasing cooling capacity levels. A cost difference of 75.6 $·year^{-1} (352.6 $·year^{-1} vs. 428.2 $·year^{-1}) is observed for a cooling capacity of 16 kW and a difference of 371.7 $·year^{-1} (2298.9 $·year^{-1} vs. 2670.6 $·year^{-1}) for a cooling capacity of 100 kW.

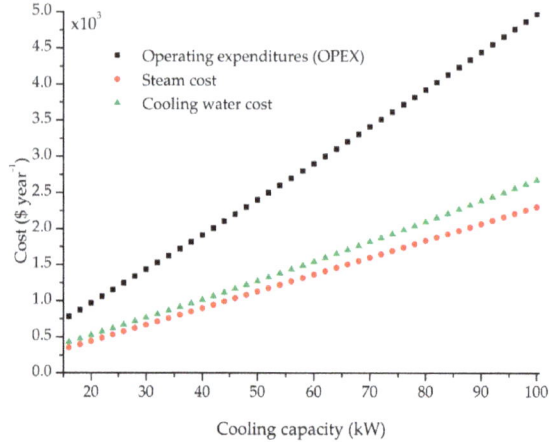

Figure 5. Optimal distribution of the operating expenditures (OPEX) versus the cooling capacity level.

Figure 6 shows the behavior of the LiBr solution concentrations (X) of the process: weak solution (X_1) and strong solutions (X_4 and X_{13} leaving the LTG and HTG, respectively; and X_{15} entering the ABS), with increasing cooling capacity levels. It can be seen that that the concentration values increase with the increase of the cooling capacity, but keep similar ratios between the concentration values in the different streams.

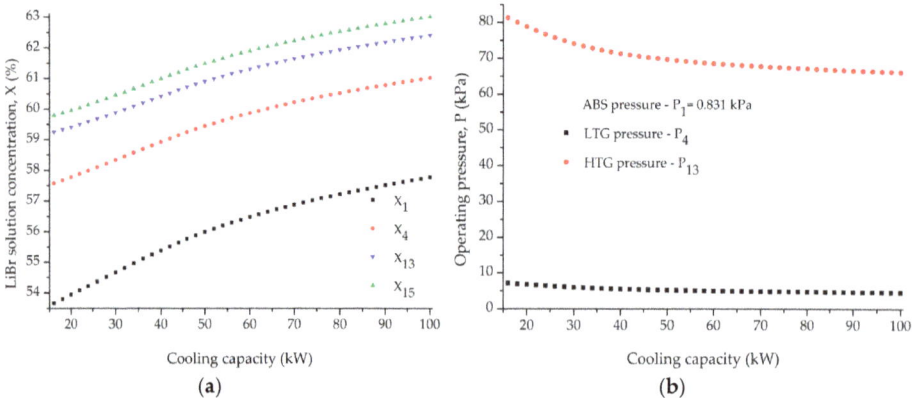

Figure 6. (a) Optimal LiBr concentration values of weak solution (X_1) and strong solutions (X_4 and X_{13} leaving the LTG and HTG, respectively; and X_{15} entering the ABS); (b) Optimal ABS (low), LTG (medium) and, HTG (high) operating pressure values, versus the cooling capacity level.

As mentioned earlier, the effectiveness factors η_{LTSHE} and η_{HTSHE} of the solution heat exchangers LTSHE and HTSHE, respectively, are considered as (free) model variables, i.e., decision variables, as opposed to other published studies, which consider these factors as (fixed) model parameters instead, usually in the range between 65% and 90%, thus always forcing their presence in the process configuration. In this work, by allowing the heat exchanger effectiveness factor to take any value, the presence or absence of the solution heat exchangers is a result of the optimization problem. First, all the solved optimization problems considered the same lower bound for η_{LTSHE} and η_{HTSHE} of 75%. The results deserve detailed discussion because they may indicate changes in the process configuration, such as the removal of one or even both solution heat exchangers in order to obtain improved solutions, in terms of total annual costs, compared to the current optimal solutions. The optimal η_{LTSHE} and η_{HTSHE} values remain constant at the imposed lower bound (75%) throughout the range of cooling capacity values.

Then, it becomes interesting to perform new optimizations while relaxing the lower bounds imposed to η_{LTSHE} and η_{HTSE} of 75%, in order to see how these bounds affect the current optimal solutions for the same range of cooling capacity values. The obtained optimization results are presented in the forthcoming discussions.

Influence of the Solution Heat Exchangers on the Optimal Solutions

The process configuration shown in Figure 1 and analyzed in the previous section—where both LTSHE and HTSHE are forced to be present—is hereafter named 'Conf. 1' and the one obtained in this subsection is referred as 'Conf. 2'. In all cases, the problem that is solved is the minimization of the TAC.

Figure 7 illustrates the optimal values of both effectiveness factors η_{LTSHE} and η_{HTSHE} obtained by considering a lower bound of 1%, which, in practical terms, is virtually zero. (Note that, in this case, a 'very small' numerical value is imposed as the lower bound, instead of zero, to prevent numerical problems that may lead to model convergence failure). As seen in Figure 7, the obtained optimal values for η_{LTSHE} result in the lower bound of η_{LTSHE}, thus indicating that the LTSHE is removed from the configuration for all the specified cooling capacity values. However, the optimal η_{HTSHE} values increase logarithmically, from 49.8% to 66.9%, with increasing cooling capacity levels in the examined range. This indicates that the heat integration between the weak and strong solutions leads to cost-effective solutions only when such integration takes place in the high-temperature region of the process through HTSHE (since LTSHE in the low-temperature region is not selected in any case).

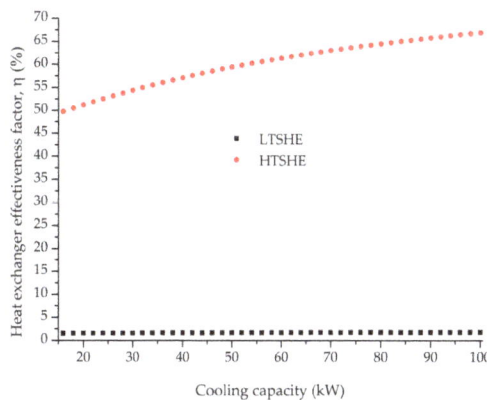

Figure 7. Optimal effectiveness factors η of the low-temperature solution heat exchanger (LTSHE) and the high-temperature solution heat exchanger (HTSHE) versus the cooling capacity when their lower bounds η_{LB} are relaxed.

Tables 1–6 compare costs, process-unit sizes, and operating conditions obtained for the two configurations corresponding to the extremes of the studied cooling capacity range, i.e., for 16 kW and 100 kW.

Table 1. Optimal costs obtained for configurations Conf. 1 and Conf. 2 for a cooling capacity of 16 kW.

Cost Item	Conf. 1	Conf. 2	Deviation (%)
TAC (M$·year^{-1})	12,794.5	11,537.2	−9.8
annCAPEX (M$·year^{-1})	12,013.6	10,684.3	−11.1
CAPEX (M$)	106,315.5	94,551.5	−11.1
EVAP	27,384.7	27,384.7	0
HTG	29,794.7	29,470.8	−1.1
LTG	29,447.1	29,195.3	−0.9
COND	3701.9	3802.2	+2.7
LTSHE	7135.8	121.2 (*)	−
HTSHE	5997.4	1911.9	−68.1
ABS	2853.9	2665.6	−6.6
OPEX (M$·year^{-1})	780.8	852.9	+9.2
Steam	352.6	405.0	+14.9
Cooling water	428.2	447.9	+4.6

(*) It is not summed in the TAC and CAPEX.

Table 2. Optimal values of heat transfer areas, heat loads, and driving forces obtained for configurations Conf. 1 and Conf. 2 for a cooling capacity of 16 kW.

	Heat Load (kW)		Heat Transfer Area (m^2)		Driving Force (°C)	
	Conf. 1	Conf. 2	Conf. 1	Conf. 2	Conf. 1	Conf. 2
EVAP	16.000	16.000	1.443	1.443	7.393	7.393
HTG	12.029	13.816	0.313	0.287	25.633	32.144
LTG	9.056	10.030	0.285	0.265	20.525	24.462
COND1	0.223	0.202	0.004	0.004	23.486	21.981
COND2	7.302	6.409	0.220	0.235	13.674	11.251
LTSHE	3.378 $\eta = 75\%$	0.047 $\eta = 1.521$	0.356	0.001	9.484	44.725
HTSHE	7.375 $\eta = 75\%$	3.197 $\eta = 49.767$	0.278	0.054	26.544	58.918
ABS	20.503	23.205	2.074	1.997	9.888	11.619

Table 3. Optimal values of operating conditions obtained for configurations Conf. 1 and Conf. 2 for a cooling capacity of 16 kW.

	Pressure (kPa)		Temperature (°C)		Solution Conc. (kg LiBr kg^{-1} sol.) × 100		Mass Flow Rate (kg·s^{-1})	
Point	Conf. 1	Conf. 2	Conf. 1	Conf. 2	Conf. 1	Conf. 2	Conf. 1	Conf. 2
1	0.813	0.813	30.944	30.967	53.668	53.681	0.085	0.058
2	7.150	5.835	30.944	30.967	53.668	53.681	0.045	0.032
3	7.150	5.835	66.918	31.659	53.668	53.681	0.045	0.032
4	7.150	5.835	78.910	76.438	57.578	58.449	0.042	0.030
5	7.150	5.835	38.298	75.638	57.578	58.449	0.042	0.030
6	0.813	0.813	38.198	42.666	57.582	59.863	0.042	0.030
7	7.150	5.835	78.910	76.438	−	−	0.003	0.003
8	7.150	5.835	39.345	35.595	−	−	0.007	0.007
9	0.813	0.813	4.005	4.005	−	−	0.007	0.007
10	0.813	0.813	4.005	4.005	−	−	0.007	0.007
11	81.299	58.161	30.944	30.967	53.668	53.681	0.040	0.026
12	81.299	58.161	117.292	89.468	53.668	53.681	0.040	0.026
13	81.299	58.161	146.074	148.517	59.245	63.889	0.037	0.022
14	81.299	58.161	55.368	89.756	59.245	63.889	0.037	0.022
15	0.813	0.813	42.516	53.893	59.787	65.401	0.037	0.022
16	81.299	58.161	146.074	148.517	−	−	0.004	0.004
17	81.299	58.161	94.023	85.209	−	−	0.004	0.004
18	7.150	5.835	39.345	35.595	−	−	0.004	0.004

Table 4. Optimal cost values obtained for configurations Conf. 1 and Conf. 2 for a cooling capacity of 100 kW.

Cost Item	Conf. 1	Conf. 2	Deviation (%)
TAC (M$·year^{-1})	35,613.9	31,338.1	−12.0
annCAPEX (M$·year^{-1})	30,644.4	26,001.7	−15.1
CAPEX (M$)	271,189.7	230,103.8	−15.1
EVAP	75,306.6	75,306.6	0
HTG	48,532.7	45,984.3	−5.3
LTG	48,642.3	45,325.2	−6.8
COND	14,649.4	13,715.8	−6.4
LTSHE	28,054.3	514.0 (*)	−
HTSHE	23,875.4	13,062.0	−45.3
ABS	32,128.8	36,709.9	+14.3
OPEX (M$·year^{-1})	4969.5	5336.4	+7.4
Steam	2298.9	2358.8	+2.6
Cooling water	2670.6	2977.6	+11.5

(*) It is not summed in the TAC and CAPEX.

Table 5. Optimal values of heat transfer areas, heat loads, and driving forces obtained for configurations Conf. 1 and Conf. 2 for a cooling capacity of 100 kW.

	Heat Load (kW)		Heat Transfer Area (m^2)		Driving Force (°C)	
	Conf. 1	Conf. 2	Conf. 1	Conf. 2	Conf. 1	Conf. 2
EVAP	100.00	100.00	9.017	9.017	7.393	7.393
HTG	78.424	80.466	2.295	1.988	22.781	26.979
LTG	56.767	61.849	2.308	1.911	15.866	20.885
COND1	1.582	1.389	0.049	0.039	12.977	14.307
COND2	45.018	40.332	3.602	3.189	5.175	5.233
LTSHE	20.039 (η = 75%)	0.316 (η = 1.794%)	2.518	0.008	7.960	38.000
HTSHE	52.188 (η = 75%)	29.361 (η = 66.910%)	1.999	0.845	26.101	34.758
ABS	131.825	138.745	7.842	8.438	16.809	16.442

Table 6. Optimal values of operating conditions obtained for configurations Conf. 1 and Conf. 2 for a cooling capacity of 100 kW.

	Pressure (kPa)		Temperature (°C)		Solution Conc. (kg LiBr kg^{-1} sol.) × 100		Mass Flow Rate (kg·s^{-1})	
Point	Conf. 1	Conf. 2	Conf. 1	Conf. 2	Conf. 1	Conf. 2	Conf. 1	Conf. 2
1	0.813	0.813	38.569	35.667	57.774	56.253	0.667	0.416
2	4.575	4.146	38.569	35.667	57.774	56.253	0.349	0.223
3	4.575	4.146	67.385	36.162	57.774	56.253	0.349	0.223
4	4.575	4.146	76.991	74.414	61.017	60.766	0.330	0.207
5	4.575	4.146	45.083	73.914	61.017	60.766	0.330	0.207
6	0.813	0.813	44.983	46.760	61.021	61.902	0.330	0.207
7	4.575	4.146	76.991	74.414	−	−	0.019	0.017
8	4.575	4.146	31.244	29.525	−	−	0.042	0.042
9	0.813	0.813	4.005	4.005	−	−	0.042	0.042
10	0.813	0.813	4.005	4.005	−	−	0.042	0.042
11	66.147	51.122	38.569	35.667	57.774	56.253	0.318	0.193
12	66.147	51.122	120.781	110.365	57.774	56.253	0.318	0.193
13	66.147	51.122	148.185	147.306	62.407	64.801	0.294	0.167
14	66.147	51.122	63.409	68.330	62.407	64.801	0.294	0.167
15	0.813	0.813	49.006	53.893	63.008	65.401	0.294	0.167
16	66.147	51.122	148.185	147.306	−	−	0.024	0.025
17	66.147	51.122	88.538	81.941	−	−	0.024	0.025
18	4.575	4.146	31.244	29.525	−	−	0.024	0.025

Figures 8–10 compare the optimal values of costs obtained for both configurations for the whole range of cooling capacity values. Figure 8a,b show that Conf. 2 has lower TAC and annCAPEX values, respectively, than Conf. 1 for all cooling capacity levels. However, Conf. 1 has slightly lower OPEX values than the OPEX values obtained for Conf. 2 (Figure 8c). The differences in TAC, annCAPEX, and OPEX values between Conf. 1 and Conf. 2 increase with increasing cooling capacity levels. As seen in Figure 8a,b and Table 1, at a cooling capacity of 16 kW, the TAC and annCAPEX values obtained for Conf. 2 are 9.8% and 11.1% lower than the values obtained for Conf. 1 (11,537.2 M\$·year^{-1} vs. 12,794.5 M\$·year^{-1}, and 10,684.3 M\$·year^{-1} vs. 12,013.6 M\$·year^{-1}, respectively). However, the OPEX in Conf. 2 is 9.2% higher than in Conf. 1 (852.9 M\$·year^{-1} vs. 780.8 M\$·year^{-1}). For a cooling capacity of 100 kW, Table 4 shows that the TAC and annCAPEX values obtained for Conf. 2 are, respectively, 12% and 15.1% lower than the values obtained for Conf. 1 (31,338.1 M\$·year^{-1} vs. 35,613.9 M\$·year^{-1}, and 26,001.7 M\$·year^{-1} vs. 30,644.4 M\$·year^{-1}, respectively). While the OPEX for Conf. 2 is 7.4% higher than for Conf. 1 (5336.4 M\$·year^{-1} vs. 4969.5 M\$·year^{-1}, respectively).

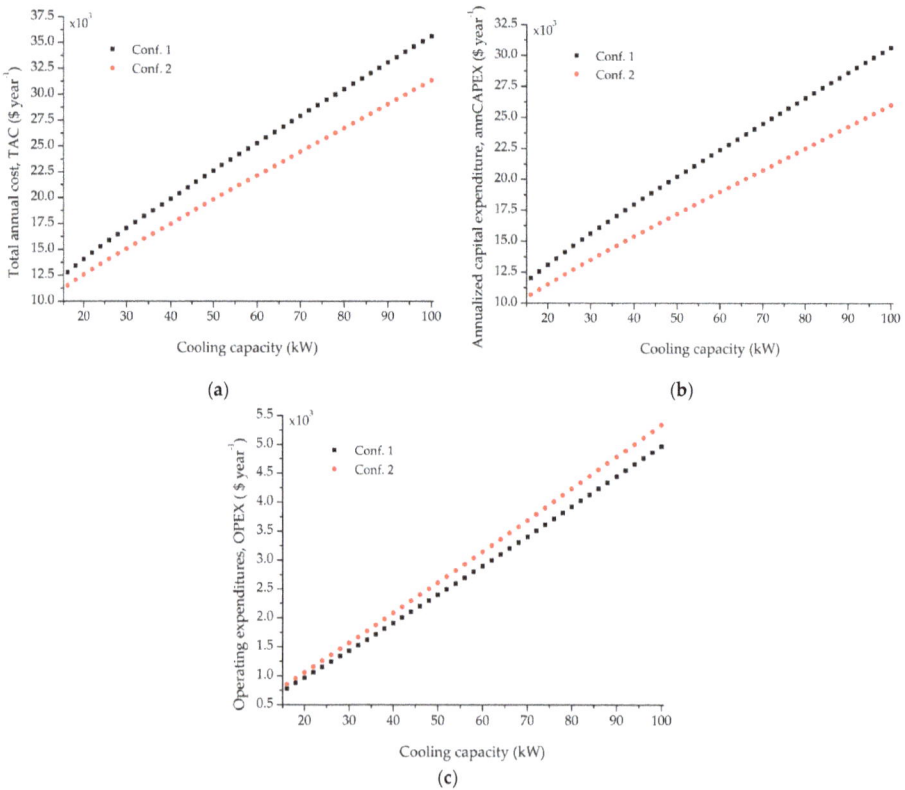

Figure 8. (**a**) Optimal total annual cost (TAC); (**b**) Optimal annualized capital expenditures (annCAPEX); (**c**) Optimal operating expenditures (OPEX) for configurations Conf. 1 and Conf. 2 as a function of the cooling capacity level.

Figure 9a compares the cost for steam (heating utility) required for different cooling capacity levels between both configurations, while Figure 9b compares the cost for cooling water requirements. It can be seen that, for all cooling capacity values, the cost for steam obtained for Conf. 2 is slightly higher than the cost obtained for Conf. 1, and that the difference remains almost constant throughout the examined range (Figure 9a). The cost for cooling water is almost the same for low capacity level values; however, for higher cooling capacity levels, the cost for Conf. 2 is greater than the cost for Conf. 1, and the difference increases with increasing cooling capacity values (Figure 9b).

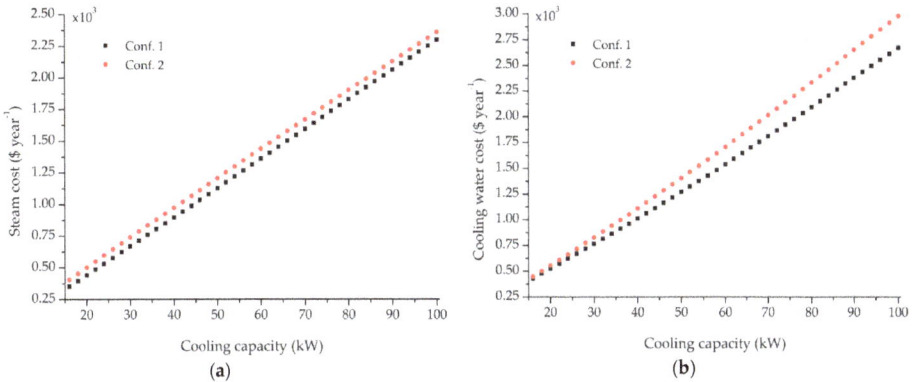

Figure 9. (a) Optimal cost values for steam requirements; (b) Optimal cost values for cooling water requirements, as a function of the cooling capacity level.

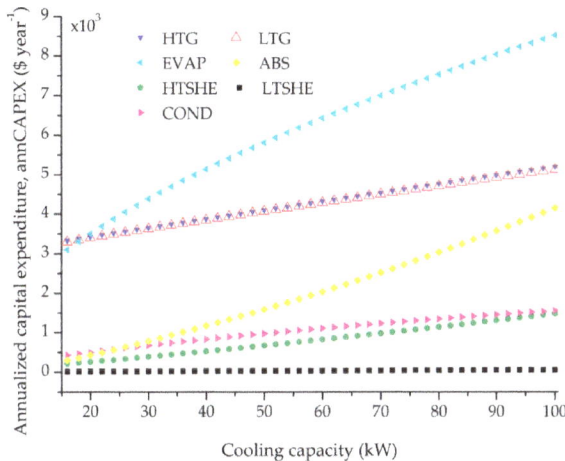

Figure 10. Optimal annualized capital expenditure (annCAPEX) values for each process unit in configuration Conf. 2 as a function of the cooling capacity level.

Figure 10 shows the investments associated with the process units obtained for Conf. 2. When comparing this figure and Figure 3 corresponding to Conf. 1, it can be seen that the trends of the individual contributions of the process units are similar for both configurations, except for LTSHE. This indicates that the elimination of LTSHE from the configuration does not modify the general trends of the investments required for the other process units as a function of the cooling capacity.

Finally, it is interesting to compare in Table 3 (16 kW) and Table 6 (100 kW) the optimal flow rate values of the weak (stream #1) and strong (stream #6) solutions for both configurations. Independently of the cooling capacity level, the optimal values of these variables obtained for Conf. 2 are significantly lower than the values obtained for Conf. 1. Moreover, all the flow rate values of the weak and strong solutions (m_1 to m_6, and m_{11} to m_{15}) obtained for Conf. 2 are comparatively lower than the values obtained for Conf. 1 (by around 30–45% depending on the particular stream considered). For 16 kW, m_1 decreases from 0.085 kg·s^{-1} to 0.058 kg·s^{-1} (a 32% decrease) and m_2 from 0.045 kg·s^{-1} to 0.032 kg·s^{-1} (a 29% decrease). However, the weak solution concentration X_1 remains virtually unchanged for 16 kW and changes by only 2.6% for 100 kW. However, the (absolute) values are different; they are 53.7% for 16 kW and 56.2% for 100 kW, in Conf. 2.

Another interesting result, from a practical point of view, is that the optimal medium and high operating pressures obtained for Conf. 2 are also significantly lower than the values obtained for Conf. 1. Table 3 shows that the medium and high pressures for Conf. 2 are 18% and 28% lower than Conf. 1, respectively, for a cooling capacity of 16 kW. Table 6 shows that these reductions are 9% and 23%, respectively, for 100 kW. However, it should be observed that, for Conf. 2 and throughout the examined range of cooling capacity values, the LiBr concentration X_{15} and temperature T_{15} of stream #15 reached the values of 65.401% and 53.893 °C, respectively, which were obtained from the model constraint that describes the crystallization line. In fact, the inequality constraints that prevent crystallization became active, thus indicating that Conf. 2 operates in a region closer to the crystallization line than Conf. 1.

Finally, in order to investigate the influence of the utility costs in the optimal solutions, the same optimization problems were solved by changing the current cost parameters. Specifically, the current cooling water and steam costs were changed to 2.95×10^{-2} \$·t^{-1} of cooling water and 84 \$·t^{-1} of steam, respectively. These numerical values are reported by Khan et al. [40] and Union Gas Limited [41], respectively. In addition, the influence of the global heat transfer coefficient values on the optimal solutions was studied. The optimization results showed that the optimal process configuration and the trends of the process variables do not vary with respect to the solutions discussed above when changes in the parameters were introduced.

5. Conclusions

This paper addressed the optimization of a double-effect H_2O-LiBr ARS through the minimization of the total annual cost for a wide range of cooling capacity values. To this end, the existing trade-offs between process configuration, sizes of the process units, and operating conditions were optimized by employing a nonlinear mathematical model, which was implemented in GAMS. Interestingly, the effectiveness factors of the solution heat exchangers, which were treated as optimization variables instead of fixed parameters, allowed us to obtain a new process configuration. The low-temperature heat exchanger is removed from the configuration throughout the examined range of cooling capacity levels, keeping only the high-temperature solution heat exchanger, indicating that the heat integration between the weak and strong LiBr solutions takes place entirely at the high-temperature zone of the process. The importance in terms of the effectiveness factor of the high-temperature solution heat exchanger increases with increasing cooling capacity levels; the sizes and operating conditions of the other process units accommodate accordingly, in order to meet the problem specifications with the minimal total annual cost. However, the improved configuration operates in a region closer to the crystallization line than the original configuration.

For a specified cooling capacity of 16 kW, the improved configuration makes it possible to reduce the total annual cost and the annualized capital expenditures by around 10% and 11%, respectively, with respect to the optimized conventional double-effect configuration, at the expense of increasing the operating expenditures by around 9%. For a cooling capacity of 100 kW, these percentages are 12%, 15%, and 7.4%, respectively. Then, the improved configuration shows better cost performances at the higher cooling capacity levels that were studied.

In future work, the proposed model will consider the variation of the heat transfer coefficients with the temperature in each process unit. Then, a superstructure-based representation embedding several candidate configurations, and thereby allowing different flow patterns, will be modeled and solved through a discrete and continuous mathematical programming model. The latter system will also include the possibility of extending the number of effects, and will make it possible to consider other heat sources.

Supplementary Materials: The following are available online at http://www.mdpi.com/2227-9717/7/1/50/s1, Figure S1: Schematic of the studied double-effect H_2O-LiBr ARS; Table S1: Parameter values for estimating process unit investment Z_k.

Author Contributions: All authors contributed to the analysis of the results and to writing the manuscript. S.F.M. developed and implemented the mathematical model of the process in GAMS, collected and analyzed data, and wrote the first draft of the manuscript. S.S.M., K.V.G., T.M. and M.C.M. provided feedback to the content and revised the final draft. M.C.M. conceived and supervised the research.

Funding: This research was funded by CONICET.

Acknowledgments: The financial support from the Consejo Nacional de Investigaciones Científicas y Técnicas (CONICET) from Argentina is gratefully acknowledged.

Conflicts of Interest: The authors declare no conflict of interest.

References

1. Garousi Farshi, L.; Mahmoudi, S.M.S.; Rosen, M.A.; Yari, M.; Amidpour, M. Exergoeconomic analysis of double effect absorption refrigeration systems. *Energy Convers. Manag.* **2013**, *65*, 13–25. [CrossRef]
2. Mazzei, M.S.; Mussati, M.C.; Mussati, S.F. NLP model-based optimal design of LiBr–H_2O absorption refrigeration systems. *Int. J. Refrig.* **2014**, *38*, 58–70. [CrossRef]
3. Kaynakli, O.; Kilic, M. Theoretical study on the effect of operating conditions on performance of absorption refrigeration system. *Energy Convers. Manag.* **2007**, *48*, 599–607. [CrossRef]
4. Avanessian, T.; Ameri, M. Energy, exergy, and economic analysis of single and double effect LiBr–H_2O absorption chillers. *Energy Build.* **2014**, *73*, 26–36. [CrossRef]
5. Mussati, S.F.; Gernaey, K.V.; Morosuk, T.; Mussati, M.C. NLP modeling for the optimization of LiBr-H_2O absorption refrigeration systems with exergy loss rate, heat transfer area, and cost as single objective functions. *Energy Convers. Manag.* **2016**, *127*, 526–544. [CrossRef]
6. Talbi, M.M.; Agnew, B. Exergy analysis: An absorption refrigerator using lithium bromide and water as the working fluids. *Appl. Therm. Eng.* **2000**, *20*, 619–630. [CrossRef]
7. Misra, R.D.; Sahoo, P.K.; Sahoo, S.; Gupta, A. Thermoeconomic optimization of a single effect water/LiBr vapour absorption refrigeration system. *Int. J. Refrig.* **2003**, *26*, 158–169. [CrossRef]
8. Kızılkan, Ö.; Şencan, A.; Kalogirou, S.A. Thermoeconomic optimization of a LiBr absorption refrigeration system. *Chem. Eng. Process. Process Intensif.* **2007**, *46*, 1376–1384. [CrossRef]
9. Rubio-Maya, C.; Pacheco-Ibarra, J.J.; Belman-Flores, J.M.; Galván-González, S.R.; Mendoza-Covarrubias, C. NLP model of a LiBr–H_2O absorption refrigeration system for the minimization of the annual operating cost. *Appl. Therm. Eng.* **2012**, *37*, 10–18. [CrossRef]
10. Srikhirin, P.; Aphornratana, S.; Chungpaibulpatana, S. A review of absorption refrigeration technologies. *Renew. Sustain. Energy Rev.* **2001**, *5*, 343–372. [CrossRef]
11. Garousi, F.; Seyed, M.; Rosen, M.A.; Yari, M. A comparative study of the performance characteristics of double-effect absorption refrigeration systems. *Int. J. Energy Res.* **2012**, *36*, 182–192. [CrossRef]
12. Garousi Farshi, L.; Seyed Mahmoudi, S.M.; Rosen, M.A. Analysis of crystallization risk in double effect absorption refrigeration systems. *Appl. Therm. Eng.* **2011**, *31*, 1712–1717. [CrossRef]
13. Arun, M.B.; Maiya, M.P.; Murthy, S.S. Performance comparison of double-effect parallel-flow and series flow water–lithium bromide absorption systems. *Appl. Therm. Eng.* **2001**, *21*, 1273–1279. [CrossRef]
14. Kaushik, S.C.; Arora, A. Energy and exergy analysis of single effect and series flow double effect water–lithium bromide absorption refrigeration systems. *Int. J. Refrig.* **2009**, *32*, 1247–1258. [CrossRef]
15. Kaynakli, O.; Saka, K.; Kaynakli, F. Energy and exergy analysis of a double effect absorption refrigeration system based on different heat sources. *Energy Convers. Manag.* **2015**, *106*, 21–30. [CrossRef]
16. Talukdar, K.; Gogoi, T.K. Exergy analysis of a combined vapor power cycle and boiler flue gas driven double effect water–LiBr absorption refrigeration system. *Energy Convers. Manag.* **2016**, *108*, 468–477. [CrossRef]
17. Misra, R.D.; Sahoo, P.K.; Gupta, A. Thermoeconomic evaluation and optimization of a double-effect H_2O/LiBr vapour-absorption refrigeration system. *Int. J. Refrig.* **2005**, *28*, 331–343. [CrossRef]

18. Bereche, R.P.; Palomino, R.G.; Nebra, S.A. Thermoeconomic analysis of a single and double-effect LiBr/H$_2$O absorption refrigeration system. *Int. J. Thermodyn.* **2009**, *12*, 89–96.

19. Gebreslassie, B.H.; Medrano, M.; Boer, D. Exergy analysis of multi-effect water–LiBr absorption systems: From half to triple effect. *Renew. Energy* **2010**, *35*, 1773–1782. [CrossRef]

20. Mussati, S.F.; Aguirre, P.A.; Scenna, N.J. Novel Configuration for a Multistage Flash-Mixer Desalination System. *Ind. Eng. Chem. Res.* **2003**, *42*, 4828–4839. [CrossRef]

21. Alasino, N.; Mussati, M.C.; Scenna, N.J.; Aguirre, P. Wastewater treatment plant synthesis and design: Combined biological nitrogen and phosphorus removal. *Ind. Eng. Chem. Res.* **2010**, *49*, 8601–8612. [CrossRef]

22. Oliva, D.G.; Francesconi, J.A.; Mussati, M.C.; Aguirre, P.A. Energy efficiency analysis of an integrated glycerin processor for PEM fuel cells: Comparison with an ethanol-based system. *Int. J. Hydrogen Energy* **2010**, *35*, 709–724. [CrossRef]

23. Oliva, D.G.; Francesconi, J.A.; Mussati, M.C.; Aguirre, P.A. Modeling, synthesis and optimization of heat exchanger networks. Application to fuel processing systems for PEM fuel cells. *Int. J. Hydrogen Energy* **2011**, *36*, 9098–9114. [CrossRef]

24. Arias, A.M.; Mussati, M.C.; Mores, P.L.; Scenna, N.J.; Caballero, J.A.; Mussati, S.F. Optimization of multi-stage membrane systems for CO$_2$ capture from flue gas. *Int. J. Greenh. Gas Control* **2016**, *53*, 371–390. [CrossRef]

25. Manassaldi, J.I.; Arias, A.M.; Scenna, N.J.; Mussati, M.C.; Mussati, S.F. A discrete and continuous mathematical model for the optimal synthesis and design of dual pressure heat recovery steam generators coupled to two steam turbines. *Energy* **2016**, *103*, 807–823. [CrossRef]

26. Gebreslassie, B.H.; Guillén-Gosálbez, G.; Jiménez, L.; Boer, D. Design of environmentally conscious absorption cooling systems via multi-objective optimization and life cycle assessment. *Appl. Energy* **2009**, *86*, 1712–1722. [CrossRef]

27. Chahartaghi, M.; Golmohammadi, H.; Shojaei, A. Performance analysis and optimization of new double effect lithium bromide–water absorption chiller with series and parallel flows. *Int. J. Refrig.* **2019**, *97*, 73–87. [CrossRef]

28. Lee, S.; Lee, J.; Lee, H.; Chung, J.; Kang, Y. Optimal design of generators for H$_2$O/LiBr absorption chiller with multi-heat sources. *Energy* **2019**, *167*, 47–59. [CrossRef]

29. Sabbagh, A.; Gómez, J. Optimal control of single stage LiBr/water absorption chiller. *Int. J. Refrig.* **2018**, 1–9. [CrossRef]

30. Lubis, A.; Jeong, J.; Giannetti, N.; Yamaguchi, S.; Saito, K.; Yabase, H.; Alhamid, M.I.; Nasruddin. Operation performance enhancement of single-double-effect absorption chiller. *Appl. Energy* **2018**, *219*, 299–311. [CrossRef]

31. Kawajir, Y.; Laird, C.; Wachter, A. Introduction to Ipopt: A Tutorial for Downloading, Installing, and Using Ipopt, 2087 ed. 2012. Available online: https://projects.coin-or.org/Ipopt (accessed on 23 November 2018).

32. Mussati, S.F.; Cignitti, S.; Mansouri, S.S.; Gernaey, K.V.; Morosuk, T.; Mussati, M.C. Configuration optimization of series flow double-effect water-lithium bromide absorption refrigeration systems by cost minimization. *Energy Convers. Manag.* **2018**, *158*, 359–372. [CrossRef]

33. Udugama, I.A.; Mansouri, S.S.; Mitic, A.; Flores-Alsina, X.; Gernaey, K.V. Perspectives on Resource Recovery from Bio-Based Production Processes: From Concept to Implementation. *Processes* **2017**, *5*, 48. [CrossRef]

34. Mansouri, S.S.; Udugama, I.A.; Cignitti, S.; Mitic, A.; Flores-Alsina, X.; Gernaey, K.V. Resource recovery from bio-based production processes: A future necessity? *Curr. Opin. Chem. Eng.* **2017**, *18*, 1–9. [CrossRef]

35. Dincer, I. *Refrigeration Systems and Applications*; Wiley: Hoboken, NJ, USA, 2003; ISBN 978-0-471-62351-9.

36. American Society of Heating, Refrigerating and Air-Conditioning Engineers. *1989 ASHRAE Handbook: Fundamentals*; ASHRAE: Atlanta, GA, USA, 1989; ISBN 978-0-910110-57-0.

37. Gilani, S.I.-u.-H.; Ahmed, M.S.M.S. Solution crystallization detection for double-effect LiBr-H$_2$O steam absorption chiller. *Energy Procedia* **2015**, *75*, 1522–1528. [CrossRef]

38. GAMS Development Corporation. *General Algebraic Modeling System (GAMS) Release 23.6.5*; GAMS Development Corporation: Washington, DC, USA, 2010.

39. Drud, A. *CONOPT 3 Solver Manual*; ARKI Consulting and Development A/S: Bagsvaerd, Denmark, 2012.

40. Khan, M.; Tahan, S.; El-Achkar, M.; Jamus, S. The study of operating an air conditioning system using Maisotsenko-Cycle. *Mater. Sci. Eng.* **2018**, *323*, 1–7. [CrossRef]
41. Union Gas Limited. Calculating the True Cost of Steam. Available online: http://members.questline.com/Article.aspx?articleID=18180&accountID=1863&nl=13848 (accessed on 23 November 2018).

processes

MDPI

Article

Supercritical CO$_2$ Transesterification of Triolein to Methyl-Oleate in a Batch Reactor: Experimental and Simulation Results

Geetanjali Yadav [1,2], Leonard A. Fabiano [1], Lindsay Soh [3], Julie Zimmerman [4], Ramkrishna Sen [2] and Warren D. Seider [1,*]

[1] Department of Chemical and Biomolecular Engineering, University of Pennsylvania, Philadelphia, PA 19104-6393, USA; yadavg@seas.upenn.edu (G.Y.); lfabiano@seas.upenn.edu (L.A.F.)
[2] Department of Biotechnology, Indian Institute of Technology Kharagpur, Kharagpur 721302, India; rksen@yahoo.com
[3] Department of Chemical and Biomolecular Engineering, Lafayette College, Easton, PA 18042, USA; lindsay.soh@gmail.com
[4] Department of Chemical and Environmental Engineering, Yale University, New Haven, CT 06511, USA; julie.zimmerman@yale.edu
[*] Correspondence: seider@seas.upenn.edu; Tel.: +1-215-898-7953

Received: 16 October 2018; Accepted: 25 December 2018; Published: 1 January 2019

Abstract: In earlier work (Silva et al., 2016; Soh et al., 2014a; Soh et al., 2015), the supercritical CO$_2$ transesterification of triolein to methyl-oleate using Nafion solid-acid catalyst and large methanol/triolein molar feed ratios was carried out. Herein, these ratios are adjusted (from 50–550) to evaluate the yield of fatty acid methyl esters in batch laboratory reactors as temperature is varied from 80–95 °C and pressure is varied from 8.0–9.65 MPa. Also, to better understand the effect of varying these operating parameters, batch reactor simulations using the Soave-Redlich-Kwong Equation of State (RK-ASPEN EOS) in ASPEN PLUS are carried-out. A single-reaction kinetic model is used and phase equilibrium is computed as the reactions proceed. Experimental data are compared with these results.

Keywords: multiphase equilibrium; RK-ASPEN; methyl-oleate; biodiesel; supercritical CO$_2$

1. Introduction

As a replacement for conventional fossil fuels to meet energy demands, a new wave of research on biodiesel production technologies has commenced for the development of alternate energy sources worldwide. These include fatty acid methyl esters (FAMEs), i.e., biodiesel, having characteristics similar to petrodiesel oil, allowing its use in compression motors without any engine modification [1]. FAMEs are commonly obtained by (1) the transesterification of vegetable oils, i.e., triglycerides (TG) of fatty acids (FAs), or (2) esterification of free fatty acids (FFA), with lower alcohols [2]. Generally, triglycerides can be classified into two groups: simple and mixed. The simple triglyceride is composed of three identical fatty acid chains, whereas fatty acid chains of a mixed triglyceride are not identical. Natural oils produced from oil-bearing crops comprise 97% of various triglycerides and 1–5% of free fatty acids (FFA). Along with simple triglycerides, vegetable oils consist of mixed triglycerides containing different fatty acid chains; e.g., C12:0 (lauric acid chain), C14:0 (myristic acid chain), C16:0 (palmitic acid chain), C18:0 (stearic acid chain), C18:1 (oleic acid chain), and C18:2 (linoleic acid chain). Their compositions are known to vary with oil sources and growth conditions [3]. Recently, alternative feedstocks such as waste/used cooking oils, and non-edible feedstocks such as jatropha, pongamia, castor and microalgal oils are used to produce biodiesel fuels, to reduce the high prices of biodiesel fuel.

Transesterification of triglycerides with homogeneous acid or base catalyst requires its neutralization and recovery from the reactor products. Increased purification and recovery steps can, eventually, affect product costs and the market. Also, the base catalyst results in the production of undesirable products due to the saponification reaction. Alternatively, heterogeneous catalysts can be separated from the liquid effluents and re-used easily [4]. Solid acid-catalyzed transesterification reactions have been explored to circumvent the problems associated with the conversion of low quality feedstocks (containing free fatty acids) to biodiesel, and thus, are preferred over base catalyzed transesterifications. Also, non-catalytic transesterifications have shown promising reaction rates for commercial application using supercritical methanol (>250 °C, 19–45 MPa) [5,6]. The partial miscibility of the oil and methanol phases at moderate temperatures and pressures hinders the rate of reaction.

Supercritical processes do not require neutralization, washing, and drying steps, allowing waste oils to be processed without these expensive pretreatment steps [5]. Supercritical carbon dioxide (Sc-CO_2) (critical point at 31 °C and 7.3 MPa) and methanol (critical point at 240 °C and 7.95 MPa) used in a single supercritical phase for the transesterification resulted in higher reaction rates and lesser time duration [6]. But, the monophasic system can suffer from high energy requirements and the need for downstream separation of glycerol from the product [7]. Operation at moderate temperatures (~80–100 °C) and pressures (8–10 MPa) in a multi-phase liquid-vapor system may allow for the same benefits without high energy burdens. Sc-CO_2 (supercritical carbon dioxide) acts as a co-solvent and can increase the rate of the reaction by eliminating or reducing the transport resistance and increasing the solubility of methanol in triolein and vice-versa [8].

In previous work Soh et al. [7], demonstrated experimentally that mixed carbon dioxide (CO_2) and methanol (MeOH) successfully transesterifies triolein into methyl-oleate at moderate pressures and temperatures below 100 °C in the presence of a heterogeneous acid catalyst, Nafion NR50. Additionally, high-pressure CO_2 was experimentally found to be effective and selective in separating algae oil triglycerides [9], with new separation approaches currently under development. Silva et al. [10] simulated a batch reactor involving six chemical species; viz., triolein, methanol, CO_2, glycerol, FAME, and water. The comparison of simulation results using the RK-ASPEN EOS (with no binary parameters) gave reasonable agreement with VLLE (vapor-liquid-liquid equilibrium) experimental results, and thus, the RK-ASPEN EOS was used in thermo-kinetic reactor model (see Section 3 below). A custom-written FORTRAN® subroutine in a USER2 block of ASPEN PLUS was used that integrates the mass balance ordinary differential equations (ODEs) and checks the multiphase equilibrium, at various time intervals, to incorporate the effect of the phase behavior on the reaction kinetics periodically using the FLSH_FLASH subroutine [10,11]. FLSH_FLASH is an ASPEN PLUS subroutine that performs only flash calculations (without reactions). Rate constants were regressed using the bulk concentrations in the experimental 50 mL, agitated reactor vessel [9].

However, since the motivation behind these reactions was to evaluate only the effect of phase behavior without assessing yields of methyl-oleate converted and a constant amount of methanol on the reaction yield, the molar ratio of methanol/triolein used was quite high (1087). Because of its impact on the FAME yield [12], these experiments have been extended to methanol/triolein molar ratios of 50, 100, 300, and 550 herein. Then, using the batch-reactor simulation model [10], the predicted FAME yields are shown to compare favorably with the experimental data. A key objective of this verification, is to show that the FAME yields can be optimized by varying the methanol/triolein molar ratio, together with operating temperature and pressure. This manuscript focuses on this verification of the laboratory data.

For the laboratory experiments (carried out at Lafayette College), corn oil and methanol at the four molar ratios were transesterified using solid-acid heterogeneous catalyst (Nafion NR50) in the presence of supercritical CO_2 at 95 °C and 9.65 MPa for 4 h in a batch reactor to yield FAME. Then, a FORTRAN® USER2 block in ASPEN PLUS V10, prepared by Silva et al. [10], was used to carry-out dynamic simulations of the batch reactor. As the mass-balances were integrated, using the kinetic model in Section 3, the FLSH_FLASH subroutine was used to compute 3-phase equilibria using the

RK-ASPEN EOS. Given the concentrations of FAME in the vapor and two liquid phases, and the phase volumes, the yield of methyl-oleate (g) was computed.

2. Materials and Methods

This section describes the chemicals, catalyst, batch reactor, and analysis methods in the experiments at Lafayette College.

(i)   Chemicals: For calibration of the supercritical fluid chromatography-mass spectroscopy (SFC-MS) unit, all oleate species standards were purchased from Sigma-Aldrich (purity $\geq$ 99%, St. Louis, MO, USA) except diolein (1,2 and 1,3 DG isomers, 2:1 isomeric ratio) from MP Biomedicals, LLC (purity $\geq$ 99%, Santa Ana, CA, USA). ACS grade methanol was obtained from J.T. Baker (Radnor, PA, USA). High Performance Liquid Chromatography (HPLC) grade heptane and ultrapure isopropanol were obtained from Alfa Aesar (Haverhill, MA, USA) and Sigma-Aldrich, Inc. Bone-dry CO_2 with a siphon tube and nitrogen gas were supplied by Airgas, Inc. (Radnor, PA, USA). Corn oil was obtained from a local market and was analyzed for fatty-acid content using standard methods (10.4 wt% C16:0, 30.8 wt% C18:1, 58.8 wt% C18:2, others in trace quantities) [13]. Note that, for experiments at low methanol/triolein molar ratios, inexpensive corn oil was purchased.

(ii) Catalyst Characteristics: Nafion NR50 was purchased from Ion Power, Inc. (New Castle, DE, USA), and stored in a desiccator [14]. For all of the experiments reported, the catalyst concentration was 0.00379 mol/L (based upon the number of active sites per μmol). To assess CO_2's effect on particle swelling, all Nafion NR50 was presoaked in methanol for at least 72 h before reaction.

(iii) Reactor and reaction conditions: Nearly all of the reactions were performed in a stainless-steel stirred reactor (Supercritical Fluid Technologies, Inc. (Newark, DE, USA), High-Pressure Reactor, 100 mL). For each reaction, the catalyst and substrates were added directly into the reactor that was then sealed and heated to the desired temperature using the built-in heating jacket and controlled by an RXTrol Jr. integrated processor (Newark, DE, USA). The reactor was then pressurized with CO_2 and stirred at 300 rpm to increase the interfacial area between triglyceride and methanol phases. Preliminary experiments indicate that this mixing speed is sufficient to minimize mass-transfer limitations within the reactor [15]. The conditions were maintained for 4 h when CO_2 was vented through a restrictor valve. After the reaction, the venting CO_2 was slowly sparged through isopropanol liquid to dissolve the reaction products. Then, this isopropanol was added to the liquids remaining in the reactor, which were dissolved in it. The resulting isopropanol was analyzed to determine concentrations of the reactor products. An internal standard was used to analytically compensate for any loss of isopropanol during sparging. All reactions were performed in at least duplicate with an initial substrate (corn oil) at loadings, depending on the methanol/triolein ratio. Note that for all loadings, the combined volumes of methanol and corn oil was 5.22 mL, giving a fixed volume of CO_2.

(iv) Analysis: Samples were analyzed by supercritical fluid chromatography-mass spectrometry (Waters® Acquity UPC^2 with Xevo TQD Triple Quadrupole Mass Spectrometer (Milford, MA, USA) with an Acquity HSS C18 column (100 Å, 1.8 μm, 2.1 mm $\times$ 100 mm) and using a 1 μL sample volume. The column was held at 45 °C with a back pressure of 1500 psi. The mobile phase consisted of CO_2 (A) and 90:10 acetonitrile: methanol (B). The elution gradient started at 15% B and increased linearly to 35% B in 3.5 min where it was held for 1 min before return to the starting conditions. The mass spectrometer was run in Atmospheric pressure chemical ionization (APCI+) mode with a desolvation temperature of 600 °C and N_2 flow rate of 1000 L/h and cone flow of 40 L/h. The APCI voltages were 3.5 kV (corona) and 50 V (cone). Each FAME was identified using its [M-H]$^+$ adduct and quantified using a calibration curve and analyzed in its linear range.

3. Multiphase Chemical Kinetics Modeling

In this section, the batch reactor model used in our ASPEN PLUS simulations is reviewed first. Then, the RK-ASPEN EOS used to calculate vapor-liquid-liquid (VLL) equilibrium is reviewed.

3.1. Batch Reactor Model

In recent research, our objective has been to take experimental data in the laboratory (to extend the data taken by Soh et al. [9,11]) and to use simulation software (ASPEN PLUS-Version 10, AspenTech, Bedford, MA, USA)—Silva et al. [10], so that at intermediate pressures, CO_2 causes triglyceride to dissolve in the methanol phase, significantly increasing the transesterification reaction rate. The additional data are needed to improve estimates of the reaction rates, especially at far smaller methanol/triglyceride ratios. Given that the triglycerides of corn oil are principally linoleic and oleic acids, triolein ($C_{57}H_{104}O_2$) was selected to represent the corn oil in the ASPEN PLUS simulations and the principal product was taken to be methyl-oleate ($C_{19}H_{36}O_2$), representing the biodiesel.

Earlier Silva et al. [10], because pure-species parameters for diglyceride and monoglyceride were unavailable, an approximate single-reaction, kinetic mechanism, was used, as shown in Figure 1. This model included just six chemical species: triolein, methanol, methyl oleate, glycerol, CO_2, and water.

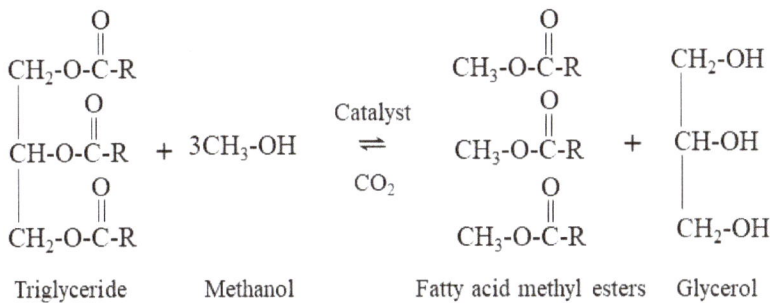

Figure 1. A single reversible transesterification reaction to convert triglycerides (TG) into biodiesel.

For this reaction, our earliest reactor model [10] was created to track the batch reactor data taken by Soh et al. [9,11], involving up to three phases (V, LI, and LII). For each phase, the mass balances were expressed:

$$\frac{d[c_j]}{dt} = v_j \times r \quad j = 1, \ldots, 4 \tag{1}$$

where j is the species counter in the reaction, and the intrinsic rate of reaction is expressed, in kmol/L·s:

$$r = k_f(c_{cat})^{n_f} c_{TG} c_{MEOH} - k_r(c_{cat})^{n_r} c_{FAME} c_{GLY} \tag{2}$$

where k_f is the forward rate constant, m^3/kmol·s, k_r is the reverse rate constant, m^3/kmol·s, c_{cat} is the Nafion catalyst concentration, n_f is the exponent of the catalyst concentration in the forward direction, n_r is the exponent of the catalyst concentration in the reverse direction, c_j is the concentration of species j, kmol/L, and v_j is the stoichiometric coefficient of species j in the reaction. In the absence of catalyst, the n coefficients are zero. Note: For species j in reaction i, the intrinsic rate of reaction is $r_j = v_j \times r$.

Prior to each time-step, when integrating mass balances (1) for each phase in a custom written FORTRAN® subroutine, in a USER2 block, the concentrations in the three phases and phase volumes are computed by minimizing G (Gibbs free energy) subject to mass-balance constraints. For this purpose, the *Gibbs flash method* in ASPEN PLUS was used as a flash convergence algorithm in the block options of the USER2 block with Redlich-Kwong Equation-of-state, RK-ASPEN, as the base method in ASPEN PLUS.

3.2. RK-ASPEN Equations-of-State

Cubic equations-of-state (EOS) were first developed roughly 130 years ago and have become the industry standard since the development of computer-aided process design in the 1970s. The cubic EOSs are named as such because they contain a cubed molar volume term (Equation (3)). Numerous variants exist, the most popular of which are the van der Waals equation, Soave-Redlich-Kwong [16], and Peng-Robinson [17]. In 1972, G. Soave replaced the $1/\sqrt{T}$ term of the original Redlich-Kwong equation with a function $\propto (T, \omega_i)$ involving the temperature and the acentric factor (the resulting equation is also known as the Soave-Redlich-Kwong equation of state; SRK EOS). Herein, Soave-Redlich-Kwong EOS will be used given as follows:

$$P = \frac{RT}{V-b} - \frac{a\,\alpha_i}{V(V+b)} \tag{3}$$

where a and b are defined as:

$$a = \sum_i \sum_j x_i x_j \left(a_i a_j\right)^{0.5} \left(1 - K_{a,ij}\right) \tag{4}$$

$$b = \sum_i \sum_j x_i x_j \left(\frac{b_i + b_j}{2}\right) \left(1 - K_{b,ij}\right) \tag{5}$$

where R is the gas constant (8.314 J/mol·K), T is temperature, V is the molar volume, P is the pressure, $K_{a,ij}$ and $K_{b,ij}$ are binary interaction parameters, and a_i and b_i are empirical parameters, calculated using Equations (6)–(9). The attractive parameter, a_i, depends on the reduced temperature ($T_{ri} = T/T_{ci}$), the critical temperature (T_{ci}) and critical pressure (P_{ci}), the accentric factor (ω_i), and an extra polar parameter (η_i). The size parameter, b_i, depends only on the critical temperature and critical pressure. γ_i is a parameter that accounts for accentricity of the molecule.

$$a_i = 0.42747 \frac{R^2 T_{Ci}^2}{P_{Ci}} \tag{6}$$

$$\alpha_i = \left[1 + \gamma_i \left(1 - T_{ri}^{0.5}\right) - \eta_i(1 - T_{ri})(0.7 - T_{ri})\right]^2 \tag{7}$$

$$\gamma_i = 0.48508 + 1.5517\omega_i - 0.15613\omega_i^2 \tag{8}$$

$$b_i = 0.08664 \frac{RT_{Ci}}{P_{Ci}} \tag{9}$$

where α_i is a dimensionless factor that becomes unity at $T = T_{ci}$. The α_i function was devised to fit the vapor pressure data of hydrocarbons and the equation does fairly well for these materials [16].

4. Results and Discussion

4.1. Experimental Data with Simulated Results

The yields of FAME obtained in: (a) experiments with corn oil were compared with those of (b) dynamic simulations of a batch reactor (100 mL). As shown in Tables 1 and 2 experiments using corn oil were carried out for four molar ratios of MeOH/TG (50, 100, 300, and 550). The first two columns show the mass (g) of corn oil and the volume (mL) of methanol fed to the batch reactor. Note that the corn oil was comprised of three principal fatty acids: palmitic acid (C16:0), oleic acid (C18:1), and linoleic acid (C18:2), with relative FAME yields of 10.4 wt%, 30.8 wt%, 58.8 wt%, respectively.

Table 1. Experimental and Simulated Methyl-Oleate Yield (g) using Methanol/TG Molar Ratios at 95 °C, 9.65 MPa.

Corn Oil (g)	MeOH (mL)	Molar Ratio MeOH/Corn Oil	Experiment 95 °C, 9.65 MPa FAME (g)	Simulation 95 °C, 9.65 MPa FAME (g)
1.551	3.52	50X	0.10	0.17
0.918	4.21	100X	0.13	0.19
0.352	4.83	300X	0.16	0.24
0.218	5	550X	0.13	0.17

Table 2. Experimental and Simulated Fractional FAME Yield using Methanol/TG Molar Ratios at 95 °C, 9.65 MPa.

Corn Oil (g)	MeOH (mL)	Molar Ratio MeOH/Triolein	Experiment 95 °C, 9.65 MPa Percent Yield [(FAME/Corn Oil) × 100]	Simulation 95 °C, 9.65 MPa Percent Yield [(FAME/Triolein) × 100]
1.551	3.52	50X	6.4	11.10
0.918	4.21	100X	14.11	20.42
0.352	4.83	300X	44.08	66.82
0.218	5	550X	61.5	79.78

When using ASPEN PLUS Version 10 with the RK-ASPEN EOS, data for CO_2, triolein, methanol, methyl-oleate and glycerol, are available in the ASPEN PLUS component library. However, corn oil, which is a complex mixture of mixed triglyceride containing palmitic, stearic, oleic, and linoleic, and other fatty acid chains, is not available in the data bank library. Therefore, we used final FAME yields as a measure to compare percent oil conversion between experiment and simulations.

In Table 1, columns 4 and 5 show the experimental FAME and simulation yields in grams. In Table 2, columns 4 and 5 show the experimental FAME and simulation yields as the FAME (grams) per gram of corn oil multiplied by 100; i.e., the percent FAME yield. Note that the experimental and simulation yields have similar trends, although some of the values are not in close agreement.

4.2. Simulated Temperature and Pressure Variations

In lieu of experimental measurements, using ASPEN PLUS, the effects of varying temperature (80 and 95 °C) and pressure (8 and 9.65 MPa) on the FAME yield in the batch reactor after 4-h were investigated at four MeOH/TG molar ratios. Table 3 shows a mix of increases and decreases with temperature and pressure. The yield of FAME increased with temperature, which is consistent with the previous work of Farobie and Matsumura [18] and Rathore and Madras [19], who reported that the conversion of oil to FAME in supercritical methanol increased with the increase in temperature from 200 to 400 °C. Methyl-oleate conversion increased with temperature and pressure in the transesterification reactor for the conversion of palm oil to biodiesel as reported by Bunyakiat et al. [20]. The increase also reduced the total batch reaction time.

As the temperature and pressure in the transesterification reactor increases, it is likely that the triglyceride and methanol mixture in the presence of supercritical CO_2 approaches the critical state. This close proximity of the otherwise partially miscible components enhances the overall solubility, thereby increasing the total yield of methyl oleate. Note that Tsai et al. [21] reported the reaction rate and FAME yield (conversion of oleic acid) increase with increasing temperature (220 to 260 °C) when other operating conditions are fixed. Kusdiana and Saka [12], also reported that increasing temperature reduces the transesterification reaction time from 3600 s to 120 s operating from 230 °C to 400 °C without using catalyst. Also, as methanol concentration increases at higher MeOH/TG molar ratios, the FAME yield increases as seen in the Table 3, consistent with the observations of

Pollardo et al. [22]. But, the increased methanol recovery and recirculation costs significantly influence the techno-economic optimization of processes to convert triolein to biodiesel.

Table 3. Simulated Fractional FAME Yield using Methanol/TG Molar Ratios.

Temp (°C)	Press (MPa)	Molar Ratio MeOH/Triolein 50× (FAME/triolein) × 100	Molar Ratio MeOH/Triolein 100× (FAME/triolein) × 100	Molar Ratio MeOH/Triolein 300× (FAME/triolein) × 100	Molar Ratio MeOH/Triolein 550× (FAME/triolein) × 100
80	8	9.09	19.34	49.45	57.38
80	9.65	14.71	16.24	42.68	52.88
95	8	6.01	11.88	32.05	35.46
95	9.65	11.11	20.42	66.83	79.79

The reaction rate constants were regressed using the bulk concentrations in the experimental 50 mL, agitated reactor vessel [9]. The relationship of the catalyst surface concentration in Equation (2) to the MeOH/triolein molar ratio deserves experimental study. As shown in Figure 2, more than 90% conversion is possible in less than 1/4 h of batch reactor time.

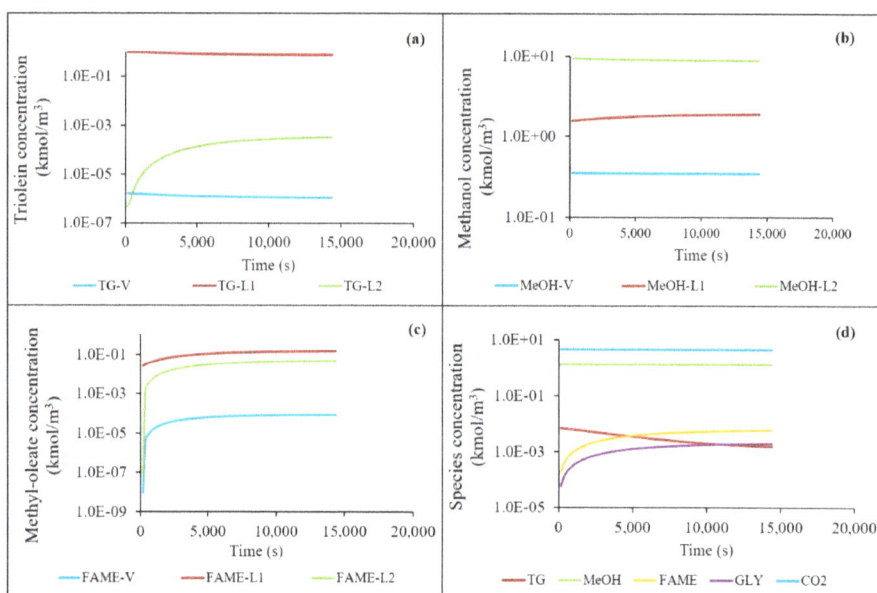

Figure 2. Concentration profiles of (**a**) Triolein, (**b**) Methanol, (**c**) FAME, and (**d**) All species, for 550X MeOH/triolein molar ratio at 9.65 MPa and 95 °C. V-Vapor, L1-Liquid 1, L2-Liquid 2 phase.

4.3. Simulated Batch Reactor Concentration Profiles

The concentration profiles for triolein, methanol, and methyl-oleate during the batch reactor simulations are shown in Figure 2. Note that for each MeOH/triolein molar ratio, the CO_2/MeOH molar ratio was fixed at 2.34. Consequently, the CO_2 concentration does not change appreciably in the three phases because CO_2 is inert and is not consumed in the reactions, as clearly seen in Figure 2d. Also, due to triolein's low vapor pressure, it is assumed that the reaction does not occur in the vapor phase. Liquid 1 is principally apolar triolein and liquid 2 is principally methanol.

Figure 2a shows the triolein to be predominant in the liquid 1 phase and somewhat increasing in concentration in liquid 2 due to the presence of supercritical CO_2 as co-solvent. Note that methanol distributes between the two liquid phases, with supercritical CO_2 increasing its solubility in the apolar phase as seen in Figure 2b. The forward reaction occurs principally in the apolar liquid phase. The reverse reaction occurs in liquid 2 with glycerol moving to liquid 2 (principally MeOH) as the batch reaction proceeds in time. Throughout the 4-h reaction time, three distinct phases (vapor and 2 liquids) exist as shown in Figure 2 for a 550 MeOH/Triolein molar ratio. FAME and glycerol are the two principal products from the transesterification reaction whose concentrations increase over the 4-h reaction time. The RK-ASPEN EOS without binary interaction parameters gave reasonable agreement with the experimental results. As the methanol critical pressure (7.95 MPa) and critical temperature (240 °C) differ significantly from the experimental reactor conditions, the mixtures are not entirely supercritical, and consequently, the reaction proceeds slower in three phases rather than a single supercritical phase.

5. Conclusions

New experimental data have been reported for the batch transesterification of corn oil to biodiesel (FAME) at lower molar ratios of methanol to oil. These data show unanticipated increases of FAME yields at higher molar ratios, suggesting the need to examine closely the catalyst surface as the MeOH/TG molar ratio decreases. The single reaction kinetic model, solved in ASPEN PLUS simulations, using multiphase equilibrium calculations with the RK-ASPEN equation-of-state and the FLSH_FLASH subroutine, confirms the trends displayed experimentally. The 1-reaction kinetic model performs well, but to obtain better agreement with the experimental results, catalyst surface behavior and additional data for the di- and mono-glyceride species, should be included which can further improve the model performance and should facilitate better the techno-economic optimization of processes to convert triglycerides to biodiesel.

Acknowledgments: The Fulbright-Kalam Climate Fellowship to Geetanjali Yadav, permitting her to visit the University of Pennsylvania for one year, is gratefully acknowledged. Also, the computational files, containing programs used in earlier algae-to-biodiesel research, and the excellent assistance/advice from Cory Silva, is appreciated. We enjoyed much assistance from the support staff at AspenTech and from the CETS IT specialist, Neal Gerrish, at the University of Pennsylvania. Junwei Xiang aided with the corn oil experiments and analysis. The SFC-MS was generously funded by the National Science Foundation under Grant No. 1626100.

Conflicts of Interest: The authors declare no conflict of interest.

References

1. Knothe, G. "Designer" Biodiesel: Optimizing Fatty Ester Composition to Improve Fuel Properties. *Energy Fuels* **2008**, *22*, 1358–1364. [CrossRef]
2. Tian, Y.; Xiang, J.; Verni, C.C.; Soh, L. Fatty Acid Methyl Ester Production via Ferric Sulfate Catalyzed Interesterification. *Biomass Bioenergy* **2018**, *115*, 82–87. [CrossRef]
3. Chang, A.-F.; Liu, Y.A. Integrated Process Modeling and Product Design of Biodiesel Manufacturing. *Ind. Eng. Chem. Res.* **2010**, *49*, 1197–1213. [CrossRef]
4. Portha, J.; Allain, F.; Coupard, V.; Dandeu, A.; Girot, E.; Schaer, E.; Falk, L. Simulation and Kinetic Study of Transesterification of Triolein to Biodiesel Using Modular Reactors. *Chem. Eng. J.* **2012**, *207–208*, 285–298. [CrossRef]
5. Kusdiana, D.; Saka, S. Kinetics of Transesterification in Rapeseed Oil to Biodiesel Fuel as Treated in Supercritical Methanol. *Fuel* **2001**, *80*, 693–698. [CrossRef]
6. Hegel, P.; Mabe, G.; Pereda, S.; Brignole, E.A. Phase Transitions in a Biodiesel Reactor Using Supercritical Methanol. *Ind. Eng. Chem. Res.* **2007**, *46*, 6360–6365. [CrossRef]
7. Maçaira, J.; Santana, A.; Costa, A.; Ramirez, E.; Larrayoz, M.A. Process Intensification Using CO_2 as Cosolvent under Supercritical Conditions Applied to the Design of Biodiesel Production. *Ind. Eng. Chem. Res.* **2014**, *53*, 3985–3995. [CrossRef]
8. Maçaira, J.; Santana, A.; Recasens, F.; Larrayoz, M.A. Biodiesel Production Using Supercritical Methanol/Carbon Dioxide Mixtures in a Continuous Reactor. *Fuel* **2011**, *90*, 2280–2288. [CrossRef]

9.   Soh, L.; Curry, J.; Beckman, E.J.; Zimmerman, J.B. Effect of System Conditions for Biodiesel Production via Transesterification Using Carbon Dioxide–Methanol Mixtures in the Presence of a Heterogeneous Catalyst. *ACS Sustain. Chem. Eng.* **2014**, *2*, 387–395. [CrossRef]

10. Beckman, E.J. Oxidation Reactions in CO_2: Academic Exercise or Future Green Processes? *Environ. Sci. Technol.* **2003**, *37*, 5289–5296. [CrossRef] [PubMed]

11. Soh, L.; Chen, C.-C.; Kwan, T.A.; Zimmerman, J.B. Role of CO_2 in Mass Transfer, Reaction Kinetics, and Interphase Partitioning for the Transesterification of Triolein in an Expanded Methanol System with Heterogeneous Acid Catalyst. *ACS Sustain. Chem. Eng.* **2015**, *3*, 2669–2677. [CrossRef]

12. Silva, C.; Soh, L.; Barberio, A.; Zimmerman, J.; Seider, W.D. Phase Equilibria of Triolein to Biodiesel Reactor Systems. *Fluid Phase Equilib.* **2016**, *409*, 171–192. [CrossRef]

13. Lepage, G.; Roy, C.C. Improved Recovery of Fatty Acid through Direct Transesterification without Prior Extraction or Purification. *J. Lipid Res.* **1984**, *25*, 1391–1396. [PubMed]

14. Soh, L.; Montazeri, M.; Haznedaroglu, B.Z.; Kelly, C.; Peccia, J.; Eckelman, M.J.; Zimmerman, J.B. Evaluating Microalgal Integrated Biorefinery Schemes: Empirical Controlled Growth Studies and Life Cycle Assessment. *Bioresour. Technol.* **2014**, *151*, 19–27. [CrossRef]

15. Soh, L.; Lane, M.K.M.; Xiang, J.; Kwan, T.A.; Zimmerman, J.B. Carbon Dioxide Mediated Transesterification of Mixed Triacylglyceride Substrates. *Energy Fuels* **2018**, *32*, 9624–9632. [CrossRef]

16. Soave, G. Equilibrium Constants from a Modified Redlich-Kwong Equation of State. *Chem. Eng. Sci.* **1972**, *27*, 1197–1203. [CrossRef]

17. Peng, D.-Y.; Robinson, D.B. A New Two-Constant Equation of State. *Ind. Eng. Chem. Fundam.* **1976**, *15*, 59–64. [CrossRef]

18. Farobie, O.; Matsumura, Y. Biodiesel Production in Supercritical Methanol Using a Novel Spiral Reactor. *Procedia Environ. Sci.* **2015**, *28*, 204–213. [CrossRef]

19. Rathore, V.; Madras, G. Synthesis of Biodiesel from Edible and Non-Edible Oils in Supercritical Alcohols and Enzymatic Synthesis in Supercritical Carbon Dioxide. *Fuel* **2007**, *86*, 2650–2659. [CrossRef]

20. Bunyakiat, K.; Makmee, S.; Sawangkeaw, R.; Ngamprasertsith, S. Continuous Production of Biodiesel via Transesterification from Vegetable Oils in Supercritical Methanol. *Energy Fuels* **2006**, *20*, 812–817. [CrossRef]

21. Tsai, Y.-T.; Lin, H.; Lee, M.-J. Biodiesel Production with Continuous Supercritical Process: Non-Catalytic Transesterification and Esterification with or without Carbon Dioxide. *Bioresour. Technol.* **2013**, *145*, 362–369. [CrossRef] [PubMed]

22. Pollardo, A.A.; Lee, H.; Lee, D.; Kim, S.; Kim, J. Effect of Supercritical Carbon Dioxide on the Enzymatic Production of Biodiesel from Waste Animal Fat Using Immobilized Candida Antarctica Lipase B Variant. *BMC Biotechnol.* **2017**, *17*, 70. [CrossRef] [PubMed]

processes

MDPI

Article

Simulation of Dual Mixed Refrigerant Natural Gas Liquefaction Processes Using a Nonsmooth Framework

Matias Vikse [1], Harry A. J. Watson [2], Truls Gundersen [1] and Paul I. Barton [2,*]

[1]   Department of Energy and Process Engineering, Norwegian University of Science and Technology (NTNU),
      7491 Trondheim, Norway; matias.vikse@ntnu.no (M.V.); truls.gundersen@ntnu.no (T.G.)
[2]   Process Systems Engineering Laboratory, Massachusetts Institute of Technology,
      Cambridge, MA 02141, USA; hwatson@alum.mit.edu
*   Correspondence: pib@mit.edu

Received: 26 September 2018; Accepted: 15 October 2018; Published: 17 October 2018

Abstract: Natural gas liquefaction is an energy intensive process where the feed is cooled from ambient temperature down to cryogenic temperatures. Different liquefaction cycles exist depending on the application, with dual mixed refrigerant processes normally considered for the large-scale production of Liquefied Natural Gas (LNG). Large temperature spans and small temperature differences in the heat exchangers make the liquefaction processes difficult to analyze. Exergetic losses from irreversible heat transfer increase exponentially with a decreasing temperature at subambient conditions. Consequently, an accurate and robust simulation tool is paramount to allow designers to make correct design decisions. However, conventional process simulators, such as Aspen Plus, suffer from significant drawbacks when modeling multistream heat exchangers. In particular, no rigorous checks exist to prevent temperature crossovers. Limited degrees of freedom and the inability to solve for stream variables other than outlet temperatures also makes such tools inflexible to use, often requiring the user to resort to a manual iterative procedure to obtain a feasible solution. In this article, a nonsmooth, multistream heat exchanger model is used to develop a simulation tool for two different dual mixed refrigerant processes. Case studies are presented for which Aspen Plus fails to obtain thermodynamically feasible solutions.

Keywords: nonsmooth modeling; process simulation; DMR liquefaction processes

---

1. Introduction

Natural gas plays a major role in the global shift towards new environmentally friendly energy sources. Low CO_2 emissions and no particle emissions upon combustion means that natural gas provides a cleaner alternative to oil and coal. However, a significant challenge with using natural gas is related to its transportation, especially over long distances. Alternative technologies for the transportation of natural gas exist, where the conventional approach is to use pipelines. However, pipeline transportation requires a large initial investment in infrastructure. Moreover, it ties the seller of the gas to a small set of customers at the receiving terminals. Excessive infrastructure and transportation costs for long distances also make pipeline gas difficult to export to the global energy market. Because of this, the recent trend has been towards natural gas liquefaction. The liquefaction of natural gas is a very energy intensive process that requires cooling to about $-162\,^\circ\text{C}$. Investments in expensive, customized, and proprietary technology are necessary. Together with high operating costs, liquefaction accounts for about 30–40% of the total cost of the Liquefied Natural Gas (LNG) chain [1]. LNG production plants are frequently divided into three main categories: base-load, peak-shaving, and small-scale plants. Single mixed refrigerant (SMR) liquefaction processes are normally considered for small-scale and peak-shaving LNG production where capital costs rather than operational costs are the main concern. In the case of base-load plants, high production volumes with accompanying high operating costs advocate more efficient designs. A popular alternative is the Dual Mixed Refrigerant (DMR) processes due to their high efficiency and flexible design. The added flexibility comes at a cost computationally, however, as DMR are significantly more complex to model, simulate, and optimize than SMR processes.

A large temperature span and small driving forces in the heat exchangers at cryogenic temperatures make LNG processes particularly difficult to analyze. The small temperature driving forces are a direct result of the exergetic losses from irreversible heat transfer increasing exponentially with a decreasing temperature at subambient conditions. As a result, even small inaccuracies in the process model at cryogenic temperatures may propagate to a significant amount of additionally required actual compression power. Furthermore, the high operating and capital costs in LNG processes favor an accurate and robust simulation tool for engineers to make the best design decisions. Nevertheless, state-of-the-art process simulation tools, such as Aspen Plus [2] and Aspen HYSYS [3], suffer from limitations in the modeling of multistream heat exchangers (MHEXs), in particular, the lack of any rigorous checks to prevent temperature crossovers and thus, the possibility that the process simulator could converge to an infeasible design [4,5].

Different modeling approaches for MHEXs that perform these checks have been proposed in the literature. A MHEX model for optimization of LNG processes was developed by Kamath et al. [4]. The model uses concepts from Pinch Analysis by treating the MHEX as a heat integration problem with no external utilities. Phase changes in the MHEX are handled using complementarity constraints and do not need to be determined a priori. Phase transitions are a known issue in modeling MHEXs as phase boundaries vary dynamically during simulation and optimization. Consequently, the location of the phase boundary as well as the phase path of a process stream cannot readily be determined a priori. Models that are capable of handling unknown phase states are therefore necessary to fully analyze LNG processes. The result is a fully equation-oriented (EO) model that was solved to local optimality for the single mixed refrigerant PRICO process. Another optimization model for MHEXs was developed by Hasan et al. [6,7] using a superstructure approach, where the MHEX is represented by a network of two-stream heat exchangers. Optimal operation is then determined by solving a heat integration (HI) problem with no external utilities. However, the model can handle phase changes in the MHEX only as long as the inlet and outlet states are known a priori. The result is a mixed-integer nonlinear program (MINLP) that is computationally expensive and requires global optimization tools to solve. Later, Rao and Karimi [8] proposed an alternative superstructure that can handle unknown phase states without including binary variables. In the superstructure, the MHEX is represented as a series of stream bundles, where each bundle is modeled as a set of two-stream heat exchangers between the hot and cold streams. Nonlinear constraints are included for the phase changes, which are modeled to occur at the endpoints of the two-stream heat exchangers. In this way, each heat exchanger operates in a single phase-regime and can be solved using a corresponding property model. In the article, property evaluations were done using a conventional process simulator (here Aspen Plus). However, tailored property models can also be added with accompanying binary variables for the phase transitions. The final model is a nonconvex NLP (or MINLP with custom property models) that is computationally expensive to solve, mostly because of the repeated property evaluations done by the process simulator, and again, requires global optimization methods. Pattison and Baldea [9] developed a MHEX model using their own pseudo-transient EO approach for modeling and simulation. The model requires no prior knowledge of the stream paths and phase boundaries. However, a relative sequence of stream temperatures must be known and fixed prior to optimization to define the enthalpy intervals for the composite curves. Temperatures are then calculated at the endpoint of each interval by introducing a nonphysical time-dependent temperature variable and then solving a system of differential-algebraic equations (DAEs). No Boolean variables or disjunctions are used for handling phase transitions. Instead, the dummy transient variable is perturbed across the phase boundary while keeping the temperature constant, and the property models are resolved using the solution from the previous time step as a starting point.

Recent literature on MHEX modeling for natural gas liquefaction has primarily focused on flowsheet optimization. The models presented in the paragraph above, require solving nonconvex optimization problems, sometimes to guaranteed global optimality, where feasible heat transfer is guaranteed at the solution through minimum approach temperature inequality constraints. Furthermore, the models only consider optimization of the simple PRICO process, which consists of a single MHEX for natural gas liquefaction. Single mixed refrigerant processes normally feature relatively simple designs with minimal auxiliary equipment, and are favored for small-scale applications or offshore production where size and capital cost are of primary concern. The large-scale production of LNG, however, demands the use of highly complex and optimized processes with additional MHEXs and refrigerants to reduce the driving forces between the hot and cold composite curves. Dual mixed refrigerant processes are examples of such designs. However, the added complexity makes DMR processes difficult to simulate and optimize with the custom MHEX models presented in the literature. Instead, these processes are normally studied using conventional process simulation tools, such as Aspen HYSYS or Aspen Plus, for simulation and a stochastic search algorithm for optimization [10].

Although the custom MHEX models focus on flowsheet optimization, process simulation remains an important step in the design and analysis of liquefaction processes. Whereas optimization ostensibly provides the designer a best known design or operation, simulation models can yield useful information on existing designs or operating points that are not necessarily optimal. Furthermore, process simulation allows for probing the behaviour of a system in the neighborhood of the current operating point, to investigate possible improvements in design that can be achieved with little effort. In addition, it can be used as a first step towards flowsheet optimization by using the simulation results as starting points for the optimizer, usually improving reliability of the convergence [11]. However, flowsheet convergence to a thermodynamically feasible operation is more challenging in simulation than optimization due to the absence of minimum temperature difference inequality constraints in the problem formulation. This is a particular issue when using conventional process simulators such as Aspen HYSYS or Aspen Plus, which frequently converge to a solution with temperature crossovers [4,5]. Furthermore, the MHEX block included in Aspen solves the overall energy balance for a single outlet temperature from the heat exchanger. Other process variables (i.e., refrigerant pressure levels and compositions) must be held fixed during simulation, making Aspen inflexible to use for LNG simulation, and often requiring the user to resort to a manual iterative procedure for locating feasible solutions.

A MHEX model that has been developed for both the simulation and optimization of LNG processes was presented by Watson et al. [5,12]. The model uses new developments in nonsmooth analysis [13] and pinch analysis concepts to solve the MHEX as a heat integration problem with no external utilities. A reformulation of the Duran and Grossmann [14] pinch location algorithm was developed to calculate the minimum temperature difference in the MHEXs and to prevent temperature crossovers during simulation. The model size is independent of the number of streams exchanging heat, which is particularly advantageous when many substreams are needed to model accurately nonlinear cooling curves [15]. Furthermore, the model includes area calculations for economic analysis. No Boolean variables or disjunctive representations are used for handling phase transitions. Instead, the model uses the nonsmooth operators max, min and mid to correctly determine the phase state, enthalpy, and temperature of the process streams. Sensitivities are computed using lexicographic directional (LD-)derivatives, which are extensions of the classical directional derivative to certain classes of nonsmooth functions that provide useful sensitivity information at nonsmooth points. This nonsmooth MHEX model has previously been used for the simulation and optimization of different single mixed refrigerant processes [11,16]. A hybrid modeling approach was used to develop the flowsheet models, where auxiliary equipment and the two-phase stream variables are solved sequentially in nested subroutines. The result is a reduced model size and increased robustness, making it possible to simulate and optimize larger and more complex SMR models. It is, therefore, interesting to investigate whether the novel modeling approach is also capable of simulating DMR processes. In this article, the nonsmooth modeling approach is used to simulate two dual mixed refrigerant processes. The first process is a relatively simple DMR process with cascading PRICO processes, whereas the second is a version of the commercial AP-DMR process [17] with single stage compression. Case studies are performed, each starting from initial points for which the process simulator Aspen Plus fails to obtain thermodynamically feasible solutions.

2. The Nonsmooth Multistream Heat Exchanger Model

The standard two-stream countercurrent heat exchanger is completely described by Equations (1)–(3), which represent the energy balance, minimum approach temperature in the heat exchanger, and total heat exchanger conductance, respectively.

$$mC_{p,H}\left(T_H^{IN} - T_H^{OUT}\right) = mC_{p,C}\left(T_C^{OUT} - T_C^{IN}\right), \tag{1}$$

$$\Delta T_{min} = \min\left\{T_H^{IN} - T_C^{OUT}, T_H^{OUT} - T_C^{IN}\right\}, \tag{2}$$

$$UA = \frac{Q}{\Delta T_{LM}}, \tag{3}$$

where $mC_{p,H/C}$ is the heat capacity flowrate for the hot (H) and cold (C) streams, respectively, UA is the heat transfer conductance, $Q \equiv mC_{p,H}\left(T_H^{IN} - T_H^{OUT}\right)$ is the total heat exchange duty, and ΔT_{LM} is the log-mean temperature difference.

The energy balance can readily be extended to the case of n_H hot and n_C cold streams as follows:

$$\sum_{i=1}^{n_H} mC_{PH,i}\left(T_{H,i}^{IN} - T_{H,i}^{OUT}\right) = \sum_{j=1}^{n_C} mC_{PC,j}\left(T_{C,j}^{OUT} - T_{C,j}^{IN}\right). \tag{4}$$

Also, the equation for the total heat exchanger conductance can be applied to the case of multistream heat exchangers by assuming vertical heat exchange between the hot and cold composite curves

$$UA = \sum_{k=1}^{K-1} \frac{\Delta Q^k}{\Delta T_{LM}^k}, \tag{5}$$

where K is the total number of enthalpy intervals and ΔQ^k is the enthalpy change of interval k.

As for the minimum approach temperature constraint, however, it cannot readily be extended to that of multistream heat exchangers. The minimum temperature difference in two-stream heat exchangers occurs at the physical endpoints of the heat exchangers, assuming constant mCp values for the streams. For multistream heat exchangers, on the other hand, the pinch point can occur at any of the stream inlet temperatures and are not necessarily situated at the physical endpoints. Instead, the approach temperature can be calculated using concepts from pinch analysis and heat integration through a pinch location algorithm. Several pinch location algorithms exist in the literature, most of which require a disjunctive optimization problem to be solved to global optimality. However, in order to avoid solving a separate optimization problem for the minimum temperature calculations, Watson et al. [5] developed a reformulation of the Duran and Grossmann model for simultaneous optimization and heat integration [14]. The reformulation solves the pinch problem through the nonsmooth equation

$$\min_{p \in P}\{EBP_C^p - EBP_H^p\} = 0, \tag{6}$$

where P is the (finite) set of candidate pinch points and $EBP_{H/C}^p$ is the enthalpy of the extended hot/cold composite curves for pinch candidate p, as defined in Watson et al. [5]. This equation can be solved using nonsmooth numerical equation solvers, and therefore, no optimization problem needs to be solved.

Multistream heat exchangers are particularly useful in natural gas liquefaction processes where the streams normally undergo phase changes. Phase changes in the heat exchanger present a commonly known modeling issue, as phase boundaries and the phases traversed in the heat exchanger are not known a priori and may change during the simulation. Instead of using Boolean variables or disjunctions to detect the correct phase behavior, the model uses the nonsmooth operators max, min and mid where the mid operator is a function mapping to its median argument [12]. This is done by partitioning each process stream into superheated (sup), two-phase (2p) and subcooled (sub) substreams whose inlet and outlet temperatures are determined by the following equations:

$$T_{\text{sup}}^{\text{in/out}} = \max\left(T_{\text{DP}}, T^{\text{IN/OUT}}\right), \tag{7}$$

$$T_{2p}^{\text{in/out}} = \text{mid}\left(T_{\text{DP}}, T_{\text{BP}}, T^{\text{IN/OUT}}\right), \tag{8}$$

$$T_{\text{sub}}^{\text{in/out}} = \min\left(T_{\text{BP}}, T^{\text{IN/OUT}}\right), \tag{9}$$

where $T^{\text{IN/OUT}}$ is the inlet or outlet temperature of the process stream, $T_{\text{sub/2p/sup}}^{\text{in/out}}$ is the corresponding inlet or outlet temperature of the substreams, and T_{DP} and T_{BP} are the dew and bubble point temperatures of the process stream. Additional stream segments may be used to improve the accuracy of the calculations, which is particularly important for the two-phase region where enthalpy and temperature vary highly nonlinearly due to phase changes. Watson et al. showed, by using the PRICO process [18] as an illustrative example, that 20 segments provided a sufficient level of accuracy for representing the two-phase region [15].

Stream temperatures in the two-phase region are calculated using successive pressure-enthalpy (PQ)-flash operations for the stream segments. As stream propertiesm and thus phase boundaries change during the simulation, the PQ-flash algorithm must be capable of handling instances of single phase flow. This is also an issue for models of auxiliary equipment in LNG processes, such as compressors and valves, which may experience instances of single phase flow during the iterations of the nonsmooth solver. For this application, a nonsmooth extension of the well-known Boston and Britt [19] flash algorithm was developed that handles instances of single phase flow without relying on post-processing methods. The flash algorithm was summarized in a three-paper series by Watson et al. [15,20] and Watson and Barton [21], and employs a mid-function for detecting the correct phase state at the outlet. The algorithm was shown to handle instances near the phase-boundaries robustly, where the conventional inside-out algorithm is prone to failure [20]. A methodology for calculating correct sensitivity information from the flash equations was presented, which allows the nonsmooth inside-out algorithm to be integrated into the flowsheet models. Rather than using a fully equation-oriented framework where the MHEX model and flash calculations are solved simultaneously, the flash calculations are embedded in the model as subroutine calls and solved with the specialized algorithm. The result is a reduced model size and increased robustness, particularly for larger and more complex models.

Solving the MHEX Model

The MHEX model requires a nonsmooth algebraic equation system to be solved. The inclusion of the nonsmooth operators max, min and mid, means that points of nondifferentiability exist where the Jacobian is undefined. A common approach for handling nonsmoothness in the literature is to formulate a smooth approximation for the nonsmooth function around the kinks. However, formulating these approximating functions is nontrivial, and sometimes, poor selection may lead to either an ill-conditioned approximation or the loss of accuracy [22].

Alternatively, extensions of the conventional derivative to certain classes of nonsmooth functions exist. There are several of these generalized derivatives described in the literature, where the Clarke Generalized Jacobian [23] for locally Lipschitz continuous functions is particularly well-known. However, the Clarke Jacobian is challenging to compute, especially for composite functions, as it follows calculus rules (e.g., the chain rule) as inclusions rather than as equations. Therefore, new interest in nonsmooth analysis has centered around a different generalized derivative known as the lexicographic (L-)derivative [24]. The L-derivative is a generalized derivative for functions satisfying the conditions for lexicographic (L)-smoothness as formulated by Nesterov [24]. It was shown later by Khan and Barton [25] that L-derivatives are as useful as elements of the Clarke Jacobian itself in many nonsmooth numerical methods. The renewed attention on the L-derivative was fueled by the development of an automatic procedure for calculating its elements for composite functions. Rather than computing elements of the L-derivative directly, Khan and Barton [13] introduced a nonsmooth extension to the directional derivative, known as the lexicographic directional (LD)-derivative. Unlike the Clarke Jacobian and L-derivatives, the LD-derivative follows calculus rules sharply, and can be computed for composite functions using a nonsmooth analog of the vector forward mode of automatic differentiation [13]. Furthermore, the L-derivative can be obtained directly from the LD-derivative. An extensive review of computing LD-derivatives and their applications is provided by Barton et al. [26].

Nonsmooth Newton-type solvers, where generalized derivative information is used in place of the function's Jacobian, exist for solving nonsmooth equation systems:

$$\mathbf{G}(\mathbf{x}^k)(\mathbf{x}^{k+1} - \mathbf{x}^k) = -\mathbf{f}(\mathbf{x}^k). \tag{10}$$

Here, $\mathbf{G}(\mathbf{x}^k)$ is an element of a generalized derivative of $\mathbf{f}$ at $\mathbf{x}^k$. Equation (10) solves for the next iterate $\mathbf{x}^{k+1}$ provided $\mathbf{G}(\mathbf{x}^k)$ is nonsingular at $\mathbf{x}^k$. However, singular generalized derivative elements may occur in the MHEX model from residuals of the form $\min\{0, y\}$ used to solve Equation (6). A Newton-type solver that is also applicable to singular generalized derivative elements is the linear programming (LP) Newton method by Facchinei et al. [27]:

$$\min_{\gamma,\mathbf{x}} \gamma$$

$$\text{s.t. } \left\| \mathbf{f}(\mathbf{x}^k) + \mathbf{G}(\mathbf{x}^k)(\mathbf{x} - \mathbf{x}^k) \right\|_\infty \leq \gamma \min \left(\left\| \mathbf{f}(\mathbf{x}^k) \right\|_\infty, \left\| \mathbf{f}(\mathbf{x}^k) \right\|_\infty^2 \right),$$

$$\left\| (\mathbf{x} - \mathbf{x}^k) \right\|_\infty \leq \gamma \left\| \mathbf{f}(\mathbf{x}^k) \right\|_\infty,$$

$$\mathbf{x} \in X, \tag{11}$$

where X is a polyhedral set (e.g., bounds) on the problem and γ is a supplementary variable that drives convergence towards the solution. The LP-Newton solves a linear program upon each iteration, where the next iterate $\mathbf{x}^{k+1}$ is given by the $\mathbf{x}$ part of the solution. Although the program in Equation (11) is also applicable for ill-conditioned generalized derivatives, previous experience with simulating single mixed refrigerant processes has shown that including a backtracking line search [28] significantly improves the step quality of the LP-Newton method at singular points [16]. Nevertheless, solving a linear program on every iteration is comparatively more expensive than computing a step with the method in Equation (10). Therefore, the model includes a hybrid solution where the LP-Newton is invoked only when the generalized derivative is either singular or ill-conditioned.

3. Simulation Cases and Results

Previously, the developed nonsmooth simulation tool was used to analyse different SMR processes under conditions for which the commercial simulator Aspen Plus failed to obtain results [16]. In particular, the additional two unknowns computed by Equations (5) and (6) add versatility, making it possible to obtain feasible operating points in cases where more than one operating parameter is unknown to the designer. This article develops simulation models for two different DMR processes. The first design constitutes a simple design where the PRICO process is used for both the warm mixed refrigerant (WMR) and the cold mixed refrigerant (CMR) cycles. Furthermore, a natural gas liquid (NGL) separator is added for the extraction of heavier hydrocarbons. The second DMR process is a version of the AP-DMR process. The main focus here is on the liquefaction part of the process, and thus, not the compression scheme. Therefore, only a single-stage compression with aftercooling is included in the model. Different case studies are presented, each solving for different sets of unknown variables. The variables considered in the analysis are the high and low pressure levels, refrigerant flowrates, inlet and outlet temperatures from the MHEXs, and refrigerant compositions for the warm and cold mixed refrigerant, as well as the MHEX specifications, i.e., the minimum temperature difference and the heat exchanger conductance. The following nomenclature is used for the parameters and unknown variables in the models:

- Pressure level of the (warm/cold) high pressure refrigerant: $P_{\text{HP(W/C)}}$.
- Pressure level of the (warm/cold) low pressure refrigerant: $P_{\text{LP(W/C)}}$.
- Inlet/outlet temperatures of the high pressure refrigerants (equal to the natural gas stream): $T_{\text{HP}}^{\text{IN/OUT}}$.
- Inlet/outlet temperatures of the low pressure refrigerant streams: $T_{\text{LP}}^{\text{IN/OUT}}$.
- Molar flowrate of the (warm/cold) refrigerants: $F_{\text{(W/C)}}$.
- Molar flowrate of component i in (warm/cold) refrigerants: $f_{\text{(W/C)},i}$.

The models were written using Julia v0.6.0 (Julia Lab – Massachusetts Institute of Technology, Cambridge, MA, USA) and run on a Dell Latitude E5470 laptop in the Ubuntu v16.10 environment with an Intel Core i7-6820HQ CPU at 2.7 GHz and 8.2 GB RAM. Simulations were done for the Peng–Robinson Equation of State where the property parameters were taken from Aspen Plus [2]. To ensure an accurate representation of the process streams, the single-phase substreams were partitioned into five stream segments for each MHEX. Similarly, two-phase substreams were represented by 20 stream segments, which was shown to be sufficiently accurate for capturing the nonlinearities in this region by Watson et al. [15] The overall flowsheet convergence tolerance was set to $\|\mathbf{y}\|_\infty < 10^{-5}$, where $\mathbf{y}$ represents the equation residuals, whereas the tolerance for the individual flash calculations was selected to be $\|\mathbf{y}\|_\infty < 10^{-8}$.

4. Example 1

The first DMR process studied in this paper is a simple configuration with cascading PRICO cycles for the warm and cold mixed refrigerants. This cycle consists of two MHEXs as well as an NGL separator for the extraction of heavier hydrocarbons. Heavier hydrocarbons freeze out at cryogenic temperatures and can cause plugging in process equipment. Furthermore, they represent a valuable commodity if sold separately. The feed gas is sent to the process at 295.15 K where it is precooled by a warm mixed refrigerant cycle consisting of ethane, propane, and n-butane. The feed gas and the refrigerants exit the heat exchanger at a temperature of $T_{\text{HP,1}}^{\text{OUT}}$. The feed gas is then sent to the NGL separator, where heavier hydrocarbons are extracted for further fractionation and/or export, before the gas enters MHEX 2 for liquefaction. The cold mixed refrigerant consists of a lighter refrigerant mixture consisting of nitrogen, methane, ethane, and propane for the liquefaction of the natural gas. Along with the feed gas, the CMR is precooled in the WMR PRICO cycle before it enters the cold heat exchanger. A process flowsheet of the DMR process is presented in Figure 1.

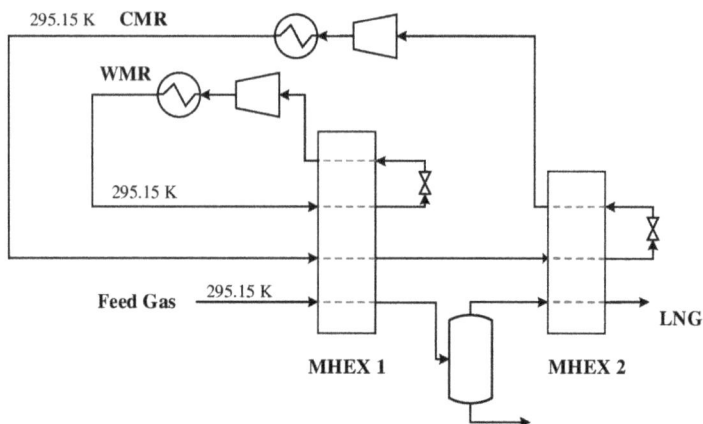

Figure 1. The Dual Mixed Refrigerant (DMR) model with cascading PRICO cycles for the warm and cold mixed refrigerant streams.

The parameter values and initial guess values for the unknown variables are provided in Table 1. The parameter values were selected such that Aspen Plus failed to converge to a feasible solution using its standard MHEX model with one equation. Essentially, the Aspen Plus model only solves the overall energy balance in Equation (4) for a single unknown temperature (chosen here as the inlet inlet temperatures to the compressors). Refrigerant compositions and pressures cannot be handled as unknown variables in these models. This results in less versatility in cases where pressure, compositions, and multiple temperatures must be adjusted to find a feasible design.

Table 1. Multistream heat exchanger (MHEX) and refrigerant stream data for Example 1. For unknown variables, the value listed is an initial guess.

Property	Value	Property	Value
η	0.8	UA_2 (MW/K)	3.0
$\Delta T_{min,1}$ (K)	3.0	$\Delta T_{min,2}$ (K)	3.0
F_W (kmol/s)	1.65	F_C (kmol/s)	1.55
P_{HPW} (MPa)	1.67	P_{HPC} (MPa)	4.30
P_{LPW} (MPa)	0.42	P_{LPC} (MPa)	0.25
$T_{HP,1}^{OUT}$ (K)	245.15	$T_{HP,2}^{OUT}$ (K)	120.15
$T_{LP,1}^{OUT}$ (K)	290.15	$T_{LP,2}^{OUT}$ (K)	240.15
Composition (mol %):			
Ethane	47.83	Nitrogen	10.00
Propane	34.17	Methane	43.80
n-Butane	18.00	Ethane	35.20
		Propane	11.00

The model contains 61 unknowns, five of which are provided by the solution of the MHEX equations. The remaining 56 variables are the temperatures of the intermediate stream segments for the single phase regions. As flash calculations are decoupled and solved separately, the model size is independent of the number of stream segments in the two-phase region. A two-equation model is used for MHEX 1 as the UA-value depends on the stream results from MHEX 2. Fixing the area prior to simulation can therefore be challenging. Therefore, the heat exchanger conductance value is instead calculated through post-processing. The simulations were carried out for a rich feed gas composition with 1.00 mol % nitrogen, 85.60 mol % methane, 4.93 mol % ethane, 3.71 mol % propane, 2.90 mol

% n-butane, 1.30 mol % i-butane, and 0.56 mol % n-pentane at a pressure of 4 MPa and flowrate of 1.0 kmol/s where both the refrigerant mixtures and the feed gas enter the precooling MHEX at a temperature of 295.15 K, as indicated in Figure 1. Two simulation cases were constructed, solving for different sets of unknown variables:

- **Case I:** P_{LPW}, P_{HPC}, $f_{W,propane}$, $T_{LP,2}^{OUT}$, UA_2.
- **Case II:** P_{HPW}, P_{LPC}, F_C, $\Delta T_{min,1}$, $\Delta T_{min,2}$.

Case I. Solved for a variable WMR composition, an unknown inlet temperature to the CMR compressor, the heat exchanger conductance in MHEX 2, as well as the low pressure level P_{LPW} and high pressure level P_{HPC} of the warm and cold mixed refrigerants, respectively. The refrigerant composition was changed in the model by varying the component molar flowrate of propane $f_{W,propane}$. A solution was obtained after four iterations and a total simulation time of 62.7 s, including initialization. The model converged to a solution with $P_{LPW} = 0.57$ MPa, $P_{HPC} = 6.53$ MPa and $T_{LP,2}^{OUT} = 242.15$ K. The design resulted in a total compression work of 21.33 MW, with heat exchanger conductance values of $UA_1 = 2.69$ MW/K and $UA_2 = 1.89$ MW/K. The work distribution of the two compressors was 16.79 MW for compressing the CMR and 4.54 MW for compressing the WMR. A new WMR composition was obtained consisting of 53.30 mol % ethane, 26.64 mol % propane and 20.06 mol % n-butane with a corresponding molar flowrate of 1.48 kmol/s. Figure 2 presents the composite curves and driving force plot for the solution.

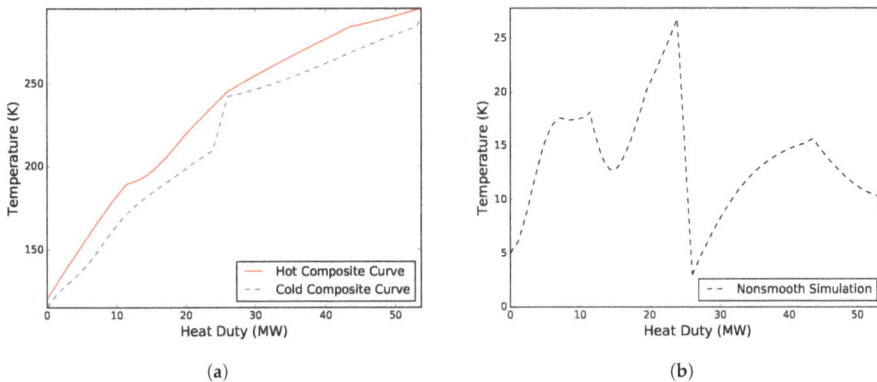

(a) (b)

Figure 2. (a) Composite curves for the feasible design in Case I. (b) The corresponding driving force plot.

Case II. Solved for the flowrate of the CMR, the minimum approach temperatures in both MHEXs, and the high pressure and low pressure levels of the warm and cold refrigerant mixtures, respectively. A solution was obtained after 13 iterations with $P_{HPW} = 1.58$ MPa, $P_{LPC} = 0.388$ MPa, $F_C = 1.952$ kmol/s, and minimum approach temperatures of 5.00 K and 3.15 K for MHEX 1 and 2, respectively. The design resulted in a total compressor work of 20.56 MW, and a heat exchanger conductance value UA_1 of 2.04 MW/K. Compressing the CMR required a total of 14.10 MW, whereas compressing the WMR required only 6.46 MW. The total simulation time, including initialization of the model, was 102.5 s. The composite curves and driving force plots for the process are presented in Figure 3. As can be seen, both solutions have a similar trend, although the cold low pressure refrigerant superheating is noticeably larger in Case I. This results in lower driving forces in MHEX 1, but also introduces a larger temperature difference at the cold end of the process, leading to a higher compression power. A discussion on superheating and its effect on design of DMR processes was made by Kim and Gundersen [29].

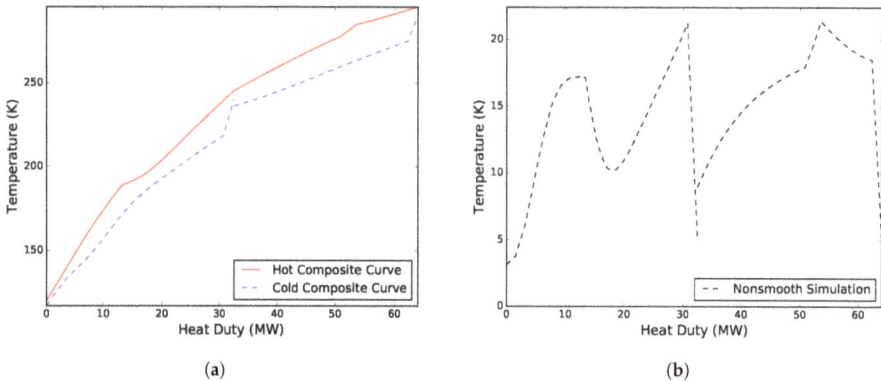

Figure 3. (**a**) Composite curves for the feasible design in Case II. (**b**) The corresponding driving force plot.

5. Example 2

A more complex DMR design is presented in Figure 4. Instead of having two cascading PRICO cycles, this design features a spiral-wound heat exchanger (SWHX) for the CMR where the refrigerant is separated after precooling to provide cooling at different temperature levels. The vapor product, which consists mainly of the light components nitrogen and methane, is liquefied and subcooled to provide cooling in the cold MHEX 3. At the same time, the liquid product is subcooled and mixed with the low pressure refrigerant from MHEX 3, where it is used to cool the feed gas in the intermediate MHEX 2. Since part of the refrigerant mixture only circulates in the warm end of the SWHX, the overall molar flowrate and thus the required heat transfer area decrease.

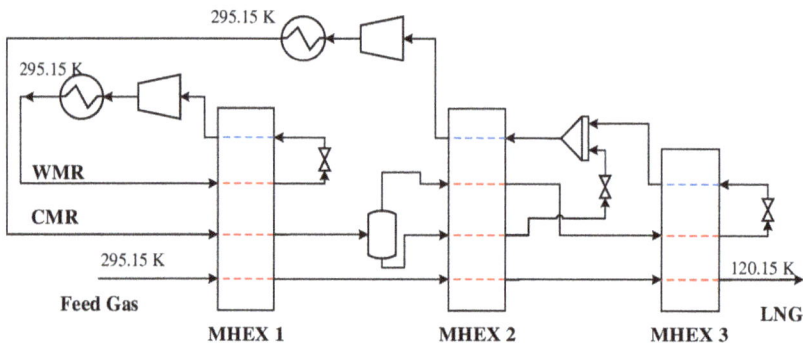

Figure 4. The DMR process with a spiral-wound heat exchanger (SWHX) for the cold mixed refrigerant.

The refrigerant streams and MHEX data for the DMR model are given in Table 2. Again, the parameter values were selected such that a solution could not be obtained using the commercial simulation tool Aspen Plus. A feed gas composition with 1.00 mol % nitrogen, 91.60 mol % methane, 4.93 mol % ethane, 1.71 mol % propane, 0.35 mol % n-butane, 0.40 mol % i-butane, and 0.01 mol % i-pentane at a pressure of 5.5 MPa and a flowrate of 1.0 kmol/s was used in the simulations.

Table 2. MHEX and refrigerant stream data for Example 2. For unknown variables, the value listed is an initial guess.

Property	Value	Property	Value
η	1.0	$\Delta T_{min,1}$ (K)	4.0
UA_3 (MW/K)	0.3	$\Delta T_{min,2}$ (K)	11.0
		$\Delta T_{min,3}$ (K)	4.0
F_W (kmol/s)	1.55	F_C (kmol/s)	1.45
P_{HPW} (MPa)	1.67	P_{HPC} (MPa)	4.85
P_{LPW} (MPa)	0.50	P_{LPC} (MPa)	0.25
$T_{HP,1}^{OUT}$ (K)	240.15	$T_{HP,2}^{OUT}$ (K)	170.15
$T_{HP,3}^{OUT}$ (K)	120.15	$T_{LP,1}^{OUT}$ (K)	280.15
$T_{LP,2}^{OUT}$ (K)	230.15	$T_{LP,3}^{OUT}$ (K)	145.15
Composition (mol %):			
Ethane	47.83	Nitrogen	7.00
Propane	34.17	Methane	41.80
n-Butane	18.00	Ethane	33.20
		Propane	18.00

The simulation model with three MHEXs consists of 96 variables and exhibits seven unknowns. Again these unknowns may be used to solve for any process stream variable such as the pressure, composition, flowrate, or temperature, as well as important MHEX data such as the minimum temperature difference and the heat exchanger conductance. Specifying the UA-value in MHEXs 1 and 2 is challenging as it depends on the solution of MHEX 3. Therefore, as for Example 1, the UA-values are calculated during post-processing to make problem specification easier. Two simulation cases were constructed solving for the following sets of unknown variables:

- **Case I:** P_{LPW}, P_{HPW}, P_{LPC}, $T_{HP,2}^{OUT}$, $T_{LP,3}^{OUT}$, F_C, $\Delta T_{min,2}$.
- **Case II:** P_{LPW}, P_{HPC}, $T_{HP,2}^{OUT}$, $T_{LP,3}^{OUT}$, $f_{W,\,ethane}$, $\Delta T_{min,2}$, UA_3.

Case I: Solved for both pressure levels of the WMR, the low pressure level and refrigerant flowrate of the CMR, the feed gas and high pressure refrigerant temperatures out of MHEX 2, the low pressure refrigerant temperature out of MHEX 3, as well as the minimum approach temperature in MHEX 2. A solution was obtained after six iterations and a total simulation time of 83.0 s with $P_{LPW} = 0.43$ MPa, $P_{HPW} = 1.62$ MPa, $P_{LPC} = 0.27$ MPa, $T_{HP,2}^{OUT} = 155.34$ K, $T_{LP,3}^{OUT} = 151.34$ K, $F_C = 1.42$ kmol/s, and $\Delta T_{min,2} = 6.68$ K. The UA-values were calculated to be 1.99 MW/K and 2.12 MW/K for MHEXs 1 and 2, respectively. The obtained feasible design resulted in a combined compression work of 14.40 MW, where 9.76 MW was needed to compress the CMR, and 4.64 MW was used to compress the WMR. Figure 5 presents the composite curves and driving force distribution in the MHEXs at the solution.

Case II: Solved for the low pressure level of the WMR, high pressure level of the CMR, the natural gas and high pressure refrigerant temperatures out of MHEX 2, the low pressure refrigerant temperature out of MHEX 3, the composition of the WMR, the minimum temperature difference in MHEX 2, and the heat exchanger conductance value for MHEX 3. The model converged after three iterations and a total simulation time of 64.2 s to a solution with $P_{LPW} = 0.44$ MPa, $P_{HPC} = 4.59$ MPa, $T_{HP,2}^{OUT} = 161.03$ K, $T_{LP,3}^{OUT} = 157.03$ K, $\Delta T_{min,2} = 7.88$ K, and $UA_3 = 0.33$ MW/K. A new WMR composition was obtained with 49.24 mol % ethane, 33.24 mol % propane, and 17.51 mol % n-butane and a total molar flowrate of 1.59 kmol/s. The feasible design required a total compression power of 14.85 MW, where 10.08 MW was spent compressing the CMR and 4.76 MW was used to compress the WMR. The heat exchanger conductance values were calculated during post-processing to be $UA_1 = 2.02$ MW/K and $UA_2 = 1.84$ MW/K, respectively. The composite curves and driving force plots for the process are presented in Figure 6.

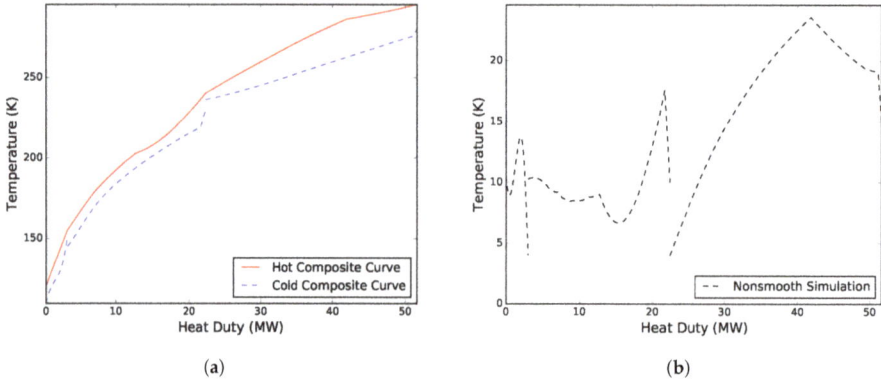

(a)   (b)

Figure 5. (**a**) Composite curves for the feasible design in Case I. (**b**) The corresponding driving force plot.

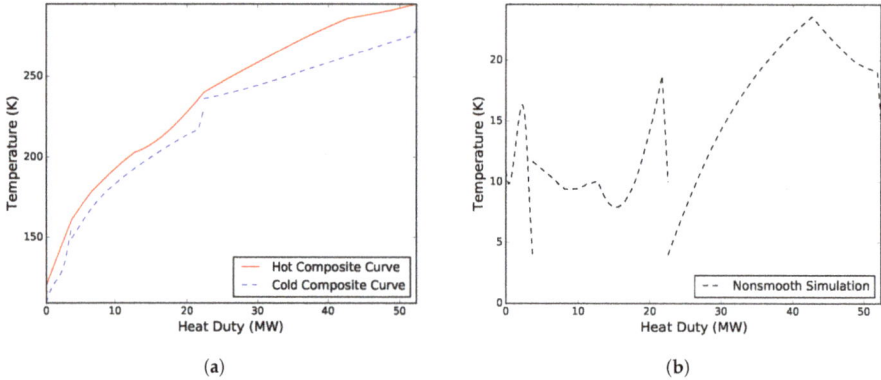

(a)   (b)

Figure 6. (**a**) Composite curves for the feasible design in Case II. (**b**) The corresponding driving force plot.

Case III: Included an NGL separator for the extraction of heavier hydrocarbons (see Figure 7). The case solved for the same set of variables as in Case I and with the same initial guesses and parameter values as given in Table 2. As for the previous examples, the model was solved from an initial guess at which Aspen Plus obtained no feasible solutions with the built-in MHEX module. Rich feed gas compositions at a reduced pressure of 4.0 MPa were used to ensure adequate separation. Simulations were carried out at three different compositions with varying methane contents (Table 3).

Table 3. Natural gas compositions for Case III.

	Compostion I:	Composition II:	Composition III:
Nitrogen (mol %)	2.00	2.00	2.00
Methane (mol %)	85.60	87.60	89.60
Ethane (mol %)	6.93	5.93	4.93
Propane (mol %)	3.71	2.71	1.71
n-Butane (mol %)	1.35	1.35	1.35
i-Butane (mol %)	0.40	0.40	0.40
i-Pentane (mol %)	0.01	0.01	0.01

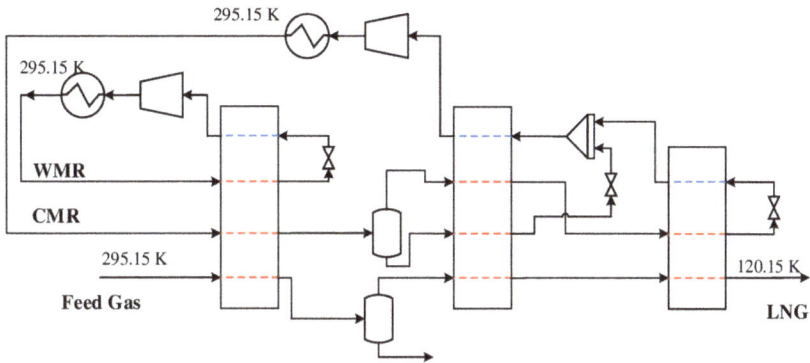

Figure 7. The DMR process in Example 2 with natural gas liquid (NGL) extraction.

Driving force distributions for the three solutions are provided in Figure 8. Solutions were obtained for all three cases within a few iterations. The first two feed gas compositions converged after seven iterations and total simulation times of 85.6 s and 86.4 s for compositions I and II, respectively. The third case converged after six iterations and a total simulation time of 82.1 s. All three solutions exhibited similar driving force profiles, with temperature differences varying mainly in the intermediate MHEX. The same trend can also be seen from the simulation results in Table 4, where the main differences between the three solutions are the UA and ΔT_{min} values for the intermediate MHEX.

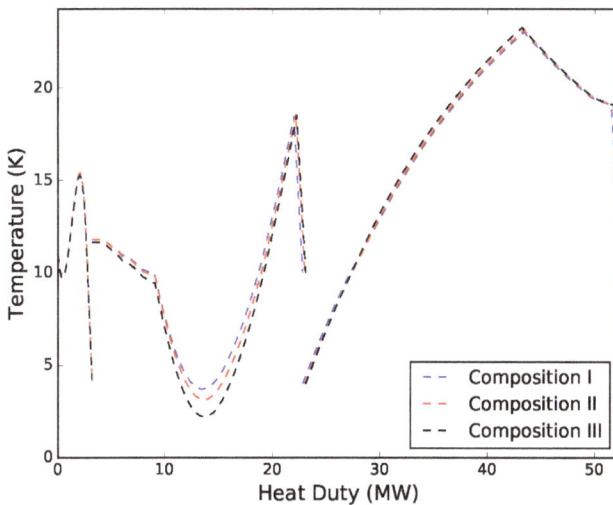

Figure 8. Driving force distributions for the DMR process with NGL extraction.

Table 4. Simulation results for the DMR process with NGL extraction.

	Composition I:	Composition II:	Composition III:
Total work (MW)	14.99	15.11	15.13
W_C (MW)	10.26	10.40	10.43
W_H (MW)	4.73	4.71	4.70
UA_1 (MW/K)	2.07	2.05	2.03
UA_2 (MW/K)	2.63	2.87	3.41
P_{LPW} (MPa)	0.43	0.43	0.43
P_{HPW} (MPa)	1.66	1.66	1.65
P_{LPC} (MPa)	0.26	0.25	0.25
$T_{HP,2}^{OUT}$ (K)	158.22	157.95	157.50
$T_{LP,3}^{OUT}$ (K)	154.22	153.95	153.50
F_C (kmol/s)	1.46	1.47	1.48
$\Delta T_{min,2}$ (K)	3.69	3.14	2.20
LNG composition (mol %):			
Nitrogen	2.08	2.05	2.03
Methane	88.04	89.24	90.61
Ethane	6.50	5.68	4.81
Propane	2.67	2.15	1.48
n-Butane	0.52	0.65	0.81
i-Butane	0.19	0.23	0.27
i-Pentane	0.00	0.00	0.00
LNG flowrate (kmol/s)	0.96	0.97	0.98

6. Conclusions

This paper successfully simulated two different dual mixed refrigerant processes using recent modeling advances in the field of nonsmooth analysis. Natural gas liquefaction processes are particularly difficult to simulate with conventional process simulators, such as Aspen Plus. Lacking rigorous checks for preventing temperature crossovers means that the user must often resort to an iterative approach to obtain feasible designs. A multistream heat exchanger model that features rigorous checks for preventing temperature crossovers during simulation was included in the models. It includes nonsmooth formulations for pinch point location and phase regime detection in the heat exchangers. As a result, it offers a compact formulation where the flowsheet models can be simulated by solving a set of nonsmooth equations, rather than solving a local or global optimization problem. All the cases successfully converged within a total simulation time of 100 s, including initialization of the models. Furthermore, each case was constructed such that Aspen Plus failed to obtain a feasible solution using the same starting point and initial guess values as the nonsmooth model. The use of additional unknowns and the possibility of varying ostensibly difficult process stream variables, such as compositions makes the tool more versatile and capable of handling even complex LNG liquefaction processes.

Author Contributions: All of the authors contributed to publishing this paper. M.V. and H.A.J.W. developed the simulation models for performing the analysis; T.G. and P.I.B. provided useful mathematical and thermodynamical insight for developing the simulation tool and analyzing the results; M.V. wrote the paper; M.V., T.G., and P.I.B. revised and polished the manuscript.

Funding: HighEFF, Statoil.

Acknowledgments: This publication was funded by HighEFF—Centre for an Energy Efficient and Competitive Industry for the Future. The authors gratefully acknowledge the financial support from the Research Council of Norway and user partners of HighEFF, an 8 year Research Centre under the FME scheme (Centre for Environment-friendly Energy Research, 257632/E20). Harry A. J. Watson acknowledges Statoil for providing funding for this project.

Conflicts of Interest: The authors declare no conflict of interest.

Nomenclature

The following abbreviations are used in this manuscript:

Roman letters

EBP	=	enthalpies of the extended composite curves (W)
f	=	component molar flowrate (mol/s)
F	=	total molar flowrate (mol/s)
G	=	element of the generalized derivative
mC_p	=	heat capacity flowrate (W/K)
n_c	=	total number of components
T	=	temperature (K)
UA	=	heat exchanger conductance (W/K)
P	=	absolute pressure (Pa)
Q	=	heat duty (W)
y	=	equation residuals
W	=	compressor work (MW)

Greek letters

ΔT_{LM}	=	log mean temperature difference (K)
ΔT_{min}	=	minimum approach temperature (K)
ΔQ	=	enthalpy change (W)
γ	=	variable in the LP-Newton method

Subscripts and superscripts

2p	=	two-phase substream
BP	=	bubble point
C	=	cold stream
DP	=	dew point
in/out	=	inlet/outlet temperature of a substream
IN/OUT	=	inlet/outlet temperature of a process stream
H	=	hot stream
HP	=	high pressure refrigerant
LP	=	low pressure refrigerant
p	=	pinch candidate
sub	=	subcooled substream
sup	=	superheated substream

References

1. Hwang, J.H.; Ku, N.K.; Roh, M.I.; Lee, K.Y. Optimal Design of Liquefaction Cycles of Liquefied Natural Gas Floating, Production, Storage, and Offloading Unit Considering Optimal Synthesis. *Ind. Eng. Chem. Res.* **2013**, *52*, 5341–5356. [CrossRef]
2. Aspen Technology Inc. *Aspen Plus v9*; Aspen Technology Inc.: Bedford, MA, USA, 2016.
3. Aspen Technology Inc. *Aspen HYSYS v9*; Aspen Technology Inc.: Bedford, MA, USA, 2016.
4. Kamath, R.S.; Biegler, L.T.; Grossmann, I.E. Modeling multistream heat exchangers with and without phase changes for simultaneous optimization and heat integration. *AIChE J.* **2012**, *58*, 190–204. [CrossRef]
5. Watson, H.A.J.; Khan, K.A.; Barton, P.I. Multistream heat exchanger modeling and design. *AIChE J.* **2015**, *61*, 3390–3403. [CrossRef]
6. Hasan, M.M.F.; Karimi, I.A.; Alfadala, H.; Grootjans, H. Modeling and simulation of main cryogenic heat exchanger in a base-load liquefied natural gas plant. *Comput. Aided Chem. Eng.* **2007**, *24*, 219–224. [CrossRef]
7. Hasan, M.M.F.; Karimi, I.A.; Alfadala, H.; Grootjans, H. Operational modeling of multistream heat exchangers with phase changes. *AIChE J.* **2009**, *55*, 150–171. [CrossRef]
8. Rao, H.N.; Karimi, I.A. A superstructure-based model for multistream heat exchanger design within flow sheet optimization. *AIChE J.* **2017**, *63*, 3764–3777. [CrossRef]
9. Pattison, R.C.; Baldea, M. Multistream heat exchangers: Equation-oriented modeling and flowsheet optimization. *AIChE J.* **2015**, *61*, 1856–1866. [CrossRef]

10. Austbø, B.; Løvseth, S.W.; Gundersen, T. Annotated bibliography—Use of optimization in LNG process design and operation. *Comput. Chem. Eng.* **2014**, *71*, 391–414. [CrossRef]
11. Watson, H.A.J.; Vikse, M.; Gundersen, T.; Barton, P.I. Optimization of single mixed-refrigerant natural gas liquefaction processes described by nondifferentiable models. *Energy* **2018**, *150*, 860–876. [CrossRef]
12. Watson, H.A.J.; Barton, P.I. Modeling phase changes in multistream heat exchangers. *Int. J. Heat Mass Transf.* **2017**, *105*, 207–219. [CrossRef]
13. Khan, K.A.; Barton, P.I. A vector forward mode of automatic differentiation for generalized derivative evaluation. *Optim. Methods Softw.* **2015**, *30*, 1185–1212. [CrossRef]
14. Duran, M.A.; Grossmann, I.E. Simultaneous optimization and heat integration of chemical processes. *AIChE J.* **1986**, *32*, 123–138. [CrossRef]
15. Watson, H.A.J.; Vikse, M.; Gundersen, T.; Barton, P.I. Reliable Flash Calculations: Part 2. Process flowsheeting with nonsmooth models and generalized derivatives. *Ind. Eng. Chem. Res.* **2017**, *56*, 14848–14864. [CrossRef]
16. Vikse, M.; Watson, H.A.J.; Gundersen, T.; Barton, P.I. Versatile Simulation Method for Complex Single Mixed Refrigerant Natural Gas Liquefaction Processes. *Ind. Eng. Chem. Res.* **2018**, *57*, 5881–5894. [CrossRef]
17. Roberts, M.J.; Agrawal, R. Dual Mixed Refrigerant Cycle for Gas Liquefaction. U.S. Patent No. 6,269,655 B1, 8 July 2001.
18. Maher, J.B.; Sudduth, J.W. Method and Apparatus for Liquefying Gases. U.S. Patent No. 3,914,949, 28 October 1975.
19. Boston, J.F.; Britt, H.I. A radically different formulation and solution of the single-stage flash problem. *Comput. Chem. Eng.* **1978**, *2*, 109–122. [CrossRef]
20. Watson, H.A.J.; Vikse, M.; Gundersen, T.; Barton, P.I. Reliable Flash Calculations: Part 1. Nonsmooth Inside-Out Algorithms. *Ind. Eng. Chem. Res.* **2017**, *56*, 960–973. [CrossRef]
21. Watson, H.A.J.; Barton, P.I. Reliable Flash Calculations: Part 3. A nonsmooth approach to density extrapolation and pseudoproperty evaluation. *Ind. Eng. Chem. Res.* **2017**, *56*, 14832–14847. [CrossRef]
22. Grossmann, I.E.; Yeomans, H.; Kravanja, Z. A rigorous disjunctive optimization model for simultaneous flowsheet optimization and heat integration. *Comput. Chem. Eng.* **1998**, *22*, 157–164. [CrossRef]
23. Clarke, F.H. *Optimization and Nonsmooth Analysis*; SIAM: Philadelphia, PA, USA, 1990.
24. Nesterov, Y. Lexicographic differentiation of nonsmooth functions. *Math. Program.* **2005**, *104*, 669–700. [CrossRef]
25. Khan, K.A.; Barton, P.I. Generalized Derivatives for Solutions of Parametric Ordinary Differential Equations with Non-differentiable Right-Hand Sides. *J. Optim. Theory Appl.* **2014**, *163*, 355–386. [CrossRef]
26. Barton, P.I.; Khan, K.A.; Stechlinski, P.; Watson, H.A.J. Computationally relevant generalized derivatives: Theory, evaluation and applications. *Optim. Methods Softw.* **2018**, *33*, 1030–1072. [CrossRef]
27. Facchinei, F.; Fischer, A.; Herrich, M. An LP-Newton method: Nonsmooth equations, KKT systems, and nonisolated solutions. *Math. Program.* **2014**, *146*, 1–36. [CrossRef]
28. Fischer, A.; Herrich, M.; Izmailov, A.F.; Solodov, M.V. A Globally Convergent LP-Newton Method. *SIAM J. Optim.* **2016**, *26*, 2012–2033. [CrossRef]
29. Kim, D.; Gundersen, T. Constraint formulations for optimisation of dual mixed refrigerant LNG processes. *Chem. Eng. Trans.* **2017**, *61*, 643–648. [CrossRef]

![processes logo] *processes*

MDPI

Article

Valorization of Shale Gas Condensate to Liquid Hydrocarbons through Catalytic Dehydrogenation and Oligomerization

Taufik Ridha, Yiru Li, Emre Gençer, Jeffrey J. Siirola, Jeffrey T. Miller, Fabio H. Ribeiro and Rakesh Agrawal *

Charles D. Davidson School of Chemical Engineering, Purdue University, West Lafayette, IN 47907, USA;
tridha@purdue.edu (T.R.); li2232@purdue.edu (Y.L.); egencer@mit.edu (E.G.); jjsiirola@gmail.com (J.J.S.);
jeffrey-t-miller@purdue.edu (J.T.M.); fabio@purdue.edu (F.H.R.)
* Correspondence: agrawalr@purdue.edu; Tel.: +1-765-494-2257

Received: 15 July 2018; Accepted: 14 August 2018; Published: 23 August 2018

Abstract: The recent shale gas boom has transformed the energy landscape of the United States. Compared to natural gas, shale resources contain a substantial amount of condensate and natural gas liquids (NGLs). Many shale basin regions located in remote areas are lacking the infrastructure to distribute the extracted NGLs to other regions—particularly the Gulf Coast, a major gas processing region. Here we present a shale gas transformation process that converts NGLs in shale resources into liquid hydrocarbons, which are easier to transport from these remote basins than NGL or its constituents. This process involves catalytic dehydrogenation followed by catalytic oligomerization. Thermodynamic process analysis shows that this process has the potential to be more energy efficient than existing NGL-to-liquid fuel (NTL) technologies. In addition, our estimated payback period for this process is within the average lifetime of shale gas wells. The proposed process holds the promise to be an energy efficient and economically attractive step to valorize condensate in remote shale basins.

Keywords: shale gas condensate; process synthesis and design; shale gas condensate-to-heavier liquids; technoeconomic analysis

1. Introduction

In order to meet the energy demands of the twenty-first century, engineers and scientists are working to develop new methods to discover, extract, and refine fossil resources including oil, coal, natural gas, shale oil, and shale gas. Recent advances in hydraulic fracturing and horizontal drilling have led to a surge in shale resource production. Similar to natural gas, methane concentration in shale gas ranges from 50% to 90%, which sets it as the major component [1,2]. However, unlike natural gas, shale gas contains higher concentrations of hydrocarbons other than methane, such as ethane, propane, butane, isobutane, and pentane. These hydrocarbons are known as condensate or natural gas liquids (NGLs), and their concentrations vary from 0% to 50% [3].

From 2006 to 2016, United States NGL production doubled from 635 million barrels to 1284 million barrels. However, not all the produced NGL can be transported to gas processing or upgrading facilities. As shown in Figure 1a–c, natural gas and hydrocarbon gas liquid (HGL) pipeline infrastructure which is used to transport NGL, and gas processing plant infrastructure are not extensive in several remote shale gas basins compared to basins that are located in historically gas producing or consuming regions such as the Gulf Coast. These remote shale gas basins constitute a large portion of United States shale resource production, shown in Figure 1d.

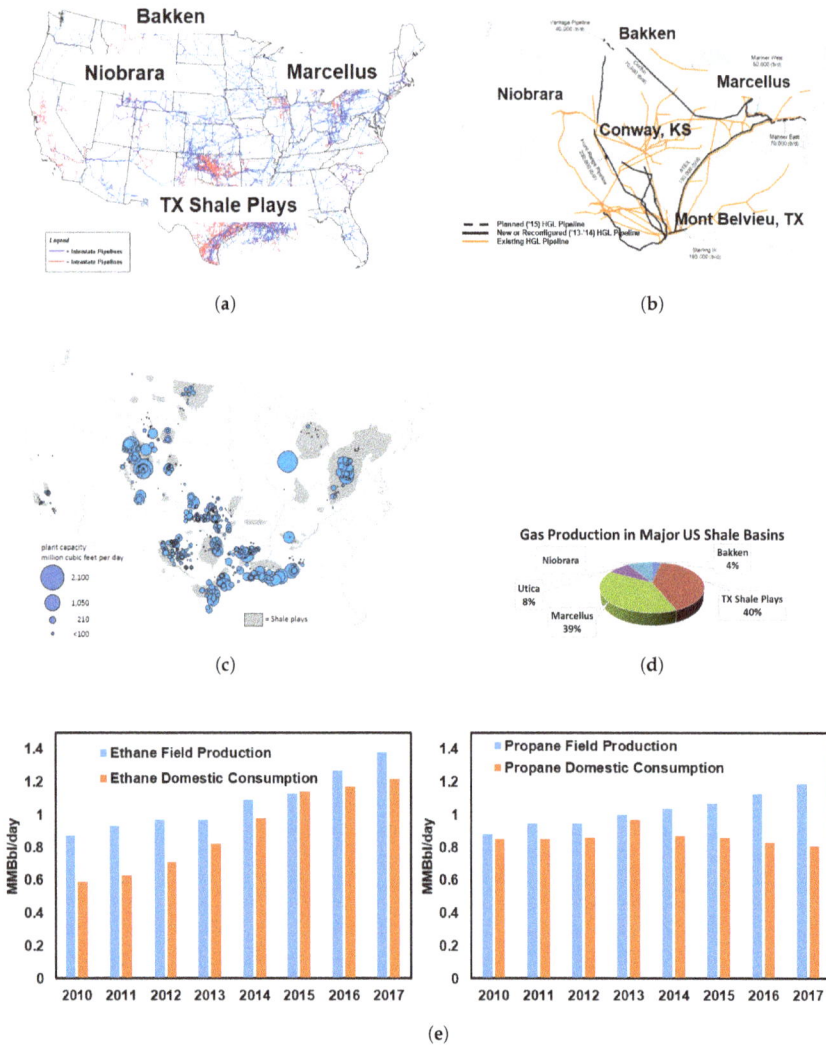

(a)

(b)

(c)

(d)

(e)

Figure 1. (**a**) United States gas transportation systems network. TX Shale Plays include Barnett, Eagle Ford, and Permian basins. Adapted from the Energy Information Agency (EIA) [4]. (**b**) Existing United States hydrocarbon gas liquid (HGL) pipeline network. Adapted from the EIA [5,6]. (**c**) Existing United States gas processing capacity. Adapted from the EIA [5,6]. (**d**) Distribution of shale gas production in the United States based on the shale basins. TX Shale Plays include Barnett, Eagle Ford, and Permian basins [4]. (**e**) United States propane and ethane production and consumption from 2010 to 2017 [7].

Currently, a substantial quantity of NGL is fed into the chemical industries. Ethane is almost exclusively used for ethylene production through steam cracking, which ultimately turns into plastics. Propane and butane are also partially used for chemical feedstocks [8]. He et al. proposed several integrated processes between gas treatment, steam cracking, and catalytic dehydrogenation, and showed the economic potential of producing ethylene and propylene from shale gas [2,9].

However, ethane crackers are highly capital-intensive facilities and take several years to build [10]. Furthermore, as shown in Figure 1e, the consumption of ethane and propane, which is mainly as feedstocks for ethylene and propylene production in the United States, is lower than their current production [7,11]. Thus, olefins such as ethylene and propylene are not reasonable target products for wellhead NGL conversion.

The United States' transportation sector is still dominated by traditional petroleum resources [12,13]. Despite increases in renewable energy and natural gas resources and advances and projected increase in light duty electric and hybrid vehicles, petroleum resources in the United States are expected to play a major role in the future, with gasoline accounting for 35% of the global transportation fuel consumption in 2040 [14,15]. Synfuel International Inc. proposed a new ethane-to-gasoline process consisting of a pyrolysis reactor followed by an ethylene reactor and oligomerization reactor to produce liquid hydrocarbons [16]. The conventional method for the gas-to-liquid (GTL) process involves the partial oxidation of natural gas to obtain synthetic gas composed of CO and H_2, followed by chain growing processes such as Fischer–Tropsch [17,18]. Another alternative to consider is the catalytic dehydrogenation of light alkanes followed by oligomerization of the olefins to form fuel range hydrocarbons.

The catalytic dehydrogenation of light alkanes has been widely studied as an alternative process for producing olefins [19–25]. However, for olefins production, there are only a few reports on process synthesis and design for the production of olefins through the oxidative and non-oxidative catalytic dehydrogenation of light alkanes [22,24,25]. UOP Oleflex is a commercially proven technology for the catalytic conversion of propane to propylene using a PtSn alloy catalyst [26]. The catalytic dehydrogenation of light alkanes can be preferred over conventional technology such as steam cracking, as it has the potential of mitigating the formation of by-products and reducing energy consumption [19,22,27]. Despite these advantages, coking is known as a major problem, which causes rapid catalyst deactivation [21]. According to our knowledge, there is a lack of use of catalytic dehydrogenation of light alkanes in the context of overall process synthesis for the transformation of NGLs to liquid hydrocarbons.

We propose a process that can upgrade shale condensate into liquid hydrocarbons via catalytic dehydrogenation followed by catalytic oligomerization. In this work, we only focus on converting ethane, propane, and butane in shale condensate into liquid fuel, and we do not consider the coupling of methane.

2. Thermodynamic Analysis of the NGL-to-Liquid Pathways

As mentioned earlier, apart from catalytic dehydrogenation followed by oligomerization, there are other routes to upgrade NGL to liquid fuel feedstocks. Alkenes or syngas are common intermediates for these routes. Taking ethane as an example, ethane can be converted to either ethylene or syngas and then upgraded to liquid fuel. Now for a comparison of different synthetic routes from NGL to liquid fuel, the energy demands of different pathways of ethane conversion are evaluated. For our current analysis, we only consider ethane to octane conversion. Figure 2 below summarizes the different pathways for the thermodynamic analysis that will be discussed.

For the "ethane–ethylene–octane" route, we consider two different dehydrogenation methods: catalytic dehydrogenation and steam cracking. For catalytic dehydrogenation, the ethane is assumed to be converted to ethylene with 100% selectivity and the conversion of ethane is 45% according to reported experimental results; for steam cracking, the conversion of ethane is 67% and selectivity towards ethylene is 81% [20,28]. The catalytic dehydrogenation reactor and steam cracker are both operated at 900 K and 3.5 bar. The dehydrogenation unit is followed by the oligomerization reactor, in which ethylene is coupled to produce octane. The oligomerization reactors are operated at 600 K. Although the coupling reaction is exothermic, the generated heat cannot be recovered to provide heat for the dehydrogenation due to the lower oligomerization operating temperature. Therefore, to compare the energy consumption, we only consider the dehydrogenation units. Through Aspen

Plus simulation, with pure ethane feed, the heat duties are 65 MJ/kmol of ethane reacted and 144 MJ/kmol of ethylene produced for the catalytic dehydrogenation reactor and 103 MJ/kmol of ethane reacted and 190 MJ/kmol of ethylene produced for the steam cracker, respectively. The actual ethane dehydrogenation reactions within the two dehydrogenation reactors are similar, and the difference in heat duty comes from the different conversion and the generation of byproducts in steam cracking. Furthermore, if we consider that the generation of high-temperature steam also demands energy input, catalytic dehydrogenation is a less-energy-intensive route for ethane conversion.

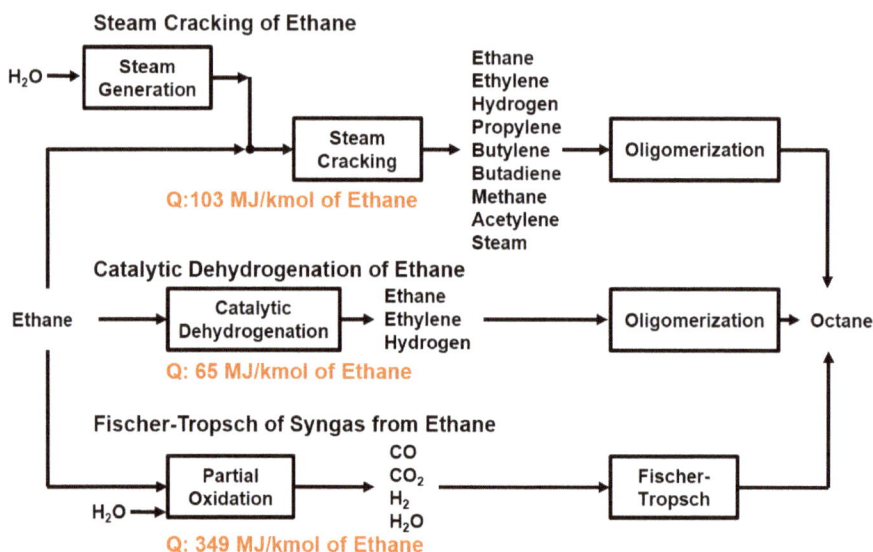

Figure 2. Three potential pathways for converting ethane to octane.

Another possible route from ethane to liquid fuel is via syngas. Ethane can be partially oxidized to syngas either by oxygen or steam and followed by a Fischer–Tropsch reactor for fuel synthesis. Considering the energy demand for air separation, we only consider ethane partial oxidation by steam. At 1000 K and 3.5 bar (the same condition as dehydrogenation), ethane and steam are reacted to produce syngas. The REQUIL reactor model in Aspen Plus was used to model the reformer or oxidation reactor. In this process, the reformer reactor consumes 349 MJ/kmol of ethane, which is higher than that of the dehydrogenation reactor. In addition, this process is counterproductive, as ethane is decomposed to carbon monoxide and hydrogen which are then later recombined to form long carbon chain molecules through Fischer–Tropsch or methanol-to-gasoline technology. Furthermore, in this process, high-temperature steam has to be generated, and the gas product from the oxidation reactor has to be compressed in order to go through the Fischer–Tropsch process. Once again, the large amount of heat generated in the Fischer–Tropsch process is at a much lower temperature than the reformer temperature, leading to a substantial degradation in the quality of heat. Therefore, among the three routes discussed, catalytic dehydrogenation followed by oligomerization is the most energy efficient method of light alkane upgrading.

3. Problem Statement

Given a shale gas condensate stream from a remote reservoir, it is desired to synthesize, simulate, and integrate an NGL-to-liquid hydrocarbons (NTL) process using catalytic dehydrogenation and oligomerization reactions and to carry out economic analysis to answer the following questions:

1. What is the necessary pretreatment of shale gas?
2. What is the correct flow sheet to achieve the NTL conversion?
3. What separation technologies are required for the process?
4. What are the economic criteria of the process and how do they compare with existing processes?
5. What is the cost differential between this process and existing GTL processes?

The following assumptions, basis, and data were used in all processes considered here:

The Bakken field is located in a remote part of North Dakota. Currently, the pipeline infrastructure is already at its full capacity, and the state's natural gas consumption is well below its shale gas production [29]. Considering the variability and decay of shale resource production, installing infrastructure for NGL distribution may not be attractive, as the payback period can easily exceed the well production lifetime [3]. Therefore, it is desirable to convert the NGL locally into liquid fuel components, as it can be refined and marketed locally and nationally through various distribution channels. A 96 million standard cubic feet per day (MMSCFD) basis feed flow rate was selected because a typical single wellhead production rate in the Bakken field ranges from 1 to 4.8 MMSCFD, and this flow rate represents a medium-scale facility that processes outputs from between 20 and 100 wells [3]. The composition of this stream is shown in Table A3 in Appendix D. Additional process assumptions shown in Table 1 are also considered.

Table 1. General process assumptions.

General Assumptions. MMSCFD: million standard cubic feet per day.
Bakken Field Shale Feed Rate: 96 MMSCFD
On-Stream Factor: 0.92
Flash Tank Pressure Drop: 0.21 bar
Heat Exchanger Pressure Drop: 0.21 bar
Ambient Temperature: 308 K
No pressure drop across the reactors
Compressor Efficiency: 0.7

4. Process Description

Shale gas requires the same conventional gas treatment as natural gas. As gas treatment is a well-known technology and UOP-ThomasRussell has an operating modular field-erected gas treatment plant with a current proven size of 200 MMSCFD, we begin with conventional shale gas treatment which consists of acid gas and water removal [15]. Depending on the nitrogen content of the raw shale gas, nitrogen removal may be necessary to meet the typical natural gas pipeline specifications, which is ≤ 4 mol % for nitrogen. In the case of the Bakken field, nitrogen removal may not be required because the region is known to produce both nitrogen-rich and nitrogen-deficient shale gas streams, and the two types of streams can be easily mixed in order to meet the pipeline specification.

Both acid gas and water removal processes are well-established and understood. Depending on the content of acid gas and water, there are various process options. Methyl diethyl amine (MDEA) absorption and triethylene glycol (TEG) absorption are the most common processes for acid gas and water removal, respectively. These processes are capable of reducing the acid gas content down to 4 ppm and the water content to 100 ppm [30]. After the shale gas is treated, it is termed dry, sweet shale gas, which can then undergo further downstream processing.

Catalytic dehydrogenation is the next step and, in this unit operation, ethane, propane, and butane undergo dehydrogenation with a catalyst that reduces selectivities toward undesirable byproducts. The dehydrogenation of ethane is an endothermic reaction, and in order to achieve a reasonable equilibrium conversion, the reaction must be performed at moderately high temperature (900–1100 K).

Hydrogen generated during dehydrogenation may need to be removed prior to the oligomerization reaction, as it can re-saturate olefins. If the oligomerization catalyst has a high hydrogen tolerance such that selectivity toward hydrogenation products is low, then hydrogen can remain in the mixture. Otherwise, hydrogen must be removed, and this separation task can be accomplished using cryogenic distillation or gas membrane separation.

After selectively dehydrogenating ethane, propane, and butane at moderately high temperature, the resulting olefins can be converted to higher molecular weight hydrocarbons through an oligomerization reaction. Catalysts for oligomerization are available, and have been used for similar applications in the past [26,31]. The product of the oligomerization reaction is a mixture of high molecular weight hydrocarbons and unconverted light alkenes. Due to a large difference in their boiling points, high molecular weight hydrocarbons can be recovered through condensation by cooling the mixture. Then, the remaining vapor, which contains unconverted light alkenes, is recycled to the inlet of the catalytic dehydrogenation reactor.

5. Process Modeling

5.1. Gas Treatment

As stated earlier, acid gas treatment and water removal are well-known processes, and the selection of the specific process depends highly on the concentration of the acid gas and water in the shale gas stream. Based on the literature, MDEA sweetening and TEG dehydration processes are suitable for the Bakken field shale gas [30]. In MDEA amine sweetening, MDEA solution is contacted with the shale gas, and carbon dioxide and hydrogen sulfide react with the amine solution. Then, the amine solution is regenerated in a stripper by releasing the acid gas from the solution. For water removal, TEG (triethylene glycol) solution is contacted with the sweet gas shale, where the water is ionically bonded with the TEG solution. The TEG solution is then recovered in a boiler by vaporizing the water. In this work, the economics and energy input of these processes are not considered as in other GTL processes. A treated natural gas stream is assumed as the feed.

5.2. Demethanizer

After gas treatment, NGL must be separated from the shale gas stream (Figure 3: 102; Figure 4: 204). As methane is not converted to liquid hydrocarbons, a high concentration of residual methane in the NGL stream from the demethanizer can possibly lead to large accumulation in downstream recycle loop. Conventionally, cryogenic distillation is used for the demethanizer. Due to the potential of relatively small-scale application of this process, membrane separation is also considered for NGL separation, which has proven to be a viable and practical option in NGL recovery from natural gas [32,33]. Considering the limitations of existing CH_4-NGL separation processes, we propose two process designs based on methane recovery of 86% and 96% in the demethanizer section, and they are labeled Process I and Process II, shown in Figures 3 and 4. For the 96% recovery demethanizer, a turbo-expander process scheme with a distillation column modeled using RadFrac in Aspen Plus was used [34]. For 85% recovery, cascade gas membrane separation was used, and cost calculation for this unit operation was based on a well-mixed membrane model. Note that the turbo-expander process scheme can also be employed for the 85% recovery, and the cascade membrane here was selected to illustrate the deployment of other separation technologies apart from distillation. The detailed schemes for these unit operations can be found in the Supplementary Information. The membrane was assumed to have a permeability of 120 barrer for C_{2+} and permselectivity of 12 for CH_4/CH_{2+} [35]. The capital cost of the membrane module was assumed to be $50/m^2$.

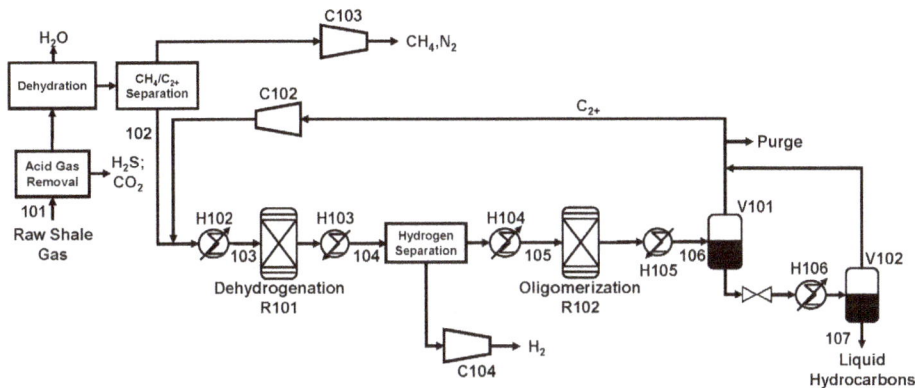

Figure 3. Process flow sheet for Process I.

5.3. Dehydrogenation

Ethane, propane, and butane can be transformed to its corresponding mono-olefins through catalytic dehydrogenation. The dehydrogenation reaction can be generalized as follows:

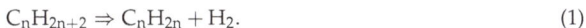

$$C_nH_{2n+2} \Rightarrow C_nH_{2n} + H_2. \tag{1}$$

The reaction is endothermic, and for light hydrocarbons, the equilibrium conversion is reasonable at high temperature ranging between 800 K and 1100 K [36]. Based on Le Chatelier's principle, lower pressure shifts the chemical equilibrium toward the product side. Hence, the reaction should be operated at low pressure. Currently, the industrial catalytic dehydrogenation of light hydrocarbons is limited to only propane and butane. Honeywell Oleflex is an example of the industrial implementation of catalytic dehydrogenation which entails the dehydrogenation of propane to propylene [26]. Using PtSn/Al$_2$O$_3$ catalyst, propane is dehydrogenated at 1.4 barg and 873 K. The dehydrogenation of ethane is usually achieved through steam cracking [26]. Ethane conversion of 45% with selectivity of 99% toward ethylene has been reported at 873 K using PtZn/SiO$_2$ catalyst [20].

Here, we assumed that through catalyst development, 95% of equilibrium conversion of ethane, propane, and butane dehydrogenation at 1073 K and 6.58 bar can be reached. Note that for dehydrogenation, 95% of the true equilibrium conversion was considered in order to account for the fact that dehydrogenation is a highly endothermic reaction and heat transfer is the rate-limiting step. In Figure 3, R101 represents the catalytic dehydrogenation reactor and 103 is the inlet stream to R101. The REQUIL reactor model in Aspen Plus was used. Three reactions (dehydrogenation of ethane, propane, and butane), and their respective temperature approaches were specified in order to adjust the equilibrium conversion. No competing reactions (e.g., hydrogenolysis of alkanes) were considered. The same modeling details for the dehydrogenation reactor were applied for Process II in Figure 4.

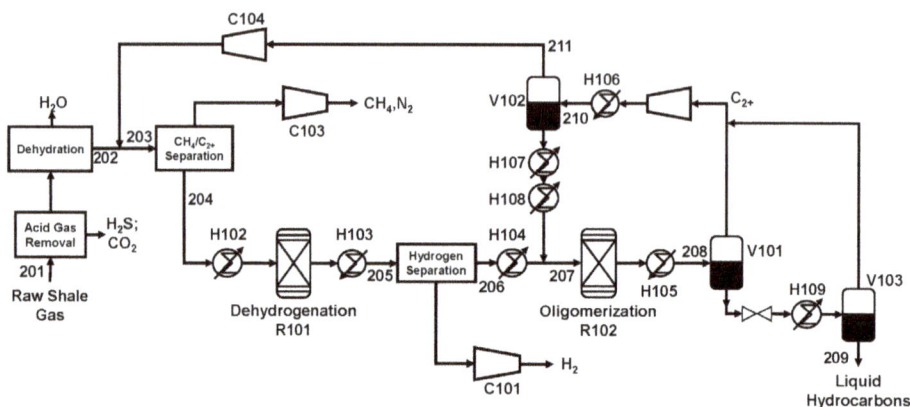

Figure 4. Process flow sheet for Process II.

5.4. Hydrogen Recovery

The product stream (Figure 3, 104; Figure 4, 205) from the dehydrogenation reactor which contains mono-olefins, hydrogen, and unconverted light alkanes is then cooled down to 500 K. Using membrane separation, hydrogen will be partially recovered. Some retained hydrogen in the retenate stream is desirable to ensure the stability of the dehydrogenation and oligomerization catalyst [20]. In Aspen Plus, the membrane was simulated using a separator and calculator block. Within the calculator block, the material balance and design equation for a well-mixed membrane were employed and the output from this block was used in the separator block to determine the purity and flow rate of the permeate and retentate streams. For sizing and economics calculation, a well-mixed membrane system and $50/m^2$ capital cost for a spiral wound membrane module were assumed [37]. The hydrogen membrane used in this work was assumed to have permeability of 250 barrer for hydrogen and selectivity of 590 and 125 for hydrogen/ethylene and hydrogen/methane, respectively [33,38]. The gas membrane was modeled as a well-mixed membrane system with a binary feed, and a polyimide membrane was used. In addition, it was assumed that the feed was a binary mixture of hydrogen and pseudo component of C_{1+}. The permselectivity of H_2/C_{1+} for this membrane here was taken to be 483. The gas membrane was designed to achieve a target of 15% mole of hydrogen in the retenate stream in order to stabilize the catalysts used in this process. The permeate purity was 83.87% mole of hydrogen. The net recovery of hydrogen through the membrane was 0.105 kmol of H_2/m^2 h. Using a single membrane configuration and setting the pressure of the permeate side at 1 bar, the hydrogen removal in the permeate was 54% and ethane slip to the permeate stream was 16%. This resulted in 15% mole of hydrogen in the retenate stream according to our simulation results.

5.5. Oligomerization

The retentate stream (Figure 3, 105) from the hydrogen membrane unit was heated to 573 K and then fed to the oligomerization reactor. In this reactor, olefins couple together to form higher molecular weight olefins. For the oligomerization of olefins , the reaction can be generalized as follows:

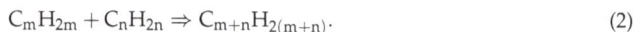

$$C_mH_{2m} + C_nH_{2n} \Rightarrow C_{m+n}H_{2(m+n)}.$$ (2)

The oligomerization reaction is exothermic and generally runs at low temperature [26]. This reaction is carried out at 573 K and 5.47 bar [26] H-ZSM-5 is commonly used for the olefin oligomerization reaction [31]. It has been reported that 90 wt % conversion to liquid has been observed from propene at 500 K and 24 bar with 88% of the liquid being C_{9+} hydrocarbons [39]. Similarly, ethylene fed with nitrogen at 773 K obtained a yield of 54.2% toward C_{5+} hydrocarbons on H-ZSM-5 [40]. Toch et al. also reported 99% ethylene conversion with 25% and 55% selectivities toward propene and gasoline, which is hydrocarbon with a carbon number ranging from five to eight, using Ni-beta zeolite at 500 K and 1.0 MPa [41,42].

In this work, we assumed that this chemical system achieves thermodynamic equilibrium at 600 K and 5.47 bar and only alkene coupling that produces a larger alkene occurs. Therefore, we only considered the C_4–C_{12} alkene oligomerization products. The selectivity to various high molecular weight alkenes are defined based on equilibrium. In Figure 3, R102 represents the oligomerization reactor. The RGIBBS reactor model in Aspen Plus was used to estimate the equilibrium composition. In addition, all paraffin molecules, methane, and hydrogen were set to be inert, indicating that they do not participate in the minimization of Gibbs free energy calculation. Note that in these coupling reactions, it is very likely for the olefins to also form both cyclic and branched molecules, but this was not considered in this study.

5.6. Liquid Hydrocarbon Recovery

After the oligomerization reactor, the final step is to recover liquid hydrocarbons and recycle the unconverted C_2 and C_3 into either of the reactors depending on whether they are olefinic or aliphatic light hydrocarbons. First, the product stream (Figure 3, 106) from the oligomerization reactor is cooled down to 275 K to condense liquid hydrocarbons. This temperature was selected because C_{9+} hydrocarbons may form into waxes and solids below 275 K. The downstream processing of the vapor stream is a crucial step in the overall separation process. This vapor stream mainly contains unconverted olefinic and aliphatic light hydrocarbons. If this vapor stream is directly recycled to the fresh feed stream of the dehydrogenation reactor and the methane recovery in the upstream CH_4/C_{2+} separator is not very high, this necessitates a very large recycle ratio. With a large recycle ratio, the feed stream entering the dehydrogenation reactor may be compositionally worse than the shale gas composition. There are several separation and recycle process configuration options to avoid a large recycle ratio, and here we consider the two following configurations:

In the first configuration, labeled Process I, the vapor stream coming out of the condenser (Figure 3, V101) after the oligomerization reactor is directly recycled to the fresh NGL stream entering the dehydrogenation reactor R101 (Process I, Figure 3). In order to avoid a large accumulation of methane, the CH_4/C_{2+} separation step must recover a large percentage of methane. For 86% and 96% methane recovery, the recycle ratios are 4.8 and 1.4, respectively, for Process I. Membrane separation can achieve 86% recovery, but it is difficult to achieve 96% recovery, which may require refrigeration and/or a multiple-stage cascade membrane system [37,43]. Thus, for Process I, the CH_4/C_{2+} separation step was designed to recover 96% of the methane in the feed.

The second configuration, labeled Process II, entails multiple recycle loops (Process II, Figure 4). By compressing and cooling the vapor stream (Figure 4, 210) to 275 K, a liquid stream containing up to 30% mono olefins of C_2, C_3, and C_4 and 40% of C_2, C_3, and C_4 alkanes is obtained, and combining this liquid stream with the feed to the oligomerization reaction results in the two recycle loops shown in Figure 4. This results in smaller recycle ratios compared to those of Process I, as the light alkenes are reacted in the oligomerization reactor. The vapor stream (Figure 4, 211) from the second condenser (Figure 4, 211) contains up to 20% methane. After compressing the vapor to 30 bar, the vapor is combined with the incoming shale gas stream (Figure 4, 202). This setup results in two loops. Each loop has a recycle ratio of less than two.

We proposed and simulated two different process designs for NGL-to-liquid fuel using Aspen Plus. The stream-data results of processes I and II are shown in Tables 2 and 3, respectively. These data were used to perform the techno-economic analysis.

Table 2. Key stream data for Process I. NGL: natural gas liquid.

Stream Name	Raw Shale Gas	Fresh NGL	Dehydrogenation Feed	Hydrogen Membrane Feed	Oligomerization Feed	Per Pass Product	Fuel-Grade Hydrocarbons	Hydrogen Rich Outlet
Stream Number	101	102	103	104	105	106	107	108
Components (Mole %)								
H_2	-	-	14.21	28.69	15.17	17.20	-	83.59
CO_2	0.57	-	-	-	-	-	-	-
CH_4	57.55	8.11	20.72	17.23	19.81	22.46	-	6.72
C_2H_6	19.89	46.92	29.54	15.80	18.18	20.61	0.01	6.16
C_2H_4	-	-	0.36	9.05	10.41	0.43	0.45	3.53
C_3H_8	11.30	30.92	12.11	3.39	4.22	4.79	-	-
C_3H_6	-	-	6.69	12.24	15.25	8.28	1.27	-
$n\text{-}C_4H_{10}$	2.82	7.84	3.35	1.34	1.67	1.90	1.65	-
$i\text{-}C_4H_{10}$	0.96	2.65	4.70	3.91	4.87	5.52	3.33	-
$n\text{-}C_5H_{12}$	0.55	1.53	0.72	0.60	0.75	0.85	6.08	-
$i\text{-}C_5H_{12}$	0.38	1.05	0.59	0.49	0.61	0.69	4.13	-
C_6H_{14}	0.22	0.61	0.20	0.16	0.20	0.23	2.82	-
C_7H_{16}	0.09	0.25	0.07	0.06	0.07	0.08	1.68	-
C_8H_{18}	0.04	0.11	0.03	0.03	0.03	0.04	0.69	-
N_2	5.20	-	-	-	-	-	0.31	-
H_2S	0.29	-	-	-	-	-	-	-
H_2O	0.15	-	-	-	-	-	-	-
C_4H_8	-	-	4.38	5.09	6.34	6.32	8.94	-
C_5H_{10}	-	-	1.75	1.46	1.82	4.24	18.63	-
C_6H_{12}	-	-	0.45	0.37	0.47	2.68	18.83	-
C_7H_{14}	-	-	0.10	0.08	0.10	1.83	15.08	-
C_8H_{16}	-	-	-	-	-	0.56	4.85	-
C_9H_{18}	-	-	-	-	-	0.59	5.21	-
$C_{10}H_{20}$	-	-	-	-	-	0.15	1.28	-
$C_{11}H_{22}$	-	-	-	-	-	0.30	2.64	-
$C_{12}H_{24}$	-	-	-	-	-	0.24	2.13	-
Total Flow (kmol/h)	4834	1733	6475	7789	6251	5514	626	1539
Temperature (K)	308	1073	1073	473	573	275	295	473
Pressure (bar)	30	7	6	6	5	5	1	1

Table 3. Key stream data for Process II.

Stream Name	Raw Shale Gas	Fresh Demethanizer Feed	Demethanizer Feed	Dehydrogenation Feed	Hydrogen Membrane Feed	Fresh Oligomerization Feed	Oligomerization Feed	Per Pass Product	Fuel-Grade Hydrocarbons	Off-Gas	Recycle Gas
Stream Number / Components (Mole %)	201	202	203	204	205	206	207	208	209	210	211
H_2	-	-	8.49	-	25.72	15.31	13.43	15.81	-	18.65	22.47
CO_2	0.57	-	-	-	-	-	-	-	-	-	-
CH_4	57.55	58.14	41.40	12.08	8.98	9.40	8.30	9.77	-	11.52	13.80
C_2H_6	19.89	20.09	22.51	41.17	17.26	18.08	16.63	19.57	-	23.03	26.48
C_2H_4	-	-	0.30	0.55	13.72	14.38	12.62	0.58	0.30	0.68	0.79
C_3H_8	11.30	11.42	9.36	18.22	3.38	4.22	4.43	5.22	-	5.98	5.98
C_3H_6	-	-	5.01	9.16	16.96	21.16	19.90	11.14	0.98	12.85	13.26
$n\text{-}C_4H_{10}$	2.82	2.85	2.30	4.47	1.07	1.34	1.87	2.20	1.56	2.12	1.39
$i\text{-}C_4H_{10}$	0.96	0.97	3.54	6.90	5.12	6.39	8.22	9.68	2.65	10.10	7.79
$n\text{-}C_5H_{12}$	0.55	0.55	0.42	0.82	0.61	0.76	1.06	1.25	7.35	0.72	0.20
$i\text{-}C_5H_{12}$	0.38	0.38	0.33	0.65	0.48	0.60	0.88	1.04	4.21	0.70	0.25
C_6H_{14}	0.22	0.22	0.14	0.27	0.20	0.25	0.27	0.32	2.91	0.08	0.01
C_7H_{16}	0.09	0.09	0.06	0.11	0.08	0.10	0.10	0.11	1.68	0.01	-
C_8H_{18}	0.04	0.04	0.03	0.05	0.04	0.05	0.04	0.05	0.69	-	-
N_2	5.20	5.25	3.27	-	-	-	-	-	0.31	-	-
H_2S	0.29	-	-	-	-	-	-	-	-	-	-
H_2O	0.15	-	-	-	-	-	-	-	-	-	-
C_4H_8	-	-	2.35	4.56	5.64	7.04	8.67	8.58	8.33	8.62	6.21
C_5H_{10}	-	-	0.46	0.90	0.67	0.83	2.65	5.80	17.67	3.68	1.22
C_6H_{12}	-	-	0.04	0.08	0.06	0.08	0.73	3.71	18.71	1.02	0.11
C_7H_{14}	-	-	-	0.01	-	-	0.17	2.56	15.49	0.23	0.01
C_8H_{16}	-	-	-	-	-	-	0.02	0.79	5.07	0.02	-
C_9H_{18}	-	-	-	-	-	-	0.01	0.85	5.52	0.01	-
$C_{10}H_{20}$	-	-	-	-	-	-	-	0.21	1.37	-	-
$C_{11}H_{22}$	-	-	-	-	-	-	-	0.43	2.85	-	-
$C_{12}H_{24}$	-	-	-	-	-	-	-	0.35	2.33	-	-
Total Flow (kmol/h)	4834	4785	7689	3951	5319	4262	4861	4130	629	3502	2903
Temperature (K)	308	323	325	1073	473	573	573	573	294	278	274
Pressure (bar)	30	29	29	5	5	4	4	4	1	1	5

6. Results and Discussion

As mentioned earlier, REQUIL and RGIBBS reactor models were used to model catalytic dehydrogenation and oligomerization reactions, respectively. For dehydrogenation, the conversions of ethane, propane, and butane per pass were 37.76%, 65.63%, and 50.16%, respectively, for Process I and Process II. In steam cracking, the molar conversion of ethane to ethylene is approximately 70% and the main by-product is a hydrogen-rich off gas [28]. Clearly, the catalytic molar conversion of ethane to ethylene is lower, but reported catalyst for the dehydrogenation of ethylene has shown to have high selectivity toward the dehydrogenation of ethane and to suppress the hydrogenolysis of ethane to methane. One of the performance metrics is the overall amount of C_{2+} being converted to C_{4+}. Equation (3) defines this metric as follows:

$$Conversion_{C_{2+}} = \frac{\sum_{i=2}^4 C_{i,in} - C_{i,out}}{\sum_{i=2}^4 C_{i,in}}, \tag{3}$$

where $C_{i,in}$ is the molar flow rate of hydrocarbons with carbon number i in the dry and sweet shale gas stream and $C_{i,out}$ is the molar flow rate of the hydrocarbons with carbon number i in the final liquid hydrocarbon, hydrogen-rich, and methane-rich streams (Figure 3: liquid hydrocarbons; H_2; CH_4, N_2). The overall C_{2+} conversion was calculated to be 76% and 72% for Processes I and II, respectively. The loss of reactants is due to the purge streams and gas membrane separation. These conversions translate to 139 and 141 BPD of liquid hydrocarbons per MMSCFD of shale gas from the Bakken field used in our simulation. Existing GTL plants using natural gas yield approximately 134 BPD per MMSCFD [44]. Both Processes I and II achieve similar yields. It is estimated that the hydrocarbon yield from syngas followed by Fischer–Tropsch is 135 bbl/MMSCF of ethane, and gasoline yield from syngas followed by methanol synthesis and methanol-to-gasoline is 111 bbl/MMSCF of ethane. The main distinctions between the two proposed processes are the process complexity, the degree of methane recovery, and their economics. Process I only possesses one recycle loop and fewer unit operations compared to Process II, which has two recycle loops and more unit operations. Demethanization in Process I cannot be achieved using existing membrane technology, while in Process II, gas membrane separation is viable for methane removal.

6.1. Energy Integration

Each process design has several process cooling and heating duties. Within the recycle loop, the recycle stream is heated to 1073 K from ambient temperature (308 K) after being combined with the fresh feed stream. The final liquid hydrocarbon stream is brought back to ambient temperature and pressure. Additionally, the dehydrogenation and olefin coupling reactions are endothermic and exothermic, respectively. Operating costs include cooling and heating duties. Integrating these duties can reduce the overall operating cost, since identifying one heat integration results in two operating cost savings, heating and cooling duties. Thermal pinch analysis can be used to determine the best heat integration in a process. The Aspen Energy Analyzer was used to determine the minimum heating and cooling duties for the two process designs considered here.

As shown in Figure 5a,b, the minimum heat duty is the horizontal gap between the cooling (blue line) and the heating curve (red line). For Process I, it was 64 MW, which is the heat of reaction for dehydrogenation. Thermal pinch results also indicated that the heat duty requirement could be reduced by 72%. For Process II, it was 65 MW, which is approximately the heat of reaction for dehydrogenation. Both minimum heat duties are equivalent to the heat of the reaction in dehydrogenation. Hence, the heat flows within the loops were being integrated except for the dehydrogenation, as it demands heat at 1073 K and no other unit operation generates heat at that temperature. The minimum cooling duty can further reduce the electricity consumption through the means of co-generation [45,46].

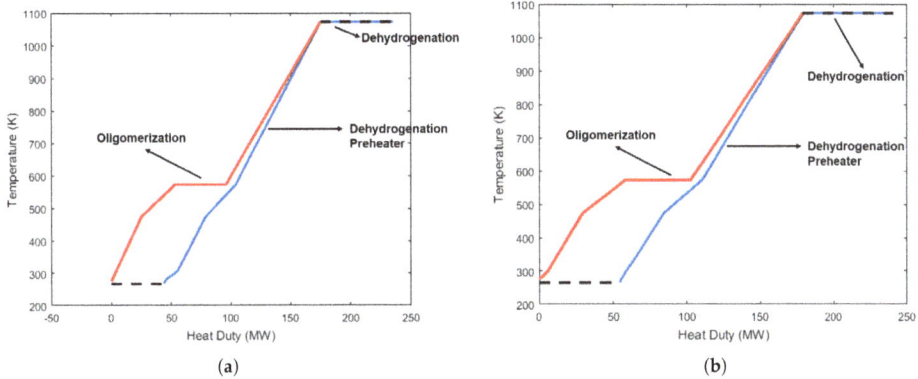

Figure 5. (**a**) Composite curve for Process I. (**b**) Composite curve for Process II.

Using the heating and cooling utilities prior to heat integration, the process thermal efficiency was calculated and shown in Table 4. For the efficiency calculation, the energy inputs were set on the basis of primary energy and the products were taken to be only liquid hydrocarbons. Hydrogen and pipeline quality gas were considered. The equation below describes this efficiency:

$$\eta = \frac{\dot{m}_{LiquidHydrocarbons}LHV_{LiquidHydrocarbons} + \dot{m}_{HydrogenRich}LHV_{HydrogenRich} + \dot{m}_{MethaneRich}LHV_{MethaneRich}}{\dot{m}_{ShaleGas}LHV_{ShaleGas} + Q_{Heat} + Q_{Electricity}}, \qquad (4)$$

where $\dot{m}_i$ is the mass flow rate of stream i, LHV_i is the lower heating value of stream i, Q_{Heat} is the total heat consumption from heat exchangers and reactors, and $Q_{Electricity}$ is the total heat consumption for electricity. These efficiencies are higher compared to GTL-FT (Fischer–Tropsch) and GTL-MTG (methanol-to-gasoline) efficiencies of 56% and 41%, respectively. Of course, GTL-FT releases a large amount of heat from the Fischer–Tropsch reactor that could be used for co-generation to improve that process efficiency. The catalytic dehydrogenation of light alkanes followed by oligomerization has the potential to be more efficient than existing technologies.

Table 4. Energy efficiency for the proposed processes and existing technologies. FT: Fischer–Tropsch; GTL: gas-to-liquid; MTG: methanol-to-gasoline.

	Energy Efficiency
Process I	0.83
Process II	0.88
GTL-FT	0.56
GTL-MTG	0.41

6.2. Economics

In order to measure the economic performance of the processes proposed in this study, an economic analysis was performed to estimate the total capital investment (TCI) and return-on-investment (ROI). Standard procedures were used to assess those economic parameters [45]. Table 5 summarizes the cost parameters that were assumed and the operating costs of both processes. Note that here we are only considering the NGL from shale gas and the resulting liquid hydrocarbon product. Hence, we are not considering the capital cost for methane gas treatment and revenue gained from methane. As shown in Table 5, the main difference between the operating costs of Process I and Process II lies in the electricity consumption.

Table 5. Key economic parameters and operating costs for Process I and II. MMSCFD: million standard cubic feet per day.

Item	Unit Cost	Process I (MMUSD)	Process II (MMUSD)
NGL in Shale Gas	$2.5/MMSCFD	32.7	32.7
Heating Utility	$4/MMBtu	6.2	6.3
Cooling Utility	$2/MMBtu	2.7	2.8
Electricity	$0.045/kWh	6.4	9.7
Liquid Hydrocarbon Sales	$1.19/gal	224	227

6.2.1. Total Capital Investment (TCI)

In order to estimate the total capital investment, two techniques were used together to estimate the capital cost of each unit operation. First, standard sizing algorithms and calculation in Aspen Economic Analyzer were used to estimate most of the unit operations. Second, a combination of cost charts, Lang's method, and estimates from various pieces of literature were used to estimate the dehydrogenation reactor and other unit operations [45,47]. Tables A1 and A2 in the Supplementary Information summarize the TCI distribution for these processes and also the technique used for each unit operation. The estimated TCIs for Process I and Process II were $251 million and $243 million, respectively. For comparison with other existing processes (i.e., GTL-FT and GTL-MTG), to produce the same amount of liquid hydrocarbons, GTL–FT costs between 300 to 525 million USD [17,18] and GTL–MTG costs approximately 1.5 billion USD [28,48]. SynFuels International Inc.'s GTL process is estimated to have TCI of $135 MMUSD for 20 MMSCFD. Using the sixth-tenth rule, the estimated capital cost for a 90 MMSCFD plant is $332 MMUSD. Figure 6 highlights the comparison of the processes in this work with other existing technologies. The TCI for the processes proposed here was at least 17% less than the alternate technologies.

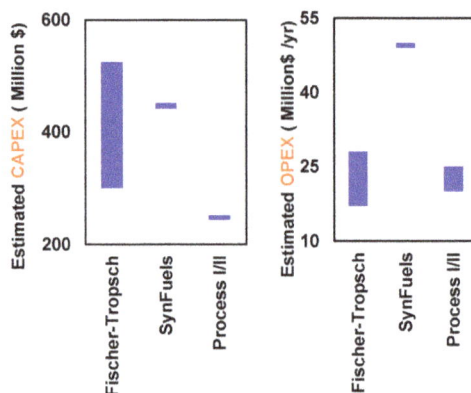

Figure 6. Comparison of total capital and operating costs from this study with the capital and operating costs of other existing technologies.

6.2.2. ROI and Payback Period

Besides the TCI, ROI is generally used to determine the economic feasibility of a plant. In order to calculate these values, the following assumptions are made: (1) linear depreciation model of five year period with 10% salvage value at the end of the period; (2) tax rate is 30% and the discount rate is 10%. Further details on the ROI evaluation can be found in Appendix B. The ROI is calculated to be 0.52 and 0.54 for Processes I and II, respectively. A process with an ROI of 0.15 or higher is considered to be lucrative. The slight difference in ROI of Processes I and II is due to the difference in the TCI of the two processes. Although Process II has a higher operating cost and a lower C2 and C3 recovery. The annual net income for this process is higher because of the lower depreciation. Therefore, this results a slightly higher ROI compared to Process I. Despite Process II having a slightly higher ROI, Process I involves a demethanizer with 95% methane recovery, which can be difficult to achieve using membrane separation technology. Note that a gas membrane system is generally deployed for gas plants of size less than 100 MMSCFD and the size of the plant considered here is 96 MMSCFD. Therefore, the ROI difference between these two processes may widen at smaller plant sizes.

The ROI values can be directly translated into payback period. The payback periods for Processes I and II are 1.9 years and 1.8 years, respectively. Considering the decline of well productivity, which can be up to 75% within three years, the payback periods of these processes are well within the lifetime of these wells [49].

7. Potential of the Proposed Processes for Modularization

Considering the economic opportunity presented by either stranded shale gas or associated shale gas, the proposed process can be deployed at modular scales. In a modular plant, the process equipment and its supporting components are mounted within a structural metal framework and each module is a self-contained process [50]. There are many factors that determine whether a process is amenable to modularization, and process complexity is one of the main factors. As stated previously, many of the existing technologies for thes conversion of natural gas liquid to liquid fuel have only been implemented at large scale. Steam cracking plants generally process up to 1500 MM lbs/year. The smallest proved GTL plant using Fischer–Tropsch process that has been proven is 14,700 bbl/day. These GTL processes mainly consist of syngas generation followed by Fischer–Tropsch or methanol synthesis with methanol-to-gasoline (MTG).

Each process described in this study is amenable to process modularization. However, the proposed process has been shown to be potentially more economically lucrative assuming high selectivity of catalytic dehydrogenation and considering the large boiling point difference between liquid hydrocarbons and light hydrocarbons. Steam cracking requires either downstream upgrading and/or separation in order to hydrogenate acetylene or remove methane. Fischer–Tropsch synthesis produces a liquid product that requires hydrogenation and hydrocracking. Therefore, the existing NTL processes clearly require more unit operations than the proposed NTL process. Although the economics of modular NTL proposed in this work have not been evaluated, this NTL process has the potential to be more economically modularized compared to other existing technologies.

8. Conclusions

Shale gas is projected to be one of the dominant forces in the future of the United States energy landscape. With a projected supply for more than one hundred years, fitting the shale gas into the United States energy landscape requires processes that can convert shale gas into different forms of energy. Shale gas utilization can vary widely from electricity production to chemical production. However, existing infrastructure and market saturation do not allow for some of its common utilizations, particularly as chemical feedstock for olefin plants. However, converting shale gas to liquid fuel can overcome limitations from existing infrastructure, as liquid fuel is transportable and easily marketed. Several large shale gas fields are located in historically non-gas-producing regions (e.g., Bakken and Niobrara basins), where infrastructure for gas distribution is limited or non-existent. Liquid hydrocarbons can be easily transported through different channels such as railways and trucks for further refining. In addition to this, the liquid fuel market is widely distributed with minimal time-variant demand. Herein, we proposed a process for the transformation of shale gas that converts the NGL in shale gas into liquids using a catalytic system that differs from the existing technologies.

There were two processes proposed in this study depending on the separation technology that is considered. Both processes entail dehydrogenation and oligomerization reactions. The main distinctions between the two processes are the separation technology used for the demethanizer and the recycle loop configurations. In terms of energy consumption, both processes have similar minimum heating and cooling duties and product yield. The main difference in energy consumption between these two processes is the electricity consumption. Based on the evaluated economic indicators, Process II is more economically attractive than Process I. In addition, it is not clear whether the demethanizer separation target in Process I can be achieved using membrane technology solely and whether Process I is amenable to modularization for wellhead applications.

Existing GTL-FT and GTL-MTG processes are estimated to be economically less attractive than the proposed processes. The total capital costs of Processes I and II are estimated to be at least 17% lower than that of the conventional GTL processes. The payback periods of Processes I and II are about two years. Clearly, the proposed processes are expected to be much more lucrative than existing technologies.

This study only considered regional or gathering scale facilities. Varying the scale of this proposed process can impact not only its economics, but also the economics and supply chain of NGL, liquid fuels, and other end-use products, especially when the entire chemical manufacturing industry is considered. It is worth assessing how this process fits into the current United States chemical manufacturing industry.

Author Contributions: Conceptualization, E.G., J.J.S., J.T.M., F.H.R. and R.A.; Formal Analysis, T.R., Y.L. and J.J.S.; Funding Acquisition, J.T.M, F.H.R., and R.A.; Investigation, T.R. and R.A.; Methodology, J.T.M, J.J.S, and R.A.; Project Administration, F.H.R. and R.A.; Writing—Original Draft, T.R.; Writing—Review & Editing, Y.L., E.G., J.J.S. and R.A.

Funding: This paper is based upon work funded in part by the National Science Foundation under Cooperative Agreement No. EEC-1647722. Any opinions, findings and conclusions or recommendations expressed in this material are those of the author(s) and do not necessarily reflect the views of the National Science Foundation.

Acknowledgments: We would like to acknowledge CISTAR (Center for Innovative and Strategic Transformation of Alkane Resources) for partial funding. In addition, we would like to acknowledge helpful and insightful discussion with David T. Allen from the McKetta Department of Chemical Engineering, University of Texas—Austin.

Conflicts of Interest: The authors declare no conflict of interest. The founding sponsors had no role in the design of the study; in the collection, analyses, or interpretation of data; in the writing of the manuscript, and in the decision to publish the results.

Abbreviations

The following abbreviations are used in this manuscript:

NGL	Natural Gas Liquids
HGL	Hydrocarbon Gas Liquids
MMSCFD	Million Standard Cubic Feet Per Day
EIA	Energy Information Agency
GTL	Gas-to-Liquids
NTL	NGL-to-Liquids
FT	Fischer–Tropsch
MTG	Methanol-to-Gasoline
TCI	Total Capital Investment
ROI	Return on Investment

Appendix A. Economic Analysis

In this work, an economic analysis was performed to evaluate the total capital investment, operating cost, return-on-investment, and break even price for crude oil. For the total capital cost investment, a combination of standard procedure from Aspen Economic Analyzer and estimates from literature along with Lang's method was used to obtain the total capital cost for each unit operation. Tables A1 and A2 summarize the capital or the installed cost for each unit operation and their methodology. The summation of all unit operation costs listed in the tables below alone does not give the total capital cost. The Aspen Plus Economic Analyzer only provides the installed cost for each unit operation, and the values obtained using Aspen Plus Economic Analyzer in the tables below are the installed costs.

Table A1. Equipment cost for unit operations in Process I.

Unit Operation	MMUSD	Method
Demethanizer Distillation Column System	1.9	Aspen Economic Analyzer
Hydrogen Membrane	1.8	Well-mixed membrane system and $50/m^2
HEX-101	0.018	Aspen Economic Analyzer
HEX-102	4.8	Aspen Economic Analyzer
HEX-103	1.37	Aspen Economic Analyzer
HEX-104	0.43	Aspen Economic Analyzer
HEX-105	0.14	Aspen Economic Analyzer
Dehydrogenation Reactor	4.6	Aspen Economic Analyzer
Oligomerization Reactor	1.8	Aspen Economic Analyzer
COMP-102	5.2	Aspen Economic Analyzer
COMP-103	0.99	Aspen Economic Analyzer
COMP-104	11.2	Aspen Economic Analyzer
V-101	0.18	Aspen Economic Analyzer
V-102	0.16	Aspen Economic Analyzer
Refrigeration	14	Aspen Economic Analyzer

Table A2. Equipment cost for unit operations in Process II.

Unit Operation	MMUSD	Method
Demethanizer Membrane System	7.3	Well-mixed membrane system and $50/m^2
Hydrogen Membrane	1.0	Well-mixed membrane system and $50/m^2
HEX-102	4.6	Aspen Economic Analyzer
HEX-103	1.3	Aspen Economic Analyzer
HEX-104	0.42	Aspen Economic Analyzer
HEX-105	0.11	Aspen Economic Analyzer
HEX-106	0.03	Aspen Economic Analyzer
HEX-107	0.02	Aspen Economic Analyzer
HEX-108	0.05	Aspen Economic Analyzer
HEX-109	0.02	Aspen Economic Analyzer
Dehydrogenation Reactor	4.7	Six Tenth Rule
Oligomerization Reactor	1.8	Aspen Economic Analyzer
COMP-101	11.2	Aspen Economic Analyzer
COMP-102	5.2	Aspen Economic Analyzer
COMP-103	0.74	Aspen Economic Analyzer
COMP-104	1.73	Aspen Economic Analyzer
Refrigeration	14	Aspen Economic Analyzer
V-101	0.18	Aspen Economic Analyzer
V-102	0.16	Aspen Economic Analyzer
V-103	0.16	Aspen Economic Analyzer

For the standard procedure from the Aspen Economic Analyzer, details can be found in the manual. Several of the unit operations (e.g., the dehydrogenation reactor, oligomerization reactor, and membranes) were estimated using literature values along with Lang's method.

Appendix B. Economic Parameters Calculation

Appendix B.1. Return-on-Investment

The equation for ROI is the following:

$$ROI = \frac{Annual\ Net\ Income(After\ Tax\ Profit)}{TCI}. \tag{A1}$$

The total capital investment is the sum of all unit operations' total capital costs. The total capital investment can be calculated by summing values in Tables A1 and A2, respectively, and multiplying this sum by Lang's factor. The second value needed to calculate the ROI is the annual net (after tax profit) cash flow. To calculate the annual net (after tax profit) cash flow, the following equation was used:

$$Annual\ Net\ Income(After\ Tax\ Profit) = (TAR - AOC - AFC - Deprec)(1 - Tax\ Rate) + Deprec, \quad \text{(A2)}$$

where TAR is the total annual revenue, AOC is the annual operating cost, AFC is the annual feedstock cost, and $Deprec$ is the depreciation. Note that the assumed selling prices for all outlet streams and feedstock costs for all raw materials are listed in Table 5. A linear depreciation model with a recovery period of five years was used to calculate the depreciation, which is given by the following:

$$Deprec = \frac{TCI - 0.1TCI}{Recovery\ Period}. \quad \text{(A3)}$$

Here, the recovery period was assumed to be five years and the salvage value was 10% of the TCI. The payback period can be calculated by taking the inverse of the return on investment.

Appendix C. CH$_4$-N$_2$/C$_{2+}$ Separation

Appendix C.1. Demethanizer

In this process configuration, a turboexpander and a Joule–Thompson valve are used to provide the refrigeration needed to liquefy the natural gas stream. Figure A1 below describes the industry standard turboexpander process employed in Process I.

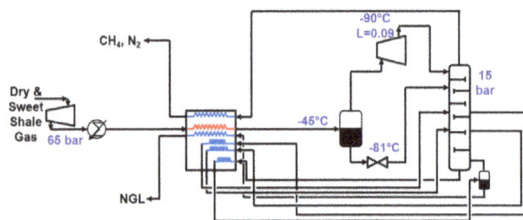

Figure A1. Turboexpander demethanizer scheme.

Appendix C.2. Cascade Gas Membrane Scheme

In this cascade gas membrane configuration, the pressure on the permeate side is atmospheric pressure and it is assumed that the pressure drop between the feed and retenate streams is negligible. The outlet pressure from every compressor is 10 bar. In order to achieve the desired 85% methane recovery, stage cuts for membranes I, II, and II were set as 1.3%, 53.4%, and 74.8%, respectively. In addition, the mole fractions of the most permeable component (in this case methane) in the retenate for membranes I, II, and III were 0.15, 0.15, and 0.39, respectively. Figure A2 describes the cascade gas membrane configuration that is employed in Process II.

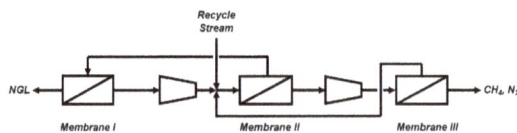

Figure A2. Cascade gas membrane demethanizer scheme.

Appendix D. Shale Gas Composition

Table A3. Composition of shale gas from the Bakken field in the United States [2].

Component	Mole Percentage-Bakken
CO_2	0.57
H_2S	0.29
H_2O	0.15
N_2	5.20
CH_4	57.55
C_2H_6	19.89
C_3H_8	11.30
$n\text{-}C_4H_{10}$	2.82
$i\text{-}C_4H_{10}$	0.96
$n\text{-}C_5H_{12}$	0.55
$i\text{-}C_5H_{12}$	0.38
C_6H_{14}	0.22
C_7H_{16}	0.09
C_8H_{18}	0.04

References

1. Ehlinger, V.M.; Gabriel, K.J.; Noureldin, M.M.B.; El-Halwagi, M.M. Process Design and Integration of Shale Gas to Methanol. *ACS Sustain. Chem. Eng.* **2014**, *2*, 30–37, doi:10.1021/sc400185b. [CrossRef]
2. He, C.; You, F. Shale Gas Processing Integrated with Ethylene Production: Novel Process Designs, Exergy Analysis, and Techno-Economic Analysis. *Ind. Eng. Chem. Res.* **2014**, *53*, 11442–11459, doi:10.1021/ie5012245. [CrossRef]
3. Ka, B.; Pe, K. Compositional variety complicates processing plans for US shale gas. *Oil Gas J.* **2009**, *107*, 50–55.
4. EIA. *Natural Gas Pipeline Network—Transporting Natural Gas in the United States*; EIA: Washington, DC, USA, 2008.
5. DeRosa, S.E. Impact of Natural Gas and Natural Gas Liquids on Chemical Manufacturing in the United States. Ph.D. Thesis, University of Texas at Austin, Austin, TX, USA, 2016.
6. Hydrocarbon Gas Liquids (HGL). *Recent Market Trends and Issues*; Technical Report; U.S. Energy Information Administration: Washington, DC, USA, 2014.
7. Energy Information Administration. *Short-Term Outlook for Hydrocarbon Gas Liquids*; Technical Report; U.S. Energy Information Administration: Washington, DC, USA, 2016.
8. He, C.; Pan, M.; Zhang, B.; Chen, Q.; You, F.; Ren, J. Monetizing shale gas to polymers under mixed uncertainty: Stochastic modeling and likelihood analysis. *AIChE J.* **2018**, *64*, 2017–2036, doi:10.1002/aic.16058. [CrossRef]
9. Gong, J.; You, F. A new superstructure optimization paradigm for process synthesis with product distribution optimization: Application to an integrated shale gas processing and chemical manufacturing process. *AIChE J.* **2018**, *64*, 123–143, doi:10.1002/aic.15882. [CrossRef]
10. *Growing U.S. HGL Production Spurs Petrochemical Industry Investment—Today in Energy*; U.S. Energy Information Administration (EIA): Washington, DC, USA, 2015.
11. Goellner, J.F.; Hamilton, B.A. Expanding the Shale Gas Infrastructure. *Chem. Eng. Progress* **2012**, 49–59.
12. Mallapragada, D.S.; Duan, G.; Agrawal, R. From shale gas to renewable energy based transportation solutions. *Energy Policy* **2014**, *67*, 499–507, doi:10.1016/j.enpol.2013.12.056. [CrossRef]
13. Mallapragada, D.S.; AgrawAL, R. *Role of Natural Gas in America's Energy Future: Focus on Transportation*; Purdue Policy Research Institute (PPRI): West Lafayette, IN, USA, 2013; p. 6.
14. *The Outlook for Energy: A View to 2040*; Technical Report; ExxonMobil: Irving, TX, USA, 2016.
15. Russell, T.H. *Changes of Cryogenic, Amine Plant and Standard Plant Concept*; Thomas Russell Co.: Tulsa, OK, USA, 2011; pp. 1–18.
16. Cantrell, J.; Bullin, J.A.; McIntyre, G.; Butts, C.; Cheatham, B. *Economic Alternative for Remote and Stranded Natural Gas and Ethane in the US*; Bryan Research and Engineering, Inc.: Bryan, TX, USA, 2016.
17. Lutz, B. New Age Gas-to-Liquid Processing. *Hydrocarb. Eng.* **2001**, *6*, 23–28.

18. Senden, M.; McEwan, M. The Shell Middle Distillates Synthesis (SMDS) Experience. In Proceedings of the 16th World Petroleum Congress, Calgary, AB, USA, 11–15 June 2000.
19. Wu, Z.; Wegener, E.C.; Tseng, H.T.; Gallagher, J.R.; Harris, J.W.; Diaz, R.E.; Ren, Y.; Ribeiro, F.H.; Miller, J.T. PdIn intermetallic alloy nanoparticles: Highly selective ethane dehydrogenation catalysts. *Catal. Sci. Technol.* **2016**, *6*, 6965–6976, doi:10.1039/C6CY00491A. [CrossRef]
20. Cybulskis, V.J.; Bukowski, B.C.; Tseng, H.T.; Gallagher, J.R.; Wu, Z.; Wegener, E.; Kropf, A.J.; Ravel, B.; Ribeiro, F.H.; Greeley, J.; et al. Zinc Promotion of Platinum for Catalytic Light Alkane Dehydrogenation: Insights into Geometric and Electronic Effects. *ACS Catal.* **2017**, *7*, 4173–4181, doi:10.1021/acscatal.6b03603. [CrossRef]
21. Sattler, J.J.H.B.; Ruiz-Martinez, J.; Santillan-Jimenez, E.; Weckhuysen, B.M. Catalytic Dehydrogenation of Light Alkanes on Metals and Metal Oxides. *Chem. Rev.* **2014**, *114*, 10613–10653, doi:10.1021/cr5002436. [CrossRef] [PubMed]
22. Baroi, C.; Gaffney, A.M.; Fushimi, R. Process economics and safety considerations for the oxidative dehydrogenation of ethane using the M1 catalyst. *Catal. Today* **2017**, *298*, 138–144, doi:10.1016/j.cattod.2017.05.041. [CrossRef]
23. Chen, K.; Bell, A.T.; Iglesia, E. Kinetics and Mechanism of Oxidative Dehydrogenation of Propane on Vanadium, Molybdenum, and Tungsten Oxides. *J. Phys. Chem. B* **2000**, *104*, 1292–1299, doi:10.1021/jp9933875. [CrossRef]
24. Wolf, D.; Dropka, N.; Smejkal, Q.; Buyevskaya, O. Oxidative dehydrogenation of propane for propylene production—Comparison of catalytic processes. *Chem. Eng. Sci.* **2001**, *56*, 713–719, doi:10.1016/S0009-2509(00)00280-3. [CrossRef]
25. Ren, T.; Patel, M.; Blok, K. Olefins from conventional and heavy feedstocks: Energy use in steam cracking and alternative processes. *Energy* **2006**, *31*, 425–451, doi:10.1016/j.energy.2005.04.001. [CrossRef]
26. Meyers, R.A. *Handbook of Petroleum Refining Processes*, 4th ed.; McGraw-Hill Education: New York, NY, USA, 2016.
27. Chauvel, A.; Lefebvre, G. *Petrochemical Processes: Technical and Economic Characteristics*; Gulf Publishing Company: Houston, TX, USA, 1989.
28. Noureldin, M.M.B.; El-Halwagi, M.M. Synthesis of C-H-O Symbiosis Networks. *AIChE J.* **2015**, *61*, 1242–1262, doi:10.1002/aic.14714. [CrossRef]
29. Ford, M.; Davis, N. *Nonmarketed Natural Gas in North Dakota Still Rising due to Higher Total Production—Today in Energy*; U.S. Energy Information Administration (EIA): Washington, DC, USA, 2014.
30. *GPSA Engineering Data Book*, 20th ed.; Gas Processors Suppliers Association: Tulsa, OK, USA, 2004.
31. Bhan, A.; Hsu, S.; Blau, G.; Caruthers, J.; Venkatasubramanian, V.; Delgass, W. Microkinetic modeling of propane aromatization over HZSM-5. *J. Catal.* **2005**, *235*, 35–51, doi:10.1016/j.jcat.2005.07.005. [CrossRef]
32. Baker, R.; Hofmann, T.; Lokhandwala, K.A. *Field Demonstration of a Membrane Process to Recover Heavy Hydrocarbons and to Remove Water from Natural Gas*; Technical Report DE-FC26-99FT40723; National Energy Technology Laboratory (NETL): Pittsburgh, PA, USA, 2007.
33. Baker, R.W. Future Directions of Membrane Gas Separation Technology. *Ind. Eng. Chem. Res.* **2002**, *41*, 1393–1411, doi10.1021/ie0108088. [CrossRef]
34. Getu, M.; Mahadzir, S.; Long, N.V.D.; Lee, M. Techno-economic analysis of potential natural gas liquid (NGL) recovery processes under variations of feed compositions. *Chem. Eng. Res. Des.* **2013**, *91*, 1272–1283, doi:10.1016/j.cherd.2013.01.015. [CrossRef]
35. Ilinitch, O.; Semin, G.; Chertova, M.; Zamaraev, K. Novel polymeric membranes for separation of hydrocarbons. *J. Membr. Sci.* **1992**, *66*, 1–8, doi:10.1016/0376-7388(92)80085-X. [CrossRef]
36. Zimmermann, H.; Walzl, R. *Ullmann's Encyclopedia of Industrial Chemistry*; Wiley-VCH Verlag GmbH & Co. KGaA: Weinheim, Germany, 2009.
37. Peters, M.S.; Timmerhaus, K.D.; West, R.E. *Plant Design and Economics for Chemical Engineers*, 5th ed.; McGraw-Hill: New York, NY, USA, 2011.
38. Al-Rabiah, A.; Timmerhaus, K.; Noble, R. Membrane Technology for Hydrogen Separation in Ethylene Plants. In Proceedings of the 6th World Congress of Chemical Engineering, Melbourne, Australia, 23–27 September 2001.
39. Wilshier, K.G.; Smart, P.; Western, R.; Mole, T.; Behrsing, T. Oligomerization of propene over H-ZSM-5 zeolite. *Appl. Catal.* **1987**, *31*, 339–359, doi:10.1016/S0166-9834(00)80701-0. [CrossRef]

40. Fernandes, D.S.; Veloso, C.O.; Henriques, C.A. Modified HZSM-5 zeolites for the conversion of ethylene into propylene and aromatics. In Proceedings of the 8th International Symposium on Acid-Base Catalysis, Rio de Janeiro, Brazil, 7–10 May 2017.

41. Toch, K.; Thybaut, J.; Marin, G. Ethene oligomerization on Ni-SiO$_2$-Al$_2$O$_3$: Experimental investigation and Single-Event MicroKinetic modeling. *App. Catal. A Gen.* **2015**, *489*, 292–304, doi:10.1016/j.apcata.2014.10.036. [CrossRef]

42. Toch, K.; Thybaut, J.W.; Arribas, M.A.; Martínez, A.; Marin, G.B. Steering linear 1-alkene, propene or gasoline yields in ethene oligomerization via the interplay between nickel and acid sites. *Chem. Eng. Sci.* **2017**, *173*, 49–59, doi:10.1016/j.ces.2017.07.025. [CrossRef]

43. Xu, J.; Agrawal, R. Membrane separation process analysis and design strategies based on thermodynamic efficiency of permeation. *Chem. Eng. Sci.* **1996**, *51*, 365–385, doi:10.1016/0009-2509(95)00262-6. [CrossRef]

44. *Gas to Liquids (GTL)*; Society of Petroleum Engineers: Richardson, TX, USA, 2015.

45. El-Halwagi, M.M. *Sustainable Design through Process Integration: Fundamentals and Applications to Industrial Pollution Prevention, Resource Conservation, and Profitability Enhancement*, 1st ed.; Butterworth-Heinemann: Amsterdam, The Netherlands; Boston, MA, USA, 2011.

46. Al-musleh, E.I.; Mallapragada, D.S.; Agrawal, R. Continuous power supply from a baseload renewable power plant. *Appl. Energy* **2014**, *122*, 83–93, doi:10.1016/j.apenergy.2014.02.015. [CrossRef]

47. Qassim, A.H.; Mathur, A.K. *Optimized CAPEX and OPEX for Acid Gas Removal Units: Design AGR without Sulphur Recovery Processes*; Society of Petroleum Engineers: Richardson, TX, USA, 2012; doi:10.2118/156096-MS. [CrossRef]

48. Helton, T.; Hindman, M. *Methanol to Gasoline: An Alternative for Liquid Fuel Production*; ExxonMobil: Houston, TX, USA, 2014.

49. Peters, E. Visualizing US Shale Oil Production. Available online: https://shaleprofile.com (accessed on 5 August 2018).

50. Yang, M.; You, F. Modular methanol manufacturing from shale gas: Techno-economic and environmental analyses of conventional large-scale production versus small-scale distributed, modular processing. *AIChE J.* **2018**, *64*, 495–510, doi:10.1002/aic.15958. [CrossRef]

![processes logo] *processes*

MDPI

Article

A Differentiable Model for Optimizing Hybridization of Industrial Process Heat Systems with Concentrating Solar Thermal Power

Matthew D. Stuber

Process Systems and Operations Research Laboratory, UTC Institute for Advanced Systems Engineering and Department of Chemical and Biomolecular Engineering, University of Connecticut, Storrs, CT 06269, USA; stuber@alum.mit.edu; Tel.: +1-860-486-3689

Received: 6 June 2018; Accepted: 19 June 2018; Published: 23 June 2018

Abstract: A dynamic model of a concentrating solar thermal array and thermal energy storage system is presented that is differentiable in the design decision variables: solar aperture area and thermal energy storage capacity. The model takes as input the geographic location of the system of interest and the corresponding discrete hourly solar insolation data, and calculates the annual thermal and economic performance of a particular design. The model is formulated for use in determining optimal hybridization strategies for industrial process heat applications using deterministic gradient-based optimization algorithms. Both convex and nonconvex problem formulations are presented. To demonstrate the practicability of the models, they were applied to four different case studies for three disparate geographic locations in the US. The corresponding optimal design problems were solved to global optimality using deterministic gradient-based optimization algorithms. The model and optimization-based analysis provide a rigorous quantitative design and investment decision-making framework for engineering design and project investment workflows.

Keywords: concentrating solar thermal; CST; concentrating solar power; CSP; parabolic trough; PTC; thermal storage; industrial process heat; hybrid solar

---

1. Introduction

Industrial process systems are enormous consumers of energy (9.23 TWh_{th}), accounting for roughly one-third of all delivered energy use (28.64 TWh_{th}) in the United States in 2017 [1]. Roughly 60% of the energy consumption in the industrial sector is from the manufacturing sector, which accounts for roughly 20% of the total US energy consumption [2]. The industrial sector has historically been the largest consumer of natural gas [1] as both a feedstock and a primary energy source. In manufacturing, industrial process heat (IPH) accounts for roughly 70% of the total process energy consumed in the US [3]. When broken down by fuel consumption, the manufacturing sector accounts for roughly 95% of all IPH demand in the US [3]. Further, since the majority of these process systems require low- or medium-temperature process heat (<250 °C) [3,4], there is a substantial opportunity to reduce fossil fuel consumption and carbon emissions by augmenting existing IPH systems with solar thermal power and designing new systems incorporating solar thermal power.

Numerous studies have been conducted over the past 40 years on the potential that solar energy can play within IPH systems [4–9]. Active interest in the topic in recent years has been motivated largely by the concern over unsustainable energy consumption, greenhouse gas emissions, and overall environmental impact of commercial and industrial sectors [10]. Attention to the feasibility of solar applications to specific industries has been made, such as: food and beverage [11], mining [12,13], agriculture [14,15], textiles [11,16–19], pulp and paper [20,21], chemicals [22,23], and building construction materials [4], among others. Recent reviews [4,6,7,9] resolve the individual industries by their similar unit operations (e.g., drying, hot water, sterilization, etc.). Results of such studies indicate general feasibility of solar in IPH applications; however, the economic viability is extremely sensitive to the application and geographic region of the project. Given the diversity of IPH demand profiles across industries and applications as well as the widely-varying solar resource performance profiles across geographic locations, a formal numerical simulation and optimization-based design approach is necessary to ensure the economic viability of hybrid solar-IPH systems.

1.1. Modeling, Simulation, and Optimization

Numerous models with varying levels of complexity and usage scenarios have been proposed for the design and simulation of solar thermal energy systems [24–37]. Specifically, the focus of this paper is on the application of modeling and simulation for mathematical optimization-based design of solar-IPH systems.

A formal mathematical optimization approach was applied by Ghobeity and Mitsos [28] to the design and operation of a solar thermal receiver and thermal energy storage (TES) system with a constant IPH demand for high-temperature steam generation (for power cycles). A sequential design approach was applied that first considers minimizing total solar cost with the solar array size and TES capacity as design decision variables and several operating decision variables related to the proposed dynamical model [28]. The model was developed for use with gradient-based local optimization solvers and utilized a total daily concentrated energy value based on a single winter day and a heliostat field efficiency calculated for a location in Cyprus [28]. The economic objective was considered as a simple linear cost model with fixed specific costs for the TES medium and solar array. The second optimization problem considered the yearly operation for the fixed design based on the winter day. Thus, for the fixed design, feasible optimal operating conditions were found with respect to 12 representative days throughout the year [28].

Modeling and simulation for hybridizing power cycles with concentrating solar thermal (CST) power was recently considered by Gunasekaran and coworkers [30] with system optimization by Ghasemi and coworkers [29] and Ayub and coworkers [36]. In [29], the focus was on optimizing the operation of the hybrid system without considering TES. In [36], economics were considered to optimize the levelized cost of electricity. However, due to the regional dependence of the solar model and economic assumptions, optimization with respect to the economics of the parabolic trough solar concentrator (PTC) array sizing were omitted. Additionally, in [36], no TES was considered. The thermo-economic model was implemented in Engineering Equation Solver (EES) and a shortcut method is used to optimize the hybrid system using intermediate variables calculated from the AspenPlus sequential-quadratic programming (SQP) optimizer.

A simplified approach to the optimal design of solar water heating systems was presented by Yan et al. [33]. The authors developed a high-level energy conservation model of the water heating system including TES as a hot water tank. The design decision variables were taken as the array size and the TES capacity and the objective was the lifecycle savings. The authors [33] proposed an approach based on diminishing marginal utility and identified optimal solutions based on the intersection of marginal energy savings and marginal embodied energy curves. Although their simplified approach was effective for the water-heating application, IPH applications and their dramatically differing energy demand profiles complicate the optimal design problem as the total utilizable energy and lifecycle savings exhibit large degrees of nonlinear sensitivity to the decision variables [33].

A hybrid solar-IPH design strategy based on heat integration was considered by Baniassadi et al. [34]. The authors considered fixed array designs determined by maximizing the solar fraction for a particular pre-determined investment and analyzed the various options for coupling solar energy with process streams. They proposed an algorithm which uses the various process stream temperatures and pinch analysis to identify the best process stream to exchange solar energy with. They utilized a commercial simulator to calculate the performance of the solar system and subsequently the economics of the retrofit strategy [34]. Although the analysis identifies some technical challenges of physically coupling the solar thermal system with the process from a heat integration perspective, it does not adequately address the fundamental design considerations of the solar system itself [34]. Further, their approach cannot be used within mathematical optimization-based design strategies as the solar performance model is only accessible as an ad-hoc black-box simulation [34].

In [31], Silva and coworkers considered a formal simulation and optimization-based design approach to hybrid solar-IPH systems. In their work, a single-tank thermocline TES device was modeled along with a PTC array for medium-temperature (100–250 °C) IPH applications. Three economic objective functions were considered: levelized cost of energy, payback time, and lifecycle savings as their base case [31]. Four design variables were considered in their work: the number of PTC modules in series, the number of PTC modules in parallel, the spacing between the rows, and the TES capacity [31]. The authors propose a derivative-free non-deterministic memetic optimization algorithm based on a modified genetic algorithm [31]. They make explicitly clear that this approach is in response to the major challenges in optimizing dynamical systems based on simulation: possible discontinuities, nonlinearity, and expensive complex objective function evaluations [31].

In [32], a hybrid solar-IPH system for zero-liquid discharge desalination (<250 °C) was considered. Similar to [31], a formal simulation and optimization-based design approach was taken using the net-present value as the economic objective. The design variables considered for the base-case were the number of PTC modules and the TES capacity. A formal robust design problem was considered as a nonconvex non-cooperative game by taking into account uncertainty in the natural gas and product markets and solving the problem formulated as a semi-infinite program [32]. Again, the authors note that due to the non-differentiable nature of the model, a derivative-free genetic algorithm-based optimization approach must be used with subsequent analysis to verify feasibility and optimality of solutions [32].

The past approaches to modeling, simulation, and optimization of solar-IPH systems have all shown favorable results using model formulations or approaches specific to the application of interest or more general formulations which pose challenges for deterministic gradient-based optimization. There exists a need for a dynamic model of a CST-TES system for designing optimal hybrid solar-IPH systems with accurate economic models for assessing technology investment risk and the economic viability of such projects. In this paper, a model is developed for low- to medium-temperature CST-TES systems for assessing the economic viability and overall project feasibility of hybrid solar-IPH systems. The model accounts for a single-axis tracking PTC array, a TES device, and utilizes high-accuracy hourly solar resource data for the United States available from an open-access database. The model is developed specifically for use with deterministic (global) optimization algorithms, and therefore accounts for convex analysis, differentiability, and the trade-offs between computational efficiency and model accuracy.

In Section 2, the hybrid solar-IPH process performance model will be developed which includes a smooth formulation, a model of a PTC with solar angle calculations for high-resolution dynamic simulation of CST performance, an economic model for assessing design feasibility and investment decision-making, and a mathematical optimization formulation. Additionally, the key theoretical results and analysis of the model are presented in Section 2. In Section 3, four numerical case studies for varying fuel costs and IPH demand profiles across three geographically-disparate locations are presented and discussed followed by the Conclusion in Section 4.

2. Model Formulations

2.1. Concentrating Solar Thermal Model

For a surface or defined geometry oriented horizontally, the performance of the solar energy collector relies on the surface tilt angle and the incidence angle of the irradiation striking the surface. Here, we consider the model of a PTC, illustrated in Figure 1 with the aperture area of a single trough given by $A_a = L_a L_m$. The total optical efficiency of the reflector is defined in the following.

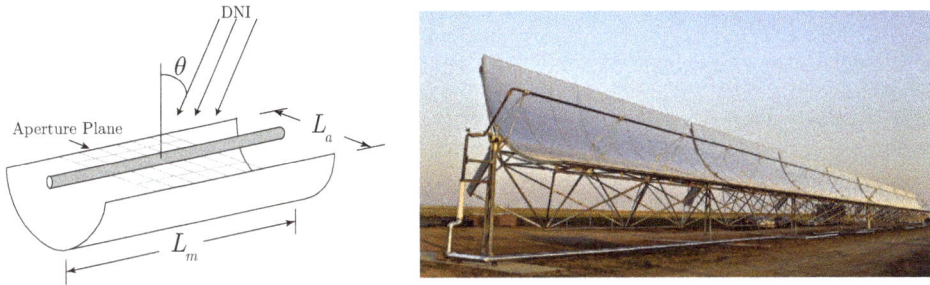

Figure 1. (left) An illustration of a parabolic trough solar concentrator (PTC) with the defined array length L_m, aperture length L_a and the aperture plane with incident solar radiation. **(right)** The large-aperture PTC (Skyfuel, Lakewood, CO, USA) powering a solar thermal desalination pilot by Stuber et al. [38].

Definition 1 (Optical Efficiency, η_{opt} [30]). *The optical efficiency of the PTC η_{opt} is given by:*

$$\eta_{opt} = \epsilon_1 \epsilon_2 \epsilon_3 \epsilon_4 \epsilon_5 \epsilon_6 r_m a_r \tau K, \tag{1}$$

where $\epsilon_i \in [0,1]$ for $i = 1, \ldots, 6$ are efficiency factors related to various components of the PTC and the incidence angle modifier K is given by the empirical relationship [39,40]:

$$K \equiv (\cos\theta + 8.84 \times 10^{-4}\theta - 5.369 \times 10^{-5}\theta^2). \tag{2}$$

Here, ϵ_1 is the shadowing factor of the receiver tube (=0 for fully-shadowed and =1 for no shadowing), ϵ_2 is the tracking error factor (=1 for no tracking error), ϵ_3 is the geometry error (=1 for no geometric or mirror alignment errors), ϵ_4 is the mirror dirt factor (=1 for perfectly clean mirrors), ϵ_5 is the dirt factor for the glass envelope enclosing the receiver tube (=1 for perfectly clean glass), and ϵ_6 are the unaccounted for losses (=1 for no unaccounted losses). The factor r_m is the reflectance of perfectly clean mirrors (=1 for 100% reflectance), a_r is the absorptance of the receiver tube itself (=1 for a perfect black-body), and τ is the transmittance of the glass envelope enclosing the receiver tube. The factor K accounts for the fact that irradiance is not normal to the PTC aperture (unless the array has two-axis tracking). The incidence angle θ follows from the solar position model in Appendix A.

With the above quantities, the specific thermal power potential (in units of kW$_{th}$/m^2) is given by

$$\hat{q} = I_d \eta_{opt}/1000,$$

with I_d the direct-normal irradiance (DNI) at a particular moment in time and geographic location (in standard units of W/m^2). Note the DNI values I_d come from experimental measurements. These values are available for specific geographic locations in the US (and some international regions) for annual time horizons with hourly resolution (i.e., 8760 discrete hourly values), via the National Renewable Energy Laboratory's (NREL) National Solar Radiation Database (NSRDB) [41]. In this work, a geographic location is specified and the typical meteorological year (TMY) data is downloaded. The numerical values of the PTC modeling parameters used in this study can be found in Table A1 of Appendix B. Finally, the solar thermal power produced (in units of kW) by a solar array is given by

$$\dot{q}_s = \hat{q}x_a, \tag{3}$$

with $x_a \in \mathbb{R}_+$ as the total aperture area of the solar array.

2.2. Process System Model

The process system of interest is represented by the block-flow diagram in Figure 2. Physically, the system consists of the solar energy collector as the source of solar thermal energy for the system, the TES, the IPH system which exists as an energy demand, and the conventional (fossil) energy source for times of inadequate solar energy.

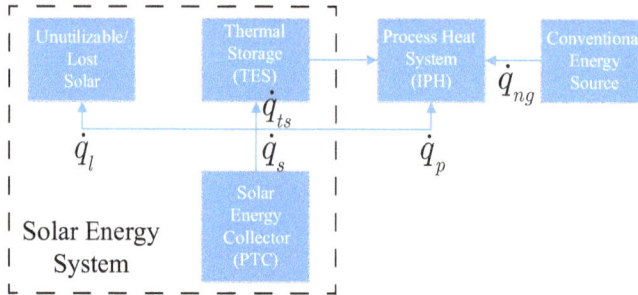

Figure 2. The block-flow diagram representation of the hybrid solar industrial process heat system being modeled.

The process system model is developed from a high-level with the following simplifying assumptions:

1. Heat losses to the environment from piping and heat exchangers is negligible.
2. Heat losses to the environment from the TES is negligible.
3. The process system requires low to medium temperature IPH (i.e., $\leq 250\ ^\circ$C).
4. Heat is always available at or above the minimum temperature as required by the IPH system.
5. The IPH system energy demand is not dependent on the state or design decisions of the solar energy system.

Under these assumptions, the process system model is developed from the overall energy balance on the entire system:

$$\dot{q}_s + \dot{q}_{ng} = \dot{q}_p + \dot{q}_{ts} + \dot{q}_l \tag{4}$$

where $\dot{q}_s$ is the instantaneous thermal power provided by the solar system, $\dot{q}_{ng}$ is the instantaneous thermal power provided by the conventional fossil-fuel (e.g., natural gas) heating system, $\dot{q}_p$ is the instantaneous thermal power demand by the industrial process (IPH), $\dot{q}_{ts}$ is the instantaneous power supply/demand of the solar thermal energy storage system, and $\dot{q}_l$ is the lost solar thermal power due to meeting the IPH requirements and reaching the maximum thermal storage capacity (i.e., the unutilizable solar energy). Note that $\dot{q}_{ts}$ can be both positive (charging state) and negative (discharging state).

The complicating details for accounting for energy in the system comes from the capacity limitations of the TES. For example, without TES, the solar array will simply deliver all its energy to the IPH system until the full power demand $\dot{q}_p$ is met. At which point, the solar array will simply de-focus, or we account for the positive difference between the supply and demand as losses or unutilizable solar $\dot{q}_l$. The negative difference between the supply and demand is simply accounted for by the conventional heating system $\dot{q}_{ng}$. However, with the existence of a TES device, the positive difference between solar thermal power supply and IPH demand can be stored and used in future times when the solar system alone cannot meet the IPH demand. This complicates the model since we assert a preference for solar energy over conventional energy sources, and hence must account for the finite capacity of the storage device and its stored capacity at each instant in time.

Rearranging Equation (4) to focus on the storage device, we get:

$$\dot{q}_{ts} + \dot{q}_a = \dot{q}_s - \dot{q}_p$$

where we have introduced $\dot{q}_a = \dot{q}_l - \dot{q}_{ng}$ for ancillary energy accounting. Physically, this quantity is used for load balancing control (i.e., balancing energy supply and demand). Equivalently, we write:

$$\frac{dq_{ts}}{dt} = \dot{q}_s - \dot{q}_p - \dot{q}_a, \quad q_{ts}(t) = 0, \ t \in [0, t^{\text{end}}],$$

assuming the TES device is empty at the start of the simulation. Applying the explicit Euler integration scheme with $h = \Delta t$ as the discrete time step-size (1 h for standard solar data), we can write:

$$q_{ts}^{i+1} = q_{ts}^i + h(\dot{q}_s^i - \dot{q}_p^i - \dot{q}_a^i), \quad q_{ts}^1 = 0. \tag{5}$$

Next, we develop the model to account for the finite capacity of the TES device. Since we know that the energy capacity of the TES device at any instant can only take values between zero and the maximum capacity, we write:

$$q_{ts}^{i+1} = \min\left\{\dot{q}_p^{\text{peak}} x_{ts}, \max\{0, q_{ts}^i + h(\dot{q}_s^i - \dot{q}_p^i)\}\right\}, \quad q_{ts}^1 = 0 \tag{6}$$

where x_{ts} is the storage capacity variable scaled in hours of peak process demand $\dot{q}_p^{\text{peak}} = \max_i \dot{q}_p^i$, determined by the process operations schedule model (known a priori). The inner max binary operation accounts for the condition that if there was excess energy available from the combined solar array and TES at the previous time, then we should consider storing it for future use. The outer min binary operation accounts for the condition that, if we are considering storing energy, we can only store at or below the maximum TES capacity. Equivalently, we can write

$$\min\left\{\dot{q}_p^{\text{peak}} x_{ts}, \max\{0, q_{ts}^i + h(\dot{q}_s^i - \dot{q}_p^i)\}\right\} \equiv \text{mid}\left\{\dot{q}_p^{\text{peak}} x_{ts}, 0, q_{ts}^i + h(\dot{q}_s^i - \dot{q}_p^i)\right\} \tag{7}$$

where the mid operator selects the median value of the three arguments. Combining Equations (5)–(7), the overall energy balance becomes:

$$
\begin{aligned}
h\dot{q}_a^i &= q_{ts}^i + h(\dot{q}_s^i - \dot{q}_p^i) - q_{ts}^{i+1} \\
&= q_{ts}^i + h(\dot{q}_s^i - \dot{q}_p^i) - \text{mid}\left\{\dot{q}_p^{\text{peak}} x_{ts}, 0, q_{ts}^i + h(\dot{q}_s^i - \dot{q}_p^i)\right\}.
\end{aligned}
\tag{8}
$$

From this model, there are three possible states from the load-balancing perspective at each discrete time point i:

1. $(\dot{q}_s^i < \dot{q}_p^i) \wedge (q_{ts}^i < h(\dot{q}_s^i - \dot{q}_p^i)) \Leftrightarrow \dot{q}_a^i < 0$

   The PTC is not providing enough energy to meet the demand of the process (TES moves into the discharging state) and there is not enough energy in the TES to meet the demand of the process over this time step. In this case, conventional energy must be supplied to the process to meet the full demand of the process.

2. $(\dot{q}_s^i > \dot{q}_p^i) \wedge (\dot{q}_p^{\text{peak}} x_{ts} - q_{ts}^i < h(\dot{q}_s^i - \dot{q}_p^i)) \Leftrightarrow \dot{q}_a^i > 0$

   The PTC is providing enough energy to meet the demand of the process (TES moves into the charging state) and there is not enough available capacity to store all excess energy in the TES device. In this case, the excess energy must be rejected.

3. $\left((\dot{q}_s^i < \dot{q}_p^i) \wedge (q_{ts}^i \geq h(\dot{q}_s^i - \dot{q}_p^i))\right) \vee \left((\dot{q}_s^i \geq \dot{q}_p^i) \wedge (\dot{q}_p^{\text{peak}} x_{ts} - q_{ts}^i \geq h(\dot{q}_s^i - \dot{q}_p^i))\right) \Leftrightarrow \dot{q}_a^i = 0$

   The TES is either in the discharging state with enough stored energy to meet the full demand, or it is in the charging state and has enough available capacity to store the solar energy in excess of the process demand.

With these relationships, we can write:

$$\dot{q}_l^i = \max\{0, \dot{q}_a^i\} \text{ and } \dot{q}_{ng}^i = -\min\{0, \dot{q}_a^i\}, \tag{9}$$

so that $\dot{q}_l$ and $\dot{q}_{ng}$ are accurately accounted for. At any given discrete time point, we have the following relationship for load balancing:

$$\dot{q}_l^i = \dot{q}_{ng}^i = 0 \lor (\dot{q}_l^i > 0 \Rightarrow \dot{q}_{ng}^i = 0) \lor (\dot{q}_{ng}^i > 0 \Rightarrow \dot{q}_l^i = 0).$$

Note that losses are accounted for here as simply the unutilizable solar power at any given time step and do not account for standard thermal losses to the environment which occur in thermal systems. As stated previously, heat losses for the TES are expected to be minimal over the timescale of charging/discharging cycles. This is valid for well-insulated systems that undergo complete or nearly-complete discharge over the course of the day. For systems that have long-term (i.e., many days or months) storage, this assumption will not be valid. Further, TES systems to be installed outdoors and above-ground may need special consideration of ambient environmental/weather conditions as large temperature swings and high winds may invalidate this assumption as significant heat transfer with the environment may occur. In all cases, some heat losses are expected during the course of operation of the physical systems which will result in an under-sized design proportional to the actual heat losses, under these assumptions.

Additionally, heat losses from piping and heat exchangers are expected to be minimal. Similarly, these assumptions are valid for well-insulated systems. To account for any of these losses, the design engineer may apply higher-fidelity models that take into account operating temperatures, heat exchanger designs, and overall heat transfer coefficients for the system(s) of concern. However, doing so at this stage significantly complicates the model from an optimization perspective and negatively impacts computational tractability. Alternatively, high-fidelity modeling may be used to estimate average losses for a particular geographic region and IPH temperature requirements. These losses can then be accounted for as an overall thermal efficiency and used to adjust the IPH demand accordingly, similar to the factor ϵ_6 in the optical efficiency model for the CST array in Equation (1). Finally, to assess the performance of a CST system with respect to its ability to augment the use of conventional energy sources for IPH, we define the annual average solar fraction:

$$SF_c = \int_0^{t_{end}} \frac{\dot{q}_s - \dot{q}_l}{\dot{q}_p} dt = 1 - \int_0^{t_{end}} \frac{\dot{q}_{ng}}{\dot{q}_p} dt.$$

In terms of the discrete values calculated from Equation (8) and the identities in Equation (9), the solar fraction can be approximated numerically using:

$$SF = \frac{\sum_i h(\dot{q}_s^i - \dot{q}_l^i)}{\sum_i h\dot{q}_p^i} = 1 - \frac{\sum_i h\dot{q}_{ng}^i}{\sum_i h\dot{q}_p^i} \tag{10}$$

which represents the fraction of total energy consumed by the process that is supplied by the solar system over the entire time interval of interest (e.g., a TMY).

Lastly, although it was assumed that $\dot{q}_p$ is not a function of the state or design decisions of the solar energy system, it need not be constant. For the purposes of studying the effects of transient load demand on the solar hybridization strategy, we will consider the model:

$$\dot{q}_p^i = \bar{q}_p \left(1 + \sigma_p \sin\left[\pi \frac{i-7}{12}\right]\right), \tag{11}$$

which assumes i is in hours ($h = 1$) with $i = 1$ corresponding to 1:00 a.m. on 1 January, $\bar{q}_p$ is the mean IPH demand, and $\sigma_p \in [0,1]$ is the fractional deviation from the mean IPH demand. For $\sigma_p = 0$, this models a constant IPH demand. For $\sigma_p > 0$, this model accounts for a dynamic ramp-up and ramp-down of demand with a maximum demand at local noon and a minimum demand at local midnight.

To use the developed models effectively within a simulation-based optimization framework, we present the following concavity result for SF.

Theorem 1. *Let $X_a, X_{ts} \subset \mathbb{R}_+$ be nonempty compact intervals. The solar fraction $SF : X_{ts} \times X_a \to \mathbb{R}$ is a concave function on its domain.*

Proof. From Equations (7)–(9) and the feasible system states, we can write

$$h\dot{q}_l^i = \max\left\{0, \dot{q}_{ts}^i + h(\dot{q}_s^i - \dot{q}_p^i) - \dot{q}_p^{\text{peak}} x_{ts}\right\}$$

for the $\dot{q}_a^i > 0$ case (or else $\dot{q}_l^i = 0$). The physical meaning of this case is that the solar system is producing more thermal power than can be used by the process or stored in the TES, and hence lost/rejected. We can see that $h(\dot{q}_s^i - \dot{q}_p^i)$ is an affine function of x_a and

$$\dot{q}_{ts}^i - \dot{q}_p^{\text{peak}} x_{ts} = \text{mid}\{\dot{q}_p^{\text{peak}} x_{ts}, 0, h(\dot{q}_s^{i-1} - \dot{q}_p^{i-1})\} - \dot{q}_p^{\text{peak}} x_{ts}$$

is a convex function of x_{ts} and x_a. Therefore,

$$\dot{q}_{ts}^i + h(\dot{q}_s^i - \dot{q}_p^i) - \dot{q}_p^{\text{peak}} x_{ts}$$

is an affine function of x_{ts} and x_a. Since $\dot{q}_l^i$ is the pointwise maximum of an affine function and a constant, it is convex. From Equation (10), we have the summation of the terms:

$$\dot{q}_s^i - \dot{q}_l^i$$

which is an affine function of x_a subtracting a convex function of x_{ts} and x_a, and hence concave. Finally, since SF in Equation (10) is a summation of concave functions (the denominator is not a function of x_{ts} or x_a), it is concave. □

2.2.1. Smooth Approximation

Since the energy balance equations from the previous section involve min and max operators, differentiability of SF as defined in Equation (10) cannot be guaranteed. Since gradient-based optimization algorithms typically require all functions to be at least once-continuously differentiable, we must develop a smooth formulation. Consider the following operator:

$$\max_\epsilon\{0, z\} \equiv \frac{1}{2}\left(z + \sqrt{z^2 + \epsilon}\right), \ z \in \mathbb{R}, 0 < \epsilon \in \mathbb{R}$$

as a smooth approximation of the binary max operator: $\max\{0, z\}$. The maximum error between the max operator and $\max_\epsilon$ is $\sqrt{\epsilon}/2$ and occurs at $z = 0$. For $|z| \gg \epsilon$, the error approaches zero. Hence, for small ϵ values, this approximation is not expected to introduce appreciable error into the model. A smooth approximation of SF, as defined in Equation (10), is given by:

$$SF_s = \frac{\sum_i(\dot{q}_s^i - \max_\epsilon\{0, \dot{q}_{ts}^i + h(\dot{q}_s^i - \dot{q}_p^i) - \dot{q}_p^{\text{peak}} x_{ts}\})}{\sum_i \dot{q}_p^i}. \tag{12}$$

Again, the motivation of the model development is for use within simulation-based (global) optimization. The following result ensures that the smooth approximation SF_s is concave and differentiable.

Theorem 2. *Let* $X_a, X_{ts} \subset \mathbb{R}_+$ *be nonempty compact intervals. Then, the smooth solar fraction* $SF_s : X_{ts} \times X_a \to \mathbb{R}$ *is concave and continuously differentiable.*

Proof. Let $g_i : \mathbb{R}^2 \to \mathbb{R} : (x_{ts}, x_a) \mapsto \dot{q}^i_{ts} + h(\dot{q}^i_s - \dot{q}^i_p) - \dot{q}^{\text{peak}}_p x_{ts}$ for some i, and $f : \mathbb{R} \to \mathbb{R} : z \mapsto \min_\epsilon \{0, z\}$. It is clear that f is convex on $\mathbb{R}$ and g_i is affine. Therefore, $\dot{q}^i_l = f \circ g_i$ is convex for every i. Further, both f and g_i are continuously differentiable on their domains. Therefore, $\dot{q}^i_l = f \circ g_i$ is continuously differentiable on $\mathbb{R}^2$. Since SF is defined as the sum of the terms

$$\dot{q}^i_s - \dot{q}^i_l,$$

with a constant denominator (with respect to x_{ts} and x_a), it is concave and continuously differentiable on $\mathbb{R}^2$. □

2.3. Economic Model

An economic analysis is at the heart of all new-technology investment decision-making. Therefore, successfully designing a solar-IPH hybridization strategy requires a thorough economic analysis to ensure an optimal venture. Often, such an analysis consists of a comparison of conventional energy prices and the levelized cost of electricity (LCOE) for electrical power systems or the levelized cost of heat (LCOH) for thermal systems. For IPH applications, the LCOH analysis is often appropriate. However, the LCOH is defined at a high-level as:

$$LCOH \equiv \frac{\text{sum of costs over lifetime}}{\text{sum of solar heat utilized}},$$

which cannot be guaranteed to be convex under appropriate assumptions. In contrast, we will develop the economic model of total lifecycle savings (LCS) which represents the total cost savings (which we wish to maximize) realized by hybridizing the IPH system with CSP.

The lifecycle savings is defined as the difference between the total lifetime cost of energy for a conventional system and that of the hybridized system, accounting for the time-value of money and energy-price inflation:

$$f^k_{LCS}(\mathbf{x}) = \sum_{i=1}^{t_{life}} \frac{SF_s(x_{ts}, x_a)C_{p,i} - C^k_{cap,i}(x_{ts}, x_a) - C_{om,i}(x_{ts}, x_a)}{(1+r)^i}, \tag{13}$$

where $\mathbf{x} = (x_{ts}, x_a)$ is simply the vector-form of our decision variables, t_{life} is the total lifetime of the project in years, r is the discount rate, $C_{p,i}$ is the cost of conventional fuel at year i, $C^k_{cap,i}$ is the annualized cost of capital including debt service at year i for financing model k, and $C_{om,i}$ is the annual operating and maintenance (O&M) cost of the CST-TES system at year i. The solar hybridization design objective then is to maximize f^k_{LCS}, and hence, it is ideal if f^k_{LCS} is concave on its domain. The optimization problem formulation will be discussed in the next section.

The cost of conventional fuel at year i is given by:

$$C_{p,i} = C_{th,0}(1 + r_{th})^{i-1} \sum_j h\dot{q}^j_p,$$

where $C_{th,0}$ is the current specific thermal energy cost and r_{th} is an inflation rate on the fuel price. The O&M cost is typically estimated under a fixed-pricing model [42] (i.e., cost per kWh of thermal energy produced).

The annualized capital expenditure and debt service of the CST-TES system at year i is given by [32]:

$$C^k_{cap,i}(x_{ts}, x_a) = \begin{cases} r_c C^k_{cap,0}(x_{ts}, x_a) \left(1 + \frac{r_c}{12}\right)^{12t_{debt}} \left(\left(1 + \frac{r_c}{12}\right)^{12t_{debt}} - 1\right)^{-1}, & i \le t_{debt} \\ 0, & i > t_{debt} \end{cases}$$

which assumes monthly-compounded interest at an annual rate of r_c, over the loan term t_{debt} (in years). The term $C_{cap,0}^k$ is the total capital expenditure (debt principal), which can be calculated using different pricing models k. The most common equipment pricing model has the form:

$$C_{cap,0}^{disc}(x_{ts}, x_a) = C_{PTC,0}x_a^{0.92} + C_{TES,0}(\dot{q}_p^{peak}x_{ts})^{0.91} \tag{14}$$

which accounts for discounts in volume pricing (economies of scale). The factors $C_{PTC,0}$ and $C_{TES,0}$ are cost parameters which can be found in Table A2. The exponents of x_a and x_{ts} and the cost parameters are derived from various vendor quotes for commercially-available and custom-built equipment. Note, this model is concave in the variables (x_{ts}, x_a). Since this term is subtracted within f_{LCS}^{disc}, concavity of f_{LCS}^{disc} cannot be guaranteed using this model, in general, which complicates solving the optimal design problem (i.e., global optimization is required).

Alternatively, a fixed-pricing (linear) model can be used:

$$C_{cap,0}^{fix}(x_{ts}, x_a) = x_a C_{PTC} + x_{ts}\dot{q}_p^{peak}C_{TES}, \tag{15}$$

where C_{PTC} is the specific installed cost of the solar array ($/m^2 aperture area) and C_{TES} is the specific installed cost of thermal energy storage ($/kWh). Taking C_{PTC} and C_{TES} as a nominal specific cost for projects at a scale relevant to the design problem will yield results that are sufficiently accurate for initial design feasibility purposes but could lack the accuracy necessary for detailed design and economic analysis. Further, without a priori knowledge of the expected scale of a design, accurate cost estimates cannot be guaranteed with this model.

The following result details the construction of the *convex envelope* of the concave discounted equipment pricing model in Equation (14). The convex envelope is the tightest convex lower-bounding function that can be constructed for any nonconvex function on a domain of interest. The importance here is that it provides a rigorous lower bound on Equation (14) over its domain of interest while providing the best-possible convex approximation of Equation (14) over that domain.

Theorem 3. *Consider the interval domain* $X \subset \mathbb{R}_+^2$ *with* $X = X_{ts} \times X_a = [x_{ts}^L, x_{ts}^U] \times [x_a^L, x_a^U]$. *The convex envelope of Equation (14) on* X *is given by*

$$C_{cap,0}^{cv}(x_{ts}, x_a) = \sum_{i \in \{a,ts\}} \left(f_i(x_i^L) + \frac{f_i(x_i^U) - f_i(x_i^L)}{x_i^U - x_i^L}(x_i - x_i^L) \right), \tag{16}$$

with $f_a(x_a) = C_{PTC,0}x_a^{0.92}$ *and* $f_b(x_{ts}) = C_{TES,0}x_{ts}^{0.91}$.

Proof. Since $C_{cap,0}^{disc}$ is separable, we can write $C_{cap,0}^{disc} = f_{ts}(x_{ts}) + f_a(x_a)$. Further, its convex envelope can be constructed as the sum of the convex envelopes of the separable factors f_{ts} and f_a [43]. The convex envelopes of these (univariate) factors are calculated simply as the secant lines between their endpoints:

$$f_i^{cv} = f_i(x_i^L) + \frac{f_i(x_i^U) - f_i(x_i^L)}{x_i^U - x_i^L}(x_i - x_i^L), \ i \in \{a, ts\}. \tag{17}$$

□

Corollary 1. *Let* $x_{ts}^L = x_a^L = 0$, $C_{PTC} = C_{PTC,0}(x_a^U)^{-0.08}$ *and* $C_{TES} = C_{TES,0}(x_{ts}^U\dot{q}_p^{peak})^{-0.09}$. *Then (15) is the convex envelope of Equation (14) on* X.

Proof. Since $x_{ts}^L = x_a^L = 0$, we have $f_{ts}(x_{ts}^L) = f_a(x_a^L) = 0$ and Equation (16) reduces to

$$C_{cap,0}^{cv} = \sum_{i \in \{a,ts\}} \left(\frac{f_i(x_i^U)}{x_i^U} x_i \right) = C_{PTC,0} \frac{(x_a^U)^{0.92}}{x_a^U} x_a + C_{TES,0} \frac{(x_{ts}^U \dot{q}_p^{peak})^{0.91}}{x_{ts}^U} x_{ts}$$

$$= C_{PTC,0}(x_a^U)^{-0.08} x_a + C_{TES,0}(x_{ts}^U \dot{q}_p^{peak})^{-0.09} x_{ts} \dot{q}_p^{peak}$$

$$= x_a C_{PTC} + x_{ts} \dot{q}_p^{peak} C_{TES}$$

$$= C_{cap,0}^{fix}.$$

□

From the above results, we have the following relationships between the models:

$$C_{cap,0}^{fix} \le C_{cap,0}^{cv} \le C_{cap,0}^{disc} \Rightarrow f_{LCS}^{disc} \le f_{LCS}^{cv} \le f_{LCS}^{fix}. \tag{18}$$

for f_{LCS}^{fix} constructed via Corollary 1.

2.4. Optimization

The solar hybridization optimal design problem is formulated as a constrained nonlinear program (NLP) using the economic objective in Equation (13) developed in the previous section. The model is formulated as

$$f_k^* = \max_{x \in \mathbb{R}^2} f_{LCS}^k(x)$$

$$\text{s.t. } g(x) \le 0 \tag{19}$$

$$x \in X = \{x \in \mathbb{R}_+^2 : x^L \le x \le x^U\}$$

where $g : \mathbb{R}^2 \to \mathbb{R}^{n_c}$ are n_c performance constraints that we may include, and X is simply a two-component interval vector with x^L and x^U as lower- and upper-bounds on the variable vector $x = (x_{ts}, x_a)$, respectively. In this work, we define only one constraint:

$$g(x) = \xi - SF_s(x_{ts}, x_a) \le 0, \ \xi \in [0,1]$$

where ξ represents some minimum fraction of the total IPH energy supply that must come from solar (e.g., a social responsibility constraint or a tax-incentive constraint). However, this general formulation can accommodate any relevant set of constraints. The parameters of the optimization model in Equation (19) can be found in Table A2.

Theorem 4. *The constrained NLP in Equation* (19) *with the inequality constraint* $g(x) = \xi - SF_s(x_{ts}, x_a) \le 0$ *and* $k \in \{cv, fix\}$ *is a convex program.*

Proof. Since SF_s is concave (Thm. 2) and $C_{cap,i}^k$ is convex for $k \in \{cv, fix\}$ and for all i (Corollary 1), f_{LCS}^k, $k \in \{cv, fix\}$ is concave by design. Further, since SF_s is concave, the function $g(x) = \xi - SF_s(x_{ts}, x_a)$ is convex. Therefore, the feasible set $\mathcal{F} = \{x \in \mathbb{R}^2 : g(x) \le 0, x^L \le x \le x^U\}$ is convex. Since Equation (19) is a concave maximization problem on a convex feasible set, it is a convex NLP. □

Note that, from Equation (18), we have the following relationship between the optimal solution values:

$$f_{disc}^* \le f_{cv}^* \le f_{fix}^*$$

for f_{LCS}^{fix} constructed via Corollary 1. Therefore, we can use the fixed-pricing model in Equation (15) for initial high-level economic feasibility studies, which will overestimate the value of the more precise discount-pricing model in Equation (14). Further, given the result of Theorem 3, we can use Equation (19) with $k = cv$ as a valid upper-bounding problem and solve the nonconvex program in Equation (19) with $k = disc$ to global optimality by applying a spatial Branch-and-Bound algorithm (see Appendix C). Therefore, for greater-detail engineering design and economic feasibility studies, a global optimum of Equation (19) with $k = disc$ can be readily obtained using the models presented herein.

3. Numerical Experiments and Results

The models presented above were implemented in the Julia programming language [44] using JuliaPro v.0.6.2.2 (Cambridge, MA, USA). Three disparate geographic locations across the US were considered as test cases for this model: Firebaugh, CA (California's Central Valley), Aurora, CO (Greater Denver Area), and Weston, MA (Greater Boston Area). The TMY hourly data for each location was obtained from the NREL NSRDB [41] (Physical Solar Model v.3). The monthly and annual average DNI values for each location can be found in Table A3. Note that the full TMY hourly dataset was used in this study. Two IPH demand profiles (i.e., constant and periodic) with equivalent total daily demand were considered for each geographic location, as were two fuel costs (i.e., commercial rate and industrial rate from [45]). For all studies, O&M cost was omitted ($C_{om,i} = 0$) to simply illustrate the trade-off between conventional fossil energy consumption cost and solar capital costs. Additionally, for each study, the fixed-pricing model in Equation (15) calculated via Corollary 1 and the discount-pricing model in Equation (14) were used.

An optimal hybridization design study was conducted for each of the three geographic locations, for each case below. All problems were solved on a personal workstation running Windows 10 v1803 operating system with an Intel Core i7-5960X CPU and 32GB of RAM. All local optimization was performed using the interior-point algorithm IPOPT [46] from within the JuMP modeling platform [47] v0.18.0 utilizing forward-mode automatic differentiation for exact derivative information. Since f_{LCS}^{fix} is concave on its domain, IPOPT is effective in solving Equation (19) with $k = fix$ (constructed using Corollary 1) to global optimality. Since f_{LCS}^{disc} is nonconvex on its domain, a global optimum cannot be guaranteed using IPOPT (or any local optimizer) alone. In this case, the problem is solved to global optimality using the the the deterministic spatial Branch-and-Bound (B&B) algorithm in the EAGO software package [48,49] v0.1.0 with a relative convergence tolerance of $\epsilon = 10^{-2}$ and the greatest relative-width midpoint branching heuristic. The B&B algorithm for maximization is presented in Algorithm A1 of Appendix C with the concept of branching illustrated in Figure A1. The concave upper-bounding problem of Equation (19) with $k = disc$ is simply given by $k = cv$. A valid upper bound is obtained at each iteration by solving the upper-bounding problem on the domain of interest using the IPOPT algorithm from JuMP. A valid lower bound is obtained by solving Equation (19) with $k = disc$ to local optimality using IPOPT from JuMP. A global optimum is provided for each case for comparison against the optimal solution of the fixed-pricing model.

3.1. Constant IPH Demand, Commercial Fuel Rate

In this study, the IPH demand was modeled using Equation (11) with $\bar{q}_p = 10^5 \text{kW}_{th}$ and $\sigma_p = 0$. Hence, we have a constant IPH demand: $\dot{q}_p^{peak} = \bar{q}_p = 10^5 \text{kW}_{th}$. The fuel cost was set at the commercial rate [45] in Table A2 and we let $\zeta = 0$.

Table 1 contains the location information and corresponding optimal solar-IPH hybridization strategies (i.e., the optimal solutions of Equation (19)) for both the linear fixed-pricing model in Equation (15) and the nonconvex discount-pricing model in Equation (14). For each location, the fixed-pricing model determines an optimal design that has a lower TES capacity and less PTC aperture area (and therefore lower annual solar fraction) than the global optimal solution. The fixed-pricing model overestimates the total lifecycle savings by 6.5% for Weston, 1.7% for Aurora, and 1.6% for Firebaugh. For initial feasibility studies, this overestimation is sufficiently small as the optimal

TES capacity and PTC aperture area are within roughly 5% of the global optimal designs for the discount-pricing model.

Table 1. The optimal solar-IPH hybridization strategies for the three geographic locations for constant IPH demand with $\bar{q}_p = 10^5$ kW$_{th}$ for the commercial fuel rate. The global optimal results for the concave capital cost model in Equation (14) are included for comparison against the linear capital cost model in Equation (15).

City, State	Lat., Long.	Time Zone	Capital Model	x_{ts}^* (h)	x_a^*(m^2)	SF_s^*	f_k^*
Weston, MA	42.37, −71.30	−5	$k = fix$ (Linear), Equation (15)	11.24	46,436.4	0.516	$2.976 M
			$k = disc$ (Concave), Equation (14)	11.72	48,229.1	0.523	$2.795 M
Aurora, CO	39.73, −104.66	−7	$k = fix$ (Linear), Equation (15)	13.85	50,618.1	0.721	$6.487 M
			$k = disc$ (Concave), Equation (14)	14.58	53,238.5	0.731	$6.377 M
Firebaugh, CA	36.85, −120.46	−8	$k = fix$ (Linear), Equation (15)	11.42	42,601.7	0.695	$7.533 M
			$k = disc$ (Concave), Equation (14)	11.72	43,615.2	0.698	$7.320 M

The monthly thermal performance profiles for the global optimal designs are shown in Figure 3. Both Aurora and Weston experience less seasonal variance in DNI than Firebaugh (see Table A3) and so the thermal performance profiles appear flatter over the year than those for Firebaugh. Even though the annual average DNI for Aurora and Firebaugh are within about 8% of each other, they differ by as much as 36% in the summer months. Hence, the optimal designs for Firebaugh are significantly smaller and provide significantly more lifecycle savings than those for Aurora. Interestingly, the optimal TES capacity for Weston and Firebaugh are very similar, whereas the PTC aperture area for Weston sits between Firebaugh and Aurora. Clearly, the widely differing DNI profiles has a dramatic influence on the optimal designs.

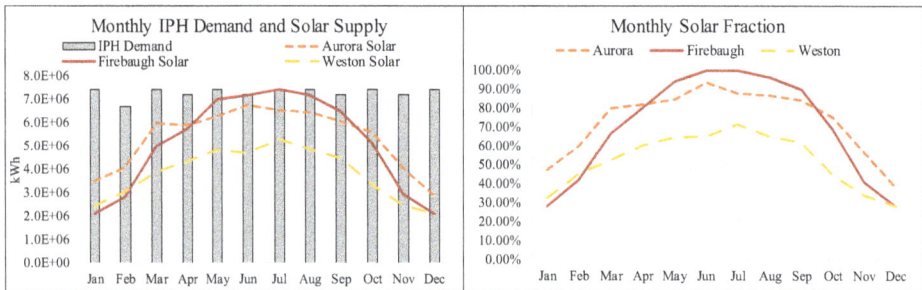

Figure 3. The thermal performance of the global optimal hybridization strategies for constant IPH demand and the commercial initial specific fuel cost.

The f_{LCS}^{fix} and SF_s functions are plotted in Figure 4 for each geographic location for comparison. Each of the surfaces are qualitatively similar but there are clear differences between the geographic regions in terms of how sharply the economics change as you move away from the optimal design. The lifecycle savings for Weston are much more sensitive to the design decisions than Aurora and Firebaugh, which exhibit broader relatively-flat regions around the optimum. This explains why the optimal economics between the fixed-pricing and discount-pricing models are within less than 2% for Aurora and Firebaugh but almost 7% for Weston. It can be concluded from this analysis that designing solar-IPH hybridization strategies for lower-DNI regions may require the use of higher-accuracy cost models (e.g., discount-pricing models) with global optimization.

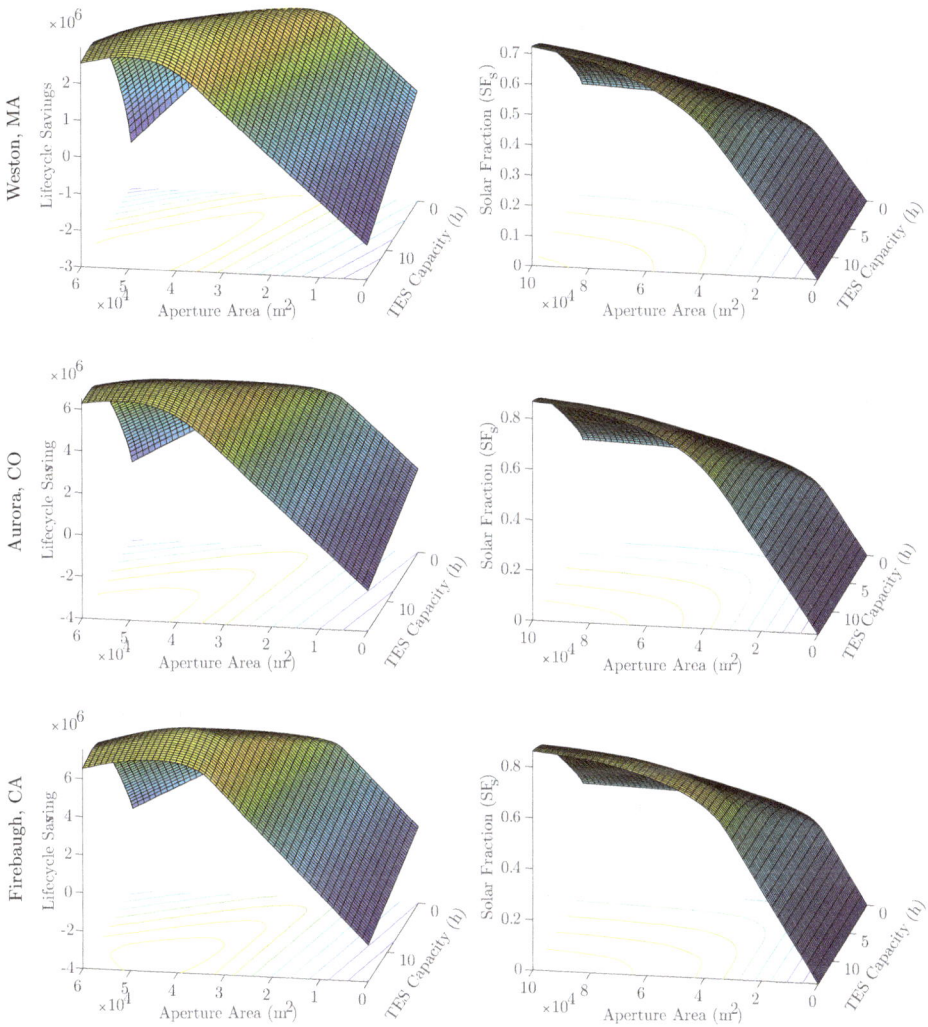

Figure 4. The objective function surfaces f_{LCS}^{fix} (**left column**) and the solar fraction surfaces SF_s (**right column**) for constant IPH demand for commercial fuel rate show similar qualitative trends between each region (rows).

3.2. Periodic IPH Demand, Commercial Fuel Rate

In this study, the IPH demand is modeled using Equation (11) with $\bar{q}_p = 10^5$ kW$_{th}$ and $\sigma_p = 0.1$. Hence, this case ramps up thermal power demand by 10% over the mean at the height of the workday ($\hat{q}_p^{peak} = 11$ MW$_{th}$) and then ramps down to 10% below the mean by midnight to simulate reduced staffing and production output overnight. We consider the commercial rate for fuel [45] and we let $\zeta = 0$.

Table 2 contains the location information and corresponding optimal solar-IPH hybridization strategies (i.e., the optimal solutions of Equation (19)) for both the linear fixed-pricing model in Equation (15) and the nonconvex discount-pricing model in Equation (14). As expected, the optimal designs have lower TES capacities than the constant IPH demand since the periodic demand is relatively synchronized with the sunrise and sunset. Further, the optimal PTC aperture areas are very similar to the constant IPH demand case since the total energy consumption between the two cases are the same. Note, the TES capacity x_{ts} is scaled against the peak IPH demand $\hat{q}_p^{peak}$. Hence, to compare the TES capacity between the periodic IPH case and the constant IPH case, we need to scale them using $\hat{q}_p^{peak}$, for each case. Thus, for the Weston fixed-pricing model and constant IPH demand, we have a TES capacity of 11.24 h $\times 10^5$ kW$_{th}$=1.124 MWh$_{th}$. For the Weston fixed-pricing model and periodic IPH demand, we have a TES capacity of 9.537 h $\times 1.1 \cdot 10^5$ kW$_{th}$=1.05 MWh$_{th}$. Thus, the TES capacity is roughly 7% larger for the constant IPH demand over the periodic IPH demand. The periodic IPH demand is clearly more economical from a solar hybridization perspective. However, without accounting for the economics of the process itself, one cannot conclude that it should be operated in such a periodic fashion to achieve overall-better economics. In other words, it may be more costly to operate the process in a periodic fashion than what is saved with an optimal solar hybrid design.

Table 2. The optimal solar-IPH hybridization strategies for the three geographic locations for periodic IPH demand with $\bar{q}_p = 10^5$ kW and $\sigma_p = 0.1$ for the commercial fuel rate. The global optimal results for the concave capital cost model in Equation (14) are included for comparison against the linear capital cost model in Equation (15).

City, State	Lat., Long.	Time Zone	Capital Model	x_{ts}^* (h)	x_a^*(m^2)	SF_s^*	f_k^*
Weston, MA	42.37,−71.30	−5	$k = fix$ (Linear), Equation (15)	9.537	46,248.8	0.507	$3.079 M
			$k = disc$ (Concave), Equation (14)	9.966	47,945.8	0.521	$2.881M
Aurora, CO	39.73,−104.66	−7	$k = fix$ (Linear), Equation (15)	11.92	50,006.8	0.706	$6.593 M
			$k = disc$ (Concave), Equation (14)	12.59	53,238.7	0.731	$6.460 M
Firebaugh, CA	36.85,−120.46	−8	$k = fix$ (Linear), Equation (15)	9.732	42,688.4	0.691	$7.641 M
			$k = disc$ (Concave), Equation (14)	9.966	43,615.2	0.698	$7.411 M

3.3. Constant IPH Demand, Industrial Fuel Rate

We consider the IPH demand modeled using Equation (11) with $\bar{q}_p = 10^5$ kW$_{th}$ and $\sigma_p = 0$ (i.e., constant demand profile). In this case, the industrial rate for fuel [45] is considered from Table A2 and we let $\xi = 0$. The objective function surfaces for both the fixed-pricing model ($k = fix$) and discount-pricing model ($k = disc$) are plotted in Figure 5. Table 3 contains the location information and corresponding optimal solar-IPH hybridization strategies (i.e., the optimal solutions of Equation (19)) for both the linear fixed-pricing model in Equation (15) and the nonconvex discount-pricing model in Equation (14).

Table 3. The optimal solar-IPH hybridization strategies for the three geographic locations for constant IPH demand with $\bar{q}_p = 10^5$ kW for the industrial fuel rate. The global optimal results for the concave capital cost model in Equation (14) are included for comparison against the linear capital cost model in Equation (15).

City, State	Lat., Long.	Time Zone	Capital Model	x_{ts}^* (h)	x_a^*(m^2)	SF_s^*	f_k^*
Weston, MA	42.37,−71.30	−5	$k = fix$ (Linear), Equation (15)	0.001	0.01	0.000	$920.2
			$k = disc$ (Concave), Equation (14)	0.001	0.01	0.000	$920.2
Aurora, CO	39.73,−104.66	−7	$k = fix$ (Linear), Equation (15)	0.001	15,720.7	0.250	$323.8 k
			$k = disc$ (Concave), Equation (14)	0.001	15,648.7	0.248	$95.59 k
Firebaugh, CA	36.85,−120.46	−8	$k = fix$ (Linear), Equation (15)	0.046	16,581.5	0.298	$698.1 k
			$k = disc$ (Concave), Equation (14)	0.001	16,424.3	0.295	$468.3 k

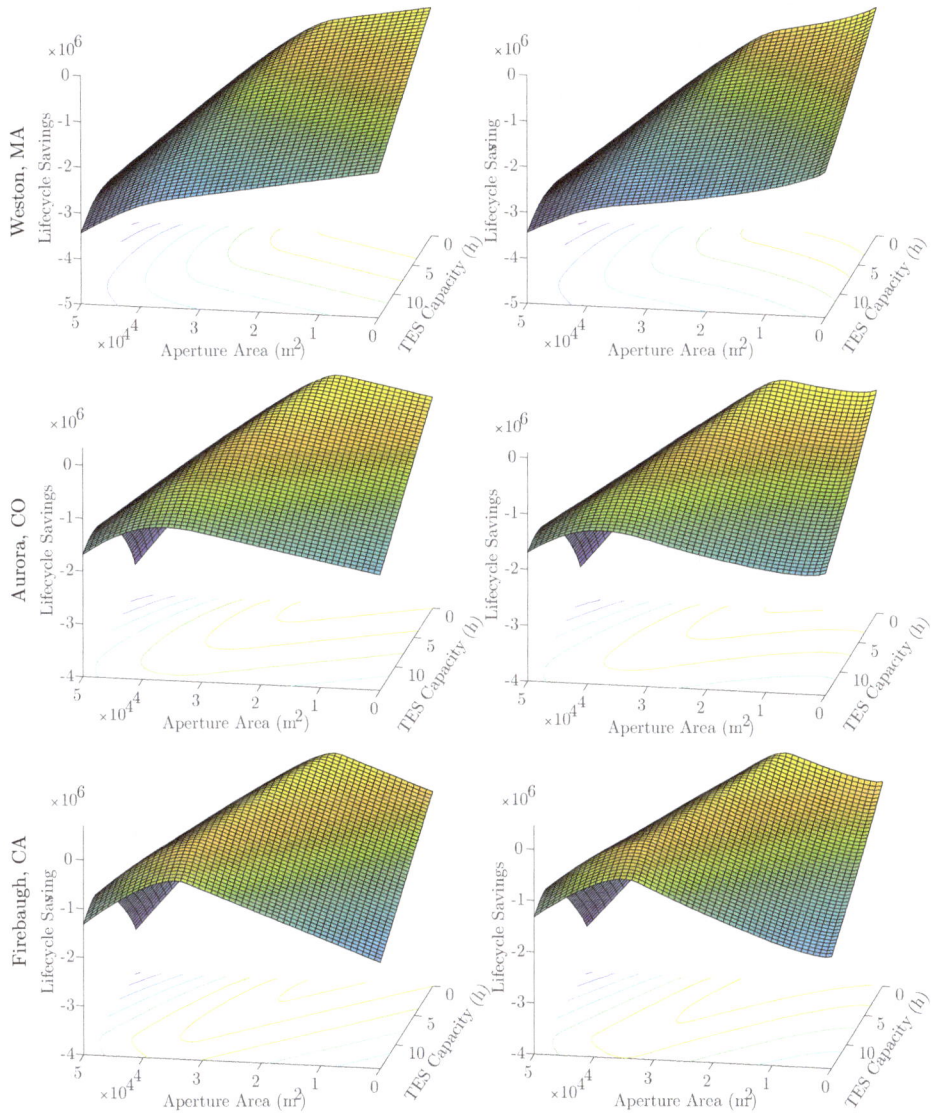

Figure 5. The objective function surfaces f_{LCS}^{fix} (**left column**) and f_{LCS}^{disc} (**right column**) for constant IPH demand for industrial fuel rates are plotted for each geographic location. The discount-pricing model exhibits nonconvexity in all cases and suboptimal local minima for Aurora and Firebaugh.

It is clear from Table 3 and Figure 5 that the lower fuel costs dramatically reduce the lifecycle savings, and therefore jeopardize the economic viability of the hybridization project. Further, the reduced fuel costs ensure that the capital cost terms have a much greater impact on the lifecycle savings. As a result, significant nonconvexity is observed for the discount-pricing model. In fact, for the Aurora and Firebaugh cases, there exists a suboptimal local minimum at $x = x^L = (x_{ts}^L, x_a^L) \approx 0$. IPOPT was applied to the nonconvex problem directly and only found the suboptimal local minima unless the domain was manually partitioned or bounded. Hence, without the application of the B&B algorithm (Algorithm A1), one may falsely conclude that solar-IPH hybridization is not economically viable. For the Weston, MA case, the fixed-pricing model concludes that the project is not economically viable, and hence, neither is the project with the discount-pricing model due to the upper-bounding relationship in Equation (18). Similar trends in the optimal solutions for the other locations are observed as for the previous cases. However, it is worth noting that for the Aurora and Firebaugh cases, the economics between the fixed-pricing and the discount-pricing models are significant. This is because significant nonconvexity is introduced for the lower fuel cost case. Hence, one can conclude that, for this low fuel cost case, the fixed-pricing model calculated via Corollary 1 over this large of a design space will not yield sufficiently accurate results. As a result, the discount-pricing model should be used with deterministic global optimization to ensure accurate results are obtained for feasibility studies.

3.4. Periodic IPH Demand, Industrial Fuel Rate

We consider the IPH demand modeled using Equation (11) with $\bar{q}_p = 10^5$ kW$_{th}$ and $\sigma_p = 0.1$ (i.e., fluctuating demand profile). In this case, the industrial rate for fuel [45] is considered from Table A2 and we let $\zeta = 0$.

Table 4 contains the location information and corresponding optimal solar-IPH hybridization strategies (i.e., the optimal solutions of Equation (19)) for both the linear fixed-pricing model in Equation (15) and the nonconvex discount-pricing model in Equation (14). The results are similar to the constant IPH demand case for the industrial fuel rate. Global optimization is required to locate the global optimal solution for the discount-pricing model case. Interestingly, the optimal solar design for the periodic IPH demand case is measurably larger in PTC aperture area over the constant IPH demand case for the same industrial fuel rate. This result differs from the commercial fuel rate cases which showed nearly identical aperture sizes between the designs. The results we see here is largely due to the very low annual solar fraction and the fact that the optimal designs call for the minimum TES capacity. Hence, the periodic IPH demand enables a greater solar fraction without adding TES capacity. However, as was identified for the constant IPH demand case, the fixed-pricing model dramatically overestimates the lifecycle savings due to the large degree of nonconvexity introduced by the discount-pricing model in the lower fuel cost case. As a result, it is still recommended for this case that the discount-pricing model is used with deterministic global optimization to ensure accurate results are obtained for feasibility studies.

Table 4. The optimal solar-IPH hybridization strategies for the three geographic locations for periodic IPH demand with $\bar{q}_p = 10^5$ kW for the industrial fuel rate. The global optimal results for the concave capital cost model in Equation (14) are included for comparison against the linear capital cost model in Equation (15).

City, State	Lat., Long.	Time Zone	Capital Model	x_{ts}^* (h)	x_a^* (m^2)	SF_s^*	f_k^*
Weston, MA	42.37,−71.30	−5	$k = fix$ (Linear), Equation (15)	0.001	0.01	0.000	$773.4
			$k = disc$ (Concave), Equation (14)	0.001	0.01	0.000	$773.4
Aurora, CO	39.73,−104.66	−7	$k = fix$ (Linear), Equation (15)	0.001	16,894.0	0.267	$350.2 k
			$k = disc$ (Concave), Equation (14)	0.001	16,879.1	0.267	$120.8 k
Firebaugh, CA	36.85,−120.46	−8	$k = fix$ (Linear), Equation (15)	0.033	17,972.1	0.320	$759.4 k
			$k = disc$ (Concave), Equation (14)	0.001	17,839.1	0.318	$529.4 k

3.5. Computational Efficiency

The solution times for each of the numerical experiments are provided in Table 5. The computational cost trade-off of using the fixed-pricing model versus the discount-pricing model is clear. In almost all cases, using the discount-pricing model requires an order-of-magnitude longer to solve the problem than the fixed-pricing model due to nonconvexity and the need for the B&B algorithm. Despite this, the solution times for the discount-pricing model are still favorable considering the model's scale with $\mathcal{O}(10^4)$ state variables. As previously mentioned, the Weston location with the lower industrial fuel rate (for both IPH demand profiles) yielded the result that the hybridization project would not be economically viable. As can be seen in Table 5, these are the only cases where solving the discount-pricing model requires roughly the same order of computational effort as solving the fixed-pricing model. Additionally, due to the relationship in Equation (18), one can immediately conclude from the solution of the fixed-pricing model that the project is not economically viable. Therefore, for the purposes of investment decision-making, an additional "economic feasibility" termination criterion can be added to Algorithm A1, whereby the algorithm simply terminates if the upper bound is below some threshold of economic viability. For the purposes of this optimal design problem, the optimal solution translates to not implementing any CSP, and hence, a termination criterion could also be added that identifies this case. With these termination criteria, the B&B algorithm (Algorithm A1) would be identical in solution time to the fixed-pricing model since it would terminate at the root node with the solution of economic infeasibility after solving only the upper-bounding problem. For such cases, $f_{LCS}^{cv} = f_{LCS}^{fix}$ if $x^L = 0$. Note, from Table A2, the lower bounds on the decision variables were not exactly zero, albeit very close with respect to their scale, which helps avoid some numerical round-off issues observed. Since $x^L \approx 0$, the results still hold.

Table 5. The solution times (in seconds) for each of the numerical experiments.

City, State	Section 3.1		Section 3.2		Section 3.3		Section 3.4	
	$k = fix$	$k = disc$	$k = fix$	$k = disc$	$k = fix$	$k = disc$	$k = fix$	$k = disc$
Weston, MA	1.09	10.4	0.632	13.3	0.103	0.42	0.103	0.39
Aurora, CO	0.70	6.69	0.495	6.40	0.187	15.5	0.208	19.4
Firebaugh, CA	0.70	10.1	1.06	9.92	0.604	12.77	0.517	12.67

4. Conclusions

A dynamic concentrating solar power performance model is presented for use with design optimization and investment decision-making problems for hybridizing IPH systems. The model utilizes high-fidelity solar resource data from the NSRDB [41] and accounts for the aperture area of a parabolic trough solar concentrator and the storage capacity of a thermal energy storage device as design decision variables. The solar performance model was constructed such that it is concave and differentiable on the domain of interest. Since determining the performance of the solar system requires a dynamic simulation (with 8760 time steps using hourly DNI data), concavity of the model on the decision domain ensures that it is efficient for use with mathematical optimization. Additionally, an economic model based on the lifecycle savings was constructed that utilizes the solar performance model as well as two capital cost models: a linear fixed-pricing model and a nonconvex discount-pricing model. The convex envelopes of the nonconvex discount-pricing model were presented, which, when used within a spatial Branch-and-Bound algorithm, enable the global solution of the economic optimization problem with the nonconvex discount-pricing model. For the higher-cost commercial fuel rate, optimization of the fixed-pricing model and the discount-pricing model show good agreement. However, for the lower-cost industrial fuel rate, significant nonconvexity complicates the problem and therefore the greater-accuracy discount-pricing model (and global optimization) is recommended for investment decision-making.

The economic model presented herein is relatively simple to illustrate the trade-off between the fuel savings and the capital cost of the project. Greater levels of detail can easily be incorporated in this model, such as taxes, incentives, capital depreciation, carbon pricing, detailed O&M, etc. Similarly, the solar thermal performance model assumes the simplifying characteristics of negligible heat losses, ideal temperature delivery, and negligible CST array shadowing. These details can be readily incorporated in this model, if necessary. The developed models may serve as the foundation for more greater-complexity model-building for systems performance and economic analyses. Specifically, the models may easily incorporate various sources of uncertainty and investment and/or thermal performance thresholds to address robust design and operations under uncertainty and new technology investment risk. However, if enhancing complexity, the user must consider how convexity/concavity and computational tractability are affected. For example, simple tax models and carbon pricing models may be incorporated as increased O&M costs without significantly affecting computational cost or convexity/concavity results. However, high-fidelity heat transfer models complicate the calculation of the annual solar fraction SF_s which substantially increase the integration cost and may introduce nonconvexity/nonconcavity. In this case, solving the optimization problem(s) will become extremely computationally expensive as convex and concave relaxations of SF_s must be introduced, ensuring many more iterations of the B&B algorithm are required to identify an optimal solution. Similarly, accounting for various sources of uncertainty may have equivalent effects. As additional modeling complexity may add significant computational burden, the models presented herein represent a balanced approach between computational cost and modeling complexity for the purposes of optimization problem tractability and sufficient accuracy of the results for engineering design and investment decision-making.

Funding: This research received no external funding.

Acknowledgments: Funding for this work was provided by the University of Connecticut.

Conflicts of Interest: The author declares no conflict of interest.

Abbreviations

The following abbreviations are used in this manuscript:

B&B	Branch-and-Bound
CSP	Concentrating Solar Power
CST	Concentrating Solar Thermal
IPH	Industrial Process Heat
LCS	Lifecycle Savings
NLP	Nonlinear Program
NREL	National Renewable Energy Laboratory
NSRDB	National Solar Radiation Database
O&M	Operations and Maintenance
PTC	Parabolic Trough Solar Concentrator
TES	Thermal Energy Storage
TMY	Typical Meteorological Year

Appendix A. Solar Position Model

Unless the CST technology of interest employs two-axis tracking, the center of the performance model is the incidence angle: the angle which the incoming solar radiation strikes the mirrors of the concentrator as measured from the normal of the concentrator aperture plane (see Figure 1). Since the sun's position in the sky is well-understood to a very high-degree of accuracy for a static global position, the incidence angle is readily calculable for every moment in time. With the incidence angle and the physical design of the CST technology, the specific thermal energy output can be calculated. The incidence angle calculation follows from [25,27,29,30,50] and the calculation of the solar position as a function of time. The model is included here for clarity and completeness.

Definition A1 (Equation of Time Factors, B, ST, TC [27]). *Due to the eccentricity of the Earth's orbit and tilt of its axis, the length of a day varies throughout the year. The variation is calculated by the* solar time *ST:*

$$B \equiv \frac{360}{365}(N - 81)$$
$$ST \equiv 9.87 \sin 2B - 7.53 \cos B - 1.5 \sin B$$

with *B* in degrees and *ST* in minutes and the integer *N* as the day number. To relate the solar time to the local time, a time correction, *TC* (in minutes), is required:

$$TC \equiv ST \pm 4[15(Time\ Zone\ (h)) - Longitude°]$$

where (+) is chosen if the location is east of the prime meridian and (−) if it is to the west.

Definition A2 (Declination Angle, *δ* [27]). *The solar declination angle δ is the angular distance of the sun's rays north (or south) of the equator:*

$$\delta \equiv 23.45° \sin B.$$

Definition A3 (Hour Angle, *γ* [27]). *The hour angle γ is the angle between the meridian of an observer and the sun. It is negative before local solar noon and positive after local solar noon. At local solar noon, γ = 0°. The hour angle (in degrees) is defined as*

$$\gamma \equiv 15(Local\ Time\ Hour + TC/60 - 12).$$

Definition A4 (Zenith Angle, *ϕ* [27]). *The solar zenith angle ϕ (in degrees) is the angle between the sun's rays and the vertical plane:*

$$\cos\phi = \sin[Latitude°]\sin\delta + \cos[Latitude°]\cos\delta\cos\gamma.$$

Definition A5 (Solar Elevation (Altitude) Angle, *α* [27]). *The solar elevation or altitude angle α (in degrees) is the angle between the sun's rays and the horizontal plane:*

$$\sin\alpha = \cos\phi \Rightarrow \phi + \alpha = \pi/2 = 90°.$$

Definition A6 (Azimuth Angle, *ζ* [27]). *The solar azimuth angle ζ is the angle between the sun's rays and due South for the Northern Hemisphere. If the sun rises and/or sets south of the East–West line (i.e., cos γ > tan δ/ tan[Latitude°]), then we have:*

$$\sin\zeta = \sin\zeta' = \frac{\cos\delta\sin\gamma}{\cos\alpha}.$$

For certain periods of the year and for certain locations, this may not be the case (i.e., cos γ ≤ tan δ/ tan[Latitude°]), and the azimuth angle becomes:

$$\zeta = \begin{cases} |\zeta'| - \pi & if\ \gamma \leq 0 \\ \pi - \zeta' & else \end{cases}$$

Definition A7 (Surface Tilt Angle, *β* [27]). *The surface tilt angle β of a horizontal surface with a N-S axis orientation and E-W tracking is given by:*

$$\tan\beta = \tan\phi\cos[90° - \zeta],\ \forall\gamma > 0$$

with the convention of −90° due East and 90° due West.

Definition A8 (Surface Incidence Angle, *θ* [25,27,50]). *The surface incidence angle θ (in degrees) is the angle between the sun's rays and the normal of the horizontal surface. For fixed N-S axis with E-W tracking, the incidence angle can be calculated as:*

$$\cos\theta = \sqrt{\cos^2\phi + \cos^2\delta\sin^2\gamma}.$$

Note, to maximize the collection of solar energy, the incidence angle must be minimized at all times. The extent to which this angle can be minimized depends on the orientation of the collector and the tracking capability of the collector (e.g., fixed non-tracking, one-axis tracking, and two-axis tracking). Other orientations and tracking capabilities are discussed further in [27].

Appendix B

Table A1. The PTC modeling parameters.

ϵ_1, Shadowing Factor	0.98
ϵ_2, Tracking Error	0.994
ϵ_3, Geometry Error	0.98
ϵ_4, Mirror Dirt Factor	$0.88/r_m \approx 0.941$
ϵ_5, Glass Receiver Envelope Dirt Factor	$(1+\epsilon_4)/2 \approx 0.971$
ϵ_6, Unaccounted Losses	0.96
r_m, Mirror Reflectance	0.935
a_m, Receiver Tube Absorptance	0.94
τ, Glass Receiver Envelope Transmittance	0.963

Table A2. The economic modeling and optimization parameters.

$C_{PTC,0}$, specific installed cost of PTC ($/m$^{1.84}$)	425.0
$C_{TES,0}$, specific installed cost of TES ($/kWh$_{th}^{0.91}$)	45.14
$C_{th,0}$, commercial initial specific fuel cost ($/kWh$_{th}$)	$7.232/293.1 \approx 0.02467$
industrial initial specific fuel cost ($/kWh$_{th}$)	$3.420/293.1 \approx 0.01167$
r, discount rate (-)	10%
r_c, capital financing APR (-)	6.5%
r_{th}, fuel cost annual inflation rate (-)	1%
t_{debt}, loan term (years)	10
t_{life}, lifetime of project (years)	30
$\mathbf{x}^L = (x_{ts}^L, x_a^L)$, variable lower bound (h, m^2)	$(0.001, 0.01)$
$\mathbf{x}^U = (x_{ts}^U, x_a^U)$, variable upper bound (commercial fuel case) (h, m^2)	$(16.0, 6 \times 10^4)$
variable upper bound (industrial fuel case) (h, m^2)	$(14.0, 5 \times 10^4)$
ζ, minimum solar fraction (-)	0

Table A3. The annual average DNI (kWh/m^2/day) for each location is resolved by month. The data are calculated from the hourly TMY values obtained from [41].

Month	Weston	Aurora	Firebaugh
Janary	3.73	4.68	3.21
Febrary	4.41	5.25	4.24
March	4.61	6.36	6.07
April	5.31	6.42	7.03
May	5.33	6.72	8.68
June	5.32	7.98	10.16
July	5.60	7.14	9.54
August	5.37	6.82	9.30
Sepember	5.24	7.21	7.70
Octorber	4.13	6.33	6.54
November	3.72	5.44	4.50
December	3.39	4.08	3.36
Ann. Avg.	4.68	6.21	6.71

Appendix C. Spatial Branch-and-Bound (B&B) Global Optimization

The spatial B&B algorithm is a deterministic global optimization algorithm that is guaranteed to find a global optimum (within some numerical tolerance ϵ) of a nonlinear program or determine that the problem is infeasible in finitely-many iterations. The algorithm systematically partitions the decision space X, with a procedure referred to as branching. Then, upper- and lower-bounds on the global optima on each partition of X are obtained by solving corresponding optimization problems (or interval analysis calculations). A summary of the algorithm is provided here for solving maximization problems as in Equation (19). Here, we present the B&B algorithm (Algorithm A1) for solving Equation (19) with $k = disc$ to global optimality. The upper- and lower-bounding problems are defined in Equations (A1) and (A2), respectively. The concept of partitioning the decision space X is illustrated in Figure A1.

Definition A9 (Upper-Bounding Problem). *Consider an interval $X^i \subset X$. The upper-bounding problem is given by:*

$$\max_{x \in \mathbb{R}^2} f_{LCS}^{cv}(x) \tag{A1}$$

$$\text{s.t. } g(x) \leq 0$$

$$x \in X^i.$$

Definition A10 (Lower-Bounding Problem). *Consider an interval $X^i \subset X$. The lower-bounding problem is given by:*

$$\max_{x \in \mathbb{R}^2} f_{LCS}^{disc}(x) \tag{A2}$$

$$\text{s.t. } g(x) \leq 0$$

$$x \in X^i.$$

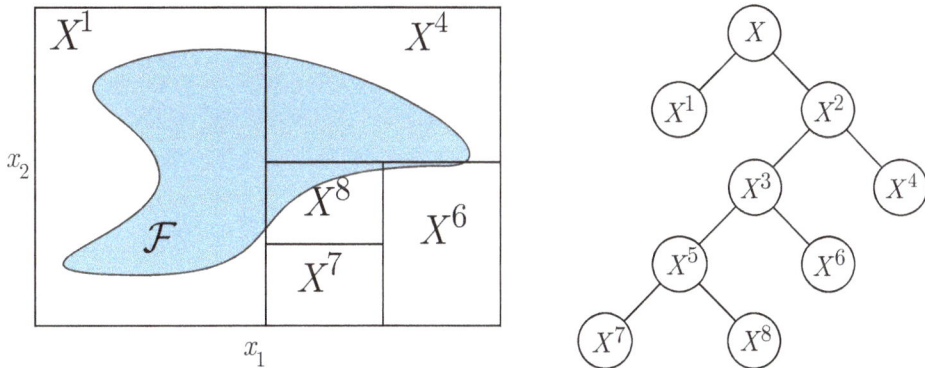

Figure A1. (**Left**) The decision space X (outer box) is shown overlaying the feasible set $\mathcal{F}$ and is partitioned into subintervals using a bisection strategy. Note that $X^7 \cap \mathcal{F} = \varnothing$ and therefore the corresponding upper- and lower-bounding problems are infeasible. (**Right**) An illustration of the B&B tree corresponding to the decision space partitioning.

Algorithm A1: Branch & Bound Maximization

$UBD, UBD_0 := +\infty, LBD := -\infty, \epsilon > 0, j := 1, S := \{X\}$
while $UBD - LBD > \epsilon$ **and** $S \neq \emptyset$ **do**

  | Select and delete X^i from S for some i
  | Construct Equation (A1) and solve it to global optimality
  | **if** *Equation* (A1) *is infeasible* **then**
  | | Set $UBD_i := -\infty$

  | **else**
  | | Set UBD_i to the optimal solution value
  | | Set $\hat{x}$ to the optimal solution
  | | **if** $\hat{x}$ *is feasible for Equation* (A2) **and** $f_{LCS}^{disc}(\hat{x}) > LBD$ **then**
  | | | set $LBD := f_{LCS}^{disc}(\hat{x}), x^* := \hat{x}$

  | Construct Equation (A2) and solve it to local optimality
  | **if** *a feasible point* $\hat{x}$ *is found* **and** $f_{LCS}^{disc}(\hat{x}) > LBD$ **then**
  | | set $LBD := f_{LCS}^{disc}(\hat{x}), x^* := \hat{x}$

  | **if** $UBD_i = -\infty$ **or** $UBD_i \leq LBD$ **then**
  | | cycle
  | Partition X^i into sets X^j and X^{j+1}
  | Set $UBD_j, UBD_{j+1} := UBD_i$
  | Add nodes X^j, X^{j+1} to S
  | Set $j := j + 2$
  | Delete from S all nodes X^i with $UBD_i \leq LBD$
  | Set $UBD := \max_{X^i \in S} UBD_i$

if $LBD = -\infty$ **then**
  | problem is infeasible

else
  | LBD is the ϵ-optimal solution value and x^* is the feasible optimal solution

---

References

1. US Energy Information Administration. *Annual Energy Outlook 2018 with Projections to 2050*; Technical Report for US Department of Energy; US Energy Information Administration: Washington, DC, USA, 2018.
2. US Department of Energy. Available online: https://www.energy.gov/eere/amo/dynamic-manufacturing-energy-sankey-tool-2010-units-trillion-btu (accessed on 15 May 2018).
3. Kurup, P.; Turchi, C. *Initial Investigation into the Potential of CSP Industrial Process Heat for the Southwest United States*; Technical Report of NREL; National Renewable Energy Laboratory: Golden, CO, USA, 2015.
4. Farjana, S.H.; Huda, N.; Mahmud, M.P.; Saidur, R. Solar process heat in industrial systems–A global review. *Renew. Sustain. Energy Rev.* **2018**, *82*, 2270–2286, doi:10.1016/j.rser.2017.08.065. [CrossRef]
5. Kalogirou, S. The potential of solar industrial process heat applications. *Appl. Energy* **2003**, *76*, 337–361, doi:10.1016/S0306-2619(02)00176-9. [CrossRef]
6. Lauterbach, C.; Schmitt, B.; Jordan, U.; Vajen, K. The potential of solar heat for industrial processes in Germany. *Renew. Sustain. Energy Rev.* **2012**, *16*, 5121–5130, doi:10.1016/j.rser.2012.04.032. [CrossRef]
7. Mekhilef, S.; Saidur, R.; Safari, A. A review on solar energy use in industries. *Renew. Sustain. Energy Rev.* **2011**, *15*, 1777–1790, doi:10.1016/j.rser.2010.12.018. [CrossRef]
8. Martinez, V.; Pujol, R.; Moia, A. Assessment of Medium Temperature Collectors for Process Heat. *Energy Proc.* **2012**, *30*, 745–754, doi:10.1016/j.egypro.2012.11.085. [CrossRef]

9.   Sharma, A.K.; Sharma, C.; Mullick, S.C.; Kandpal, T.C.   Solar industrial process heating: A review. *Renew. Sustain. Energy Rev.* **2017**, *78*, 124–137, doi:10.1016/j.rser.2017.04.079. [CrossRef]
10. Panwar, N.L.; Kaushik, S.C.; Kothari, S.   Role of renewable energy sources in environmental protection: A review. *Renew. Sustain. Energy Rev.* **2011**, *15*, 1513–1524, doi:10.1016/j.rser.2010.11.037. [CrossRef]
11. Ramos, C.; Ramirez, R.; Beltran, J. Potential Assessment in Mexico for Solar Process Heat Applications in Food and Textile Industries. *Energy Proc.* **2014**, *49*, 1879–1884, doi:10.1016/j.egypro.2014.03.199. [CrossRef]
12. Murray, J. Aluminum Production Using High-Temperature Solar Process Heat. *Sol. Energy* **1999**, *66*, 133–142, doi:10.1016/s0038-092x(99)00011-0. [CrossRef]
13. Hockaday, S.; Dinter, F.; Harms, T.   Opportunities for concentrated solar thermal heat in the minerals processing industry.   In Proceedings of 4th Southern African Solar Energy Conference (SASEC) 2016, Stellenbosch, South Africa, 31 October–2 November 2016.
14. Mekhilef, S.; Faramarzi, S.Z.; Saidur, R.; Salam, Z.   The application of solar technologies for sustainable development of agricultural sector.   *Renew. Sustain. Energy Rev.* **2013**, *18*, 583–594, doi:10.1016/j.rser.2012.10.049. [CrossRef]
15. Hussain, M.I.; Lee, G.H.   Utilization of Solar Energy in Agricultural Machinery Engineering: A Review. *J. Biosyst. Eng.* **2015**, *40*, 186–192, doi:10.5307/jbe.2015.40.3.186. [CrossRef]
16. Beesing, M.E. Textile Drying Using Solar Process Steam. In Proccedings of the Energy Conference ASME 1979 International Gas Turbine Conference and Exhibit and Solar, San Diego, CA, USA, 12–15 March 1979, doi:10.1115/79-sol-23. [CrossRef]
17. Sharma, A.K.; Sharma, C.; Mullick, S.C.; Kandpal, T.C.   GHG mitigation potential of solar industrial process heating in producing cotton based textiles in India.   *J. Clean. Prod.* **2017**, *145*, 74–84, doi:10.1016/j.jclepro.2016.12.161. [CrossRef]
18. Sharma, A.K.; Sharma, C.; Mullick, S.C.; Kandpal, T.C. Incentives for promotion of solar industrial process heating in India: a case of cotton-based textile industry. *Clean Technol. Environ. Policy* **2018**, *20*, 813–823, doi:10.1007/s10098-018-1499-1. [CrossRef]
19. Frey, P.; Fischer, S.; Drück, H.; Jakob, K. Monitoring Results of a Solar Process Heat System Installed at a Textile Company in Southern Germany, *Energy Proc.* **2015**, *70*, 615–620. doi:10.1016/j.egypro.2015.02.168. [CrossRef]
20. Sharma, A.K.; Sharma, C.; Mullick, S.C.; Kandpal, T.C.   Potential of Solar Energy Utilization for Process Heating in Paper Industry in India: A Preliminary Assessment.   *Energy Proc.* **2015**, *79*, 284–289, doi:10.1016/j.egypro.2015.11.486. [CrossRef]
21. Sharma, A.K.; Sharma, C.; Mullick, S.C.; Kandpal, T.C. Carbon mitigation potential of solar industrial process heating: paper industry in India. *J. Clean. Prod.* **2016**, *112*, 1683–1691, doi:10.1016/j.jclepro.2015.04.093. [CrossRef]
22. Meier, A.; Bonaldi, E.; Cella, G.M.; Lipinski, W.; Wuillemin, D.   Solar chemical reactor technology for industrial production of lime. *Sol. Energy* **2006**, *80*, 1355–1362, doi:10.1016/j.solener.2005.05.017. [CrossRef]
23. Romero, M.; Steinfeld, A. Concentrating solar thermal power and thermochemical fuels. *Energy Environ. Sci.* **2012**, *5*, 9234–9245, doi:10.1039/c2ee21275g. [CrossRef]
24. Kalogirou, S.; Lloyd, S.; Ward, J. Modelling, optimisation and performance evaluation of a parabolic trough solar collector steam generation system. *Sol. Energy* **1997**, *60*, 49–59, doi:10.1016/s0038-092x(96)00131-4. [CrossRef]
25. Duffie, J.A.; Beckman, W.A. *In Solar Engineering of Thermal Processes*; John Wiley & Sons: Hoboken, NJ, USA, 2013.
26. Kalogirou, S.A. A detailed thermal model of a parabolic trough collector receiver. *Energy* **2012**, *48*, 298–306, doi:10.1016/j.energy.2012.06.023. [CrossRef]
27. Kalogirou, S.A. *Solar Energy Engineering Processes and Systems*, 2nd ed.; Academic Press: Oxford, UK, 2014.
28. Ghobeity, A.; Mitsos, A.   Optimal design and operation of a solar energy receiver and storage. *J. Sol. Energy Eng.* **2012**, *134*. doi:10.1115/1.4006402. [CrossRef]
29. Ghasemi, H.; Sheu, E.; Tizzanini, A.; Paci, M.; Mitsos, A.   Hybrid solar-geothermal power generation: Optimal retrofitting. *Appl. Energy* **2014**, *131*, 158–170, doi:10.1016/j.apenergy.2014.06.010. [CrossRef]
30. Gunasekaran, S.; Mancini, N.; El-Khaja, R.; Sheu, E.; Mitsos, A. Solar-geothermal hybridization of advanced zero emissions power cycle. *Energy* **2014**, *65*, 152–165, doi:10.1016/j.energy.2013.12.021. [CrossRef]

31. Silva, R.; Berenguel, M.; Perez, M.; Fernandez-Garcia, A. Thermo-Economic Design Optimization of Parabolic Trough Solar Plants for Industrial Process Heat Applications with Memetic Algorithms. *Appl. Energy* **2014**, *113*, 603–614, doi:10.1016/j.apenergy.2013.08.017. [CrossRef]

32. Stuber, M.D. Optimal Design of Fossil-Solar Hybrid Thermal Desalination for Saline Agricultural Drainage Water Reuse. *Renew. Energy* **2016**, *89*, 552–563, doi:10.1016/j.renene.2015.12.025. [CrossRef]

33. Yan, C.; Wang, S.; Ma, Z.; Shi, W. A simplified method for optimal design of solar water heating systems based on life-cycle energy analysis. *Renew. Energy* **2015**, *74*, 271–278, doi:10.1016/j.renene.2014.08.021. [CrossRef]

34. Baniassadi, A.; Momen, M.; Amidpour, M. A new method for optimization of Solar Heat Integration and solar fraction targeting in low temperature process industries. *Energy* **2015**, *90*, 1674–1681, doi:10.1016/j.energy.2015.06.128. [CrossRef]

35. Frasquet, M. SHIPcal: Solar heat for industrial processes online calculator. *Energy Proc.* **2016**, *91*, 611–619, doi:10.1016/j.egypro.2016.06.213. [CrossRef]

36. Ayub, M.; Mitsos, A.; Ghasemi, H. Thermo-economic analysis of a hybrid solar-binary geothermal power plant. *Energy* **2015**, *87*, 326–335, doi:10.1016/j.energy.2015.04.106. [CrossRef]

37. Al-Aboosi, F.Y.; El-Halwagi, M.M. An Integrated Approach to Water-Energy Nexus in Shale-Gas Production. *Processes* **2018**, *6*, 52, doi:10.3390/pr6050052. [CrossRef]

38. Stuber, M.D.; Sullivan, C.; Kirk, S.A.; Farrand, J.A.; Schillaci, P.V.; Fojtasek, B.D.; Mandell, A.H. Pilot demonstration of concentrated solar-powered desalination of subsurface agricultural drainage water and other brackish groundwater sources. *Desalination* **2015**, *355*, 186–196, doi:10.1016/j.desal.2014.10.037. [CrossRef]

39. Dudley, V.E.; Kolb, G.J.; Mahoney, A.R.; Mancini, T.R.; Matthews, C.W.; Sloan, M.; Kearney, D. *Test Results: SEGS LS-2 Solar Collector*; Technical Report of Sandia National Labs.; National Technical Information Service: Albuquerque, NM, USA, 1994.

40. Forristall, R. *Heat Transfer Analysis and Modeling of a Parabolic Trough Solar Receiver Implemented in Engineering Equation Solver*; Technical Report of NREL; National Renewable Energy Laboratory: Golden, CO, USA, 2003.

41. National Renewable Energy Laboratory. NSRDB: National Solar Radiation Database. Available online: https://nsrdb.nrel.gov/nsrdb-viewer (accessed on 21 May 2018).

42. International Renewable Energy Agency. *Renewable Energy Technologies: Cost Analysis Seris—Concentrating Solar Power*; Technical Report of IRENA; International Renewable Energy Agency: Abu Dhabi, UAE, 2012.

43. Falk, J.E.; Soland, R.M. An Algorithm for Separable Nonconvex Programming Problems. *Manag. Sci.* **1969**, *15*, 550–569, doi:10.1287/mnsc.15.9.550. [CrossRef]

44. Bezanson, J.; Edelman, A.; Karpinski, S.; Shah, V.B. Julia: A Fresh Approach to Numerical Computing. *SIAM Rev.* **2017**, *59*, 65–98. doi:10.1137/141000671. [CrossRef]

45. US Energy Information Administration. Available online: https://www.eia.gov/dnav/ng/NG_PRI_SUM_DCU_NUS_M.htm (accessed on 16 April 2018).

46. Wächter, A.; Biegler. On the implementation of an interior-point filter line-search algorithm for large-scale nonlinear Programming. *Math. Program.* **2006**, *106*, 25–57, doi:10.1007/s10107-004-0559-y. [CrossRef]

47. Dunning, I.; Huchette, J.; Lubin, M. JuMP: A modeling language for mathematical optimization. *SIAM Rev.* **2017**, *59*, 295–320, doi:10.1137/15M1020575. [CrossRef]

48. Wilhelm, M.; Stuber, M.D. Easy Advanced Global Optimization (EAGO): An Open-Source Platform for Robust and Global Optimization in Julia. In Proccedings of the AIChE Annual Meeting 2017 Minneapolis, Minneapolis, MN, USA, 31 October 2017.

49. Wilhelm, M.; Stuber, M.D. EAGO: Easy Advanced Global Optimization Julia Package. Available online: https://github.com/PSORLab/EAGO.jl (accessed on 1 May 2018).

50. Kreith, F.; Kreider, J.F. *Principles of Solar Engineering*; Hemisphere Publishing Corporation: Washington, DC, USA, 1978.

processes

MDPI

Article

An Integrated Approach to Water-Energy Nexus in Shale-Gas Production

Fadhil Y. Al-Aboosi [1,2] and Mahmoud M. El-Halwagi [1,*]

[1]   The Artie McFerrin Department of Chemical Engineering, Texas A&M University,
    College Station, TX 77843-3122, USA; alaboosi@tamu.edu
[2]   Department of Energy Engineering, Baghdad University, Baghdad 10071, Iraq
*   Correspondence: El-Halwagi@tamu.edu; Tel.: +1(979)845-3484

Received: 18 April 2018; Accepted: 4 May 2018; Published: 8 May 2018

Abstract: Shale gas production is associated with significant usage of fresh water and discharge of wastewater. Consequently, there is a necessity to create proper management strategies for water resources in shale gas production and to integrate conventional energy sources (e.g., shale gas) with renewables (e.g., solar energy). The objective of this study is to develop a design framework for integrating water and energy systems including multiple energy sources, the cogeneration process and desalination technologies in treating wastewater and providing fresh water for shale gas production. Solar energy is included to provide thermal power directly to a multi-effect distillation plant (MED) exclusively (to be more feasible economically) or indirect supply through a thermal energy storage system. Thus, MED is driven by direct or indirect solar energy and excess or direct cogeneration process heat. The proposed thermal energy storage along with the fossil fuel boiler will allow for the dual-purpose system to operate at steady-state by managing the dynamic variability of solar energy. Additionally, electric production is considered to supply a reverse osmosis plant (RO) without connecting to the local electric grid. A multi-period mixed integer nonlinear program (MINLP) is developed and applied to discretize the operation period to track the diurnal fluctuations of solar energy. The solution of the optimization program determines the optimal mix of solar energy, thermal storage and fossil fuel to attain the maximum annual profit of the entire system. A case study is solved for water treatment and energy management for Eagle Ford Basin in Texas.

Keywords: cogeneration; process integration; solar energy; thermal storage; desalination; optimization

---

1. Introduction

Recently, major discoveries of shale gas reserves have led to substantial growth in production. For instance, the U.S. production of shale gas has increased from 2 trillion ft^3 in 2007 to 17 trillion ft^3 in 2016 with an estimated cumulative production of more than 400 trillion ft^3 over the next two decades [1]. Consequently, there are tremendous monetization opportunities to convert shale gas into value-added chemicals and fuels such as methanol, olefins, aromatics and liquid transportation fuels [2–9]. A major challenge to a more sustainable growth of shale gas production is the need to address natural resource, environmental] and safety issues [10,11]. Specifically, the excessive usage of fresh water and discharge of wastewater constitute major problems. Hydraulic fracturing and horizontal drilling are the essential technologies to extract natural gas from shale rock. Water plays a significant role in shale gas production through mixing millions of gallons of water with sand, chemicals, corrosion inhibitors, surfactants, flow improvers, friction reducers and other constituents to produce fracturing fluid. Under the high pressure, the fracturing fluid is injected into the wellbore to make cracks within the rock layers to increase the production [12,13]. Large quantities of water are used in the fracturing and related process [14]. The typical annual water consumption per well for hydraulic fracturing ranges between 1000 and 30,000 m^3, leading to substantial amounts of water usage. For instance, the annual water usage in shale gas production is estimated to be about 120 MM m^3. In the Eagle Ford Shale Play, the annual water use is 18 MM m^3 for 1040 wells [15]. Wastewater associated with shale gas production is discharged in two forms: flowback water (which is released over several weeks following production) and produced water (which is the long-term wastewater) [14,16]. Treatment of shale gas wastewater followed by recycling and reuse can provide major economic and environmental benefits [12–17]. Regrettably, a small fraction of the shale-gas wastewater is recycled. A recent study [18] reported that in 2014, less than 10% of the roughly 80,000 wells in the U.S. used recycled water after proper treatment. Lira-Barragán et al. [18] developed a mathematical programming model for the combination of water networks in the shale gas site by taking into consideration the requirement of water, the uncertainty of used and flowback water, and the optimal size of treatment units, storage systems and disposals. Gao and You [12] addressed the shale-gas water problem as a mixed integer linear fractional programming (MILFP) problem to maximize the profit per unit of freshwater consumption. Yang et al. [14] developed a two-stage mixed integer nonlinear programming (MINLP) model for shale gas formations with the uncertainty of water availability. Several approaches may be used for treatment and management of shale gas wastewater [13–20]. These approaches include conventional technologies such as multi-effect distillation and reverse osmosis. Additionally, emerging technologies such as membrane distillation may be used to exploit excess heat from flared gases, compression stations and other on-site sources and to provide a modular system with high levels of salt rejection [16,21–29]. Additionally, renewable energy (such as solar) may be utilized to enhance the sustainability of the system. Therefore, it is important to consider the water management problem for shale gas production via a water-energy nexus framework.

This work is aimed at developing a new systematic approach to the design, operation, integration and optimization of a dual-purpose system, which integrates solar energy and fossil fuels to produce electricity and desalinated water while treating shale-gas wastewater. In addition to fossil fuels, a concentrated solar power field, a thermal storage system, conventional steam generators and the cogeneration process are coupled with two water treatment plants: reverse osmosis (RO) and multiple-effect distillation (MED). A multi-period mixed integer nonlinear program (MINLP) formulation is developed to account for the diurnal fluctuations of solar energy. The solution of the mixed integer nonlinear program (MINLP) determines the optimal mix of solar energy, thermal storage and fossil fuel and the details of wastewater treatment and water recycling.

2. Problem Statement

Consider a shale-gas production site with the following known information:

- Flowrate and characteristics of produced and flared shale gas.
- Demand for fresh water (flowrate and quality).
- Flowrate and characteristics of flowback and produced wastewater.

The site is not connected to an external power grid.
It is desired to systematically design an integrated system that:

- Treats the wastewater for on-site recycling/reuse.
- Uses solar energy and fossil fuels to provide the needed electric and thermal power needs.
- Satisfies technical, economic and environmental requirements.

Given are:

- Flowrate and composition of shale gas (sold and flared).
- Flowrate and purity needs for fresh water.
- Total volumetric flow of wastewater (flow-back and produced water) of shale gas play.
- Flowrate of flared gases that may be used in the cogeneration process.

- Electric energy requirement for RO and MED (kWh_e/m^3).
- Thermal energy requirement for MED (kWh_t/m^3).

To solve the problem, the following questions should be addressed:

- What is the maximum annual profit of the whole system for producing desalinated water and electricity for the various percentage contributions of RO and MED in the total desalinated water production?
- What is the minimum total annual cost of the entire system?
- What is the economic feasibility of the system?
- What is the optimal mix of solar energy, thermal storage and fossil fuel for the MED plant and the entire system?
- What is the optimal design and integration of the system?
- What are the optimal values of the design and operating variables of the system (e.g., minimum area of a solar collector, maximum capacity of a thermal storage system, etc.)?
- What is the feasible range of the percentage contribution of RO and MED in the total desalinated water production?

The superstructure integrates the primary components of solar energy and fossil fuels to produce electricity and desalinated water, as shown in Figure 1:

- To achieve a steady supply of thermal power to the whole system, solar energy (as direct solar thermal power), fossil fuel (shale gas, flared gas) and thermal energy storage (as indirect solar thermal power) are used.
- Solar energy is used as a source of heat to provide thermal power directly to the MED plant exclusively (to be more economically feasible), while the surplus thermal power is stored.
- A two-stage turbine is used to enhance the cogeneration process efficiency.

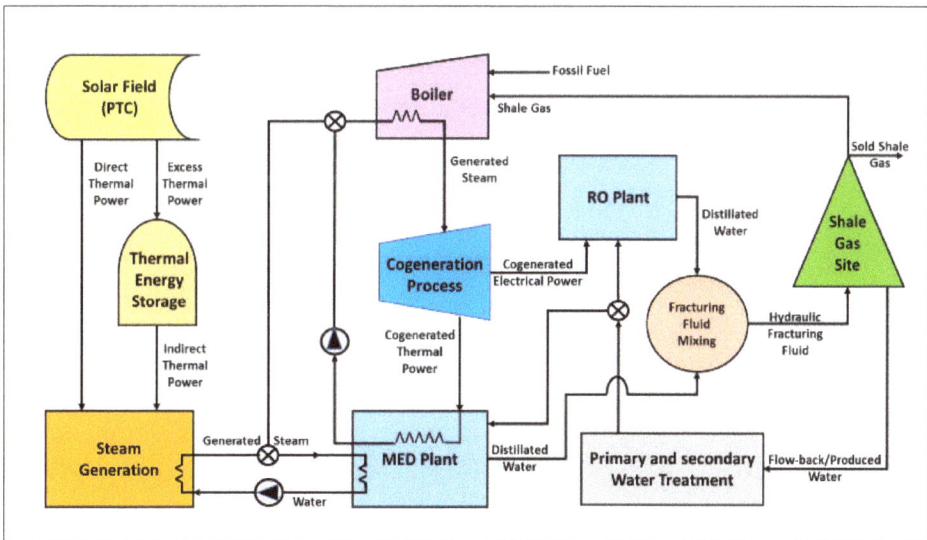

Figure 1. Proposed superstructure representation.

3. Approach

A hierarchical design is proposed to efficiently address the water-energy nexus problem. Figure 2 demonstrates the main steps of the approach. The first step is to gather the required data for the system, then to select and formulate the appropriate models that describe the major system components. Once the preceding steps are achieved, the computational optimization is applied to the integrated system to maximize the annual profit of the system that produces a specific level of desalinated water and electricity. In treating wastewater, focus is given to the management of flowback and produced shale gas wastewater. To decompose the optimization problem, the percentage contribution of RO and MED to treating wastewater is iteratively discretized. It is worth noting that the proposed discretization approach offers significant reduction in the complexity of solving the optimization problem. Such decomposition leads to computational efficiency. Similar approaches have been proposed earlier in the literature for other applications [23,30,31]. For each discretization, the RO and MED systems are designed separately because their treatment tasks for the iterative discretization are known. Consequently, the thermal and electric loads are calculated. Next, a multi-period mixed-integer nonlinear program (MINLP) is solved using the software LINGO to optimize the power mix for each period. Upon identification of the solar load, the solar area and storage capacity are calculated, and the total annual profit of the system is calculated. The procedure is repeated for the selected discretizations of the fractional contribution of RO and MED. The results are compared, and the maximum-profit solution is selected.

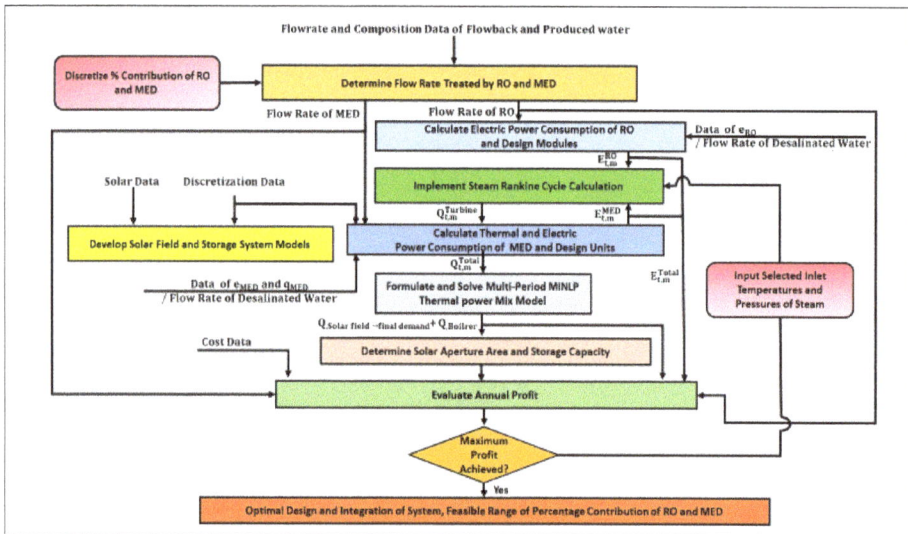

Figure 2. Proposed approach.

4. Modeling the Building Blocks

The performance models for MED and RO have been taken from the literature [32–36]. For the solar system, a parabolic trough collector was selected. The modeling of the solar system was based on literature models and data [37–40] as described in this section. The solar thermal power (per unit length of a collector) produced by the solar field when the direct normal irradiance (DNI) strikes the collector aperture plane is given by the following expression:

$$Q_{sun \to collector}(W/m) = DNI \cdot cos\theta \cdot W_c \qquad (1)$$

where *DNI* (W/m²) is the direct normal irradiance, θ is the solar incidence angle and W_c (m) is the width of the collector aperture.

For the north-south orientation, the incidence angle is calculated as follows:

$$cos\theta = \sqrt{cos^2\theta_z + cos^2\delta \cdot sin^2\omega} \tag{2}$$

where θ_z is the solar zenith angle, δ is the declination and ω is the hour angle.

To calculate the thermal power (per unit length of a collector) absorbed by the receiver tube of a collector loop, the influences of the optical losses can be taken into consideration by inserting four parameters into the equation given by the following expression:

$$Q_{collector \rightarrow reciever}(W/m) = DNI \cdot cos\theta \cdot W_c \cdot \eta_{opt} \cdot K(\theta) \cdot F_f \cdot R_{SL} \cdot O_{EL} \tag{3}$$

where η_{opt} is the peak optical efficiency of a collector, $K(\theta)$ is the incidence angle modifier, F_f is the soiling factor (mirror cleanliness), R_{SL} is the row shadow loss and O_{EL} is the optical end loss.

The peak optical efficiency of a collector when the incidence angle on the aperture plane is 0° is:

$$\eta_{opt} = \rho \cdot \gamma \cdot \tau \cdot \alpha \big|_{\theta=0°} \tag{4}$$

where ρ is the reflectivity, γ is the intercept factor, τ is the glass transmissivity and α is the absorptivity of the receiver pipe.

The incidence angle modifier for an LS-3 collector is given by:

$$K(\theta) = 1 - 2.23073 \times 10^{-4} \cdot \theta - 1.1 \times 10^{-4} \cdot \theta^2 + 3.18596 \times 10^{-6} \cdot \theta^3$$
$$-4.85509 \times 10^{-8} \cdot \theta^4 \quad 0° \le \theta \le 80°$$
$$K(\theta) = 0 \quad \theta > 80° \tag{5}$$

The row shadow factor is:

$$R_{SL} = min\left[max\left(0.0, \frac{L_{spacing}}{W_c} \cdot \frac{cos\theta_z}{cos\theta}\right); 1.0\right] \tag{6}$$

where $L_{spacing}$ (m) is the length of spacing between troughs.

The optical end loss is:

$$O_{EL} = 1 - \frac{f \cdot tan\theta}{L_{SCA}} \tag{7}$$

where f is focal length of the collectors (m) and L_{SCA} is the length of a single collector assembly (m).

The total thermal power (per unit length of a collector) loss from a collector represents the combination of the radiative heat loss from the receiver pipe to the ambient environment ($Q_{reciever \rightarrow ambient}$) and convective and conductive heat losses from the receiver pipe to its outer glass pipe ($Q_{receiver \rightarrow glass}$) and is calculated by the following expression:

$$Q_{collector \rightarrow ambient}(W/m) = U_{rec} \cdot \pi \cdot d_o \cdot (T_{rec} - T_{amb}) \tag{8}$$

where U_{rec} (W/m²$_{rec}$·K) is the overall heat transfer coefficient of a receiver pipe, d_o(m) is the outer diameter of a receiver pipe, T_{rec}(K) is the mean receiver pipe temperature and T_{amb}(K) is the ambient air temperature.

The overall heat transfer coefficient of a collector is found experimentally depending on the receiver pipe temperature, and it can be given in the second-order polynomial equation:

$$U_{rec} = a + b(T_{rec} - T_{amb}) + c(T_{rec} - T_{amb})^2 \tag{9}$$

where the *a*, *b* and *c* coefficients have been calculated experimentally for the LS-3 collector and have been reported in the literature [37].

The thermal power (per unit length of a collector) that transferred from a collector to a fluid is given in the following expression [41]:

$$Q_{collector \rightarrow fluid}(W/m) = Q_{collector \rightarrow receiver} - Q_{collector \rightarrow ambient} \tag{10}$$

The thermal power (per unit length of a collector) loss from the headers (pipes) is given in the following expression [42]:

$$Q_{LFP}(W/m) = 0.0583 \cdot W_c \cdot (T_{rec} - T_{amb}) \tag{11}$$

The thermal power (per unit length of a collector) loss from the expansion tank (vessel) is given in the following expression [42]:

$$Q_{,LFV}(W/m) = 0.0497 \cdot W_c \cdot (T_{rec} - T_{amb}) \tag{12}$$

The useful thermal power (per unit length of a collector) produced by the solar field is given by the following expression, which represents the sum of Equations (10)–(12):

$$Q_{solar\ field \rightarrow final\ demand}(W/m) = Q_{collector \rightarrow receiver} - Q_{collector \rightarrow ambient} - Q_{LFP} - Q_{LFV} \tag{13}$$

The inlet thermal power of the thermal storage is given in the following expression:

$$Q_{in} = m_{ms} \cdot C_{P,ms} \cdot (T_{HT} - T_{CT}) = \eta_{EX} \cdot m_{oil} \cdot C_{P,oil} \cdot (\Delta T) \tag{14}$$

The expression of the discharge process (outlet thermal power) is given by:

$$Q_{out} = m_{oil} \cdot C_{P,oil} \cdot (\Delta T) = \eta_{EX} \cdot m_{ms} \cdot C_{P,ms} \cdot (T_{HT} - T_{CT}) \tag{15}$$

where m_{ms} is the molten salt flow rate (kg/s), ($C_{P,ms} = 1443 + 0.172\ T_{ms}$) is the specific heat of the molten salt (J/kg·°C), T_{ms} is the temperature (°C) of the molten salt, T_{HT} is the hot tank temperature (°C), T_{CT} is the cold tank temperature (°C), η_{EX} is the efficiency of the heat exchanger, m_{oil} is the oil mass flowrate (kg/s) and ΔT is the difference between the inlet and outlet of the oil.

$$Q_{TES} = Q_{acc} + Q_{in} - Q_{out} - Q_{loss} \tag{16}$$

where Q_{acc} is the accumulated thermal power in the tank from preceding iterations and Q_{loss} is the thermal power loss (kW/m²) of the cold and heat tanks, and it is given in the following empirical equation [43]:

$$Q_{loss} = 0.00017 \cdot T_{ms} + 0.012 \tag{17}$$

where T_{ms} is the temperature (°C) of the molten salt in the hot and in the cold tanks.

The optimal values of the Rankine cycle parameters of the cogeneration process can be satisfied by formulating the entire cycle as an optimization problem. Thus, there is a necessity to obtain suitable correlations of the thermodynamic properties that can be used in the optimization formulations. In the thermodynamic calculations of the Rankine cycle, mathematical equations are used to replace the steam tables because they could easily be incorporated into the optimization formulations. However, available correlations for steam tables are complicated (e.g., nonlinear, nonconvex function), and it is hard to insert them into the optimization task. Consequently, a new set of thermodynamic correlations has been developed in the literature [44] to estimate the properties of steam, and they can be incorporated easily into the optimization formulation and cogeneration design. The isentropic efficiency of the steam turbine can be obtained from the turbine hardware model, which was developed by Mavromatis and

Kokossis [45], to show the efficiency variation with the load, the turbine size and operating conditions, as in the following correlation:

$$\eta_{is} = \frac{6}{5 \cdot B} \left(1 - \frac{3.41443 \cdot 10^6 \cdot A}{\Delta h_{is} \cdot m^{max}} \right) \left(1 - \frac{m^{max}}{6 \cdot \dot{m}} \right) \tag{18}$$

where $\dot{m}$ is the inlet turbine steam flowrate (Ib/h), m^{max} is the maximum mass flowrate of a turbine (Ib/h) and A and B are parameters that depend on the inlet saturation temperature (°F) and the type of turbine as in the following correlations:

$$A = a_o + a_1 \cdot T_{sat} \tag{19}$$

$$B = a_2 + a_3 \cdot T_{sat} \tag{20}$$

where a_o, a_1, a_2, a_3 are the correlation constants and can be found in the literature [46].

5. Optimization Formulation

Because of the diurnal nature of solar energy, a multi-period approach is adopted. The annual operation is discretized into a number of operational periods (e.g., monthly). The index m refers to the operational period. For each operational period, an average meteorological day is used to represent the solar intensity data. In turn, the meteorological day is discretized into a number of sub-periods (e.g., 24 h) where the index t is used to designate a sub-period. Two water-treatment technologies are used: multi-effect distillation (MED) and reverse osmosis (RO). MED consumes mostly thermal energy and some electric energy, which are respectively given by the specific requirements: q_{MED} (kWh$_t$/m^3) and e_{MED} (kWh$_e$/m^3). RO requires electric energy, which is represented by the following specific energy consumption term: e_{RO} (kWh$_e$/m^3).

For each sub-period t, the thermal power needs for water treatment are obtained directly from the combustion of fossil fuels ($Q_{t,m}^{Fossil}$), directly from a solar thermal collector ($Q_{t,m}^{Direct,SC}$), indirectly from solar energy through thermal storage ($Q_{t,m}^{Out\_Stored\_SC}$) and from steam leaving the cogeneration turbine ($Q_{t,m}^{Turbine}$). Hence,

$$Q_{t,m}^{Total} = Q_{t,m}^{Fossil} + Q_{t,m}^{Direct,SC} + Q_{t,m}^{Out\_Stored\_SC} + Q_{t,m}^{Turbine} \quad \forall t, \forall m \tag{21}$$

where:

$$Q_{t,m}^{Total} = F_{t,m}^{MED} q_{MED} \quad \forall t, \forall m \tag{22}$$

The electric power provided by the cogeneration turbine is given by:

$$E_{t,m}^{Total} = F_{t,m}^{RO} e_{RO} + F_{t,m}^{MED} e_{MED} \quad \forall t, \forall m \tag{23}$$

The thermal power captured by the solar collector ($Q_{t,m}^{SC}$) is directly used ($Q_{t,m}^{Direct,SC}$) or is stored ($Q_{t,m}^{In\_Stored-SC}$) for subsequent usage, i.e.,

$$Q_{t,m}^{SC} = Q_{t,m}^{Direct,SC} + Q_{t,m}^{In\_Stored-SC} \quad \forall t, \forall m \tag{24}$$

Over a sub-period, t, the thermal power balance for the thermal storage unit is given by:

$$Q_{t,m}^{Stored-SC} = Q_{t-1,m}^{Stored-SC} + Q_{t,m}^{In\_Stored-SC} - Q_{t,m}^{Out\_Stored-SC} - Q_{t,m}^{Stored-Loss} \quad \forall t, \forall m \tag{25}$$

Such collected energy is a function of the solar-radiation intensity ($Solar\_Radiation_{t,m}$) and the effective surface area of the solar collector (A^{SC}).

Although each period requires a certain area of the solar collector, the design value (which is also used for capital cost estimation) is the largest of all needed areas, i.e.,

$$A_{t,m}^{SC} \leq A_{Design}^{SC} \quad \forall t, \forall m \tag{26}$$

The cogeneration turbine is modelled through a performance function (e.g., isentropic expansion with an efficiency) that combines inlet and outlet steam conditions and relates the produced power to heat.

$$\Omega_{t,m}^{Turbine}(D_{t,m}^{Tutbine}, O_{t,m}^{Turbine}, Steam_{t,m}^{In}, Steam_{t,m}^{Out}, Power_{t,m}^{Out}) = 0 \quad \forall t, \forall m \tag{27}$$

The objective function seeks to maximize the profit for the water-energy nexus system:

Maximize annual profit = annual value of treated water + annual value of avoided cost of discharging wastewater − cost of fossil fuels − total annualized cost of solar collection system − total annualized cost of solar storage system − total annualized cost of cogeneration system − total annualized cost of MED system − total annualized cost of RO system:

$$
\begin{aligned}
\text{Maximum annual profit} = {} & \sum_m \sum_t \left(v_{t,m}^{RO} F_{t,m}^{RO} + v_{t,m}^{MED} F_{t,m}^{MED} \right) + c^{Waste} W_w - \\
& \sum_m \sum_t \left(c_{t,m}^{Fossil} F_{t,m}^{Fossil} \right) - AFC^{SC} - \sum_m \sum_t OPEX_{t,m}^{SC} - AFC^{SC\_Storage} - \\
& \sum_m \sum_t OPEX_{t,m}^{SC\_Storage} - AFC^{Cogen} - \sum_m \sum_t OPEX_{t,m}^{Cogen} - AFC^{MED} - \\
& \sum_m \sum_t OPEX_{t,m}^{MED} - AFC^{RO} - \sum_m \sum_t OPEX_{t,m}^{RO}
\end{aligned}
\tag{28}
$$

It is worth noting that the economic objective function can be altered to include sustainability and safety metrics by using the sustainability and safety weighted return on investment metrics [47,48].

6. Case Study

To demonstrate the viability of the proposed approach for solution strategies, a case study will be solved based on the Eagle Ford Shale Play, which is located in south Texas. A dual-purpose system that integrates solar energy and fossil fuels for producing electricity and fresh water has been considered. The optimal design, operation and integration of the system will be found through this case study, which requires particular input data for each unit of the entire system. As mentioned earlier, this system includes the concentrated solar power field, a thermal storage system, conventional steam generators and a cogeneration process into two water treatment plants, a reverse osmosis plant (RO) and a multiple-effect distillation plant (MED).

7. Flowback/Produced Water of Shale Gas Play

In order to supply a specific amount of flow-back and produced water (FPW) from a shale play to a desalination plant, the calculation of an FPW flow average for many years is an appropriate option to avoid the uncertainty in the amount of FPW. Specifically, we know that wastewater of shale play is typically subjected to heavy regulation and should be stored in containers so that these containers can be utilized to get a constant flow approximately. Additionally, a large number of wells in a shale play can contribute to making the flow rate of FPW approximately constant because when the FPW production of one well starts declining, another well will start its production and compensate a drop of production in other wells.

The value of flowback and produced water returned from shale gas formations to the surface in the Eagle Ford Basin is estimated to be 151.22×10^6 m^3 [49] for 10 plays since the early 2000s until 2015. Table 2 summarizes the costs of RO and MED. Additional data can be obtained from the literature [50–52]. The techno-economic data for RO and MED are reported in Table 1.

Table 1. Techno-economic data for RO and MED [32,50].

Technology	Thermal Energy Consumption (kWht/m³ Desalinated Water)	Electric Energy Consumption (kWhe/m³ Desalinated Water)	Annualized Fixed Cost (AFC) ($/year)	Operating Cost ($/m³ seawater)	Water Recovery (m³ Desalinated Water/m³ Feed Seawater)	Value of Desalinated Water ($/m³ Desalinated Water)	Outlet Salt Content (ppm)
RO	-	4	$2.0 \times 10^6 + 1166.$ (flowrate of seawater, m³/day)$^{0.8}$	0.18	0.55	0.88	200
MED	65	2	$13.0 \times 10^6 + 2227.$ (flowrate of seawater, m³/day)$^{0.7}$	0.24	0.65	0.82	80

8. Solar Energy

The solar data are summarized in Appendix A. Table 2 summarizes the main cost data for the solar collectors.

Table 2. The direct capital cost of parabolic trough collector items [53,54].

Item	Receivers	Mirrors	Concentrator Structure	Concentrator Erection	Drive	Piping
Cost $/m²	43	40	47	14	13	10

Item	Electronic and Control	Header Piping	Civil Works	Spares, HTF, Freight	Contingency	Structures and Improvement
Cost $/m²	14	7	18	17	11	7

The total fixed capital cost of the solar field ($) is the sum of the heat collection element (HCE), mirror, support structure, drive, piping, civil work, structures and improvements, as follows:

$$FCI_{SF} = C_{SF} \cdot A_{SF} \tag{29}$$

where C_{SF} is the solar field cost per area unit ($241/m²) and A_{SF} is the solar field aperture area (m²).

The thermal storage system is assumed to be an indirect two-tank type, which uses the binary solar salt (sodium and potassium nitrate) as a storage material with the following fixed capital cost estimation ($):

$$FCI_{TES} = C_{TES} \cdot SC \cdot Q_{solar\ field \rightarrow final\ demand} \tag{30}$$

where C_{TES} is the thermal storage system cost per thermal energy unit ($27.18/kWh), SC is the number of storage capacity hours (h) and $Q_{solar\ field \rightarrow final\ demand}$ is the useful thermal power produced by the solar field (kW).

The fixed capital cost estimation of a steam generator system ($) is calculated as:

$$FCI_{SG} = C_{SG} \cdot Q_{solar\ field \rightarrow final\ demand} \tag{31}$$

where C_{SG} is the steam generator system cost per thermal power unit ($/kW$_t$).

The fixed capital cost of a boiler ($), which is assumed to a water-tube boiler fueled with gas or oil, is estimated as follows [44]:

$$FCI_B = 3 \cdot N_p \cdot N_T \cdot Q_{Boiler}^{0.77} \tag{32}$$

where Q_{Boiler} is the amount of thermal power (BTU/h) transferred to the steam and equal to $(Q_{Boiler}/\eta_{boiler})$, η_{boiler} is the efficiency of a boiler and N_p is a factor to account for the operation pressure, and it is given by: $N_p = 7 \times 10^{-4} \cdot P_g + 0.6$; P_g is the gauge pressure (psi) of a boiler; N_T is a factor accounting for the superheat temperature and is given by: $N_T = 1.5 \times 10^{-6} \cdot T_{SH}^2 + 1.13 \times 10^{-3} \cdot T_{SH} + 1$; T_{SH} is the superheat temperature (°F), $T_{SH} = T^{in} - T_{sat}^{in}$; T^{in} is the temperature at the inlet of a turbine; T_{sat}^{in} is the saturation temperature at the inlet of a turbine.

The fixed capital cost of a turbine ($), which is assumed to be a non-condensing turbine, is estimated as follows [44]:

$$FCI_T = 475 \cdot E_T \tag{33}$$

where E_T is the turbine shaft power output (BTU/h); $E_{,T} = m \cdot \left(h^{in} - h_{act}^{out}\right)$.

9. Flared Gas

The shale gas production from the Eagle Ford wells can be used as a fuel for the cogeneration process. Furthermore, the flared gas can be used also as a fuel source for the cogeneration process as it will contribute to saving a considerable amount of shale gas along with diminishing CO_2 emissions accompanying the flared gas. In the Eagle Ford fields, 4.4 billion cubic feet of gas were flared in 2013, which represented around 13% of the gas in the formation [55].

10. Total Cost

The annual fixed cost (AFC) ($/year) of the system is determined as follows:

$$AFC = [(FCI_{SF} + FCI_{TES} + FCI_{SG} + FCI_B + FCI_T + FCI_{PST})/N] + AFC_{RO} + AFC_{MED} \tag{34}$$

The operation and maintenance cost ($/h) of the solar field, cogeneration process, thermal storage system, administration and operations is estimated as follows, based on data given by [53,54]:

$$OC_{OM} = C_{OM} \cdot \left(Q_{solar\ field \to final\ demand} + Q_{Boiler}\right) \tag{35}$$

where C_{OM} is the operation and maintenance cost per thermal power unit ($0.0203/kWh).

The type and amount of the selected fuel are necessary to estimate the cost of fuel ($/h), and it is formulated as follows:

$$OC_F = C_F \cdot Q_B \cdot 3413 \times 10^{-6} \tag{36}$$

where C_F is the fuel cost ($/MMBTU), Q_B is the amount of thermal power (BTU/h) that equals $(Q_{Boiler}/\eta_{boiler})$ and η_{boiler} is the efficiency of a boiler.

The annual operating cost (AOC) ($/year) is determined as follows:

$$AOC = a_Y \cdot (OC_{OM} + OC_F) \tag{37}$$

where a_Y is the annual operation time (h/year).

The annual income ($/year) is the sum of the total desalinated water production value and the savings value of a reduction in the cost of transportation, fresh water acquisition and disposal:

$$
\begin{aligned}
\text{Annual income} = \ & a_Y \cdot \{(0.88 \cdot \text{flowrate of desalinated water from } RO, \text{m}^3/\text{hr} \\
& + \ 0.82 \cdot \text{flowrate of desalinated water from } MED, \text{m}^3/\text{hr}) \\
& + \ [(C_F W + C_D S + C_T R) \cdot \text{total flowrate of desalinated water from } (RO, MED)]/0.11924\}
\end{aligned} \tag{38}
$$

where C_{FW} is the fresh water cost per volume unit (0.24$/bbl), C_{DS} is the disposal cost per volume unit (0.05$/bbl) and C_{TR} is the transportation cost per volume unit (0.89$/bbl).

The net profit represents the sum of the total desalinated water production value and the saving value of a reduction in the cost of transportation, fresh water acquisition and disposal. The treatment process of flowback and produced water in a shale gas site can contribute effectively to saving money for each barrel of flowback and produced water that should be transported by truck and disposed. Table 3 shows the cost of transportation, fresh water acquisition, primary/secondary treatment and disposal depending on the characteristics of a water treatment plant with a capacity of 2380 barrels/day in Eagle Ford Basin [56].

Table 3. Cost of transportation, fresh water, treatment and disposal of FPW.

Type	($/barrel)
Fresh water	0.24
Disposal (deep well + landfill)	0.05
Primary and secondary treatment	0.34
Transportation	0.89

11. Results and Discussion

A detailed performance model of the parabolic trough was applied to the case study to determine the useful thermal power (per unit length of a collector) produced by the solar field. The calculations of the solar field have been carried out depending on the monthly average of hourly direct solar irradiance, hourly ambient temperature and hourly incidence angle. Moreover, the characteristics of the LS-3 collector were adopted, and all types of thermal losses (convection, conduction, radiation) are considered for the entire solar field. The hourly variations in the useful thermal power for 12 months were obtained, as shown in Figure 3.

The obtained results showed that the gained thermal power in the months of January, February, November and December is less than the other eight months of the year due to low DNI and the high cosine effect. However, the four months that have the lowest value of useful thermal power still have significant potential to provide thermal power to the system. Selecting the solar irradiance at around 500 W/m^2 at the design point to calculate the total area of the collectors can give a great chance for these four months to contribute efficiently to supplying sufficient thermal power, despite a low value of average direct normal irradiance in the region selected as a case study. In the same direction, the eight months that have a higher DNI can be exploited to provide direct thermal power to MED and surplus thermal power to a thermal storage system. Indeed, the optimal area of collectors and storage system capacity are based on the minimum total annual cost of the entire system that can be obtained through an optimization solution.

The monthly distribution of the optimal thermal power mix for the MED plant and the entire system has been determined for the different percentage contributions of RO and MED in the total desalinated water production. The optimal thermal power mix for the MED plant includes the direct thermal power of the solar field, the indirect thermal power of the thermal storage system, the surplus thermal power of the cogeneration system and the direct thermal power from the combustion of fossil fuels. The monthly distribution varies over the year due to the availability of DNI and the variability of the incident angle, as shown in Figures 4–6.

The solution of the case study introduces two scenarios to the optimal operation for MED in accordance with the availability of solar energy regardless of the percentage contribution of MED. The first scenario is for the months of January, February, November and December and shows that it favors the harnessing of direct solar thermal power during the diurnal hours and utilizing fossil fuel in the early hours of the day and in the evening. However, stored solar thermal power can be contributed from 1–2 h only because of the lack of solar energy in these months, as illustrated in Figure 7, adapted from [57].

The second scenario is for the months of April, March, May, June, July, August, December and October and shows sharply diminishing fossil fuel use up to 2 h only. Typically, direct solar thermal power is exploited in the middle of the day, while stored solar thermal power is dispatched in the early hours and in the evening, as shown in Figure 8, adapted from [43]. In future work, the previous two scenarios can be applied to the entire system in the case of integrating solar energy into cogeneration process.

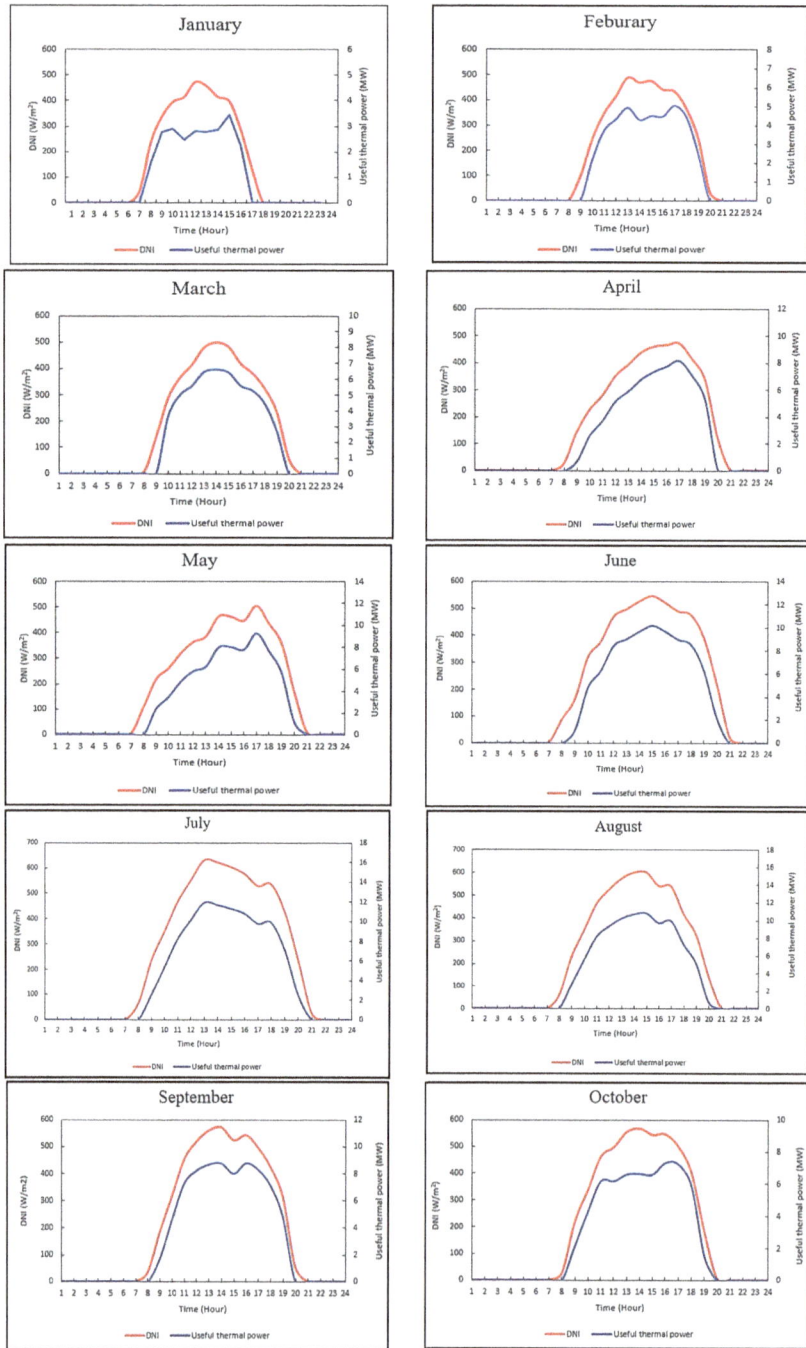

Figure 3. *Cont.*

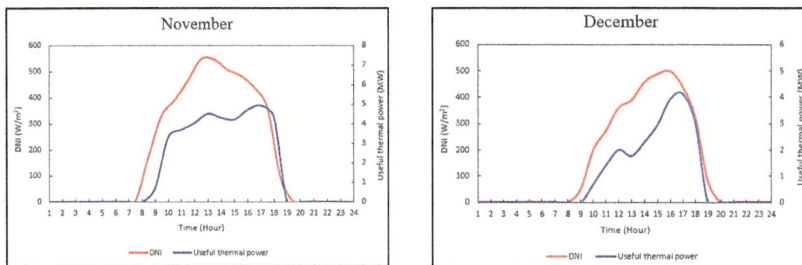

Figure 3. Monthly average of hourly direct normal irradiance (DNI) and useful thermal power.

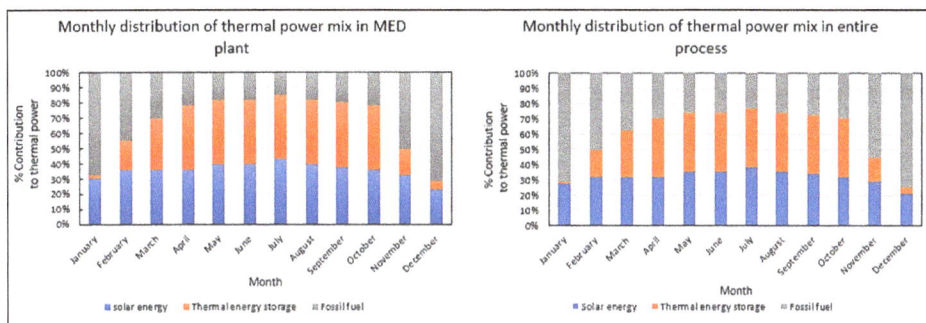

Figure 4. Optimal thermal power mix for MED plant and the entire system with 30% *RO* 70% *MED*.

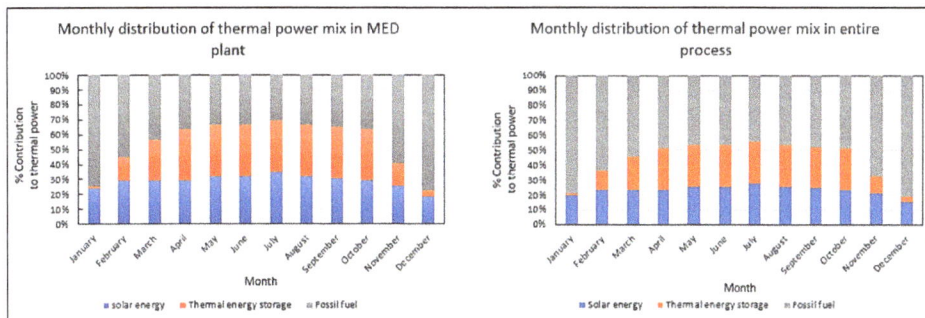

Figure 5. Optimal thermal power mix for the MED plant and the entire system with 60% *RO* 40% *MED*.

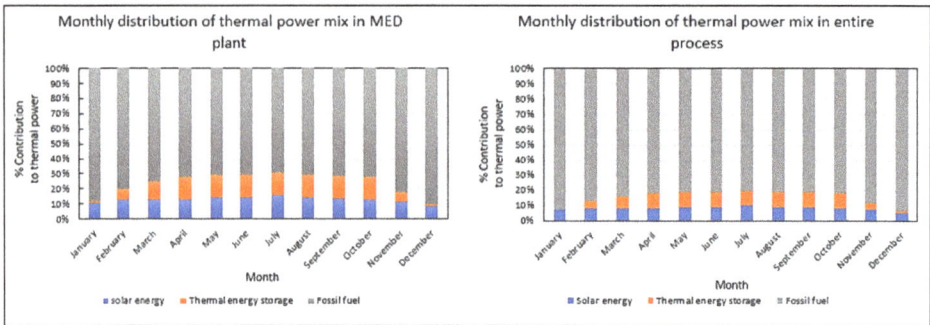

Figure 6. Optimal thermal power mix for the MED plant and the entire system with 80% *RO* 20% *MED*.

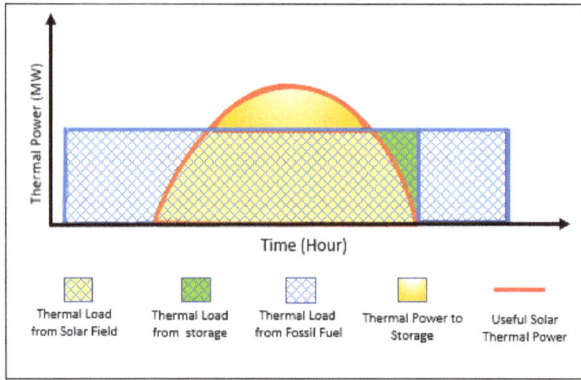

Figure 7. Optimal operation for MED during January, February, November and December.

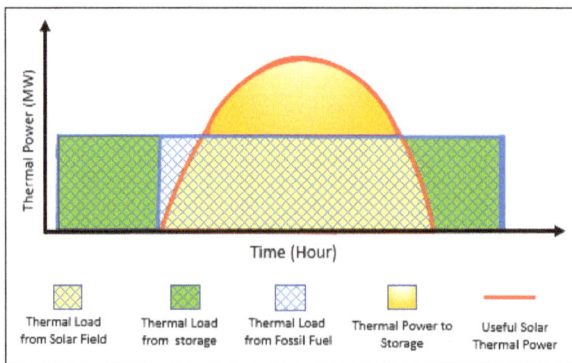

Figure 8. Optimal operation for MED during April, March, May, June, July, August, December and October.

It is observed that the total annual cost of the system as mentioned in the previous section can be reduced by increasing the percentage contribution of RO over MED, but it requires consuming a great amount of fossil fuel. More consumption of fossil fuel causes serious environmental impacts due to emitting a massive amount of CO_2. From the case study, sustaining fossil fuel resources and diminishing the emissions of greenhouse gas require enhancing the percentage contribution of MED in the system based on solar energy as a provider for a high percentage of the thermal power. Figure 9 offers an obvious comparison between the economic and environmental aspects of the system through the different percentage contributions of RO and MED in the total desalinated water production. Reconciliation of economic and environmental objectives can be achieved using a sustainability weighted return on investment calculation [47,48].

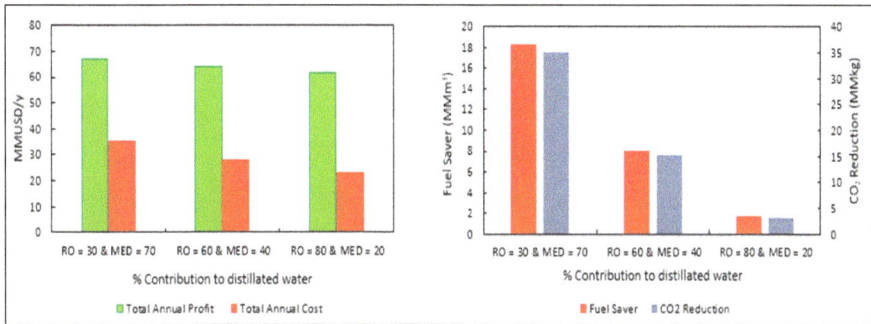

Figure 9. Comparison between the economic and environmental aspects.

The case study shows that in the Eagle Ford fields, 4.4 billion cubic feet of gas were flared in 2013, which represented around 13% of the gas in the formation [55]. Therefore, this significant amount of flared gas can be exploited as a major source of energy for the system or sharing shale gas in a specific percentage as a minor source of energy, and the results of the different percentage contribution of flared gas are shown in Table 4.

Table 4. Technical and economic results for the system.

Percentage Contribution * (%)	Percentage Contribution ** (%)	Total Annual Cost (MM $/year)	Annual Net (After Tax) Profit (MM $/year)	ROI (%)	Payback Period (year)
30 *RO* 70 *MED*	0.0	35.3	50.4	14.9	5.9
30 *RO* 70 *MED*	50	35.1	50.6	14.96	5.6
30 *RO* 70 *MED*	100	34.8	50.8	15	5.5
60 *RO* 40 *MED*	0.0	28.1	48.8	17.2	4.9
60 *RO* 40 *MED*	50	27.8	49	17	4.8
60 *RO* 40 *MED*	100	27.5	49.2	17.3	4.8
80 *RO* 20 *MED*	0.0	23.5	47.7	19.1	4.4
80 *RO* 20 *MED*	50	23.2	47.9	19.2	4.3
80 *RO* 20 *MED*	100	22.8	48.1	19.3	4.3

* The percentage contribution of RO and MED plants in the total desalinated water production; ** the percentage contribution of flared gas as a source of energy.

12. Conclusions

A water-energy nexus framework has been used to address water management in shale gas production. The following key elements have been integrated: solar energy, fossil fuel, cogeneration process, MED and RO. A hierarchical approach and a multi-period MINLP have been developed and solved to find the optimal mix of solar energy, thermal storage and fossil fuel and the optimal usage of water treatment technologies. A case study for Eagle Ford Basin in Texas has been solved to show the applicability of the proposed approach. The system has been analyzed according to the technical, economic and environmental aspects. The multi-period method has been applied to discretize the operational period to track the diurnal fluctuations of solar energy. The percentage utilization of water treatment technologies has been iteratively discretized. Once the solution of the mixed integer nonlinear program (MINLP) was applied to each discretization, the optimal mix of solar energy, thermal storage and fossil fuel, the optimal values of the design and operating variables of the system (e.g., minimum area of a solar collector, maximum capacity of the thermal storage system, etc.) have been determined. The results show the system's economic and environmental merits using a water-energy nexus framework and enabling effective water management strategies while incorporating renewable energy.

Author Contributions: This paper is a collaborative research between the two authors. The general problem definition, methodology, optimization approach, and solution method were developed via discussions between the two authors. Al-Aboosi carried out the computational aspects of the optimization program and the techno-economic analysis for the case study under the supervision of El-Halwagi. Both authors contributed to writing and editing of the manuscript.

Conflicts of Interest: The authors declare no conflict of interest.

Nomenclature

a_0, a_1, a_2, a_3	Correlation constants
a, b and c	Coefficients for the LS-3 collector
AFC^{MED}	Annualized fixed capital cost of the multi-effect desalination
AFC^{RO}	Annualized fixed capital cost of the reverse osmosis
AFC^{SC}	Annualized fixed capital cost of the solar collector
AFC^{cogen}	Annualized fixed capital cost of the cogeneration system
A^{SC}	Effective surface area of the solar collector
A_{SF}	Solar field aperture area
AFC	Total annual fixed cost
AOC	Total annual operating cost
A and B	Parameters that depend on the type of the turbine
bbl	Barrel
c^{Waste}	Value of avoided cost of discharging wastewater
$c_{t,m}^{Fossil}$	Value of fossil fuel
C_{DS}	Disposal cost per volume unit
C_F	Fuel cost per thermal power unit
C_{FW}	Fresh water cost per volume unit
C_{OM}	Operation and maintenance cost per thermal power unit
C_{PST}	Primary and secondary treatment cost per volume unit
C_{SF}	Solar field cost per area unit
C_{SG}	Steam generator system cost per thermal power unit
C_{TES}	Thermal storage system cost per thermal power unit
C_{TR}	Transportation cost per volume unit
Cp_{ms}	Specific heat of the molten salt
Cp_{oil}	Specific heat of oil
do	Outer diameter of the receiver pipe
$D_{t,m}^{Tutbine}$	Design variable of the turbine
DNI	Direct normal irradiance
e_{MED}	Electric energy requirements of MED
e_{RO}	Electric energy requirements of RO
E_T	Turbine shaft power output
$E_{t,m}^{Total}$	Electric energy provided by the cogeneration turbine
ft^3	Cubic feet
f	Focal length of the collectors
FCI_B	Fixed capital cost of a boiler
FCI_{PST}	Fixed capital cost of the primary and secondary treatment

FCI_{SF}	Total fixed capital cost of the solar field
FCI_{SG}	Fixed capital cost estimation of the steam generator system
FCI_T	Fixed capital cost of the turbine
FCI_{TES}	Fixed capital cost of the thermal storage system
FCI_{Total}	Total fixed capital cost
F_f	Soiling factor (mirror cleanliness)
FPW	Flowback and produced water
$F_{t,m}^{Fossil}$	Volumetric flow rate of fossil fuel
$F_{t,m}^{MED}$	Volumetric flow rate of desalinated water from MED
$F_{t,m}^{RO}$	Volumetric flow rate of desalinated water from RO
h_{act}^{out}	Actual outlet enthalpy of the turbine
h^{in}	Inlet enthalpy of the steam
h_{is}^{out}	Outlet isentropic enthalpy
HCE	Sum of heat collection element
$K(\theta)$	Incidence angle modifier
L_{SCA}	Length of a single collector assembly
$L_{spacing}$	Length of spacing between troughs
$\dot{m}$	Inlet turbine steam flowrate
m^{max}	Maximum mass flowrate of the turbine
m_{ms}	Mass flow rate of molten salt
m_{oil}	Mass flowrate of oil
MED	Multi-effect distillation plant
MINLP	Mixed integer nonlinear program
MM	Million
N_P	Factor to account for the operation pressure of the boiler
N_T	Factor accounting for the superheat temperature of the boiler
N	Service life of the property in years
NSRDB	National Solar Radiation Data Base
OC_{OM}	Operation and maintenance cost
O_{EL}	Optical end loss
OC_F	Cost of fuel
$O_{t,m}^{Turbine}$	Operation variable of the turbine
$OPEX_{t,m}^{MED}$	Annualized operational expenditure of MED
$OPEX_{t,m}^{RO}$	Annualized operational expenditure of RO
$OPEX_{t,m}^{SC}$	Annualized operational expenditure of the solar collector
$OPEX_{t,m}^{SC-storage}$	Annualized operational expenditure of the thermal storage system
$OPEX_{t,m}^{cogen}$	Annualized operational expenditure of the cogeneration system
P_g	Gauge pressure of the boiler
PTC	Parabolic trough collector
q_{MED}	Thermal energy requirements of MED
Q_{Boiler}	Thermal power output of the boiler rate
Q_{LFP}	Thermal power that loss from the headers (pipes)
Q_{LFV}	Thermal power that loss from the expansion tank (vessel)
Q_{TES}	Net thermal power inside the tank
Q_{in}	Inlet thermal power
Q_B	Amount of thermal power that produced by the boiler
Q_{acc}	Accumulated thermal power in the tank from preceding iterations
$Q_{collector \rightarrow ambient}$	Total thermal power that loss from a collector to ambient

$Q_{collector \to fluid}$	Thermal power that transferred from a collector to a fluid
$Q_{collector \to reciever}$	Thermal power that absorbed by the receiver tube of a collector loop
Q_{out}	Outlet thermal power
$Q_{solar\ field \to final\ demand}$	Useful thermal power that produced by the solar field
$Q_{sun \to collector}$	Solar thermal power that produced by the solar field
Q_{loss}	Thermal power loss
$Q_{t,m}^{Direct,SC}$	Direct thermal power from the solar thermal collector
$Q_{t,m}^{Fossil}$	Direct thermal power from the combustion of fossil fuels
$Q_{t,m}^{In\_Stored-SC}$	Inlet thermal power of the thermal storage system
$Q_{t,m}^{Out\_Stored-SC}$	Indirect thermal from solar energy through the thermal storage system
$Q_{t,m}^{SC}$	Thermal power captured by the solar collector
$Q_{t,m}^{Stored-Loss}$	Loss thermal power of the thermal storage system
$Q_{t,m}^{Stored-SC}$	Thermal power stored in the thermal storage system
$Q_{t,m}^{Total}$	Total thermal power needs for water treatment
$Q_{t,m}^{Turbine}$	Thermal power from steam leaving the cogeneration turbine
$Q_{t-1,m}^{Stored-SC}$	Thermal power stored from previous iterations
R_{SL}	Row shadow loss
RO	Reverse osmosis plant
ROI	Return on investment
SC	Number of storage capacity hours
T_{CT}	Cold tank temperature
T_{HT}	Hot tank temperature
T_{SH}	Superheat temperature
T_{amb}	Ambient air temperature
T_{in}	Temperature at the inlet of the turbine
T_{ms}	Temperature of the molten salt
T_{rec}	Mean receiver pipe temperature
T_{sat}^{in}	Saturation temperature at the inlet of a turbine
U_{rec}	Overall heat transfer coefficient of the receiver pipe
W_c	Width of the collector aperture
W_w	Volumetric flow rate of discharging wastewater
W	Watt

Subscript and Superscript Symbols

ac	Actual
acc	Accumulated
amb	Ambient
B	Boiler
c	Collector aperture
Cogen	Cogeneration process
CT	Cold tank
DS	Disposal
EL	End loss
f	Factor
F	Fuel
FW	Freshwater
g	Gauge
HT	Hot tank
is	Isentropic
LFP	Loss from pipes
LFV	Loss from vessel

m	Time period (month)
MED	Multi-effect distillation plant
ms	Molten salt
OM	Operation and maintenance
P	Pressure
PST	Primary and secondary treatment
rec	Receiver
RO	Reverse osmosis plant
sat	Saturation
SC	Solar collector
SCA	Single collector assembly
SF	Solar field
SG	Steam generator
SH	Superheat
SL	Shadow loss
t	Time period (h)
T	Turbine
TES	Thermal energy storage
TR	Transportation
w	wastewater

Greek Symbols

η_{boiler}	Efficiency of the boiler
η_{is}	Isentropic efficiency of the steam turbine
a_Y	Annual operation time
$\Omega_{t,m}^{Turbine}$	Vector set of the turbine
$v_{t,m}^{MED}$	Value of produced water from MED
$v_{t,m}^{RO}$	Value of produced water from RO
$\forall m$	For every month (operational period)
$\forall t$	For every hour (sub- period)
Δh_{is}	Isentropic enthalpy change
η_{opt}	Peak optical efficiency of a collector
θ	Solar incidence angle
θz	Solar zenith angle
γ	Intercept factor
δ	Declination
ΔT	Difference between inlet and outlet of the oil
ρ	Reflectivity
τ	Glass transmissivity
ω	Hour angle
α	Absorptivity of the receiver pipe

Appendix A. Solar Data for the Case Study

The solar data for Eagle Ford Shale Play as extracted from National Solar Radiation Data Base (NSRDB) are shown in Tables A1–A4 to represent:

- Average hourly dry bulb temperature (°C)
- Average hourly wet bulb temperature (°C)
- Average hourly direct solar irradiance (W/m^2)
- Average hourly solar incidence angle (degree).

The solar beam radiation is 500 (W/m^2) at the design point.

Table A1. Average hourly dry bulb temperature (°C)

Hour \ Month	January	February	March	April	May	June	July	August	September	October	November	December
0.5	7.1	8.1	13.4	17.3	20.9	23.6	13.4	25.1	24.1	18.9	13.1	8.2
1.5	6.6	7.71	13.0	16.9	20.4	23.3	13.0	24.5	23.6	18.2	12.6	7.7
2.5	6.1	7.24	12.6	16.4	19.9	23.1	12.6	24.0	23.2	17.4	12.3	7.36
3.5	6.0	6.98	12.3	16.2	19.6	23.0	12.3	23.6	22.9	17.1	11.6	7.11
4.5	5.9	6.74	12.0	16.0	19.3	22.8	12.0	23.2	22.6	16.8	11.4	7.13
5.5	5.9	6.49	11.7	15.8	19.0	22.8	11.7	22.8	22.4	16.5	11.3	6.96
6.5	5.5	7.37	12.6	16.8	20.1	23.3	12.6	24.2	22.4	17.9	10.9	7.03
7.5	5.4	8.28	13.5	17.8	21.2	24.6	13.5	25.6	23.7	19.3	11.8	7.21
8.5	7.7	9.20	14.5	18.8	22.3	26.0	14.5	27.0	25.6	20.6	14.0	9.10
9.5	10	11.1	16.2	20.1	23.4	27.3	16.2	28.5	27.0	22.1	16.3	11.0
10.5	12	13.0	17.9	21.4	24.5	28.4	17.9	30.1	28.2	23.6	18.0	12.8
11.5	13	14.9	19.6	22.7	25.6	29.4	19.6	31.6	29.4	25.2	19.3	14.1
12.5	14	15.7	20.5	23.5	26.2	30.4	20.5	32.4	30.3	25.8	20.3	15.1
13.5	15	16.6	21.4	24.4	26.8	31.3	21.4	33.3	30.7	26.5	21.1	16.0
14.5	15	17.5	22.3	25.2	27.5	31.4	22.3	34.1	31.0	27.2	21.3	16.4
15.5	16	17.0	21.7	24.8	27.4	31.7	21.7	33.5	31.2	26.5	21.2	16.5
16.5	15	16.5	21.2	24.4	27.4	31.2	21.2	32.9	31.0	25.8	20.5	16.0
17.5	13	16.1	20.7	23.9	27.3	30.4	20.7	32.3	30.2	25.1	19.0	14.4
18.5	12	14.6	19.1	22.5	26.1	29.0	19.1	30.9	28.8	24.0	17.3	12.7
19.5	10.9	13.21	17.5	21.2	24.95	27.64	17.5	29.53	27.76	22.88	15.84	11.2
20.5	9.73	11.77	16.0	19.8	23.7	26.47	16.0	28.10	26.68	21.75	14.63	10.3
21.5	8.63	10.79	15.3	19.2	23.0	25.44	15.3	27.30	25.93	21.00	13.95	9.77
22.5	7.91	9.825	14.5	18.5	22.3	24.75	14.5	26.46	25.36	20.25	13.45	9.55
23.5	7.56	8.846	13.8	17.7	21.5	24.0	13.8	25.6	24.7	19.6	13.30	9.31

Table A2. Average hourly wet bulb temperature (°C).

Month / Hour	January	February	March	April	May	June	July	August	September	October	November	December
0.5	5.7	6.3	9.85	15.3	18.5	21.6	22.9	22.0	21.5	16.3	11.4	6.41
1.5	5.4	6.0	9.69	15.1	18.3	21.5	22.8	22.0	21.3	15.9	11.1	6.03
2.5	4.9	5.7	9.52	14.9	18.0	21.4	22.7	21.9	21.2	15.4	10.8	5.75
3.5	4.9	5.5	9.43	14.7	17.8	21.4	22.7	21.8	21.0	15.1	10.2	5.55
4.5	4.8	5.3	9.35	14.6	17.6	21.4	22.6	21.6	20.9	14.9	10.1	5.56
5.5	4.8	5.0	9.21	14.5	17.4	21.4	22.6	21.4	20.8	14.6	10.0	5.40
6.5	4.5	5.7	9.64	15.1	18.1	21.7	22.9	22.0	20.8	15.6	9.78	5.44
7.5	4.3	6.3	10.0	15.7	18.8	22.2	23.3	22.6	21.4	16.4	10.3	5.60
8.5	6.1	7.0	10.4	16.3	19.4	22.6	23.4	23.1	22.0	17.2	11.6	6.99
9.5	7.5	8.0	11.3	17.0	19.8	22.7	23.6	23.4	22.2	17.8	12.7	8.08
10.5	8.4	8.9	12.0	17.6	20.1	22.8	23.6	23.4	22.2	18.3	13.3	8.90
11.5	9.1	9.6	12.5	18.1	20.4	23.0	23.5	23.3	22.1	18.7	13.8	9.42
12.5	9.5	10	12.7	18.4	20.7	23.2	23.5	23.3	22.3	18.8	14.0	9.82
13.5	10	10	12.9	18.6	21.0	23.2	23.5	23.2	22.2	18.9	14.2	10.1
14.5	10	10	13.0	18.8	21.2	22.9	23.5	23.0	22.1	19.0	14.1	10.3
15.5	10	10	12.8	18.5	21.1	22.9	23.4	22.8	22.0	18.7	14.1	10.2
16.5	9.8	10	12.5	18.3	20.9	22.8	23.3	22.6	22.0	18.5	13.8	10.0
17.5	9.2	9.8	12.2	18.1	20.7	22.7	23.3	22.3	22.0	18.2	13.3	9.39
18.5	8.6	9.4	11.9	17.6	20.5	22.4	23.4	22.4	21.8	18.0	12.7	8.72
19.5	8.0	8.9	11.4	17.1	20.2	22.3	23.4	22.4	21.8	17.7	12.2	8.13
20.5	7.4	8.3	10.8	16.5	19.8	22.1	23.2	22.1	21.6	17.3	11.7	7.78
21.5	6.9	7.9	10.6	16.3	19.5	22.0	23.2	22.2	21.6	17.1	11.4	7.50
22.5	6.4	7.4	10.3	16.0	19.2	21.9	23.1	22.1	21.6	16.9	11.3	7.37
23.5	6.1	6.8	9.91	15.5	18.7	21.7	22.9	21.9	21.6	16.7	11.4	7.30

Table A3. Average hourly direct solar irradiance (W/m^2)

Hour \ Month	January	February	March	April	May	June	July	August	September	October	November	December
0.5	0	0	0	0	0	0	0	0	0	0	0	0
1.5	0	0	0	0	0	0	0	0	0	0	0	0
2.5	0	0	0	0	0	0	0	0	0	0	0	0
3.5	0	0	0	0	0	0	0	0	0	0	0	0
4.5	0	0	0	0	0	0	0	0	0	0	0	0
5.5	0	0	0	0	5.1	3.8	1	0.0	0	0	0	0
6.5	48	95	9.6	26	109	86	65	57	34	26	1.8	49
7.5	240	244	140	145	216	164	236	229	184	221	171	199
8.5	339	346	287	228	258	319	350	347	315	337	328	272
9.5	396	413	365	281	318	377	467	463	450	460	388	359
10.5	415	487	413	352	362	470	550	524	516	497	462	389
11.5	473	468	478	394	383	496	630	573	557	553	545	459
12.5	457	474	498	439	462	526	621	599	569	566	544	489
13.5	415	440	481	461	460	545	603	600	521	542	504	499
14.5	397	433	417	467	445	520	576	540	540	544	481	440
15.5	283	365	380	473	503	489	529	539	493	498	437	323
16.5	128	246	323	414	434	475	536	417	422	401	361	80
17.5	0.4	32	234	338	356	389	427	323	311	181	93	0
18.5	0	0	54	119	166	217	234	140	53	3.6	0	0
19.5	0	0	0	0.1	7.2	21	24	4.3	0	0	0	0
20.5	0	0	0	0	0	0	0	0	0	0	0	0
21.5	0	0	0	0	0	0	0	0	0	0	0	0
22.5	0	0	0	0	0	0	0	0	0	0	0	0
23.5	0	0	0	0	0	0	0	0	0	0	0	0

Table A4. Average hourly solar incidence angle (degree).

Hour \ Month	January	February	March	April	May	June	July	August	September	October	November	December
0.5	0	0	0	0	0	0	0	0	0	0	0	0
1.5	0	0	0	0	0	0	0	0	0	0	0	0
2.5	0	0	0	0	0	0	0	0	0	0	0	0
3.5	0	0	0	0	0	0	0	0	0	0	0	0
4.5	0	0	0	0	0	0	0	0	0	0	0	0
5.5	0	0	0	0	0	0	0	0	0	0	0	0
6.5	0	0	0	6.04	16.1	20.2	19.2	11.1	0	0	0	0
7.5	0	4.33	7.10	2.51	9.26	13.4	12.3	5.49	4.95	16.1	23.4	0
8.5	30.6	23.6	14.3	4.99	2.85	6.99	5.77	2.49	11.8	23.4	31.8	34.4
9.5	37.8	30.5	20.7	10.9	2.76	1.52	1.14	7.13	18.0	29.8	38.5	41.4
10.5	43.8	36.3	26.1	15.6	7.01	2.69	4.28	11.8	22.9	35.0	44.0	47.1
11.5	48.2	40.6	30.0	18.7	9.73	5.40	7.20	14.9	26.3	38.4	47.6	51.1
12.5	50.2	42.7	31.8	20.0	10.7	6.44	8.40	16.2	27.5	39.3	48.5	52.6
13.5	49.5	42.1	31.0	19.0	9.79	5.70	7.78	15.3	26.2	37.5	46.5	51.1
14.5	46.1	39.0	27.8	16.0	7.06	3.20	5.40	12.7	22.8	33.4	42.2	47.2
15.5	40.7	34.0	22.9	11.5	2.79	0.83	1.58	8.49	17.8	27.8	36.2	41.4
16.5	34.0	27.7	16.9	5.74	2.83	6.15	3.82	3.07	11.7	21.1	29.2	34.4
17.5	21.1	20.4	9.96	2.65	9.23	12.4	10.1	3.61	4.78	11.5	0	0
18.5	0	0	0.17	8.06	16.1	19.2	16.9	10.5	0.99	0	0	0
19.5	0	0	0	0	0	0	0	0	0	0	0	0
20.5	0	0	0	0	0	0	0	0	0	0	0	0
21.5	0	0	0	0	0	0	0	0	0	0	0	0
22.5	0	0	0	0	0	0	0	0	0	0	0	0
23.5	0	0	0	0	0	0	0	0	0	0	0	0

References

1. Zhang, C.; El-Halwagi, M.M. Estimate the Capital Cost of Shale-Gas Monetization Projects. *Chem. Eng. Prog.* **2017**, *113*, 28–32.
2. Al-Douri, A.; Sengupta, D.; El-Halwagi, M.M. Shale Gas Monetization—A Review of Downstream Processing to Chemicals and Fuels. *J. Nat. Gas Sci. Eng.* **2017**, *45*, 436–455.
3. Ortiz-Espinoza, P.A.; Jiménez-Gutiérreza, A.; Nourledin, M.; El-Halwagi, M.M. Design, Simulation and Techno-Economic Analysis of Two Processes for the Conversion of Shale Gas to Ethylene. *Comp. Chem. Eng.* **2017**, *107*, 237–246. [CrossRef]
4. Pérez-Uresti, S.I.; Adrián-Mendiola, J.M.; El-Halwagi, M.M.; Jiménez-Gutiérrez, A. Techno-Economic Assessment of Benzene Production from Shale Gas. *Processes* **2017**, *5*, 33. [CrossRef]
5. Jasper, S.; El-Halwagi, M.M. A Techno-Economic Comparison of Two Methanol-to-Propylene Processes. *Processes* **2015**, *3*, 684–698. [CrossRef]
6. Julián-Durán, L.; Ortiz-Espinoza, A.P.; El-Halwagi, M.M.; Jiménez-Gutiérrez, A. Techno-economic assessment and environmental impact of shale gas alternatives to methanol. *ACS Sustain. Chem. Eng.* **2014**, *2*, 2338–2344. [CrossRef]
7. Salkuyeh, Y.K.; Adams, T.A., II. A novel polygeneration process to co-produce ethylene and electricity from shale gas with zero CO_2 emissions via methane oxidative coupling. *Energy Convers. Manag.* **2015**, *92*, 406–420. [CrossRef]
8. Ehlinger, M.V.; Gabriel, K.J.; Noureldin, M.M.B.; El-Halwagi, M.M. Process Design and Integration of Shale Gas to Methanol. *ACS Sustain. Chem. Eng.* **2014**, *2*, 30–37. [CrossRef]
9. Salkuyeh, Y.K.; Adams, T.A., II. Shale gas for the petrochemical industry: Incorporation of novel technologies. In *Computer Aided Chemical Engineering*; Elsevier: New York, NY, USA, 2014; Volume 34, pp. 603–608.
10. Hasaneen, R.; El-Halwagi, M.M. Integrated Process and Microeconomic Analyses to Enable Effective Environmental Policy for Shale Gas in the United States. *Clean Technol. Environ. Policy* **2017**, *19*, 1775–1789. [CrossRef]
11. Arredondo-Ramírez, K.; Ponce-Ortega, J.M.; El-Halwagi, M.M. Optimal Planning and Infrastructure Development of Shale Gas. *Energy Convers. Manag.* **2016**, *119*, 91–100. [CrossRef]
12. Gao, J.; You, F. Optimal design and operations of supply chain networks for water management in shale gas production: MILFP model and algorithms for the water-energy nexus. *AIChE J.* **2015**, *61*, 1184–1208. [CrossRef]
13. Carrero-Parreño, A.; Onishi, V.C.; Salcedo-Díaz, R.; Ruiz-Femenia, R.; Fraga, E.S.; Caballero, J.A.; Reyes-Labarta, J.A. Optimal Pretreatment System of Flowback Water from Shale Gas Production. *Ind. Eng. Chem. Res.* **2017**, *56*, 4386–4398. [CrossRef]
14. Yang, L.; Grossmann, I.E.; Manno, J. Optimization models for shale gas water management. *AIChE J.* **2014**, *60*, 3490–3501. [CrossRef]
15. Nicot, J.-P.; Scanlon, B.R. Water Use for Shale-Gas Production in Texas. *U.S. Environ. Sci. Technol.* **2012**, *46*, 3580–3586. [CrossRef] [PubMed]
16. Elsayed, A.N.; Barrufet, M.A.; Eljack, F.T.; El-Halwagi, M.M. Optimal Design of Thermal Membrane Distillation Systems for the Treatment of Shale Gas Flowback Water. *Int. J. Membr. Sci. Technol.* **2015**, *2*, 1–9.
17. Boschee, P. Produced and Flowback Water Recycling and Reuse: Economics, Limitations, and Technology. *Oil Gas Facil.* **2014**, *3*, 16–21. [CrossRef]
18. Lira-Barragán, L.; Ponce-Ortega, J.M.; Guillén-Gosálbez, G.; El-Halwagi, M.M. Optimal Water Management under Uncertainty for Shale Gas Production. *Ind. Eng. Chem. Res.* **2016**, *55*, 1322–1335. [CrossRef]
19. Chen, H.; Carter, K.E. Water usage for natural gas production through hydraulic fracturing in the United States from 2008 to 2014. *J. Environ. Manag.* **2016**, *170*, 152–159. [CrossRef] [PubMed]
20. Shaffer, D.L.; Chavez, L.H.A.; Ben-Sasson, M.; Castrillón, S.R.; Yip, N.Y.; Elimelech, M. Desalination and reuse of high-salinity shale gas produced water: Drivers, technologies, and future directions. *Environ. Sci. Technol.* **2013**, *47*, 9569–9583. [CrossRef] [PubMed]
21. Lokare, O.R.; Tavakkoli, S.; Rodriguez, G.; Khanna, V.; Vidic, R.D. Integrating membrane distillation with waste heat from natural gas compressor stations for produced water treatment in Pennsylvania. *Desalination* **2017**, *413*, 144–153. [CrossRef]

22. Tavakkoli, S.; Lokare, O.R.; Vidic, R.D.; Khanna, V. A techno-economic assessment of membrane distillation for treatment of Marcellus shale produced water. *Desalination* **2017**, *416*, 24–34. [CrossRef]
23. Bamufleh, H.; Abdelhady, F.; Baaqeel, H.M.; El-Halwagi, M.M. Optimization of Multi-Effect Distillation with Brine Treatment via Membrane Distillation and Process Heat Integration. *Desalination* **2017**, *408*, 110–118. [CrossRef]
24. Gabriel, K.; El-Halwagi, M.M.; Linke, P. Optimization Across Water-Energy Nexus for Integrating Heat, Power, and Water for Industrial Processes Coupled with Hybrid Thermal-Membrane Desalination. *Ind. Eng. Chem. Res.* **2016**, *55*, 3442–3466. [CrossRef]
25. Elsayed, A.N.; Barrufet, M.A.; El-Halwagi, M.M. An Integrated Approach for Incorporating Thermal Membrane Distillation in Treating Water in Heavy Oil Recovery using SAGD. *J. Unconv. Oil Gas Resour.* **2015**, *12*, 6–14. [CrossRef]
26. González-Bravo, R.; Nápoles-Rivera, F.; Ponce-Ortega, J.M.; Nyapathi, M.; Elsayed, N.A.; El-Halwagi, M.M. Synthesis of Optimal Thermal Membrane Distillation Networks. *AIChE J.* **2015**, *61*, 448–463. [CrossRef]
27. González-Bravo, R.; Elsayed, N.A.; Ponce-Ortega, J.M.; Nápoles-Rivera, F.; Serna-González, M.; El-Halwagi, M.M. Optimal Design of Thermal Membrane Distillation Systems with Heat Integration with Process Plants. *Appl. Therm. Eng.* **2014**, *75*, 154–166. [CrossRef]
28. Elsayed, A.N.; Barrufet, M.A.; El-Halwagi, M.M. Integration of Thermal Membrane Distillation Networks with Processing Facilities. *Ind. Eng. Chem. Res.* **2014**, *53*, 5284–5298. [CrossRef]
29. Tovar-Facio, J.; Eljack, F.; Ponce-Ortega, J.M.; El-Halwagi, M.M. Optimal Design of Multiplant Cogeneration Systems with Uncertain Flaring and Venting. *ACS Sustain. Chem. Eng.* **2016**, *5*, 675–688. [CrossRef]
30. Pham, V.; Laird, C.; El-Halwagi, M.M. Convex Hull Discretization Approach to the Global Optimization of Pooling Problems. *Ind. Eng. Chem. Res.* **2009**, *48*, 1973–1979. [CrossRef]
31. Gabriel, F.; El-Halwagi, M.M. Simultaneous Synthesis of Waste Interception and Material Reuse Networks: Problem Reformulation for Global Optimization. *Environ. Prog.* **2005**, *24*, 171–180. [CrossRef]
32. El-Halwagi, M.M. *Sustainable Design through Process Integration: Fundamentals and Applications to Industrial Pollution Prevention, Resource Conservation, and Profitability Enhancement*, 2nd ed.; IChemE/Elsevier: New York, NY, USA, 2017.
33. Khor, S.C.; Foo, D.C.Y.; El-Halwagi, M.M.; Tan, R.R.; Shah, N. A Superstructure Optimization Approach for Membrane Separation-Based Water Regeneration Network Synthesis with Detailed Nonlinear Mechanistic Reverse Osmosis Model. *Ind. Eng. Chem. Res.* **2011**, *50*, 13444–13456. [CrossRef]
34. Alnouri, S.; Linke, P.; El-Halwagi, M.M. Synthesis of Industrial Park Water Reuse Networks Considering Treatment Systems and Merged Connectivity Options. *Comp. Chem. Eng.* **2016**, *91*, 289–306. [CrossRef]
35. El-Halwagi, M.A.; Manousiouthakis, V.; El-Halwagi, M.M. Analysis and Simulation of Hollwo Fiber Reverse Osmosis Modules. *Sep. Sci. Technol.* **1996**, *31*, 2505–2529. [CrossRef]
36. El-Halwagi, M.M. Synthesis of Optimal Reverse-Osmosis Networks for Waste Reduction. *AIChE J.* **1992**, *38*, 1185–1198. [CrossRef]
37. Goswami, D.Y.; Kreith, F. *Energy Conversion*; CRC Press: Boca Raton, FL, USA, 2007.
38. Mittelman, G.; Epstein, M. A novel power block for CSP systems. *Sol. Energy* **2010**, *84*, 1761–1771. [CrossRef]
39. Eck, M.; Hirsch, T.; Feldhoff, J.F.; Kretschmann, D.; Dersch, J.; Morales, A.G.; Gonzalez-Martinez, L.; Bachelier, C.; Platzer, W.; Riffelmann, K.-J. Guidelines for CSP yield analysis–optical losses of line focusing systems; definitions, sensitivity analysis and modeling approaches. *Energy Procedia* **2014**, *49*, 1318–1327. [CrossRef]
40. Channiwala, S.; Ekbote, A. A generalized model to estimate field size for solar-only parabolic trough plant. In Proceedings of the 3rd Southern African Solar Energy Conference, Skukuza, South Africa, 11–13 May 2015.
41. Lovegrove, K.; Stein, W. *Concentrating Solar Power Technology: Principles, Developments and Applications*; Elsevier: New York, NY, USA, 2012.
42. Quaschning, V.; Kistner, R.; Ortmanns, W. Influence of direct normal irradiance variation on the optimal parabolic trough field size: A problem solved with technical and economical simulations. *J. Sol. Energy Eng.* **2002**, *124*, 160–164. [CrossRef]
43. Herrmann, U.; Kearney, D.W. Survey of thermal energy storage for parabolic trough power plants. *J. Sol. Energy Eng.* **2002**, *124*, 145–152. [CrossRef]
44. Al-Azri, N.; Al-Thubaiti, M.; El-Halwagi, M.M. An Algorithmic Approach to the Optimization of Process Cogeneration. *J. Clean Technol. Environ. Policy* **2009**, *11*, 329–338. [CrossRef]

45. Mavromatis, S.; Kokossis, A. Conceptual optimisation of utility networks for operational variations—I. Targets and level optimisation. *Chem. Eng. Sci.* **1998**, *53*, 1585–1608. [CrossRef]
46. Bamufleh, S.H.; Ponce-Ortega, J.M.; El-Halwagi, M.M. Multi-objective optimization of process cogeneration systems with economic, environmental, and social tradeoffs. *Clean Technol. Environ. Policy* **2013**, *15*, 185–197. [CrossRef]
47. El-Halwagi, M.M. A Return on Investment Metric for Incorporating Sustainability in Process Integration and Improvement Projects. *Clean Technol. Environ. Policy* **2017**, *19*, 611–617. [CrossRef]
48. Guillen-Cuevas, K.; Ortiz-Espinoza, A.P.; Ozinan, E.; Jiménez-Gutiérrez, A.; Kazantzis, N.K.; El-Halwagi, M.M. Incorporation of Safety and Sustainability in Conceptual Design via A Return on Investment Metric. *ACS Sustain. Chem. Eng.* **2018**, *6*, 1411–1416. [CrossRef]
49. Kondash, J.A.; Albright, E.; Vengosh, A. Quantity of flowback and produced waters from unconventional oil and gas exploration. *Sci. Total Environ.* **2017**, *574*, 314–321. [CrossRef] [PubMed]
50. Atilhan, S.; Linke, P.; Abdel-Wahab, A.; El-Halwagi, M.M. A Systems Integration Approach to the Design of Regional Water Desalination and Supply Networks. *Int. J. Process Syst. Eng.* **2011**, *1*, 125–135. [CrossRef]
51. Ghaffour, N.; Missimer, T.M.; Amy, G.L. Technical review and evaluation of the economics of water desalination: Current and future challenges for better water supply sustainability. *Desalination* **2013**, *309*, 197–207. [CrossRef]
52. Mezher, T.; Fath, H.; Abbas, Z.; Khaled, A. Techno-economic assessment and environmental impacts of desalination technologies. *Desalination* **2011**, *266*, 263–273. [CrossRef]
53. National Renewable Energy Laboratory. *Assessment of Parabolic Trough and Power Tower Solar Technology Cost and Performance Forecasts*; DIANE Publishing: Collingdale, PA, USA, 2003.
54. Price, H. A parabolic trough solar power plant simulation model. In Proceedings of the ASME International Solar Energy Conference, Kohala Coast, HI, USA, 15–18 March 2003.
55. Horwitt, D.; Sumi, L. *Up in Flames: US Shale Oil Boom Comes at Expense of Wasted Natural Gas, Increased CO₂*; Earthworks: Washington, DC, USA, 2014.
56. RPSEA. Advanced Treatment of Shale Gas Fracturing Water to Produce Re-Use or Discharge Quality Water. 2015. Retrieved November, 2017. Available online: https://www.rpsea.org/node/222 (accessed on 18 Novemeber 2017).
57. Giuliano, S.; Buck, R.; Eguiguren, S. Analysis of solar-thermal power plants with thermal energy storage and solar-hybrid operation strategy. *J. Sol. Energy Eng.* **2011**, *133*, 031007. [CrossRef]

Article

Building Block-Based Synthesis and Intensification of Work-Heat Exchanger Networks (WHENS)

Jianping Li, Salih Emre Demirel and M. M. Faruque Hasan *

Artie McFerrin Department of Chemical Engineering, Texas A&M University, College Station, TX 77843-3122, USA; ljptamu@tamu.edu (J.L.); emredemirel@tamu.edu (S.E.D.)
* Correspondence: hasan@tamu.edu; Tel.: +1-979-862-1449

Received: 16 November 2018; Accepted: 1 January 2019; Published: 7 January 2019

Abstract: We provide a new method to represent all potential flowsheet configurations for the superstructure-based simultaneous synthesis of work and heat exchanger networks (WHENS). The new representation is based on only two fundamental elements of abstract building blocks. The first design element is the block interior that is used to represent splitting, mixing, utility cooling, and utility heating of individual streams. The second design element is the shared boundaries between adjacent blocks that permit inter-stream heat and work transfer and integration. A semi-restricted boundary represents expansion/compression of streams connected to either common (integrated) or dedicated (utility) shafts. A completely restricted boundary with a temperature gradient across it represents inter-stream heat integration. The blocks interact with each other via mass and energy flows through the boundaries when assembled in a two-dimensional grid-like superstructure. Through observation and examples from literature, we illustrate that our building block-based WHENS superstructure contains numerous candidate flowsheet configurations for simultaneous heat and work integration. This approach does not require the specification of work and heat integration stages. Intensified designs, such as multi-stream heat exchangers with varying pressures, are also included. We formulate a mixed-integer non-linear (MINLP) optimization model for WHENS with minimum total annual cost and demonstrate the capability of the proposed synthesis approach through a case study on liquefied energy chain. The concept of building blocks is found to be general enough to be used in possible discovery of non-intuitive process flowsheets involving heat and work exchangers.

Keywords: WHENS; work and heat integration; building blocks; superstructure; MINLP

---

1. Introduction

Heat and work are used as the primary energy utilities in most chemical process plants. Both heat and work are interchangeable, and it is imperative that we consider them together when we perform energy integration. In this regard, the work and heat exchanger network synthesis (WHENS) is a class of design problems that aims to simultaneously optimize heat and work exchangers and their networks [1–3]. WHENS improves energy efficiency and brings economic benefits to energy systems [4]. Significant research has been done in the past in heat exchanger network synthesis (HENS) [5]. Work exchange network synthesis (WENS) [6–11] has also gained attention in recent years. However, WHENS problems are more challenging compared to individual HENS and work exchange network (WEN) problems [2]. Fu and Gundersen [12] defined a WHENS problem as follows: "Given a set of process streams with supply and target states (temperature and pressure), as well as utilities for power, heating and cooling; design a work and heat exchange network of heat transfer equipment such as heat exchangers, evaporators and condensers, as well as pressure-changing equipment such as compressors, expanders, pumps and valves, in such a way that the exergy consumption is minimized or the exergy production is maximized". Apart from exergy, other objectives of WHENS may include cost minimization, utility reduction, and equipment reduction.

An indicative list of recent contributions in WHENS research is provided in Table 1. These contributions can be broadly classified into pinch technology-based graphical approaches and mathematical programming-based optimization approaches. Pinch analysis relies on fundamental thermodynamic insights and involves appropriate placement [13] and grand composite curves [14–16]. Though significant progress has been made in terms of theoretical development [17,18] and methodological advances [19,20], there are several limitations of pinch analysis. Firstly, this approach is time-consuming when applied to systems involving many process streams [2]. Secondly, the stream identity (hot/cold, high/low-pressure) and the starting and final states of each process stream must be specified *a priori*. Mathematical programming-based optimization approaches, e.g., refs. [3,21–24], overcome some of these limitations. However, they require a suitable representation of all candidate network configurations. This can be done by developing a superstructure, which is a giant flowsheet incorporating many alternative configurations [25–27]. To this end, a comprehensive but intelligent representation of the superstructure is critical to include as many network configurations and flowsheet candidates as possible, while keeping the corresponding mathematical program computationally tractable [28].

There exist several superstructure representations in the WHENS literature [23], e.g., state-space representation [29], multi-stage superstructure [3,24] and representation involving heuristics [30]. However, these superstructures suffer from several fundamental limitations. Firstly, one needs to pre-postulate all equipment configurations in the superstructure based on existing knowledge of unit operations, engineering experience and heuristics. If one excludes the best configuration as one of the alternatives in the superstructure, then it will be never discovered. Given the complexity, interchangeability and trade-offs between work and heat exchange networks, this inability to incorporate innovation could sometimes result in inferior solutions. With increasing competitions and awareness for energy sustainability, there is a need for incorporating novel and "out-of-the-box" solutions when solving a WHENS problem.

Secondly, pathways leading to novel intensified designs are neglected in classic superstructure representations. Process intensification refers to significant reduction of equipment sizes, waste generation, and increase of productivity [31]. New opportunities could arise in WHENS through incorporating process intensification principles. It can bring about new technologies which are smaller, cleaner, safer, and more energy-efficient [32–34]. To this end, the goal of WHENS and process intensification are often complementary to each other. For example, one could use a multi-stream heat exchanger (MHEX) instead of two-stream exchangers that would reduce the number of equipment and, at the same time, would improve the overall performance of a work-heat exchange network. Few works considered incorporating MHEXs in heat integration. Hasan et al. exploited a stagewise superstructure to find the optimal heat exchanger network (HEN) that best represents the operational of MHEXs using historical data [35]. Rao and Karimi addressed MHEXs based on a single-stage superstructure consisting of two-stream exchangers [36].

To summarize, a key challenge in WHENS is to systematically discover and screen both existing and novel, classic and intensified alternative configurations. Superstructure provides an excellent means to automatically generate many network configurations, but the traditional superstructures could still miss innovative solutions due to a lack of representation. To this end, Hasan and co-workers have recently put forward a novel superstructure representation using abstract building blocks for systematic process synthesis and intensification [37–40]. With a generic block representation, there is no need to pre-specify the stream and equipment identities and flowsheet configurations. Streams can intermittently change their identities as needed. Classical and intensified equipment are configured automatically. Furthermore, there is no need to specify any work and heat integration stages. Therefore, the block representation could potentially avoid the above limitations when applied to WHENS.

Table 1. An indicative list of recent contributions in WHENS literature.

Reference	Approach	Application/Case Studies
Wechsung, Aspelund, Gundersen, Barton (2011) [29]	Combination of pinch analysis, exergy analysis, and optimization to find heat exchanger network (HEN) with minimal irreversibility by varying pressure levels of process streams	An offshore natural gas liquefaction process
Razib, Hasan, Karimi (2012) [7]	First formalization of an optimization-based systematic work exchange network (WEN) synthesis problem	Integration among high-pressure and low-pressure streams
Dong, Yu, Zhang (2014) [21]	Superstructure optimization for heat, mass and pressure exchange networks with exergoeconomic analysis	Wastewater distribution network in a petroleum refining process
Onishi, Ravagnani, Caballero (2014a) [41]	Superstructure optimization for HEN design with pressure recovery	Cryogenic process design
Onishi, Ravagnani, Caballero (2014b) [22]	MINLP-based WHENS using a multi-stage superstructure for optimal pressure recovery of process gaseous streams	Integration among high-pressure and low-pressure streams
Fu and Gundersen (2015a,b,c,d) [14–16,42]	Graphical methodology for HEN design including compressors or expanders to minimize exergy consumption above or below ambient temperature	Integration of process streams with supply and target states
Huang and Karimi (2016) [3]	MINLP-based approach to synthesize WHENS for optimized selection of end-heaters and end-coolers to meet the desired temperature targets	Integration among high-pressure and low-pressure streams and a transport chain for stranded natural gas
Fu, Gundersen (2016) [13]	Correct integration of both compressors and expanders in HEN to minimize exergy consumption	Integration of process streams with the same supply and target temperatures
Fu, Gundersen (2016c) [12]	Graphical methodology using thermodynamic insights for WHENS	CO_2 capture processes
Onishi, Ravagnani, Caballero (2017) [43]	Multi-objective optimization of WHENS using a multi-stage superstructure	Integration among process streams based on economic and environmental criteria
Zhuang, Liu, Liu, Du (2017) [44]	Synthesis of direct work exchange network (WEN) in adiabatic process involving heat integration based on transshipment model	Integration of high-pressure and low-pressure streams in a chemical plant
Nair, Rao, Karimi (2018) [24]	MINLP-based general WHENS framework considering stream temperature, pressure and/or phase changes without *a priori* classification of stream identity	C3 splitting and offshore liquefied natural gas (LNG) processes

In this work, we formalize and employ the concept of abstract building blocks to represent all alternative configurations within a superstructure for synthesis problems involving simultaneous work and heat integration (WHENS). The remaining of the article is structured as follows. First, we elaborate the representation of work and heat exchange networks using building blocks in Section 2. Next, we present a mixed-integer nonlinear formulation (MINLP) for WHENS in Section 3. We demonstrate the applicability of our approach with a case study on WHENS in Section 4. Finally, we present some concluding remarks.

2. A Building Block Representation of WHENS

WHENS is more complex than HENS and WENS. HENS involves several specified hot and cold streams with initial and final temperatures. A hot stream undergoes successive cooling either using a cold utility (e.g., cooling water or a refrigerant) in coolers or through exchanging heat with one or more cold streams using heat exchangers. Similarly, a cold stream undergoes successive heating either through using a hot utility (e.g., steam) or through directly integrating heat with one or more hot streams. Heat can be also recovered from hot streams using a working fluid which then transfers that heat to cold streams. Similar to HENS, WENS involves high-pressure and low-pressure streams with specified flow rates and initial and final pressure ratings. A high-pressure stream undergoes successive release in pressure through valves or expanders. A low-pressure stream undergoes successive compression using movers such as pumps and compressors. If the movers are dedicated to individual streams and use single shafts, then they need utility (e.g., electricity, steam turbine). However, if an expander and a compressor share a common shaft, then the shaft work generated by the expander is integrated with the compressor. Thus, an integration of work is achieved. Unlike HENS and WENS, WHENS involves process streams that might undergo both temperature and pressure changes (sometimes in multiple stages) to achieve the target temperature and pressure ratings. Therefore, WHENS involves more than two types of streams, which can be initially (i) hot and high-pressure; (ii) hot and low-pressure; (iii) cold and high-pressure; (iv) cold and low-pressure; and (v) neutral (e.g., a refrigerant circulating through multiple equipment in a refrigeration cycle). Furthermore, the interchangeability of work and heat is often reflected in an intermittent change of stream identities. For example, an initially hot and high-pressure stream may become a cold and low-pressure stream after excessive expansion. Similarly, an initially cold stream can later become a hot stream through compression.

The goal in WHENS is to identify the optimal unit operations and equipment sizing involving mixing, splitting, cooling, heating, pressurizing, and depressurizing (note that inter-stream mixing is not allowed). In this section, we first describe how we can create representations for different types of unit operations in WHENS by using only two fundamental design elements of abstract building blocks, namely a block interior and a block boundary. We then discuss the details of building blocks and the construction of a block-based WHENS superstructure.

2.1. Elements of Building Block Representation

The new representation is based on the concept of "abstract building blocks" originally proposed by Hasan and co-workers for general process design, integration and intensification [37–39]. Each building block has two fundamental design elements. The first design element is the block interior that is used to represent splitting, mixing, utility cooling and utility heating of individual streams (Figure 1a). Each block interior is assigned with a temperature, a pressure, a composition, and a phase. The second design element is the shared boundaries between two adjacent blocks that permit inter-stream heat and work transfer and integration. Specifically, each block has four boundaries (left, right, top and bottom). Each of these boundaries can be one of the three types: unrestricted, semi-restricted, and completely restricted (Figure 1b). An unrestricted boundary is assigned when the two blocks sharing this boundary have the same pressure and composition (temperatures and phases can be different). A semi-restricted boundary represents expansion/compression of a stream while leaving a block to another with a different pressure. The pressurizing/depressurizing is done through either common (integrated) or dedicated (utility) shafts. (Please note that, in the original representation of Demirel et al. [37], a semi-restricted boundary assumes a more general task as an interface for mass/heat/energy transfer. In WHENS, we need it only to represent work transfer, which simplifies the model). A completely restricted boundary between two blocks with a temperature gradient represents inter-stream heat integration using a common heat exchanger (e.g., shell and tube exchanger with the cold fluid in the shell-side and the hot fluid in the tube-side).

2.2. Equipment Representation

Using the basic concepts of a building block as described above, we can represent more intricate and complex processes. To do this, we need to orient multiple blocks in a two-dimensional grid-like structure. These blocks will interact with each other via mass and energy flows through various boundaries, and automatically generate many alternative equipment and flowsheet configurations.

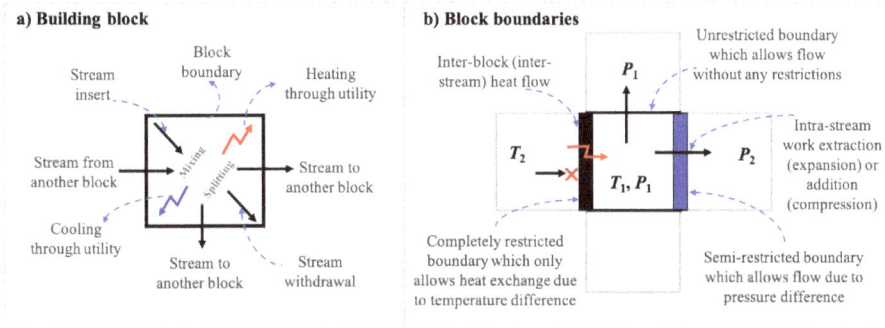

Figure 1. Elements of abstract building blocks: (**a**) block interior (**b**) block boundary.

For example, Figure 2a shows a block representation and its corresponding flowsheet configuration of an operation involving a single stream undergoing a pressure change. The block representation is given by two blocks $B_{1,1}$ and $B_{1,2}$, separated by a semi-restricted (blue) boundary. (From now on, $B_{i,j}$ will represent a block placed in row i and column j). The stream with pressure P_1 is fed into block $B_{1,1}$ and goes through the boundary to achieve the target pressure P_2. Depending on the inlet and outlet pressures, this boundary is assigned with an expander/valve (when $P_1 \geq P_2$) or a compressor (when $P_1 \leq P_2$). In Figure 2b, two high-pressure streams, (HP_1 and HP_2) are integrated with a low-pressure stream (LP_1). These two high-pressure streams pass through two expanders Exp_1 and Exp_2 respectively to achieve the desired pressure. The pressure of stream LP_1 is increased to the target pressure after two compressors (Com_1 and Com_2). The corresponding block representation involves 3×3 building blocks. Feed HP_1 is supplied into block $B_{1,1}$ and withdrawn as product in block $B_{1,2}$ while feed HP_2 is supplied into block $B_{1,2}$ and withdrawn as product in block $B_{2,2}$. The low-pressure stream LP_1 is fed into block $B_{1,3}$ and withdrawn from block $B_{3,2}$. Please note that there are four semi-restricted boundaries in this block representation, which include the right boundaries of block $B_{1,1}$, $B_{2,1}$ and $B_{3,2}$ and the bottom boundary of block $B_{1,3}$. Specifically, the right boundaries of block $B_{1,1}$ and block $B_{2,1}$ are assigned with expanders. The bottom boundary of block $B_{1,3}$ and the right boundary of $B_{3,2}$ are assigned with compressors. The expander at right boundary of block $B_{1,1}$ and the compressor at the right boundary of block $B_{3,2}$ are sharing shaft 1 (both are marked as blue). The expander at right boundary of block $B_{2,1}$ and the compressor at bottom boundary of block $B_{1,3}$ are sharing shaft 2 (both are marked as red). For illustration purpose, the boundary between block $B_{2,3}$ and block $B_{3,3}$ is specified as unrestricted boundary, where the pressure $P_{2,3}$ and $P_{3,3}$ are the same.

In the case of heat exchange (Figure 2c), a cold stream C_1 is integrated with a hot stream H_1 through a heat exchanger. Two equivalent block representations are presented. On involves a representation with two blocks. The left block allows the entering of cold stream C_1 and produces the product stream with the desired temperature. Hot stream H_1 enters the right block and is withdrawn from the same block. This heat exchanger is represented through a completely restricted boundary between block $B_{1,1}$ and $B_{1,2}$. The energy flow is transferred from block $B_{1,2}$ to block $B_{1,1}$. Another representation involves more blocks but better captures the relation of temperature change. Cold stream C_1 enters block $B_{1,1}$ with inlet temperature as $T_{1,1}$ and flows into block $B_{1,2}$ with target temperature as $T_{2,1}$. Hot stream H_1 enters block $B_{1,2}$ with inlet temperature as $T_{1,2}$ and is withdrawn from block $B_{2,2}$ with temperature as $T_{2,2}$. The right boundary of block $B_{2,1}$ is a completely restricted boundary. As shown in Figure 2d, a block with multiple completely restricted boundary can represent an MHEX. The cold stream C_1 enters block $B_{1,2}$ and takes heat from hot stream H_1 in block $B_{1,1}$ and from hot stream H_2 in block $B_{1,3}$.

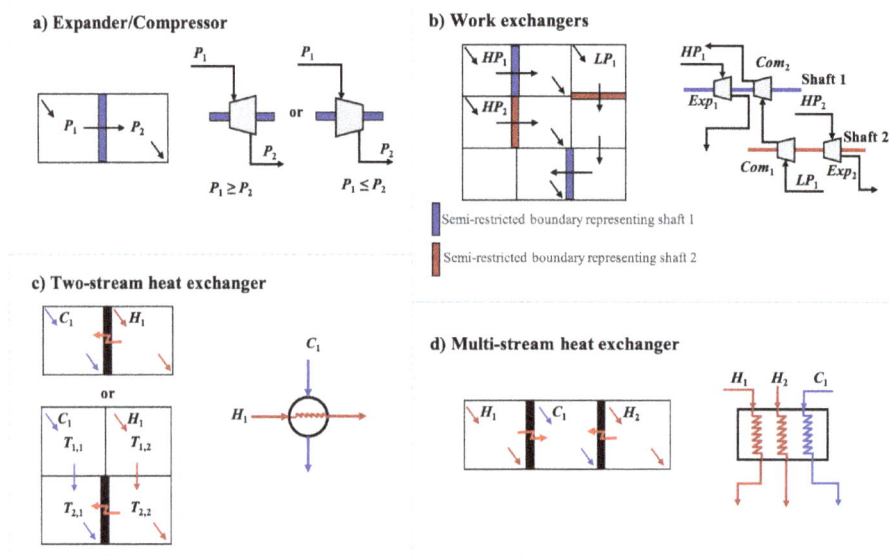

Figure 2. Equipment representations using building blocks for work and heat exchanger network: (**a**) Expander/compressor; (**b**) Work-exchanger shafts for work integration; (**c**) Two-stream exchanger for heat integration; (**d**) Multi-stream heat exchanger (MHEX).

2.3. Flowsheet Representation

As we add more blocks in the 2-D grid assembly, we enlarge the space for representing more and more equipment and flowsheet alternatives in a single structure. The versatility of the block representation can be seen in Figure 3, where blocks are used to represent WHENS superstructures taken from a range of literature problems such as separation system for propane and propylene [24], liquefied energy chain [24,29], general work and heat integration process [13], and single mixed refrigerant (SMR) process [45]. For instance, the separation system involving three process streams S_1, S_2 and S_3 is represented by a block representation with $i = 5$ and $j = 4$ to involve all connectivities for work and heat integration. S_1 enters block $B_{2,1}$ and flows through a valve before entering block $B_{3,1}$. S_2 with varying identity is supplied into block $B_{2,3}$ and is compressed at the bottom boundary of block $B_{2,3}$. The heat transfer happens at right boundary of block $B_{3,1}$ between stream S_1 and stream S_2, right boundary of $B_{2,2}$ between cold side of S_2 and hot side of S_2, right boundary of $B_{4,3}$ between stream S_2 and S_3 and right boundary of $B_{5,4}$ between stream S_2 and stream S_3.

A process with multi-stream heat exchanger is shown in Figure 3d. The natural gas (NG) feed enters block $B_{2,1}$ and flows into the block $B_{3,1}$ with its right boundary as completely restricted boundary. The NG stream goes through a valve assigned on the bottom boundary of block $B_{3,1}$ before entering block $B_{5,1}$ with a flash boundary. The details of these separation boundaries can be found in Li et al. [38] Refrigerant flow enters block $B_{2,2}$ and undergoes sequential compression and cooling in block $B_{1,2}$, $B_{1,3}$, $B_{1,4}$ and $B_{2,4}$. The outlet stream from block $B_{3,4}$ serves as a hot stream, supplying heat to the same stream after valve operation at the right boundary of block $B_{4,2}$. The MHEX is represented by a block, i.e., $B_{3,2}$ with two completely restricted boundary. Based on the representation approach, we develop the corresponding MINLP model for WHENS, which is discussed in the next section.

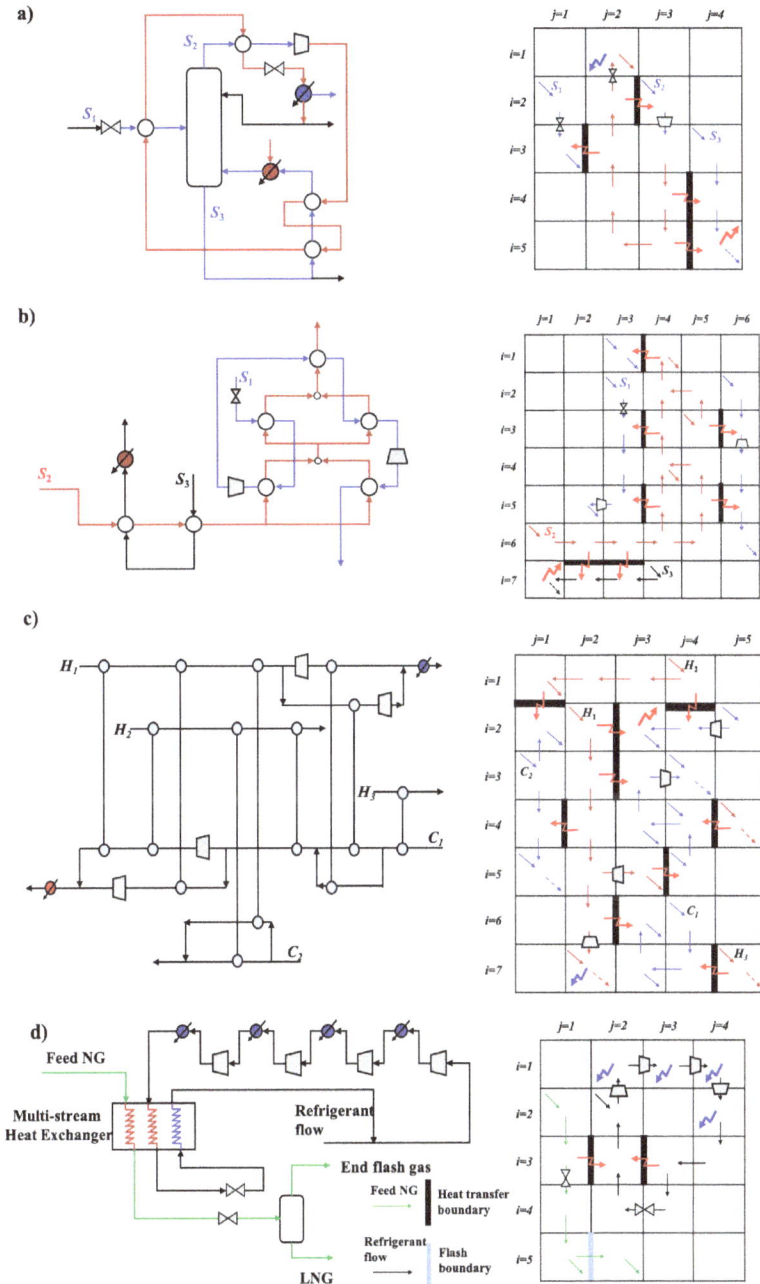

Figure 3. Various flowsheets and networks representations for work and heat integration in WHENS: (**a**) Work and heat exchange network for a separation system. (**b**) Work and heat exchange network for liquefied energy chain. (**c**) Work and heat exchange network with three hot streams and two cold streams. (**d**) Work and heat exchange network for single mixed refrigerant (SMR) process.

2.4. Block Superstructure for WHENS

As shown in Figures 2 and 3, the block representation indicates towards a unified approach for WHENS while accounting for the interplay of pressure and temperature. As we infer more, a generalized two-dimensional grid-like orientation of building blocks can be used to contain numerous flowsheet configurations for simultaneous heat and work integration. To this end, our block-based WHENS superstructure is shown in Figure 4. This representation consists of building blocks arranged in a grid with I number of rows and J number of columns. Feed f with component flowrate as $M_{i,j,k,f}$ and product streams p with component flowrate as $H_{i,j,k,p}$ are potentially supplied into or withdrawn from block $B_{i,j}$. Each block has temperature and pressure attributes as $T_{i,j}$ and $P_{i,j}$. These blocks are connected to each other through adjacent connecting streams $F_{i,j,k,d}$ and jump connecting streams $J_{i,j,i',j',k}$ (see black arrows and gray arrows in Figure 4a respectively).

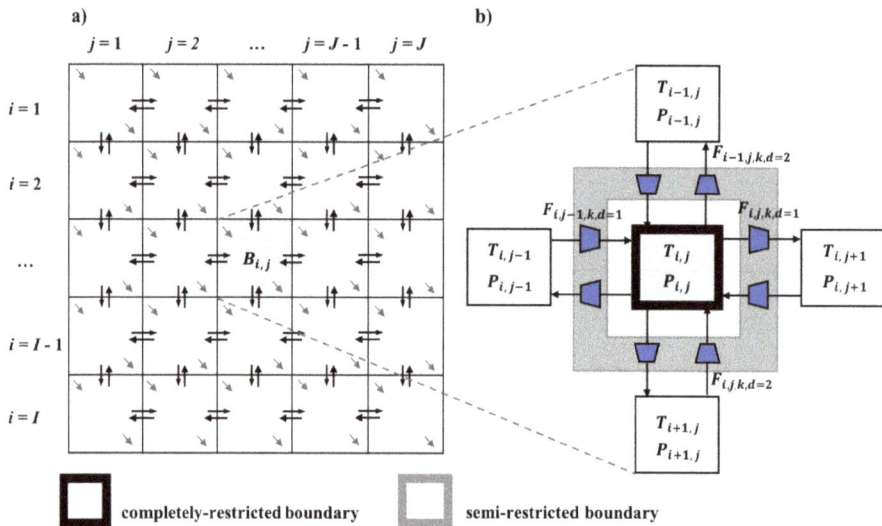

Figure 4. A general superstructure representation using building blocks for work and heat integration in WHENS: (**a**) General block representation; (**b**) Interaction of blocks through boundaries and connecting flows.

Each adjacent connecting stream has both positive and negative components as $FP_{i,j,k,d}$ and $FN_{i,j,k,d}$ to allow more network alternatives. $FP_{i,j,k,d}$ designates the flow from block $B_{i,j}$ to block $B_{i,j+1}$ in horizontal direction ($d = 1$) or the flow from block $B_{i+1,j}$ to block $B_{i,j}$ in vertical direction ($d = 2$). $FN_{i,j,k,d}$ designates the flow from block $B_{i,j+1}$ to block $B_{i,j}$ in horizontal direction ($d = 1$) or the flow from block $B_{i,j}$ to block $B_{i+1,j}$ in vertical direction ($d = 2$). Besides, jump flow from block $B_{i,j}$ to block $B_{i',j'}$ with component flowrate as $J_{i,j,i',j',k}$ is introduced to avoid unnecessary intermediate blocks for transferring material and energy flow, where i' and j' designate the row and column position of a different block in the block superstructure.

Adjacent blocks are separated via unrestricted, semi-restricted or completely restricted boundary. Both unrestricted and semi-restricted boundaries allow mass and energy flow. The thermodynamic driving force at these boundaries enables the changes of temperature and pressure. Unrestricted boundary allows mass and energy flow while ensuring the inlet pressure equal to outlet pressure, i.e., the block pressure $P_{i,j}$ in adjacent blocks separated by unrestricted boundary are the same. Semi-restricted boundary allows pressure change between adjacent blocks and hence indicates the existence of a pressure exchanger, i.e., compressors, expanders, or valves (shown in Figure 4b gray box). Completely restricted boundary prohibits mass flow while allowing energy flow (shown in Figure 4b black box). The existence of completely restricted boundary indicates a heat exchanger between two streams in the adjacent blocks. When a block includes more than one completely restricted boundaries, this block can be regarded as an MHEX. The inlet pressure for pressure exchangers are $P_{i,j}$ when the adjacent connecting streams across the semi-restricted boundary are outlet flow from block $B_{i,j}$. The outlet pressure for these pressure exchangers are $P_{i,j+1}$ in horizontal direction and $P_{i+1,j}$ in vertical direction when $F_{i,j,k,d}$ is coming out from block $B_{i,j}$. The block temperature $T_{i,j}$ is the temperature of outlet streams from block $B_{i,j}$. With the block temperature $T_{i,j}$ and pressure $P_{i,j}$, the unit enthalpy $EH_{i,j,k}$ in block $B_{i,j}$ for component k can be determined. In addition to the heat transfer happening at completely restricted boundary, each block also allows external utility stream to supply extra heat duty $Q_{i,j}^h$ or cold duty $Q_{i,j}^c$.

3. MINLP Model for WHENS

We now present an MINLP model for WHENS using building block-based superstructure. The overall problem is described as follows. When given a set $FS = \{f|f = 1,...,|FS|\}$ of inlet process streams with temperature and pressure specifications as T_f^{feed} and P_f^{feed}, and a set $PS = \{p|p = 1,...,|PS|\}$ of outlet process streams with target temperature and pressure ranges as $[T_p^{min}, T_p^{max}]$ and $[P_p^{min}, P_p^{max}]$, respectively, synthesize the optimal work and heat exchanger network that minimizes the total annual cost. The MINLP model will involve block material and energy balances, flow directions, work calculations, phase relations, boundary and task assignments, and logic constraints. The known flowrates of inlet process streams is designated as F_f^{feed}. The objective is to synthesize a work and heat exchanger network that captures the interplay of pressure and temperature to minimize the total annual cost. The set $D = \{d|d = 1,2\}$ designates the flow alignment. The flow alignment $d = 1$ when the stream is flowing in the horizontal direction, i.e., from block $B_{i,j}$ to $B_{i,j+1}$; $d = 2$ when the stream is flowing in the vertical direction, i.e., from block $B_{i,j}$ to $B_{i+1,j}$. The temperature range and flowrate range for all connecting flows including direct connecting flow and jump connecting flow is set as $[T^{min}, T^{max}]$ and $[FL, FU]$ respectively. The assumptions for this work are continuous steady-state operation, adiabatic expansion/compression, and linear relation of stream enthalpy with pressure and temperature. With these, we now describe the MINLP model for WHENS based on block superstructure.

3.1. Block Material Balance

The generic material balance is imposed on each individual block. The inlet flow for component k at block $B_{i,j}$ includes horizontal inlet flow $F_{i,j-1,k,1}$, vertical inlet flow $F_{i-1,j,k,2}$, external feed stream $M_{i,j,k}^f$, and inlet stream via jump flow $J_{i,j,k}^f$. The outlet flow for component k at block $B_{i,j}$ includes the horizontal outlet flow $F_{i,j,k,1}$, vertical outlet flow $F_{i,j,k,2}$, external product stream $H_{i,j,k}^p$ and outlet stream via jump flow $J_{i,j,k}^p$.

$$F_{i,j-1,k,1} - F_{i,j,k,1} + F_{i-1,j,k,2} - F_{i,j,k,2} + M_{i,j,k}^f - H_{i,j,k}^p + J_{i,j,k}^f - J_{i,j,k}^p = 0 \quad i \in I, j \in J, k \in K \quad (1)$$

We set $F_{i=I,j,k,1} = F_{i,j=J,k,2} = 0$ to avoid other interactions between the superstructure and the environment except those through external feeds and products. External feed stream of component k, $M_{i,j,k}^f$, collect the component flowrate, $M_{i,j,k,f}$, from all available feed f. External product stream $H_{i,j,k}^p$ is the summation of component flowrate, $H_{i,j,k,p}$, from all possible product stream p in block $B_{i,j}$. Similarly, $J_{i,j,k}^f$ and $J_{i,j,k}^p$ are determined from the jump connecting flow $J_{i,j,i',j',k}$. These are achieved through the following constraints.

$$M_{i,j,k}^f = \sum_{f \in FS} M_{i,j,k,f} \quad i \in I, j \in J, k \in K \quad (2)$$

$$H_{i,j,k}^{p} = \sum_{p \in PS} H_{i,j,k,p} \quad i \in I, j \in J, k \in K \tag{3}$$

$$J_{i,j,k}^{f} = \sum_{(i',j') \in LN} J_{i',j',i,j,k} \quad i \in I, j \in J, k \in K \tag{4}$$

$$J_{i,j,k}^{p} = \sum_{(i',j') \in LN} J_{i,j,i',j',k} \quad i \in I, j \in J, k \in K \tag{5}$$

Here the set LN collects all jump connectivities from block $B_{i,j}$ to block $B_{i',j'}$.

We define a feed fraction variable $0 \leq z_{i,j,f}^{feedfrac} \leq 1$ for feed stream f in block $B_{i,j}$. Therefore, the component flowrate through feed f into block $B_{i,j}$, $M_{i,j,k,f}$ can be determined as follows:

$$M_{i,j,k,f} = F_{f}^{feed} y_{k,f}^{feed} z_{i,j,f}^{feedfrac}, \quad i \in I, j \in J, k \in K, f \in FS \tag{6}$$

$$0 \leq \sum_{i \in I} \sum_{j \in J} z_{i,j,f}^{feedfrac} \leq 1, \quad f \in FS \tag{7}$$

where F_{f}^{feed} is the maximum available amount of feed stream f. $y_{k,f}^{feed}$ is composition of component k in feed f. Summation of these feed fraction variables will be less than 1 if the overall feed amount is less than F_{f}^{feed}.

The flowrate range for product stream p is restricted from minimum product flowrate D_{p}^{L} to maximum product flowrate D_{p}^{U}.

$$D_{p}^{L} \leq \sum_{i \in I} \sum_{j \in J} \sum_{k \in K} H_{i,j,k,p} \leq D_{p}^{U}, \quad p \in PS \tag{8}$$

In general, the flowrate range for product stream p is equal to inlet flowrate, indicating $D_{p}^{L} = D_{p}^{U}$.

3.2. Flow Directions

We consider the connecting stream $F_{i,j,k,d}$ between adjacent blocks as a bidirectional flow. Its positive and negative components are $FP_{i,j,k,d}$ and $FN_{i,j,k,d}$ respectively. The selection of flow direction is achieved through the following binary variable:

$$z_{i,j,d}^{Plus} = \begin{cases} 1 & \text{if } F_{i,j,k,d} \text{ is from block } B_{i,j} \text{ to } B_{i,j+1} \ (d=1) \text{ or from block } B_{i,j} \text{ to } B_{i+1,j} \ (d=2) \\ 0 & \text{otherwise} \end{cases}$$

The following constraints ensure that only one component of this connecting stream is activated:

$$F_{i,j,k,d} = FP_{i,j,k,d} - FN_{i,j,k,d} \quad i \in I, j \in J, k \in K, d \in D \tag{9}$$

$$FP_{i,j,k,d} \leq FU z_{i,j,d}^{Plus}, \quad i \in I, j \in J, k \in K, d \in D \tag{10}$$

$$FN_{i,j,k,d} \leq FU(1 - z_{i,j,d}^{Plus}), \quad i \in I, j \in J, k \in K, d \in D \tag{11}$$

Besides, with these stream components, the overall inlet flow is the summation of all incoming streams into block $B_{i,j}$:

$$\phi_{i,j,k} = FP_{i,j-1,k,1} + FN_{i,j,k,1} + FP_{i-1,j,k,2} + FN_{i,j,k,2} + J_{i,j,k}^{f} + \sum_{f} M_{i,j,k,f} \quad i \in I, j \in J, k \in K \tag{12}$$

3.3. Block Energy Balance

The block energy balance includes stream enthalpy, feed enthalpy, product enthalpy, external heating/cooling, work energy associated with expansion/compression and contacting energy flow across the block boundary. Then the steady-state energy balance for block $B_{i,j}$ is formulated as follows:

$$EB_{i,j}^{in} - EB_{i,j}^{out} + EM_{i,j} - EP_{i,j} + Q_{i,j} + W_{i,j} + EF_{i,j-1,1} + EF_{i-1,j,2} - EF_{i,j,1} - EF_{i,j,2} = 0 \quad i \in I, j \in J \quad (13)$$

where $EB_{i,j}^{in}$ and $EB_{i,j}^{out}$ are inlet enthalpy and outlet enthalpy streams to block $B_{i,j}$ respectively. $EM_{i,j}$ is the overall stream enthalpy carried by feed streams. $EP_{i,j}$ is the overall product enthalpy carried by product streams. $Q_{i,j}$ is the enthalpy supplied by utility streams into block $B_{i,j}$. $W_{i,j}$ is the work energy supplied through compression operation or expansion operation for block $B_{i,j}$ respectively. Besides, $EJ_{i,j}$ is the stream enthalpy carried along with the jump connecting streams. $EF_{i,j,d}$ represents the energy flow going through the completely restricted boundary and indicates the amount of heat transfer between adjacent streams. These energy flow variables are shown in Figure 5.

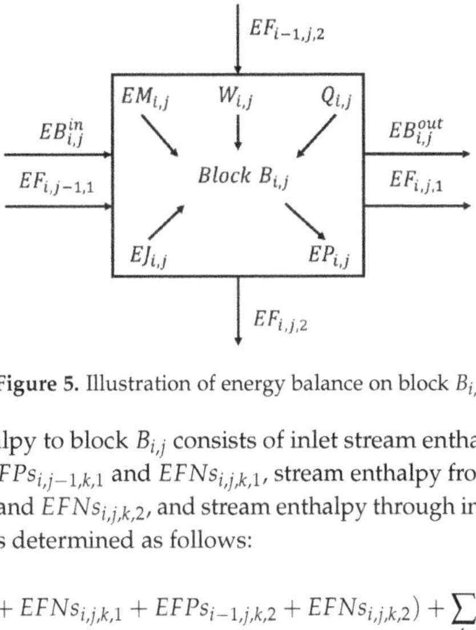

Figure 5. Illustration of energy balance on block $B_{i,j}$.

The inlet stream enthalpy to block $B_{i,j}$ consists of inlet stream enthalpy from adjacent blocks in horizontal directions, i.e., $EFPs_{i,j-1,k,1}$ and $EFNs_{i,j,k,1}$, stream enthalpy from adjacent blocks in vertical directions, i.e., $EFPs_{i-1,j,k,2}$ and $EFNs_{i,j,k,2}$, and stream enthalpy through inlet jump connecting streams, i.e., $EJ_{i',j',i,j,k}$. Hence, $EB_{i,j}^{in}$ is determined as follows:

$$EB_{i,j}^{in} = \sum_{k}(EFPs_{i,j-1,k,1} + EFNs_{i,j,k,1} + EFPs_{i-1,j,k,2} + EFNs_{i,j,k,2}) + \sum_{k}\sum_{(i,j)\in LN} EJ_{i',j',i,j,k} \quad \forall i,j \quad (14)$$

Similarly, the outlet stream enthalpy to block $B_{i,j}$, $EB_{i,j}^{out}$ is determined as follows:

$$EB_{i,j}^{out} = \sum_{k}(EFPs_{i,j,k,1} + EFPs_{i,j,k,2} + EFNs_{i-1,j,k,2} + EFNs_{i,j-1,k,1} + \sum_{(i',j')\in LN} EJ_{i,j,i',j',k}) \quad \forall i,j \quad (15)$$

These stream enthalpies are determined based on the flowrate and the unit enthalpy. For outlet streams from block $B_{i,j}$ in flow alignment d, the initial unit enthalpy for these streams are the block enthalpy $EH_{i,j,k}$. For the inlet stream to block $B_{i,j}$ in horizontal direction with flow alignment $d = 1$, the initial unit enthalpy is the unit enthalpy $EH_{i,j+1,k}$ in block $B_{i,j+1}$. For the inlet stream to block $B_{i,j}$ in vertical direction with flow alignment $d = 2$, the initial unit enthalpy is the enthalpy $EH_{i+1,j,k}$ in block $B_{i+1,j}$.

$$EFPs_{i,j,k,d} = FP_{i,j,k,d}EH_{i,j,k} \quad i \in I, j \in J, k \in K, d \in D \quad (16)$$

$$EFNs_{i,j,k,1} = FN_{i,j,k,d}EH_{i,j+1,k} \quad i \in I, j \in J, k \in K, d \in D \quad (17)$$

$$EFNs_{i,j,k,2} = FN_{i,j,k,d}EH_{i+1,j,k} \quad i \in I, j \in J, k \in K, d \in D \quad (18)$$

The feed stream enthalpy in block $B_{i,j}$ is based on the flowrate of feed streams and stream enthalpy of feed streams EF_f.

$$EM_{i,j} = \sum_{k,f} M_{i,j,k,f} EF_f \quad i \in I, j \in J, k \in K, f \in FS \tag{19}$$

EF_f is a parameter determined by the feed temperature and pressure.

The product stream enthalpy from block $B_{i,j}$ is based on the flowrate of product streams and stream enthalpy in this block.

$$EP_{i,j} = \sum_{k,p} H_{i,j,k,p} EH_{i,j,k} \quad i \in I, j \in J, (k,p) \in kps \tag{20}$$

The utility enthalpy term consists of hot utility, $Q_{i,j}^h$, and cold utility $Q_{i,j}^c$, which are supplied into block $B_{i,j}$.

$$Q_{i,j} = Q_{i,j}^h - Q_{i,j}^c \quad i \in I, j \in J \tag{21}$$

To obtain the fixed cost of heaters and coolers, we define the following two binary variables:

$$z_{i,j}^{hot} = \begin{cases} 1 & \text{if block } B_{i,j} \text{ involves hot utility} \\ 0 & \text{otherwise} \end{cases}$$

$$z_{i,j}^{cod} = \begin{cases} 1 & \text{if block } B_{i,j} \text{ involves cold utility} \\ 0 & \text{otherwise} \end{cases}$$

It is straightforward to relate heat duty of heaters and coolers with these two binary variables:

$$Q_{i,j}^h \leq EU z_{i,j}^{hot}, \quad Q_{i,j}^c \leq EU z_{i,j}^{cod} \quad i \in I, j \in J \tag{22}$$

Here EU is the upper bound of stream enthalpy.

The work energy is determined by the amount of work added into or taken out of block $B_{i,j}$, which are denoted as $W_{i,j}^{com}$ for compression and $W_{i,j}^{exp}$ for expansion, respectively. The calculation of $W_{i,j}^{com}$ and $W_{i,j}^{exp}$ is explained later in this Section 3.8.

$$W_{i,j} = W_{i,j}^{com} - W_{i,j}^{exp} \quad i \in I, j \in J \tag{23}$$

The stream enthalpy across the completely restricted boundary is either in the positive direction ($z_{i,j,d}^{Plus} = 1$) or in the negative direction ($z_{i,j,d}^{Plus} = 0$).

$$EFP_{i,j,d} \leq EU z_{i,j,d}^{fplus} \quad i \in I, j \in J, d \in D \tag{24}$$

$$EFN_{i,j,d} \leq EU(1 - z_{i,j,d}^{fplus}) \quad i \in I, j \in J, d \in D \tag{25}$$

3.4. Product Stream Assignments and Logical Constraints

We define binary variables for each product stream p at block $B_{i,j}$ to determine whether they are active in $B_{i,j}$ or not:

$$z_{i,j,p}^{product} = \begin{cases} 1 & \text{if product stream } p \text{ is withdrawn from block } B_{i,j} \\ 0 & \text{otherwise} \end{cases}$$

The identification of block as product block is achieved through the following logical relation, which involves product binary variable.

$$\sum_{k \in K} P_{i,j,k,p} \leq D_p^U z_{i,j,p}^{product} \quad i \in I, j \in J, p \in PS \tag{26}$$

For each block, there are at most one type of product stream present in block $B_{i,j}$. The logic proposition is illustrated as follows:

$$\sum_{p \in PS} z_{i,j,p}^{product} \leq 1 \quad i \in I, j \in J \tag{27}$$

At least one stream for product p appears in the block superstructure.

$$\sum_{i \in I}\sum_{j \in J} z_{i,j,p}^{product} \geq 1 \quad p \in PS \tag{28}$$

The temperature range for block with product stream p is $\left[T_p^{min}, T_p^{max}\right]$.

$$T_p^{min} z_{i,j,p}^{product} + T^{min}\left(1 - z_{i,j,p}^{product}\right) \leq T_{i,j} \leq T_p^{max} z_{i,j,p}^{product} + T^{max}\left(1 - z_{i,j,p}^{product}\right) \quad i \in I, j \in J, p \in PS \tag{29}$$

Likewise, the pressure range for product block is $\left[P_p^{min}, P_p^{max}\right]$ as follows.

$$P_p^{min} z_{i,j,p}^{product} + P^{min}\left(1 - z_{i,j,p}^{product}\right) \leq P_{i,j} \leq P_p^{max} z_{i,j,p}^{product} + P^{max}\left(1 - z_{i,j,p}^{product}\right) \quad i \in I, j \in J, p \in PS \tag{30}$$

We impose the following constraint to tighten the bounds of block pressure $P_{i,j}$ for blocks not involving product streams. This constraint states that if the block includes component k, then the block pressure $P_{i,j}$ is larger than the minimum product pressure P_p^{min}.

$$P_{i,j} \geq P^{min}\left(1 - z_{i,j,k}^{mix}\right) + P_p^{min} z_{i,j,k}^{mix} \quad k, p \in kps(k, p) \tag{31}$$

Here, the set $kps(k, p)$ specifies the type of component k in product p.

Similarly, if the block includes component k, then the block temperature $T_{i,j}$ is correspondingly bounded above the minimum product temperature T_p^{min}.

$$T_{i,j} \geq T^{min}\left(1 - z_{i,j,k}^{mix}\right) + T_p^{min} z_{i,j,k}^{mix} \quad k, p \in kps(k, p) \tag{32}$$

In WHENS problem, we assume mixing of different streams. Hence, we define the following binary variable to decide which component is allowed to exist in block $B_{i,j}$.

$$z_{i,j,k}^{mix} = \begin{cases} 1 & \text{if component } k \text{ exists in block } B_{i,j} \\ 0 & \text{otherwise} \end{cases}$$

The following constraints ensure that at most one component is allowed in block $B_{i,j}$ and all other inlet component streams are prohibited from entering this block.

$$\phi_{i,j,k} \leq \phi_{i,j,k}^{up} z_{i,j,k}^{mix} \quad i \in I, j \in J, k \in K \tag{33}$$

$$\sum_k z_{i,j,k}^{mix} \leq 1 \quad i \in I, j \in J \tag{34}$$

If the block $B_{i,j}$ supplies product stream p, the required component k should exists in this block. Similarly, if the block $B_{i,j}$ takes feed stream f, the feed components k are inside the same block.

$$z_{i,j,k}^{mix} \geq z_{i,j,p}^{p} \quad i \in I, j \in J, (k, p) \in kps(k, p) \tag{35}$$

$$z_{i,j,k}^{mix} \geq z_{i,j,f}^{feed,frac} \quad i \in I, j \in J, (k, f) \in kfs(k, f) \tag{36}$$

Here the set $kfs(k, f)$ specifies all components k in feed f.

The existence of component k in block $B_{i,j}$ facilitate the tightening the bounds of block temperature $T_{i,j}$.

$$\sum_k T_k^{mink} z_{i,j,k}^{mix} \le T_{i,j} \le \sum_k T_k^{maxk} z_{i,j,k}^{mix} \quad i \in I, j \in J \tag{37}$$

Here T_k^{maxk} and T_k^{mink} are maximum and minimum temperature in the system with $T_k^{maxk} = \max(\max_{f \in FS} T_f^{feed}, \max_{p \in pS} T_p^{max})$ and $T_k^{mink} = \min(\min_{f \in FS} T_f^{feed}, \min_{p \in PS} T_p^{min})$

3.5. Boundary Assignment

The block boundaries are assigned as either unrestricted, semi-restricted or completely restricted boundary. If there is no pressure change across adjacent blocks through their connecting streams, then the inter-block boundary is unrestricted. If there is pressure change across a boundary, then the boundary is semi-restricted. If there is no mass allowed to flow between adjacent blocks, then the inter-block boundary is identified as completely restricted boundary.

$$z_{i,j,d}^{ur} = \begin{cases} 1 & \text{If boundary between } B_{i,j} \text{ and } B_{i,j+1} \text{ for } d = 1 \text{ (between } B_{i,j} \text{ and } B_{i+1,j} \text{ for } d = 2) \\ & \text{is unrestricted} \\ 0 & \text{Otherwise} \end{cases}$$

$$z_{i,j,d}^{sr} = \begin{cases} 1 & \text{If boundary between } B_{i,j} \text{ and } B_{i,j+1} \text{ for } d = 1 \text{ (between } B_{i,j} \text{ and } B_{i+1,j} \text{ for } d = 2) \\ & \text{is semi-restricted} \\ 0 & \text{Otherwise} \end{cases}$$

$$z_{i,j,d}^{cr} = \begin{cases} 1 & \text{If boundary between } B_{i,j} \text{ and } B_{i,j+1} \text{ for } d = 1 \text{ (between } B_{i,j} \text{ and } B_{i+1,j} \text{ for } d = 2) \\ & \text{is completely restricted} \\ 0 & \text{Otherwise} \end{cases}$$

Only one type of the boundaries is activated between two adjacent blocks.

$$z_{i,j,d}^{ur} + z_{i,j,d}^{sr} + z_{i,j,d}^{cr} = 1 \quad i \in I, j \in J, d \in D \tag{38}$$

Mass flow is prohibited while energy flow is allowed across a completely restricted boundary between adjacent blocks.

$$FP_{i,j,k,d} + FN_{i,j,k,d} \le FU(1 - z_{i,j,d}^{cr}) \quad i \in I, j \in J, d \in D \tag{39}$$

$$EFP_{i,j,d} + EFN_{i,j,d} \le EUz_{i,j,d}^{cr} \quad i \in I, j \in J, d \in D \tag{40}$$

3.6. Phase Relation and Stream Enthalpies

Each block has phase assignment according to the components existing in it, temperature, and pressure condition. We define binary variables for liquid phase and gas phase in block $B_{i,j}$. This phase relations are adapted from the work of Nair et al. [24].

$$z_{i,j}^l = \begin{cases} 1 & \text{If liquid phase exists in block } B_{i,j} \\ 0 & \text{Otherwise} \end{cases}$$

$$z_{i,j}^v = \begin{cases} 1 & \text{If gas phase exists in block } B_{i,j} \\ 0 & \text{Otherwise} \end{cases}$$

Besides the following 0–1 continuous variable is defined for the two-phase zone in block $B_{i,j}$.

$$z_{i,j}^{lv} = \begin{cases} 1 & \text{If gas phase exists in block } B_{i,j} \\ \dfrac{T_{i,j} - T_{i,j}^{BP}}{T_{i,j}^{DP} - T_{i,j}^{BP}} & \text{if gas and liquid phase coexist in block } B_{i,j} \\ 0 & \text{If liquid phase exists in block } B_{i,j} \end{cases}$$

$$z_{i,j}^l + z_{i,j}^{lv} \leq 1 \quad i \in I, j \in J \tag{41}$$

$$z_{i,j}^v \leq z_{i,j}^{lv} \quad i \in I, j \in J \tag{42}$$

The enthalpy expression for liquid phase and gas phase is linearly dependent on the block temperature $T_{i,j}$ and block pressure $P_{i,j}$.

$$H_{i,j,k}^l = a_k^l T_{i,j} + b_k^l P_{i,j} + c_k^l \quad i \in I, j \in J, k \in k \tag{43}$$

$$H_{i,j,k}^v = a_k^v T_{i,j} + b_k^v P_{i,j} + c_k^v \quad i \in I, j \in J, k \in k \tag{44}$$

Here a_k^l, b_k^l, c_k^l, a_k^v, b_k^v and c_k^v are parameters used for determining the enthalpy of liquid and gas phase.

The enthalpy expression for two-phase region is approximated as the linear segment between enthalpy at bubble point $T_{i,j}^{BP}$ and that at dew point $T_{i,j}^{DP}$.

$$H_{i,j,k}^{lv} = z_{i,j}^{lv}(H_{i,j,k}^v - H_{i,j,k}^l) + H_{i,j,k}^l \quad i \in I, j \in J, k \in k \tag{45}$$

The bubble point and dew point for component k in block $B_{i,j}$ are linearly dependent on the block pressure $P_{i,j}$.

$$BP_{i,j,k} = a_k^b P_{i,j} + b_k^b \quad i \in I, j \in J, k \in k \tag{46}$$

$$DP_{i,j,k} = a_k^d P_{i,j} + b_k^d \quad i \in I, j \in J, k \in k \tag{47}$$

Here a_k^b, b_k^b, a_k^d, and b_k^d are parameters for determining bubble point and dew point.

The general bubble point and dew point temperatures are then assigned to block bubble temperature $T_{i,j}^{BP}$ and block dew temperature $T_{i,j}^{DP}$, if component k exists in block $B_{i,j}$.

$$BP_{i,j,k} - T^{max}(1 - z_{i,j,k}^{mix}) \leq T_{i,j}^{BP} \leq BP_{i,j,k} + T^{max}(1 - z_{i,j,k}^{mix}) \quad i \in I, j \in J, k \in k \tag{48}$$

$$DP_{i,j,k} - T^{max}(1 - z_{i,j,k}^{mix}) \leq T_{i,j}^{DP} \leq DP_{i,j,k} + T^{max}(1 - z_{i,j,k}^{mix}) \quad i \in I, j \in J, k \in k \tag{49}$$

The definition of $z_{i,j}^{lv}$ is achieved through the following constraint.

$$T^{min} z_{i,j}^l + (1 - z_{i,j}^{lv} - z_{i,j}^l)BP_{i,j,k} + z_{i,j}^{lv} DP_{i,j,k} \leq T_{i,j} \leq T^{max} z_{i,j}^v + (1 - z_{i,j}^{lv})BP_{i,j,k} + DP_{i,j,k}(z_{i,j}^{lv} - z_{i,j}^v))$$
$$i \in I, j \in J, k \in k \tag{50}$$

When $z_{i,j}^l = 0$ and $z_{i,j}^v = 0$, the above constraint is reduced to $z_{i,j}^{lv} = (T_{i,j} - T_{i,j}^{BP})/(T_{i,j}^{DP} - T_{i,j}^{BP})$.

Block temperature $T_{i,j}$ is related with $z_{i,j}^l$ and $z_{i,j}^v$ respectively through the following two constraints.

$$T_{i,j}^{BP} - T^{max} z_{i,j}^l \leq T_{i,j} \leq T_{i,j}^{BP} + T^{max}(1 - z_{i,j}^l) \quad i \in I, j \in J \tag{51}$$

$$T_{i,j}^{DP} - T^{max}(1 - z_{i,j}^v) \leq T_{i,j} \leq T_{i,j}^{DP} + T^{max} z_{i,j}^v \quad i \in I, j \in J \tag{52}$$

Similarly, the obtained enthalpy expressions $H^l_{i,j,k}$, $H^v_{i,j,k}$, and $H^{lv}_{i,j,k}$ map to the block enthalpy $EH_{i,j,k}$ via $z^l_{i,j}$, $z^v_{i,j}$, $z^{lv}_{i,j}$ and $z^{mix}_{i,j,k}$ through the following big-M constraints.

$$H^l_{i,j,k} - EH^{max}_{i,j,k}(1 - z^{mix}_{i,j,k}) \leq EH_{i,j,k} \leq H^v_{i,j,k} + EH^{max}_{i,j,k}(1 - z^{mix}_{i,j,k}) \quad i \in I, j \in J, k \in K \tag{53}$$

$$H^v_{i,j,k} - dH^{max}_{DPT,k}(1 - z^v_{i,j}) \leq EH_{i,j,k} \leq H^l_{i,j,k} + dH^{max}_{BPT,k}(1 - z^l_{i,j}) \quad i \in I, j \in J, k \in K \tag{54}$$

$$H^{lv}_{i,j,k} - (H^l_{i,j,k} - H^{min}_k)z^l_{i,j} \leq EH_{i,j,k} \leq H^{lv}_{i,j,k} + (H^{max}_k - H^v_{i,j,k})z^v_{i,j} \quad i \in I, j \in J, k \in K \tag{55}$$

Here $EH^{max}_{i,j,k}$, $dH^{max}_{DPT,k}$, $dH^{max}_{BPT,k}$, H^{min}_k and H^{max}_k are appropriate big-M values.

3.7. Heat Transfer Boundary Modeling

Instead of following the conventional heat integration [46], we propose a heat transfer boundary model. The representation for model describes the heat transfer across a wall (completely restricted boundary). The block that supplies the energy flow is the heat source, while the block that takes the energy flow is the heat sink. Hence, there is no need to assign binary variable for determining the identity of process streams in each block since they are automatically determined by the heat transfer direction. Besides, no stage number for heat integration need to be specified in advance.

The total amount of heat duty $EFtot_{i,j,d}$ exchanged at boundary between block $B_{i,j}$ and $B_{i,j+1}$ in horizontal direction (or between block $B_{i,j}$ and $B_{i+1,j}$ in vertical direction) is determined as follows:

$$EFtot_{i,j,d} = EFP_{i,j,d} + EFN_{i,j,d} \quad i \in I, j \in J, d \in D \tag{56}$$

The inlet temperature to block $B_{i,j}$ is designated $T^{in}_{i,j}$. The bound of $T^{in}_{i,j}$ can be tightened to be $[T^{mink}_k, T^{maxk}_k]$ if component k exists in block $B_{i,j}$. This is achieved through the following constraint:

$$T_{i,j} - (T^{max} - T^{min})(1 - z^{mix}_{i,j,k}) \leq T^{in}_{i,j} \leq T_{i,j} + (T^{max} - T^{min})(1 - z^{mix}_{i,j,k}) \quad i \in I, j \in J, k \in K \tag{57}$$

The inlet temperature to block $B_{i,j}$, $T^{in}_{i,j}$, is equal to the temperature of overall inlet streams after mixing effect. The $T^{in}_{i,j}$ is obtained through the following energy balance at the inlet port of the block $B_{i,j}$. Since compression or expansion operation also contribute to the temperature change at the inlet part of each block, $W^{comp}_{i,j}$ and $W^{exp}_{i,j}$ are included into the energy balance at the inlet port.

$$W^{comp}_{i,j} - W^{exp}_{i,j} + EB^{in}_{i,j} + EM_{i,j} = EFP^{in}_{i,j-1,1} + EFP^{in}_{i-1,j,2} + \sum_d EFN^{in}_{i,j,d} + EM^{in}_{i,j} + EJ^{in}_{i,j} \quad i \in I, j \in J \tag{58}$$

The inlet stream enthalpy terms on the right-hand side of the above equation is determined based on the inlet temperature of the destinate block $T^{in}_{i,j}$, the pressure of the destinate block $P_{i,j}$ (since $W^{comp}_{i,j}$ and $W^{exp}_{i,j}$ already contributes to the pressure change).

$$EFP^{in}_{i,j,1} = \sum_k FP_{i,j,k,1}\{H^l_k(T^{in}_{i,j+1}, P_{i,j+1})z^l_{i,j} + H^v_k(T^{in}_{i,j+1}, P_{i,j+1})z^v_{i,j}$$
$$+ z^{lv}_{i,j}(1 - z^l_{i,j} - z^v_{i,j})H^{lv}_k(T^{in}_{i,j+1}, P_{i,j+1})\} \quad i \in I, j \in J \tag{59}$$

$$EFN^{in}_{i,j,1} = \sum_k FN_{i,j,k,1}\{H^l_k(T^{in}_{i,j}, P_{i,j})z^l_{i,j+1} + H^v_k(T^{in}_{i,j}, P_{i,j})z^v_{i,j}$$
$$+ z^{lv}_{i,j+1}(1 - z^l_{i,j+1} - z^v_{i,j+1})H^{lv}_k(T^{in}_{i,j}, P_{i,j})\} \quad i \in I, j \in J \tag{60}$$

$$EFP_{i,j,2}^{in} = \sum_k FP_{i,j,k,2}\{H_k^l(T_{i+1,j}^{in}, P_{i+1,j})z_{i,j}^l + H_k^v(T_{i+1,j}^{in}, P_{i+1,j})z_{i,j}^v$$
$$+ z_{i,j}^{lv}(1 - z_{i,j}^l - z_{i,j}^v)H_k^{lv}(T_{i+1,j}^{in}, P_{i+1,j})\} \quad i \in I, j \in J \tag{61}$$

$$EFN_{i,j,2}^{in} = \sum_k FN_{i,j,k,2}\{H_k^l(T_{i+1,j}^{in}, P_{i+1,j})z_{i+1,j}^l + H_k^v(T_{i,j}^{in}, P_{i,j})z_{i+1,j}^v$$
$$+ z_{i+1,j}^{lv}(1 - z_{i+1,j}^l - z_{i+1,j}^v)H_k^{lv}(T_{i,j}^{in}, P_{i,j})\} \quad i \in I, j \in J \tag{62}$$

$$EJ_{i,j,i',j'}^{in} = \sum_k JF_{i,j,i',j',k}\{H_k^l(T_{i',j'}^{in}, P_{i',j'})z_{i,j}^l + H_k^v(T_{i',j'}^{in}, P_{i',j'})z_{i,j}^v$$
$$+ z_{i,j}^{lv}(1 - z_{i,j}^l - z_{i,j}^v)H_k^{lv}(T_{i',j'}^{in}, P_{i',j'})\} \quad i \in I, j \in J \tag{63}$$

$$EM_{i,j}^{in} = \sum_{(k,fs)\in kfs} M_{i,j,k,f}\{H_k^l(T_{i,j}^{in}, P_{i,j})(1 - zph_k^{feed}) + H_k^v(T_{i,j}^{in}, P_{i,j})zph_k^{feed}\} \quad i \in I, j \in J \tag{64}$$

The phase of the above streams, $EFP_{i,j,1}^{in}$, $EFN_{i,j,1}^{in}$, $EFP_{i,j,2}^{in}$, $EFN_{i,j,2}^{in}$, $EJ_{i,j,i',j'}^{in}$ and $EM_{i,j}^{in}$, are the same as the phase of block where these streams originate from. Parameter zph_k^{feed} describes the phase of feed streams, which is equal to one if feed enters the system as gas and equal to zero if the feed enters the system as liquid.

When the energy flow direction in horizontal direction is from block $B_{i,j}$ to block $B_{i,j+1}$ ($z_{i,j,1}^{plus} = 1$), the inlet and outlet approach temperatures for process streams in adjacent blocks separated by completely restricted boundary are determined as follows:

$$dtin_{i,j,1}^{plus} \leq T_{i,j}^{in} - T_{i,j+1} + dt^{up}(2 - z_{i,j,1}^{plus} - z_{i,j,1}^{cr}) \quad i \in I, j \in J \tag{65}$$

$$dtout_{i,j,1}^{plus} \leq T_{i,j} - T_{i,j+1}^{in} + dt^{up}(2 - z_{i,j,1}^{fplus} - z_{i,j,1}^{cr}) \quad i \in I, j \in J \tag{66}$$

Here when energy flow direction is in positive direction $z_{i,j,1}^{plus} = 1$ and $z_{i,j,1}^{cr} = 1$. The inlet approach temperature and outlet approach temperature are $dtin_{i,j,1}^{plus} = Tin_{i,j} - T_{i,j+1}$ and $dtout_{i,j,1}^{plus} = T_{i,j} - T_{i,j+1}^{in}$.

Similarly, the approach temperatures for process streams exchanging heat in vertical direction ($d = 2$) are determined as follows:

$$dtin_{i,j,2}^{plus} \leq T_{i,j}^{in} - T_{i+1,j} + dt_{i,j,2}^{up}(2 - z_{i,j,2}^{plus} - z_{i,j,2}^{cr}) \quad i \in I, j \in J \tag{67}$$

$$dtout_{i,j,d}^{plus} \leq T_{i,j} - T_{i+1,j}^{in} + dt^{up}(2 - z_{i,j,2}^{plus} - z_{i,j,2}^{cr}) \quad i \in I, j \in J \tag{68}$$

When the energy flow direction in horizontal direction is from block $B_{i,j+1}$ to block $B_{i,j}$, the inlet and outlet approach temperatures are obtained with the following two relations:

$$dtin^{neg}(i,j,1) \leq T_{i,j+1}^{in} - T_{i,j} + dt_{i,j,1}^{up}(1 + z_{i,j,1}^{plus} - z_{i,j,1}^{cr}) \quad i \in I, j \in J \tag{69}$$

$$dtout_{i,j,1}^{neg} \leq T_{i,j+1} - T_{i,j}^{in} + dt_{i,j,1}^{up}(1 + z_{i,j,1}^{plus} - z_{i,j,1}^{cr}) \quad i \in I, j \in J \tag{70}$$

Similarly, in vertical direction ($d = 2$) with energy flow direction from block $B_{i+1,j}$ to block $B_{i,j}$, the inlet and outlet approach temperatures are obtained with the following two relations:

$$dtin_{i,j,2}^{neg} \leq T_{i+1,j}^{in} - T_{i,j} + dt^{up}(1 + z_{i,j,2}^{plus} - z_{i,j,2}^{cr}) \quad i \in I, j \in J \tag{71}$$

$$dtout_{i,j,2}^{neg} \leq T_{i+1,j} - T_{i,j}^{in} + dt^{up}(1 + z_{i,j,2}^{plus} - z_{i,j,2}^{cr}) \quad i \in I, j \in J \tag{72}$$

Similarly, approach temperatures for blocks with hot utility and cold utility are given by:

$$dtin_{i,j}^{CU} \leq T_{i,j}^{in} - TCUOUT + dt^{up}(1 - z_{i,j}^{cod}) \quad i \in I, j \in J \tag{73}$$

$$dtout_{i,j}^{CU} \leq T_{i,j} - TCUIN + dt^{up}(1 - z_{i,j}^{cod}) \quad i \in I, j \in J \tag{74}$$

$$dtin_{i,j}^{HU} \leq THUOUT - T_{i,j}^{in} + dt^{up}(1 - z_{i,j}^{hot}) \quad i \in I, j \in J \tag{75}$$

$$dtout_{i,j}^{HU} \leq THUIN - T_{i,j} + dt^{up}(1 - z_{i,j}^{hot}) \quad i \in I, j \in J \tag{76}$$

Parameters $TCUOUT$ and $TCUIN$ are outlet and inlet temperature of cold utility while $THUOUT$ and $THUIN$ are outlet and inlet temperature of hot utility.

Only one heat duty at heat transfer boundary and one approach temperature variable are required for determining the heat exchanger area. These are ensured through the following inequalities:

$$dtin_{i,j,d} \leq dtin_{i,j,d}^{plus} \quad dtin_{i,j,d} \leq dtin_{i,j,d}^{neg} \quad i \in I, j \in J, d \in D \tag{77}$$

$$dtout_{i,j,d} \leq dtout_{i,j,d}^{plus} \quad dtout_{i,j,d} \leq dtout_{i,j,d}^{neg} \quad i \in I, j \in J, d \in D \tag{78}$$

With the approach temperature and heat duty of heat exchangers, we determine the heat exchange areas as follows:

$$A_{i,j,d} = EFtot_{i,j,d} / U^{hx}(dtin_{i,j,d}dtout_{i,j,d}(dtin_{i,j,d} + dtout_{i,j,d})/2)^{1/3} \quad i \in I, j \in J, d \in D \tag{79}$$

$$A_{i,j}^{HU} = Q_{i,j}^{h} / U^{HU}(dtin_{i,j}^{HU}dtout_{i,j}^{HU}(dtin_{i,j}^{HU} + dtout_{i,j}^{HU})/2)^{1/3} \quad i \in I, j \in J \tag{80}$$

$$A_{i,j}^{CU} = Q_{i,j}^{c} / U^{CU}(dtin_{i,j}^{CU}dtout_{i,j}^{CU}(dtin_{i,j}^{CU} + dtout_{i,j}^{CU})/2)^{1/3} \quad i \in I, j \in J \tag{81}$$

Here, $A_{i,j,d}$, $A_{i,j}^{HU}$ and $A_{i,j}^{CU}$ represent heat exchange area between process streams, between process stream in block $B_{i,j}$ and hot utility, between process stream in block $B_{i,j}$ and cold utility respectively. We use Chen's approximation to calculate the logarithmic mean temperature difference in area calculations. U^{hx}, U^{HU}, U^{CU} are overall heat transfer coefficient at heat exchangers, heaters, and coolers.

3.8. Work Calculation

A semi-restricted boundary can be assigned with either an expander, a compressor, or a valve. These assignments are indicated through the following binary variables:

$$z_{i,j,d,m}^{sre} = \begin{cases} 1 & \text{If turbine on SSTC } m \text{ exists on right or bottom boundary } (d = 1 \text{ or } d = 2) \text{ of block } B_{i,j} \\ 0 & \text{Otherwise} \end{cases}$$

$$z_{i,j,d,m}^{src} = \begin{cases} 1 & \text{If compressor on SSTC } m \text{ exists on right or bottom boundary } (d = 1 \text{ or } d = 2) \text{ of block } B_{i,j} \\ 0 & \text{Otherwise} \end{cases}$$

$$z_{i,j,d}^{srv} = \begin{cases} 1 & \text{If valve exists on right or bottom boundary } (d = 1 \text{ or } d = 2) \text{ of block } B_{i,j} \\ 0 & \text{Otherwise} \end{cases}$$

Only one of these pieces of pressure-changing equipment is allowed on semi-restricted boundary.

$$z_{i,j,d}^{sr} = \sum_m z_{i,j,d,m}^{sre} + \sum_m z_{i,j,d,m}^{src} + z_{i,j,d}^{srv} \quad i \in I, j \in J, d \in D \tag{82}$$

For the right or bottom boundary of block $B_{i,j}$ ($d = 1$ or $d = 2$), there can be a situation that $z_{i,j,d}^{srv} = 1$ while the flow associated with the boundary $F_{i,j,k,d} = 0$. This does not indicate an existence of valve operation at the boundary but suggests that the temperature and pressure relation at the boundary is relaxed. This avoids the happening of infeasibility. Since $z^{srv}_{i,j,d}$ is also not related with the cost function, its value has no influence on the objective value.

The existence of shaft m is indicated through the following binary variable:

$$z^{xm}_m = \begin{cases} 1 & \text{If shaft } m \text{ exists} \\ 0 & \text{Otherwise} \end{cases}$$

If an SSTC exists, there is at least one compressor or turbine. To avoid symmetric solution, we prefer shaft with lower index.

$$z^{xm}_m \leq \sum_i \sum_j (z^{sre}_{i,j,1,m} + z^{src}_{i,j,1,m} + z^{sre}_{i,j,2,m} + z^{src}_{i,j,2,m}) \leq NU_m z^{xm}_m \quad i \in I, j \in J, m \in M \tag{83}$$

$$z^{xm}_m \geq z^{xm}_{m+1} \quad m \in M \tag{84}$$

On each shaft, there exists a motor or a generator. These are represented with binary variables:

$$z^{gen}_m = \begin{cases} 1 & \text{If shaft } m \text{ involves a motor} \\ 0 & \text{Otherwise} \end{cases}$$

$$z^{mot}_m = \begin{cases} 1 & \text{If shaft } m \text{ involves a generator} \\ 0 & \text{Otherwise} \end{cases}$$

If a shaft does not exist, then the motor and generator on this shaft also do not exist. If a generator or a motor exists on a shaft, then there is at least one semi-restricted boundary assigned with turbines or compressors on this shaft.

$$z^{gen}_m + z^{mot}_m \leq z^{xm}_m \quad m \in M \tag{85}$$

$$z^{gen}_m \leq \sum_i \sum_j (z^{sre}_{i,j,1,m} + z^{sre}_{i,j,2,m}) \quad i \in I, j \in J, m \in M \tag{86}$$

$$z^{mot}_m \leq \sum_i \sum_j (z^{src}_{i,j,1,m} + z^{sre}_{i,j,2,m}) \quad i \in I, j \in J, m \in M \tag{87}$$

We define the positive variable $PR^F_{i,j,d}$ to designate the pressure ratio between the block $B_{i,j+1}$ and $B_{i,j}$ for flow alignment $d = 1$ or between the block $B_{i+1,j}$ and $B_{i,j}$ for flow alignment $d = 2$. The calculation of $PR^F_{i,j,d}$ is activated when the boundary of block $B_{i,j}$ is semi-restricted or unrestricted at the corresponding flow alignment d ($z^{sr}_{i,j,d} = 1$ or $z^{ur}_{i,j,d} = 1$). The pressure ratio is taken as 1 to avoid the calculation of the pressure ratio if this boundary is not semi-restricted. In horizontal direction, the pressure ratio is determined as follows:

$$\frac{P_{i,j+1}}{P_{i,j}} - PR^{up}(1 - z^{sr}_{i,j,1}) \leq PR^F_{i,j,1} \leq \frac{P_{i,j+1}}{P_{i,j}} + PR^{up}(1 - z^{sr}_{i,j,1}) \quad i \in I, j \in J \tag{88}$$

$$\frac{P_{i,j+1}}{P_{i,j}} - PR^{up}(1 - z^{ur}_{i,j,1}) \leq PR^F_{i,j,1} \leq \frac{P_{i,j+1}}{P_{i,j}} + PR^{up}(1 - z^{ur}_{i,j,1}) \quad i \in I, j \in J \tag{89}$$

$$1 - PR^{up}z^{sr}_{i,j,1} \leq PR^F_{i,j,1} \leq 1 + PR^{up}z^{sr}_{i,j,1} \quad i \in I, j \in J \tag{90}$$

Here, PR^{up} is taken as the maximum pressure ratio, which is determined as P^{max}/P^{min}. Similarly, in vertical direction, the pressure ratio is determined as follows:

$$\frac{P_{i+1,j}}{P_{i,j}} - PR^{up}(1 - z^{sr}_{i,j,2}) \leq PR^F_{i,j,2} \leq \frac{P_{i+1,j}}{P_{i,j}} + PR^{up}(1 - z^{sr}_{i,j,2}) \quad i \in I, j \in J \tag{91}$$

$$\frac{P_{i+1,j}}{P_{i,j}} - PR^{up}(1 - z_{i,j,2}^{ur}) \le PR_{i,j,2}^F \le \frac{P_{i+1,j}}{P_{i,j}} + PR^{up}(1 - z_{i,j,2}^{ur}) \quad i \in I, j \in J \tag{92}$$

$$1 - PR^{up} z_{i,j,2}^{sr} \le PR_{i,j,2}^F \le 1 + PR^{up} z_{i,j,2}^{sr} \quad i \in I, j \in J \tag{93}$$

For feed stream f, the pressure ratio is taken as the ratio between block pressure $P_{i,j}$ and parameter P_f^{feed} for feed pressure .

$$PR_{i,j,f}^{feed} = \frac{P_{i,j}}{P_f^{feed}} \quad i \in I, j \in J, f \in FS \tag{94}$$

The work term $W_{i,j}$ consists of compression work term $W_{i,j}^{com}$ and expansion work term $W_{i,j}^{exp}$. Both $W_{i,j}^{com}$ and $W_{i,j}^{exp}$ consist of work components for direct connecting streams ($W_{i,j,d}^{comp,FP}$ for positive component, $W_{i,j,d}^{comp,FN}$ for negative component), feed streams($W_{i,j,f}^{comp,FS}$), and jump connecting streams ($W_{i',j',i,j}^{comp,Jf}$). Accordingly,

$$W_{i,j}^{com} = \sum_{d \in D}(W_{i,j,d}^{comp,FP} + W_{i,j,d}^{comp,FN}) + \sum_{f \in FS} W_{i,j,f}^{comp,FS} + \sum_{(i',j') \in LN(i,j,i',j')} W_{i',j',i,j}^{comp,Jf}, \quad i \in I, j \in J \tag{95}$$

$$W_{i,j}^{exp} = \sum_{d \in D}(W_{i,j,d}^{exp,FP} + W_{i,j,d}^{exp,FN}) + \sum_{f \in FS} W_{i,j,f}^{exp,FS} + \sum_{(i',j') \in LN(i,j,i',j')} W_{i,j,i',j'}^{comp,Jp}, \quad i \in I, j \in J \tag{96}$$

From these pressure ratio definitions, we calculate the isentropic work on direct connecting streams, feed streams and jump connecting streams. In the horizontal direction, the inlet isentropic work is determined as follows:

$$\eta W_{i,j,1}^{comp,FP} - W_{i,j,1}^{exp,FP}/\eta = \sum_{k \in K} FP_{i,j-1,k,1} T_{i,j-1,1}^s R_{gas} \frac{\gamma}{\gamma-1}\{(PR_{i,j-1,1}^F)^{\frac{\gamma-1}{\gamma}} - 1\} \quad i \in I, j \in J \tag{97}$$

$$\eta W_{i,j,1}^{comp,FN} - W_{i,j,1}^{exp,FN}/\eta = \sum_{k \in K} FN_{i,j,k,1} T_{i,j,1}^s R_{gas} \frac{\gamma}{\gamma-1}\{(\frac{1}{PR_{i,j,1}^F})^{\frac{\gamma-1}{\gamma}} - 1\} \quad i \in I, j \in J \tag{98}$$

Here R_{gas} is the gas constant and γ is the adiabatic compression coefficient. η is the adiabatic compression efficiency. Similarly, the isentropic work for a vertical entering stream is calculated as follows:

$$\eta W_{i,j,2}^{comp,FP} - W_{i,j,2}^{exp,FP}/\eta = \sum_{k \in K} FP_{i-1,j,k,2} T_{i-1,j,2}^s R_{gas} \frac{\gamma}{\gamma-1}\{(PR_{i-1,j,2}^F)^{\frac{\gamma-1}{\gamma}} - 1\} \quad i \in I, j \in J \tag{99}$$

$$\eta W_{i,j,2}^{comp,FN} - W_{i,j,2}^{exp,FN}/\eta = \sum_{k \in K} FN_{i,j,k,2} T_{i,j,2}^s R_{gas} \frac{\gamma}{\gamma-1}\{(\frac{1}{PR_{i,j,2}^F})^{\frac{\gamma-1}{\gamma}} - 1\} \quad i \in I, j \in J \tag{100}$$

The work terms related to feed streams and jump connecting streams are calculated in a similar way:

$$\eta W_{i,j,f}^{comp,FS} - W_{i,j,f}^{exp,FS}/\eta = \sum_{k \in K} M_{i,j,k,f} T_f^{feed} R_{gas} \frac{1}{n_{fs}}\{(PR_{i,j,f}^{feed})^{n_{fs}} - 1\} \quad i \in I, j \in J, f \in FS \tag{101}$$

$$\eta W_{i,j,i',j'}^{comp,JF} - W_{i,j,i',j'}^{exp,JF}/\eta = J_{i,j,i',j'}^T T_{i,j} R_{gas} \frac{\gamma}{\gamma-1}\{(\frac{P_{i',j'}}{P_{i,j}})^{\frac{\gamma-1}{\gamma}} - 1\} \quad (i,j,i',j) \in LN(i,j,i',j') \tag{102}$$

Here n_{fs} is the adiabatic compression coefficient.

These work components for direct connecting streams are related with the boundary type and the type of pressure exchangers assigned on semi-restricted boundary.

$$W_{i,j,1}^{comp,FN} \leq W^{comp,max} \sum_m z_{i,j,1,m}^{sre} \quad W_{i,j,1}^{exp,FN} \leq W^{exp,max} \sum_m z_{i,j,1,m}^{src} + z_{i,j,1}^{srv} \quad i \in I, j \in J \tag{103}$$

$$W_{i,j,1}^{comp,FN} \leq W^{comp,max} z_{i,j,1}^{sr} \quad W_{i,j,1}^{exp,FN} \leq W^{exp,max} z_{i,j,1}^{sr} \quad i \in I, j \in J \tag{104}$$

$$W_{i,j,1}^{comp,FP} \leq W^{exp,max} \sum_m z_{i,j-1,1,m}^{src} \quad W_{i,j,1}^{exp,FP} \leq W^{exp,max} \sum_m z_{i,j-1,1,m}^{sre} + +z_{i,j-1,1}^{srv} \quad i \in I, j \in J \tag{105}$$

$$W_{i,j,1}^{comp,FP} \leq W^{exp,max} z_{i,j-1,1}^{sr} \quad W_{i,j,1}^{exp,FP} \leq W^{comp,max} z_{i,j-1,1}^{sr} \quad i \in I, j \in J \tag{106}$$

$$W_{i,j,2}^{comp,FN} \leq W_{i,j}^{comp,max} \sum_m z_{i,j,2,m}^{src} \quad W_{i,j,2}^{exp,FN} \leq W_{i,j}^{comp,max} \sum_m z_{i,j,2,m}^{sre} + +z_{i,j,2}^{srv} \quad i \in I, j \in J \tag{107}$$

$$W_{i,j,2}^{comp,FN} \leq W^{comp,max} z_{i,j,2}^{sr} \quad W_{i,j,2}^{exp,FN} \leq W^{exp,max} z_{i,j,2}^{sr} \quad i \in I, j \in J \tag{108}$$

$$W_{i,j,2}^{comp,FP} \leq W^{comp,max} \sum_m z_{i-1,j,2,m}^{src} \quad W_{i,j,2}^{exp,FP} \leq W^{exp,max} \sum_m z_{i-1,j,2,m}^{sre} + z_{i-1,j,2}^{srv} \quad i \in I, j \in J \tag{109}$$

$$W_{i,j,2}^{comp,FP} \leq W^{comp,max} z_{i-1,j,2}^{sr} \quad W_{i,j,2}^{exp,FP} \leq W^{exp,max} z_{i-1,j,2}^{sr} \quad i \in I, j \in J \tag{110}$$

The compression work energy $W_{i,j}^{comp}$ and expansion energy $W_{i,j}^{exp}$ are determined as follows:

$$\begin{aligned} W_{i,j}^{comp} = W_{i,j,1}^{comp,FN} + W_{i,j,1}^{comp,FP} + W_{i,j,2}^{comp,FN} + W_{i,j,2}^{comp,FP} + \sum_{fs \in FS} W_{i,j,fs}^{comp,FS} \\ + \sum_{(i',j') \in LN(i,j,i',j')} W_{i',j',i,j}^{comp,Jf}, \quad i \in I, j \in J \end{aligned} \tag{111}$$

$$\begin{aligned} W_{i,j}^{exp} = W_{i,j,1}^{exp,FN} + W_{i,j,1}^{exp,FP} + W_{i,j,2}^{exp,FN} + W_{i,j,2}^{exp,FP} + \sum_{fs \in FS} W_{i,j,fs}^{exp,FS} \\ + \sum_{(i',j') \in LN(i,j,i',j')} W_{i',j',i,j}^{exp,Jf}, \quad i \in I, j \in J \end{aligned} \tag{112}$$

The following shaft balance distributes the energy generated by turbines to compressors on the same shaft. Additional energy is transferred to motors to generate electricity. If the energy supply is not enough, electricity is consumed to activate motors.

$$\begin{aligned} wH_m + \sum_{i,j} W_{i,j,1}^{exp,FN} z_{i,j,1,m}^{sre} + \sum_{i,j} W_{i,j,1}^{exp,FP} z_{i,j-1,1,m}^{sre} + \sum_{i,j} W_{i,j,2}^{exp,FN} z_{i,j,2,m}^{sre} + \sum_{i,j} W_{i,j,2}^{exp,FP} z_{i-1,j,1,m}^{sre} = \\ \sum_{i,j} W_{i,j,1}^{comp,FN} z_{i,j,1,m}^{src} + \sum_{i,j} W_{i,j,1}^{comp,FP} z_{i,j-1,1,m}^{src} + \sum_{i,j} W_{i,j,2}^{comp,FN} z_{i,j,2,m}^{src} + \sum_{i,j} W_{i,j,2}^{comp,FP} z_{i-1,j,2,m}^{src} + WG_m \end{aligned} \tag{113}$$

$$z_m^{mot} wH^L \leq WH_m \leq z_m^{mot} wH^U \tag{114}$$

$$z_m^{gen} wG^L \leq WG_m \leq z_m^{gen} wG^U \tag{115}$$

wH^U and wG^U designate the maximum capacity of motors and generators, respectively. Please note that valves do not contribute to work energy. Hence, whenever the semi-restricted boundary is assigned with a valve, then that work energy term is ignored.

Additional constraints ensure that the block assigned with turbines or compressors should involve with gas phase.

$$\sum_m (z_{i,j,1,m}^{sre} + z_{i,j,1,m}^{src}) + \sum_m (z_{i,j,2,m}^{sre} + z_{i,j,2,m}^{src}) \leq z_{i,j}^v \tag{116}$$

3.9. Objective Function

We consider an economic objective similar to Nair et al. [24] as follows:

$$
\begin{aligned}
\min \quad TAC = \gamma_a \Bigg(&\sum_m \alpha_m^{gen}(CF_m^{gen} z_m^{gen} + CP_m^{gen} WG_m^{\beta_m^G}) + \sum_m \alpha_m^{mot}(CF_m^{mot} z_m^{mot} + CP_m^{mot} WH_m^{\beta_m^G}) \\
&+ \sum_{i,j,d} \alpha^c (CF^c \sum_m z_{i,j,d,m}^{src} + \sum_m CP^c z_{i,j,d,m}^{src}(W_{i,j,d}^{comp,FP})^{\beta_m^c} + \sum_m CP^c z_{i,j,d,m}^{src}(W_{i,j,d}^{comp,FN})^{\beta_m^c}) \\
&+ \sum_{i,j,d} \alpha^e (CF^e \sum_m z_{i,j,d,m}^{sre} + \sum_m CP^e z_{i,j,d,m}^{sre}(W_{i,j,d}^{exp,FP})^{\beta_m^e} + \sum_m CP^e z_{i,j,d,m}^{sre}(W_{i,j,d}^{exp,FN})^{\beta_m^e}) \\
&+ \sum_{i,j,d} \alpha^{hx} (CF^{hx} z_{i,j,d}^{cr} + CP^{hx} A_{i,j,d}) + \sum_{i,j} \alpha^{hot}(CF^{hot} z_{i,j}^{hot} + CP^{hot} A_{i,j}^{HU}) \\
&+ \sum_{i,j} \alpha^{cod}(CF^{cod} z_{i,j}^{cod} + CP^{cod} A_{i,j}^{CU})) + \sum_{i,j}(CC^{HU} Q_{i,j}^h + CC^{CU} Q_{i,j}^c) \\
&+ \sum_m (CC^E WH_m - Rev^G WG_m) \Bigg)
\end{aligned}
\tag{117}
$$

Here γ_a is the annualized factor. This objective function aims at minimizing total annual cost (*TAC*). This *TAC* mainly involves capital costs including those of generator, motor, compressor, expander, and heat exchangers, as well as operating cost including costs of running motors, utility consumption. Besides, the electricity generated by generators bring revenue. The parameter *CF* is the fixed cost for different equipment. *CP* is the appropriate cost coefficient for associated equipment. *CC* is the unit cost for utilities. Rev^G is the price of electricity while parameter α is the cost factor.

4. WHENS Case Study on Liquefied Natural Gas (LNG)-Based Cryogenic Energy Chain

The above MINLP model is applied to a WHENS problem related to a liquefied energy chain reported by Nair et al. [24]. Liquefaction is an energy-intensive process that converts natural gas (NG) into liquid form for economic and safe transportation [47–49]. The overall procedure for solving the case study is illustrated in Figure 6.

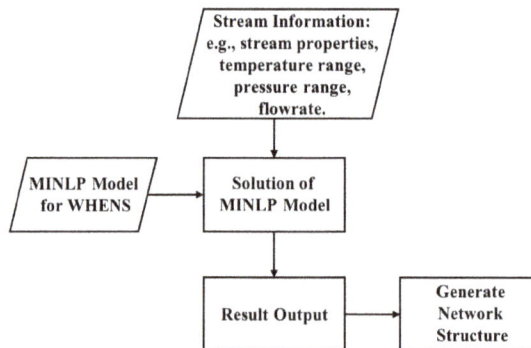

Figure 6. Procedure for block-based WHENS.

The stream information such as stream properties, temperature range, pressure range, and flowrate is directed to the block-based MINLP model for WHENS problem. This MINLP is solved using commercial solvers and results in block configurations. The block configurations are then converted to classic work and heat exchange networks. The procedure for this conversion can be found in Demirel et al. [37] and Li et al. [38] There are four process streams, including liquid inert nitrogen (S_1), liquefied natural gas or LNG (S_2), liquid carbon dioxide (S_3), and the propane pre-cooled mixed refrigerant (C3MR) (S_4) and one external stream as hot utility (HU). Among these process streams, part of S_1 also serves as cold utility stream (CU). The information is provided in Table 2.

Table 2. Specification for process streams and utility streams (HU: hot utility; CU: cold utility).

Specification/Parameter	S_1	S_2	S_3	S_4	HU	CU
Feed pressure, P_f^{feed} (MPa)	10	10	6	-	-	-
Feed temperature, T_f^{feed} (K)	103.45	319.80 (298.15)	221.12	-	383.15	93.15
Target pressure, $P_p^{min} = P_p^{max}$ (MPa)	0.1	10	6	-	-	-
Target pressure range, $T_p^{min} = T_p^{max}$ (K)	-	104.75 (113.15)	293.15	-	383.15	93.15
Flowrate, F_f^{feed} (kg/s)	1.2	1	2.46	-	-	-
Molecular weight, MW (kg/kmol)	28	19	44	23.82	28	18

Stream property information is provided in Table 3 and include bubble point, dew point, enthalpy calculation for streams in liquid and gas phases. Since we assume these variables are linearly dependent on both system temperature and pressure, the following linear coefficients are sufficient to capture the thermodynamic relations. The unit for bubble point and dew point are K while the units for liquid enthalpy and gas enthalpy are kJ/kg.

Table 3. Specification for stream properties.

Stream	a_k^b	b_k^b	a_k^d	b_k^d	a_k^l	b_k^l	c_k^l	a_k^v	b_k^v	c_k^v
Nitrogen	10.284	93.947	10.284	93.948	2.495	-0.57	-625.05	1.15	-2.38	-342.2
Natural gas	0	197.35	0	265.15	3.51	0	0	3.46	0	123.77
Carbon dioxide	-	-	-	-	2.318	0	0	-	-	-

The equipment considered are compressors, expanders, motors, generators, and heat exchangers. The cost coefficients for them are reported in the Table 4. Please note that the additional amount of energy brought by generator can be converted into electricity, contributing to the revenue gaining.

Table 4. Cost coefficients for equipment and utilities (*CF*: fixed cost for different equipment; *CP*: appropriate cost coefficient for associated equipment; *CC*: unit cost for utilities).

	Capital Cost (K $)				Operating Cost
	CF	CP	β	α	CC or Rev
Compressor	184.12	2.4×10^{-5}	2.988	2.5	-
Expander	29.20	0.4872	1	2.5	-
Motor	-1.1	2.1	0.6	4	455.04 ($/(KW·a))
Generator	-1.1	2.1	0.6	4	455.04 ($/(KW·a))
Heat exchanger	27.05	0.5027	0.8003	3.5	337 ($/(KW·a))
HU	-	-	-	-	337 ($/(KW·a))
CU	-	-	-	-	1000 ($/(KW·a))

$\gamma_a = 0.18, \gamma = 1.51, dt^{min} = 4$ K, $U^L = U^V = U^{LV} = 0.1$ m^{2}K/KW, $U^{HU} = U^{CU} = 1$ m^{2}K/KW

We select a 3×3 block superstructure. To facilitate the solution, we reduce the number of binary variables by prohibiting the use of valves. The heat transfer boundaries are only allowed in the horizontal direction. To reduce the number of non-linear terms, we fix all jump connecting streams to be zero. This restricts the number of process alternatives but helps to demonstrate the capability of the proposed approach. This case study is solved using solver ANTIGONE 24.4.3 developed by Misener et al. [50] in GAMS 24.4 (2015 version and developed by GAMS Development Corporation in Fairfax, VA, USA) on a Dell OptiPlex 9020 computer (Intel 8 Core i7-4770 CPU 3.4 GHz, 15.5 GB memory) running Springdale Linux. We consider two different cases of this case study to show the capability of the proposed approach. Case 1 involves liquefaction of NG using available process streams. Case 2 achieves the liquefaction of NG using C3MR.

Solver SCIP (representing Solving Constraint Integer Programs) [51] is used for initializing the proposed model. This model is solved within 2 h with optimal total annual cost as 0.696 MM\$/year with total capital cost of 0.225 MM\$/year and operating cost of 0.471 MM\$/year. The block configuration and the equivalent process flowsheet are shown in Figure 7. Stream S_3 flows into block $B_{3,1}$ and is cooled using the HU. Stream S_3 is distributed in block $B_{3,2}$, block $B_{3,3}$ and block $B_{2,3}$ with feed fractions of 0.34, 0.18, and 0.32, respectively. A CU is supplied into block $B_{3,2}$ to partially cool down the stream S_1. The vertical outlet flow of block $B_{3,3}$ is integrated with stream S_2, where S_2 serves as the hot stream and S_1 in block $B_{2,3}$ serves as the cold stream. Please note that the identity of these two streams were not postulated in advance. This was identified by the heat transfer direction reported in the solution. For instance, the energy flow at the right boundary of block $B_{2,2}$ is from block $B_{2,2}$ to block $B_{2,3}$. Hence, the stream in block $B_{2,2}$ is a hot stream while the stream in block $B_{2,3}$ is a cold stream. The additional amount of heat transferred from block $B_{2,2}$ is compensated by a CU in block $B_{2,3}$. After heat integration, stream S_2 is withdrawn in block $B_{2,2}$ and stream S_1 is withdrawn in block $B_{1,3}$ after the expansion at the bottom boundary of block $B_{1,3}$. The relevant block temperature and block pressure can be found in Figure 7a.

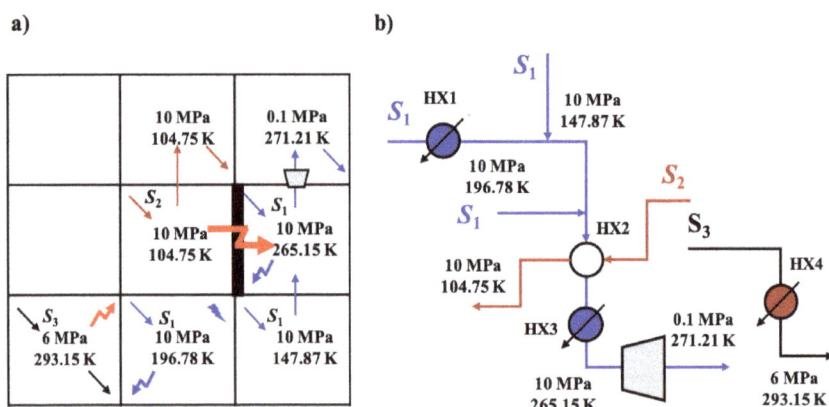

Figure 7. Resultant integrated work and heat exchanger network for the liquefied energy chain (case 1): (a) Bock representation; (b) Equivalent WHEN structure.

The corresponding process flowsheet is shown in Figure 7b. Streams S_1 and S_2 are integrated through heat exchanger HX2 with heat duty of 862.61 KW. Stream S_3 is not involved in either heat integration or work integration. A heater with heat duty as 410.74 KW helps stream S_3 to achieve the design target. The heat duty of coolers on stream S_1 are 10.42 KW, 381.36 KW for cooler HX1 and HX3, respectively. Through this case study on liquefied energy chain, we show that the proposed approach could enable both heat and work integration opportunities. These integration alternatives reduced the energy consumption and total annual cost, resulting in significant energy intensification within the network structure. The identity of process streams was determined simultaneously together with network generation. Besides, compared with classic superstructure representation approach, no information on work and heat exchange stage number was required.

Next, we consider a conceptual design problem where the goal is to obtain a liquefaction process for a NG stream from an initial gaseous condition of 1 atm and 298.15 K to a final condition of saturated LNG at 1 atm (which corresponds to a final temperature below 113.15 K). Interestingly, this process synthesis problem can be formulated as a WHENS problem, where we have two process streams, namely the NG (hot stream) and C3MR. The refrigerant can be considered as a neutral stream or a circulating working fluid with the same initial and final conditions. Any feasible process configuration for this design problem would involve heat exchangers, and the temperatures indicate that cryogenic cooling is necessary. However, consider that cooling water at the ambient temperature is the only CU that is available for the process. To this end, work exchangers (compressors and expanders) would be required. The refrigerant would undergo a cycle involving multiple alternating heat and work exchangers to first achieve a cryogenic temperature through expansion at which heat can be gained from the NG stream, and then achieve a high temperature through compression at which heat can be released to the cooling water. This is what happens in a typical refrigeration cycle.

While such an answer is well-known for this problem, this is to illustrate the possibility of discovering non-intuitive process flowsheets involving heat and work exchangers using the building block approach. Consider a naive designer who does not have any prior knowledge of how a refrigeration cycle works, or how a stream should be liquefied at cryogenic conditions when cooling utilities are only available at the ambient condition. The designer can still obtain the same solution, as it is already embedded in the general block-based WHENS superstructure (see Figure 8).

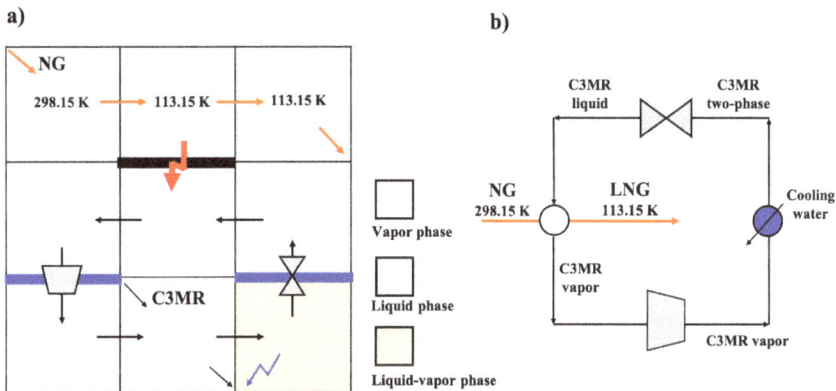

Figure 8. Manifestation of a refrigeration cycle using building blocks for cryogenic liquefaction process: (**a**) block representation, and (**b**) equivalent WHEN structure.

The NG stream enters the process as a gaseous stream in block $B_{1,1}$ and finally exits as a saturated liquid from block $B_{1,3}$. Before exiting, NG exchanges heat with the refrigerant C3MR through the bottom boundary of block $B_{1,2}$. C3MR cycles through the blocks placed in rows 2 and 3. Starting from block $B_{3,2}$, the refrigerant C3MR enters as a vapor stream and is cooled into a two-phase mixture in a utility cooler using cooling water as the CU. The outlet stream from block $B_{3,3}$ is expanded which results in a liquid C3MR with a reduced temperature when entering the block $B_{2,3}$ through a valve placed at the top semi-restricted boundary of block $B_{3,3}$. The horizontal outlet stream of block $B_{2,2}$ is vaporized using the heat released by NG through the completely restricted boundary between the blocks $B_{1,2}$ and $B_{2,2}$. The C3MR vapor after the heat exchanger is compressed across a semi-restricted boundary assigned on the bottom of block $B_{2,1}$ before it is finally withdrawn in block $B_{3,2}$. The seamless entrance and exit of C3MR within the same block $B_{3,2}$ suggests the existence of a cycle. The corresponding process flowsheet is shown in Figure 8b.

5. Conclusions

We presented a method to automatically generate numerous alternative configurations for the synthesis of integrated work and heat exchange networks using building blocks. The block representation is abstract, and it requires a transformation to obtain classic unit operations and flowsheet configurations. However, it is also analogous in a sense that the transformations from blocks-to-flowsheets and from flowsheets-to-blocks are systematic. The benefits of block-based representation over a unit operations-based representation is that the former uses only two fundamental design elements, namely the block interior and the block boundary. Alternative arrangement of flows to block interior and assignment of work/heat transfer phenomena to block boundaries give rise to alternative networks for systematic WHENS. The heat and work transfer models are general such that they do not depend on the postulation of stream identities. Besides, there is no need to specify any stagewise integration. With this representation approach, we formulated an MINLP model for WHENS. Using a case study, we demonstrated the capability of the block-based approach for WHENS. We also considered the possibility of discovering non-intuitive process flowsheets involving heat and work exchangers. Specifically, the case of NG liquefaction indicated that a designer could generate the design of a refrigeration cycle using the building block approach, even when designers have no prior knowledge of how a refrigeration cycle works, or how a stream should be liquefied at cryogenic conditions when cooling utilities are only available at ambient conditions. This provides hints towards the potential of our approach for the discovery of novel processes through process synthesis. However, further research is needed to extend the simultaneous process synthesis along with work and heat integration to more complex scenarios.

Author Contributions: J.L., S.E.D., and M.M.F.H. conceived the ideas, models and prepared the manuscript.

Funding: The authors gratefully acknowledge financial support from the U.S. National Science Foundation (NSF CBET-1606027) and the DOE/RAPID NNMI Institute.

Conflicts of Interest: The authors declare no conflict of interest.

References

1. Yu, H.; Fu, C.; Vikse, M.; Gundersen, T. Work and heat integration—A new field in process synthesis and PSE. *AIChE J.* **2018**. [CrossRef]
2. Fu, C.; Vikse, M.; Gundersen, T. Work and heat integration: An emerging research area. *Energy* **2018**, *158*, 796–806. [CrossRef]
3. Huang, K.; Karimi, I. Work-heat exchanger network synthesis (WHENS). *Energy* **2016**, *113*, 1006–1017. [CrossRef]
4. Subramanian, A.; Gundersen, T.; Adams, T.A., II. Modeling and simulation of energy systems: A Review. *Processes* **2018**, *6*, 238. [CrossRef]
5. Furman, K.C.; Sahinidis, N.V. A critical review and annotated bibliography for heat exchanger network synthesis in the 20th century. *Ind. Eng. Chem. Res.* **2002**, *41*, 2335–2370. [CrossRef]
6. Huang, Y.; Fan, L. Analysis of a work exchanger network. *Ind. Eng. Chem. Res.* **1996**, *35*, 3528–3538. [CrossRef]
7. Razib, M.; Hasan, M.M.F.; Karimi, I.A. Preliminary synthesis of work exchange networks. *Comput. Chem. Eng.* **2012**, *37*, 262–277. [CrossRef]
8. Liu, G.; Zhou, H.; Shen, R.; Feng, X. A graphical method for integrating work exchange network. *Appl. Energy* **2014**, *114*, 588–599. [CrossRef]
9. Zhuang, Y.; Liu, L.; Zhang, L.; Du, J. Upgraded graphical method for the synthesis of direct work exchanger networks. *Ind. Eng. Chem. Res.* **2017**, *56*, 14304–14315. [CrossRef]
10. Zhuang, Y.; Liu, L.; Du, J. Direct Work Exchange Networks Synthesis of Isothermal Process Based on Superstructure Method. *Chem. Eng. Trans.* **2017**, *61*, 133–138.
11. Amini-Rankouhi, A.; Huang, Y. Prediction of maximum recoverable mechanical energy via work integration: A thermodynamic modeling and analysis approach. *AIChE J.* **2017**, *63*, 4814–4826. [CrossRef]
12. Fu, C.; Gundersen, T. Heat and work integration: Fundamental insights and applications to carbon dioxide capture processes. *Energy Convers. Manag.* **2016**, *121*, 36–48. [CrossRef]
13. Fu, C.; Gundersen, T. Correct integration of compressors and expanders in above ambient heat exchanger networks. *Energy* **2016**, *116*, 1282–1293. [CrossRef]
14. Fu, C.; Gundersen, T. Integrating compressors into heat exchanger networks above ambient temperature. *AIChE J.* **2015**, *61*, 3770–3785. [CrossRef]

15. Fu, C.; Gundersen, T. Integrating expanders into heat exchanger networks above ambient temperature. *AIChE J.* **2015**, *61*, 3404–3422. [CrossRef]

16. Fu, C.; Gundersen, T. Sub-ambient heat exchanger network design including compressors. *Chem. Eng. Sci.* **2015**, *137*, 631–645. [CrossRef]

17. Aspelund, A.; Berstad, D.O.; Gundersen, T. An extended pinch analysis and design procedure utilizing pressure based exergy for subambient cooling. *Appl. Therm. Eng.* **2007**, *27*, 2633–2649. [CrossRef]

18. Gundersen, T.; Berstad, D.O.; Aspelund, A. Extending pinch analysis and process integration into pressure and fluid phase considerations. *Chem. Eng. Trans.* **2009**, *18*, 33–38.

19. Kansha, Y.; Tsuru, N.; Sato, K.; Fushimi, C.; Tsutsumi, A. Self-heat recuperation technology for energy saving in chemical processes. *Ind. Eng. Chem. Res.* **2009**, *48*, 7682–7686. [CrossRef]

20. Tsutsumi, A.; Kansha, Y. Thermodynamic mechanism of self-heat recuperative and self-heat recovery heat circulation system for a continuous heating and cooling gas cycle process. *Chem. Eng. Trans.* **2017**, *61*, 1759–1764.

21. Dong, R.; Yu, Y.; Zhang, Z. Simultaneous optimization of integrated heat, mass and pressure exchange network using exergoeconomic method. *Appl. Energy* **2014**, *136*, 1098–1109. [CrossRef]

22. Onishi, V.C.; Ravagnani, M.A.; Caballero, J.A. Simultaneous synthesis of work exchange networks with heat integration. *Chem. Eng. Sci.* **2014**, *112*, 87–107. [CrossRef]

23. Vikse, M.; Fu, C.; Barton, P.I.; Gundersen, T. Towards the use of mathematical optimization for work and heat exchange networks. *Chem. Eng. Trans.* **2017**, *61*, 1351–1356.

24. Nair, S.K.; Nagesh Rao, H.; Karimi, I.A. Framework for work-heat exchange network synthesis (WHENS). *AIChE J.* **2018**, *61*, 871–876. [CrossRef]

25. Floudas, C.A.; Ciric, A.R.; Grossmann, I.E. Automatic synthesis of optimum heat exchanger network configurations. *AIChE J.* **1986**, *32*, 276–290. [CrossRef]

26. Yeomans, H.; Grossmann, I.E. A systematic modeling framework of superstructure optimization in process synthesis. *Comput. Chem. Eng.* **1999**, *23*, 709–731. [CrossRef]

27. Chen, Q.; Grossmann, I. Recent developments and challenges in optimization-based process synthesis. *Annu. Rev. Chem. Biomol. Eng.* **2017**, *8*, 249–283. [CrossRef] [PubMed]

28. Demirel, S.E.; Li, J.; Hasan, M.M.F. Systematic process intensification. *Curr. Opin. Chem. Eng.* **2019**, in press.

29. Wechsung, A.; Aspelund, A.; Gundersen, T.; Barton, P.I. Synthesis of heat exchanger networks at subambient conditions with compression and expansion of process streams. *AIChE J.* **2011**, *57*, 2090–2108. [CrossRef]

30. Uv, P.M. Optimal Design of Heat Exchanger Networks with Pressure Changes. Master's Thesis, NTNU, Trondheim, Norway, 2016.

31. Stankiewicz, A.I.; Moulijn, J.A. Process intensification: Transforming chemical engineering. *Chem. Eng. Prog.* **2000**, *96*, 22–34.

32. Reay, D.; Ramshaw, C.; Harvey, A. *Process Intensification: Engineering for Efficiency, Sustainability and Flexibility*; Butterworth-Heinemann: Oxford, UK, 2013.

33. Tian, Y.; Demirel, S.E.; Hasan, M.M.F.; Pistikopoulos, E.N. An Overview of Process Systems Engineering Approaches for Process Intensification: State of the Art. *Chem. Eng. Process.-Process Intensif.* **2018**, *133*, 160–210. [CrossRef]

34. Demirel, S.E.; Li, J.; Hasan, M.M.F. A General Framework for Process Synthesis, Integration and Intensification. *Comput. Aided Chem. Eng.* **2018**, *44*, 445–450.

35. Hasan, M.M.F.; Karimi, I.A.; Alfadala, H.E.; Grootjans, H. Operational modeling of multistream heat exchangers with phase changes. *AIChE J.* **2009**, *55*, 150–171. [CrossRef]

36. Nagesh Rao, H.; Karimi, I.A. A superstructure-based model for multistream heat exchanger design within flow sheet optimization. *AIChE J.* **2017**, *63*, 3764–3777. [CrossRef]

37. Demirel, S.E.; Li, J.; Hasan, M.M.F. Systematic process intensification using building blocks. *Comput. Chem. Eng.* **2017**, *105*, 2–38. [CrossRef]

38. Li, J.; Demirel, S.E.; Hasan, M.M.F. Process synthesis using block superstructure with automated flowsheet generation and optimization. *AIChE J.* **2018**, *64*, 3082–3100. [CrossRef]

39. Li, J.; Demirel, S.E.; Hasan, M.M.F. Process Integration Using Block Superstructure. *Ind. Eng. Chem. Res.* **2018**, *57*, 4377–4398. [CrossRef]

40. Li, J.; Demirel, S.E.; Hasan, M.M.F. Fuel Gas Network Synthesis Using Block Superstructure. *Processes* **2018**, *6*, 23. [CrossRef]

41. Onishi, V.C.; Ravagnani, M.A.; Caballero, J.A. Simultaneous synthesis of heat exchanger networks with pressure recovery: optimal integration between heat and work. *AIChE J.* **2014**, *60*, 893–908. [CrossRef]
42. Fu, C.; Gundersen, T. Sub-ambient heat exchanger network design including expanders. *Chem. Eng. Sci.* **2015**, *138*, 712–729. [CrossRef]
43. Onishi, V.C.; Ravagnani, M.A.; Jiménez, L.; Caballero, J.A. Multi-objective synthesis of work and heat exchange networks: Optimal balance between economic and environmental performance. *Energy Convers. Manag.* **2017**, *140*, 192–202. [CrossRef]
44. Zhuang, Y.; Liu, L.; Liu, Q.; Du, J. Step-wise synthesis of work exchange networks involving heat integration based on the transshipment model. *Chin. J. Chem. Eng.* **2017**, *25*, 1052–1060. [CrossRef]
45. Qadeer, K.; Qyyum, M.A.; Lee, M. Krill-herd-based investigation for energy saving opportunities in offshore LNG processes. *Ind. Eng. Chem. Res.* **2018**, *57*, 14162–14172. [CrossRef]
46. Yee, T.F.; Grossmann, I.E. Simultaneous optimization models for heat integration—II. Heat exchanger network synthesis. *Comput. Chem. Eng.* **1990**, *14*, 1165–1184. [CrossRef]
47. Lim, W.; Choi, K.; Moon, I. Current status and perspectives of liquefied natural gas (LNG) plant design. *Ind. Eng. Chem. Res.* **2013**, *52*, 3065–3088. [CrossRef]
48. Vikse, M.; Watson, H.; Gundersen, T.; Barton, P. Simulation of Dual Mixed Refrigerant Natural Gas Liquefaction Processes Using a Nonsmooth Framework. *Processes* **2018**, *6*, 193. [CrossRef]
49. Kazda, K.; Li, X. Approximating nonlinear relationships for optimal operation of natural gas transport networks. *Processes* **2018**, *6*, 198. [CrossRef]
50. Misener, R.; Floudas, C.A. ANTIGONE: Algorithms for continuous/integer global optimization of nonlinear equations. *J. Glob. Optim.* **2014**, *59*, 503–526. [CrossRef]
51. Achterberg, T. SCIP: Solving constraint integer programs. *Math. Program, Comput,* **2009**, *1*, 1–41. [CrossRef]

processes

Article

Development of a Dynamic Model and Control System for Load-Following Studies of Supercritical Pulverized Coal Power Plants

Parikshit Sarda [1], Elijah Hedrick [1], Katherine Reynolds [1], Debangsu Bhattacharyya [1,*], Stephen E. Zitney [2] and Benjamin Omell [3]

[1] Department of Chemical and Biomedical Engineering, West Virginia University, 395 Evansdale Drive, Morgantown, WV 26506-6070, USA; pss0007@mix.wvu.edu (P.S.); ebhedrick@mix.wvu.edu (E.H.); kgreynolds@mix.wvu.edu (K.R.)
[2] National Energy Technology Laboratory, 3610 Collins Ferry Road, Morgantown, WV 26507, USA; Steve.Zitney@NETL.DOE.GOV
[3] National Energy Technology Laboratory, 626 Cochrans Mill Road, Pittsburgh, PA 15236, USA; Benjamin.Omell@netl.doe.gov
* Correspondence: debangsu.bhattacharyya@mail.wvu.edu; Tel.: +304-293-9335

Received: 8 October 2018; Accepted: 14 November 2018; Published: 17 November 2018

Abstract: Traditional energy production plants are increasingly forced to cycle their load and operate under low-load conditions in response to growth in intermittent renewable generation. A plant-wide dynamic model of a supercritical pulverized coal (SCPC) power plant has been developed in the Aspen Plus Dynamics® (APD) software environment and the impact of advanced control strategies on the transient responses of the key variables to load-following operation and disturbances can be studied. Models of various key unit operations, such as the steam turbine, are developed in Aspen Custom Modeler® (ACM) and integrated in the APD environment. A coordinated control system (CCS) is developed above the regulatory control layer. Three control configurations are evaluated for the control of the main steam; the reheat steam temperature is also controlled. For studying servo control performance of the CCS, the load is decreased from 100% to 40% at a ramp rate of 3% load per min. The impact of a disturbance due to a change in the coal feed composition is also studied. The CCS is found to yield satisfactory performance for both servo control and disturbance rejection.

Keywords: dynamic modeling; process control; load-following; supercritical pulverized coal (SCPC); cycling; time-delay; smith predictor

1. Introduction

Due to the increased penetration of renewables into the electric grid, traditional thermal power plants are being forced to cycle their load and operate under low-load condition to meet changing load demands. However, these plants were designed for neither frequent cycling nor sustained low-load operation. Load-following and part-load operation can lead to considerable losses in efficiency, adverse impacts on plant health, and increases in emissions. To reduce the undesired effects of load-following and part-load operation, advanced control strategies can be helpful for maintaining key controlled variables in their desired range. For developing advanced controllers and studying their performance, a dynamic model of the plant is necessary. Since the model needs to run reasonably fast and achieve desired accuracy, the trade-off between model fidelity and computational expense is an important consideration. For supercritical power plants, an additional computational difficulty is the high degree of nonlinearity in steam properties, especially when the plant transitions between the supercritical and subcritical regimes during load-following.

While there is a large body of literature on dynamic modeling and control of subcritical pulverized coal plants [1], there are fewer studies on supercritical pulverized coal (SCPC) plants. Existing literature on dynamic modeling of SCPC plants can be largely grouped into two categories—those that have focused on individual equipment items and those that have focused on plant-wide model development. The key equipment items that affect the dynamics under load-following operation are those in the boiler section, steam turbine section, and feedwater heater (FWH) section. The existing literature focused on these individual sections is discussed first, followed by a discussion on the literature focused on the plant-wide dynamic model development including plant-wide control.

For supercritical boilers, a number of models with varying scopes have been described in the literature. A "non-equal fragmented model" that captures heat and mass transport characteristics along the height of the water wall has been developed [2]. A model for calculating the heat flux distribution and 3D temperature distribution in a supercritical boiler has also been reported [3]. These studies [2,3] focused only on the furnace combustion and water-wall section. However, it should be noted that the dynamics of the overall boiler depend on the other components of the boiler since the boiler feedwater (BFW) passes through the economizer before going to the water wall. The economizer dynamics, in turn, depend on the dynamics of the superheaters, attemperators and reheaters since the flue gas passes through these sections before entering the economizer. Furthermore, the BFW gets heated in the FWHs before being fed to the economizer. Therefore, due to the pathways of the flue gas and the BFW/steam, all boiler components and some upstream and downstream components must be simulated together. A dynamic model of a 600 MW supercritical plant was developed and used for studying start-up and dynamic behavior [4]. This model included the economizer, superheater, water circulation pump, and water storage tank. The air flow rate was assumed to be sufficient for complete combustion. Thus, no combustion control system was developed. In power plants, the dynamics of the air side can have considerable impact on the dynamics of water/steam-side components, especially during startup and load-following; therefore, consideration of the air-side dynamics is desired. In this work, all sections of the boiler, including air-side control, are modeled with due consideration of the configuration of a typical SCPC plant.

Several models of the steam turbines (ST) are available in the literature. A nonlinear ST model based on approximations of fundamental equations has been developed [5]. This model used identical turbine models for all turbines stages and was validated with steady-state heat balance data from a commercial turbine unit. However, the operation of the governing stage can be different than other stages. Furthermore, models of the back-end condensing stages should account for the presence of moisture. Another nonlinear mathematical model of a ST for a 440 MW power plant was developed to predict the transient behaviors of the turbine system where the model parameters were estimated by a genetic algorithm [6]. Although this model considered separate sub-models for the high-pressure (HP) and intermediate-pressure (IP) sections, where flow was homogeneous, and later stages of the low-pressure (LP) section, where flow was considered to be heterogeneous in order to detect the presence of moisture it also assumed a constant pressure ratio using a first-order transfer function instead of calculating the actual pressure profile. The turbine efficiency was also assumed to be constant. For ST models intended for load-following studies, three aspects should be captured. First, the models of the governing stage, other non-condensing stages, and the condensing stages should be developed such that they can capture the differences in the performance characteristics of these stages. Second, for the non-condensing stages, the efficiency change under load-following operation should be included because of the sliding-pressure operation in SCPC plants and the large variation in the inlet temperature profile that may occur during load-following and low-load operation. Third, the assumption of a fixed number of condensing stages may not always be valid under load-following operation, especially under low-load condition where the reheat temperature may not be maintained at the desired value. Thus, the model should be capable of detecting moisture, if present, in the stages that may be non-condensing under nominal condition. If moisture is present, then the moisture fraction is the variable of interest at the outlet rather than the temperature, which remains constant at the saturation temperature for a given pressure. This change in variables resulting from phase transition can lead to computational issues in a simultaneous equation solver and must be handled effectively in a dynamic model. In this work, conducted in Aspen Plus Dynamics® (APD), pressure and enthalpy are calculated at the ST outlet rather than pressure and temperature (non-condensing) or pressure and vapor fraction (condensing). Then, properties calls are made to obtain the vapor fraction and temperature given the pressure and enthalpy. The changed set of variables is found to yield a model that can be solved reliably if a ST stage transitions back-and-forth between pure vapor phase and two-phase operation.

The FWHs play a key role in achieving high power plant efficiency. An optimal configuration of the FWH network was proposed for increasing the efficiency of a coal-fired power plant [7]. This work considered only high-pressure FWHs as a part of regenerative heating but neglected low-pressure FWHs. For evaluating the performance of FWHs, a nonlinear correlation among the terminal temperature difference, drain cooler approach, and temperature rise was developed as a function of load [8]. However, the extraction flow rate was considered as self-regulating, whereby there is no control valve and the steam flow adjusts itself by a thermal equilibrium process. In another study, a thermodynamic optimization of the heat transfer in the FWHs and exhaust flue gas heat recovery system of SCPC plants was proposed to increase plant efficiency and reduce CO_2 emissions [9]. That work was based on the assumptions of constant pressure ratio in the turbine stages, constant turbine efficiency, constant drain cooling approach, and constant temperature rise. It should be noted that under sliding-pressure operation during load-following, extraction pressure can change considerably leading to control limitations and changes in the condensation temperature of steam, which, in turn, affects the dynamics of the other sections. In this paper, a model of the entire FWH network including a regulatory control layer is developed and included in the plant-wide model so that its impact during load-following operation can be studied.

Works on plant-wide dynamic models and control of SCPC plants in the existing literature are very few in comparison to the works on coal-fired subcritical power plants [10,11]. Recently a comprehensive review of dynamic modeling of thermal power plants was provided [12]. A dynamic model of an SCPC power plant developed in the process simulation software Apros® was used to investigate operational flexibility and transient behavior. Under sliding-pressure operation, the load was decreased from 100% to 27.5% in six steps in 185 min, a ramp rate of 0.4%/min [13]. Energy utilization in a 660 MW SCPC power plant under load-following condition has been studied using a model developed in the GSE software [14]. The GSE model was used to study energy-saving opportunities during load-following by considering a typical coal consumption rate. A 50% load change under sliding-pressure mode was obtained with a maximum ramp rate of 0.5%/min. The ramp rates considered in both these studies are far below the cycling demands and current industrial practices of about 3–8% load change per minute [15]. In addition, none of these studies included the industry-standard coordinated control system (CCS) that is essential to study the dynamics of operating plants. Recently, another dynamic model of an SCPC power plant was developed in Apros® and validated against steady-state and transient plant data [16]. The ramp rates studied in this paper are in an industrially acceptable range of 3–8% load per minute. However, few details about the control configuration, except for the load control and main steam temperature control, were provided. Also, no disturbance rejection studies were conducted. During rapid load-following operation, careful consideration must be provided for not only the dynamics of the main steam temperature but also the dynamics of the reheat steam temperature, since they affect both the plant efficiency as well as the extent of condensation in the last few turbine stages, which affects ST health. Furthermore, developing the plant-wide control system requires simultaneous consideration of the FWH section, boiler section, and the ST section due to strong interactions among these sections. The CCS including the FWH control is necessary for plant-wide control. In this paper, a CCS is designed and its performance for load-following is studied. It should be noted that the CCS presented here does not represent or reproduce that of any vendor or any power plant but was developed by the authors based on the information available in the open literature and control requirements under load-following operation.

In this work, first a steady-state model of an SCPC power plant is developed. The configuration and nominal operating conditions of the plant are similar to Case B12B (for a 550 MWe net SCPC plant using Illinois No. 6 coal) from the cost and performance baseline study by the National Energy Technology Laboratory (NETL) [17]. The steady-state model is developed using Aspen Plus® (AP) and ACM then converted to a pressure-driven APD model, where the regulatory control layer and CCS are developed. Tight control of the main steam temperature is desired under load-following condition, since a lower temperature leads to losses in efficiency, and a higher temperature can lead to damage in the superheater tubes in the boiler and the leading stage(s) of the turbine. While the fire side of the boiler has very fast dynamics, the steam-side dynamics are comparatively slower due to the considerable thermal holdup in the boiler tubes. Due to considerable time delay, tight control of the steam temperature under load-following operation becomes challenging. For the main steam temperature control, a Smith predictor for time-delay systems is developed and implemented as part of the overall CCS in addition to traditional strategies in power plants for steam temperature control. For evaluating the performance of the CCS, the plant load is ramped down from 100% to 40% at 3% load change/min under sliding-pressure operation. The remaining sections of this paper are arranged as follows. Section 2 provides details of the SCPC power plant configuration and dynamic process sub-models. Section 3 describes the design of the control system. Section 4 provides the simulation results followed by the conclusions in Section 5.

2. Process Sub-Models

2.1. Plant Configuration

The SCPC power plant configuration presented consists of a once-through steam boiler with flue gas treatment and a supercritical steam cycle with single steam reheater. There are four main sections: the feedwater treatment and heating sections, the supercritical boiler section that includes air fans as well as the air preheater, the ST section, and the flue gas treatment section, including some consideration for acid gas recovery (AGR). The configuration of the plant is shown in Figure 1, as adapted from the NETL study [17]. The referenced configuration also includes CO_2 capture, but a detailed model of that section is not included in the current study. Nevertheless, the steam extraction for the AGR section was modeled to correctly characterize the power produced in the ST; these extraction flows were assumed to change proportionally with load. Another important note is that the coal feed in Figure 1 is located after the coal pulverizers, which were not considered as part of this study. It should also be noted here that the double-ended arrows indicate extracted steam flowing for use as a heating medium and the then-cooled effluent returned to the surface condenser in the ST section.

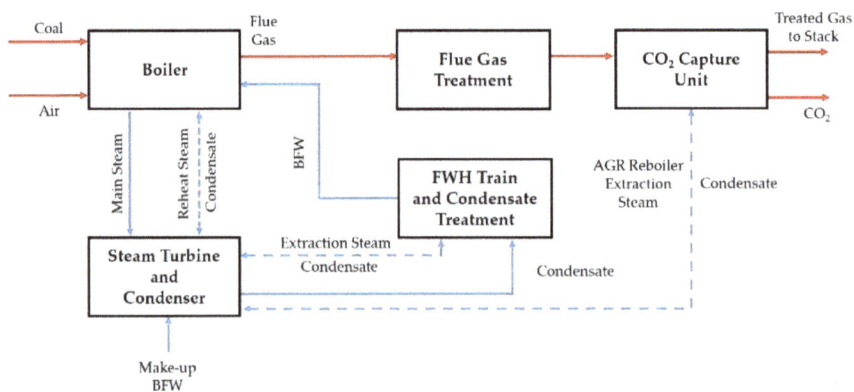

Figure 1. Supercritical Pulverized Coal (SCPC) Power Plant Block Flow Diagram.

In the boiler, pulverized coal is combusted producing hot flue gas. The boiler section consists of various components including an economizer, water wall, separator, reheater, multiple superheaters, two-stage attemperation for the main steam, and one-stage attemperation for the reheated steam. In the steam cycle, the supercritical steam at 593.3 °C and 241.2 bar is sent to a HP turbine, where it is expanded to 47 bar in three stages. The expanded steam is then returned to the boiler where it is reheated to 593.3 °C, before it is sent to a three-stage IP turbine and subsequently to the five LP turbines. To enhance the overall power cycle efficiency, steam is extracted from the turbines for feedwater heating.

The dynamic SCPC model in the APD software was generated by first developing a valid pressure-flow network in the steady-state SCPC model in the AP software. This modeling task required connecting the pressure nodes in the SCPC plant through flow nodes that relate pressure drop with volumetric flow rate [18]. In dynamic simulations, specification of equipment sizes, their geometries, and orientations are crucial for capturing the transient behavior of the system. Volumetric holdup in equipment items affects the rate of accumulation, which is one of the key factors that determines the transient response [18]. Each of the vessels was sized based on its steady-state operating conditions, and these geometrical details were used in the APD model; in dynamic simulation this allows for the dynamics to be captured relative to the nominal condition and provides the most logical basis for equipment design. The dynamic SCPC model operating at base load was shown to be in good agreement with the steady-state results from the NETL baseline study [17].

Specific component lists with appropriate physical property packages were assigned to the individual sections of the plant as necessary in order to accurately capture the interactions in each section based on local components and conditions. This helped to minimize the zero-flow components in specific streams and equipment models, thereby improving solver convergence properties and reducing computational time.

2.2. Feedwater Heaters

Figure 2 shows the layout of the feedwater pretreatment and heating section of the SCPC plant with, one deaerator (DA) and seven total exchangers consisting of five FWHs and two drain coolers (DCs) [17]. The main difference between the FWHs and the DCs is that in the FWHs heating is accomplished primarily using the latent heat from the extracted steam, whereas in the DCs the sensible heat of the condensate from the FWHs is used for heating the feedwater. Extracted steam from HP Stage 1 and HP Stage 2 is fed to FWH 1 and FWH 2, respectively, with an extraction from IP Stage 1 fed to FWH 3. The condensate from these three FWHs is sent to DC 1 and subsequently to the deaerator. In the deaerator, extracted steam from IP Stage 2 is used for removing dissolved oxygen. Extracted steam from LP Stage 1 and LP Stage 2 is fed to FWH 4 and FWH 5, respectively. The condensate from these two FWHs is sent to DC 2 and subsequently to the surface condenser.

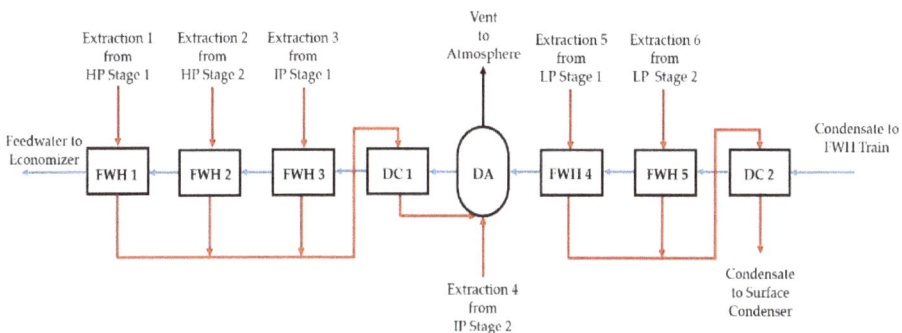

Figure 2. Feedwater Pretreatment and Heating Section Block Flow Diagram.

Aspen Exchanger Design and Rating (EDR) was used to size each of the FWHs as a shell-and-tube heat exchanger based on its steady-state operating conditions. Aspen EDR sizes heat exchangers based on a constrained optimization, accounting for the process conditions within an economic framework [19]. Sizing information for the FWHs including the volumes and metal masses of the shells and tube bundles was used in the APD models. For more information on the Aspen library models, interested readers are referred to several resources available in the public domain [18,20–24].

2.3. Simplified Boiler Model

As noted above, gas-side dynamics of SCPC boilers are very fast in comparison to the water/steam side. Additionally, the flue gas has a low density and heat capacity in comparison to the water/steam in SCPC plants. For comparison, the ratio of the product of specific enthalpy and density (characterizing the thermal holdup) for water/steam to gas is more than 500 under conditions at the water-wall of the boiler. Therefore, in this work, gas-side dynamics have been neglected, and the gas side is assumed to be instantaneous. The ultimate analysis of the coal provided in Table 1 is the same as that for the Illinois No. 6 coal in the NETL baseline study [17].

Table 1. Ultimate Analysis of Coal Feed to the Boiler.

Component	Weight %
H_2O	11.12
C	63.75
H_2	4.50
N_2	1.25
Cl	0.29
S	2.51
O_2	6.88
Ash	9.7

The following sections of the boiler are modeled with due consideration of thermal and volumetric holdup: economizer, water wall, primary superheater, platen superheater, finishing superheater, and reheater. Typical inlet and outlet temperatures of the water and flue gas in these sections are estimated based on the NETL study [17], information available in the open literature [25], and the results of an energy balance.

The flue gas exiting the boiler section is sent to the flue gas desulfurization (FGD) unit. Since this work primarily focuses on the dynamics of the front end of the power plant, models of back end sections like the flue gas treatment section are very simple. A simple stoichiometric reactor with 98% conversion of SO_2 was developed for the FGD section where the SO_2 in the flue gas reacts with lime slurry to form calcium sulfite that is then oxidized with air to form gypsum. The flue gas finally leaves the system via the carbon capture unit.

2.4. Fan Models

The primary air (PA) and forced draft (FD) fans are used for providing air to the pulverizers and burners in the boiler, respectively. During load-following operation of the plant, changes in these air flow rates affect the energetics in the boiler and the auxiliary power requirements. Therefore, the control system needs to be designed appropriately. For large power plants, the PA and FD fans are typically operated by variable frequency drives (VFDs) that modulate the fan speed to obtain the desired flow rate. Since fan curves that represent the head and power with respect to flow rate at various revolutions per minute (RPMs) are not currently available for the FD and PA fans corresponding to this work, an approximate method is developed. A family of curves available in the open literature [26] for similar sized fans is scaled to match the desired range of head and flow. Then, a quadratic function between the head and flow is regressed to the family of curves simultaneously where each regression coefficient is considered to be a linear function of RPM.

2.5. Steam Turbine

Three separate ST models were considered to capture the operating characteristics of the various stages of the ST:

1. Leading (Governing) Stage
2. High-Pressure (HP), Intermediate-Pressure (IP), and Low-Pressure (LP) Stages
3. Final Stage before Condenser

Figure 3 shows the layout of the turbine section of the SCPC plant [17]. The main steam from the finishing superheater of the boiler is throttled and fed to the governing stage of the HP turbine. There are three physical stages in the HP section. Extractions 1 and 2 from the first and second stages of the HP turbine section are sent to FWHs 1 and 2, respectively. After the HP section, steam is heated to 593 °C under the nominal condition by returning it to the boiler through a single reheater followed by attemperation. The reheated steam is sent to the IP section of the turbine that comprises two physical stages. Extractions 3 and 4 from the first and second IP stages, respectively, are sent to FWH 3 and the deaerator, respectively. After the IP turbines there are auxiliary extractions connected to various reboilers and a single turbine for auxiliary equipment, and the steam goes to the LP section that comprises five physical stages, with two extractions to FWHs 4 and 5, after LP stages 1 and 2, respectively. The effluent steam from the final LP stage is then fed to a surface condenser where it is condensed with cooling water (CW). The condenser is integrated with a hotwell from which the FWH pump returns water to the feedwater treatment and heating section.

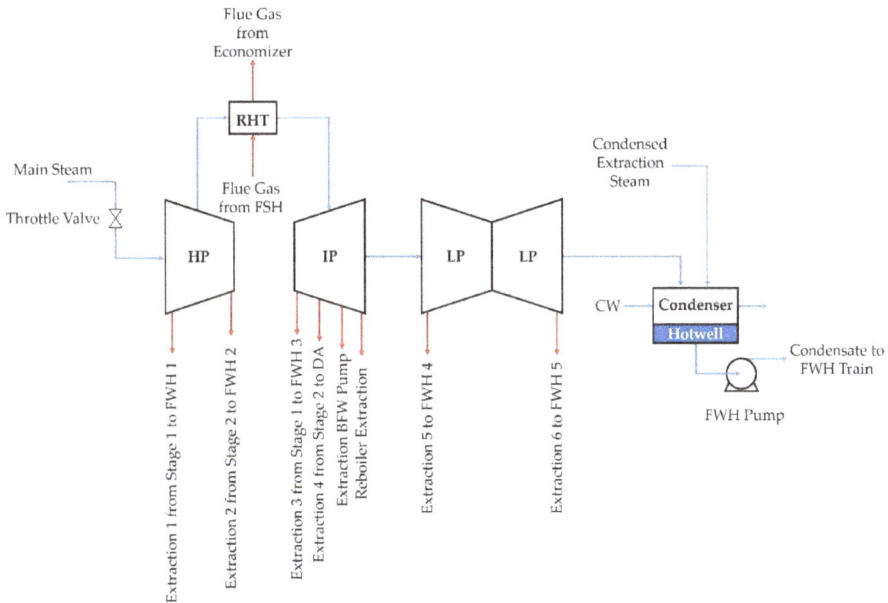

Figure 3. Turbine Section Block Flow Diagram.

Each of the turbine stage sub-models described below was developed in ACM. These custom models were then compiled into library blocks and used in the steady-state model in AP. The same blocks were used in APD to build a model for the entire ST, including extractions and auxiliary equipment items. The isentropic power generated by any given turbine stage without condensation is given by Equation (1), where the power for the condensing stage is shown later.

$$\mathcal{P} = \eta_{is} M \Delta h_{is} \tag{1}$$

Turbine dynamics are fast and, therefore, have been neglected in this model. Turbine dynamics can be important during plant startup/shutdown, but those operations are not considered in this work.

2.5.1. Leading Stage

The leading stage model was considered separately for two reasons: capturing the change in the leading stage efficiency under extremely nonlinear property changes and considering the control setups for fixed-pressure operation [27]. For fixed-pressure operation, separate governing valves are used in the control of separate arcs of admission into the turbine. Here, an array of four instantiations of the model are used to represent the true leading stage at the nominal condition. In each model instance, the flow, M, through the stage is calculated by Equation (2), where the flow parameter C_{flow} is designed for the nominal load. The efficiency, η_{is}, of the leading stage is then calculated using Equations (3) and (4). The equations used for this model are adapted from the work of Liese [27].

$$M = C_{flow} \frac{P_{in}}{\sqrt{T_{in}}} \sqrt{\frac{\gamma}{\gamma - 1} \left[\left(\frac{P_{out}}{P_{in}} \right)^{\frac{2}{\gamma}} - \left(\frac{P_{out}}{P_{in}} \right)^{\frac{\gamma+1}{\gamma}} \right]} \tag{2}$$

$$V_o = \frac{44.72}{W} \sqrt{(1 - R)(h_{out} - h_{outi})} \tag{3}$$

$$\eta_{is} = 2 \left(\frac{V_{rbl}}{V_o} \right) \left[\left(\sqrt{1 - R} - \frac{V_{rbl}}{V_o} \right) + \sqrt{\left(\sqrt{1 - R} - \frac{V_{rbl}}{V_o} \right)^2 + R} \right] \tag{4}$$

2.5.2. HP, IP, and LP Stages

The model for all the stages between the leading and last stage models is a thermodynamic stage-by-stage model [28]. Here, a model that represents a single thermodynamic stage was developed and then repeated as needed to represent the boundaries of each section with extractions placed at the appropriate pressure levels. The HP, IP, and LP sections are comprised of seven, fourteen, and seven thermodynamic stages, respectively.

For the thermodynamic stages, the isentropic efficiency, η_{is}, is correlated with the specific shaft speed, N_s, as seen in Equations (5)–(7). The other important parameter in these calculations is the isentropic head parameter, k_{is}, which is used for calculating the total enthalpy change as shown in Equation (8). Here, the calculation is made for each stage at the nominal operating condition via Equation (9) and then the value remains fixed during dynamic simulation. These equations are adapted from the work of Lozza [28].

$$\Delta ef = -0.0049(ln|N_s + 0.001|)^3 - 0.031(ln|N_s + 0.001|)^2 - 0.060(ln|N_s + 0.001|) - 0.022 \tag{5}$$

$$ef = 0.072(ln|N_s + 0.001|)^3 + 0.020(ln|N_s + 0.001|)^2 - 0.010(ln|N_s + 0.001|) + 0.89 \tag{6}$$

$$\eta_{is} = ef(1 - \Delta ef) - 0.87(1 - f_v) \tag{7}$$

$$\Delta h_{is} = \frac{k_{is} u_{mv}^2}{2W} \tag{8}$$

$$k_{is} = 2.2 + 8.9e^{-43N_s} \tag{9}$$

Accounting for moisture is essential in calculating the actual power produced by a given stage. The existence of moisture also significantly affects the efficiency of the stage. As noted earlier, if moisture is present, then the model needs to calculate moisture fraction as opposed to temperature, which becomes fixed. A logic-based approach for detection of moisture based on the dew point calculation and subsequent structural changes, if moisture is present, does lead to convergence issues during load-following since APD uses an equation-oriented approach. The change in solving from pressure and temperature to pressure and vapor fraction yields a discontinuous system about the dew point, creating the convergence problems. One way of avoiding the logic-based approach is to change the variables from temperature or vapor fraction to enthalpy. Thus, while pressure and temperature fully define the system in absence of condensation, and pressure and moisture fraction fully define the system under condensation, pressure, and enthalpy fully define the system for both presence and absence of condensation. This change in variables avoids issues with structural changes in an equation-oriented framework.

2.5.3. Final Stage

The modeling of the final turbine stage is important since this stage typically operates under a choked flow condition, and it has different performance characteristics than other stages. In addition, due to condensation in this stage under typical operating conditions, the stage efficiency calculation needs to be modified [27]. The Stodola equation (Equation (10)) is considered to represent the flow through this stage in the presence of condensation. The exit pressure of the last stage is constrained to the pressure of the surface condenser, which is again affected by the cooling duty of the condenser. Equations (11) and (12) show the calculation of the end-line end-point and the used-energy end-point enthalpies; these enthalpies correspond to the calculation of the efficiency for this stage by accounting for the generation of moisture. Then efficiency is calculated from Equations (13) and (14). The power produced by the condensing stage is shown by Equation (15) as a function of the real enthalpy drop accounting for the presence of moisture. Here, these equations are adapted from Liese [27], as follows:

$$M = C_{flow} \frac{P_{in}}{\sqrt{T_{in}}} \sqrt{1 - \left(\frac{P_{out}}{P_{in}}\right)^2} \tag{10}$$

$$h_{elep} = h_{in} + (h_{is} - h_{in})\eta_{dry}f_v(1 - 0.65f_l) \tag{11}$$

$$h_{ueep} = h_{elep} + TEL\eta_{dry}f_v(1 - 0.65f_l) \tag{12}$$

$$\Delta h_r = h_{in} - h_{ueep} \tag{13}$$

$$\eta_{is} = \frac{\Delta h_r}{\Delta h_{is}} \tag{14}$$

$$P_{cond} = M\Delta h_r \tag{15}$$

3. Control System Design

As mentioned earlier, the water-side of the SCPC system is a time-delay system that makes the design of the control system challenging. In addition, steam properties and heat transfer characteristics are highly nonlinear as the system transitions from the supercritical to subcritical region or vice versa during load-following. Furthermore, the highly complex configuration of the FWHs, coupled with the sliding-pressure operation that changes the pressure of the steam extractions, leads to considerable further challenges in the control system design. In SCPC power plants, a coordinated control system is usually used for load-following [29]. The CCS is implemented as the supervisory layer that exploits the regulatory control as degrees of freedom to achieve the control objectives during load following.

3.1. Regulatory Control Layer

The regulatory control layer is developed using the minimum amount of control needed for dynamic convergence. It consists of 16 single-loop feedback controllers and 13 cascade control loops, where proportional-integral-derivative (PID) controllers are used. A few of these controllers are discussed in detail below. Under the nominal condition of the SCPC plant, phase separation does not take place in the separator that is located between the water wall and primary superheater. Therefore, under the nominal condition, the inventory on the water side is controlled in the hotwell and in the deaerator. The deaerator level is maintained by manipulating the incoming BFW flow rate while the hotwell level is maintained by manipulating the demineralized water flow rate to the hotwell, under the assumption that a condensate storage tank can be neglected.

As mentioned earlier, maintaining the reheat temperature while following the load is important. A lower temperature than desired will lead to efficiency losses and higher condensation in the LP turbine section. A higher temperature than desired will lead to damage in the superheater tubes and turbine. Typically, at the vertical downpass of the boiler, the flue gas is split between the reheater and primary superheater. The flue gas split fraction going to the reheater section can be modulated by a damper to help control the reheat temperature. In addition, the BFW flow rate to the attemperator, located after the reheater, also assists in the reheat temperature control.

3.2. Supervisory Control Layer

As noted before, the typical supervisory control layer for SCPC plants is the CCS, which helps to follow the load with due consideration of the synergies between the boiler and turbine and interactions among the manipulated and controlled variables, as laid out briefly in the literature [29]. Figure 4 shows the overall CCS master diagram that has been developed in this work and implemented in APD. While the required coal flow rate for a desired power output can be calculated based on the calorific value of the coal and the overall system efficiency, the system efficiency changes under load-following operation. Therefore, the heat rate correction is considered while calculating the trim to the boiler master and the turbine master inputs. It should be noted here that there are various functions and ratios implemented in the control systems described below. For simplicity, these functions are represented only in the form *f(x)* in the control diagrams for main steam temperature control in Section 3.4 and BFW flow control in Section 3.5.

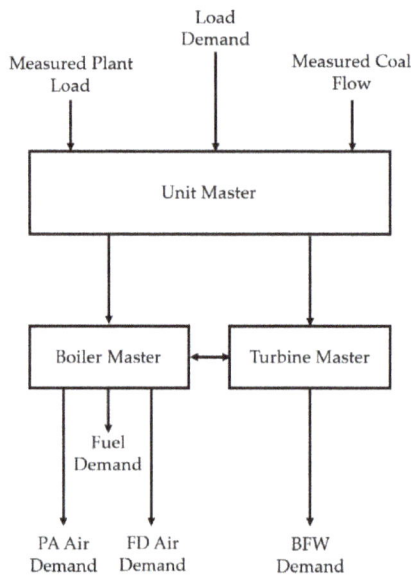

Figure 4. Coordinated Control System (CCS) Master Diagram.

3.3. Air Flow Rate Control

Figures 5 and 6 show the control diagrams for the two fans that supply air to the boiler: the forced draft (FD) fan and the primary air (PA) fan, respectively. In the SCPC plant, the PA fan supplies air to the pulverizers transporting coal to the burners. Here, the air through the pulverizers is accounted for to accurately model the system interactions even though the pulverizers are not explicitly modeled.

Based on the output signal of the boiler master controller, set points for air flow for the PA and FD fans are calculated. The corresponding set points for fan speeds in RPM are sent to the respective fan VFDs that modulate the frequencies to obtain the desired RPMs, based on the performance curves explained in Section 2.4. The VFD control is represented by a simple PID controller. For the FD fan, a trim is provided based on the oxygen concentration in the boiler outlet flue gas. Proper control of the excess oxygen is crucial in that, if the excess oxygen drops too low, incomplete combustion might result leading to a process safety risk; if it becomes higher than needed, the higher heat loss through the exiting flue gas from the system would reduce the boiler efficiency.

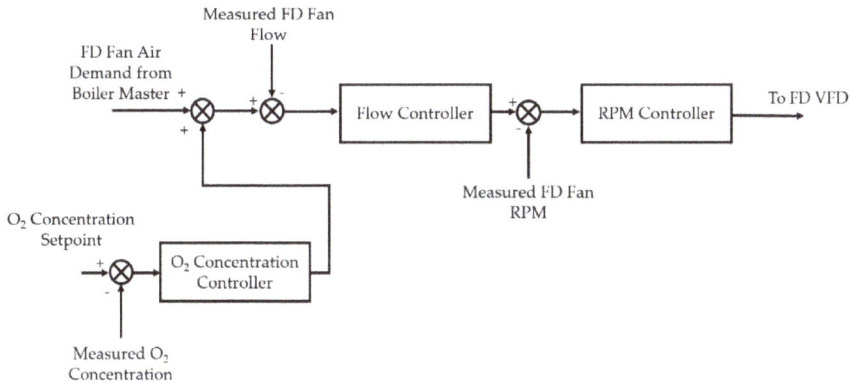

Figure 5. Forced Draft (FD) Fan Air Control Scheme.

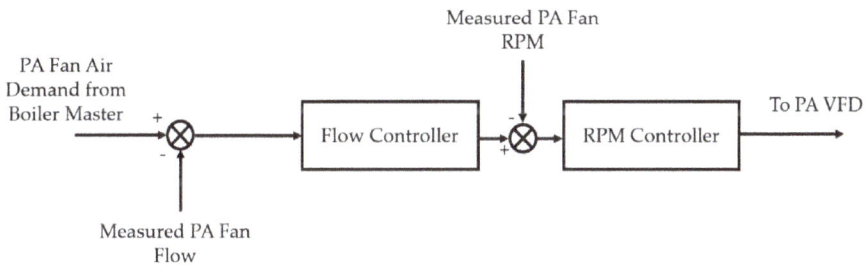

Figure 6. Primary Air (PA) Fan Air Control Scheme.

3.4. Main Steam Temperature Control

As mentioned before, tight control of the main steam temperature is desired for maintaining efficiency during load following. Furthermore, large deviations in the main steam temperature can lead to undesired creep and fatigue in the boiler tubes and turbine components. Temperature of the main steam is controlled by attemperation using BFW spray at two locations: the first immediately before the platen superheater (SH) and the second immediately before the finishing SH as shown in Figure 7. Here, the second attemperator plays the key role in controlling the main steam temperature by regulating the spray flow, while the first attemperator assists by ensuring that the second attemperator spray is within a set range of operation, leaving room for changes in response to disturbances or fast load changes.

Three configurations for main steam temperature control are investigated here, where the manipulated variable is the BFW flow rate injected into Attemperator 2. Configuration 1 and Configuration 2 are typical control strategies reported in literature [10]. Configuration 3 is the strategy proposed in this work, with comparisons of the three strategies to follow in the results.

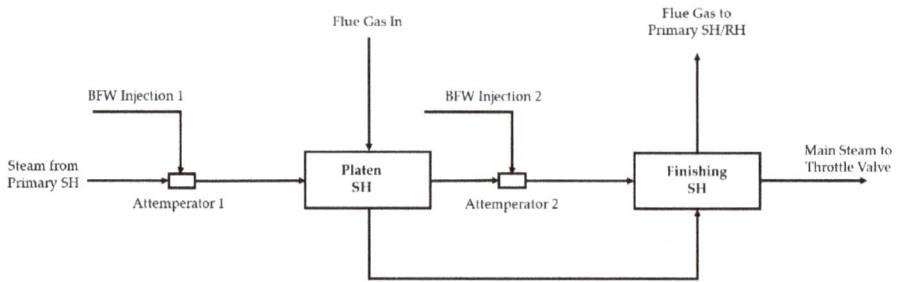

Figure 7. Schematic of High-Pressure Steam Attemperation.

3.4.1. Configuration 1

Configuration 1, shown in Figure 8, consists of a simple feedback loop with a feedforward correction based on the steam flow rate [10]. As discussed before, large excursions in the main steam temperature should be avoided. However, there are considerable nonlinearities in the steam properties especially during transitions between the supercritical and subcritical regions. A gain-scheduled controller is used in Configuration 1 to help improve control for this nonlinear system. The feedforward term helps to improve the disturbance rejection characteristics of the loop.

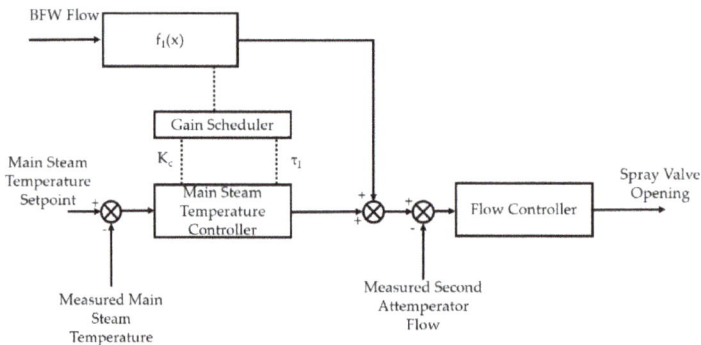

Figure 8. Configuration 1 Control Scheme for Main Steam Temperature.

3.4.2. Configuration 2

In Configuration 1, temperature of the main steam is controlled without any consideration of the intermediate steam temperature immediately after Attemperator 2. The temperature of this intermediate steam responds faster to changes in the BFW spray flow rate in comparison to the main steam temperature, which lags due to the thermal and volumetric holdup of the finishing SH. In Configuration 2, the intermediate steam temperature controller manipulates the BFW injection flow rate to Attemperator 2 as shown in Figure 9 [30]. The PID controller that is used for the main steam temperature control generates the set point for the intermediate steam temperature controller. It should be noted that this configuration does not consider any feedforward correction.

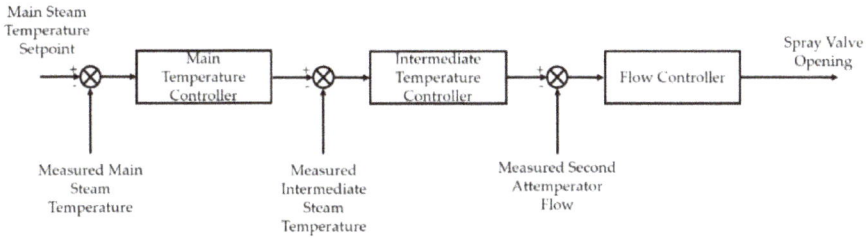

Figure 9. Configuration 2 Control Scheme for Main Steam Temperature.

3.4.3. Configuration 3

As noted before, there is significant time-delay in the water/steam side of the SCPC plant. For closed-loop stability of such systems, a smaller gain has to be used in the PID controller leading to sluggish response that is undesired for main steam temperature control. One classic approach for control of a time-delay system is the Smith predictor [31]. For designing the Smith Predictor, the finishing SH is represented as a first-order process with time-delay as follows [31]:

$$y(s) = \frac{K_c e^{-\theta s}}{\tau_s + 1} \times u(s) \tag{16}$$

A minor feedback loop is introduced in the conventional feedback structure, along with a feedforward compensation. The block diagram for the Configuration 3 control scheme developed in this study is shown in Figure 10.

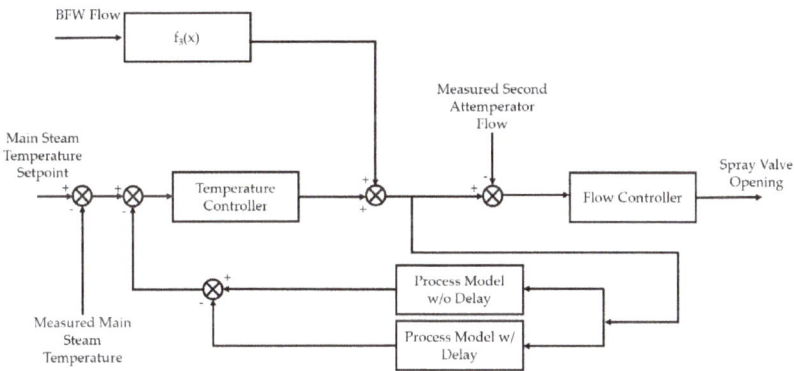

Figure 10. Configuration 3 (Smith Predictor) Control Scheme for Main Steam Temperature.

3.5. BFW Flow Control

Figure 11 represents the control diagram for the BFW flow control. The BFW flow plays a key role in achieving sliding-pressure operation and ensuring that the temperature constraints at various locations of the boiler can be satisfied. The BFW flow rate set point is load-dependent and corrected via the enthalpy at the water wall (WW) outlet in the boiler and by the degree of attemperation as shown in Figure 11 [32]. A trim is also provided based on the water wall outlet enthalpy, that can be calculated based on the water wall outlet temperature and pressure. The CCS determines the load-dependent set point for the BFW controller based on the turbine master output signal. The trim, which is based on the opening of the main steam Attemperator 1 valve, ensures that the Attemperator 1 valve opening retains sufficient gain to move this system in response to sudden load changes.

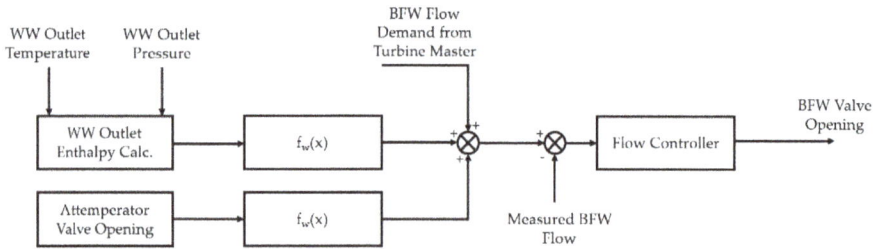

Figure 11. Boiler Feedwater (BFW) Flow Control Scheme.

4. Results

Table 2 compares the results of the simulation at full-load condition for the SCPC plant-wide dynamic model developed in this study and the steady-state NETL baseline study [17].

Table 2. Steady-State Validation for Full Load.

Parameter	Unit	NETL Baseline Study	SCPC Model	Error
Coal Flow Rate	tonne/h	225	228	1.53%
Gross Power	MW	641	620	−3.28%
Net Power	MW	550	532	−3.21%
Heat Rate	kJ/kWh	11,086	11,629	4.90%
Main Steam Pressure	MPa	24.2	24.1	−0.37%
Main Steam Temperature	°C	593	593	0.00%
Main Steam Flow Rate	tonne/h	2003	2027	1.19%

Using the CCS detailed above, transient studies were conducted on the response of the SCPC plant to ramp changes in power demand (load). Here, the studies were conducted for a load decrease from 100% to 40% over 20 min, corresponding to a ramp rate of 3% load per min. This ramp rate is within an acceptable range of power industry ramp rates while maintaining all key operating variables within allowable deviations from their set points. A near-perfect tracking of the load was accomplished (not shown here). During these studies, each of the configurations detailed above were used in turn, and in the following results their responses are either shown explicitly or deemed similar to one another and discussed as such.

Figure 12 represents the response of the BFW flow rate and the main steam pressure to the 60% ramp down in load starting at time equal to 1 hr. The BFW flowrate and main steam pressure decrease by approximately 63% and 62%, respectively. These responses are hardly affected by the main steam temperature control figurations. The main steam pressure slides from 242 bar to 93 bar, corresponding to a ramp rate of 7.5 bar per min.

Figure 13 depicts the response of main steam temperature to the 60% reduction in load for each of the control configurations detailed above. A ±10 °C band shown in Figure 13 is considered to be the acceptable range. Both Configurations 1 and 3 lead to main steam temperatures that are well within the band; however, Configuration 2 results in a large undershoot that is unacceptable because of boiler efficiency losses, ST efficiency losses, added thermal stresses on the reheater, and added condensation in the trailing LP stages, leading to damage to the ST. The large undershoot of about 18 °C Configuration 2 comes from a lack of accounting for the dead time in the system, leading to a controller that is out of sync with the system dynamics and thereby has to catch up with the system. Configuration 3 provides the best control performance, limiting the maximum deviation in the main steam temperature to about 7 °C and resulting in a settling time of about 15 min following the end of the ramp down in load. Configuration 3 is the best performer, due to the characterization of the dead time by the Smith predictor, though the control performance of Configuration 1 is similar.

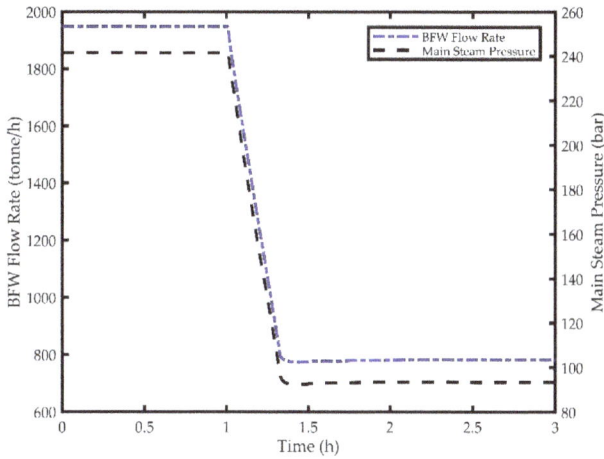

Figure 12. Response of BFW Flowrate and Main Steam Pressure to a 60% Reduction in Load.

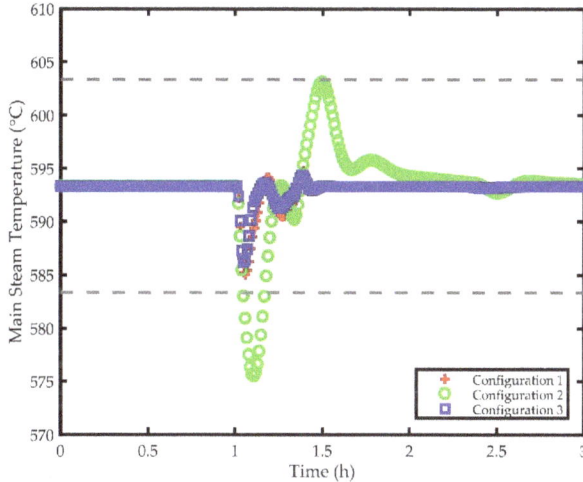

Figure 13. Transients of Main Steam Temperature for Different Control Strategies.

Figure 14 shows the response of using attemperation after the reheater to control the temperature of the reheated steam returning to the IP turbine. Impacts of the three control configurations, that were similar to the main steam temperature control, are shown in Figure 14, with the bands representing deviations of ±10 °C. It can be seen here that the reheat temperature could be brought back to the original set point by each of the configurations considered. Though Configuration 3 has slightly higher overshoot than Configurations 1 and 2, it has faster settling time and lower oscillation. The performance of Configuration 1 is found to be the worst. However, the performance of each configuration is acceptable for controlling the reheat steam temperature.

Finally, Figure 15 represents how the oxygen concentration in the boiler flue gas outlet responds to the 60% ramp decrease in load. Here again, the configuration used for main steam temperature control has no effect on the response of the oxygen concentration so only one plot is shown. It can be seen in Figure 15 that the maximum deviation in oxygen concentration is within ±5%.

The composition of coal fed to a power plant can change considerably. The CCS should be designed for rejecting this disturbance efficiently while maintaining a set load. The base case composition of Illinois No. 6 coal shown in Table 1 is changed as shown in Table 3 for this transient study, corresponding to 2.59% reduction in the calorific value of the coal feed. This change is centered around expected deviations in coal composition, even when considering coal of a similar grade or from the same mine. Here, it can be observed that the calorific value of the coal can deviate over a range of feeds, a disturbance that the CCS must be able to handle.

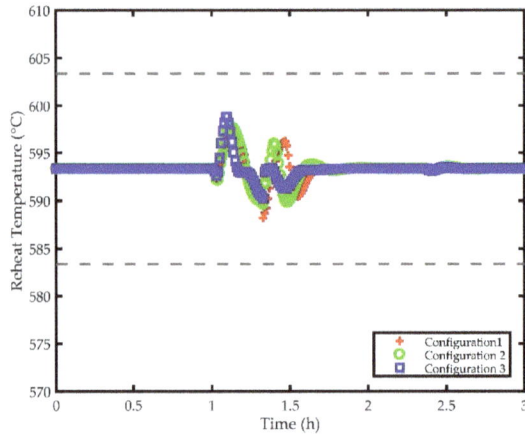

Figure 14. Response of Reheat Temperature for Different Control Strategies.

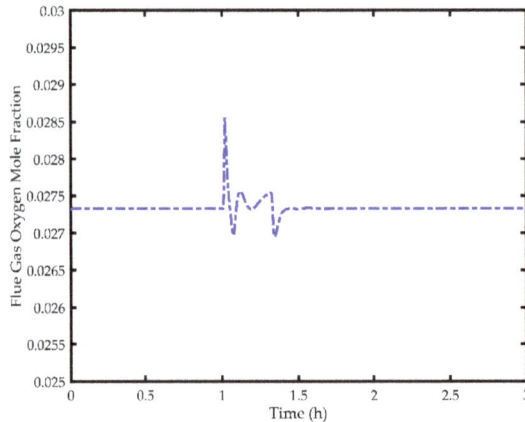

Figure 15. Response of the Boiler Outlet Oxygen Mole Fraction to a 60% Reduction in Load.

Table 3. Comparison of Coal Compositions for Disturbance Rejection Study.

	Ultimate Coal Analysis	
	Base Case	Changed
H_2O	11.12	13.18
C	63.75	59.36
H_2	4.5	5.18
N_2	1.25	1.49
Cl	0.29	0.29
S	2.51	2.88
O_2	6.88	7.92
Ash	9.7	9.7

Figure 16 shows the transients in load and coal flow for the change in coal feed composition at time equal to 1 h. Here, because of the lower calorific value of the new coal, the load drops by approximately 0.4%, leading to an increase in the coal feed to compensate. Note that the results are only shown here for using Configuration 3 to control the main steam and reheat steam temperatures, given similarities across the results for the three control configurations.

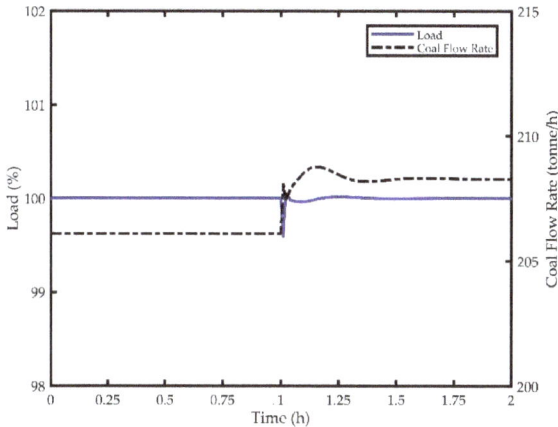

Figure 16. Disturbance Rejection Results for Load and Coal Flow.

Figure 17 shows the transients in the main steam temperature and flue gas oxygen concentration in response to the disturbance in coal composition. Here, again, a $\pm 10\,^\circ\text{C}$ band is set on the main steam temperature. It is observed that Configuration 2 has lower undershoot (about 8 °C) than Configuration 1 but has higher overshoot than Configuration 1 (about 5 °C). Configuration 3 results in considerably lower under/overshoot with a maximum deviation of about 5 °C. Configuration 3 also results in a settling time that is more than 20 min faster compared to other configurations. Irrespective of the configuration for steam temperature control, the oxygen concentration remains relatively constant at its setpoint as shown in Figure 17.

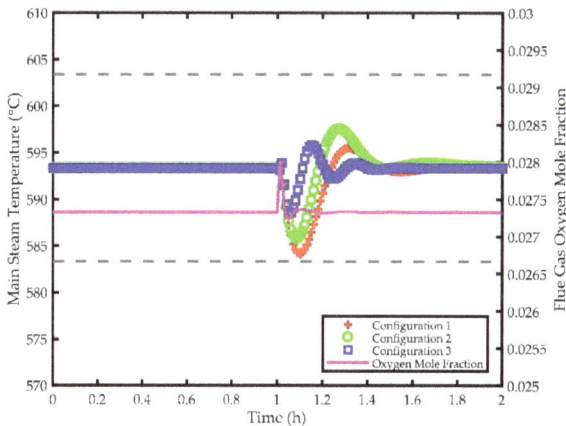

Figure 17. Disturbance Rejection Results for Main Steam Temperature and Flue Gas Oxygen Concentration.

5. Conclusions

A plant-wide dynamic model of a supercritical pulverized coal power plant with CO_2 capture is developed in this work. A coordinated control strategy is designed and its performance is studied for both servo control and disturbance rejection. Sliding-pressure operation is considered while ramping down the load from 100% to 40% at a ramp rate of 3% load per min. As the main and reheat steam temperatures must be controlled tightly, performance of three control configurations is evaluated. For main steam temperature control Configuration 3, which includes the Smith predictor for handling time delays on the water/steam-side, results in maximum deviation of about 7 °C in the main steam temperature and a settling time of about 15 min following the end of the ramp change. Configuration 1 also provides similar control performance. However, Configuration 2 results in a much poorer tracking control performance with the maximum deviation of about 18 °C in the main steam temperature and much longer settling time. For the reheat temperature control, Configuration 3 has the best performance and Configuration 1 has the worst performance while performances of all three configurations are acceptable.

Performance of the control system is evaluated for a disturbance in the coal feed composition. The reduction in the calorific value of the coal feed could be rejected very efficiently by the coordinated control strategy. Maximum deviation in the load is found to be about 0.4% when the coal calorific value was changed by about 2.6%. Maximum deviation in the main steam temperature is found to be about 9 °C, 8 °C, and 5 °C for Configuration 1, Configuration 2, and Configuration 3, respectively. Settling time of Configuration 3 is found to be faster by more than 20 min in comparison to the other configurations. It should be noted that no feedforward input is considered for this disturbance. Overall, Configuration 3 with the Smith predictor is found to provide the best performance for main and reheat steam temperature control for both tracking and disturbance rejection scenarios. Our future work in this area will focus on development of higher fidelity models for various plant components and development of model-based control strategies for further improvement in control performance.

Author Contributions: All authors contributed equally.

Funding: PS, EH, KR, DB gratefully acknowledge financial support from National Energy Technology Laboratory through DE-FE0025912 contract for site support services.

Acknowledgments: The authors acknowledge valuable discussion with Eric Liese from NETL, Morgantown while developing the ST model.

Disclaimer: This report was prepared as an account of work sponsored by an agency of the United States Government. Neither the United States Government nor any agency thereof, nor any of their employees, makes any warranty, express or implied, or assumes any legal liability or responsibility for the accuracy, completeness, or usefulness of any information, apparatus, product, or process disclosed, or represents that its use would not infringe privately owned rights. Reference herein to any specific commercial product, process, or service by trade name, trademark, manufacturer, or otherwise does not necessarily constitute or imply its endorsement, recommendation, or favoring by the United States Government or any agency thereof. The views and opinions of authors expressed herein do not necessarily state or reflect those of the United States Government or any agency thereof.

Conflicts of Interest: The authors declare no conflict of interest.

Notation

Variable	Name
h_{is}	turbine isentropic outlet specific molar enthalpy
k_{is}	turbine isentropic head factor
u_{mv}	turbine mean peripheral exit velocity
W	molecular weight
N_s	turbine specific shaft speed
ef	turbine efficiency parameter
η_{is}	turbine isentropic efficiency
f_v	turbine outlet vapor fraction
M	turbine inlet mass flow rate
C_{flow}	turbine flow coefficient
P_{in}	turbine stage inlet pressure
P_{out}	turbine stage outlet pressure
T_{in}	turbine inlet temperature
h_{out}	turbine actual stage outlet enthalpy
h_{outi}	turbine isentropic stage outlet enthalpy
h_{elep}	turbine end line end point molar enthalpy
h_{in}	inlet molar enthalpy
η_{dry}	turbine "dry" efficiency
f_l	turbine outlet liquid fraction
h_{ueep}	turbine used energy end point molar enthalpy
TEL	turbine total exhaust losses
Δh_r	turbine enthalpy drop
γ	turbine heat capacity ratio
V_o	turbine current operating velocity
V_{rbl}	turbine design velocity
R	turbine blade reaction
f_v	turbine outlet vapor fraction
$\mathcal{P}$	turbine power produced
$y(s)$	main steam temperature (output)
K_c	process gain
θ	system time delay
τ_s	system time constant
$u(s)$	BFW injection flow rate (input)
$\mathcal{P}_{cond}$	power produced, condensing stage
t	time

References

1. Adam, E.J.; Marchetti, J.L. Dynamic simulation of large boilers with natural recirculation. *Comput. Chem. Eng.* **1999**, *23*, 1031–1040. [CrossRef]
2. Wang, J.; Zhang, Y.; Li, Y.; Huang, S. A non-equal fragment model of a water-wall in a supercritical boiler. *J. Energy Inst.* **2015**, *88*, 143–150. [CrossRef]
3. Shu, Z.; Zixue, L.; Yanxiang, D.; Huaichun, Z. Development of a distributed-parameter model for the evaporation system in a supercritical W-shaped boiler. *Appl. Therm. Eng.* **2014**, *62*, 123–132. [CrossRef]
4. Deng, K.; Yang, C.; Chen, H.; Zhou, N.; Huang, S. Start-Up and dynamic processes simulation of supercritical once-through boiler. *Appl. Therm. Eng.* **2017**, *115*, 937–946. [CrossRef]
5. Ray, A. Dynamic modelling of power plant turbines for controller design. *Appl. Math. Modell.* **1980**, *4*, 109–112. [CrossRef]
6. Chaibakhsh, A.; Ghaffari, A. Steam turbine model. *Simul. Modell. Pract. Theory* **2008**, *16*, 1145–1162. [CrossRef]

7. Devandiran, E.; Shaisundaram, V.S.; Ganesh, P.S.; Vivek, S. Influence of Feed Water Heaters on the Performance of Coal Fired Power Plants. *Int. J. Latest Technol. Eng. Manag. Appl. Sci.* **2016**, *5*, 115–119.
8. Almedilla, J.R.; Pabilona, L.L.; Villanueva, E.P. Performance Evaluation and Off Design Analysis of the HP and LP Feed Water Heaters on a 3 × 135 MW Coal Fired Power Plant. *J. Appl. Mech. Eng.* **2018**, *7*, 1–14. [CrossRef]
9. Espatolero, S.; Romeo, L.M.; Cortés, C. Efficiency improvement strategies for the feedwater heaters network designing in supercritical coal-fired power plants. *Appl. Therm. Eng.* **2014**, *73*, 449–460. [CrossRef]
10. Chen, C.; Zhou, Z.; Bollas, G.M. Dynamic modeling, simulation and optimization of a subcritical steam power plant. Part I: Plant model and regulatory control. *Energy Convers. Manag.* **2017**, *145*, 324–334. [CrossRef]
11. Flynn, D. *Thermal Power Plant Simulation and Control*; The Institution of Engineering and Technology: London, UK, 2003; ISBN 978-0-85296-419-4.
12. Alobaid, F.; Mertens, N.; Starkloff, R.; Lanz, T.; Heinze, C.; Epple, B. Progress in dynamic simulation of thermal power plants. *Prog. Energy Combust. Sci.* **2017**, *59*, 79–162. [CrossRef]
13. Starkloff, R.; Alobaid, F.; Karner, K.; Epple, B.; Schmitz, M.; Boehm, F. Development and validation of a dynamic simulation model for a large coal-fired power plant. *Appl. Therm. Eng.* **2015**, *91*, 496–506. [CrossRef]
14. Wang, C.; Liu, M.; Li, B.; Liu, Y.; Yan, J. Thermodynamic analysis on the transient cycling of coal-fired power plants: Simulation study of a 660 MW supercritical unit. *Energy* **2017**, *122*, 505–527. [CrossRef]
15. Lindsay, J.; Dragoon, K. *Summary Report on Coal Plant Dynamic Performance Capability*; Renewable Northwest Project: Portland, OR, USA, 2010; pp. 4–7.
16. Hentschel, J.; Zindler, H.; Spliethoff, H. Modelling and transient simulation of a supercritical coal-fired power plant: Dynamic response to extended secondary control power output. *Energy* **2017**, *137*, 927–940. [CrossRef]
17. Fout, T.; Zoelle, A.; Keairns, D.; Pinkerton, L.L.; Turner, M.J.; Woods, M.; Kuehn, N.; Shah, V.; Chou, V. *Cost and Performance Baseline for Fossil Energy Plants Volume 1a: Bituminous Coal (PC) and Natural Gas to Electricity*; Technical Report; U.S. Department of Energy: Pittsburgh, PA, USA, 2015.
18. Turton, R.; Bailie, R.C.; Whiting, W.B.; Shaeiwitz, J.A.; Bhattacharyya, D. *Analysis, Synthesis, and Design of Chemical Processes*, Prentice Hall International Series in the Physical and Chemical Engineering Sciences; 4th ed.; Prentice Hall: Upper Saddle River, NJ, USA, 2012; ISBN 978-0-13-261812-0.
19. Aspen Exchanger Design & Rating. Available online: https://www.aspentech.com/en/products/engineering/aspen-exchanger-design-and-rating (accessed on 1 January 2018).
20. AspenTech Aspen Plus 12.1 User Guide 06/03. Available online:. (accessed on 30 October 2018).
21. Adams, T.A., II. *Learn Aspen Plus in 24 Hours*, 1st ed.; McGraw-Hill Education: New York, NY, USA, 2017; ISBN 978-1-260-11645-8.
22. Schefflan, R. *Teach Yourself the Basics of Apsen Plus*, 2nd ed.; Wiley-AIChE: Hoboken, NJ, USA, 2016; ISBN 978-1-118-98059-0.
23. Jana, A.K. *Process Simulation and Control Using Aspen*, 2nd ed.; PHI Learning: New Delhi, India, 2012; ISBN 978-81-203-4568-3.
24. Al-Malah, K.I.M. *Aspen Plus: Chemical Engineering Applications*; Wiley: Hoboken, NJ, USA, 2016; ISBN 978-1-119-13123-6.
25. Shulka, A. 660 MW Supercritical Boiler. Available online: https://www.slideshare.net/AshvaniShukla/660-mw-supercritical-boiler (accessed on 21 June 2018).
26. The Basics of Fan Performance Tables, Fan Curves, System Resistance Curves and Fan Laws (FA/100-99). Available online: http://www.greenheck.com/library/articles/10 (accessed on 1 January 2017).
27. Liese, E. Modeling of a Steam Turbine Including Partial Arc Admission for Use in a Process Simulation Software Environment. *J. Eng. Gas Turbines Power* **2014**, *136*, 112605. [CrossRef]
28. Lozza, G. Bottoming Steam Cycles for Combined Gas Steam Power Plants: A Theoretical Estimation of Steam Turbine Performance and Cycle Analysis. In Proceedings of the 1990 ASME Cogen Turbo, New Orleans, LA, USA, 27 August 1990; pp. 83–92.
29. Wu, X.; Shen, J.; Li, Y.; Lee, K.Y. Steam power plant configuration, design, and control. *Wiley Interdisc. Rev. Energy Environ.* **2015**, *4*, 537–563. [CrossRef]

30. Draganescu, M.; Guo, S.; Wojcik, J.; Wang, J.; Liu, X.; Hou, G.; Xue, Y.; Gao, Q. Generalized Predictive Control for superheated steam temperature regulation in a supercritical coal-fired power plant. *CSEE J. Power Energy Syst.* **2015**, *1*, 69–77. [CrossRef]
31. Ogunnaike, B.A.; Ray, H.W. *Process Dynamics, Modeling and Control*; Oxford University Press: Oxford, UK, 1994.
32. Dong, L.; Ming, Z.; Yong-jun, L.; Ru-jia, S. Discussion of Ultra-Supercritical Units Feed Water Control Strategy. *Procedia Eng.* **2011**, *15*, 828–833. [CrossRef]

![processes logo] *processes*

MDPI

Article

Waste Fuel Combustion: Dynamic Modeling and Control

Nathan Zimmerman [1,*], Konstantinos Kyprianidis [1] and Carl-Fredrik Lindberg [1,2]

[1]   Mälardalen University, Box 883, 721 23 Västerås, Sweden; konstantinos.kyprianidis@mdh.se (K.K.);
      carl-fredrik.lindberg@se.abb.com (C.-F.L.)
[2]   ABB Force Measurement, Tvärleden 2, 721 36 Västerås, Sweden
[*] Correspondence: nathan.zimmerman@mdh.se; Tel.: +46-21-10-7335

Received: 15 October 2018; Accepted: 9 November 2018; Published: 13 November 2018

Abstract: The focus of this study is to present the adherent transients that accompany the combustion of waste derived fuels. This is accomplished, in large, by developing a dynamic model of the process, which can then be used for control purposes. Traditional control measures typically applied in the heat and power industry, i.e., PI (proportional-integral) controllers, might not be robust enough to handle the the accompanied transients associated with new fuels. Therefore, model predictive control is introduced as a means to achieve better combustion stability under transient conditions. The transient behavior of refuse derived fuel is addressed by developing a dynamic modeling library. Within the library, there are two models. The first is for assessing the performance of the heat exchangers to provide operational assistance for maintenance scheduling. The second model is of a circulating fluidized bed block, which includes combustion and steam (thermal) networks. The library has been validated using data from a 160 MW industrial installation located in Västerås, Sweden. The model can predict, with satisfactory accuracy, the boiler bed and riser temperatures, live steam temperature, and boiler load. This has been achieved by using process sensors for the feed-in streams. Based on this model three different control schemes are presented: a PI control scheme, model predictive control with feedforward, and model predictive control without feedforward. The model predictive control with feedforward has proven to give the best performance as it can maintain stable temperature profiles throughout the process when a measured disturbance is initiated. Furthermore, the implemented control incorporates the introduction of a soft-sensor for measuring the minimum fluidization velocity to maintain a consistent level of fluidization in the boiler for deterring bed material agglomeration.

Keywords: circulating fluidized bed boiler; refuse derived fuel; waste to energy; dynamic modeling; process control

---

1. Introduction

Circulating fluidized bed (CFB) boilers have been gaining ground as a viable standard when it comes to thermally treating municipal waste. Despite the complex inherent nature of refuse derived fuels (RDF), waste can be thermally treated to produced heat and power. This approach can in turn impact the environment in a positive way by reducing the need for antiquated techniques such as land filling, as well as alleviating the need for coal and oil fired boilers. The complexities of RDF are matched by the advantages of CFB boilers, which can accommodate a high degree in the fluctuation of fuel composition, high combustion efficiency, lower emissions, and have a rather quick response to load changes [1–3].

The complications that can arise during the combustion of RDF are necessary to predict or better yet avoid in order for large scale industrial CFBs to provide heat and power at a consistent rate. For this reason, a multi-functional dynamic modeling library has been developed to aid in maintaining a high level of operation by detecting and/or preventing any unwanted phenomenon due to the transient behavior of RDF. The implementation of more advanced control has the ability to reduce the magnitude of potential peaks and dips in boiler temperature profiles. This paper addresses the issues by developing a dynamic model of RDF combustion. Showing that a model based on first principal modeling techniques can be used to design and implement a model predictive control strategy for better control of the boiler temperature profiles. The short literature review that follows dictates this.

1.1. Modelling and Control

Modeling industrial applications is a standard practice which allows one to firstly determine the feasibility of any new industrial installation, but secondly to provide insight into how the process will operate under different operating conditions. For example, coal, biomass, and RDF all have unique inherent characteristics, i.e., heating value (HV). As well as each one being able to present a different series of problematic outcomes during the combustion process, i.e., efficiency and emissions. The focus of this paper is on a CFB fired with RDF. However, previously published works on the combustion of coal and biomass are to be presented first. Because these practices, in many cases, are the corner-stone for developing applicable modeling techniques with respect to RDF combustion.

1.1.1. Modeling

The aforementioned attributes of CFB combustion have been tested through modeling schemes in order to further understand the process of combustion and the associated generation of pollutants. In a review for the combustion of coal in CFBs, Basu [4] outlined three levels of consideration for predictive models and their possible applications:

- Level I: The 1D approach, the boiler is considered to be either a plug flow or stirred tank and mass and energy balances are used to describe the phenomena occurring.
- Level II: The boiler region is split into core and annulus regions, quasi-2D model, to incorporate the changes in solid and gas temperatures and concentrations to achieve a higher resolution.
- Level III: The combustion process is based on the Navier–Stokes equation, a 3D model, and chemical kinetics and physical processes are incorporated in detail.

At the 1D level, in an approach presented by Gungor et al. [3], they developed and validated a model for coal combustion, by splitting the boiler bed and riser into two separate regions. In this approach, they were able to predict the temperature, carbon and oxygen concentrations, and emissions in both regions. Huang et al. [5] express the need for a dynamic model of the combustion process in CFBs in order to more accurately describe and quantify the process due to its transient behavior. They presented two dynamic modeling approaches for the combustion of coal in the riser of a CFB, which predict the dynamic response of the boiler by determining the phase shift and pressure drops at different locations in the riser. In a review of modeling techniques for a single fuel particle by Jiang et al. [6], it was determined that the drying and combustion modeling of biomass are similar to that of coal, but that the devolitilization models differ due to chemical and physical characteristics. A number of publications present combustion models that focus on splitting the boiler into two distinct regions, the boiler bed (dense region) and the boiler riser (dilute region), and further distinguish the dilute region into two sections, e.g., the core-annulus approach (quasi-2D modeling). This approach has been thoroughly reviewed by Huilin et al. [7] and Pallarés et al. [8]. The same approach has also been used in detail to describe the hydrodynamics and particle velocity behaviors in CFBs. Pugsley and Berruti [9] developed a model for predicting boiler hydrodynamics, where a similar goal was also achieved by Davidson [10]. A framework for identifying parameters associated with hydrodynamics has been presented by Xu and Gao [11]. The effects that different operating conditions have on the hydrodynamics can be found in the work presented by Wang et al. [12], where efforts on identifying the fluidization state in CFBs have been presented by Cai et al. [13]. The different types of modeling approaches presented above highlight the achievements when considering the combustion of coal, where such methods can be applied to the combustion of biomass and waste.

In order to incorporate the ever increasing popularity of biomass fuels, Wang et al. [14] presented a 1D model for evaluating the combustion behavior and ash characteristics for several different types of biomass fuels and RDF. They found that the yield of particle burnout values are comparable between biomass and coal fuels, where particle size and moisture content are important factors on burnout. A 2D modelling approached, for biomass combustion, was developed by Gungor [15], which was used to analyze the effects that different operational parameters have on boiler temperature profiles and emissions. By adapting coal-fired combustion models, Saastamoinen [16] was able to develop a simplified model suitable for biomass fuels with a varying size, moisture content, and shape, but was only used to compliment computational fluid dynamics calculations. Several publications take into consideration the adherent problems that accompany the combustion of biomass and waste fuels. Liu et al. [17] investigated the the impact of bed material size and agglomeration tendencies at different gas velocities. The affects of fuel composition on emissions have been investigated by Demirbas [18] and Krzywanski et al. [19]. The influence of fuel properties on heat exchanger fouling have been thoroughly investigated by Tang et al. [20], Lindberg et al. [21], and Pettersson et al. [22]. In regards to emissions, Desroches-Ducarne et al. [23] presented a model for determining the gaseous pollutants that can form during the combustion of waste derived fuels by adopting reaction rates from coal combustion.

When it comes to the temperature behavior, heat transfer, and combustion characteristics in CFBs, there is a lack of information in the literature when pertaining to RDF fired CFBs—specifically, by using a dynamic model to capture the transient temperature behavior and heat transfer associated with a complex fuel source. Despite rigorous fuel preparation methods, the composition of RDF can vary from one hour to the next and this has a direct impact on the operating temperatures within the boiler. If not controlled properly, it can lead to the aforementioned issues presented earlier.

1.1.2. Control

Historically, the fuel of choice in the heat and power industry can be attributed to the use of oil and coal. Managing the boiler output could in a large part be easily controlled by adjusting the the fuel feeding rate via simple proportional-integral control loops. With the trend of transitioning further away from fossil fuels into renewable energy sources, more robust control schemes are needed to combat fuels that can experience a large variance in composition in a short amount of time. Alamoodi and Daoutidis [24] state that, in power generation, there is a strong nonlinear relationship between the boiler, superheaters, and steam turbine. These complex highly nonlinear systems complicate control strategies, and conventional proportional-integral (PI) and proportional-integral-derivative (PID) control schemes are incapable of achieving satisfactory performance [25]. With model predictive control (MPC), a prediction on the future outputs, in a multiple input multiple output (MIMO) system, over a finite prediction horizon can be made. Based on the past values of the manipulated variables, the calculated future values of manipulated variables, and the past values of any disturbances (feedforward variable). The performance of the MPC algorithm is directly linked to the model's accuracy in simulating the process's dynamic behavior [26].

Several groups have been working towards control strategies within the heat and power industry. Hadavand et al. [27] developed a coal combustion model based on mass and heat transfer processes using mass and energy balances, which incorporate process dynamics such as gas and particle interactions, to develop a state space dynamic model to control the boiler bed temperature. By using multiple inputs and multiple outputs, Alamoodi and Daoutidis [24] developed a nonlinear decoupling controller to control the superheated steam temperature in a coal fired plant allowing for frequent changes in the power demand. Ji et al. [26] presented three control schemes with Linear Parameter Varying MPC in a coal fired CFB with favorable results. Sun et al. [25] showed that MPC produced more favorable results when compared to PI control schemes in a coal fired bubbling fluidized bed boiler. Havlena and Findejs [28] applied an MPC approach to better control coal-fired boilers. With a selection of control variables, in particular the disposition of the air and fuel flows, the boiler efficiency could be increased while reducing NO_x. Kortela and Jämsä-Jounela [29] presented a MPC strategy for a biomass fired grate boiler, which incorporated fuel soft-sensors to estimate the variation in the fuel's moisture content.

1.2. Aim and Contribution

Presented in this paper is a dynamic modeling library for CFBs that has been constructed in Modelica programming language, version 3.4, developed by the Modelica Association, Linköping, Sweden. The boiler model has been validated using online measurements from a 160 MW installation at Mälarenergi in Västerås, Sweden. The model's functionality is that it can predict the boiler bed, riser, live steam temperature profiles, and load. This has in turn allowed for the potential to model the onslaught of bed material agglomeration, through the continuation of work by Zimmerman et al. [30], by calculating the minimum fluidization velocity in the boiler. To combat the fluctuating degree in RDF heating value, and in turn have a higher consistency in boiler bed, riser, and steam temperatures. Two different multivariate model predictive control schemes are introduced and are compared with the more traditional PI controlled scheme. In all three control schemes, the same set-points were used on the boiler bed, riser, steam temperatures and load, as well as the minimum fluidization velocity. The MPC approaches incorporated a linearized model based on the physical model constructed using subspace identification.

Based on the knowledge gaps presented in the previous section, the following two questions are defined as the contributions of this paper:

1. Can a dynamic modeling library for the the combustion of waste fuel, based on first principle modeling techniques, be used to improve RDF fired CFB plant operations?
2. To what degree does implementing feedforward MPC on a waste-fired CFB compare to that of MPC without feedforward and traditional PI-control schemes?

2. Circulating Fluidized Bed Boilers

Circulating fluidized bed boilers are a unique type of boiler that have the ability to combust problematic fuels with a varying composition and characteristic makeup of density, moisture content, heating value, and ash content [2,31], due to the high heat capacity of the bed material. The system uses sand as a fluidizing medium in order to create a more uniform distribution when fuel is fed in. A degree of solid reflux of bed material (sand, fuel, char) occurs within the boiler providing a better temperature distribution. The illustration in Figure 1 is that of a CFB boiler and its constituent sections.

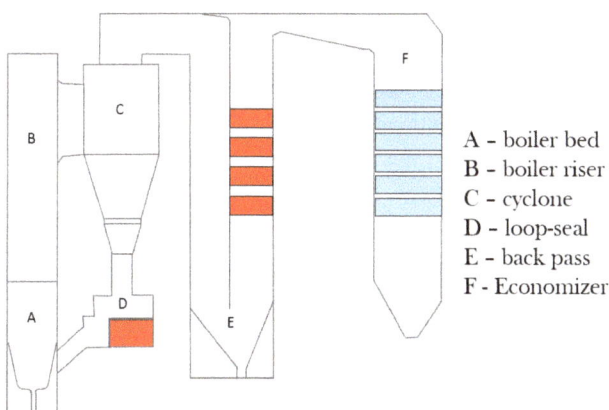

A – boiler bed
B – boiler riser
C – cyclone
D – loop-seal
E – back pass
F - Economizer

Figure 1. Circulating fluidized bed boiler.

The primary air source entering the boiler bed allows for the mixing of bed material and fuel to begin. Fuel enters the boiler bed where it is engulfed with hot non-combustible solids (sand and ash), which heats and drys the fuel. This is followed by the devolitilization step, which releases a wide range of condensible and non-condensible gases. The devolitilization stage is proceeded by the combustion of char, the devolitilized mass of the fuel, or can also overlap with the devolitilization step. The secondary air enters just above the boiler bed and this is to promote a better unification and mixing of solids and theoretically a more uniform temperature profile [2]. The residual carbon is burned until the particle size is small enough to be transported by the fluidizing gas into the riser. The amount of carbon utilized in the process is a function of the fuel's heating value. Therefore, lower combustion losses can be achieved with a more reactive fuel or with longer residence times from larger boiler dimensions. However, it has been determined by Xiao et al. [32], who with prolonged residence times a portion of the unburnt carbon, can become inert and will be discharged along with the fly ash. It has also been shown by Gungor [3] that the carbon content can be reduced when the superficial velocity is decreased and as the bed temperature rises, or when the air-to-fuel ratio is lowered along with the bed temperature. However, the former is achieved due to a reduction in the fuel feeding rate and therefore a reduction in carbon content of the fuel being injected into the boiler. If large particles escape the riser, the circulating aspect of the CFB allows for unburnt char and carbon to be recirculated back to the boiler via the cyclone and loopseal in order to be further utilized. The size, density, and moisture content of the fuel source plays a big roll in how quickly each of these stages occur as well as affects temperature profile variation in the boiler. The boiler is also comprised of evaporator tubes, which allows for a fraction of the combustion heat to be absorbed, and is further utilized in the convection sections after the cyclone. The top of the boiler is connected to a cyclone (gas-solid separator) and this allows for the majority of particles (bed material and unburnt char) to be separated from the flue gas to then be returned to the boiler bed.

To date, it appears that CFB technology is second to grate-fired boilers when it comes to producing heat and power via waste/biomass. In a review of developed regions using MSW as a fuel stock, Lu et al. [33] determined from a report by The International Solid Waste Association [34] that 4.5% of the implemented technologies are from fluidized bed boilers, 87.9% are of grate boilers, and 7.6% are of type "other". Lu et al. also determined the highest penetration of fluidized bed technology, as a percentage of different incinerator types in a given region, are in China (28.9%), Sweden (20.1%), U.S. (19.6%), and Japan (11.8%). The possible reasons for a lower utilization is that fluidized bed boilers have a higher capital/operation cost and are more difficult process to control, when compared to grate-fired boilers [35,36].

Refuse Derived Fuels

Refuse derived fuels come from the process of gathering and sorting municipal solid waste (MSW). MSW is an unavoidable by-product of human nature and over the past several decades has led to negative environmental impacts due to land filling [37]. The reason it is imperative to first sort MSW is that it is comprised of organics, paper, plastics, metals, glass, combustible, and non-combustible materials. The sorting process initially starts with a shredding process that breaks the incoming material into credit card size pieces. This is proceeded by the removal of metals, to be recycled, and then an air classifier is used to allow for other heavy materials to be separated. The separation of the unwanted fractions also increases the effective energy of the by product, where Redemann et al. [38] found that they had a heating value increase of $6\,MJ\,kg^{-1}$ post sorting. From the sorting process, the goal is to accomplish:

- Create a more uniform size distribution in the final RDF product,
- Sort out the recyclable fraction, i.e., metal, glass, and plastics,
- Sort out the non-combustible fraction, i.e., ceramics and rocks.

Post sorting, the heterogeneous high calorific by-product, can officially be labeled as RDF, and it can be further classified into biomass- and fossil-based fractions. The fossil-based portion is what in essence leads to RDF having such a high lower heating value, and from the literature can vary between 12.9 to 25.1 MJ kg^{-1} [14,31,39]. This variability can be interpreted as quite high. However, it is not too far off from that of biomass. With technological advancements in boiler design and regulations against land filling, waste incineration has become a popular alternative for producing heat and power in Northern Europe, despite its complex nature.

The concept of MSW incineration is by no means a new concept for producing heat and power. According to Lu et al. [33], as of 2015, there are 1179 plants in operation with an approximate capacity of 700,000 tonnes per day, where the EU-27 is second to China, and accounts for 40% and 30% of the plants operating and global capacity, respectively. Ripa et al. [40] heed caution to increasing the use of MSW as a fuel source, and that it should not decrease the capability of obtaining and recycling materials from MSW. Furthermore, a holistic approach is necessary in order to connect local waste management authorities and the different ministries responsible for policies on waste management and energy, in order to explore the full potential of waste to energy [41]. Global implications have been observed, for instance, China's decision to restrict the importation of certain wastes have made waste disposal problematic for some countries who depend upon high volumes of imported goods [42]. There are also environmental concerns that accompany MSW incineration. An increased complication of dioxin and furans can be accompanied with RDF fuels due to plastics. However, these levels can be managed, along with traditional concentrations of NO_x and SO_x, by operating the boiler within optimal temperature ranges and flue gas cleaning. The benefits that CFB combustion present allows for such a complex fuel to be utilized due to its unique design of allowing for unburnt char to be recirculated back into the boiler bed to be completely combusted. However, due to the biomass-based portion of RDF, the moisture content can fluctuate. This can lead to unforeseen and unwanted bed temperature fluctuations and therefore a difficult system to control.

3. Model Description and Methodology

3.1. Process Parameters

The model proposed has been modeled after a 160 MW industrial installation at Mälarenergi in Västerås, Sweden. The unit provides roughly half of the district heating needs of the surround municipality by utilizing up to 60 t h^{-1} of RDF. The boiler is 60 m long, 32 m wide, and 55 m tall. There is an on-site state-of-the-art sorting facility for processing waste from the local municipality and from abroad, and has the capacity to sort 480,000 tonnes of unsorted waste per year. After the sorting process, the fuel is fed and stored in four fuel bunkers that have a combined capacity to provide the boiler with several days of operation. There are five sets of heat exchangers incorporated into block 6: economizer, evaporator, superheater 1 (SH1), superheater 2 (SH2), superheater 3 (SH3). The purpose for having three superheaters is to achieve a steam temperature of 470 °C in SH3. SH1 and SH2 are located in the 2nd draft of the back pass and are convective, and SH3 is located in the loopseal, and is a superheated fluidized bed.

3.2. Description of Model and Methodology

CFB models can generally be characterized into two main approaches: steady-state and dynamic modeling. The steady-state approach is where the variables controlling the system do not change over time. This means that any observed behavior in a system will continue to be as such, and that any change in state variables over time will equal zero. In a dynamic approach, the state variables are changing with respect to time, and this makes it possible to monitor transitions and events occurring over time. The model presented in this paper is characterized as a dynamic model. The reason this approach has been chosen is because the transient behavior of the changing fuel composition has a direct effect on the boiler operating temperature. Looking at process information, the boiler bed and riser temperatures can fluctuate significantly over the course of daily operation.

The model has been developed using Modelica programming language. This offers a unique opportunity to model multi-domain dynamic systems by solving ordinary differential equations and algebraic differential equations, by employing a solver for both differential and algebraic equations. Using this tool also allows for one to use an object-oriented modeling approach. Blocks for each necessary component can be developed and connected together; the CFB library constructed is illustrated in Figure 2. Process signals are incorporated as input into the model, and are represented in Table 1. From these input parameters, Equations (1) and (2) were used to calculate the mass and energy balances, respectively.

$$\frac{d(M_{i,j})}{dt} = \Sigma \dot{m}_{in,i,j} - \Sigma \dot{m}_{out,i,j} \tag{1}$$

$$\frac{d(M_{i,j}H_{i,j})}{dt} = \Sigma \dot{m}_{in,i,j} H_{in,i,j} - \Sigma \dot{m}_{out,i,j} H_{out,i,j} + \alpha Q_{H,i} - Q_{c,i}, \tag{2}$$

where the mass flow, $\dot{m}$ (kg s^{-1}), is used to calculate the mass, M (kg), in each control volume i, i.e., the boiler bed and riser, cyclone, loop-seal, backpass, and economizer. Within each control volume, the subscript j denotes the difference in species, i.e., solids or gas. The enthalpy, H (kJ kg^{-1}), α is the percentage of combustion occurring in boiler bed, Q_H is the amount of heat (kW) released from the fuel, and Q_c is the amount of heat transferred (kW) from convection. In Equation (3), the enthalpies for each species j in the given control volume i are calculated by

$$\frac{d(H_{i,j})}{dt} = cp_{i,j} \frac{d(T_i)}{dt}. \tag{3}$$

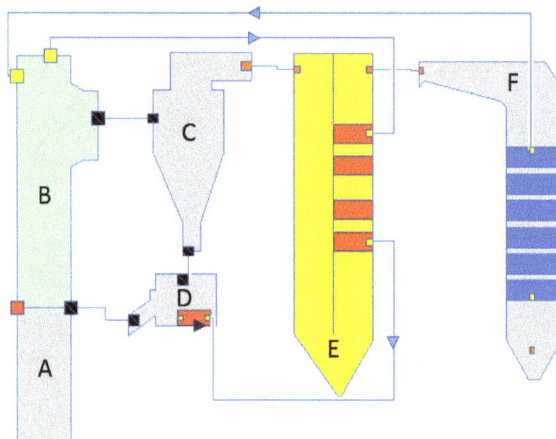

Figure 2. Object-oriented approach, CFB (Circulating fluidized bed) in Dymola, boiler bed and riser (**A,B**), cyclone (**C**), hot solid re-circulation via SH3 in the loop-seal (**D**), SH1 and SH2 (**E**), and the economizer (**F**).

Table 1. Process signals used as input into the model.

	Temperature (°C)	Mass Flow Rate (kg s^{-1})
fuel	-	x
primary air	x	x
secondary air	x	x
flue gas recirculation	x	x
feed water	x	x

The enthalpy, H (kJ kg^{-1}), is calculated by using the corresponding specific heat, cp (kJ kg^{-1} K^{-1}), and temperatures, T (k),for each control volume i and evolved species j.

4. Validation

The achieved simulation results for the boiler bed, riser (temperature at the top of the boiler), and live steam temperatures are illustrated in Figure 3, for an evaluation period during full load operation. The fourth plot represents the four fuel feeders into the boiler. The fuel feeders feed fuel into the boiler laterally, but it can be seen that, during a good portion of the time, the rate of each feeder is not consistent with the next. This can lead to an instability, not captured by the model, of the temperature profiles, despite assistance from the fluidizing medium. It can be observed that the simulation profiles (solid-lines) do follow the process sensor values (dash-lines). Despite the fact that the values from Table 1 are incorporated, continuous information on the fuel's HV and composition are not available. The volumetric flow rate for the fuel is given from process information as m^3 h^{-1}, which was converted to the mass flow rate kg s^{-1} by using an average bulk density of approximately 200 kg m^{-3}. The heating value of the fuel is unknown, so, for the purposes of this work, an energy balance on all heat exchanger surfaces has been carried out to back calculate the heating value using available process data. Consequently, a higher relative error is believed to be because of a significant deviation in the fuel's HV because a constant bulk density is assumed. Therefore, the trends in the temperature change are captured, but there are deviations in the magnitudes. This can be quantified by looking at the percent error between the simulation and actual measurements. For the simulated model results, the percent error are between 0 & 3%, but, around days 6 & 10, there is a higher percent error of up to 5%. With respect to control, the model's level of accuracy in capturing trends is satisfactory for implementing different control schemes, and will be discussed in detail in later sections.

Offline Process Monitoring

It was previously mentioned how there are five sets of heat exchangers throughout the process. The main objective of the superheaters are to use energy from the flue gas in SH1 and SH2, and hot solids in SH3 to achieve a steam temperature of approximately 470 °C. Due to the problematic ash composition from RDF combustion, there is an inherit risk of fouling on the heat exchanger surfaces. One way to avoid fouling is through periodic soot-blowing, which will help in removing any "loose" buildup on the heat exchanger surfaces, which is easily done in the back pass section of the thermal network, SH1 and SH2, but not in SH3 or the evaporating tubes because the latter are in constant contact with high temperature bed material and ash.

By following a similar practice as that of Sandberg [43], the overall heat transfer coefficient, U (kW m^{-2} K^{-1}), from January 2015 to January 2018 has been analyzed. During this period, there have been three scheduled shutdowns for maintenance. By looking at the overall heat transfer coefficient before and after maintenance, it has been determined that there is an amount of fouling occurring. For these three years, the U values corresponding to when the boiler was operating at a full load are considered. Illustrated in Figure 4, there are four distinct periods, before and after maintenance for each year. There are distinguished jumps between before and after maintenance, and there are corresponding negative slopes preceding the maintenance period. This phenomenon is not observed during the fourth period of SH3, and is attributed to not having data for 2018 in order to continue the analysis. The overall negative trend for each period represents a decrease in heat exchanger performance by quantifying a reduction in heat transfer. Thus, it is possible that more attention is needed during maintenance or that maintenance should be scheduled earlier, when the evaporator and SH3 are considered.

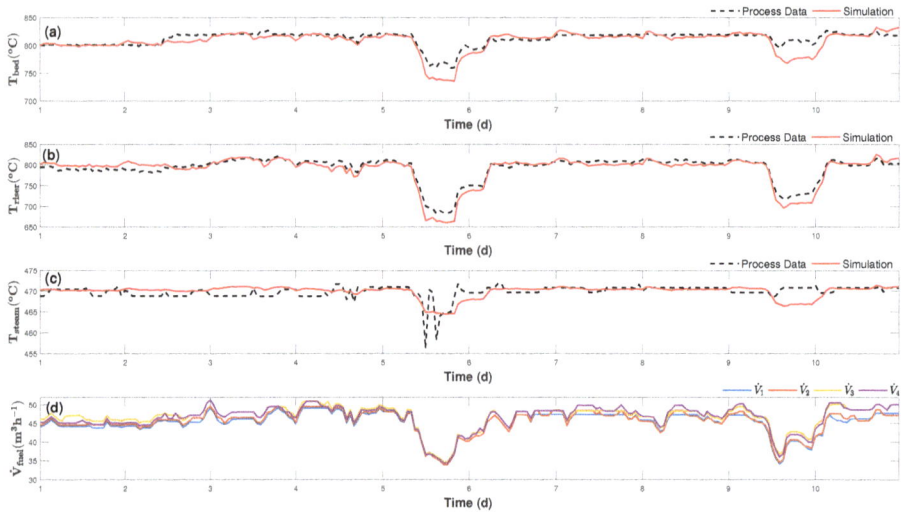

Figure 3. Simulation and actual temperature profiles for the boiler bed (**a**); riser (**b**); steam (**c**); and the volumetric flow rate measurement for the four fuel feeders (**d**).

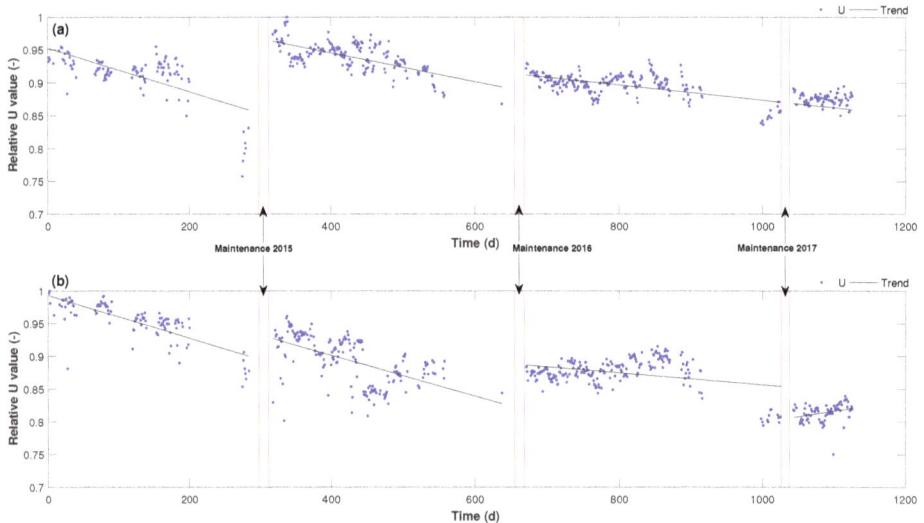

Figure 4. Normalized overall heat transfer coefficient for the evaporator (**a**) and SH3 (**b**) at full load. The red lines indicate the duration of maintenance.

5. Results and Discussion

5.1. Agglomeration

Mechanisms for the formation of agglomerates have been thoroughly reviewed by Yan et al. [44] during multi-waste incineration, Lin et al. [45] during straw incineration, and Skrivfars et al. [46] during biomass incineration. The main factors that can lead to agglomeration are temperature fluctuations in the boiler, inconsistent fluidization, and fuel composition. The content of Alkali and alkali earth metals in the fuel, such as K, Na, Ca, Mg can increase agglomeration potential significantly, and if Cl is present this can facilitate the progression even further. Furthermore, modes for agglomeration prediction have been presented by Gatterning [47], agglomeration detection methods have been presented by Bartels et al. [48,49], and the mitigation of agglomeration have been presented by Morris et al. [50]. Proven methods for predicting agglomeration, controlled agglomeration test as well as fuel and ash analysis, work only for fuels that are relatively homogeneous. Since these tests are done at sampling intervals and not in real time, an online detection method is needed. Agglomeration indexes have also been used by Visser [51] and Vamvuka et al. [52], but these indexes rely on a fuel composition that is relatively consistent. The above-mentioned methods should be used cautiously as they neglect fuel-ash-bed material interactions and therefore can only give insight into agglomeration prediction. Another option for reducing the tendency of agglomeration is to avoid using sand as the bed material, and, instead, replace it with less reactive materials [53]. When agglomerates form, they disrupt the boiler dynamics and it is from this perspective that the changes in the boiler velocity profiles can be observed, specifically the minimum fluidization velocity, U_{mf} (m s^{-1}), to detect agglomeration. Within the physical model, it is possible to calculate U_{mf}, Equation (4), by using a derivative form of the Ergun eqution [54] expressed in [55]:

$$\frac{1.75}{\epsilon_{mf}^3 \phi} \cdot \left(\frac{d_p U_{mf} \rho_g}{\mu_g}\right)^2 + \frac{150(1-\epsilon_{mf})}{\epsilon_{mf}^3 \phi^2} \cdot \frac{d_p U_{mf} \rho_g}{\mu_g} = \frac{d_p^3 \rho_g (\rho_p - \rho_g) g}{\mu_g^2}, \tag{4}$$

where the particle diameter d_p is taken to be 500 μm, sand density $\rho_p = 1600$ kg m^{-3}, assuming that the sand is round by using $\phi = 0.86$, voidage at minimum fluidization, $\epsilon_{mf} = 0.44$ from [55], gravitational constant, g (m s^{-2}), and the gas density, ρ_g (kg m^{-3}) and viscosity, μ_g (kg m^{-1} s^{-1}) are calculated internally.

During the evaluation period illustrated in Figure 3, a campaign was led to take daily samples of the bed material and recirculated fines within the boiler. Illustrated in Figure 5a is a breakdown of the size distribution from the sampled bed material.

It can be observed that there is a large change, from day-to-day, when compared to virgin sand. A sample of virgin sand shows that it primarily consists of particles less than 0.4 mm. From the model's prediction of the minimum fluidization, a correlation between U_{mf} and agglomeration has been detected and is illustrated in Figure 6. From day one, it can be observed that there is a good correlation between the particle size decreasing along with a reduction in the minimum fluidization. The size of the bed material particles sampled then begin to increase along with an increase in minimum fluidization up to days 4 and 5, where both proceed to decrease and increase in similar trends up to day 11. Since roughly three quarters of the virgin sand is comprised of having a diameter of less than 0.4 mm, the change in particle size diameter between 0.4 and 0.63 mm was observed. The idea is that agglomeration has occurred when the size distribution has increased above 0.4 mm due to the binding of bed material. It was also discovered that recirculated fines greater than 0.63 mm correlate to U_{mf}. In comparing Figure 5a to the models prediction of U_{mf}, Figure 6, there is a noticeable trend between agglomeration and increased U_{mf}. This is because, as there is an increase in the particle size of the bed material, there is also an accompanied increase in the boiler's velocity in order to achieve an adequate level fluidization. Further analysis on the samples taken, illustrated in Figure 5b, shows that the major elemental components that lead to agglomeration have a significant presence in the bed material and fly ash. For reference, the percent composition distribution in virgin sand is presented, which is comprised of 91% Si. The bed material shows a significant increase in the content of Ca and Na and over 60% of the fly ash is composed of Ca and Al, which are heavy proponents of agglomeration.

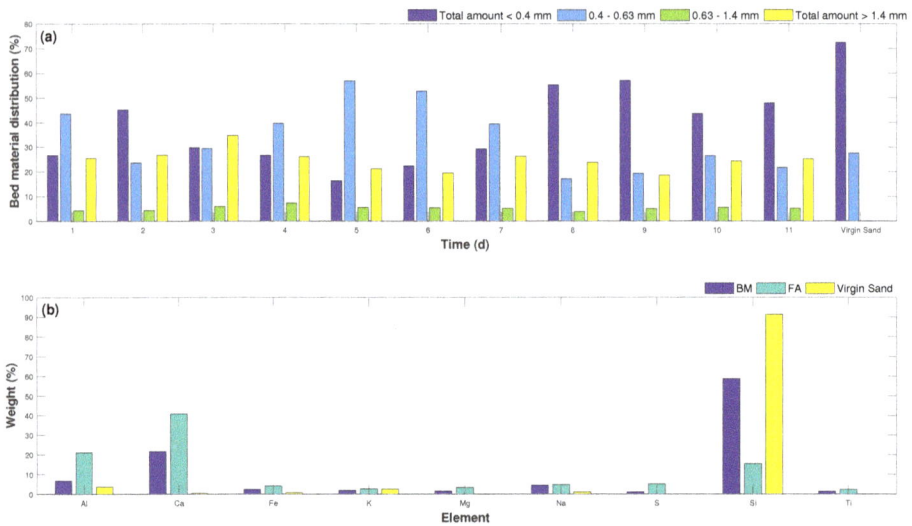

Figure 5. The percent in size distributions for the measured bed material (**a**), and the average percent composition of bed material (BM) and fly ash (FA), in comparison to virgin sand (**b**).

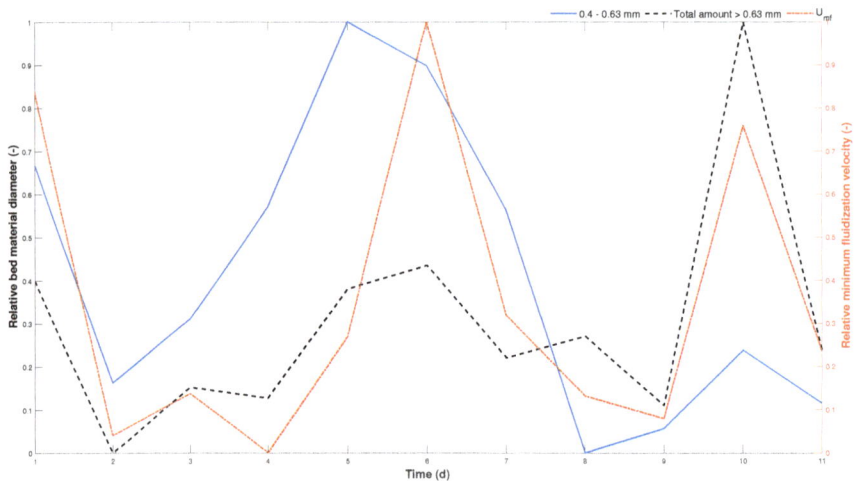

Figure 6. Detection of bed agglomeration: correlation between boiler minimum fluidization and increase of bed material (BM) and recirculated fines (RF) diameter, normalized values.

From Figure 4, it was observed that the further away from a given maintenance period, the slope of the heat transfer coefficient represents a negative trend. This negative trend can also be attributed to the percent composition distribution in Figure 5b as these elements can also lead to fouling. By examining Figures 5a and 6, it can be observed that there is a driving force for agglomeration when there is an increase in U_{mf} throughout the evaluation period. In Figure 5b, it can be observed that approximately 15%, 60%, and 90% of the fly ash, bed material, and sand, respectively, are comprised of Si. It is intuitive that a larger fraction goes to Si when considering virgin sand. The higher percent in the bed material, with respect to other elements, can be attributed to ash melts coating the sand particles, hence agglomeration is occurring. The driving forces behind ash melts can be attributed to hot regions within the boiler and/or a lack of proper fluidization. It should also be noted that a fraction of the Si reported could be attributed to glass in the fuel, but this value is not known.

5.2. Control

The agglomeration of bed material is prominent in waste fired CFBs, as was described in the previous section. However, since it has been shown that the minimum fluidization can be of value in detecting agglomeration, it can be used as a soft-sensor to assist in better control. This is due to the fact that U_{mf} is calculated within the model. For this reason, it can be used as a control variable along with other designated control variables in a control scheme. In order to replicate the control scheme currently being used in the plant, a PI control scheme has been implemented. It will then be used as a base case comparison to MPC control. Noise has been added to the measured outputs, as well as low-pass filters, in order to further replicate the real process. The PI control was executed by implementing set-points on calculated boiler outputs: boiler bed temperature, boiler riser temperature, steam temperature, minimum fluidization, and the load. The control variables for the respective measured outputs are the tertiary air, secondary air, fuel, primary air, and feed water, respectively.

The measured outputs and corresponding control variables, at a 30-s sampling interval, are illustrated in Figure 7, where a step change in the fuel's heating value has been initiated. The step change serves as an unmeasured disturbance, and the five control variables in each PI controller signal actuators in order to meet the required set-points in the measured outputs. It can be observed that, when there is a step change in the heating value, the measured outputs trend towards the desired set-point values. When tuning the controllers there was a trade-off between maintaining performance and noise sensitivity and can be observed by a noisy control variable response in the tertiary air flow in Figure 7. The tertiary air is actually the flue gas from downstream being recirculated back into the system. The reason for this is that it has a high impact in controlling the boiler bed temperature by reducing the amount of available oxygen. Hence, it is reasonable to have more excitation in this control variable because the bed temperature will affect all other temperatures downstream.

Model Predictive Control

When it comes to handling complex systems with MIMO, it was previously discussed how conventional PI control schemes are lacking in robustness. Model predictive control techniques have been gaining ground in process industries as a way to better improve plant operations over that which can be achieved by using a series of PI controllers. One of the underlying concepts in MPC is to predict future process behavior by minimizing the difference between the predicted process response and the reference points with a prediction model [56]. MPCs have the ability to address different control strategies, where, at a sampling interval, the problem is solved and updated up to a set prediction in the time horizon [57]. The ability to add weights and constraints on control variables and measurable outputs makes MPC a very powerful tool in large process industries.

The MPC approach was implemented by estimating a discrete-time model from the validated dynamic model by using subspace identification (SID), where a 30 s sampling time was used. Model identification methods, like SID, identify the system matrices and system order by a non-iterative method to determine a state-space linear time invariant model [58]. By using Numerical Subspace State-Space System Identification (N4SID), it was possible to identify a linear relationship between the inputs and outputs. In turn, this reduces the complexity of the dynamic model into an adaptable "lighter" model for implementing control. Studies for MPC approaches based on a linearized plant model around an operating point have been achieved by Khani et al. [59] and Al Seyab et al. [60] with favorable results. Therefore, with the developed validated model, it is possible to develop a linearized model, to be used for model predictive control, to, in turn, be used to control the dynamic model around multiple operating points.

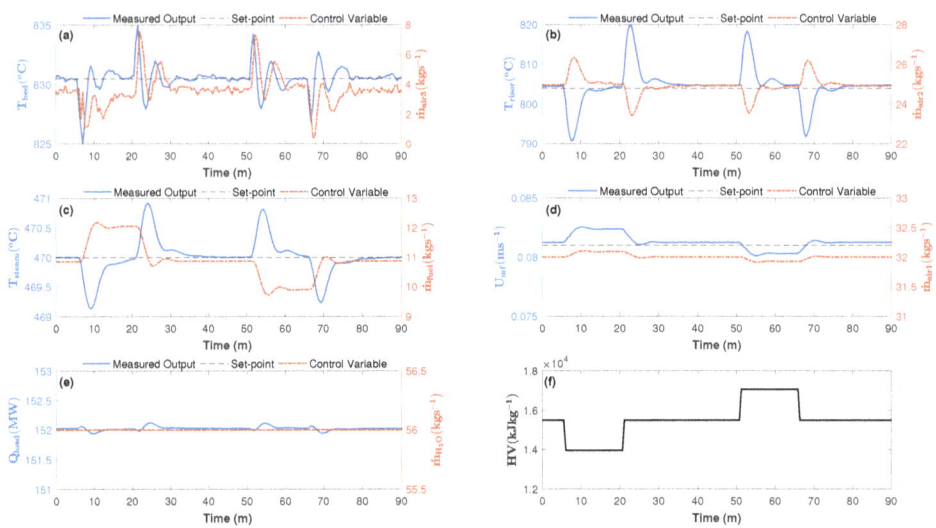

Figure 7. Measured output response to a step-change in the fuel's heating value (**f**). Simulated using PI (proportional-integral control) controllers with set-points on the measured outputs in conjunction with the following control variable relationships: bed temperature—tertiary air flow rate (**a**); riser temperature—secondary air flow rate (**b**); steam temperature—fuel mass flow rate (**c**); minimum fluidization—primary air flow rate (**d**); and load—feedwater flow rate (**e**).

It was essential to first generate adequate input signals of a sufficient excitation in both amplitude and frequency for the control variables, which do not cause the measured output signals to deviate too far from normal operating conditions [61]. Ten sets of input signals were generated, and sequentially each data set was used as input into the physical model to acquire the outputs. The next step was to detrend the data by removing the mean values to eliminate any bias in the trend. N4SID was then used to create the state-space model, where several model orders were tested and it was determined that a second order model provided sufficient results, illustrated in Figure 8.

The estimated model was then used to design and develop a feedforward MPC (FFMPC) controller, and an MPC controller without feedforward (NFFMPC). A step change was implemented in the fuel's heating value, the same as in the PI-control case illustrated in Figure 7, to test the response of the control schemes. Taking this value to be known allows for the implementation of a FFMPC controller, or, in other words, the heating value is considered to be a known measurable disturbance. In the MPC approach, without feedforward, the fuel's heating value is considered as an unmeasured disturbance, i.e., there is no cognizant knowledge of the fuel's heating value. For the MPC structure, there are five manipulated variables, same as in the PI controllers: fuel mass flow, feed water, primary, secondary, and tertiary air mass flow rates. There is also one measured disturbance, the fuel's heating value. There are five measured outputs: boiler bed, flue gas, steam temperatures, and load. The fifth measured output is a soft-sensor measurement of the minimum fluidization taken from the dynamic model. The MPC controller was connected directly with the described dynamic model, and noise has been added to the measured outputs to exhibit a more realistic sensor reading. These signals are then used as feedback into the MPC. The MPC in turn uses its internal state-space model and signals a change to the actuators controlling the manipulated variables into the plant.

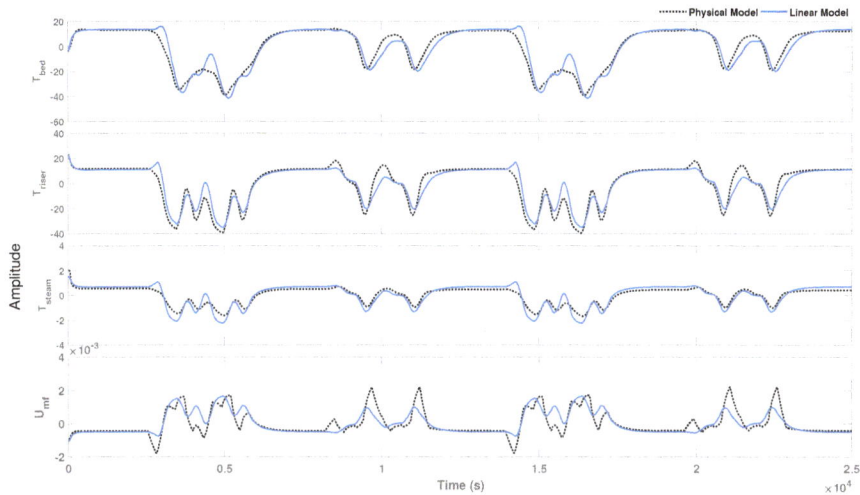

Figure 8. Simulated response comparison showing the linear model is within an expectable range to be used in controller design.

Illustrated in Figures 9 and 10 are the measured outputs and control variables respectively from the two different MPC schemes and the PI-control scheme. The same heating value disturbance is taken as input into the models. It can be observed that both MPC approaches are capable of combating a high fluctuation in the quality of the fuel. The bed, flue gas, steam, minimum fluidization, and load profiles are operating around their set-points. However, it is noticeable that when FFMPC control is implemented that it offers better control in minimizing the swings in measured outputs while also producing a more consistent control variable operating point. The reduction in temperature spikes, that FFMPC provides, means that it is achievable to operate a waste fired boiler at a more efficient level. This is because the combustion in the boiler will be more stable, which will provide a more uniform temperature profile throughout the boiler and there would be a foreseeable reduction in potential hot zones.

To quantify the performance of each control scheme the standard deviation for each measured output (MO) and control variable (CV) were calculated, observed in Table 2. For the MOs, i.e., the boiler bed, riser and steam temperatures, and the plant load, it can be observed that FFMPC offers a much smaller deviation from the set-points. This means that FFMPC can provide a more stable temperature in the boiler during operation, and therefore help to deter unfavorable emissions. The CV actuators, i.e., the fuel, air, and feed water mass flow rates, operate at a more consistent frequency, and are therefore not changing as often, leading to smoother operation. The response in the measured outputs, when speaking about FFMPC, is reflected in the control variables. There is a noticeable degree of an earlier reaction time for the actuators when FFMPC is implemented. Because both MPC approaches incorporate a state-space model of the validated process, it can be observed that the CV actuators are able to handle a disturbance more robustly than the PI case. This is because the state-space model is able to handle multiple inputs and control multiple outputs simultaneously, whereas with the PI case the CVs and MOs were coupled together. A step-response for one of the MO would show that it is not clear as to which CV to control because more than one CV would actually adjust to this step change. Therefore, better control has been obtained in both MPC cases, which both provide more stable operation conditions.

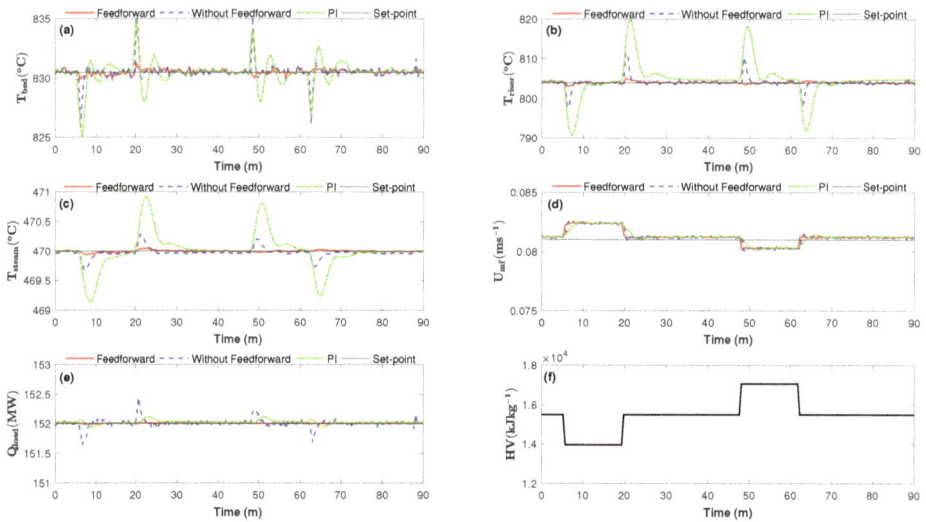

Figure 9. Measured output response to a step change in the fuel's heating value (**f**) to the bed temperature (**a**); riser temperature (**b**); steam temperature (**c**); minimum fluidization (**d**); and load (**e**). The figures represent feedforward and without feedforward MPC control, with reference PI control, and corresponding set-points.

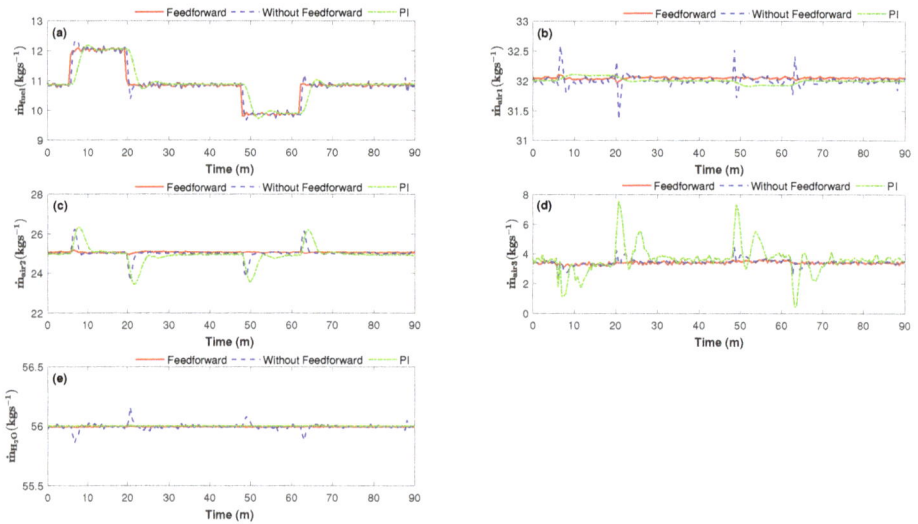

Figure 10. Control variable responses for the fuel mass flow (**a**); primary air flow (**b**); secondary air flow (**c**); tertiary air flow (**d**); and feed water flow rate (**e**). Illustrating feedforward and without feedforward MPC (model predictive control), with reference PI control.

Table 2. The standard deviation in the measured output performance and control variables in all three cases.

MO/CV	FFMPC	NFFMPC	PI
T_{bed} (°C)	0.222	0.788	1.0877
T_{riser} (°C)	0.265	1.336	4.331
T_{steam} (°C)	0.015	0.071	0.298
Q_{load} (MW)	0.0039	0.074	0.029
U_{mf} (m s^{-1})	0.0006	0.0006	0.0006
$\dot{m}_{fuel}$ (kg s^{-1})	0.611	0.626	0.596
$\dot{m}_{air1}$ (kg s^{-1})	0.016	0.105	0.0461
$\dot{m}_{air2}$ (kg s^{-1})	0.026	0.259	0.433
$\dot{m}_{air3}$ (kg s^{-1})	0.103	0.217	0.974
$\dot{m}_{H2O}$ (kg s^{-1})	0.0016	0.026	0.0002

MO = measured output; CV = control variable; FFMPC = feedforward model predictive control; NFFMPC = model predictive control without feedforward; PI = proportional-integral control.

6. Conclusions

In this study, a method for modeling and controlling the transient behavior associated with waste fuel combustion has been presented. The model can simulate the boiler bed temperature, riser temperature, live steam temperature, and load profiles by using process sensors. The validated model has then been used to show its capability in detecting the agglomeration of bed material, and the boiler's performance based on heat transfer in the thermal network. In respects to control, three cases have been presented that adequately keep the boiler temperature profiles and fluidization velocity at a more consistent operating point, and maintains an adequate live steam temperature and boiler load. Great detail has been put into presenting the difficulties that come with modeling the combustion of waste because of the severity of its heterogeneous composition. It has also been demonstrated that, with a validated dynamic model, it is possible to improve a boiler's performance with or without information on the fuel's heating value, i.e., with feedforward or without feedforward model predictive control. Both of which have been shown to outperform the PI-control case. Model predictive control with feedforward shows a significant potential in reducing temperature swings in the boiler as well as maintaining a more consistent fluidization velocity for deterring agglomeration. The model predictive control approach provides a solid platform, and flexibility, for constraining states and inputs, and produces a better response than PI-control when transient conditions are imminent.

Author Contributions: The manuscript was conceptualized by all three authors, where N.Z. developed the models, analyzed results, and wrote the manuscript, K.K. contributed in model methodology and validation and manuscript review and editing, and C.-F.L. contributed in control methodology and implementation, and manuscript review and editing.

Funding: The funding for this research is supported by The Swedish Knowledge Foundation (KSS-20120276) and The Future Energy Research Profile at Mälardalen University, Västerås, Sweden.

Acknowledgments: The authors want to acknowledge Mälarenergi AB, Västerås, Sweden, for their support in data acquisition and process information.

Conflicts of Interest: The authors declare no conflict of interest.

References

1. Van Caneghem, J.; Brems, A.; Lievens, P.; Block, C.; Billen, P.; Vermeulen, I.; Dewil, R.; Baeyens, J.; Vandecasteele, C. Fluidized bed waste incinerators: Design, operational and environmental issues. *Prog. Energy Combust. Sci.* **2012**, *38*, 551–582. [CrossRef]
2. Basu, P.; Fraser, S.A. *Circulating Fluidized Bed Boilers: Design, Operation and Maintenance*; Springer International Publishing: Cham, Switzerland, 2015. [CrossRef]
3. Gungor, A. One dimensional numerical simulation of small scale CFB combustors. *Energy Convers. Manag.* **2009**, *50*, 711–722. [CrossRef]

4. Basu, P. Combustion of coal in circulating fluidized-bed boilers: A review. *Chem. Eng. Sci.* **1999**, *54*, 5547–5557. [CrossRef]
5. Huang, Y.; Turton, R.; Park, J.; Famouri, P.; Boyle, E.J. Dynamic model of the riser in circulating fluidized bed. *Powder Technol.* **2006**, *163*, 23–31. [CrossRef]
6. Jiang, X.; Chen, D.; Ma, Z.; Yan, J. Models for the combustion of single solid fuel particles in fluidized beds: A review. *Renew. Sustain. Energy Rev.* **2017**, *68*, 410–431. [CrossRef]
7. Huilin, L.; Guangbo, Z.; Rushan, B.; Yongjin, C.; Gidaspow, D. A coal combustion model for circulating fluidized bed boilers. *Fuel* **2000**, *79*, 165–172. [CrossRef]
8. Pallarès, D.; Johnsson, F. Macroscopic modelling of fluid dynamics in large-scale circulating fluidized beds. *Prog. Energy Combust. Sci.* **2006**, *32*, 539–569. [CrossRef]
9. Pugsley, T.S.; Berruti, F. A predictive hydrodynamic model for circulating fluidized bed risers. *Powder Technol.* **1996**, *89*, 57–69. [CrossRef]
10. Davidson, J.F. Circulating fluidised bed hydrodynamics. *Powder Technol.* **2000**, *113*, 249–260. [CrossRef]
11. Xu, G.; Gao, S. Necessary parameters for specifying the hydrodynamics of circulating fluidized bed risers—A review and reiteration. *Powder Technol.* **2003**, *137*, 63–76. [CrossRef]
12. Wang, C.; Zhu, J.; Li, C.; Barghi, S. Detailed measurements of particle velocity and solids flux in a high density circulating fluidized bed riser. *Chem. Eng. Sci.* **2014**, *114*, 9–20. [CrossRef]
13. Cai, R.; Zhang, H.; Zhang, M.; Yang, H.; Lyu, J.; Yue, G. Development and application of the design principle of fluidization state specification in CFB coal combustion. *Fuel Process. Technol.* **2018**. [CrossRef]
14. Wang, G.; Silva, R.B.; Azevedo, J.L.T.; Martins-Dias, S.; Costa, M. Evaluation of the combustion behaviour and ash characteristics of biomass waste derived fuels, pine and coal in a drop tube furnace. *Fuel* **2014**, *117*, 809–824. [CrossRef]
15. Gungor, A. Two-dimensional biomass combustion modeling of CFB. *Fuel* **2008**, *87*, 1453–1468. [CrossRef]
16. Saastamoinen, J. Simplified model for calculation of devolatilization in fluidized beds. *Fuel* **2006**, *85*, 2388–2395. [CrossRef]
17. Liu, Z.S.; Peng, T.H.; Lin, C.L. Effects of bed material size distribution, operating conditions and agglomeration phenomenon on heavy metal emission in fluidized bed combustion process. *Waste Manag.* **2012**, *32*, 417–425. [CrossRef] [PubMed]
18. Demirbas, A. Combustion characteristics of different biomass fuels. *Prog. Energy Combust. Sci.* **2004**, *30*, 219–230. [CrossRef]
19. Krzywanski, J.; Rajczyk, R.; Bednarek, M.; Wesolowska, M.; Nowak, W. Gas emissions from a large scale circulating fluidized bed boilers burning lignite and biomass. *Fuel Process. Technol.* **2013**, *116*, 27–34. [CrossRef]
20. Tang, Z.; Chen, X.; Liu, D.; Zhuang, Y.; Ye, M.; Sheng, H.; Xu, S. Experimental investigation of ash deposits on convection heating surfaces of a circulating fluidized bed municipal solid waste incinerator. *J. Environ. Sci.* **2016**, *48*, 1–10. [CrossRef] [PubMed]
21. Lindberg, D.; Backman, R.; Chartrand, P.; Hupa, M. Towards a comprehensive thermodynamic database for ash-forming elements in biomass and waste combustion: Current situation and future developments. *Fuel Process. Technol.* **2013**, *105*, 129–141. [CrossRef]
22. Pettersson, A.; Niklasson, F.; Moradian, F. Reduced bed temperature in a commercial waste to energy boiler—Impact on ash and deposit formation. *Fuel Process. Technol.* **2013**, *105*, 28–36. [CrossRef]
23. Desroches-Ducarne, E.; Dolignier, J.; Marty, E.; Martin, G.; Delfosse, L. Modelling of gaseous pollutants emissions in circulating fluidized bed combustion of municipal refuse. *Fuel* **1998**, *77*, 1399–1410. [CrossRef]
24. Alamoodi, N.; Daoutidis, P. Nonlinear control of coal-fired steam power plants. *Control Eng. Pract.* **2017**, *60*, 63–75. [CrossRef]
25. Sun, L.; Li, D.; Lee, K.Y. Enhanced decentralized PI control for fluidized bed combustor via advanced disturbance observer. *Control Eng. Pract.* **2015**, *42*, 128–139. [CrossRef]
26. Ji, G.; Huang, J.; Zhang, K.; Zhu, Y.; Lin, W.; Ji, T.; Zhou, S.; Yao, B. Identification and predictive control for a circulation fluidized bed boiler. *Knowl.-Based Syst.* **2013**, *45*, 62–75. [CrossRef]
27. Hadavand, A.; Jalali, A.A.; Famouri, P. An innovative bed temperature-oriented modeling and robust control of a circulating fluidized bed combustor. *Chem. Eng. J.* **2008**, *140*, 497–508. [CrossRef]
28. Havlena, V.; Findejs, J. Application of model predictive control to advanced combustion control. *Control Eng. Pract.* **2005**, *13*, 671–680. [CrossRef]

29. Kortela, J.; Jämsä-Jounela, S.L. Modeling and model predictive control of the BioPower combined heat and power (CHP) plant. *Int. J. Electr. Power Energy Syst.* **2015**, *65*, 453–462. [CrossRef]
30. Zimmerman, N.; Kyprianidis, K.; Lindberg, C.F. Agglomeration Detection in Circulating Fluidized Bed Boilers Using Refuse Derived Fuels. In Proceedings of the 2016 9th EUROSIM Congress on Modelling and Simulation, Oulu, Finland, 12–16 September 2016; pp. 123–128. [CrossRef]
31. Hernandez-Atonal, F.D.; Ryu, C.; Sharifi, V.N.; Swithenbank, J. Combustion of refuse-derived fuel in a fluidised bed. *Chem. Eng. Sci.* **2007**, *62*, 627–635. [CrossRef]
32. Xiao, X.; Yang, H.; Zhang, H.; Lu, J.; Yue, G. Research on Carbon Content in Fly Ash from Circulating Fluidized Bed Boilers. *Energy Fuels* **2005**, *19*, 1520–1525. [CrossRef]
33. Lu, J.W.; Zhang, S.; Hai, J.; Lei, M. Status and perspectives of municipal solid waste incineration in China: A comparison with developed regions. *Waste Manag.* **2017**, *69*, 170–186. [CrossRef] [PubMed]
34. ISWA. *Waste-to-Energy State-of-the-Art-Report*; Technical Report; ISWA: Copenhagen, Denmark, 2013.
35. Yin, C.; Li, S. Advancing grate-firing for greater environmental impacts and efficiency for decentralized biomass/wastes combustion. *Energy Procedia* **2017**, *120*, 373–379. [CrossRef]
36. Beyene, H.D.; Werkneh, A.A.; Ambaye, T.G. Current updates on waste to energy (WtE) technologies: A review. *Renew. Energy Focus* **2018**, *24*, 1–11. [CrossRef]
37. Manfredi, S.; Tonini, D.; Christensen, T.H. Contribution of individual waste fractions to the environmental impacts from landfilling of municipal solid waste. *Waste Manag.* **2010**, *30*, 433–440. [CrossRef] [PubMed]
38. Redemann, K.; Hartge, E.U.; Werther, J. Ash management in circulating fluidized bed combustors. *Fuel* **2008**, *87*, 3669–3680. [CrossRef]
39. Sever Akdag, A.; Atımtay, A.; Sanin, F. Comparison of fuel value and combustion characteristics of two different RDF samples. *Waste Manag.* **2016**, *47*, 217–224. [CrossRef] [PubMed]
40. Ripa, M.; Fiorentino, G.; Giani, H.; Clausen, A.; Ulgiati, S. Refuse recovered biomass fuel from municipal solid waste. A life cycle assessment. *Appl. Energy* **2016**, *186*, 211–225. [CrossRef]
41. Malinauskaite, J.; Jouhara, H.; Czajczyńska, D.; Stanchev, P.; Katsou, E.; Rostkowski, P.; Thorne, R.J.; Colón, J.; Ponsá, S.; Al-Mansour, F.; et al. Municipal solid waste management and waste-to-energy in the context of a circular economy and energy recycling in Europe. *Energy* **2017**, *141*, 2013–2044. [CrossRef]
42. Perrot, J.F.; Subiantoro, A. Municipal waste management strategy review and waste-to-energy potentials in New Zealand. *Sustainability* **2018**, *10*, 3114. [CrossRef]
43. Sandberg, J. Fouling in Biomass Fired Boilers. Ph.D. Thesis, Mälardalen University, Västerås, Sweden, 2011.
44. Yan, R.; Liang, D.T.; Laursen, K.; Li, Y.; Tsen, L.; Tay, J.H. Formation of bed agglomeration in a fluidized multi-waste incinerator. *Fuel* **2003**, *82*, 843–851. [CrossRef]
45. Lin, W.; Dam-Johansen, K.; Frandsen, F. Agglomeration in bio-fuel fired fluidized bed combustors. *Chem. Eng. J.* **2003**, *96*, 171–185. [CrossRef]
46. Skrivfars, B.J.; Zevenhoven, M.; Backman, R.; Ohman, M. *Effect of Fuel Quality on the Bed Agglomeration Tendency in A Biomass Fired Fluidised Bed Boiler*; Therm. Eng. Res. Assoc. Rep. No. Varmeforsk B8–803; Värmeforsk: Stockholm, Sweden, 2000.
47. Gatternig, B. Predicting Agglomeration in Biomass Fired Fluidized Beds. Ph.D. Thesis, University of Erlangen–Nuremberg, Erlangen, Germany, 2015.
48. Bartels, M.; Nijenhuis, J.; Kapteijn, F.; van Ommen, J.R. Case studies for selective agglomeration detection in fluidized beds: Application of a new screening methodology. *Powder Technol.* **2010**, *203*, 148–166. [CrossRef]
49. Bartels, M.; Nijenhuis, J.; Kapteijn, F.; van Ommen, J.R. Detection of agglomeration and gradual particle size changes in circulating fluidized beds. *Powder Technol.* **2010**, *202*, 24–38. [CrossRef]
50. Morris, J.D.; Daood, S.S.; Chilton, S.; Nimmo, W. Mechanisms and mitigation of agglomeration during fluidized bed combustion of biomass: A review. *Fuel* **2018**. [CrossRef]
51. Visser, H.J.M. *The Influence of Fuel Composition on Agglomeration Behaviour in Fluidised-Bed Combustion*; Duurzame Energy; ECN Biomass: Delft, The Netherland, 2004; p. 44.
52. Vamvuka, D.; Zografos, D.; Alevizos, G. Control methods for mitigating biomass ash-related problems in fluidized beds. *Bioresour. Technol.* **2008**, *99*, 3534–3544. [CrossRef] [PubMed]
53. Corcoran, A.; Knutsson, P.; Lind, F.; Thunman, H. Comparing the structural development of sand and rock ilmenite during long-term exposure in a biomass fired 12 MWthCFB-boiler. *Fuel Process. Technol.* **2018**. [CrossRef]
54. Ergun, S. Fluid flow through packed columns. *Chem. Eng. Prog.* **1952**. [CrossRef]

55. *Fluidization Engineering*, 2nd ed.; Butterworth Publishers: Waltham, MA, USA, 1991; p. 491.
56. Kozak, S. State-of-the-art in control engineering. *J. Electr. Syst. Inf. Technol.* **2014**, *1*, 1–9. [CrossRef]
57. Sultana, W.R.; Sahoo, S.K.; Sukchai, S.; Yamuna, S.; Venkatesh, D. A review on state of art development of model predictive control for renewable energy applications. *Renew. Sustain. Energy Rev.* **2017**, *76*, 391–406. [CrossRef]
58. Garg, A.; Corbett, B.; Mhaskar, P.; Hu, G.; Flores-Cerrillo, J. Subspace-based model identification of a hydrogen plant startup dynamics. *Comput. Chem. Eng.* **2017**, *106*, 183–190. [CrossRef]
59. Khani, F.; Haeri, M. Robust model predictive control of nonlinear processes represented by Wiener or Hammerstein models. *Chem. Eng. Sci.* **2015**, *129*, 223–231. [CrossRef]
60. Al Seyab, R.; Cao, Y. Nonlinear model predictive control for the ALSTOM gasifier. *J. Process Control* **2006**, *16*, 795–808. [CrossRef]
61. Van Overschee, P.; De Moor, B. *Subspace Identification for Linear Systems*; Springer: Boston, MA, USA, 1996. [CrossRef]

processes

MDPI

Article

Diagnostics-Oriented Modelling of Micro Gas Turbines for Fleet Monitoring and Maintenance Optimization

Moksadur Rahman *, Valentina Zaccaria, Xin Zhao and Konstantinos Kyprianidis

School of Business, Society and Engineering, Mälardalen University, Västerås 72123, Sweden; valentina.zaccaria@mdh.se (V.Z.); xin.zhao@mdh.se (X.Z.); konstantinos.kyprianidis@mdh.se (K.K.)
* Correspondence: moksadur.rahman@mdh.se; Tel.: +46-(0)21-10-1594

Received: 14 October 2018; Accepted: 31 October 2018; Published: 2 November 2018

Abstract: The market for the small-scale micro gas turbine is expected to grow rapidly in the coming years. Especially, utilization of commercial off-the-shelf components is rapidly reducing the cost of ownership and maintenance, which is paving the way for vast adoption of such units. However, to meet the high-reliability requirements of power generators, there is an acute need of a real-time monitoring system that will be able to detect faults and performance degradation, and thus allow preventive maintenance of these units to decrease downtime. In this paper, a micro gas turbine based combined heat and power system is modelled and used for development of physics-based diagnostic approaches. Different diagnostic schemes for performance monitoring of micro gas turbines are investigated.

Keywords: micro gas turbine; modelling; diagnostics, gas path analysis, analysis by synthesis

1. Introduction

According to multiple sources, the global market for micro gas turbine (MGT) will experience an expeditious growth in the coming years [1–3].The power generation segment of its product portfolio is expected to contribute to a bulk portion of this growth. In particular, the combined heat and power (CHP) configuration of this energy generator are grabbing much attention from both the industry and the policy makers [4,5]. Especially in the context of European Unions'(EU) initiatives against climate change, these micro-CHP units could play a vital role in achieving both short- and long-term emission reduction targets.

MGTs are gas turbines combined with high speed generators whose electrical output can range between few kilowatts and few hundreds kilowatts. They offer a number of benefits compared to other technologies for distributed heat and power generation, including compact size, lightweight, fewer moving parts, lower maintenance needs, lower noise and vibration, high reliability, higher fuel flexibility, lower emission levels, potential for low cost mass production, and potential for integration with others decentralised energy generators [6–9]. On the other hand, the main drawbacks of MGTs are low electrical efficiency, high research and development cost, and high ownership cost at the moment [5,7]. However, numerous development activities are under-way by academia, industry, and policy makers to overcome the aforementioned challenges.

To bring the cost of ownership down, multiple vendors are offering micro-CHP units that are being developed by utilizing commercial off-the-shelf (COTS) components from automotive turbocharger industry. The inclusion of mature turbocharger technology offers not only cost reduction, but also high reliability and robustness that the industry achieved through decades of continuous improvement. On the contrary, turbochargers are not optimised for MGT operation due to the trade-off between design point efficiency and cost of manufacturing. For this reason, the MGT market is still considered to be a niche market, and the cost for development of turbo-machinery optimised for MGT operation can only be justified with mass production. Hence, modification of automotive turbocharger is often preferred to improve design point efficiency.

In the context of a power sector with high share of intermittent energy sources, MGTs should offer high reliability and availability to ensure the security of supply. Subsequently, an on-line condition-based monitoring and fault detection system is necessary. MGT runtime can be extended by early detection of faults and performance degradation. Thus, it will be possible to plan maintenance activity long before a breakdown occurs. Eventually, this will improve the availability and lower the maintenance cost. Another important aspect of future MGT market that emphasises the need of a condition-based monitoring and fault detection system is the ownership structure. Traditionally, gas turbines are owned by utilities and large companies. However, MGT-based CHP units could also be owned by private persons and small and medium enterprises (SME). Hence, a service-oriented approach is vital, where the service provider (i.e., technology provider or system installer or independent service provider) might need to manage a large fleet of MGT units. This is where the concept of fleet level monitoring and diagnostics can play an important role. The service provider will be responsible for monitoring the MGT fleet and planning maintenance based on the engine health conditions and severity of deterioration. Hence, an integrated approach for MGT fleet monitoring and diagnostics is necessary to foster the adoption of this promising technology. Since the MGT technology is still in the early stage of commercialization phase, sufficient data on degraded and faulty operations are lacking, which rules out the use of data-driven diagnostics approaches for the moment. Moreover, data-driven approaches perform worse for cases that fall outside of the training dataset, which could be a limitation considering the wide operating flexibility that is expected from MGT units. Therefore, physics-based modelling for MGT diagnostics appears necessary to overcome the aforementioned challenges.

Realizing its potential, the exploitation of a physics-based model for MGT fleet monitoring and diagnostics is investigated in this paper. A diagnostic-oriented model for an MGT based CHP unit is developed. An in-house Fortran-based gas turbine modelling tool named EnVironmental Assessment (EVA) is used for model development [10,11]. Model adaptation techniques for individual MGT units in a fleet are also discussed in detail. Subsequently, a multi-level fault detection and isolation methods are presented along with a detailed fault diagnostics scheme that can identify the location and magnitude of different component faults. Finally, several simulation trials are performed to test the presented fault diagnostics scheme.

2. Review on Performance Based Gas Path Diagnostics

The performance of gas turbines deteriorates over time, leading to reduced output capacity and thermal efficiency that in turn result in reduced profitability and increased emissions [12]. Generally, any performance deterioration in a gas turbine can be linked with performance deterioration of one or more gas path components. The performance of these components deteriorate over time due to various degradation mechanisms such as fouling, erosion, corrosion, internal liner surface cracking, increase in tip and seal clearance, foreign object damage, plugging of the injector and the cooling holes, etc. [13–15]. The rate at which these deterioration mechanisms take place could be different depending on the manufacturing tolerance, engine operating conditions, operating regime, i.e., part or full load, start-stop cycles, and fuel type and quality. Deterioration generally causes deviations in the component performance parameters, i.e., efficiency, pressure ratio, flow capacity, and others, which in turn lead to deviations in the gas path measurable parameters such as temperatures, pressures, speed, and flow rates. Using the gas path measurable parameters to detect the change in component health parameters forms the foundation for performance-based gas path analysis (GPA) methods for gas turbine diagnostics. Two of the well-known variants of this method are physics-based and data-driven GPA. Numerous comprehensive review articles explore existing GPA approaches (both physics-based and data-driven) and their relative performance [16–20]. In these articles, the comparative pros and cons of different approaches are examined based on attributes such as reliability, accuracy, model complexity, computational efficiency, the ability to cope with noise and bias, and number of measurements required for diagnostics. The findings of these reviews can be summarized by stating that there is no single approach that outperforms the others in all the attributes; rather they are complementary and each has its own benefits and drawbacks. Hence, hybrid schemes that combine both physics-based and data-driven approaches should be preferred. As a starting point for MGT diagnostics and in light of the limitations previously discussed, the focus of this study is limited to a physics-based approach.

Physics-based GPA approaches for gas turbines diagnostics have been widely studied by the research community over the years. As the name suggests, these approaches explicitly rely on the physics-based models of gas turbines. The models are based on mathematical and thermodynamic equations that principally correlate gas path measurable parameters with component performance parameters. The has approach developed much since Urban pioneered it in 1967 [21]. Urban [22,23] and others [24–26] further investigated the approach, which is widely referred to as linear GPA in the literature. In linear GPA, unknown variations in components performance parameters are computed from known variations in measurable parameters by using a set of linear equations. The equations are derived by linearising the non-linear equations that link components performance parameters with measurable parameters, around a specific steady-state operating point. Being conceptually simple and computationally light, linear GPA offers numerous benefits such as fault isolation and quantification and multiple faults diagnostics. On the other hand, the method has multiple limitations. It requires many relevant measurements for fault diagnostics, which can be quite rare in commercial units. Due to the assumption of linearity, the method shows instability and large inaccuracy under higher level of deteriorations. Moreover, it is unable to deal with sensor noise and bias.

To cope with the non-linear behaviour of the gas turbines and improve the accuracy of GPA, non-linear GPA was introduced by House [27] and Esher [28]. The method was further improved by many others. Unlike linear GPA, in non-linear GPA, the full equations are treated directly without any linearisation. To deal with the engine to engine variations, an adaptive approach of non-linear GPA was examined by Stamatis et al. [29]. The author introduced modification factors to the health parameters to take care of the individual engine variation that are computed through an optimisation procedure. Li [30] developed a two step approach for linear and non-linear adaptive GPA that can detect both single and simultaneously occurring multiple faults. Not long ago, Larsson [31] presented a systematic design procedure to construct a fault detection and isolation system by using complex non-linear models. In a more recent work, Liu [32] proposed a dynamic tracking filter incorporating state observer to detect the variation of six performance parameters of three gas path components by using four measurement parameters. However, the usage of linear state observer resulted in reduced accuracy in fault detection capability of this approach; hence, a non-linear state observer is required. In another work, Kang [33] suggested a compressor map adaptation technique to enhance the accuracy of performance based diagnostics of a heavy-duty gas turbine. Comparison of different diagnostics approaches are studied in Koskoletos et al. [34]. The authors performed a comparative analysis among: (1) probabilistic neural network (PNN); (2) k-nearest neighbours; (3) optimization; (4) combinatorial; (5) adaptive 2X2; and (6) combination of PNN and adaptive 2X2 method. They concluded that Methods 3–6 can be used for component fault magnitude estimation and prognostic purpose.

Previous research efforts have made valuable contributions in improving performance-based gas path diagnostics methods. However, most of this work is focused on large scale industrial gas turbines; only a few studies are focused on micro gas turbine [35–38]. Performance-based diagnostics of MGT by employing GPA poses numerous challenges. To keep the cost of ownership down, MGTs include only few measurements of gas path parameters. Moreover, some of these measurements are used for control purpose, meaning that they cannot be utilized for diagnostics. Due to the high manufacturing tolerances, engine components show wide variations in performance, which could lead to deviations in measured parameters that are comparable to fault conditions. Hence, model tuning is essential for individual engines before employing any fault diagnostics approach.

3. Methodology

To develop a model-based diagnostics scheme for an MGT fleet, a diagnostics-oriented model of a nominal MGT unit is developed and presented in this article. The developed model is validated against the measurements from the performance test of a commercial MGT unit. Sequentially, a scheme for model tuning is employed to account for engine to engine variations. Finally, a diagnostics scheme is proposed and tested with the help of simulation studies.

3.1. Gas Path Modelling of the Micro Gas Turbine

The operating principle of an MGT is identical to large-scale open cycle gas turbines. Both operate on the well-known Brayton cycle. As shown in Figure 1, a typical MGT with CHP configuration consists of a compressor, a recuperator, combustor, a turbine, a high-speed generator, and an exhaust recovery heat exchanger. The air is compressed in the compressor and then preheated by the exhausts in the recuperator before being further heated by burning fuel in the combustor. The high temperature working fluid is then expanded in the turbine that operates the compressor and the high-speed generator. The remaining exhaust heat is recovered by water in the recovery heat exchanger.

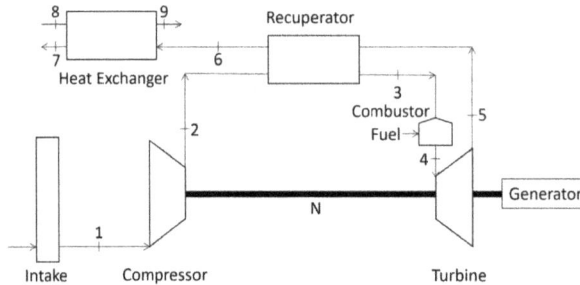

Figure 1. Layout of the MGT in CHP configuration (MGT: Micro Gas Turbine; CHP: Combined Heat and Power).

The MGT unit under study in this paper is the EnerTwin Micro-CHP that is marketed by Micro Turbine Technology (MTT) B.V. It is a single-shaft, radial, recuperated gas turbine manufactured in CHP configuration. The micro-CHP unit have a capacity of 3.2 kW of electric and 16 kW thermal output.

A modular modelling technique is employed here to develop the MGT model. All the gas path components, i.e., compressor, recuperator, combustor and turbine including duct and rotating shaft are modelled and integrated just like the way they are connected in reality. As mentioned previously, the MGT model is developed by using an in-house Fortran-based gas turbine modelling tool called EVA. It is a multidisciplinary conceptual design tool that comprises various modules incorporating substantial detail within a wide range of disciplines, i.e., gas turbine performance, aerodynamic and mechanical design, emissions prediction, and environmental impact. In this work, the gas turbine performance analysis module of the tool is utilized.

The compressor and turbine models are based on their corresponding characteristic maps provided by the manufacturer and standard mass and energy balance equations. The characteristics maps provide correlations between pressure ratio (PR), shaft speed (N), mass flow rate ($\dot{m}$), and isentropic efficiency (η_{is}), as shown in Equation (1).

$$\left(\frac{\dot{m}\cdot\sqrt{T_{in}}}{P_{in}},\eta_{is}\right) = f\left(PR,\frac{N}{\sqrt{T_{in}}}\right) \tag{1}$$

where T_{in} and P_{in} refer to the inlet temperature and pressure respectively. For a known shaft speed and pressure ratio, corresponding mass flow rate and isentropic efficiency are extracted from the characteristic maps. The extracted values are then used in well-known thermodynamic equations to calculate parameters related to compressor and turbine. For example, compressor outlet temperature (T_2), pressure (P_2), and compression work ($W_{comp.}$) are calculated utilizing Equations (4), (7) and (9).

Starting by assuming isentropic compression and expansion in the compressor and turbine ,respectively, the Gibbs equation takes the following form:

$$\underbrace{S_{out} - S_{in}}_{=0} = \int_{T_{in}}^{T_{out,is.}} \frac{C_p}{T} \cdot dT - R_g \cdot \ln\left(\frac{P_{out}}{P_{in}}\right) \tag{2}$$

Here, S refers to the entropy, C_p is the specific heat capacity and R_g is the universal gas constant. Defining entropy function as in Equation (3),

$$\phi(T) = \int_{T_{ref.}}^{T} \frac{C_p}{T} \cdot dT \tag{3}$$

Equation (4) is derived:

$$\phi(T_{out,is.}) = \phi(T_{in}) + R_g \cdot \ln\left(\frac{P_{out}}{P_{in}}\right) \tag{4}$$

Here, $\phi(T_{in})$ and $\phi(T_{out,is})$ are the temperature dependent entropy functions at the inlet and outlet of the component (i.e., compressor or turbine).

However, the isentropic assumptions are not applicable to real compression and expansion processes which have inherent losses due to compressor and turbine inefficiencies. To account for these losses, isentropic efficiencies for compressor ($\eta_{comp.,is.}$) and turbine ($\eta_{turb.,is.}$) are defined as in Equations (5) and (6).

$$\eta_{comp.,is.} = \frac{h(T_{2,is.}) - h(T_1)}{h(T_2) - h(T_1)} \tag{5}$$

$$\eta_{turb.,is.} = \frac{h(T_5) - h(T_4)}{h(T_{5,is.}) - h(T_4)} \tag{6}$$

Subsequently, compressor and turbine pressure ratios are defined by Equations (7) and (8).

$$PR_{comp.} = \frac{P_2}{P_1} \tag{7}$$

$$PR_{turb.} = \frac{P_4}{P_5} \tag{8}$$

Finally, compressor and turbine works are calculated using Equations (9) and (10),

$$W_{comp.} = \dot{m}_1 (h_2 - h_1) \tag{9}$$

$$W_{turb.} = \dot{m}_4 (h_4 - h_5) \tag{10}$$

Here, $\phi(T_1)$ and $\phi(T_2)$ are the temperature dependent entropy functions at the inlet and outlet of the compressor, and R_g is the universal gas constant.

The recuperator is modelled as a counter-current plate type heat exchanger. Heat exchanger's key performance parameters related to heat transfer and pressure drop are used for this purpose. The thermal effectiveness (ϵ) is used as heat transfer performance parameter, while relative pressure drop in the air-side (ΔP_{23}) and gas-side (ΔP_{56}) are introduced as pressure drop performance parameters.

A heat flux-based definition of effectiveness is considered instead of a temperature-based one to include the influence of the gas composition on the specific heat capacity, as shown in Equation (11).

$$\epsilon = \frac{Q_{act.}}{Q_{max.}} \tag{11}$$

$$\Delta P_{23} = \frac{P_3 - P_2}{P_2} \tag{12}$$

$$\Delta P_{56} = \frac{P_6 - P_5}{P_5} \tag{13}$$

where $Q_{act.}$ and $Q_{max.}$ refer to actual heat transfer and maximum possible heat transfer in the recuperator. The maximum possible heat transfer is achieved when the fluid with minimum heat capacity rate undergoes the maximum temperature difference available present in the exchanger, which is the difference in the entering temperatures for the hot and cold fluids.

The combustor performance is given in terms of combustion efficiency ($\eta_{comb.}$) and relative pressure drop (ΔP_{34}). The combustor efficiency can be computed by Equation (14), while relative pressure drop can be computed using an equation similar to Equation (12). Using these parameters, fuel to air ratio (FAR) and pressure at exit of the combustor (P_4) are determined. Finally, energy balance is applied for the combustor to estimate the enthalpy at the combustor outlet that in-turn is used to determine the temperature. For simplicity, a constant lower heating value (LHV) is used in Equation (14).

$$\eta_{comb.} = \frac{\dot{m}_3 (h_3 - h_2)}{\dot{m}_f \cdot LHV} \tag{14}$$

Compressor, turbine, and generator are mounted on the same shaft; hence, the shaft mechanical efficiency is defined as in Equation (15).

$$\eta_{shaft} = \frac{W_{comp.} + W_{gen.}}{W_{turb.}} \tag{15}$$

Pressure losses in the ducts are also taken into account by using an equation similar to Equation (12).

3.1.1. Matching Scheme for Gas Path Modelling

The steady state operating points of the gas turbine are obtained by matching the compressor and turbine. This is done by superimposing turbine map on the compressor map while mass flow and energy continuity are maintained. The serial nested loops method is used where initial guesses are continuously updated until all the residuals error terms, corresponding to components mass and energy balance, reach predefined accuracy. To do this, a Jacobian matrix is built where each element of the matrix is the sensitivity ratio between each output (or target) and state ($\delta Y / \delta X$). The Jacobian is used to compute and minimize the residuals between model outputs and targets. In normal conditions, the Jacobian consists of the following pairs of outputs and states in the rows and columns as shown in Table 1.

For example, the mass continuity in the compressor needs to be satisfied, which means that the mass flow calculated from the compressor map needs to match the compressor inlet mass flow. The speed factor $N_{rel.}$, defined as $N/N_{des.}$, is then varied until the residual between the two mass flow values ($\dot{m}_{comp.,res.}$) is below a predefined defined threshold.

Table 1. Jacobian matrix used for modelling (PWX: shaft power).

Outputs \ States	$N_{rel.}$	$\dot{m}_{int.,corr.,rel.}$	$\beta_{comp.}$	$\beta_{turb.}$	$Q_{hex.,air,nond.}$	$Q_{hex.,gas,nond.}$	PWX	$\dot{m}_{corr.,f,rel.}$
$Torque_{res.}$	0.8978	0.0000	0.7910	0.8590	0.1371	0.0000	0.0001	0.1106
$\dot{m}_{comp.,res.}$	2.0321	0.9998	0.9557	0.0000	0.0000	0.0000	0.0000	0.0000
$\dot{m}_{turb.,res.}$	0.4596	0.0000	0.9285	0.0951	0.1412	0.0000	0.0000	0.1102
$Q_{hex.,air,res.}$	0.4589	0.0000	0.0633	0.2508	0.4263	0.0000	0.0000	0.4334
$Q_{hex.,gas,res.}$	0.4589	0.0000	0.0633	0.2508	0.5737	1.0000	0.0000	0.4334
$\dot{m}_{noz.,res.}$	7.7099	0.0000	1.6493	7.3678	0.0106	0.4625	0.0000	0.0330
$N_{rel,res.}$	1.0000	0.0000	0.0000	0.0000	0.0000	0.0000	0.0000	0.0000
$T_{5,res.}$	0.7408	0.0000	0.6401	0.1407	0.3219	0.0000	0.0000	0.2432

3.1.2. Modified Matching Scheme for Adaptation

For model tuning and gas path diagnostics purpose, correcting factors for the component performance parameters i.e., efficiencies, flow capacities, and effectiveness were included in the MGT model. These factors are included as state variables along with their corresponding target variables. Then, a new Jacobian matrix is built, and new residuals are generated between output variables and target values (Table 2) to achieve a solution.

Table 2. Target and state pairs used in the Jacobian matrix for diagnostics (FC: flow capacity).

Targets	States
$Torque_{res.}$	$N_{rel.}$
$\dot{m}_{comp.,res.}$	$\dot{m}_{int.,corr.,rel.}$
$\dot{m}_{turb.,res.}$	$\beta_{comp.}$
$Q_{hex.,air,res.}$	$\beta_{turb.}$
$Q_{hex.,gas,res.}$	$Q_{hex.,air,nond.}$
$\dot{m}_{noz.,res.}$	$Q_{hex.,gas,nond.}$
$N_{rel,res.}$	PWX
$T_{5,res.}$	$\dot{m}_{corr.,f,rel.}$
$T_{2,res.}$	$\Delta\eta_{comp.}$
$P_{2,res.}$	$\Delta FC_{comp.}$
$P_{5,res.}$	$\Delta\eta_{turb.}$
$\dot{m}_{1,res.}$	$\Delta FC_{turb.}$
$T_{3,res.}$	$\Delta\epsilon_{rec.}$

In this case, the standard matching scheme is modified to fit the required state variables. The correcting factor on the flow capacity for compressor and turbine is varied to satisfy the mass flow continuity in these two components. Once the flow capacity is fixed, the beta lines in the compressor and turbine maps are varied to match the desired exit pressure/temperature. The correction factor on the compressor efficiency is varied to match the torque on the shaft, while the speed factor becomes the variable that minimizes the residual between produced power and load demand.

3.2. Scheme for the Model Tuning

As noted previously, due to the high manufacturing tolerances, MGT components show wide variations in performance characteristics which result in engine to engine performance deviations. Hence, a baseline or nominal model to represent all the MGTs in a fleet may lead to inaccuracy in diagnostics. To reduce the model plant mismatch for healthy engines, model tuning is performed by following a tuning scheme, as shown in Figure 2. The scheme is followed to get individual models for each of the MGTs in the fleet. It is important to note here that the model tuning need to be performed for healthy engines operating in nominal conditions.

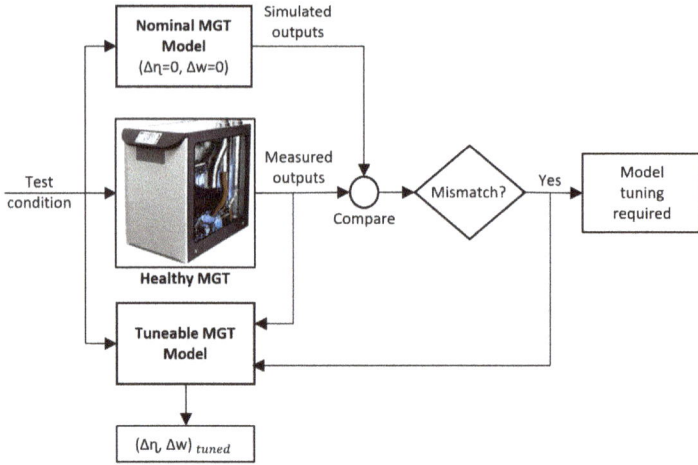

Figure 2. Scheme for model tuning.

According to the tuning scheme, the MGT under consideration is operated in test conditions and data are collected. After that, the nominal MGT model is simulated with same test conditions and the results are compared with the measured data. The measured data are fed to the tuneable MGT model which is used to back-calculate the performance deltas to match the measured data from the healthy MGT. The performance deltas are the deviations in components performance parameters, i.e., efficiency and flow capacity of the turbine and compressor, effectiveness and pressure drop of the recuperator. These calculated performance deltas are then used to modify the nominal model to achieve the tuned model.

3.3. Scheme for the Diagnostics

Typically, diagnostics schemes for gas turbines are based on a multi-level approach that includes data pre-processing, threshold monitoring, sensor fault detection, engine fault detection and isolation, and fault identification [39–41]. However, here the focus will be only on fault identification, meaning assessment of fault location and magnitude. The proposed scheme for the diagnostics of individual MGT unit is presented in Figure 3.

The fault diagnostics scheme that is carried out here is based on a modification of the matching scheme previously discussed, where the Jacobian matrix includes additional states (the performance modification deltas) and additional targets (gas path measurements). In addition, a further step founded upon a signature-based algorithm is used to isolate and quantify the detected faults. A nominal model of representative average engine is used to create fault signatures by simulating different component faults that stored in a signature database. The faults are simulated by assuming associated performance deltas and using these deltas in the MGT model for fault simulation. An inventory of faults that are used to build an example signature database is listed in Table 3. The signature database also includes multiple faults that assumed to be occurred at the same time.

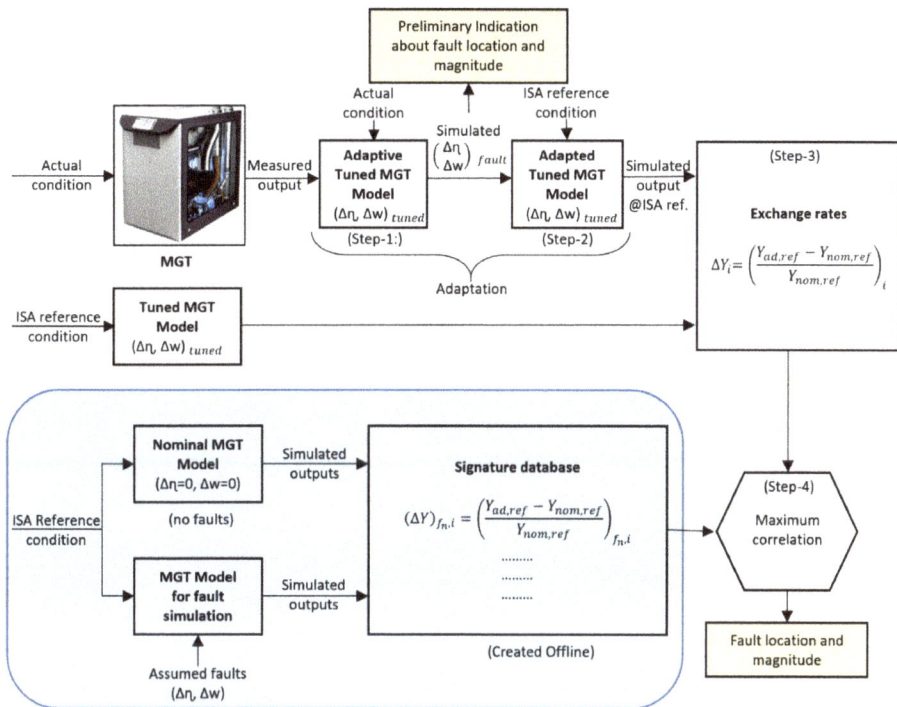

Figure 3. Scheme for fault identification (ISA: International Standard Atmosphere).

Table 3. Considered fault inventory.

Component Name	Fault Description
Compressor	• 1% drop in isentropic efficiency ($\Delta\eta_{comp.}$) • 1% drop in flow capacity ($\Delta FC_{comp.}$)
Turbine	• 1% drop in isentropic efficiency ($\Delta\eta_{turb.}$) • 1% drop in flow capacity ($\Delta FC_{turb.}$)
Recuperator	• 1% drop in effectiveness ($\Delta\epsilon_{rec.}$)
Duct	• 1% flow leakage from compressor outlet duct
Shaft/Bearings	• 1% drop in shaft efficiency due to increased bearing loss

The creation of fault database is something that is performed off-line, whereas the rest of the scheme is performed close to real-time. The first step of the real-time diagnostics scheme is based on a modification of the Jacobian matrix and is called analysis by synthesis (AnSyn). During this step, the measurements from an engine under operation are used to calculate any deviation in performance deltas by simulating the adaptive tuned model of the engine in actual operating conditions. The detected deltas can provide a good indication of the fault location and magnitude for single or multiple faults in compressor, turbine and recuperator as listed in Table 3. However, this step cannot detect other faults such as flow leakages, shaft loss, etc. Hence, the signature-based algorithm is also applied here. In the next step, the computed performance deltas are used as inputs in the adapted tuned model at ISA (International Standard Atmosphere) reference conditions, and exchange rates are calculated. The exchange rates are measurement deviations that are converted to ISA reference conditions. Subsequently, a correlation function, as shown in Equation (9), is used to find the correlations between engine exchange rates and the signatures from the database. The correlation function calculates the Pearson Product-Moment Correlation Coefficient (PPMCC) for two sets of values, in this case signatures from measurements and database.

$$Corr(x,y) = \frac{\sum_{i=1}^{k}(x_{n,i} - \bar{x}_n)(y - \bar{y}_i)}{\sqrt{\sum_{i=1}^{k}(x_{n,i} - \bar{x}_n)^2}\sqrt{\sum_{i=1}^{k}(y - \bar{y}_i)^2}} \tag{16}$$

Here, $Corr(x,y)$ is referred as the correlation coefficient. In the correlation function, subscript i is used to denote the series of sensor measurements where k is the total number of available measurements from the engine. x_n refers to the signature resulted by the fault in nth component and y refers to the exchange rates at a given operating point.

The maximum correlations give the location of the fault. To get the magnitude of the faults, Equation (17) is solved in an iterative way to determine the coefficient estimates c_m that give the magnitude of the corresponding faults.

$$Y(x) = \sum_{i=1}^{l} c_m X_m \tag{17}$$

Here, X_m and Y are the vectors that consists of signatures and exchange rates for specific faults which are indicated by the maximum correlation. Subscript l refers to the number of faults and m corresponds to a specific fault. Linear regression is employed to solve the above equation and determine the magnitude for single and multiple faults.

To prove the effectiveness of the proposed diagnostics scheme, in this work, the developed MGT model is used to generate measurements related to different faults.

4. Results and Discussion

Here, the findings from the gas path modelling and diagnostics, and their inferences are elaborated in detail. At first, the modelling error are presented against the performance test results of a commercial MGT unit. Thereafter, the proposed diagnostics scheme is demonstrated by formulating different case studies, which is complemented by sensitivity studies for different measurement uncertainties, i.e., sensor noise and bias.

4.1. Gas Path Modelling

In Table 4, the model outputs at nominal load are compared with the corresponding values from performance test results of a commercial scale MGT unit. As it can be seen, the simulated model have acceptable accuracy at the nominal load. It is important to note that the diagnostics scheme tested in this paper is applied only at nominal load.

Table 4. Modelling error against performance test result at the nominal load.

Parameters	Modelling Error (%)
PWX (W)	0.02
P_2 (kPa)	−0.26
T_3 (K)	−0.01
N (RPM)	0.00
T_5 (K)	0.00

The model outputs for three other off-design points at part-load are also compared with the performance test results. The comparison results are presented in Figure 4 as percentage error. It can be observed that the speed (N) and the turbine outlet temperature (T_5) give zero error. This is in-line with the matching scheme described in Section 3.1.1, where N and T_5 are used as target variables. The error in shaft power (PWX) becomes positive and then negative as the power moves from nominal value to part-load. The shaft power was not available directly from the performance test, and it was calculated from the electrical power output. Assumptions were made about auxiliary power consumption, generator efficiency, and inverter efficiency, these can be contributed to the mismatch between simulated and experimental data.

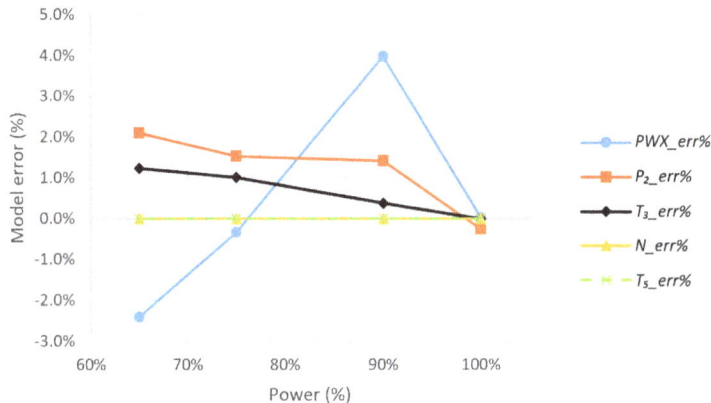

Figure 4. Comparison between model and performance test results for off-design operation (PWX: shaft power).

The error in the compressor outlet pressure (P_2) increases as power decreases. The compressor and turbine maps used here are the maps supplied by the turbocharger manufacturer; hence, they are not corrected for the modifications performed to the turbine and compressor. This might explain some of the errors in off-design operating points including P_2. Finally, the recuperator effectiveness was assumed to be constant over entire operating range. This could potentially be responsible for error in recuperator cold side outlet temperature (T_3). Overall, the results demonstrate a sufficient agreement between the simulation results with the performance test results for off-design operating points.

4.2. Diagnostics

To demonstrate the diagnostics scheme described in Section 3.3, two sets of case studies were formulated: one with only single faults occurring one at a time, and the other with multiple faults occurring concurrently. For the first set (S1 to S7), faults listed in Table 3 are considered, but the fault magnitudes are assumed to be 1.5% instead of 1% as used for the signature database. The results from the AnSyn step and the signature-based algorithm are presented below.

In Figure 5, the location and magnitude of all the single faults except the flow leakage and the shaft loss can be detected in the AnSyn step. This can be explained by looking at the matching scheme used for gas path component diagnostics as elaborated in Section 3.1.2. The modified Jacobian matrix (Table 2) includes performance deltas and their corresponding measurement pairs for the compressor, the turbine and the recuperator. However, there are no target-state pairs considered for the flow leakage and the shaft loss. This necessarily means that these faults will be detected during the AnSyn as multiple performance deviations. In these cases, in the AnSyn an equivalent fault is created by distributing the fault effects among other performance deltas that are included in the matching scheme. Alongside, the flow leakage and the shaft loss only affect turbine and recuperator deltas. This is because the GPA measurements corresponding to the compressor deltas are not affected by these two faults. However, due to the measurement uncertainties, compressor deltas are also expected to be affected marginally in reality.

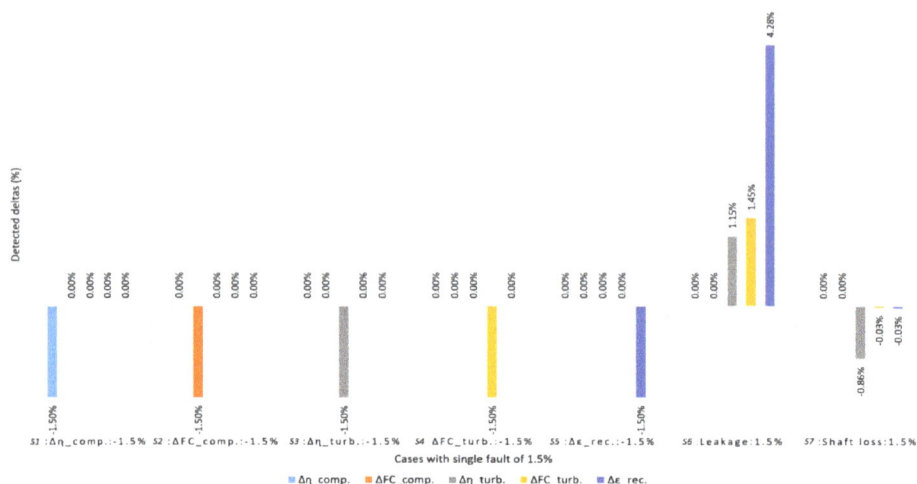

Figure 5. Fault location and severity for cases with single faults detected by AnSyn (AnSyn: analysis by synthesis).

It should be noted here that the AnSyn detects general performance deviations in all fault cases, without identifying the cause of the deviation. In some cases, deviations in efficiency and flow capacity can be directly related to a specific fault (e.g., compressor fouling or turbine erosion or recuperator fouling), while other faults such as flow leakage or additional shaft loss cannot be directly linked to the results from the AnSyn. Hence, a second step based on signature correlation is necessary. In reality, occurring faults will always have effect on more than one delta (e.g., compressor fouling reduces both efficiency and flow), but all the faults will correspond to deviations in the five performance parameters here presented, making AnSyn effective for fault detection. For consequent fault isolation and identification, without the need of increasing the number of required measurements, a second layer of fault diagnostics is applied. This second layer is the signature-based algorithm that includes correlation and regression analysis for fault localization and magnitude quantification respectively. Table 5 shows the correlation coefficients between exchange rates and signatures for case studies with single fault. It is found that, for all the cases, the maximum correlation coefficient always leads to correct location of the fault and thus is placed diagonally in the table. One observation from this study is that the faults corresponding to turbine isentropic efficiency loss and shaft loss have very close correlation coefficients, as red numbers in Table 5. This is quite obvious, since for a fixed turbine pressure ratio, both parameters highly depend on the ratio between actual isentropic enthalpy drop across turbine. However, a better decision about the fault location can be made by merging results from AnSyn and correlation steps.

Table 5. Correlation coefficients for cases with single fault.

Cases	Correlation between Exchange Rates and Signatures for Cases with Single Fault						
	$-1\% \, \Delta\eta_{comp.}$	$-1\% \, \Delta FC_{comp.}$	$-1\% \, \Delta\eta_{turb.}$	$-1\% \, \Delta FC_{turb.}$	$-1\% \, \Delta\epsilon_{rec.}$	1% Leakage	1% Shaft Loss
S1	**1.000**	0.693	0.783	0.496	−0.754	0.129	0.790
S2	0.694	**1.000**	0.863	0.505	−0.815	0.134	0.869
S3	0.784	0.863	**1.000**	0.247	−0.994	0.476	0.997
S4	0.497	0.506	0.248	**1.000**	−0.145	−0.733	0.268
S5	−0.754	−0.814	−0.994	−0.144	**1.000**	−0.565	−0.992
S6	0.127	0.135	0.477	−0.733	−0.566	**1.000**	0.459
S7	0.790	0.869	0.997	0.267	−0.992	0.457	**1.000**

Once the location of the fault is known, linear regression is applied to get the magnitude of the corresponding fault. Fault magnitudes estimated by the linear regression for the single faults cases are listed in Table 6. It is observed that linear regression resulted in faults magnitude that is very close to the actual simulated faults.

Table 6. Fault magnitudes for cases with single fault.

Cases	Fault Magnitude	Detected Fault Magnitude Using	
		AnSyn	Regression
S1	−1.500	−1.500	−1.511
S2	−1.500	−1.500	−1.506
S3	−1.500	−1.500	−1.509
S4	−1.500	−1.500	−1.503
S5	−1.500	−1.500	−1.500
S6	1.500	-	1.506
S7	1.500	-	1.508

The second set of case studies (M1 to M7) that includes multiple faults are listed in Table 7. The magnitudes of the faults are chosen between 1% and 1.5% so that the combination is different from the signature database where all the faults are 1% in magnitude.

Table 7. List of case studies for multiple fault.

Case Identifier	Fault Number	Fault Location and Magnitude
M1	Fault-1: Fault-2:	$\Delta\eta_{comp.} = -1.5\%$ $\Delta FC_{comp.} = -1.0\%$
M2	Fault-1: Fault-2:	$\Delta FC_{comp.} = -1.5\%$ $\Delta FC_{turb.} = -1.0\%$
M3	Fault-1: Fault-2:	$\Delta\eta_{comp.} = -1.5\%$ $\Delta\eta_{turb.} = -1.5\%$
M4	Fault-1: Fault-2:	$\Delta FC_{comp.} = -1.0\%$ $\Delta\eta_{turb.} = -1.5\%$
M5	Fault-1: Fault-2: Fault-3:	$\Delta\eta_{comp.} = -1.0\%$ $\Delta FC_{comp.} = -1.0\%$ $\Delta\epsilon_{rec.} = -1.5\%$
M6	Fault-1: Fault-2:	$Flow\,leakage = 1.0\%$ $\Delta FC_{comp.} = -1.5\%$
M7	Fault-1: Fault-2:	$Shaft\,loss = 1.5\%$ $\Delta\eta_{comp.} = -1.5\%$

Results from the AnSyn for simultaneously occurring multiple faults are displayed in Figure 6. As expected, AnSyn can identify correct fault locations and magnitudes for all cases except those including flow leakage and shaft loss. However, when these faults are combined with compressor's faults, corresponding compressors deltas can be correctly identified by the AnSyn.

The correlation coefficients between exchange rates and signatures for the above cases are reported in Table 8. For each case, this step gives maximum correlation for corresponding signatures from the database that reveals the location of the faults correctly.

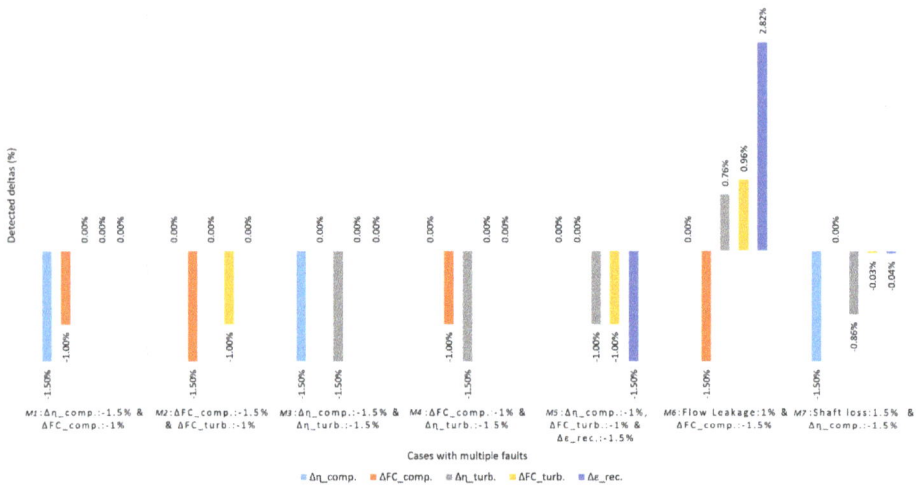

Figure 6. Preliminary estimation of fault location and severity for cases with multiple faults by AnSyn.

Table 9 summarizes the estimated magnitudes of faults for different cases with multiple faults by regression analysis and the comparison with AnSyn. As can be seen, the regression can give good indication of faults magnitude with some level of error, but the AnSyn performs better. The accuracy can be further improved by including more measurements in the multiple linear regression. However, a limited number of measurements are available in reality for such analysis. Additionally, if a sensor fault occurs and one or more measurements need to be removed from the scheme, the accuracy of the regression can even deteriorate. Sensor faults will also reduce the fault detectability by the AnSyn, since corresponding performance delta for the removed measurement also need to be removed from the matching scheme, as presented in Table 2.

Overall, the results presented until now show that the proposed diagnostics scheme can be used to detect the location and magnitude of different component faults with acceptable accuracy. Combining results from AnSyn with the signature-based algorithm increases the confidence on the final outcome of the proposed scheme. However, the above analysis thus far does not include any measurement uncertainty (i.e., sensor noise and bias), which can be quite common in reality. Hence, the influence of measurement uncertainty on the proposed fault diagnostics scheme is assessed in the following section.

First, the influence of measurement uncertainty on the AnSyn is examined by performing a sensitivity analysis. Here, the measurements corresponding to each of the performance deltas, as listed in Table 2, are varied within measurement uncertainty range. The measurement uncertainties used in this paper are obtained from the literature [42,43]. It is considered here that $P5$ measurement is available through a differential pressure sensor across recuperator.

Figure 7 shows sensitivity analysis of measurement uncertainties on all five performance deltas. Here, case study "S1" i.e., 1.5% drop in compressor isentropic efficiency is considered for this analysis. As can be seen in Figure 7, the compressor isentropic efficiency is only affected by P_2 and T_2 measurements. However, there is very strong linkage between P_2 and compressor flow capacity, and P_5 and recuperator effectiveness. Hence, measurement uncertainties in P_2 and P_5 can give misleading indication of faults related to compressor flow capacity and recuperator effectiveness, respectively; although their magnitudes are not as prominent as the fault. Other measurement uncertainties have negligible influence on different performance deltas. Therefore, it can be summarized from the sensitivity analysis that the influence of measurement uncertainties is limited and will have the effect of a reduced accuracy in fault magnitude estimation. Moreover, the sensor data can be filtered before using it for diagnostics purpose to decrease false alarm due to sensor noise related uncertainties.

Table 8. Correlation coefficients for cases with multiple faults.

Cases	Correlation between Exchange Rates and Signatures for Cases with Multiple Faults						
	−1% Δη_comp. & −1% ΔFC_comp.	−1% ΔFC_comp. & −1% ΔFC_turb.	−1% Δη_comp. & −1% Δη_turb.	−1% ΔFC_comp. & −1% Δη_turb.	−1% Δη_comp., −1% ΔFC_turb. & −1% Δε_rec.	1% Leakage & −1% ΔFC_comp.	1% Shaft Loss & −1% Δη_comp.
M1	**0.997**	0.658	0.946	0.871	0.877	0.437	0.971
M2	0.745	**0.991**	0.606	0.617	0.935	−0.093	0.629
M3	0.952	0.522	**1.000**	0.970	0.832	0.653	0.995
M4	0.890	0.500	0.972	**0.999**	0.810	0.722	0.946
M5	0.827	0.961	0.723	0.705	**0.983**	0.002	0.748
M6	0.562	−0.071	0.737	0.800	0.315	**0.988**	0.682
M7	0.972	0.553	0.995	0.944	0.846	0.595	**1.000**

Table 9. Fault magnitudes for cases with multiple faults.

Cases	Fault Number	Fault Magnitude	Detected Fault Magnitude Using	
			AnSyn	Regression
M1	Fault-1:	−1.500	−1.500	−1.486
	Fault-2:	−1.000	−1.000	−0.973
M2	Fault-1:	−1.500	−1.500	−1.262
	Fault-2:	−1.000	−1.000	−0.958
M3	Fault-1:	−1.500	−1.500	−1.520
	Fault-2:	−1.500	−1.500	−1.522
M4	Fault-1:	−1.000	−1.000	−0.948
	Fault-2:	−1.500	−1.500	−1.505
M5	Fault-1:	−1.000	−1.000	−0.984
	Fault-2:	−1.000	−1.000	−1.002
	Fault-3:	−1.500	−1.500	−0.993
M6	Fault-1:	1.000	-	0.999
	Fault-2:	−1.500	−1.500	−1.527
M7	Fault-1:	1.500	-	1.525
	Fault-2:	−1.500	−1.500	−1.517

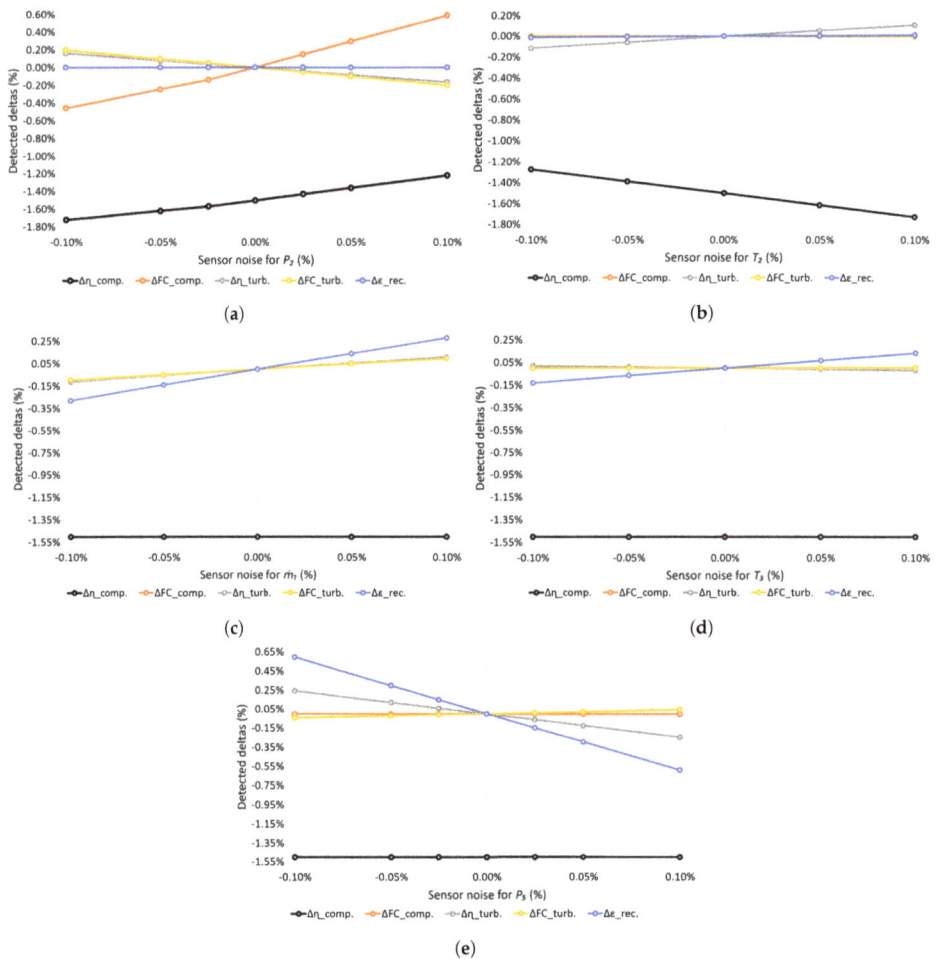

Figure 7. Sensitivity of AnSyn against measurement uncertainties in: (**a**) P_2; (**b**) T_2; (**c**) $\dot{m}_1$; (**d**) T_3; and (**e**) P_5.

Thereafter, the effect of measurement uncertainties on the correlation step is assessed. As in the previous section, the case study "S1", i.e., 1.5% drop in compressor isentropic efficiency is considered. For each of the sensors, maximum deviation in the measurement is assumed. As can be seen in Table 10, the measurement uncertainties has influence on the correlation coefficient. However, the influence is negligible and still faults are identified correctly by giving maximum correlation coefficient for corresponding fault location.

Table 10. Correlation coefficients for different measurement uncertainties.

Cases	Correlation between Exchange Rates and Signatures for Different Measurement Uncertainties				
	$-1\%~\Delta\eta_{comp.}$	$-1\%~\Delta FC_{comp.}$	$-1\%~\Delta\eta_{turb.}$	$-1\%~\Delta FC_{turb.}$	$-1\%~\Delta\epsilon_{rec.}$
P_2–0.1%	**0.996**	0.692	0.776	0.447	−0.750
T_2–0.2%	**0.983**	0.780	0.885	0.453	−0.860
$\dot{m}_1$–0.2%	**0.989**	0.738	0.767	0.615	−0.722
T_3–0.2%	**0.989**	0.620	0.686	0.541	−0.650
P_5–0.01%	**1.000**	0.689	0.784	0.493	−0.754

Eventually, the influence of measurement uncertainties on the fault magnitude detection by using linear regression is examined with the help of sensitivity analysis. The sensor measurements are varied by introducing different level of uncertainties and the corresponding fault magnitudes are calculated for case study $S1$. The result of the analysis is summarized in Figure 8. The most influencing measurement uncertainty in this case corresponds to T_2 and T_3. Despite the high sensitivity for some measurements, the regression can still provide fault magnitude with acceptable accuracy.

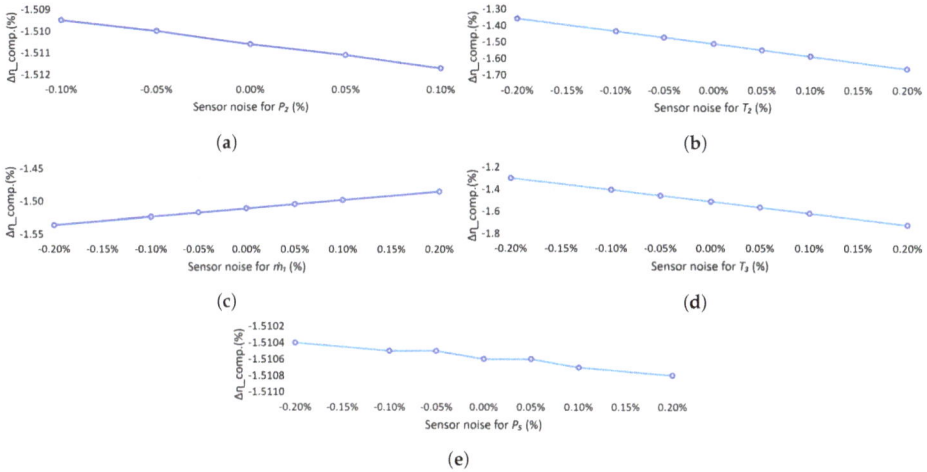

Figure 8. Sensitivity of fault magnitude detection using linear regression against different measurement uncertainties in: (a) P_2; (b) T_2; (c) $\dot{m}_1$; (d) T_3; and (e) P_5.

5. Conclusions

In this article, a multi-layer approach for monitoring and diagnostics of MGT fleet is investigated. A diagnostics-oriented model based on gas path analysis lies at the core of this approach. Subsequently, a model tuning approach is proposed to account for engine to engine variation as a result of production scatter. The two layers of diagnostics approach that are investigated in this paper include: (1) analysis by synthesis (AnSyn); and (2) signature based algorithm to detect the location and magnitude of different component faults. To perform AnSyn, performance deltas corresponding to each component fault are introduced in the physics-based model, where each delta is associated with a measurement of the engine. The performance deltas are then calculated by using real-time measurements from the engine that gives both location and magnitude of the fault. Due to the lack of measurements, not all the faults can be included in the AnSyn step. Moreover, to improve the robustness of the diagnostics approach, a signature based algorithm is applied as the second layer. Fault signatures were generated with the engine model and compared with the simulated faulty engine data. Correlation function and linear regression are applied to get the location and magnitude of the faults, respectively. Finally, results from both layers are merged together. The proposed diagnostics scheme was tested by formulating case studies corresponding to single and multiple faults. Furthermore, sensitivity studies were performed for different measurement uncertainties (i.e., sensor noise and bias) to evaluate the robustness of the scheme against measurement uncertainties. The result shows that the proposed diagnostics approach performs satisfactorily even under measurement uncertainties.

Overall, magnitude of triple faults seems hard to detect by signature based algorithm, given that the number of measurements available for the analysis is limited to five. The accuracy can be improved by including more measurements in the analysis. In case of sensor faults, the corresponding measurements need to be removed from the matching scheme for AnSyn along with their associated performance deltas. This will unavoidably reduce the detectability of the corresponding fault by AnSyn. At the same time, the signature based algorithm might result in reduced accuracy or false alarm.

Author Contributions: M.R. and V.Z. outlined the paper and designed the case studies; M.R. performed the simulations, analysed the results and wrote the paper; X.Z. contributed through EVA code modification; and K.K. and V.Z. supervised the work.

Funding: This research was funded by European Commission under Horizon 2020 programme, grant number 723523.

Acknowledgments: The authors gratefully acknowledge the financial support from European Commission under Horizon 2020 programme, SPIRE-02-2016 through FUDIPO project (http://fudipo.eu/). The authors also thank Mikael Stenfelt of Malardalen University and Mark Oostveen of MTT BV for their continuous support.

Conflicts of Interest: The authors declare no conflict of interest.

Abbreviations

The following abbreviations are used in this manuscript:

MGT   Micro Gas Turbine
COTS   Commercial Of The Shelve
CHP   Combined Heat and Power
EU   European Union
SME   Small and Medium Enterprises
EVA   EnVironmental Assessment
GPA   Gas Path Analysis
FAR   Fuel to Air Ratio
LHV   Lower Heating Value
ISA   International Standard Atmosphere
PPMCC   Pearson Product-Moment Correlation Coefficient

References

1. Zampilli, M.; Bidini, G.; Laranci, P.; D'Amico, M.; Bartocci, P.; Fantozzi, F. Biomass microturbine based EFGT and IPRP cycles: Environmental impact analysis and comparison. In Proceedings of the Asme Turbo Expo: Turbine Technical Conference and Exposition, Charlotte, NC, USA, 26–30 June 2017; Volume 3, pp. 1–9. [CrossRef]
2. European Turbine Network (ETN). *R & D Recommendation Report 2016—For the Next Generation of Gas Turbines*; Technical Report; European Turbine Network (ETN): Brussels, Belgium, 2017.
3. Owens, B. *The Rise of Distributed Power*; Technical Report; General Electric Company: Boston, MA, USA, 2014, doi:10.1093/0195139690.001.0001.
4. European Parliament. *EU Startegy on Heating and Cooling*; 2016/2058(INI); European Parliament: Brussels, Belgium, 2016.
5. European Turbine Network (ETN). *Micro Gas Turbine Technology Summary: Research and Development for European Collaboration*; Technical Report; European Turbine Network (ETN): Brussels, Belgium, 2017.
6. Gopisetty, S.; Treffinger, P. Generic Combined Heat and Power (CHP) Model for the Concept Phase of Energy Planning Process. *Energies* **2016**, *10*, 11. [CrossRef]
7. Rahman, M.; Malmquist, A. Modeling and Simulation of an Externally Fired Micro-Gas Turbine for Standalone Polygeneration Application. *J. Eng. Gas Turbines Power* **2016**, *138*, 112301. [CrossRef]
8. Kim, S.; Chun Kim, K. Performance Analysis of Biogas-Fueled SOFC/MGT Hybrid Power System in Busan, Republic of Korea. *Proceedings* **2018**, *2*, 605. [CrossRef]
9. Kalathakis, C.; Aretakis, N.; Mathioudakis, K. Solar Hybrid Micro Gas Turbine Based on Turbocharger. *Appl. Syst. Innov.* **2018**, *1*, 27. [CrossRef]
10. Kyprianidis, K.G.; Quintero, R.F.C.; Pascovici, D.S.; Ogaji, S.O.T.; Pilidis, P.; Kalfas, A.I. EVA: A Tool for EnVironmental Assessment of Novel Propulsion Cycles. In Proceedings of the ASME Turbo Expo 2008: Power for Land, Sea, and Air, Berlin, Germany, 9–13 June 2008; pp. 547–556.
11. Kyprianidis, K.G.; Dahlquist, E. On the trade-off between aviation NOx and energy efficiency. *Appl. Energy* **2017**, *185*, 1506–1516. [CrossRef]
12. Razak, A.M.Y. *Industrial Gas Turbines: Peformance and Operability*; Woodhead Publishing Limited, Abington Hall: Abington, UK, 2007; p. 9.

13. Kurz, R.; Brun, K.; Wollie, M. Degradation effects on industrial gas turbines. *J. Eng. Gas Turbines Power* **2009**, *131*, 62401. [CrossRef]
14. Kurz, R. Gas Turbine Degradation. In Proceedings of the 43rd Turbomachinery & 30th Pump Users Symposia, Houston, TX, USA, 23–25 September 2014; pp. 1–36.
15. Tahan, M.; Tsoutsanis, E.; Muhammad, M.; Abdul Karim, Z.A. Performance-based health monitoring, diagnostics and prognostics for condition-based maintenance of gas turbines: A review. *Appl. Energy* **2017**, *198*, 122–144. [CrossRef]
16. Li, Y.G. Performance-analysis-based gas turbine diagnostics: A review. *Proc. Inst. Mech. Eng. Part A J. Power Energy* **2002**, *216*, 363–377. [CrossRef]
17. Volponi, A.J.; Doel, D.L.; Provost, M.J.; Grodent, M.; Navez, A.; Mathioudakis, K.; Romessis, C.; Stamatis, A.; Singh, R. *Gas Turbine Condition Monitoring and Fault Diagnostics*; Lecture Series 2003-1; Von Karman Institute for Fluid Dynamics: Brussels, Belgium, 2003.
18. Marinai, L.; Probert, D.; Singh, R. Prospects for aero gas-turbine diagnostics: A review. *Appl. Energy* **2004**, *79*, 109–126. [CrossRef]
19. Stamatis, A.G. Evaluation of gas path analysis methods for gas turbine diagnosis. *J. Mech. Sci. Technol.* **2011**, *25*, 469–477. [CrossRef]
20. Fentaye, A.D.; Gilani, S.I.U.H.; Baheta, A.T. Gas turbine gas path diagnostics: A review. In Proceedings of the 3rd International Conference on Mechanical Engineering Research (ICMER 2015), Kuantan, Malaysia, 18–19 August 2015; Volume 74, p. 5.
21. Urban, L.A. *Gas Turbine Engine Parameter Interrelationships*, 1st ed.; Hamilton Standard Division of United Aircraft Corporation: Windsor Locks, CT, USA, 1967.
22. Urban, L.A. Parameter Selection for Multiple Fault Diagnostics of Gas Turbine Engines. *J. Eng. Power* **1975**, *97*, 225–230. [CrossRef]
23. Urban, L.A.; Volponi, A.J. *Mathematical Methods of Relative Engine Performance Diagnostics*; Technical Paper; Society of Automotive Engineers (SAE): Warrendale, PA, USA, 1992, doi:10.4271/922048.
24. Passalacque, J. Description of automatic gas turbine engine trends diagnostic system. In Proceedings of the 1st National Research Council of Canada (NRCC) Conference on Gas Turbines Operations and Maintenance: Ottawa, ON, Canada, 1974.
25. Stamatis, A.; Mathioudakis, K.; Papailiou, K.; Berios, G. Jet engine fault detection with discrete operating points gas path analysis. *J. Propuls. Power* **1991**, *7*, 1043–1048. [CrossRef]
26. Simani, S.; Patton, R.J.; Daley, S.; Pike, A. Identification and fault diagnosis of an industrial gas turbine prototype model. In Proceedings of the 39th IEEE Conference on Decision and Control, Sydney, Australia, 12–15 December 2000; Volume 3, pp. 2615–2620.
27. House, P. Gas Path Analysis Techniques Applied to a Turboshaft Engine. Ph.D. Thesis, Cranfield University, Cranfield, Bedfordshire, UK, 1992.
28. Esher, P.C. Pythia: An Object-Oriented Gas Path Analysis Computer Program for General Applications. Ph.D. Thesis, Cranfield University, Cranfield, UK, 1995.
29. Stamatis, A.; Mathioudakis, K.; Papailiou, K.D. Adaptive Simulation of Gas Turbine Performance. *J. Eng. Gas Turbines Power* **1990**, *112*, 168–175. [CrossRef]
30. Li, Y.G. Gas Turbine Performance and Health Status Estimation Using Adaptive Gas Path Analysis. *J. Eng. Gas Turbines Power* **2010**, *132*, 41701–41709. [CrossRef]
31. Larsson, E.; Åslund, J.; Frisk, E.; Eriksson, L. Gas Turbine Modeling for Diagnosis and Control. *J. Eng. Gas Turbines Power* **2014**, *136*, 71601–71617. [CrossRef]
32. Liu, Y. Design of Fault Detection System for a Heavy Duty Gas Turbine with State Observer and Tracking Filter. In Proceedings of the ASME Turbo Expo 2017: Turbomachinery Technical Conference and Exposition, Charlotte, NC, USA, 26–30 June 2017; p. V006T05A017.
33. Kang, D.W.; Kim, T.S. Model-based performance diagnostics of heavy-duty gas turbines using compressor map adaptation. *Appl. Energy* **2018**, *212*, 1345–1359. [CrossRef]
34. Koskoletos, O.A.; Aretakis, N.; Alexiou, A.; Romesis, C.; Mathioudakis, K. Evaluation of Aircraft Engine Diagnostic Methods through ProDiMES. In Proceedings of the ASME Turbo Expo 2018: Turbomachinery Technical Conference and Exposition, Oslo, Norway, 11–15 June 2018; p. V006T05A023.

35. Davison, C.R.; Birk, A.M. Automated fault diagnosis of a micro turbine with comparison to a neural network technique. In Proceedings of the ASME Turbo Expo 2006: Power for Land, Sea, and Air, Barcelona, Spain, 8–11 May 2006; pp. 795–804.

36. Davison, C.R.; Birk, A.M. Prediction of time until failure for a micro turbine with unspecified faults. In Proceedings of the ASME Turbo Expo 2006: Power for Land, Sea, and Air, Barcelona, Spain, 8–11 May 2006; pp. 805–814.

37. Mahmood, M.; Martini, A.; Traverso, A.; Bianchi, E. Model Based Diagnostics of AE-T100 Micro Gas Turbine. In Proceedings of the ASME Turbo Expo 2016: Turbomachinery Technical Conference and Exposition, Seoul, Korea, 13–17 June 2016; pp. V006T05A021.

38. Kim, M.J.; Kim, J.H.; Kim, T.S. The effects of internal leakage on the performance of a micro gas turbine. *Appl. Energy* **2018**, *212*, 175–184. [CrossRef]

39. Zaccaria, V.; Stenfelt, M.; Aslanidou, I.; Kyprianidis, K.G. Fleet Monitoring and Diagnostics Framework Based on Digital Twin of Aero-Engines. In Proceedings of the ASME Turbo Expo 2018: Turbomachinery Technical Conference and Exposition, Oslo, Norway, 11–15 June 2018; p. V006T05A021.

40. Ogaji, S.O.T.O. Advanced Gas-Path Fault Diagnostics for Stationary Gas Turbines. Ph.D. Thesis, Cranfield University, Cranfield, UK, 2003.

41. Aslanidou, I.; Zaccaria, V.; Rahman, M.; Oostveen, M.; Olsson, T.; Kyprianidis, K.G. Towards an Integrated Approach for Micro Gas Turbine Fleet Monitoring, Control, and Diagnostics. In Proceedings of the Global Power and Propulsion Society (GPPS) Forum 2018, Zurich, Switzerland, 10–22 January 2018.

42. Badami, M.; Giovanni Ferrero, M.; Portoraro, A. Dynamic parsimonious model and experimental validation of a gas microturbine at part-load conditions. *Appl. Therm. Eng.* **2015**, *75*, 14–23. [CrossRef]

43. Mohtar, H.; Chesse, P.; Chalet, D. Describing uncertainties encountered during laboratory turbocharger compressor tests. *Exp. Tech.* **2012**, *36*, 53–61. [CrossRef]

processes

Article

Dynamic Modeling and Control of an Integrated Reformer-Membrane-Fuel Cell System

Pravin P. S., Ravindra D. Gudi * and Sharad Bhartiya *

Department of Chemical Engineering, Indian Institute of Technology Bombay, Mumbai 400076, India; pravin_ps@iitb.ac.in
* Correspondence: ravigudi@iitb.ac.in (R.D.G.); bhartiya@che.iitb.ac.in (S.B.);
  Tel.: +91-(22)-2576-7231 (R.D.G.); +91-(22)-2576-7225 (S.B.)

Received: 30 July 2018; Accepted: 11 September 2018; Published: 17 September 2018

Abstract: Owing to the pollution free nature, higher efficiency and noise free operation, fuel cells have been identified as ideal energy sources for the future. To avoid direct storage of hydrogen due to safety considerations, storing hydrocarbon fuel such as methane and suitably reforming in situ for hydrogen production offers merit for further investigation. Separating the resulting hydrogen in the reformate using membrane separation can directly feed pure gas to the anode side of fuel cell for power generation. Despite the numerous works reported in literature on the dynamic and steady state modeling and analysis of reformers, membrane separation units and fuel cell systems, there has been limited work on an analysis of the integrated system consisting of all the three components. This study focuses on the mathematical modeling and analysis of the integrated reformer, membrane, fuel cell system from first principles in a dynamic framework. A multi loop control strategy is developed and implemented on the mathematical model of the integrated system in which appropriate controllers based on the system dynamics are designed to examine and study the overall closed loop performance to achieve rapidly fluctuating target power demand and rejection of reformer feed and fuel cell coolant temperature disturbances.

Keywords: auto thermal reformer; palladium membrane hydrogen separation; polymer electrolyte membrane fuel cell (PEMFC); multi-loop control

1. Introduction

The main focus of emergent hydrogen economy for the last few decades has been on the use of hydrogen fuel cells for stationary and portable applications. Almost all practical fuel cells available in the current market use hydrogen or hydrogen-rich hydrocarbons as the fuel [1]. While designing a fuel cell system, the primary challenge of a systems engineer is to make sure that the desired quantity of fuel is delivered at the appropriate time. In the case of fuel cells driven by hydrogen, the prevailing method to deliver hydrogen entails storing of the fuel in appropriately designed high pressure tanks that can withstand high pressures of the order of 350 to 700 bars [1]. Large volume vessels are generally needed for hydrogen storage because of its low volumetric energy density [1]. For automotive applications, while this demanding requirement of large volume might have a lesser impact on heavy duty vehicles, it can be a crucial issue for light duty vehicles. A key issue for portable applications is the non-availability of hydrogen fueling infrastructure for refueling purposes. Furthermore, hydrogen when exposed to air can easily catch fire and can lead to explosion thereby underlining the high risk of direct hydrogen storage. The need to avoid direct storage of hydrogen triggered the development of integrated fuel processing system that converts hydrogen rich hydrocarbons directly to a hydrogen fuel stream. It is well known that hydrogen production by reforming natural gas is an economically viable alternative since infrastructure for natural gas supply already exists.

In the current work, an auto thermal reformer that incorporates both endothermic steam reforming and exothermic partial oxidation is used. Hence, the hydrocarbon fuel viz. natural gas reacts with oxygen from air and steam over a catalyst in a fixed bed reactor at suitable operating conditions to obtain appropriate hydrogen yield along with other gases such as CO, CO_2, O_2, unconverted CH_4, H_2O (steam) and the inert N_2 [2]. The advantage of auto thermal reforming over steam reforming include self-sustained reaction by using the heat generated by the combustion reaction for the prevalent endothermic reactions. Among the methods for separation of gases at the reformer exit, dense palladium membrane has been employed in the current work for separation of H_2 from mixture of other gases [3]. Palladium membranes have the advantage of high hydrogen permeability as well as fast hydrogen absorption and transport kinetics. Another advantage is its excellent thermal stability at an operating temperature range of 300–850 °C [4]. These behaviors are important from a rapid dynamic response viewpoint when used in an integrated system. Finally, a low temperature polymer electrolyte membrane fuel cell (PEMFC) is used for electrical power generation that occurs in response to the electrochemical reaction of hydrogen and air as source of oxygen.

To overcome the deficiencies of direct hydrogen storage, this work explores the potential of an integrated natural gas reformer fuel cell system with membrane-based gas separation/purification using the dynamic models for each. The frequency of start/stop cycles as well as the weight and size of the integrated system are a few of the vital factors to be considered while designing the power supply system [5]. While significant works have been carried out on the dynamic behavior of the individual subsystems viz. reformer, membrane and fuel cell, only a handful of them have directed their research on integration of these subunits [5,6]. The integration of the downstream fuel cell system with the upstream fuel processing system poses several challenges from a dynamics and control viewpoint. Firstly, the response time of a fuel cell is at least an order of magnitude faster than the fuel processing system (auto thermal reformer). Also, the variation in power demands, depending on the duty cycle of the electrical load in the application, needs to be adhered by the fuel cell and this should trigger appropriate decisions at the reformer level. Secondly, due to slow reformer dynamics, lead-like behavior in control mechanisms could be anticipated. Also, the integrated system needs to accommodate line-pack associated aspects for hydrogen. Hence, in this work, we focus on examining the feasibility of the integrated reformer fuel cell system with a membrane separation unit for power applications. Use of hydrocarbon fuels with solid oxide fuel cell via internal reforming have been reported [7,8]. In Ref [9], syngas has been used with molten carbonate fuel cell for power generation. Hydrocarbon as the primary source of hydrogen in PEMFC is challenging since PEMFC cannot tolerate impurities. The authors in Ref [6] discuss a detailed design and control framework of a bio-ethanol steam reformer integrated with PEMFC using Aspen HYSYS simulation software. Protection of the fuel cell stack due to hydrogen/oxygen starvation, higher system efficiency (i.e., maximum ethanol to hydrogen conversion) were a few of the objectives considered by the authors while designing the controllers. The authors in Ref [5] investigated and compared high temperature and low temperature PEMFCs with different fuel processors using glycerol as the fuel. Ref [10] examined the dynamic electrochemical model of a PEMFC along with a methanol reformer and a power conditioning unit. This latter work used a simplified transfer function-based fuel cell model and did not discuss the possibility of integration of the units. Out of the very few works on the integrated reformer fuel cell system reported in the literature, Ref [11,12] explored the dynamic system interaction using liquid methanol as the fuel. However the authors did not consider separation of the hydrogen gas exiting from the reformer output.

The main contribution of this paper is exploring the integrated reformer membrane fuel cell system for dynamic performance under realistic scenarios. An appropriate control structure is designed based on four single loop controllers for overall system operation. The integrated system is evaluated for tracking the target power profile for a representative power application as well as disturbance rejection. The results show the potential of the integrated system for power generation applications and highlight the need for auxiliary power supply during cold startup of the reformer.

The rest of the paper is organized as follows. A brief description of the fuel processing, fuel purification, power generation subsystems as well as the overall integrated system along with its mathematical modeling are discussed in the next Section. In Section 3, we present three case studies: the first being an open loop simulation for start-up of the integrated reformer membrane fuel cell system. Subsequently, the relative gain array analysis is presented to choose the input-output pairings for carrying out control relevant studies. The second case study involves the integrated system used for a generic power application with suitably designed controllers and the final case study evaluates disturbance rejection by considering uncertainty in both the inlet CH_4 feed concentration to the reformer as well as coolant temperature in fuel cell. This is followed by simulation results and discussions. Conclusions are presented in Section 4.

2. System Description

To understand the behavior of the integrated reformer membrane fuel cell system as a whole, detailed dynamic modeling of the individual subsystems is necessary. This will help identify possible challenges for achieving smooth, demand-driven production and delay-free delivery of hydrogen fuel [6]. The basic block diagram showing the integrated system is depicted in Figure 1. The input to the auto thermal reformer consists of appropriate proportion of methane, air and steam thoroughly mixed and preheated to reforming temperatures during start of the reaction. Once the reactions are initiated, the preheater can be removed from the loop as the heat liberated by the exothermic reactions can be further used by the endothermic reactions. The reforming reaction of methane with air and steam results in products such as H_2, CO and CO_2 along with some unreacted CH_4, O_2, N_2 and steam. Two heat exchangers, one at the reformer exit and other at the fuel cell anode inlet side, are used to maintain the gas temperatures at their target values and is assumed to be under perfect control. The gas mixture exiting the reformer is then separated to obtain pure hydrogen by using the palladium membrane based separation unit. The pure hydrogen gas separated from the gas mixtures is then fed to the control valve 2 inlet through a check valve or non-return valve (NRV). The hydrogen is then used by the fuel cell to generate power. Four controllers, one to regulate hydrogen flow at the fuel cell anode inlet side, second to regulate methane flow at the reformer inlet side, third to regulate the coolant flow at the PEMFC coolant circulation side and fourth to regulate the ratio of air to H_2 in PEMFC are implemented to achieve effective operation of the integrated system. The three major components namely, fuel processing subsystem, fuel purification subsystem and power generation subsystem are discussed in detail in the subsequent sections.

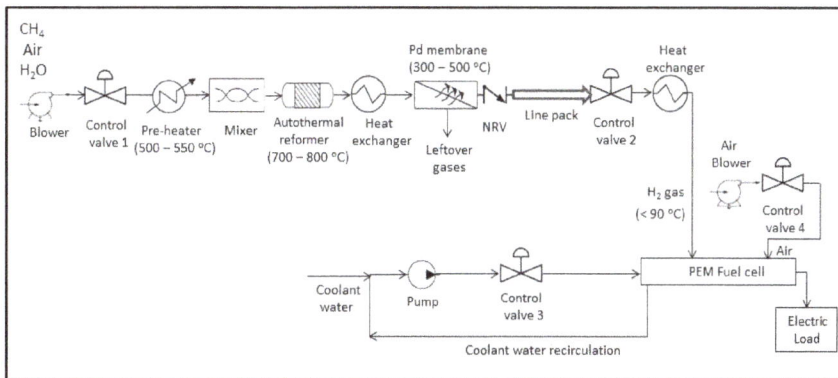

Figure 1. Block diagram showing the integrated reformer membrane fuel cell system. (NRV: non-return valve ; PEM: polymer electrolyte membrane)

2.1. Fuel Processing Subsystem: Auto Thermal Reformer

An auto thermal reformer employing methane as the fuel is chosen as the fuel processing subsystem in the current study. The upstream part of the reactor is dominated by the highly exothermic oxidation reaction whereas endothermic steam reforming reactions dominates the downstream part. A feed stream consisting of steam to carbon molar ratio of 1:1 and oxygen to carbon molar ratio of 0.45:1 is considered. Many reactions such as steam reforming, water gas shift, total combustion, partial oxidation, partial combustion, dry reforming, boudouard reaction and decomposition reaction are likely to occur in an auto thermal reformer [2]. Among this possible set of reactions, only the dominant reactions with significant rates are considered in this study in order to reduce the complexity of the mathematical model. Reactions in Equations (1) and (2) represent the steam reforming reaction while Equations (3) and (4) represent the water gas shift and total combustion reactions respectively [2].

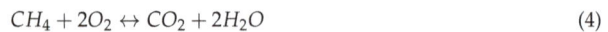

$$CH_4 + H_2O \leftrightarrow CO + 3H_2 \tag{1}$$

$$CH_4 + 2H_2O \leftrightarrow CO_2 + 4H_2 \tag{2}$$

$$CO + H_2O \leftrightarrow CO_2 + H_2 \tag{3}$$

$$CH_4 + 2O_2 \leftrightarrow CO_2 + 2H_2O \tag{4}$$

Mathematical Model of Auto Thermal Reformer

Auto thermal reforming has been widely studied and experimental validation has been reported in the literature [2,13–15]. The authors in [13] performed a steady state analysis of hydrogen production from methane using an auto thermal reformer with a dual catalyst bed configuration, but did not discuss the dynamic behavior of the system. Extensive experiments were conducted on an in-house developed auto thermal reformer by the authors in [14] and the experimental results were used to validate the mathematical model developed. Ref [15] analyzed the steady state modeling aspects of a miniaturized methanol reformer for fuel cell powered mobile applications but did not study the dynamic behavior of the system. In this work, a one-dimensional dynamic model of the auto thermal reformer is selected [2]. Energy and mass balances are applied both in gas and solid phases to obtain the mathematical model of the reformer. The reformer consists of a cylindrical reactor of 0.2 m in length and with Nickel as the catalyst having a density of 1870 kg m^{-3}. Both the gas feed as well as catalyst temperature is maintained at a value of 542 °C. The reformer operating pressure is chosen to be 1.5 atm with a Gas Hourly Space Velocity (GHSV) of 3071 h^{-1} implying a residence time of 1.17 s and a gas mass flow velocity of 0.15 kg m^{-2} s^{-1}. The schematic showing the flow diagram of the modeled region in an auto thermal reformer is shown in Figure 2. The auto thermal reformer model involves seven main species (CH_4, O_2, CO, CO_2, H_2, H_2O) including one inert component (N_2 from air). For more particulars on the operating conditions, kinetic parameters, constants, kinetic reaction model and gas properties examined in the model, the reader is referred to Ref [2].

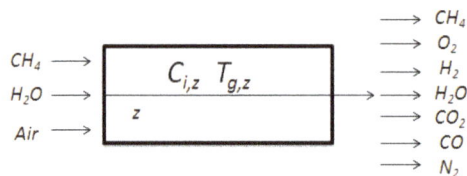

Figure 2. Flow diagram of modeled region in an auto thermal reformer.

While modeling the reformer subsystem, it is assumed that the gas behavior is ideal and the reforming operation is adiabatic in nature. Thermal dispersion in the axial direction is modeled and negligible radial gradients are assumed. Temperature gradient in the catalyst particles is also ignored assuming the particle size and bed porosity to be uniform.

Mass balance for each species in the gas phase is given by:

$$\epsilon_b \frac{\partial C_i}{\partial t} + \frac{\partial (uC_i)}{\partial z} + k_{g,i}a_v(C_i - C_{i,s}) = \epsilon_b D_z \frac{\partial^2 C_i}{\partial z^2} \tag{5}$$

$$i = CH_4, O_2, CO, CO_2, H_2, H_2O, N_2$$

Mass balance for the solid phase is given by:

$$k_{g,i}a_v(C_i - C_{i,s}) = (1 - \epsilon_b)\rho_{cat}r_i \tag{6}$$

The first term on the left hand side of Equation (5) represents the mass accumulation term of each component while the second term indicates convective diffusion. Mass transfer from gas to solid is represented by the third term and is evaluated using Equation (6). The term on the right hand side of Equation (5) gives the axial dispersion of components.

Energy balance on the gas phase is given by:

$$\epsilon_b \rho_f C_{pg} \frac{\partial T_g}{\partial t} + u\rho_f C_{pg} \frac{\partial T_g}{\partial z} = h_f a_v(T_s - T_g) + \lambda_z^f \frac{\partial^2 T_g}{\partial z^2} \tag{7}$$

The first term on the left hand side of Equation (7) represents the heat accumulation term while the second term indicates convective heat transfer. The first term on the right hand side represents the heat transfer from gas to solid and the second term indicates the conductive heat transfer.

Energy balance on the solid phase is given by:

$$\rho_{bed} C_{p,bed} \frac{\partial T_s}{\partial t} + h_f a_v(T_s - T_g) =$$
$$\rho_{cat}(1 - \epsilon_b) \sum -\Delta H_{rxn,j} \eta_j R_j \tag{8}$$

The term on the right hand side of Equation (8) involves the heat of reaction term multiplied by the catalyst density and packed bed porosity.

The pressure drop across the reformer bed is given by Ergun equation and can be expressed as:

$$\frac{\partial P_g}{\partial z} = -K_D u - K_V u^2 \tag{9}$$

Boundary Conditions are as follows:

At the reformer inlet $z = 0$

$$C_i = C_{i,o}; \quad T_g = T_{g,o}; \quad T_s = T_{s,o}; \quad P_g = P_{g,o} \tag{10}$$

At the reformer exit $z = L$

$$\frac{\partial C_i}{\partial z} = 0; \quad \frac{\partial T_g}{\partial z} = 0; \quad \frac{\partial T_s}{\partial z} = 0 \tag{11}$$

Initial conditions are as follows:

$$C_i = C_{i,o}; \quad T_g = T_{g,o}; \quad T_s = T_{s,o} \tag{12}$$

The reformer model equations Equations (5)–(12) provide a dynamic description of the concentration of six species taking part in the reforming reaction along with the temperature of the gas and solid phase.

2.2. Fuel Purification Subsystem: Palladium Based Membrane Separation

Efficient operation of a low temperature PEMFC inevitably demands purified hydrogen gas at the anode inlet. The efficiency and lifetime of the fuel cell strongly depends on the rate of poisoning of the catalyst due to the presence of impurities in the feed. Palladium alloy-based membrane separation is reported to be an attractive method for the purification of hydrogen gas from mixture of other gases [16]. Another alternative method for hydrogen gas purification is the pressure swing adsorption (PSA) that works according to the species molecular characteristics and affinity for an adsorbent material [17,18]. However due to size constraints, this method may not be suitable for portable applications and hence not considered for this particular study. The authors in [19] has developed a mathematical model of the palladium-based membrane separation for hydrogen gas separation and has experimentally validated the model. Irrespective of whether the separation is required for portable or stationary applications, palladium-based membrane separation can be efficiently exploited for the gas purification process. Palladium membrane has the added advantage of having low permeability for other gases compared to hydrogen gas which effectively qualifies it for gas separation/purification [3,4].

Mathematical Model of Palladium Membrane Separation

According to Sievert's law [16], permeation of gas occurs through the palladium membrane based on the difference in partial pressures of hydrogen at both sides of the membrane. The molar flow rate of hydrogen permeating through the palladium membrane according to Sievert's law is given by [16]

$$M_{H_2, perm} = (1 - \theta_{mem}) \left(\frac{A_{mem}}{L_{mem}} \right) P_e [C_{H_2}^n - C_{H_2, m}^n] (RT_{gr})^n \tag{13}$$

Variable θ_{mem} stands for the membrane poisoning caused by the adsorption of CO on the membrane surface which can partially prevent or obstruct the hydrogen permeation and $(1 - \theta_{mem})$ denotes the fraction of membrane actually available for the hydrogen gas separation. The membrane permeability P_e is a function of temperature which can be represented by the Arrhenius law as

$$P_e = P_{e_0} exp \left(\frac{-E_a}{RT_{gr}} \right) \tag{14}$$

where P_{e_0}, E_a, R and T_{gr} represents permeability pre-exponential factor, activation energy for hydrogen permeability, gas constant and gas temperature at the exit of heat exchanger 1 respectively. More hydrogen crossover can be facilitated than the predicted behavior by the Sievert's law due to the presence of defects in the membrane layers. However, due to this, other gas species can also pass across the membrane allowing the overall flux to be the sum of hydrogen flux given by Sievert's law and the flux of other gases through the defects. This aspect has not been pursued in the present work.

2.3. Power Generation Subsystem: PEMFC

Hydrogen gas exiting from the fuel processing subsystem after undergoing necessary purification is fed to the anode side of the PEMFC to generate electricity. Using a blower, air from the atmosphere is fed to the cathode side of the PEMFC. In this study, hydrogen is considered to be the limiting reagent while air is considered as the excess reagent. Although the stoichiometric ratio of the inlet molar flow rate of hydrogen to oxygen is 2, the molar flow rate of air is chosen to be always 5 times greater than the molar flow rate of hydrogen fed to the fuel cell. This is achieved by employing a ratio controller that senses the flow rate of hydrogen and adjusts the control valve 4 position to allow requisite flow of air to enter the fuel cell cathode. As soon as the reactants are directed to the active catalyst sites through the fuel cell electrodes, electrochemical reactions initiate and load current is generated. The electrochemical reactions that occur in a PEMFC are shown in Table 1. A single cell has an open circuit voltage of around 1 V and under load, it drops down to 0.6 to 0.7 V. Real world applications demand electricity of the order of several tens or hundreds of volts. To achieve these high voltage values, individual fuel cells are stacked in series so that the aggregate voltage fits any specified voltage requirement. To maintain acceptable conductivity, the polymer membrane requires to be hydrated with water. For achieving this, the fuel cell must operate at temperatures below 90 °C.

Table 1. Electrochemical reactions in PEMFC (PEMFC: polymer electrolyte membrane fuel cell)

Electrode	Reactions
Anode	$H_2 \leftrightarrow 2H^+ + 2e^-$
Cathode	$\frac{1}{2}O_2 + 2H^+ + 2e^- \leftrightarrow H_2O$
Overall	$H_2 + \frac{1}{2}O_2 \leftrightarrow H_2O$

Mathematical Model of PEMFC

Numerous studies in fuel cell dynamics have been reported in the literature [5,20,21]. The study of the dynamic operation of a fuel cell is crucial for both portable and stationary applications, where power demand varies, and the fuel cell does not generally operate at the designed optimal steady-state [20]. Considerable amount of work using fuel cell system dynamics have been reported in which authors in Ref [5] studied the case of a high temperature PEMFC with and without a water gas shift reactor used for carbon monoxide removal. In Ref [21], the dynamics of the fuel cell using an air compressor as source for oxygen/air has been studied by assuming direct storage of hydrogen. The model equations for PEMFC considered in this paper involves an along the channel model adopted from Ref [20]. The schematic showing the flow diagram of the modeled region in a PEMFC is shown in Figure 3 [22]. The mathematical model includes set of ordinary differential equations comprising of consumption equations for each species (H_2 at the anode and air at the cathode), balance of water and temperature of gas at both the anode and cathode side as well as the coolant temperature. The temperature variation of the polymer membrane consists of a partial differential equation while the Nernst equation, which is an algebraic equation, describes the voltage obtained from the PEMFC.

For more details on the amount of water molecules transported by each proton, membrane conductivity, various assumptions and parameter values used in the model, the reader is referred to the references [20,23]. Apart from the energy balance equation, all the remaining equations are considered to be at quasi steady state.

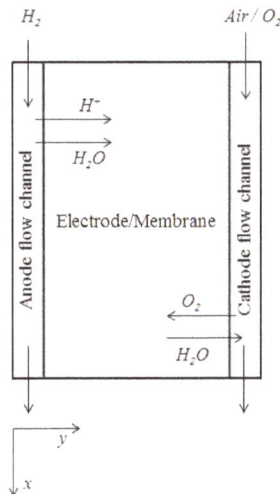

Figure 3. Flow diagram of modeled region in a PEMFC (PEMFC: polymer electrolyte membrane fuel cell) [22].

Equations for consumption of a single phase species i along the channel length is given by

$$\frac{dM_i(x)}{dx} = -hN_i(x) \tag{15}$$

$$i = H_2, O_2, N_2$$

where,

$$N_{H_2}(x) = \frac{I(x)}{2F} \quad N_{O_2}(x) = \frac{I(x)}{4F} \quad N_{N_2}(x) = 0 \tag{16}$$

Balance of water in liquid form is influenced by the effect of evaporation and condensation which is given by

$$\frac{dM_{w,k}^l(x)}{dx} = \frac{k_c hd}{RT_k(x)} \left\{ \frac{M_{w,k}^v(x)}{\sum_i^{N_k} M_{i,k}} P_k - P_w^{sat}(T_k) \right\} \tag{17}$$

$$k = a, c$$

It is significant to note that Equation (17) is used only when liquid water exists at the anode or cathode and/or partial pressure of water in vapour form exceeds its saturation pressure.

At the anode side, equation for water vapor balance is given by

$$\frac{dM_{w,a}^v(x)}{dx} = -\frac{dM_{w,a}^l(x)}{dx} - \frac{h\alpha(x)}{F} I(x) \tag{18}$$

The first term on the right hand side of Equation (18) represents the evaporation/condensation rate of liquid water while the second term indicates the net amount of water migrated across the membrane.

At the cathode side, equation for water vapor balance is given by

$$\frac{dM_{w,c}^v(x)}{dx} = -\frac{dM_{w,c}^l(x)}{dx} + \frac{h\alpha(x)}{F} I(x) + \frac{h}{2F} I(x) \tag{19}$$

The electrochemical reaction at the cathode side of the fuel cell generates some amount of water vapor and is represented by the last term of Equation (19). Appropriate heat removal and humidification are essential for the membrane to be suitably hydrated and conductive which in turn lowers the ohmic losses in the cell. The humidification of membrane is characterised by the term alpha (α) given by Equation (20) which denotes the ratio of water molecules per proton. Equations (18) and (19) contain the term α which indicates the net amount of water migrated across the polymer membrane. Equation (21) indicates the effect of humidity or water content on the conductivity of the membrane. Reduction in membrane humidity can in turn reduce the membrane conductivity and can further cause a reduction in the cell voltage (see Equation (31)). The ratio of water molecules per proton, α, is expressed as

$$\alpha = n_d - \frac{F}{I(x)} D \frac{dc_w}{dy} - c_w \frac{k_p}{\mu} \frac{F}{I(x)} \frac{dp_w}{dy} \tag{20}$$

The expression for conductivity of the membrane is given as

$$\sigma_m = \left(0.00514 \frac{M_{m,dry}}{\rho_{m,dry}} c_{w,a} - 0.00326 \right) exp \left(\frac{1}{303} - \frac{1}{T_s} \right) \tag{21}$$

The expressions for $c_{w,a}$ and $c_{w,c}$ are given by

$$c_{w,k} = \frac{\rho_{m,dry}}{M_{m,dry}}(0.043 + 17.8a_k - 39.8a_k^2 + 36.0a_k^3) \quad \text{for } a_k \leq 1 \tag{22}$$

$$c_{w,k} = \frac{\rho_{m,dry}}{M_{m,dry}}[(14 + 1.4(a_k - 1))] \quad \text{for } a_k > 1 \tag{23}$$

$$k = a, c$$

The activities of water in the anode and cathode are defined as,

$$a_a = \frac{x_{w,a}P}{P_{w,a}^{sat}} = \left(\frac{M_{w,a}^v}{M_{w,a}^v + M_{H_2}}\right)\frac{P}{P_{w,a}^{sat}} \tag{24}$$

$$a_c = \frac{x_{w,c}P}{P_{w,c}^{sat}} = \left(\frac{M_{w,c}^v}{M_{w,c}^v + M_{O_2} + M_{N_2}}\right)\frac{P}{P_{w,c}^{sat}} \tag{25}$$

The equation for the variation in anode and cathode gas temperatures are given by

$$\frac{dT_k(x)}{dx} = \frac{U_g A_g\{T_{fc}(x) - T_k(x)\}}{\sum_i C_{p,i}M_i(x)}, \quad k = a, c \tag{26}$$

The variation in temperature of coolant circulated is expressed as

$$\frac{dT_{cool}(x)}{dx} = \frac{U_w A_{cool}\{T_{fc}(x) - T_{cool}(x)\}}{C_{p,w}M_{cool}} \tag{27}$$

The spatial and time dependency of solid temperature is determined by the energy balance equation given by

$$\begin{aligned}
\rho_s C_{p,s}\frac{\partial T_{fc}}{\partial t} &= k_s\frac{\partial^2 T_{fc}}{\partial x^2} + \frac{U_g A_g}{f}(T_a + T_c - 2T_{fc}) \\
&+ \frac{U_w A_{cool}}{f}(T_{cool} - T_{fc}) - \frac{e}{f}\left(\frac{\Delta H}{2F} + V_{cell}\right)I(x) \\
&+ \frac{\Delta H_{vap}(T_{fc})}{f}\left(\frac{dM_{w,a}^l(x)}{dx} + \frac{dM_{w,c}^l(x)}{dx}\right)
\end{aligned} \tag{28}$$

The first term on the right hand side of Equation (28) represents the heat transfer by conduction, second term indicates the heat transfer to the fuel and oxidant flows, third term gives the heat transfer to the coolant channels, fourth term represents the heat generation by the reactions and the last term indicates the heat of evaporation/condensation.

The boundary conditions given by Equation (29) indicates the heat energy lost to the surroundings from both the edges of the membrane by convection.

$$\begin{aligned}
k_s\frac{\partial T_{fc}}{\partial x}\Big|_{x=0} &= U_c(T_{fc} - T_{inf}) \\
k_s\frac{\partial T_{fc}}{\partial x}\Big|_{x=l} &= -U_c(T_{fc} - T_{inf})
\end{aligned} \tag{29}$$

The cell voltage attained from the PEMFC given by the Nernst equation is expressed as

$$V_{cell} = V_{oc} - \eta(x) - \frac{I(x)t_m}{\sigma_m(x)} \tag{30}$$

The overall output voltage obtained from the PEMFC considering the losses is given by

$$V_{cell} = V_{oc}^0 + \frac{RT_{fc}}{n_e F} \ln\left(\frac{P_{H_2}P_{O_2}^{0.5}}{P_{H_2O}}\right) - \frac{RT_{fc}}{F} \ln\left(\frac{I(x)}{i_o P_{O_2}(x)}\right)$$
$$- \frac{I(x)t_m}{\sigma_m(x)} \tag{31}$$

The third term of Equation (31) represents the activation over potential and last term represents the ohmic over potential. Concentration over potential is not taken into account in this particular mathematical model.

2.4. Integrated Reformer Membrane Fuel Cell System

The integrated system essentially comprises of a combination of the three subsystems viz. reformer, membrane and fuel cell to form a single unit. Present work discusses the detailed mathematical modeling and control studies of the integrated system with a multi-loop control strategy along with a few case studies. As can be noticed from Figure 1, control valve 1 is installed at the inlet side of the reformer to facilitate flow of stored hydrocarbon fuel to the reactor. A blower is installed ahead of the control valve 1 in order to maintain the feed side upstream pressure greater than the downstream pressure. The molar flow rate of gas species through the control valve 1 is given by [24].

$$M_{CH_4} = \frac{94.8 F_p C_{v1} P_{in} Y f_1}{3600 W_{CH_4}} \sqrt{\frac{X_1 W_{CH_4}}{T_{in} Z}} \tag{32}$$

where $f_1 \in [0, 1]$, Y and Z represents the control valve 1 opening, expansion and compressibility factors respectively. X_1 is the ratio of pressure drop across the control valve 1 to the absolute inlet pressure given by

$$X_1 = \frac{P_{in} - P_g}{P_{in}} \tag{33}$$

The velocity u of gas species entering the reformer is a function of the molar flow rate according to the equation given by

$$u = \frac{M_{CH_4} W_{CH_4}}{\rho_{CH_4} A_1} \tag{34}$$

Furthermore as seen from Figure 1, control valve 2 is installed at the anode inlet side of the PEMFC to facilitate flow of hydrogen gas to the anode side.

$$V\frac{dC_{H_2,m}}{dt} = M_{H_2,perm} - M_{H_2}(x = 0) \tag{35}$$

where, $M_{H_2}(x = 0)$ is given by [24]

$$M_{H_2}(x = 0) = \frac{94.8 F_p C_{v2} P_m Y f_2}{3600 W_{H_2}} \sqrt{\frac{X_2 W_{H_2}}{T_{gr} Z}} \tag{36}$$

where, pressure in the line pack P_m is calculated based on the ideal gas law given by

$$P_m = C_{H_2,m} R T_{gr} \tag{37}$$

X_2 is the ratio of pressure drop across the control valve 2 to the absolute inlet pressure given by

$$X_2 = \frac{P_m - P_a}{P_m} \tag{38}$$

Pressure drop across the lines are neglected in this study.

A dynamic mathematical model for the variation in molar concentration of hydrogen in the line pack before the control valve 2 is obtained by applying simple mass balance as can be seen in Equation (35). The change in moles of hydrogen in the line pack is given by $C_{H_2,m}$. $M_{H_2,perm}$ represents the molar flow rate of hydrogen permeating through the membrane and $M_{H_2}(x = 0)$ depicts the molar flow rate of hydrogen to the PEMFC anode inlet through control valve 2. The reformer is considered cylindrical in shape with a volume of 5.65×10^{-4} m^3. The palladium membrane considered is having a volume of 6.4×10^{-7} m^3. The volume occupied by the fuel cell stack containing 375 cells is calculated to be 0.01547 m^3. The line pack connecting pipe is considered to have a volume of 2.5×10^{-4} m^3. Considering all these volumes, the total volume comes to around 0.016 m^3. Assuming the total volume occupied by all the other ancillary components like the control valves, heat exchangers, mixers etc. to be around 0.02 m^3, then the total volume of the integrated power system would be approximately equal to 0.036 m^3.

The third loop involves control valve 3 which facilitates the circulation of coolant to and from the fuel cell in order to maintain the average solid temperature of the PEMFC at a pre-defined acceptable value. The spatial and time dependence of solid temperature of the PEMFC is given by a partial differential equation already discussed in Equation (28). Fourth loop involves a ratio controller that manipulates the control valve 4 position based on the hydrogen flow rate, in order to feed appropriate amount of air to the fuel cell cathode through an air blower.

2.5. Numerical Solution of the Integrated System

In the case of an auto thermal reformer, the mathematical model consists of set of partial differential equations for species concentrations and temperatures (Equations (5)–(12)). Considering seven gas species concentrations along with gas and solid temperatures, the resulting set of partial differential equations are converted to a set of nine ordinary differential equations using method of lines that discretize the spatial derivatives by finite differencing. In the case of palladium membrane, the molar flow rate of hydrogen permeating through the membrane is expressed by an algebraic equation (Equation (13)) while the change in molar concentration of H_2 in the line pack is given by an ordinary differential equation (Equation (35)). The mathematical model for PEMFC involves mass balances given by a set of ordinary differential equations (Equations (15)–(27), energy balance described by a partial differential equation (Equation (28)) and the output voltage given by an algebraic equation (Equation (31)). The partial differential equation is converted to set of ordinary differential equations using method of lines. Since the mass balance equations in PEMFC consist of quasi steady state equations, but involve time dependent quantities, they are solved separately along the space after time simulation of the partial differential equations over short time interval of 0.1 s. The reformer length is divided into a uniform grid of 100 intervals while the PEMFC flow channel is divided into a uniform grid of 400 intervals. The data used for base case and the parameters used for simulation of the integrated system are given in Appendix A and B respectively.

Using the ode15s differential equation solver available in MATLAB environment 2014b, the resulting sets of equations are solved simultaneously to study the dynamic and steady state behavior of the integrated system.

2.6. Controllability Analysis and Choice of Pairing

The ability to maintain the process outputs within specified bounds by appropriately adjusting the manipulated variables characterizes the controllability of a process [25]. To perform the controllability analysis of the integrated system, the dynamic model of the system is considered with three inputs:

molar flow rate of CH_4 to the reformer inlet through control valve 1 (u_1), molar flow rate of H_2 to the fuel cell anode through control valve 2 (u_2) and molar flow rate of coolant to the fuel cell through control valve 3 (u_3). The molar concentration of H_2 in the line pack (or H_2 pressure in the line pack assuming ideal gas law) (y_1), power demanded by the fuel cell output load (y_2) and average solid temperature of the fuel cell (y_3) are selected as outputs. Relative gain array (RGA) is used to select the input-output pairings and to study their interactions [25].

Remark 1. *Since the dynamics of PEMFC are significantly faster than the reformer dynamics, it takes considerable amount of time for the fuel processing subsystem during cold start to produce hydrogen to run the fuel cell thereby making the overall system dynamics much slower [12]. Hence, to ensure acceptable performance of the output electrical load powered by the integrated system, backup power sources like batteries or super capacitors that have a much faster response time is desirable. The choice of such relevant ancillary power sources and their impact on the overall system dynamics will be a subject matter of a future paper.*

Remark 2. *In the designed system, although a small amount of hydrogen inventory still needs to be maintained in the line pack to meet transients, the hydrogen stored is in gaseous form and is in relatively small amounts. This is a safer storage option in contrast with the case where liquid / gaseous hydrogen is directly stored in large volumes and high pressure composite vessels for vehicular fuel cell applications.*

3. Case Studies: Dynamic Analysis and Control of the Integrated System

3.1. Case Study 1: Open Loop Simulation for Start-up of Integrated Reformer-Membrane-Fuel Cell System

Without incorporating controllers into the loop, an open loop start-up simulation is carried out first to analyze the dynamic nature of the integrated reformer membrane fuel cell system. The flow rate of input fuel species to the reformer is determined by the percentage opening of the control valve 1 which is maintained at a fixed position of 50% throughout the simulation. As already discussed in the case of auto thermal reformer, a steam to carbon molar ratio of 1:1 and oxygen to carbon molar ratio of 0.45:1 is considered for the simulation studies. This ensures maximum methane conversion thereby increasing the hydrogen yield at the reformer output. It can also be seen that the percentage conversion of methane is around 99.58% revealing that the hydrocarbon fuel is almost completely converted into useful energy. During start-up of the plant, it is assumed that the gas pipeline connecting the NRV to the control valve 2 is already filled with some amount of hydrogen so as to maintain a line pack pressure of 1 atm. It is also assumed that the fuel cell anode and cathode sides are completely purged with inert gas making the initial concentration of gas species present to be ideally zero. The steady state molar concentration profiles of the species in the reformer after a time period of 2000 s are shown in Figure 4a where the molar concentration of hydrogen increases along the reformer length as the reaction progresses and settles to a value of 21.5 mol m^{-3}. The change in molar concentration of water along the reformer length at steady state can also be seen in Figure 4a. It can be inferred that the molar concentration of water starts decreasing near the inlet due to the consumption of water in the steam reforming and water gas shift reactions. However, at the reformer exit, water is seen to be slightly increasing. In the water gas shift reaction, CO reacts with steam to produce CO_2 and H_2. When larger quantities of H_2 and CO_2 are available, the equilibrium shifts in favour of the reverse reaction resulting in an increase in quantity of water along the reformer length (see Figure 4a). It is assumed that the inlet gas feed temperature to the reformer as well as the initial catalyst temperature is maintained at a value of 815 K during the start and throughout the simulation. Because of the relatively low operating pressure of 1.5 atm and a gas hourly space velocity (GHSV) of 0.15 kg m^{-2} s^{-1}, the oxidation reaction is considerably slower than the endothermic steam reforming reactions. Due to this, the gas and catalyst temperatures decrease during the initial part of the reformer as shown in Figure 4b. Once the reactions picks up, the temperature of both gas as well as catalyst increases sharply. The gas temperature profile is found to be nearly same as the catalyst temperature profile, because of the effective transport of heat from the catalyst to the gas during the operation.

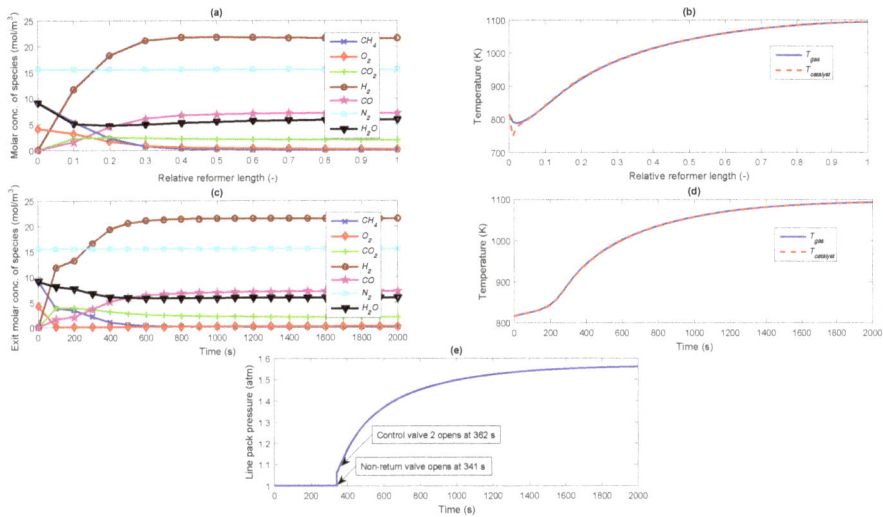

Figure 4. (**a**) Steady state profile for molar concentration of species; (**b**) Steady state profile for gas and catalyst temperatures; (**c**) Dynamic profile for molar concentration of species at reformer exit; (**d**) Dynamic profile for gas and catalyst temperatures at reformer exit; (**e**) Pressure in the line pack.

The dynamic profiles of the exit molar concentration of each species taking part in the reforming reactions are illustrated in Figure 4c. It can be noticed from the profile that the molar concentrations of the species settle to their steady state values in approximately 600 s. The exit molar concentration of H_2 and CO increase while that for all the other species decrease with time as they are consumed in the reforming reactions. N_2, which is an inert component does not take part in any of the reactions and thus exhibits a constant profile. As the exothermic reactions dominates the endothermic reactions, the gas temperature and catalyst temperature profiles at the reformer exit are found to be increasing with time as seen in Figure 4d and settle at a value of approximately 1090 K.

As already discussed, before the plant start-up, some amount of hydrogen is assumed to be present between the pipeline connecting NRV to the control valve 2 to maintain a pressure of 1 atm. The NRV after the membrane separation unit (see Figure 1) restricts the flow of hydrogen further from the reformer exit through the membrane to the control valve 2 inlet until the upstream pressure (membrane permeate side) exceeds the downstream pressure (line pack pressure maintained at 1 atm initially). As can be seen from Figure 4e, the NRV remains closed till a time period of 341 s, after which it opens to allow flow of hydrogen to build up the line pack pressure beyond 1 atm. As can be seen from Figure 4e, the plant is designed in such a manner that as the line pack pressure increases and becomes 1.1 times greater than the downstream pressure (1 atm), the control valve 2 is opened to 50% (at 362 s as seen from Figure 4e). The initially closed control valve 2 allows the molar concentration of H_2 in the line pack to build up (in turn causing a build up in the line pack pressure) as shown in Figure 5a. In response to the sudden opening of the control valve 2, the profiles for line pack pressure and molar concentration of H_2 in the line pack starts increasing as can be seen from Figures 4e and 5a respectively.

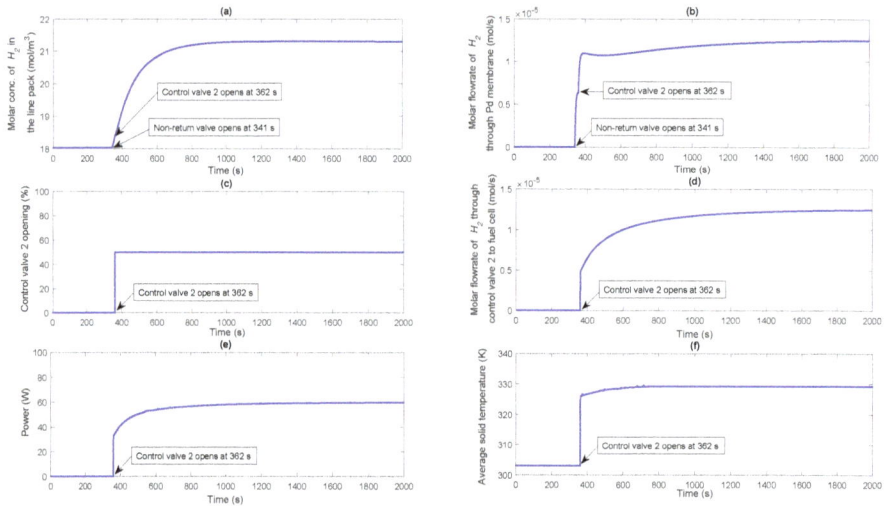

Figure 5. (**a**) Molar concentration of H_2 in the line pack; (**b**) Molar flow rate of H_2 through Pd membrane; (**c**) Percentage opening of the control valve 2; (**d**) Molar flow rate of H_2 through control valve 2 to fuel cell; (**e**) Power generated by a single fuel cell; (**f**) Average solid temperature of fuel cell.

According to the Sievert's law given by Equation (13), the molar flow rate of hydrogen permeating through the palladium membrane represented by $M_{H_2,perm}$ increases as soon as the NRV opens at 341 s (see Figure 5b). At 362 s, when control valve 2 opens, a small disturbance is expected in the molar flow rate of hydrogen as can be understood from Sievert's law given by Equation (13). In response to opening of the control valve to 50% at 362 s (see Figure 5c), pressurized H_2 from the line pack starts flowing through the control valve 2 to the fuel cell anode, the molar flow rate of which can be seen from Figure 5d. As the molar flow rate of H_2 to the fuel cell anode increases in response to opening of the control valve 2, the electrochemical reactions commence, which in turn increases the power generated by the fuel cell as shown in Figure 5e. From the start up simulation, it can be observed that the dynamics of reformer-membrane-fuel cell is sluggish. This is because of lack of availability of H_2 for 362 s, when the control valve 2 opens to 50% once the line pack pressure exceeds 1.1 times the downstream fuel cell pressure. However, as can be seen from Figure 5b,d, the flow rate from the membrane to the line pack limits the flow rate of H_2 to the fuel cell. Thus, the fuel cell dynamics are slower due to the limitation of the hydrogen produced by the reformer. The dynamics associated with shut-down of the integrated system has not been studied in the current work. The mathematical modeling of the ancillary components like heaters, mixers etc. are not taken into account for this particular study for simplification. Due to the dominating exothermic reactions occurring in the auto thermal reformer, there is always a possibility for cogeneration which can enhance the overall efficiency of the plant. The heat extracted by the coolant from the PEM fuel cell can also be used for heat transfer to the fuel. This will be a subject matter for our future work. The average solid temperature of the fuel cell increases and settles to a value of approximately 330 K as can be seen from Figure 5f, which is based on a constant coolant circulation flow rate of 0.055 mol s^{-1}.

Relative Gain Array (RGA) Analysis

The controlled variables chosen in this particular study are the molar concentration of H_2 in the line pack (or H_2 pressure in the line pack assuming ideal gas law) (y_1), power demanded by the output electrical load (y_2) and average solid temperature of the fuel cell (y_3). For achieving satisfactory operation of the integrated reformer membrane fuel cell assembly, the available manipulated variables selected are the molar flow rate of CH_4 to the reformer inlet through control valve 1 (u_1), molar flow rate of H_2 to the fuel cell anode through control valve 2 (u_2) and molar flow rate of coolant to the fuel cell through control valve 3 (u_3). For designing appropriate single input single output (SISO) controllers to form a multi-loop control strategy, suitable input-output pairings need to be decided for which the RGA analysis is carried out. By performing step changes in the manipulated variables and observing the changes in output variables of the integrated system, steady state model of the system in deviation form is obtained as,

$$\begin{bmatrix} \Delta y_1 \\ \Delta y_2 \\ \Delta y_3 \end{bmatrix} = \begin{bmatrix} 1.75 & 0.35 & 0 \\ 0.225 & 1 & 0.13 \\ 0.789 & 1.764 & -0.45 \end{bmatrix} \begin{bmatrix} \Delta u_1 \\ \Delta u_2 \\ \Delta u_3 \end{bmatrix} \tag{39}$$

Given the steady state gain matrix K of the system, the steady-state RGA matrix for the selected inputs and outputs can be obtained by Equation (40).

$$(RGA)_{y_1,y_2,y_3,u_1,u_2,u_3} = K \otimes (K^{-1})^T = \begin{bmatrix} 1.0638 & -0.0638 & 0 \\ -0.0317 & 0.7047 & 0.3270 \\ -0.0321 & 0.3591 & 0.6730 \end{bmatrix} \tag{40}$$

It can be noticed that the RGA for diagonal pairings $(u_1 - y_1)$, $(u_2 - y_2)$ and $(u_3 - y_3)$ are 1.0638, 0.7047 and 0.6730 respectively which is closer to 1 indicating stronger interaction. This pairing seems appropriate from a process point of view because the molar flow rate of CH_4 to the reformer inlet through control valve 1 (u_1) has a direct effect on the amount of H_2 produced at the reformer output which in-turn has an effect on the molar concentration of H_2 in the line pack (or H_2 pressure in the line pack assuming ideal gas law) (y_1). Similarly, the molar flow rate of H_2 to fuel cell anode through control valve 2 (u_2) has a direct effect on the power demanded by the output electrical load (y_2). Also, the molar flow rate of coolant circulated through the fuel cell (u_3) has a direct effect on the average solid temperature of fuel cell (y_3). As can be seen from the steady-state RGA matrix, the RGA for off-diagonal pairings are negative and/or are far way from 1 indicating lesser interaction. A process flow diagram with selected input-output pairings and with appropriate control loops is shown in Figure 6.

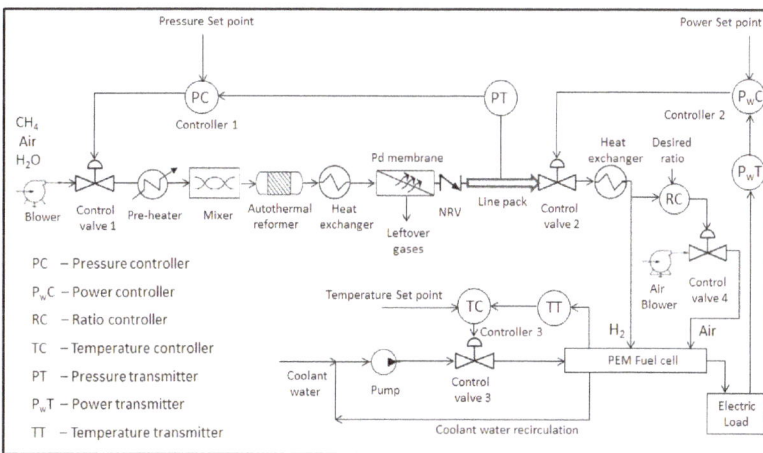

Figure 6. Process flow diagram with control loops for the integrated system.

3.2. Case Study 2: Integrated System Delivering Target Power Demand

For validating the capability of employing the integrated system for a realistic application, an electrical load of about 27 kW is considered, which corresponds to a light motor vehicle [26], powered by an integrated reformer membrane fuel cell system. Taking into account the transmission and DC/AC converter losses, the maximum power requested is assumed to be 30 kW. In this particular case study, the PEM fuel cell modeled can generate a maximum current density of 1.5319 A cm^{-2} at a cell voltage of 0.53 V based on the model parameters used. As the electrode area per cell is assumed to be 100 cm^2, a single fuel cell can generate a maximum power of approximately 0.08 kW. In order for the fuel cell to generate a power of 30 kW, a fuel cell stack consisting around 375 such individual cells connected in series is necessary. Modeling a fuel cell stack containing these many cells is beyond the scope of this paper. For the purpose of simulating the fuel cell stack, it is assumed here that each cell in the stack behaves identically. Thus, the overall stack voltage will be given by the sum of 375 individual cell voltages given by Equation (31). The average molar flow rate of hydrogen for the fuel cell system is 1.3×10^{-5} mol s^{-1} or 4.68×10^{-2} mol h^{-1}. This corresponds to a methane flow rate of 15.12 mol h^{-1}. Assuming that methane is stored at 200 atm pressure and a temperature of 298 K and using ideal gas law, the corresponding volume of methane is 1.8 litres. The output electrical load power demand is assumed to exhibit slower dynamics for the first 65 s and extremely faster dynamics during rest of the time. In this case study, controller 1 tries to regulate the molar concentration of H_2 in the line pack to 21.315 mol m^{-3} irrespective of variation in the power requested by the fuel cell. This is achieved by manipulating the control valve 1 position thereby varying the molar flow rate of fuel to the reformer inlet. In other words, controller 1 controls the hydrogen pressure in the line pack by manipulating the methane flow rate to the reformer inlet. Line pack pressure is calculated from the molar concentration of hydrogen present in the line pack using the ideal gas law as given by Equation (37). Inlet pressure at the anode and cathode sides are fixed at 1 atm. Line pack pressure control ensures presence of sufficient hydrogen always to prevent hydrogen starvation, which can permanently damage the fuel cell stack. The molar concentration of H_2 in the line pack is chosen as the controlled variable keeping the molar flow rate of fuel from the fuel storage to the reformer inlet through the control valve 1 as the manipulated variable. In addition to this, controller 2 is designed to take necessary action to supply requisite amount of H_2 to fuel cell anode in order to achieve the desired target power. Output power demanded by the fuel cell stack is identified as the controlled variable while the molar flow rate of H_2 through the control valve 2 to the fuel cell anode is selected as the manipulated variable. Third loop involves controller 3 and is designed to maintain a fixed average solid temperature of fuel cell by manipulating the molar flow rate of coolant circulated. As obvious, average solid temperature is selected as the controlled variable while molar flow rate of coolant is selected as the corresponding manipulated variable. Finally, fourth loop involves a ratio controller that regulates the ratio of air to hydrogen to be fed to the cathode and anode inlets of the fuel cell respectively. The tuning parameters for the four controllers are listed in Table 2.

Table 2. Design parameters for various control loops.

Loop	Controller	K_p	K_I	K_D	Ratio
Pressure controller	PID	0.3	0.0001	0.4	-
Power controller	P	0.1	-	-	-
Temperature controller	PID	−0.1	−0.0001	−0.05	
Air to H_2 ratio controller	Ratio	-	-	-	5

Based on the power demanded by the fuel cell load, the controller 2 manipulates the control valve 2 position as shown in Figure 7a. Opening of the control valve 2 lets an appropriate amount of H_2 to flow to the fuel cell anode, the molar flow rate of which can be seen from Figure 7b. Due to the feedback mechanism, controller 1 takes necessary action to manipulate the percentage opening of control valve 1 as shown in Figure 7c.

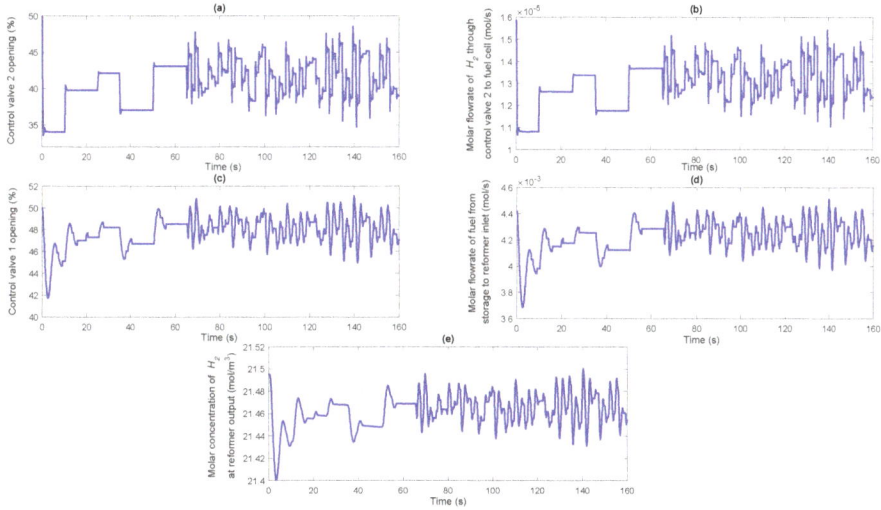

Figure 7. (a) Percentage opening of the control valve 2; (b) Molar flow rate of H_2 through control valve 2 to fuel cell; (c) Percentage opening of the control valve 1; (d) Molar flow rate of fuel from storage to reformer inlet; (e) Molar concentration of H_2 at reformer output.

Proportionately, the molar flow rate of stored fuel coming as input to the reformer starts varying as illustrated in Figure 7d. In response to this, the molar concentration of H_2 exiting from the reformer also starts varying as given in Figure 7e. According to Sievert's law given by Equation (13), in response to variation in the molar concentration of H_2 coming from the reformer, the molar flow rate of H_2 flowing through the Pd membrane starts varying as is obvious from Figure 8a. The controller 1 takes necessary action to maintain a constant H_2 molar concentration in the line pack between NRV and control valve 2 as shown in Figure 8b. As can be seen from the figure, the dashed line indicates the desired target molar concentration to be maintained while the solid line indicates the actual concentration in the line pack. As already discussed, the variation in the molar flow rate of H_2 to the fuel cell anode changes the power generated by the fuel cell which can be seen from Figure 8c. The dashed line shows the desired target power demanded by the electrical load while the solid line indicates the actual power supplied by the PEMFC. The profile showing the variation in the line pack pressure is given in Figure 8d. The controller 3 takes care of maintaining the average solid temperature of the fuel cell at a desirable value irrespective of variations in the power demanded by the fuel cell electric load. The manipulation of coolant flow rate by varying the control valve 3 position by the controller 3 can be observed from Figure 8e. The corresponding profile for the average solid temperature of the PEMFC is shown in Figure 8f. It is observed that the designed controllers with appropriately chosen controller parameters were able to control the H_2 molar concentration in the line pack, fuel cell power demand and the average solid temperature of fuel cell with relatively short settling times and with negligible offset. Both slow and fast varying load scenarios have been simulated in the present case study for the closed loop system for power control and observed that the designed controller could tracks the set point with a settling time of the order of 0.4 s.

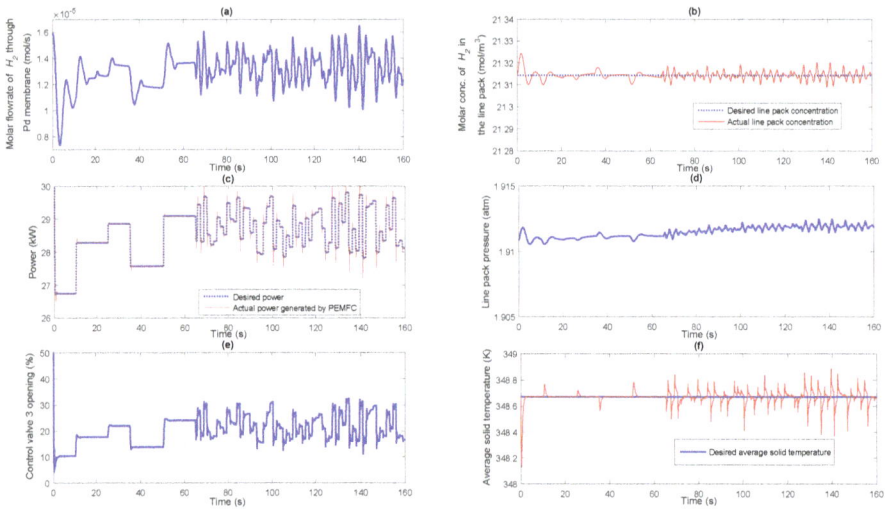

Figure 8. (**a**) Molar flow rate of H_2 through Pd membrane; (**b**) Performance of controller 1 maintaining a fixed molar concentration of H_2 in the line pack; (**c**) Performance of controller 2 tracking the target power profile; (**d**) Pressure in the line pack; (**e**) Percentage opening of the control valve 3; (**f**) Performance of controller 3 maintaining a fixed average solid temperature of fuel cell.

3.3. Case Study 3: Disturbance in both the Reformer Inlet CH_4 Feed Concentration and Coolant Temperature

The ability of the designed controller to reject the unknown disturbances occurring in the integrated reformer membrane fuel cell system is analyzed in this case study. In the previous case study, it was assumed that the inlet CH_4 concentration and coolant temperature are fixed and are not affected by any external disturbances. In this particular case study, disturbance in the form of step change is introduced in both the inlet CH_4 concentration as well as coolant temperature as shown in Figure 9a,b respectively. The control structure discussed here is also similar to the one discussed in the previous case study. Controller 1 tries to maintain a fixed molar concentration of H_2 in the line pack by manipulating the flow rate of fuel entering the reformer while controller 2 make sure that the power demanded by the electrical load is delivered by the fuel cell at the appropriate time. Controller 3 maintains a fixed average solid temperature in the fuel cell irrespective of changes in the output power demand as well as variations in the coolant flow rate and temperature. Based on the power demanded by the fuel cell electrical load, controller 2 manipulates the control valve 2 as seen from Figure 9c. Based on the opening of control valve 2, the requisite amount of H_2 flows to the fuel cell anode, the molar flow rate of which is shown in Figure 9d. Once the H_2 concentration in the line pack gets disturbed, controller 1 takes necessary action to manipulate the opening of control valve 1 so as to maintain the line pack H_2 concentration at a constant value. The change in control valve 1 opening based on the command from the controller 1 can be seen from Figure 9e. Based on the control valve 1 opening, appropriate amount of stored fuel is supplied as input to the auto thermal reformer, the molar flow rate of which can be seen in Figure 9f. Because of the presence of impurities in CH_4 concentration at the inlet, controller 1 takes additional action so as to compensate for the unknown disturbance happening at the input side of the plant. As the fuel inlet to the reformer varies, the molar concentration of H_2 at the reformer output also starts varying as shown in Figure 10a. Accordingly, the molar flow rate of H_2 passing through the Pd membrane also starts varying as given in Figure 10b. The profile showing the desired and the actual molar concentration of H_2 in the line pack is shown in Figure 10c. It can be seen from Figure 10d that the power demanded by the electrical load and actual power delivered by the fuel cell. Because of the presence of disturbance in the coolant flow rate as well as changes in the power demanded, the average solid temperature starts varying from the required set point value. A significant deviation in the fuel cell power compared to the desired power at the initial time period (first 5 s) can be observed from Figure 10d. This is due to the disturbance in the coolant inlet temperature from its nominal value of 340 K to 345 K, as observed from Figure 9b. This in turn resulted in loss of power, due to which, controller 3 increases the coolant flow rate which restores the temperature of the fuel cell and consequently the power (see Figure 10d,e). Based on the command from the controller 3, the average solid temperature sticks to the pre-defined set value as can be seen from Figure 10f. It is observed that the designed controllers with appropriately chosen controller parameters were able to control the parameters and track the set point even in the presence of unexpected disturbances happening in the plant. Due to the presence of disturbances, the deviations of molar concentration of hydrogen in the line pack as well as the average solid temperature from the set point is more compared to the scenario when disturbances were absent.

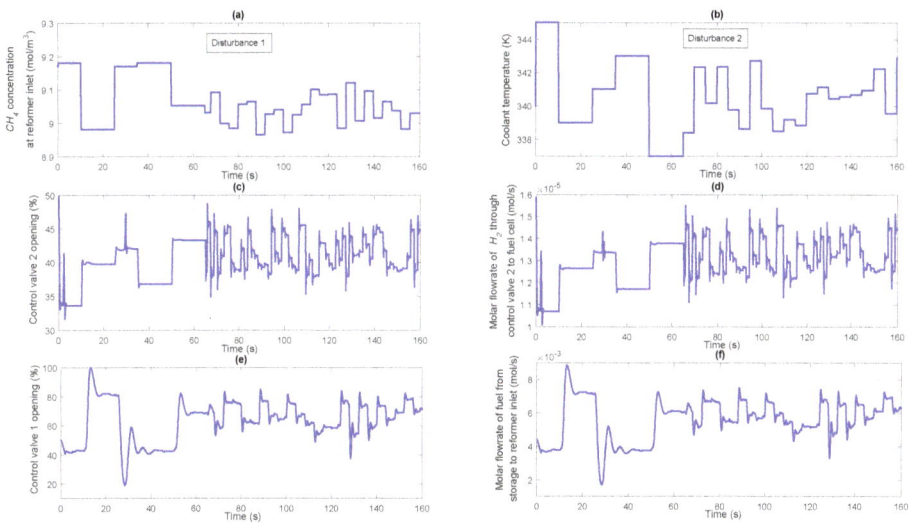

Figure 9. (**a**) Disturbance in the concentration of CH_4 at reformer inlet; (**b**) Disturbance in the coolant temperature; (**c**) Percentage opening of the control valve 2; (**d**) Molar flow rate of H_2 through control valve 2 to fuel cell; (**e**) Percentage opening of the control valve 1; (**f**) Molar flow rate of fuel from storage to reformer inlet.

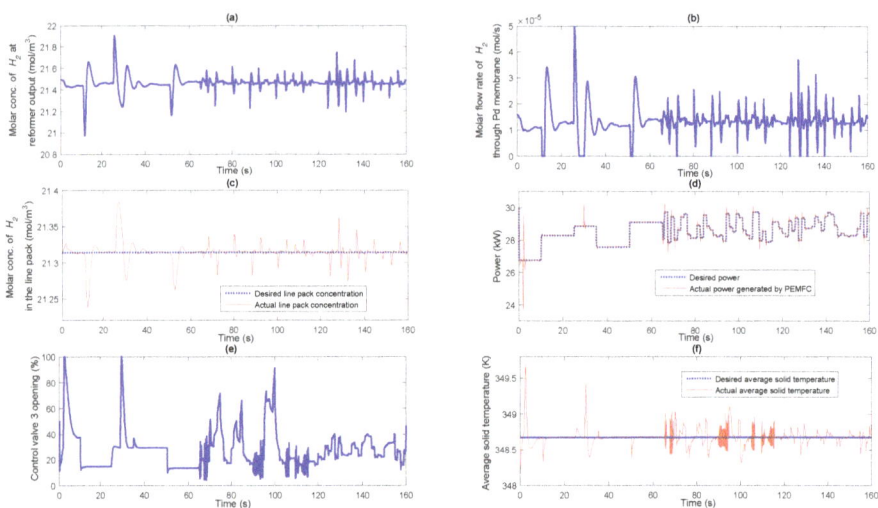

Figure 10. (**a**) Molar concentration of H_2 at reformer output; (**b**) Molar flow rate of H_2 through Pd membrane; (**c**) Performance of controller 1 maintaining a fixed molar concentration of H_2 in the line pack; (**d**) Performance of controller 2 tracking the target power profile; (**e**) Percentage opening of the control valve 3; (**f**) Performance of controller 3 maintaining a fixed average solid temperature of fuel cell.

4. Conclusions

A system level mathematical model of an integrated reformer membrane fuel cell system has been developed in order to assess the feasibility of the integrated system for stationary and portable power applications. Such an integrated system promotes safety and leverages existing natural gas infrastructure. The dynamic analysis of the open loop system revealed significant difference in the dynamics related to the upstream reformer as compared with the fuel cell. This behavior points the need for an auxiliary power system during startup of the integrated system which is outside the scope of this work. A multi loop control strategy is implemented on to the integrated system wherein power demanded by the fuel cell load, molar concentration of H_2 in the line pack and average solid temperature of the fuel cell are chosen as the controlled variables for the two control loops. Relevant controllers are designed appropriately to control the system which included set point tracking and disturbance rejection studies. The simulation results show faster settling times with negligible offset proving the effectiveness of the designed controller.

Author Contributions: All of the authors contributed to publishing this article. The collection of materials and summarization of this article was done by P.P.S. and the conceptual ideas, methodology and guidance for the research were provided by R.D.G. and S.B. All the coding and programming works was done by P.P.S.

Funding: This research was funded by the Dept. of Science & Technology, Govt. of India, Grant 13DST057.

Conflicts of Interest: The authors declare no conflict of interest.

Nomenclature

α	ratio of water molecules per proton (molecules proton^{-1})
ΔH	heat of reaction (kJ mol^{-1})
ϵ_b	packing bed porosity
η_j	effectiveness factor of reaction j

η	over potential (V)
λ_z^f	effective thermal conductivity $(W(mK)^{-1})$
μ	water viscosity $(g\,cm^{-1}s^{-1})$
ρ_{bed}	density of the catalyst bed $(kg\,m^{-3})$
ρ_{cat}	density of the catalyst pellet $(kg\,m^{-3})$
ρ_f	density of the fluid $(kg\,m^{-3})$
ρ_s	density of the solid $(kg\,m^{-3})$
$\rho_{m,dry}$	density of dry membrane of fuel cell $(g\,cm^{-3})$
σ_m	ionic conductivity of the fuel cell membrane $(Ohm^{-1}\,cm^{-1})$
θ_{mem}	Pd membrane poisoning caused by the adsorption of CO
A	heat exchange area per unit length (cm)
A_1	area of the fuel flow pipe to the reformer inlet (m^2)
a_k	activity of water in stream k
a_v	external catalyst surface area per unit volume of catalyst bed (m^2m^{-3})
A_{mem}	Pd membrane surface area (m^2)
C_i	molar concentration of species i in the gas phase $(mol\,m^{-3})$
C_p	molar heat capacity $(J(g\,mol\,K)^{-1})$
c_w	concentration of water in the membrane of fuel cell $(mol\,cm^{-3})$
$C_{H_2,m}$	molar concentration of H_2 in the line pack $(mol\,m^{-3})$
$C_{i,s}$	molar concentration of species i in the solid phase $(mol\,m^{-3})$
$C_{p,bed}$	specific heat of the catalyst bed $(J\,(kg\,K)^{-1})$
C_{pg}	specific heat of the fluid $(J\,(kg\,K)^{-1})$
C_{v1}	control valve 1 flow coefficient
C_{v2}	control valve 2 flow coefficient
$c_{w,k}$	concentration of water at k interface of the membrane of fuel cell $(mol\,cm^{-3})$
D	diffusion coefficient of water in membrane of fuel cell $(cm^2\,s^{-1})$
d	fuel cell channel height (cm)
D_z	axial dispersion coefficient $(m^2\,s^{-1})$
e	membrane area per unit length (cm)
E_a	activation energy for H_2 permeability $(J\,mol^{-1})$
F	Faraday's constant $(C\,mol^{-1})$
f	cross-section of solid in fuel cell (cm^2)
f_1	opening of control valve 1 (%)
f_2	opening of control valve 2 (%)
F_p	piping geometry factor
h	fuel cell channel width (cm)
h_f	gas to solid heat transfer coefficient $(W\,m^{-2}s^{-1})$
I	current density $(A\,cm^{-2})$
i_o	exchange current density $(A\,cm^{-2})$
k	heat conduction coefficient $(W\,(cmK)^{-1})$
k_c	condensation rate constant (s^{-1})
K_D	parameter corresponding to the viscous loss term $(Pa\,s\,m^{-2})$
K_V	parameter corresponding to the kinetic loss term $(Pa\,s^2\,m^{-3})$
$k_{g,i}$	gas to solid mass transfer coefficient of component i $(m\,s^{-1})$
L_{mem}	Pd membrane thickness (m)
M	molar flow rate $(mol\,s^{-1})$
$M_{m,dry}$	equivalent weight of dry membrane of fuel cell $(g\,mol^{-1})$
n	pressure exponent
n_e	number of electrons taking part in charge reactions
N_i	molar flux of species i $(mol\,s^{-1}\,cm^{-2})$
P	pressure (atm)
$P_{w,k}^{sat}$	vapor pressure of water in k^{th} channel
P_e	permeability of Pd membrane $(mol\,(msPa^n)^{-1})$

$P_{e,o}$ permeability pre-exponential factor (mol $(msPa^n)^{-1}$)
P_{in} methane fuel storage pressure (atm)
R gas constant (J $(mol K)^{-1}$)
r_i rate of consumption or formation of species i (mol $(kg_{cat}s)^{-1}$)
R_j rate of reaction j (mol $(kg_{cat}s)^{-1}$)
t time (s)
T_g reformer gas temperature (K)
t_m fuel cell membrane thickness (cm)
T_{fc} fuel cell solid temperature (K)
T_{gr} gas temperature at the exit of heat exchanger 1 (K)
T_{inf} ambient temperature (K)
T_{in} methane fuel storage temperature (K)
U convective heat transfer coefficient (W $cm^{-2} K^{-1}$)
u superficial gas flow velocity (m s^{-1})
V volume of the line pack (m^3)
V_{cell} cell voltage (V)
V_{oc} open circuit voltage (V)
W molecular weight (kg mol^{-1})
X ratio of pressure drop to the absolute inlet pressure
x direction along the fuel cell channel length (cm)
Y expansion factor
Z compressibility factor
z axial dimension (m)

Suffixes

a anode
c cathode
g gas
m line pack
s solid
sat saturation
w water

Appendix A. Data for the Base Case

Parameter	Values
C_{CH_4} (mol m^{-3})	9.1705
C_{O_2} (mol m^{-3})	4.1273
C_{CO_2} (mol m^{-3})	0.0001
C_{H_2O} (mol m^{-3})	9.1705
C_{H_2} (mol m^{-3})	0.0001
C_{CO} (mol m^{-3})	0.0001
C_{N_2} (mol m^{-3})	15.5287
T_g (K)	815
T_s (K)	815
$C_{H_2,m}$ (mol m^{-3})	0.001
M_{H_2} (mol s^{-1})	1.14×10^{-5}
M_{O_2} (mol s^{-1})	5.7×10^{-6}
$M_{w,a}^l$ (mol s^{-1})	0
$M_{w,c}^l$ (mol s^{-1})	0
$M_{w,a}^v$ (mol s^{-1})	9.8555×10^{-6}
$M_{w,c}^v$ (mol s^{-1})	4.9278×10^{-6}
T_a (K)	353
T_c (K)	353
T_{cool} (K)	340
T_{fc} (K)	342

Appendix B. Parameter Values Used for Simulation

Parameter	Values
ϵ_b	0.4
z (m)	0.2
D_z (m^2 s^{-1})	2.91×10^{-4}
a_v (m^{-1})	1200
ρ_{cat} (kg m^{-3})	1870
ρ_{bed} (kg m^{-3})	1122
$C_{p,bed}$ (J kg^{-1} K^{-1})	850
$\Delta H_{rxn,1}$ (kJ mol^{-1})	206.2
$\Delta H_{rxn,2}$ (kJ mol^{-1})	164.9
$\Delta H_{rxn,3}$ (kJ mol^{-1})	−41.1
$\Delta H_{rxn,4}$ (kJ mol^{-1})	−802.7
η_1 (K)	0.07
η_2 (K)	0.06
η_3 (K)	0.7
η_4 (K)	0.05
K_D (K)	$6.539 \ times 10^3$
θ_{mem}	0.2
A_{mem} (m^2)	0.0064
L_{mem} (m)	0.0001
R (J mol^{-1} K^{-1})	8.314
n	0.67
$P_{e,o}$ (mol m^{-1} s^{-1} Pa^{-n})	0.4
E_a (J mol^{-1})	8000
x (cm)	10
h (cm)	0.1
d (cm)	0.1
F (C mol^{-1})	96484.69
k_c (s^{-1})	100
P_a (atm)	1
P_c (atm)	1
U_g (W cm^{-2} K^{-1})	0.025
A_g (cm)	0.4
U_w (W cm^{-2} K^{-1})	0.025
A_{cool} (cm)	0.4
$C_{p,w}$ (J g^{-1} mol^{-1} K^{-1})	75.38
M_{cool} (mol s^{-1})	5.5556×10^{-2}
ρ_s (g cm^{-3})	2
$C_{p,s}$ (J g^{-1} mol^{-1} K^{-1})	1
k_s (W cm^{-1} K^{-1})	0.005
e (cm)	0.1
U_c (W cm^{-2} K^{-1})	0.025
T_{inf} (K)	350
V_{oc} (V)	1.1
t_m (cm)	0.01275
n_e	2
i_o (A cm^{-2})	0.01
F_p	0.667
C_{v1}	0.67
Y	1
W_{CH_4} (kg mol^{-1})	0.016
Z	1
P_{in} (atm)	2
V (m^3)	0.00025
C_{v2}	0.002
W_{H_2} (kg mol^{-1})	0.002

References

1. Qi, A.; Peppley, B.; Karan, K. Integrated fuel processors for fuel cell application: A review. *Fuel Process. Technol.* **2007**, *88*, 3–22. [CrossRef]
2. Halabi, M.; de Croon, M.; van der Schaaf, J.; Cobden, P.; Schouten, J. Modeling and analysis of autothermal reforming of methane to hydrogen in a fixed bed reformer. *Chem. Eng. J.* **2008**, *137*, 568–578. [CrossRef]
3. Iwuchukwu, I.J.; Sheth, A. Mathematical modeling of high temperature and high-pressure dense membrane separation of hydrogen from gasification. *Chem. Eng. Process. Process. Intensif.* **2008**, *47*, 1292–1304. [CrossRef]

4. Okazaki, J.; Ikeda, T.; Tanaka, D.A.P.; Sato, K.; Suzuki, T.M.; Mizukami, F. An investigation of thermal stability of thin palladium–silver alloy membranes for high temperature hydrogen separation. *J. Membr. Sci.* **2011**, *366*, 212–219. [CrossRef]

5. Authayanun, S.; Mamlouk, M.; Scott, K.; Arpornwichanop, A. Comparison of high-temperature and low-temperature polymer electrolyte membrane fuel cell systems with glycerol reforming process for stationary applications. *Appl. Energy* **2013**, *109*, 192–201. [CrossRef]

6. Basualdo, M.; Feroldi, D.; Outbib, R. PEM Fuel Cells with Bio-Ethanol Processor Systems: A Multidisciplinary Study of Modelling, Simulation, Fault Diagnosis and Advanced Control. In *Green Energy and Technology*; Springer: London, UK, 2011.

7. Lorenzo, G.D.; Corigliano, O.; Faro, M.L.; Frontera, P.; Antonucci, P.; Zignani, S.; Trocino, S.; Mirandola, F.; Aricò, A.; Fragiacomo, P. Thermoelectric characterization of an intermediate temperature solid oxide fuel cell system directly fed by dry biogas. *Energy Convers. Manag.* **2016**, *127*, 90–102. [CrossRef]

8. Kupecki, J.; Motylinski, K.; Milewski, J. Dynamic analysis of direct internal reforming in a sofc stack with electrolyte-supported cells using a quasi-1d model. *Appl. Energy* **2018**, *227*, 198–205. [CrossRef]

9. Lorenzo, G.D.; Milewski, J.; Fragiacomo, P. Theoretical and experimental investigation of syngas-fueled molten carbonate fuel cell for assessment of its performance. *Int. J. Hydrogen Energy* **2017**, *42*, 28816–28828. [CrossRef]

10. El-Sharkh, M.; Rahman, A.; Alam, M.; Byrne, P.; Sakla, A.; Thomas, T. A dynamic model for a stand-alone pem fuel cell power plant for residential applications. *J. Power Sources* **2004**, *138*, 199–204. [CrossRef]

11. Ipsakis, D.; Voutetakis, S.; Seferlis, P.; Papadopoulou, S.; Stoukides, M. Modeling and analysis of an integrated power system based on methanol autothermal reforming. In Proceedings of the 17th Mediterranean Conference on Control and Automation, Thessaloniki, Greece, 24–26 June 2009; pp. 1421–1426.

12. Stamps, A.T.; Gatzke, E.P. Dynamic modeling of a methanol reformer- pemfc stack system for analysis and design. *J. Power Sources* **2006**, *161*, 356–370. [CrossRef]

13. Patcharavorachot, Y.; Wasuleewan, M.; Assabumrungrat, S.; Arpornwichanop, A. Analysis of hydrogen production from methane autothermal reformer with a dual catalyst-bed configuration. *Theor. Found. Chem. Eng.* **2012**, *46*, 658–665. [CrossRef]

14. Ding, O.; Chan, S. Autothermal reforming of methane gas-modelling and experimental validation. *Int. J. Hydrogen Energy* **2008**, *33*, 633–643. [CrossRef]

15. Vadlamudi, V.K.; Palanki, S. Modeling and analysis of miniaturized methanol reformer for fuel cell powered mobile applications. *Int. J. Hydrogen Energy* **2011**, *36*, 3364–3370. [CrossRef]

16. Pinacci, P.; Drago, F. Influence of the support on permeation of palladium composite membranes in presence of sweep gas. *Catal. Today* **2012**, *193*, 186–193. [CrossRef]

17. Doong, S.; Yang, R. Hydrogen purification by the multibed pressure swing adsorption process. *React. Polym. Ion Exch. Sorbents* **1987**, *6*, 7–13, [CrossRef]

18. Canevese, S.; Marco, A.D.; Murrai, D.; Prandoni, V. Modelling and control of a psa reactor for hydrogen purification. *IFAC Proceed. Vol.* **2007**, *40*, 99–104. [CrossRef]

19. Bhargav, A. Model Development and Validation of Palladium-Based Membranes for Hydrogen Separation in Pem Fuel Cell Systems. Ph.D. Thesis, University of Maryland, College Park, MD, USA, 2010.

20. Golbert, J.; Lewin, D.R.; Model-based control of fuel cells: (1) Regulatory control. *J. Power Sources* **2004**, *135*, 135–151. [CrossRef]

21. Pukrushpan, J.T.; Stefanopoulou, A.G.; Peng, H. Modeling and control for pem fuel cell stack system. In Proceedings of the 2002 American Control Conference (IEEE Cat. No.CH37301), Anchorage, AK, USA, 8–10 May 2002; Volume 4, pp. 3117–3122.

22. Nguyen, T.; White, R. A water and heat management model for proton-exchange-membrane fuel cells. *J. Electrochem. Soc.* **1993**, *140*, 2178–2186. [CrossRef]

23. Methekar, R.; Prasad, V.; Gudi, R. Dynamic analysis and linear control strategies for proton exchange membrane fuel cell using a distributed parameter model. *J. Power Sources* **2007**, *165*, 152–170. [CrossRef]

24. Perry, R.H.; Green, D.W.; Maloney, J.O. *Perry's Chemical Engineers Handbook*, 7th ed.; The McGraw-Hill Companies Inc.: New York, NY, USA, 1999.

25. Chatrattanawet, N.; Skogestad, S.; Arpornwichanop, A. Control structure design and dynamic modeling for a solid oxide fuel cell with direct internal reforming of methane. *Chem. Eng. Res. Des.* **2015**, *98*, 202–211. [CrossRef]

26. Sarkar, A.; Banerjee, R. Net energy analysis of hydrogen storage options. *Int. J. Hydrog Energy* **2005**, *30*, 867–877. [CrossRef]

processes

MDPI

Article

Improving Flexibility and Energy Efficiency of Post-Combustion CO₂ Capture Plants Using Economic Model Predictive Control

Benjamin Decardi-Nelson, Su Liu and Jinfeng Liu *

Department of Chemical & Materials Engineering, University of Alberta, Edmonton, AB T6G 1H9, Canada; decardin@ualberta.ca (B.D.-N.); su7@ualberta.ca (S.L.)
* Correspondence: jinfeng@ualberta.ca; Tel.: +1-780-492-1317

Received: 30 July 2018; Accepted: 17 August 2018; Published: 21 August 2018

Abstract: To reduce CO_2 emissions from power plants, electricity companies have diversified their generation sources. Fossil fuels, however, still remain an integral energy generation source as they are more reliable compared to the renewable energy sources. This diversification as well as changing electricity demand could hinder effective economical operation of an amine-based post-combustion CO_2 capture (PCC) plant attached to the power plant to reduce CO_2 emissions. This is as a result of large fluctuations in the flue gas flow rate and unavailability of steam from the power plant. To tackle this problem, efficient control algorithms are necessary. In this work, tracking and economic model predictive controllers are applied to a PCC plant and their economic performance is compared under different scenarios. The results show that economic model predictive control has a potential to improve the economic performance and energy efficiency of the amine-based PCC process up to 6% and 7%, respectively, over conventional model predictive control.

Keywords: optimal control; post-combustion CO_2 capture; energy efficiency; time-varying operation

1. Introduction

The reliance on fossil fuels, especially coal for electricity, is one of the major causes of increased amount of anthropogenic carbon dioxide in the atmosphere and climate change. At the moment, almost 40% of the world's energy is produced from combustion of coal [1]. Although renewable energies such as those obtained from wind or solar or biomass can be helpful in reducing CO_2 emissions, they are not mature enough to fully take over. Therefore, for these fossil fuels to keep contributing to the global energy mix, carbon capture and storage (CCS) must be implemented. CCS involves capturing carbon dioxide from large point sources such as power plants, transporting and storing it in deep geological or oceanic wells for a long time in a supercritical state [2]. Among the various options for CO_2 capture, post-combustion CO_2 capture (PCC) using reactive solvents is considered the most mature and viable option as it can easily be retrofitted to existing power plants. However, a major setback to the realization of this technology is its high regeneration heat requirement. Studies have shown that a PCC plant attached to a power plant reduces the power plant's efficiency from about 40% to 30% [3].

Figure 1 shows an overview of a fossil-fueled power plant with CO_2 capture. In a typical fossil-fueled power plant with CO_2 capture, the fuel is combusted in air to generate flue gas, which is used to generate high pressure steam in the steam generator. The steam is then used to turn a turbine leading to electricity generation. At the same time, part of the steam is sent to the PCC plant for regeneration of the captured CO_2. This results in a reduction of the power plant's efficiency. The flue gas (containing about 15 mol % CO_2) on the other hand is channeled to the PCC plant for treatment before it is released into the atmosphere. The power plant's ability to meet the varying demands of the end users while capturing CO_2 is a major concern. This could result in large fluctuations in the flue gas flow rate and the availability of the steam from the power plant to the the capture plant. These fluctuations as well as issues such as complex/nonlinear system dynamics could make attainment of the operational objectives and control of the PCC plant difficult. These challenges have motivated the research on the development of more flexible and economical control for PCC plants.

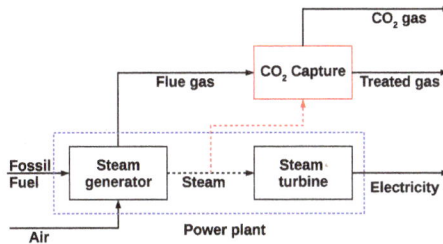

Figure 1. Overview of electricity generation with post-combustion CO_2 capture.

Ziaii et al. [4] designed a ratio controller to maintain a 90% CO_2 capture efficiency in the absorber. They considered a disturbance in the steam availability to the reboiler and used the liquid to gas flow ratio as the manipulated variable. Lin et al. [5] presented two different PI control strategies to maintain a desired absorption efficiency. They used the lean solvent flow rate and the reboiler heat input as the manipulated variables.

Advanced control algorithms such as Model Predictive Control (MPC) have also been applied to the control of PCC plants. Bedelbayev et al. [6] presented an MPC scheme for the absorber in standalone. Panahi and Skogestad [7,8] implemented an MPC scheme with two controlled variables—the CO_2 recovery and the reboiler temperature. In the context of scheduling and control, He et al. [9] studied the flexibility of a post-combustion CO_2 capture process using MPC. They conducted this study by investigating key process variables such as CO_2 absorption efficiency and CO_2 composition in the product stream in both closed-loop and open-loop. They observed that, under high frequency disturbances in the flue gas flow rate, a well-tuned controller is needed to avoid oscillations in the absorption efficiency. Bankole et al. [10] used a two-level control strategy consisting of an upper-level scheduler and a lower-level supervisory controller to handle the flexible operation of a load-following power plant with CO_2 capture under three different capture scenarios. MacDowell and Shah [11] also evaluated the operation of a PCC plant under four different scenarios—load following, solvent storage, exhaust gas by-pass and time-varying solvent regeneration—using multi-period optimization.

An emerging MPC formulation that has a potential to determine the optimal operating policy directly in the control layer is Economic Model Predictive Control (EMPC). In EMPC, real-time process optimization and feedback control are unified [12]. This means that, in EMPC, an economic cost function or objective is directly optimized instead of tracking a predetermined set-point [13,14]. To date, EMPC has had some successful applications in different processes [15–19]. In the studies of Zeng et al. [15] and Liu et al. [16], for example, EMPC was applied to improve the effluent quality of a waste water treatment process and the operation of an oil sands primary separation vessel, respectively. Idris et al. [17] applied EMPC to a continuous catalytic distillation process whileMendoza-Serrano et al. [18] and Touretzky et al. [19] applied EMPC to temperature control in buildings. In these applications, EMPC showed a potential to significantly improve the performance of the classical MPC. However, it should be pointed out that the reported EMPC designs and their benefits are usually process specific. The general case where EMPC has a clear edge over classical MPC remains to be understood.

In this study, EMPC was applied to a PCC plant attached to a load-following power plant and its performance was compared with classical tracking MPC from different aspects. In particular, the operation of the PCC plant under time-varying flue gas flow rate and steam price was considered.

The performance of tracking MPC and EMPC under two different carbon tax policies are investigated namely: without tax-free emission limit and with tax-free emission limit. The rest of the paper is organized as follows: in Section 2, a detailed PCC plant model based on first principles is presented. In Section 3, the designs of both MPC and EMPC as well as their numerical implementation are presented. In Section 4, a systematic investigation of the performance of both MPC and EMPC from different aspects including time-varying conditions and uncertainty is conducted.

2. Model Development

A typical PCC plant is comprised of two major operating units namely, absorption and desorption columns, where reactive separation takes place. A schematic diagram of the PCC plant considered in this work is shown in Figure 2. The process starts in the absorption column where the flue gas (containing high amount of CO_2) from the power plant is contacted with the lean solvent—solvent containing low amount of CO_2 (in this work, state-of-the-art 5 M Monoethanolamine (MEA) is used as solvent)—in a counterclockwise way. The flue gas leaves the absorption column as treated gas with a low amount of CO_2. The solvent with a high amount of CO_2 (rich solvent) exits the absorption column and goes through the lean-rich heat exchanger where it trades heat with the lean solvent exiting the reboiler. The rich solvent enters the desorption unit and contacts hot vapor from the reboiler. In the desorption column, the acid gas (CO_2) is stripped from the rich solvent. Gas with a high concentration of CO_2 (90–99%) exits the desorption column from the top while the solvent moves out of the desorption column to the reboiler. In the reboiler, the solvent is heated to a temperature of about 120 °C. The gas exits the reboiler to the desorption column while the lean solvent is recycled to the absorption column for re-absorption. In this work, the lean solvent flow rate and the steam from the power plant are the manipulated variables. The control objective is to reduce CO_2 emissions and heat consumption at a reasonable economic cost. The disturbances are the flue gas flow rate and steam from the power plant.

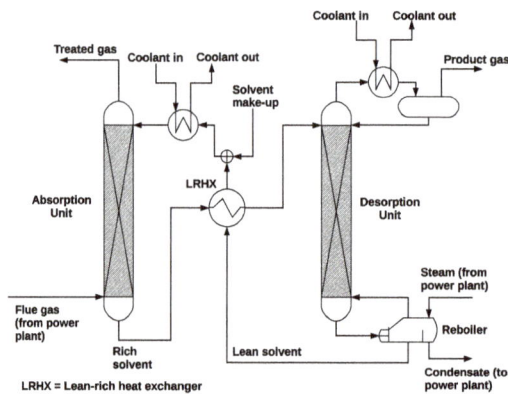

Figure 2. Schematic diagram of a full cycle amine-based post-combustion capture plant.

Several studies on dynamic modelling of the PCC process exist in literature. The model used in this work is based on the work of Harun et al. [20]. However, the lean-rich heat exchanger was modeled using the log-mean temperature approach. In addition, the buffer tank and column heat losses were not considered in this work. The details of the model used in this study are summarized in this section.

2.1. Absorption and Desorption Units

The model formulations for the absorption and desorption units are similar. The differences are the direction of reactions and the presence of reboiler and condenser in the desorption unit.

2.1.1. Modeling Assumptions

The assumptions used in the modeling of the PCC plant are:

- Well mixed bulk and liquid phases. Each stage therefore behaves like a continuously stirred tank reactor (CSTR) with no spacial variation in properties.
- Reactions occur only in the liquid film and the influence of the reaction on mass transfer is described using enhancement factor.
- Mass and heat transfer are described by the two film theory [21].
- Pressure drop in the two columns is linear.
- No heat losses to the surrounding area.

2.1.2. Balance Equations

Equations (1)–(4) describe the mass and heat balances occurring in the two columns. These equations are written along the axis of the columns to determine the axial concentration and temperature profiles. Since the assumption that a stage in both columns is well mixed has been made, there are no radial changes in temperature and concentration in each column:

$$\frac{dC_{Li}}{dt} = \frac{4F_L}{\pi D_c^2}\frac{\partial C_{Li}}{\partial l} + (N_i a^l), \quad i = CO_2, MEA, H_2O, N_2, \tag{1}$$

$$\frac{dC_{Gi}}{dt} = -\frac{4F_G}{\pi D_c^2}\frac{\partial C_{Gi}}{\partial l} - (N_i a^l), \quad i = CO_2, MEA, H_2O, N_2, \tag{2}$$

$$\frac{dT_L}{dt} = \frac{4F_L}{\pi D_c^2}\frac{\partial T_L}{\partial l} + \frac{(Q_L a^l)}{\sum_{i=1}^{n} C_{Li} C_{p,i}}, \tag{3}$$

$$\frac{dT_G}{dt} = -\frac{4F_G}{\pi D_c^2}\frac{\partial T_G}{\partial l} + \frac{(Q_G a^l)}{\sum_{i=1}^{n} C_{Gi} C_{p,i}}. \tag{4}$$

In Equations (1)–(4) above, C_i is the phase (subscripts L and G are liquid phase and gas phase respectively) concentration of component i in $kmol/m^3$, F is the phase volumetric flow in m^3/s, D_c is the cross-sectional area of the column in m, N_i is the mass transfer rate in $kmol/m^2s$, T is temperature in K, l is the length of the column in m, C_p is the heat capacity in $kJ/kmol$, Q is the heat transfer rate in kJ/m^2s and a^l is the interfacial area in m^2/m^3. In the absorption unit, F_L is a manipulated variable to control the CO_2 emission, whereas F_G is a disturbance from the power plant.

2.1.3. Heat and Mass Transfer Rates

The mass transfer rates in the columns are determined using Equation (5):

$$N_i = K_{Gi} P_t (y_i - y_i^*), \quad i = CO_2, MEA, H_2O, \tag{5}$$

where K_{Gi} is the overall mass transfer coefficient in $kmol/m^3 Pa$, P_t is the stage pressure in Pa, and y_i and y_i^* are the bulk and equilibrium gas phase component mole fractions. The overall gas phase mass transfer coefficients are used to compute the mass transfer rates between the two phases. This is convenient as it avoids the determination of interfacial concentrations. In this work, the mass transfer rate of N_2 is assumed to be zero or negligible.

The interface equilibrium H_2O and MEA concentrations are determined from their saturated vapor pressures as shown in Equation (6), whereas that of CO_2 is determined using Henry's law as shown in Equation (7):

$$y_i^* P_t = x_i \gamma_i P_i^v, \quad i = H_2O, MEA, \tag{6}$$

$$y_i^* P_t = He_i C_i^* \gamma_i, \qquad i = CO_2, \tag{7}$$

where γ_i is the activity coefficient of component i, He_i is the Henry's constant and P_i^v is the saturated vapor pressure of component i in Pa.

The overall mass transfer coefficients are determined using Equation (8) below:

$$\frac{1}{K_{Gi}} = \frac{1}{k_{Gi}} + \frac{He}{k_{Li}E_f}, \qquad i = CO_2, MEA, H_2O, \tag{8}$$

where k_{Gi} and k_{Gi} are the gas and liquid mass transfer coefficients of component i, respectively, in kmol/m²s and E_f is the enhancement factor.

The mass transfer coefficients and interfacial area are estimated using Onda's correlations [22]. The enhancement factor approach is used to determine the influence of the reactions on the rates of mass transfer. The pseudo-first order enhancement factor during absorption is calculated using Equation (9) determined by van Kravelen and Hoftijzer [23]:

$$E = \frac{\sqrt{k_i D_i C_j}}{k_{Li}}, \qquad i = CO_2, j = MEA, \tag{9}$$

where k_i is the second order rate constant in m³/kmol and D_i is the liquid phase diffusivity of component i in m²/s. In the case of desorption, the enhancement factor is calculated using Equation (10) [24]:

$$E = 1 + \frac{(\frac{D_{MEACOO^-}}{D_{CO_2}})\sqrt{K_{eq}C_{MEA}^B}}{(1+2(\frac{D_{MEACOO^-}}{D_{MEA}})\sqrt{K_{eq}C_{CO_2}^B})(\sqrt{C_{CO_2}^B}+\sqrt{C_{CO_2}^I})}, \tag{10}$$

where D_{MEACOO^-}, D_{CO_2} and D_{MEA} are the liquid phase diffusivities of carbamate ion, CO_2 and MEA respectively in m²/s, $C_{CO_2}^B$ and C_{MEA}^B are the bulk liquid phase concentrations of CO_2 and MEA respectively in kmol/m³, $C_{CO_2}^I$ is the concentration of CO_2 at the gas–liquid interface in kmol/m³, and K_{eq} is the overall equilibrium rate constant in m³/kmol.

The heat transfer rates across the interface for gas and liquid phases are determined using Equations (11) and (12), respectively. The heat of reaction is accounted for in the liquid heat transfer rate computation:

$$Q_G = h_{GL}(T_L - T_G), \tag{11}$$

$$Q_L = h_{GL}(T_L - T_G) + \Delta H_{rxn} N_{CO_2} + \Delta H_{vap} N_{H_2O}, \tag{12}$$

where h_{GL} is the gas–liquid heat transfer coefficient in kJ/m²s, ΔH_{rxn} is the heat of reaction between MEA and CO_2 in kJ/m³s and ΔH_{vap} is the heat of vaporization of H_2O in kJ/m³s. The gas–liquid heat transfer coefficient was determined using the Chilton–Colburn analogy [25].

2.2. Heat Exchanger Model

The heat exchanger, especially the lean-rich heat exchanger, is a crucial component of the post combustion capture plant. It seeks to make the process heat efficient by taking heat from the hot lean solvent (source) to the cool rich solvent (sink) exiting the absorber. Since there is no movement of mass from one stream to the other and no accumulation is assumed, the mass balances are not considered in the heat exchanger model.

2.2.1. Energy Balance

The energy balance equations are shown in Equations (13) and (14):

$$\frac{dT_{tube}}{dt} = \frac{\dot{V}_{tube}}{V_{tube}}(T_{tube,in} - T_{tube,out}) + \dot{Q}\frac{1}{\hat{C}p_{tube}\rho_{tube}V_{tube}}, \tag{13}$$

$$\frac{dT_{shell}}{dt} = \frac{\dot{V}_{shell}}{V_{shell}}(T_{shell,in} - T_{shell,out}) + \dot{Q}\frac{1}{\hat{C}p_{shell}\rho_{shell}V_{shell}}, \tag{14}$$

where T is the temperature in K, $\dot{V}$ is the volumetric flow rate in m^3/s, V is the volume in m^3, $\dot{Q}$ is the heat transfer rate in kJ/s, $\check{C}p$ is the average molar liquid heat capacity in kJ/kmol, ρ is the average molar density in kmol/m^3 and ΔT_{LMTD} is the log-mean temperature difference between the hot side and the cold side of the heat exchanger in K. Subscripts *tube*, *shell*, *in* and *out* represent the tube-side, shell-side, inlets and outlets of the heat exchanger, respectively.

The log-mean temperature difference between the shell-side and tube-side was determined using Equation (15) below:

$$\dot{Q} = UA\Delta T_{LMTD}, \tag{15}$$

$$\Delta T_{LMTD} = \frac{\Delta T_1 - \Delta T_2}{ln(\Delta T_1/\Delta T_2)}, \tag{16}$$

where ΔT_1 and ΔT_2 are equal to $(T_{shell,in} - T_{tube,out})$ and $(T_{shell,out} - T_{tube,in})$, respectively.

2.3. Reboiler Model

The reboiler is the most heat-intensive part of the plant. In this unit, the rich solvent is heated to break the chemical bonds between CO_2 and MEA. H_2O, CO_2 and MEA are vaporized in the process and channeled to the bottom of the desorption unit. The solvent containing low amount of the acid gas (CO_2) exits the unit as lean solvent.

2.3.1. Material Balance

The material balance around the reboiler is shown in Equation (17):

$$\frac{dM_i}{dt} = F_{in}x_{i,in} - Vy_{i,out} - Lx_{i,out}, \quad i = CO_2, MEA, H_2O, \tag{17}$$

where M_i is the mass holdup of component i in kmol, F is the molar flow rate in kmol/s, V and L are the vapor and liquid flow rates, respectively, in kmol/s. The subscripts *in* and *out* denote inlet and outlet streams, respectively. In this study, we assumed that the liquid level and the reboiler pressure does not change. Thus, there is no vapour or liquid holdup.

2.3.2. Energy Balance

The energy balance equation is shown in Equation (18):

$$\rho C_p V \frac{dT_{reb}}{dt} = F_{in}x_{in} - VH_{v,out} - LH_{L,out} + Q_{reb}, \tag{18}$$

where T_{reb} is the temperature in K, ρ is the density in kmol/m^3, C_p is the molar heat capacity in kJ/kmol, V is the holdup volume in m^3, H is the enthalpy in kJ and Q_{reb} is the heat input in KJ/s. Q_{reb} is used as a manipulated variable to control the reboiler temperature.

2.4. Physical and Chemical Properties

The accuracy of process models hinges on accurate predictions of the physical and chemical properties of the system. The non-ideal liquid phase behaviour of the CO_2-MEA-H_2O system was modeled using the electrolyte Non-Random Two Liquids (eNRTL) thermodynamic model. The liquid phase activity coefficients were obtained from Aspen Properties (version 9.0, Aspen Technology, Inc., Bedford, Massachusetts 01730, USA) and implemented via look-up tables. The gas phase behaviour was modeled as an ideal phase. The heat and mass transfer coefficients of the two phases are dependent on several property calculations such as diffusivity, viscosity, specific heat capacity, density and several other correlations. These correlations as well as the kinetic parameters for the CO_2-MEA-H_2O system used in this work can be found in the work of Harun [26].

2.5. Model Discretization

The partial differential equations in the column equations are converted to ordinary differential equations using the method of lines. This results in the conversion of the derivatives with respect to the length of the column being discretized into five stages, thus rendering the partial differential equations as ordinary differential equations. After discretization, the model described in this section can be written in a compact form as shown in the system of Differential Algebraic Equations (DAEs) below:

$$\dot{x} = f(x, z, u, p), \tag{19}$$

$$g(x, z, u, p) = 0, \tag{20}$$

$$y = h(x, z, u, p), \tag{21}$$

where $x \in \mathbb{R}^{103}$ denotes the differential states, $z \in \mathbb{R}^7$ denotes the algebraic states, $u = [F_L, Q_{reb}]$ denotes the manipulated input solvent flow rate in L/s and reboiler heat input in MW, $p = [F_G]$ denotes the uncontrolled input flue gas flow rate, $y = [y_{CO_2}, T_{reb}]$ denotes the outputs CO_2 flow rate in the treated gas in kg/h and reboiler temperature in K. The system of DAEs described in Equations (19) – (21) will be used in the MPC and EMPC designs.

3. Control Problem Formulation and Design

In this work, two control objectives are considered. The first is reduction of the amount of CO_2 released into the atmosphere. The second control objective is the minimization of the heat consumed at the regeneration section of the plant. In the case of MPC, these objectives have been translated to tracking the optimal CO_2 emission and reboiler temperature. Hard constraints are imposed to ensure that the temperature of the solvent in the reboiler does not go beyond the point where the solvent degrades (120 °C). The designed controller must achieve these two control objectives while rejecting disturbances in the form of fluctuations in the flue gas flow rate originating from the power plant and time-varying steam price due to the time-varying electricity demand.

3.1. Conventional Set-Point Tracking MPC

To operate chemical plants in a time-varying manner, process economic optimization and control are usually decomposed into a hierarchical structure or layer. In this work, the conventional set-point tracking MPC takes a two-step approach, namely: steady-state optimization (SSO) and dynamic tracking. First, a steady-state optimization is performed to determine the optimal operating point for MPC. The optimization problem is as follows:

Optimization problem 1. Steady-state optimization problem

$$\min_{x_s, z_s, u_s, \epsilon} \quad l = \alpha_1 u_{2,s} + \alpha_2 \epsilon, \tag{22}$$

$$\text{s.t.} \quad f(x_s, z_s, u_s, p) = 0, \tag{23}$$

$$g(x_s, z_s, u_s, p) = 0, \tag{24}$$

$$y_s = h(x_s, z_s, u_s, p), \tag{25}$$

$$y_{1,s} - y_{limit} \leq \epsilon, \tag{26}$$

$$\epsilon \geq 0, \tag{27}$$

$$385.15 \leq y_{2,s} \leq 393.15, \tag{28}$$

$$0.1 \leq u_{1,s} \leq 2.0, \tag{29}$$

$$0.05 \leq u_{2,s} \leq 0.5, \tag{30}$$

where y_{limit} is the threshold value of CO_2 above which cost will be incurred in kg/h, ϵ is a variable which represents the amount of CO_2 released above the threshold value in kg/h, and α_1 indicates the price of steam from the power plant, whereas α_2 is the carbon price/tax.

Equation (22) is the economic cost function, Equations (23)–(25) are the steady-state of Equations (19)–(21) described in Section 2 in compact form, and Equations (26) and (27) form a soft constraint on the CO_2 emissions. Equations (28)–(30) are hard constraints on the reboiler temperature, solvent flow rate and reboiler heat input, respectively. The constraint on the reboiler temperature is to ensure that the temperature does not fall below the minimum CO_2 regeneration temperature or go beyond the degradation temperature of the solvent.

MPC is an optimal control strategy that has gained popularity in the chemical process industry. The MPC controller is designed to track a predetermined optimal steady-state value. The cost function is quadratic and penalizes the deviations of the state and input trajectories from the set-point. The MPC used in this work takes the form shown below:

Optimization problem 2. Tracking MPC optimization problem

$$\min_{u(t)\in S(\Delta)} \int_{t_k}^{t_k+N_p} ((y_s - \tilde{y}(t))^T Q(y_s - \tilde{y}(t)) + (u_s - u(t))^T R(u_s - u(t)))dt, \tag{31}$$

$$\text{s.t.} \quad \dot{\tilde{x}} = f(\tilde{x}(t), \tilde{z}(t), u(t), p(t)), \tag{32}$$

$$g(\tilde{x}(t), \tilde{z}(t), u(t), p(t)) = 0, \tag{33}$$

$$\tilde{y} = h(\tilde{x}(t), \tilde{z}(t), u(t), p(t)), \tag{34}$$

$$\tilde{x}(t_k) = x(t_k), \tag{35}$$

$$385.15 \le \tilde{y}_2(t) \le 393.15, \tag{36}$$

$$0.1 \le u_1(t) \le 2.0, \tag{37}$$

$$0.05 \le u_2(t) \le 0.5, \tag{38}$$

where $S(\Delta)$ is the family of piece-wise constant functions with a sampling time Δ, N_p is the prediction horizon, $\tilde{x}$, $\tilde{y}$ and $\tilde{z}$ denote the predicted differential, output and algebraic states, respectively. Q and R are the weighting matrices on the outputs and control actions, respectively. Their determination and values are described in the results section.

In Optimization problem 2, Equation (31) is the cost function; Equations (32)–(34) represent the PCC model; Equation (35) is the initial condition of the system at time instant t_k; and Equations (36)–(38) are the same as the constraints in Optimization problem 1. At the next sampling time, Optimization problem 2 is solved again. This forms an implicit feedback control law.

3.2. Economic Model Predictive Control

EMPC integrates steady-state optimization and dynamic tracking by directly optimizing economic performance in the dynamic control problem. The difference between MPC and EMPC is the cost function. While MPC uses a quadratic cost function and require set-point updating, EMPC uses a general economic cost function and does not require any set-points.

The EMPC at time instant t_k is formulated as the optimization problem below:

Optimization problem 3. EMPC optimization problem

$$\min_{u(t)\in S(\Delta)} \int_{t_k}^{t_k+N_p} (\alpha_1(t)u_2(t) + \alpha_2\epsilon(t))dt, \tag{39}$$

$$\text{s.t.} \quad \dot{\tilde{x}} = f(\tilde{x}(t), \tilde{z}(t), u(t), p(t)), \tag{40}$$

$$g(\tilde{x}(t), \tilde{z}(t), u(t), p(t)) = 0, \tag{41}$$

$$\tilde{y} = h(\tilde{x}(t), \tilde{z}(t), u(t), p(t)), \tag{42}$$

$$\tilde{x}(t_k) = x(t_k), \tag{43}$$

$$\tilde{y}_1(t) - y_{limit} \leq \epsilon(t), \tag{44}$$

$$\epsilon(t) \geq 0, \tag{45}$$

$$385.15 \leq \tilde{y}_2(t) \leq 393.15, \tag{46}$$

$$0.1 \leq u_1(t) \leq 2.0, \tag{47}$$

$$0.05 \leq u_2(t) \leq 0.5. \tag{48}$$

In Optimization problem 3 above, Equation (39) is the cost function that minimizes the accumulated economic cost over the prediction horizon of N_p sampling periods. At the next sampling time, Optimization problem 3 is solved again.

3.3. Implementation

The resulting optimization problems are highly nonlinear with algebraic equations, thus the simultaneous approach to dynamic optimization is used in this work. This is because the simultaneous approach is more robust to problems with such properties compared to the sequential approach [27]. This approach has been presented in Biegler [28] and is summarized below:

- The continuous time differential and algebraic model equations are discretized by approximating the state and control profiles by a family of polynomials on finite elements. This involves dividing the control horizon into a number of finite elements with the size of each element corresponding to one sampling time. Within each element, the state and input profiles are approximated by a family of polynomials. In this work, Radau orthogonal polynomials in Lagrange form is used.
- The dynamic optimization problem is then formulated as a large scale nonlinear programming (NLP) problem.
- The NLP problem is solved using a computationally efficient solver that exploits the sparsity in the resulting matrix.

The formulation of the optimization problem is presented below:

Optimization problem 4. Reformulated optimization problem.

$$\min_{\tilde{x}_{i,k}, \tilde{z}_{i,k}, u_{i,k}} \sum_{i=0}^{N-1} l_e(\tilde{x}_i, \tilde{z}_i, u_i, p_i), \tag{49}$$

$$\text{s.t.} \quad \sum_{j=1}^{K} \ell_j(\tau_k)\tilde{x}_{i,j} - h_i f(\tilde{x}_{i,k}, \tilde{z}_{i,k}, u_{i,k}, p_i) = 0 \quad i = 1, ..., N \quad k = 1, ..., K, \tag{50}$$

$$g(\tilde{x}_{i,k}, \tilde{z}_{i,k}, u_{i,k}, p_i) = 0 \quad i = 1, ..., N \quad k = 1, ..., K, \tag{51}$$

$$\tilde{x}_{i+1,0} = \sum_{j=0}^{K} \ell_j(1)\tilde{x}_{i,j} \quad i = 1, ..., N-1, \tag{52}$$

$$u_{i,k} = u_{i,0} \quad i = 1, ..., N-1 \quad k = 1, ..., K, \tag{53}$$

$$u_{i,k} \in \mathbb{U} \quad i = 1, ..., N-1 \quad k = 1, ..., K, \tag{54}$$

$$x_{i,k} \in \mathbb{X} \quad i = 1, ..., N \quad k = 1, ..., K, \tag{55}$$

where

$$\ell_j(\tau) = \prod_{k=1,k\neq j}^{K} \frac{\tau - \tau_k}{\tau_j - \tau_k}. \tag{56}$$

In Optimization problem 4 above, subscripts i and k represent the i-th finite element in the prediction horizon and k-th collocation point in each finite element, respectively. $\ell_j(\tau_k)$ represents the Lagrange interpolation profile in a finite element, h is the length of a finite element and K is the number of collocation points in an element. τ is a root in the interpolating polynomial. Equation (49) is the objective function to be minimized, Equation (50) is an implicit formulation of the approximated model, Equation (51) is the algebraic equations, Equation (52) is a constraint that ensures that the beginning of an element is equal to the end of the previous element i.e., continuity constraint, Equation (53) ensures that the inputs are constant within an element similar to a zero-order hold (ZOH) fashion, and Equations (54) and (55) are inputs and state constraints. The output equation and soft constraints can also be added to this formulation. However, it has been omitted for simplicity. It should be noted that, in this formulation, the current state of the plant is enforced using the state constraints at $i = 0$ as the current state or measurement from the plant is not required in the formulation.

The model as well as the optimization problems (steady-state, EMPC and MPC) were implemented in Python (version 2.7) and casADi (version 3.4.0)—a software framework to facilitate the implementation and solution to optimal control problems using automatic differentiation [29]. The dynamic optimization problems (MPC and EMPC) were formulated as Optimization problem 4 and solved using the nonlinear programming (NLP) solver, IPOPT [30].

4. Simulations, Results and Discussion

In this section, the results of this study are presented. First, the determination of the tuning parameters of MPC is presented. Following this, the case of time-varying operation of the PCC plant is considered. Within this case, a suitable update strategy for MPC is determined and the performance of both controllers with and without carbon tax-free limit is investigated. Finally, the performance of the controllers under uncertainty in the flue gas flow rate and CO_2 concentration are presented and briefly discussed.

4.1. Simulation Setup

The prediction and control horizons for both MPC and EMPC was set as 360 min with 10 min being the sampling time (N in Optimization problem 4 is 36). The number of collocation points within each element (K in Optimization problem 4 was fixed at 3). No terminal constraints or costs were used in this work at the expense of using a sufficiently long horizon. The PCC plant configuration and flue gas condition are shown in Table 1.

Table 1. PCC plant configuration and flue gas condition. Adapted from [20]. PCC: Post-combustion CO_2 Capture.

Property	Value
Packing properties (Absorber and desorber)	
Column internal diameter m), D_c	0.43
Packing height (m)	6.1
Packing type	IMTP #40
Nominal packing size (m)	0.038
Specific packing area (m^2/m^3)	143.9
Flue gas condition	
Temperature (K)	319.7
Volumetric flow rate (m^3/s)	0.0832
CO_2 mole fraction	0.15
N_2 mole fraction	0.80
MEA mole fraction	0.00
H_2O mole fraction	0.05
Lean-rich heat exchanger	
Volume of tube side (m^3), V_{tube}	0.016
Volume of shell side (m^3), V_{shell}	0.205
Overall heat transfer coefficient (J/Ks), UA	1899.949

As can be observed from Table 1, the plant configuration used in this study is a pilot plant. The reason for this choice stems from the fact that several authors have used this configuration in the study of flexible operation and control of the PCC process [9,20]. Again, this configuration has been used to validate the absorber and desorber models on which this work is based on [20]. The scale of the plant is not expected to affect the outcomes of this work. This is because the primary goal of this work is to compare the ease of flexible operation and performance (economic and energy efficiency) improvements of EMPC over conventional MPC when applied to the PCC process. Thus, the study conducted is relative and not absolute. The results of this study can therefore be used irrespective of the size of the plant under study.

4.1.1. Time-Varying Steam Price and Flue Gas Flow Rate

In a typical power plant, the operation is periodic everyday (24 h). The demand for electricity is usually low at dawn and rises to a peak in the middle of the day and then drops to a low at night. A consequence of changing electricity demand is that the flue gas flow rate fluctuates in a similar manner affecting the operation of the attached post combustion capture plant. Again, during the high peak periods, steam, which is required in the regeneration section, may be unavailable since it will be needed for electricity generation. It is therefore important to determine how a deployed controller in the PCC plant reacts to such changes. The selected MPC controller was compared to EMPC in terms of economic performance. To assess the performance of EMPC and MPC for this kind of operation, the 24-h operation was scaled to 5 h. Figure 3 shows the shape of a periodic disturbance for both the steam price and the flue gas flow rate. The demand for electricity is low in the first hour which translates into a low flue gas flow rate. It then ramps up for half an hour and remains constant for 2 h before ramping down. In addition, as the demand for electricity increases, a decision has to be made whether to use the steam to generate more electricity or channel it to the PCC plant for CO_2 regeneration. This decision is accounted for as an increase in the price of steam. The nominal value for the price of steam was taken from the study of He et al. [9].

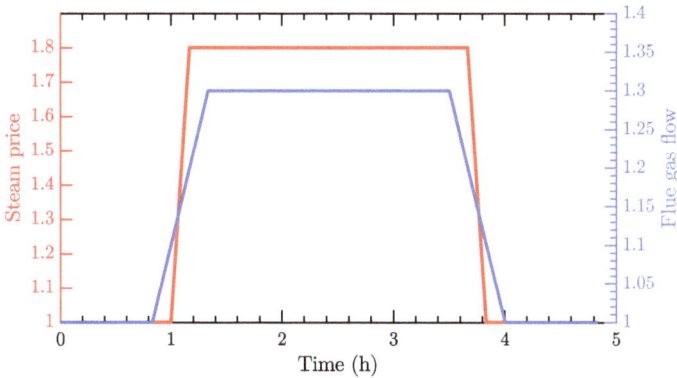

Figure 3. Scaled hourly changes in steam price (Nominal value: $0.01/kg) and flue gas flow rate (Nominal value: 0.0832 m^3/s). Price (red), Flue gas flow (blue).

4.1.2. Tracking MPC Parameter Tuning

To ensure fair comparison of the two controllers, the MPC need to be well tuned. Several MPC tuning techniques exist. However, in this work, the MPC was tuned such that the closed-loop trajectories and economic performance of MPC was as close as possible to that of EMPC. This was achieved using two step tests. The first step test corresponds to a state where the plant transitions from a peak electricity demand period (steam price is 1.8 and flue gas flow is 1.3) to a low electricity demand period (steam price is 1.0 and flue gas flow is 1.0). The second step test is the reverse of the first step test where the plant transitions from a low electricity demand period to a peak electricity demand period. Three MPCs with different tuning parameters (as shown in Table 2) and EMPC were designed and tested on the step tests. The resulting closed-loop trajectories are shown in Figures 4 and 5.

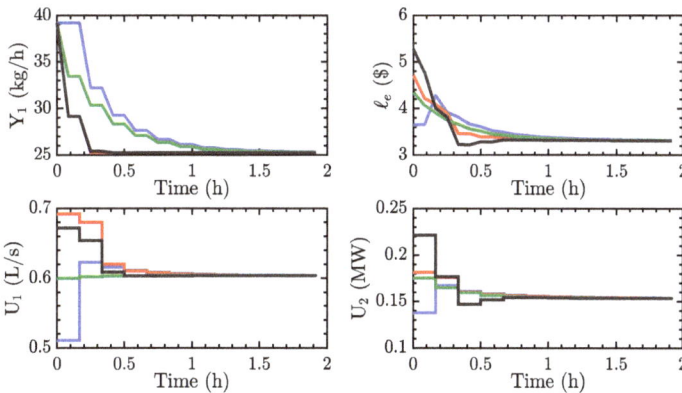

Figure 4. Closed-loop trajectories of EMPC and different MPC tuning parameters starting from a higher economic cost to a lower one. EMPC (blue), MPC I (red), MPC II (green), MPC III (black). The average economic performances are 4.700, 4.768, 4.713, 4.692 for EMPC, MPC I, MPC II, MPC III, respectively; (**top-left**) mass flow rate of CO_2 in the treated flue gas; (**top-right**) economic performance of controller; (**bottom-left**) volumetric flow rate of solvent; (**bottom-right**) heat input to reboiler. EMPC: Economic Model Predictive Control

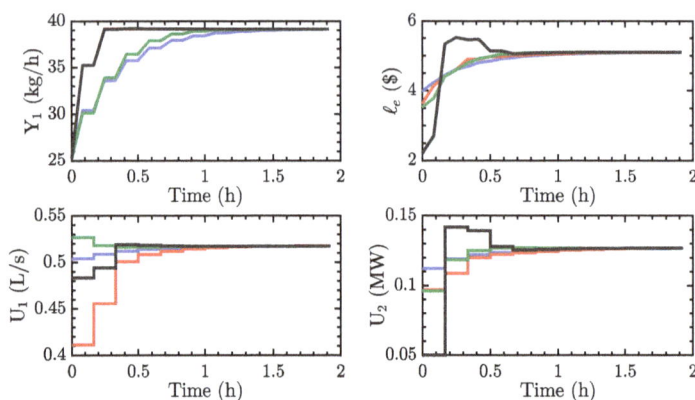

Figure 5. Closed-loop trajectories of EMPC and different MPC tuning parameters starting from a lower economic cost to a higher one. EMPC (blue), MPC I (red), MPC II (green), MPC III (black). The average economic performances are 3.654, 3.668, 3.650, 3.685 for EMPC, MPC I, MPC II, MPC III, respectively; (**top-left**) mass flow rate of CO_2 in the treated flue gas; (**top-right**) economic performance of controller; (**bottom-left**) volumetric flow rate of solvent; (**bottom-right**) heat input to reboiler.

Based on Figures 4 and 5, MPC II was chosen for further tests. The weighting matrices are shown below. These parameters imply weakly tracking the CO_2 emission as compared to the reboiler temperature and putting less weights on the inputs. The weights on the inputs are to ensure that the MPC does not make excessive control actions as compared to that of the EMPC:

$$Q = \begin{bmatrix} a & 0 \\ 0 & b \end{bmatrix}, \quad R = \begin{bmatrix} c & 0 \\ 0 & d \end{bmatrix}$$

Table 2. Tuning parameters for the three different MPCs tested. MPC: Model Predictive Control.

MPC	a	b	c	d
I	0.0001	0.01	0	0
II	0.0001	0.01	0.05	0.05
III	1.0	0.01	0.05	0.05

4.2. MPC Set-Point Update Strategy

In this work, the set-point update strategy is based on the frequency at which the steady-state optimization (SSO) of is carried out. Thus, the optimal steady-state operating point or set-point is determined at regular time intervals taking into consideration all information affecting the plant. Table 3 shows different MPC set-point update strategies, and their impact on the economic performance and heat duty compared to that of EMPC. For example, in strategy 1, the set-point is kept constant throughout the operation of the plant irrespective of the changes in flue gas flow rate and price of steam. In strategies 2–4, on the other hand, the SSO is computed and the set-point of MPC is updated at regular time intervals according to the update frequency. It can be observed in Table 3 that not updating the set-point is the worst strategy economically. This is because, when the set-point is kept constant throughout the operation of the PCC plant, any fluctuation in the flue gas flow rate is rejected at the expense of high operating costs. This is shown in Figure 6. This gets better as the frequency of the updates increase. Updating the set-point every sampling time (10 min) is the best option, however, it still leaves an economic and heat duty gap between MPC and EMPC. This could be because the SSO layer does not take into account future economic information resulting in some economic losses and ineffective use of steam for regeneration.

Table 3. Different MPC set-point update strategies and percent decrease in economics and heat duty. Heat duty is defined as the ratio of the amount of reboiler heat input to the mass of CO_2 absorbed.

Strategy	Description	Avg. Cost (%)	Avg. Heat Duty (%)
1	No update/Fixed operating point	5.83	6.52
2	Update every hour	2.02	4.27
3	Update every 30 min	0.79	2.99
4	Update every 10 min (sampling time)	0.03	2.32

4.3. Operation under Different Carbon Tax Policies

In this set of simulations, the two operating scenarios are considered. In the first scenario, a tax-free emission limit is imposed. Emission beyond this limit leads to economic penalties. In the second scenario, charges are applied to any CO_2 emitted. These scenarios were implemented as soft constraints in Optimization problems 1 and 3. Considering the scale of the PCC plant used in this study, y_{limit} in Equations (26) and (45) was fixed at a value of 10.00 kg/h and 0.00 kg/h for the first and second scenarios, respectively. The horizon (prediction and control) for both MPC and EMPC is fixed at 360 min with a sampling time of 10 min (this implies N = 36). An MPC-SSO structure is used with MPC set-point update frequency of 1 h. The weighting matrices of the MPC are the same as the ones presented in Section 4.1. All information (flue gas flow rate and price of steam) is available to both controllers. In the case of MPC, the SSO uses the instantaneous information on the flue gas flow rate and the steam price to conduct steady-state economic optimization and updates the set-point of MPC every hour while in the case of EMPC, future information is used to determine the optimal input to the PCC plant at every sampling time. The MPC uses future information on the flue gas flow rate to determine the optimal input to the PCC plant to reach the target set-point every sampling time. The simulation time is 5 h with time-varying steam price and flue gas flow rate as presented in Figure 3.

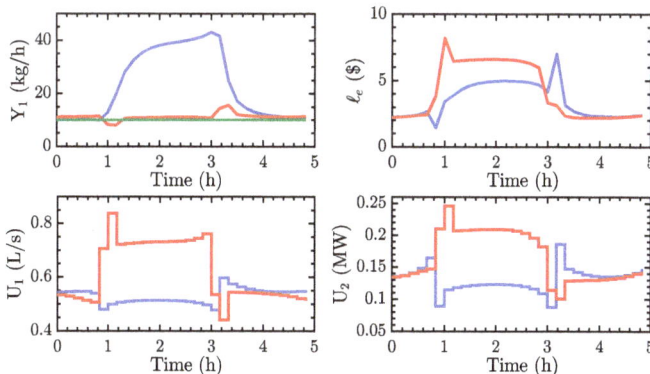

Figure 6. CO_2 emission, input and economic cost trajectories of the PCC plant with carbon tax set at $50.00 per tonne CO_2. The set-point of MPC is not updated. EMPC (blue), MPC (red), Emission limit (green); (**top-left**) mass flow rate of CO_2 in the treated flue gas; (**top-right**) economic performance of controller; (**bottom-left**) volumetric flow rate of solvent; (**bottom-right**) heat input to reboiler. PCC: Post-combustion CO_2 Capture.

4.3.1. Carbon Tax with Tax-Free Emission Limit

Four sets of simulations are conducted in this scenario. The settings remained the same in each of the simulations with the exception of the carbon tax, which varied from simulation to simulation. The carbon tax considered are $50.00, $100.00, $150.00 and $200.00 per tonne CO_2. The goal is to determine the operating strategy, compare the economic performance and the influence of the carbon tax on the performance of the two controllers under scenario one.

Figures 7 and 8 show the CO_2 emission, input and economic cost trajectories for the PCC process for each controller under scenario one at carbon tax of $50.00 and $200.00, respectively. At a carbon tax of $50.00 per tonne CO_2 (Figure 7), both EMPC and MPC-SSO schemes either stayed on or above the emission limit. This is because there is no incentive for capturing more CO_2 and operating below the emission limit. In addition, the effect of not considering the process economics every sampling time for MPC can be seen in the CO_2 emission trajectories of the system. EMPC starts to quickly violate the emission limit as a result of the increase in flue gas flow rate and heat price while MPC sluggishly does that. The sluggishness could be because MPC tries to reject the ramp change in the flue gas flow rate until the set-point is updated by the SSO. Again, in the last hour when the flue gas flow rate and heat price decrease, EMPC rapidly decreases the emissions compared to that of MPC. Although MPC had information about the ramp increase and decrease in flue gas flow rate, it had no idea of the change in steam price, hence the sluggishness. This information became available to MPC only after the set-point of MPC was updated by the SSO. However, in the case of EMPC, both heat price and flue gas flow rate were considered. Again, in Figure 8, both controllers tried to stay above the emission limit—however, with much less emissions compared to when the carbon tax is $50.00 per tonne CO_2. This is because there is much more weight now on emissions compared to the cost of steam. It can also be observed that the CO_2 emission trajectories of both MPC and EMPC are different. MPC operates well below the emission limit and uses much more resources without any economic gains. Another reason is that, at a carbon tax of $200.00, the control dynamics change from slow tracking of CO_2 to fast tracking. However, MPC had already been tuned for a single case and needed re-tuning. This resulted in the increase in both the average economic cost and heat duty (Figures 9 and 10).

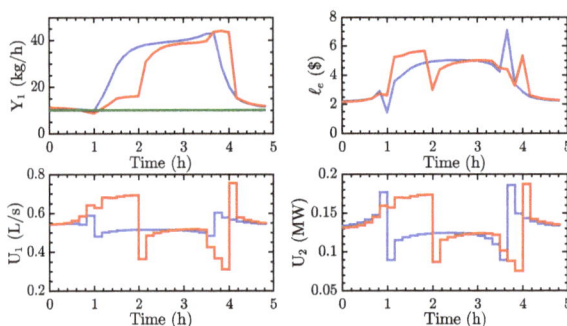

Figure 7. CO_2 emission, input and economic cost trajectories of the PCC plant for scenario 1 with carbon tax set at $50.00 per tonne CO_2. EMPC (blue), MPC (red), Emission limit (green); (**top-left**) mass flow rate of CO_2 in the treated flue gas; (**top-right**) economic performance of controller; (**bottom-left**) volumetric flow rate of solvent; (**bottom-right**) heat input to reboiler.

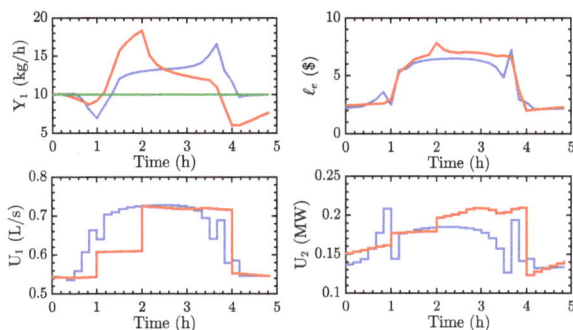

Figure 8. CO_2 emission, input and economic cost trajectories of the PCC plant for scenario 1 with carbon tax set at $200.00 per tonne CO_2. EMPC (blue), MPC (red), Emission limit (green); (**top-left**) mass flow rate of CO_2 in the treated flue gas; (**top-right**) economic performance of controller; (**bottom-left**) volumetric flow rate of solvent; (**bottom-right**) heat input to reboiler.

The average economic performances and heat duties of MPC and EMPC under scenario one and different carbon tax is shown in Figures 9 and 10. As can be observed in Figure 9, the economic performance gap between MPC and EMPC remained fairly constant as the carbon tax increases until $150.00, where the gap widened. Similarly, the gap between the average heat duty remained the same until $150.00 where it increases. One possible explanation for this is that, in this scenario, when the PCC plant operates on the emission limit, the carbon tax is not charged. Only the heat input is optimized. The dynamics of the economics therefore changes from heat input to CO_2 emissions and back to heat input. This shifting dynamics could have resulted in the increased gap between MPC and EMPC when the CO_2 price is high. In addition, one key observation is that the gap between MPC and EMPC is smaller for the economic cost compared to the heat duty plot. This implies that though the economics seem to be close, the heat input is not efficiently used in the case of MPC compared to EMPC. Thus, to ensure a better economic performance and heat duty of MPC, re-tuning or increased update frequency is probably necessary. However, determining the optimal tuning parameters of MPC is not always obvious and is a major challenge in MPC design. EMPC on the other hand is able to determine the optimal operating strategy without requiring an update in the controller settings and change in set-point update strategy. This makes EMPC well suited for flexible operation of the PCC process while being energy efficient.

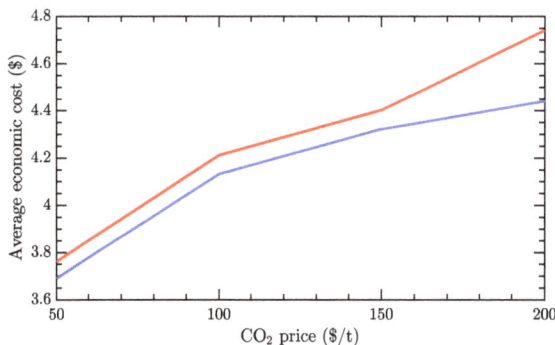

Figure 9. Average economic performance of the controllers under scenario 1 at different CO_2 prices. EMPC (blue), MPC (red).

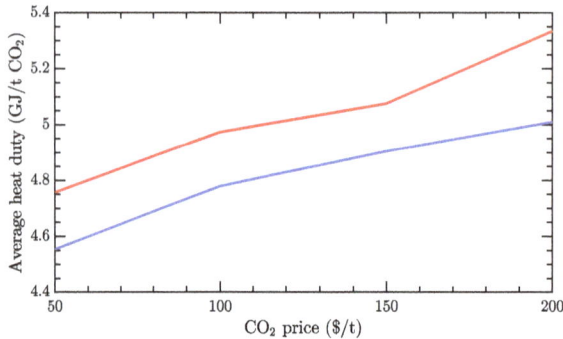

Figure 10. Average heat duty of the controllers under scenario 1 at different CO_2 prices. EMPC (blue), MPC (red).

4.3.2. Carbon Tax without Tax-Free Emission Limit

The simulations conducted in this section and their purpose is similar to that of scenario one. The controller settings therefore remain the same. The only difference between the two scenarios is how the carbon tax is enforced. In this scenario, charges are applied to any CO_2 released into the atmosphere. In Figures 11 and 12, the CO_2 emission trajectory for both controllers remain fairly the same in the first three hours and begin to differ slightly in the last two hours. However, this difference is not significant enough to cause much differences in the performance metrics (economic cost and heat duty) when the carbon tax increases. This can be observed in Figures 13 and 14 where the gaps in the average economic cost and average heat duty are almost constant. Similar to the results in scenario one, the gap in the heat duty between MPC and EMPC is wider compared to their economic performance. Thus, under this scenario, changes in the carbon tax does not affect the performance of the controllers. This could be because, throughout the operation, the cost of CO_2 emission dominates the decision making as compared to the case of scenario one where, when the PCC plant operates on the emission limit, CO_2 emissions are not penalized. The difference in economic performance of both controllers can therefore be solely attributed to the timescale separation between MPC and SSO.

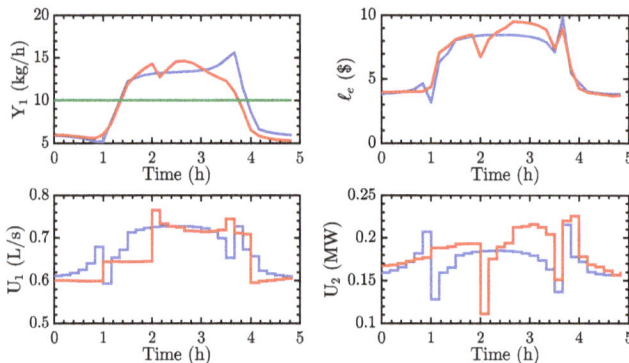

Figure 11. CO_2 emission, input and economic cost trajectories of the PCC plant for scenario 2 with carbon tax set at $50.00 per tonne CO_2. EMPC (blue), MPC (red), Emission limit (green); (**top-left**) mass flow rate of CO_2 in the treated flue gas; (**top-right**) economic performance of controller; (**bottom-left**) volumetric flow rate of solvent; (**bottom-right**) heat input to reboiler.

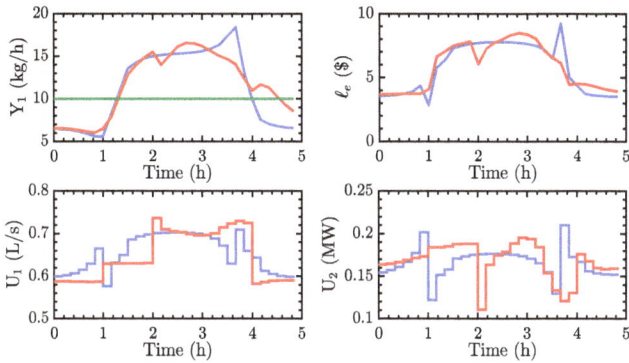

Figure 12. CO_2 emission, input and economic cost trajectories of the PCC plant for scenario 2 with carbon tax set at $200.00 per tonne CO_2. EMPC (blue), MPC (red), Emission limit (green); (**top-left**) mass flow rate of CO_2 in the treated flue gas; (**top-right**) economic performance of controller; (**bottom-left**) volumetric flow rate of solvent; (**bottom-right**) heat input to reboiler.

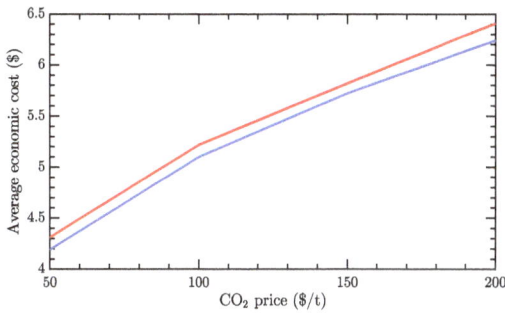

Figure 13. Average economic performance of the controllers under scenario 2 at different CO_2 prices. EMPC (blue), MPC (red).

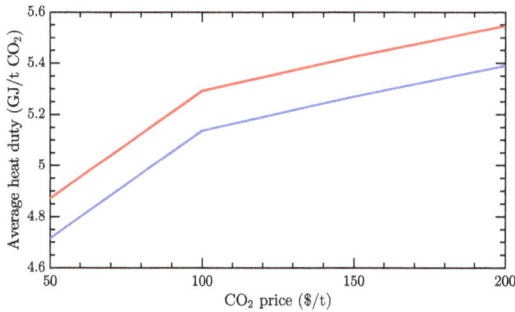

Figure 14. Average heat duty of the controllers under scenario 2 at different CO_2 prices. EMPC (blue), MPC (red).

4.4. Operation under Uncertainty

The presence of disturbances is a problem in many practical MPC applications. This could result in performance degradation or infeasibility. In this section, we subject the PCC process to disturbances in the flue gas flow rate and CO_2 concentration. This is to test the susceptibility of the controllers to uncertainties. The uncertainties were obtained by adding Gaussian white noise to the flue gas flow rate and flue gas concentration, respectively. The simulation for each uncertainty was repeated many times, each time with a different random number seed value (varied from 1 to 10). Only one of plots for each uncertainty experiment has been presented. The controller design for both MPC and EMPC remain the same as the previous sections.

Figures 15 and 16 show the trajectories of CO_2 emission, economic cost and inputs for both controllers under uncertainty in flue gas flow rate and CO_2 concentration, respectively. As can be observed in the figures, both controllers showed the satisfactory performance with very little performance degradation. Thus, under small uncertain disturbances, both controllers perform the same when there is uncertainty in the flue gas flow rate and concentration. One reason for this observation could be that the PCC plant is insensitive to small disturbances. However, due to numerical challenges encountered at high disturbances, further tests were not conducted. Uncertainty in the steam price was not considered because it is only considered in the cost function and does not affect the process dynamics. Thus, both controllers are expected to be affected in the same way.

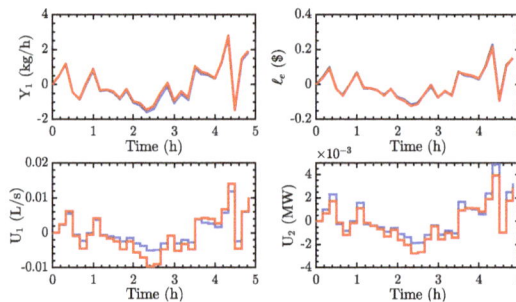

Figure 15. CO_2 emission, inputs and economic cost trajectories of the PCC plant under uncertainty in flue gas flow rate. Nominal value =1.0. Uncertainty generated by adding Gaussian white noise with zero mean and standard deviation $\sigma = 0.1$, to the flue gas flow rate. EMPC (blue), MPC (red); **(top-left)** mass flow rate of CO_2 in the treated flue gas; **(top-right)** economic performance of controller; **(bottom-left)** volumetric flow rate of solvent; **(bottom-right)** heat input to reboiler.

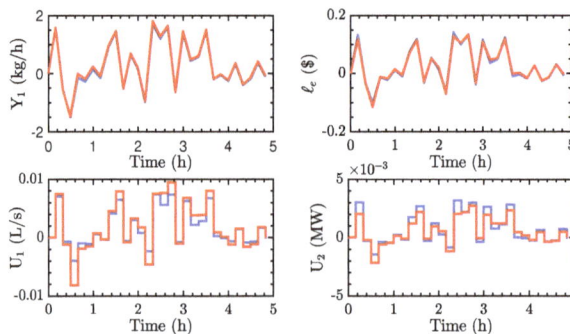

Figure 16. CO_2 emission, inputs and economic cost trajectories of the PCC plant under uncertainty in flue gas CO_2 concentration. Uncertainty generated by adding Gaussian white noise with zero mean and standard deviation $\sigma = 0.06$, to the flue gas CO_2 concentration. EMPC (blue), MPC (red); **(top-left)** mass flow rate of CO_2 in the treated flue gas; **(top-right)** economic performance of controller; **(bottom-left)** volumetric flow rate of solvent; **(bottom-right)** heat input to reboiler.

5. Conclusions

Attaching an amine-based PCC plant to a power plant reduces the overall efficiency of the power plant especially in the wake of frequent changes in electricity demand and diversification of electricity generation sources. However, operating the PCC plant in a flexible manner can reduce the impact of attaching a PCC plant to a power plant. This study investigates the operation of a PCC process under time-varying disturbances in flue gas flow rate and steam price using EMPC. The economic cost and heat duty of the operation of the PCC process under EMPC was compared to that of tracking MPC. Two scenarios, namely with and without emission limits, were also investigated. This was achieved by first presenting a first principle model of the PCC process—followed by formulation and design of both controllers and the solution to the resulting optimization problems using the simultaneous approach for dynamic optimization.

The results show that operating the PCC plant at a constant set-point leads to significant economic losses and inefficient utilization of the steam from the power plant. Specifically, operating the PCC plant in a time-varying fashion using EMPC improves the operation of the process up to about 6% in terms of economic efficiency and 7% in terms of heat duty or energy efficiency. In addition, a nearly constant gap between the average economic cost and heat duty of the process operation at different carbon prices using EMPC and MPC was observed with the heat duty having a larger gap compared to the economic cost. However, under scenario one, the gap further increased at a CO_2 price of $200.00 per tonne, whereas the gap remained fairly the same under scenario two. Although this gap can be reduced by increasing the frequency of updating the set-point of MPC and/or re-tuning MPC altogether, obtaining the optimal tuning parameters of MPC is not always obvious. EMPC, on the other hand, was able to determine the appropriate control policy depending on the scenario without further tuning or updating. This makes it suitable for flexible operation of the PCC process while being energy efficient. Finally, it was observed in this study that both controllers are robust to small uncertain disturbances in the flue gas flow rate and concentration. The insight gained from this study is useful in an era where our energy demands need to be met while ensuring that emissions are at a minimum.

Author Contributions: All authors contributed to the conceptualization of this work, simulation design and analysis of results. B.D.-N. and S.L. conducted the simulations and prepared the manuscript. J.L. supervised this research and reviewed the manuscript.

Funding: This research was funded by Natural Sciences and Engineering Research Council of Canada (NSERC) grant number RGPIN 435767.

Conflicts of Interest: The authors declare no conflict of interest.

Abbreviations

The following abbreviations are used in this manuscript:

CCS	Carbon Capture and Storage
CSTR	Continuously Stirred Tank Reactor
EMPC	Economic Model Predictive Control
eNTRL	electrolyte Non-Random Two Liquids
DAEs	Differential Algebriac Equations
MEA	Monoethanolamine
MPC	Model Predictive Control
NLP	Nonlinear Programming
ODEs	Ordinary Differential Equations
PCC	Post-combustion CO_2 Capture
PDEs	Partial Differential Equations
SSO	Steady-State Optimization
ZOH	Zero-Order Hold

References

1. Miller, B.G. *Fossil Fuel Emission Control Technologies: Stationary Heat and Power Systems*, 1st ed.; Elsevier: Waltham, MA, USA, 2015.
2. Metz, B.; Intergovernmental Panel on Climate Change (Eds.) *IPCC Special Report on Carbon Dioxide Capture and Storage*; Cambridge University Press: Cambridge, UK, 2005.
3. Davison, J.; Mancuso, L.; Ferrari, N. Costs of CO_2 Capture Technologies in Coal Fired Power and Hydrogen Plants. *Energy Procedia* **2014**, *63*, 7598–7607. [CrossRef]
4. Ziaii, S.; Rochelle, G.T.; Edgar, T.F. Dynamic Modeling to Minimize Energy Use for CO_2 Capture in Power Plants by Aqueous Monoethanolamine. *Ind. Eng. Chem. Res.* **2009**, *48*, 6105–6111. [CrossRef]
5. Lin, Y.J.; Wong, D.S.H.; Jang, S.S.; Ou, J.J. Control Strategies for Flexible Operation of Power Plant with CO_2 Capture Plant. *AIChE J.* **2012**, *58*, 2697–2704. [CrossRef]
6. Bedelbayev, A.; Greer, T.; Lie, B. Model Based Control of Absorption Tower for CO_2 Capturing. In Proceedings of the 49th International Conference of Scandinavian Simulation Society (SIMS 2008), Oslo, Norway, 7–8 October 2008.
7. Panahi, M.; Skogestad, S. Economically Efficient Operation of CO_2 Capturing Process Part I: Self-Optimizing Procedure for Selecting the Best Controlled Variables. *Chem. Eng. Process.* **2011**, *50*, 247–253. [CrossRef]
8. Panahi, M.; Skogestad, S. Economically Efficient Operation of CO_2 Capturing Process. Part II. Design of Control Layer. *Chem. Eng. Process.* **2012**, *52*, 112–124. [CrossRef]
9. He, Z.; Sahraei, M.H.; Ricardez-Sandoval, L.A. Flexible Operation and Simultaneous Scheduling and Control of a CO_2 Capture Plant Using Model Predictive Control. *Int. J. Greenh. Gas Control* **2016**, *48*, 300–311. [CrossRef]
10. Bankole, T.; Jones, D.; Bhattacharyya, D.; Turton, R.; Zitney, S.E. Optimal Scheduling and Its Lyapunov Stability for Advanced Load-Following Energy Plants with CO_2 Capture. *Comput. Chem. Eng.* **2018**, *109*, 30–47. [CrossRef]
11. Mac Dowell, N.; Shah, N. The Multi-Period Optimisation of an Amine-Based CO_2 Capture Process Integrated with a Super-Critical Coal-Fired Power Station for Flexible Operation. *Comput. Chem. Eng.* **2015**, *74*, 169–183. [CrossRef]
12. Ellis, M.; Durand, H.; Christofides, P.D. A Tutorial Review of Economic Model Predictive Control Methods. *J. Process Control* **2014**, *24*, 1156–1178. [CrossRef]
13. Angeli, D.; Amrit, R.; Rawlings, J.B. On Average Performance and Stability of Economic Model Predictive Control. *IEEE Trans. Autom. Control* **2012**, *57*, 1615–1626. [CrossRef]
14. Liu, S.; Liu, J. Economic Model Predictive Control with Extended Horizon. *Automatica* **2016**, *73*, 180–192. [CrossRef]
15. Zeng, J.; Liu, J. Economic Model Predictive Control of Wastewater Treatment Processes. *Ind. Eng. Chem. Res.* **2015**, *54*, 5710–5721. [CrossRef]
16. Liu, S.; Zhang, J.; Liu, J. Economic MPC with Terminal Cost and Application to an Oilsand Primary Separation Vessel. *Chem. Eng. Sci.* **2015**, *136*, 27–37. [CrossRef]
17. Idris, E.A.; Engell, S. Economics-Based NMPC Strategies for the Operation and Control of a Continuous Catalytic Distillation Process. *J. Process Control* **2012**, *22*, 1832–1843. [CrossRef]
18. Mendoza-Serrano, D.I.; Chmielewski, D.J. Smart Grid Coordination in Building HVAC Systems: EMPC and the Impact of Forecasting. *J. Process Control* **2014**, *24*, 1301–1310. [CrossRef]
19. Touretzky, C.R.; Baldea, M. Integrating Scheduling and Control for Economic MPC of Buildings with Energy Storage. *J. Process Control* **2014**, *24*, 1292–1300. [CrossRef]
20. Harun, N.; Nittaya, T.; Douglas, P.L.; Croiset, E.; Ricardez-Sandoval, L.A. Dynamic Simulation of MEA Absorption Process for CO_2 Capture from Power Plants. *Int. J. Greenh. Gas Control* **2012**, *10*, 295–309. [CrossRef]
21. Whitman, W.G. The Two Film Theory of Gas Absorption. *Int. J. Heat Mass Trans.* **1962**, *5*, 429–433. [CrossRef]
22. Onda, K.; Takeuchi, H.; Okumoto, Y. Mass Transfer Coefficients between Gas and Liquid Phases in Packed Columns. *J. Chem. Eng. Jpn.* **1968**, *1*, 56–62. [CrossRef]
23. Danckwerts, P. The Reaction of CO_2 with Ethanolamines. *Chem. Eng. Sci.* **1979**, *34*, 443–446. [CrossRef]
24. Tobiesen, F.A.; Juliussen, O.; Svendsen, H.F. Experimental Validation of a Rigorous Desorber Model for CO_2 Post-Combustion Capture. *Chem. Eng. Sci.* **2008**, *63*, 2641–2656. [CrossRef]

25. Chilton, T.H.; Colburn, A.P. Mass Transfer (Absorption) Coefficients Prediction from Data on Heat Transfer and Fluid Friction. *Ind. Eng. Chem.* **1934**, *26*, 1183–1187. [CrossRef]
26. Harun, N. Dynamic Simulation of MEA Absorption Process for CO_2 Capture from Power Plants. Ph.D. Thesis, University of Waterloo, Waterloo, ON, USA, 2012.
27. Amrit, R.; Rawlings, J.B.; Biegler, L.T. Optimizing Process Economics Online Using Model Predictive Control. *Comput. Chem. Eng.* **2013**, *58*, 334–343. [CrossRef]
28. Biegler, L.T. *Nonlinear Programming: Concepts, Algorithms, and Applications to Chemical Processes*; Society for Industrial and Applied Mathematics: Philadelphia, PA, USA, 2010.
29. Andersson, J.A.E.; Gillis, J.; Horn, G.; Rawlings, J.B.; Diehl, M. CasADi—ASoftware Framework for Nonlinear Optimization and Optimal Control. *Math. Program. Comput.* **2018**. [CrossRef]
30. Wächter, A.; Biegler, L.T. On the Implementation of an Interior-Point Filter Line-Search Algorithm for Large-Scale Nonlinear Programming. *Math. Program.* **2006**, *106*, 25–57. [CrossRef]

![processes logo] *processes*

MDPI

Article

Dynamic Optimization of a Subcritical Steam Power Plant Under Time-Varying Power Load

Chen Chen and George M. Bollas *

Department of Chemical & Biomolecular Engineering, University of Connecticut, 191 Auditorium Road, Unit 3222, Storrs, CT 06269, USA; chen.chen@uconn.edu
* Correspondence: george.bollas@uconn.edu; Tel.: +1-860-486-6037

Received: 2 June 2018; Accepted: 21 July 2018; Published: 3 August 2018

Abstract: The increasing variability in power plant load in response to a wildly uncertain electricity market and the need to to mitigate CO_2 emissions, lead power plant operators to explore advanced options for efficiency optimization. Model-based, system-scale dynamic simulation and optimization are useful tools in this effort and are the subjects of the work presented here. In prior work, a dynamic model validated against steady-state data from a 605 MW subcritical power plant was presented. This power plant model was used as a test-bed for dynamic simulations, in which the coal load was regulated to satisfy a varying power demand. Plant-level control regulated the plant load to match an anticipated trajectory of the power demand. The efficiency of the power plant's operation at varying loads was optimized through a supervisory control architecture that performs set point optimization on the regulatory controllers. Dynamic optimization problems were formulated to search for optimal time-varying input trajectories that satisfy operability and safety constraints during the transition between plant states. An improvement in time-averaged efficiency of up to 1.8% points was shown to be feasible with corresponding savings in coal consumption of 184.8 tons/day and a carbon footprint decrease of 0.035 kg/kWh.

Keywords: power plants; supervisory control; dynamic simulation; dynamic optimization

1. Introduction

The excessive emissions of CO_2 from fossil-fueled power plants contribute to the greenhouse effect and global warming. Increasing the efficiency of power generation cycles and integration with CO_2 capture units are nowadays accepted as the most promising short-term approaches to reducing CO_2 emissions while we transition to renewable and carbon free energy sources [1,2]. Efficiency improvements can be achieved through the optimization of power plant operating strategies or through modification of plant design. For instance, new fossil-fueled power plants use a combination of steam and gas turbines to generate electricity, resulting in thermal efficiencies as high as 61% [3]. Moreover, modern coal-fired Rankine cycle systems can achieve efficiencies as high as 47% using ultra-supercritical boilers [4]. For instance, the commercial power plant of Lünen (Germany) burns low-sulfur hard coal [5] at an efficiency of up to 46% in a 750 MW ultra-supercritical once-through boiler, operating at steam conditions of 600 °C and 280 bar [6]. Nonetheless, subcritical coal-fired steam power plants that operate on the principle of the Rankine Cycle still supply more than one-third of the electricity demand in the US [7]. Subcritical power plants operating at pressures lower than 220 bar have a nominal efficiency of 37% [8]. Compared to supercritical and ultra-supercritical plants, the more common subcritical plants are advantageous in terms of lower installation costs, operating and maintenance experience [5,7]. Therefore, optimization of the efficiency of subcritical power plants is the first realistic step in our efforts to reduce CO_2 emissions from the power sector.

Due to seasonal and daily fluctuations in power demand and new deployment programs focused on renewable energy, dynamic simulations and optimization are required for power plants in order to respond to the resulting time-varying power demand. The contribution of electricity generation from renewable energy sources throughout the world will expand from the current 21% to 29.8% by 2040 [7]. The impact of this increase in penetration of renewable sources has been explored by many researchers. For example, Shah et al. [9] showed that the higher penetration of large-scale photovoltaic plants in the power grid will lead to significant variation in the power flow across the grid and unstable power generation profiles for the balancing conventional plants. The work by Edmunds et al. [10] showed that today's power plants are subject to more intense ramping operations due to the increasing variable renewable penetration. Critz et al. [11] focused on the challenges arising from the inability to accurately forecast renewable power generation. Correspondingly, Eser et al. [12] showed that the high penetration of renewable energy sources will result in an increase in periodic start-ups of thermal power plants. Thus, simulation of the dynamic behavior of the integrated electricity sector and, in particular, the dynamicity of the fossil-fueled power plants, which will provide the balanced power (between renewables input and market demand), is increasingly of interest to improve productivity and stability and reduce cost and emissions.

The efficiency of conventional fossil-fueled power plants that are based on the Rankine Cycle mostly depends on the steam temperature and pressure [4], with the majority of previous work on efficiency optimization of these plants focusing on steady-state analyses. The work by Fu et al. [13] showed an average efficiency increase of 0.1% points for every increment of 8 °C in boiler feedwater temperature, every decrement of 4.5 °C in flue gas temperature and every increment of 10 bar in main steam pressure, compared to a reference case with an efficiency of 45.5%. Sanpasertparnich and Aroonwilas [14] presented potential efficiency improvements of up to 8.88% points for subcritical coal-fired power plants. They identified the preheated air temperature, main steam temperature and pressure of stream extracted from the high-, intermediate-, and low-pressure turbines (HP, IP and LP, respectively), as being the most critical variables in the optimization of power plant performance. In work by Tzolakis et al. [15], an absolute net efficiency gain of 0.55% was shown to be feasible by reducing the mass flow rate of the steam exiting the HP turbines and increasing the mass flow rate of the steam exiting the IP and LP turbines. These significant efforts in the area of steady-state optimization of power plants paved the way for future work on dynamic optimization. Moreover, advancements in process modeling tools, such as Dymola [16] and gPROMS [17], have made it easier to simulate these processes dynamically. For instance, Chen et al. [18] developed a Dymola [16] dynamic model of a combined cycle power plant integrated with chemical-looping combustion, with the combustion process optimized in gPROMS [17] to maximize the power plant efficiency. Franke et al. [19] presented a model-based, dynamic optimization framework exploiting the Modelica language [20] to improve power plant performance. Their work illustrated the efficiency benefits of applying offline optimization results to online power plant operations. Lind and Sallberg [21] used modern acausal simulation and optimization tools to optimize the start-up procedure of a combined cycle power plant. Their analysis showed that the thermal stress in the heat recovery steam generator is the major constraint limiting the rapid start-up of the gas turbines to full load.

One practical approach to improve the efficiency of existing fossil-fueled power plants is to deploy supervisory control schemes targeted to efficiency optimization. Supervisory control architectures are often used to perform tasks of process optimization without changing the plant infrastructure and design. Skogestad [22] presented a systematic procedure for designing advanced control structures at the supervision level for complex chemical plants. The critical first steps in designing a supervisor logic are to define the operational and economic objectives and the available degrees of freedom. Common degrees of freedom include the set points of the regulatory controllers, system boundaries not controlled and system parameters tuned to a particular operating scheme. For instance, Lestage et al. [23] presented a linear supervisory control design for constrained real-time optimization of an ore grinding plant in which they optimized the set points of the local controllers to maximize throughput. Baillie and Bollas [24] presented the key steps in the development of a high-fidelity model for a chiller plant which was used in supervisory resilient control architectures for plant optimization under fault scenarios by

Mittal et al. [25]. Obviously, supervisory control is a promising approach for the efficiency optimization of power plants, wherein there exists a large number of regulatory controllers that must be maintained for safety and performance reasons. In one such effort, Sáez et al. [26] developed a supervisory algorithm based on adaptive predictive control to optimize the operation of the gas turbine of a combined cycle power plant in Chile. They showed the potential of 3% fuel consumption savings by manipulating variables such as the fuel flow, air flow and steam flow. Ponce et al. [27] presented a dynamic simulator of an integrated solar combined cycle power plant, incorporating a supervisory control strategy. Fuel savings of 1.7–3.7% were shown to be feasible through the manipulation of set points of the regulatory controllers of the steam pressure, gas turbine power and steam turbine power. These efforts focused mostly on the optimization of a few power plant components instead of solving a problem that maximizes the power plant efficiency by using all or most of the degrees of freedom. In this work, the optimization problem serving the supervisory controller deals with the integrated coal-fired steam power plant.

In prior work [28], a power plant model was developed and validated against steady-state data from a fossil-fueled subcritical power plant with a reheat, regenerative cycle [29]. The power plant modeled exhibited a full-load power generation of 605 MW at efficiency of 38.7%. The modeled power plant operates with nominal steam turbine conditions of 174 bar and 538 °C, generated by the combustion of bituminous B coal. Conventional proportional-integral-derivative (PID) controllers were incorporated into the system model. Dynamic simulation of the power plant operating with step changes in fuel load showed that the controllers are robust in maintaining the controlled variables at set point. In this work, open-source data of time-varying power demand along with its forecast from the New England area are used to study this plant under realistic operating conditions [30]. A fuel load controller is implemented to meet the time-varying power demand, and controllers are added to adjust the air flow and water flow for a time-varying load. Supervisor control strategies are applied for static and dynamic optimization of the power plant's efficiency. This optimization is accomplished by manipulating the set points of the regulatory controllers of the temperature of the superheated steam and preheated air, and the mass flow rates of steam extracted from the steam turbines. Steady-state and dynamic optimization results are compared and discussed in an effort to explore the value proposition of each. Previous work on power plant optimization focused on steady-state simulation and optimization. The main contribution of this work is to use dynamic optimization with embedded plant-level control to optimize the transient operation of power plants. This can enable plant operators to operate power plants efficiently at variable load demands which becomes increasingly important with the higher grid-penetration of renewables.

2. Power Plant Studied and Plant Model

The power plant studied and simulated in prior work [28] was the fossil fuel-fired subcritical power plant, shown in Figure 1, with operating conditions at full load, as reported by Singer [29]. The plant employs a reheat, regenerative cycle to produce 605 MW of electricity by burning fossil-fuel, with nominal turbine conditions of 174 bar and 538 °C steam. The combustion of bituminous B coal [5] with preheated air produces hot flue gas that evaporates and superheats water. The feedwater is converted to high temperature superheated steam through a series of heat exchange steps in the boiler, including the economizer, evaporator, reheater, and superheater. The superheated steam produced in the boiler is expanded in a series of high-pressure (HP), intermediate-pressure (IP) and low-pressure (LP) turbines connected to a generator to convert the heat to mechanical torque and produce electricity. The steam exiting the last LP turbine is condensed in the condenser. The condensate is preheated in four heat exchange steps, including a deaerator and three water preheaters, which are supplied with steam streams extracted from the HP, IP and LP turbines. Three pumps, namely the condensate booster pump, condensate pump and boiler feed pump, are used to re-circulate the water after being condensed in the condenser. The preheated condensate re-enters the boiler under high pressure and closes the loop.

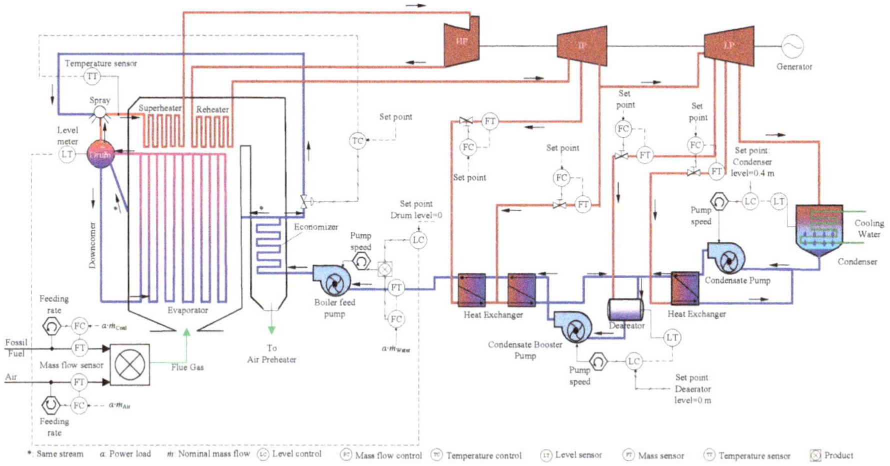

Figure 1. Reheat regenerative cycle: 605 MW subcritical-pressure fossil power plant with a control system design [29].

The plant in Figure 1 was simulated with a dynamic power plant model developed in Dymola [16] using the Modelon ThermalPower library [31]. The Modelica language used in Dymola is a non-proprietary, object-oriented, equation-based language for the modeling of complex physical systems [20] that is well-suited for the objective of large-scale dynamic simulation of power plants. A comprehensive list of the operating data of the power plant was provided in prior work, and the plant model was provided as Supporting Information to that work [28]. The model of the plant operating at full load was shown to be in excellent agreement with steady-state data from the reference power plant. Figure 1 also shows the design of the control system of that plant, including controllers for safety regulation (marked with black solid lines), plant-level controllers (marked with blue dashed lines) and controllers for plant optimization (marked with red dotted lines). The regulatory control system, including the controllers of superheated steam temperature and of the water levels in the drum, condenser and deaerator, was discussed in detail in prior work [28]. This regulatory control system was tuned using bump tests and the dynamic responses of the model were assessed qualitatively in terms of robustness and plant stability. The multilayer control scheme designed in this work and the controllers required to meet the time-varying power load are discussed in the following text.

3. Power Plant Under Time-Varying Power Demand

Extensive studies of the power demand and its forecasting have resulted in excellent models of the power demand per market sector, such as gray-box prediction models, to forecast real-time electricity demand with an error less than 8% [32,33]. The forecasted power demand is typically used by utility companies to predict the grid load and maintain service reliability. In this work, the data of power demand (along with its forecast) in the New England area were used. In particular, the data from 17 April 2016 were used as a realistic sample of power demand fluctuations [30]. The duration for the temporal forecasted power demand studied was 24 h. To meet the full power load of the reference power plant (605 MW), the ISO New England data (maximum value is 18,000 MW) was uniformly scaled-down, as shown in Figure 2a. The underlying assumption in this normalization was that the power demand from one power plant is proportional to the total power consumed. Therefore, it was considered that a fraction of the total power demand (scaled by a constant factor) and its daily fluctuation needed to be met by one power plant. The reality with renewable inputs in the grid is, as mentioned, a more abruptly fluctuating load for the power plant. It is thus anticipated that the efficiency gains from the analysis presented herein are a lower bound to the potential efficiency gains when renewable energy becomes a more dominant contribution to the electric grid.

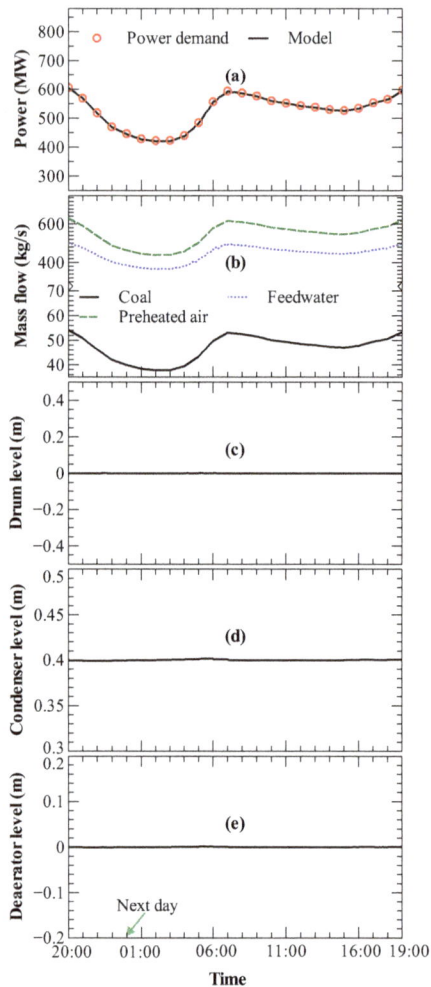

Figure 2. Dynamic performance of the power plant model: (**a**) power demand and power generated by the plant model; (**b**) mass flow rates of coal, preheated air and feedwater; (**c**) water level in the drum; (**d**) water level in the condenser; (**e**) water level in the deaerator.

In prior work [28], the power plant model was validated dynamically, showing fast responses to sudden changes in coal load. The regulatory control system incorporated in the power plant model was shown to be robust in maintaining controlled variables at set points. Here, plant-level controllers were added to the plant model, as shown in Figure 1, to adjust the coal load, preheated air flow and feedwater flow so that the plant met the time-varying power load of Figure 2a. The mass flow rate of feedwater ($\dot{m}_{FW}$) circulating in the plant and the mass flow rate of preheated air ($\dot{m}_{Air}$) mixed with the fuel were assumed to be proportional to the power load [34]. The mass flow rates of feedwater and preheated air were set to adjust with the plant load, by multiplying the nominal $\dot{m}_{Air}$ and $\dot{m}_{FW}$ by the temporal power load change ratio. The mass flow rate of coal was adjusted by a PID fuel load controller to match the temporal power demand. Table 1 presents the tuning parameters of the feed-forward control of water and air feed rates and the PID controller of fuel load. The measurement for the fuel load controller was the power generation (P), the manipulated variable was the coal mass flow ($\dot{m}_{Coal}$), and the set point was the temporal profile of the normalized power demand of the New England area [30]. These new controllers were tuned following standard methodologies discussed elsewhere [28]. The plant time scale and response times to load changes were studied and shown to be in the order of seconds and always less than a minute. Figure 2a shows that the power generated by the plant model matched the power demand of the normalized New England area data [30]. Figure 2b shows the transient responses of $\dot{m}_{Coal}$, $\dot{m}_{Air}$ and $\dot{m}_{FW}$ to the dynamically varying power demands of Figure 2a. Figure 2c–e show that the safety-critical regulated variables (water levels in the drum, condenser and deaerator) were robustly controlled and exhibited negligible oscillations. The dynamic performance of the plant model over the entire 24-hour period suggests that the model provides a robust test-bed of the plant physics and its controls and thus, it was used in the following experiments for steady-state and dynamic optimization. Figure 2 also shows that power demand was rapidly and accurately matched by the feed-forward controller, as tuned with the settings of Table 1.

Table 1. Controllers for the power plant in response to a time-varying power load *.

Feed-Forward Control: Air and Feedwater Controllers				
Controlled variables	$\dot{m}_{Air}$		$\dot{m}_{FW}$	
PID (Proportional-Integral-Derivative) Control: Fuel Load Controller				
Controlled variables	Manipulated variables	K_p	K_i	K_d
P	$\dot{m}_{Coal}$	1×10^{-8}	1×10^{-10}	1×10^{-6}

* P: power generation; $\dot{m}_{Air}$: mass flow rate of air; $\dot{m}_{FW}$: mass flow rate of feedwater; $\dot{m}_{Coal}$: mass flow rate of coal; K_p: coefficient of the proportional term; K_i: coefficient of the integral term; K_d: coefficient of the derivative term.

4. Optimization of an Integrated Power Plant

4.1. Objective and Optimization Variables

The objective of plant-level optimization is to maximize the efficiency of the power plant while operating at steady-state or to optimize the integral of the efficiency over time if the power plant is operating in a transient fashion. Another objective for plant-level optimization can be to reduce the plant's settling time, but, as shown previously [28], the time scale of the plant studied here is small, making settling time reduction a secondary concern. This was accomplished by calculating the optimal set points for the regulatory controllers without violating the operability and safety constraints. The plant efficiency was calculated as [29]

$$\eta = \frac{P_{ST} - P_{Pumps}}{\dot{m}_{Coal} LHV_{Coal}}, \tag{1}$$

where η is the efficiency of the plant; $\dot{m}_{Coal}$ is the mass flow rate of coal; LHV_{Coal} is the lower heating value of coal; P_{ST} is the power generated by steam turbines; and P_{Pumps} is the power consumed by pumps. Here, high-volatile bituminous B coal with an average LHV of 28 MJ/kg was used [5]. Other auxiliary energy losses were not not considered in Equation (1), as previous work has shown that auxiliary efficiency losses are small, often in the order of ~2 MW for coal-fired steam cycles for plant sizes similar the one studied here [29,35]. As discussed in the introduction, the power plant efficiency of Equation (1) can be improved by manipulating several plant variables. Table 2 summarizes the optimization variables, range of variability and efficiency improvements achieved in relevant previous work. In the majority of previous analyses [13–15,36–44], plant efficiency optimization was performed by manipulating the temperature of superheated steam (T_{SH}). For example, Xiong et al. [41] showed that the higher superheated steam temperature increases the power generated by the HP turbine, improving cycle efficiency. Several other variables have been explored in the literature in regard to their capability to improve plant efficiency. Sanpasertparnich et al. [14,36] presented the impact of preheated air temperature (T_{Air}) on power plant efficiency. Tzolakis et al. [15,42] optimized the plant efficiency at full load, by manipulating the mass flow extracted from steam turbines ($\dot{m}_{ST}$, which included $\dot{m}_{HP}$, $\dot{m}_{IP}$ and $\dot{m}_{LP}$). Other optimization variables, such as the moisture content of coal [45], the mass flow rate of feedwater [42], the isentropic efficiency of turbines [41], the temperature of flue gas exiting the boiler [43], and the pressure of steam extracted from turbines [14,36] require changes in the existing infrastructure and were not considered here. In summary, the common plant efficiency optimization variables T_{SH}, T_{Air}, and $\dot{m}_{ST}$ were chosen in this work. For the purpose of illustration, two optimization cases were considered. Case Study I presents plant optimization by manipulating T_{SH} and T_{Air} within an operation horizon of 24 h. Case Study II presents plant optimization by manipulating $\dot{m}_{ST}$ with an operation horizon of 4 h. This separation of optimization variables was done to allow for easy comparison with the trends reported in the literature and presented in Table 2. The results of each optimization problem are discussed in detail in the following section.

4.2. Supervisory Control

The control system of a plant is usually divided into several layers, typically separated by different time scale requirements and objectives. The control architecture includes regulatory control (seconds), supervisory control (minutes), local optimization (hours), site-wide optimization (days) and scheduling (weeks) [22]. Supervisory control can be designed to manipulate regulatory control set points and the remaining degrees of freedom of the plant (if any) to optimize the plant's efficiency within constraints imposed by the local controllers [46]. The critical first steps in designing a supervisor logic are to define the operational and economic objectives and the available degrees of freedom. Common degrees of freedom include the set points of the regulatory controllers, uncontrolled system boundaries and system parameters tuned to a particular operating scheme.

Figure 3 illustrates a scheme for such a supervisory control strategy for the power plant studied. The control system includes the supervisory control, regulatory control and plant level control. The regulatory control structure includes optimization controllers (marked as red dotted lines in Figure 1), which are the regulatory controllers used for plant optimization and safety controllers (marked as black solid lines in Figure 1), which regulate the level of water in the drum, condenser and deaerator. The main function of the supervisory control is to update the set points of the optimization controllers ($\mathbf{y}_O^{sp}$) to maximize the plant's efficiency as shown by Equation (1). The plant level controllers (marked with blue dashed lines in Figure 1) adjust the mass flow rates of coal, preheated air and feedwater according to the market power demand ($\mathbf{y}_P^{sp}$). The set points of the control system include the set points of the safety controllers, plant-level controllers and optimization controllers, i.e., $\mathbf{y}^{sp} = \{\mathbf{y}_S^{sp}, \mathbf{y}_P^{sp}, \mathbf{y}_O^{sp}\}$. These controllers manipulate the control inputs ($\mathbf{u} = \{\mathbf{u}_S, \mathbf{u}_P, \mathbf{u}_O\}$) to maintain the controlled variables at their set points. In principle, one should consider the disturbance ($\mathbf{\omega}^x$) and measurement noise ($\mathbf{\omega}^y$), which are responsible for a difference ($\mathbf{e}$) between the model ($\mathbf{y}^{pred}$) and power plant outputs ($\mathbf{y}^{meas}$). An estimator could update the model's parameters ($\hat{\theta}$) and filter plant states ($\mathbf{x}$) to eliminate this model-plant mismatch. In this work, disturbance and measurement noise were not considered, mostly to simplify the analysis, as the efficiency benefits are not affected by them (although the robustness of the supervisor will be). Therefore, $\mathbf{\omega}^x$ and $\mathbf{\omega}^y$ were considered negligible, and data filtering and state estimation (blocks in gray in Figure 3) are not discussed. The supervisory control updates the optimal $\mathbf{y}_O^{sp}$ according to an objective function maximizing Equation (1) in a formulation that includes the system model equations as discussed in the following text.

Table 2. Review of power plant optimization efforts and respective variables *.

Ref.	Δη (%)	T_{SH} (°C)	$\dot{m}_{IP}/\dot{m}_{HP}/\dot{m}_{LP1}/\dot{m}_{LP2}/\dot{m}_{LP}$ (kg/s)	T_{Air} (°C)	$\dot{m}_{FW}$ (kg/s)	p_{SH} (bar)	β (%)	Others
						Optimization Variables		
[14,36]	7.8	[530,600]	[3.9,19.6]/ [6.2,43.3]/ [15.1,28.7]/ [11.1,42.1]	[166,190]	[250,350]	[166,190]	[11.1,17.6]	
[15]	0.55		[0,30.8]/ [0,51.2]/ [0,21.1]/ [0,0.94]					
[37]	0.41		[16,26]/ [14,24]/ [12.6,24]/ [34,57]					
[38,39]	2.8	[600,625]		[35,275]	[400,475]	[20,30]		Excess air ∈ [0,25%], T_{RH} ∈ [580,620]
[40]	2	[550,700]				[230,350]		η_{ST} ∈ [0.75,0.87], $p_{HP/IP/LP}$(bar) ∈ [60/9/0.0356, 80/25.5/2.68]
[13]	5.9	[487,1076]				[230,350]		
[41]	2.5	[535,545]				[150,450]		η_{ST} ∈ [0.8,0.95]
[42]	1.3	[485,537]			[45,57]			
[43]	0.79	[115,278]			[21,38.4]			T_{FG}(°C) ∈ [85,125]
[44]	3.5	[460,530]				[64,110]		p_{CON}(bar) ∈ [0.01,0.05]

* T_{SH}: temperature of superheated steam; $\dot{m}_{HP}$: mass flow rates of steam extracted from high-pressure turbine; $\dot{m}_{IP}$: mass flow rates of steam extracted from intermediate-pressure turbine; $\dot{m}_{LP}$: mass flow rates of steam extracted from low-pressure turbine; p_{SH}: pressure of the superheated steam; T_{Air}: temperature of the preheated air; $\dot{m}_{FW}$: mass flow rate of feedwater; β: coal moisture content; η_{ST}: isentropic efficiency of steam turbines; T_{RH}: temperature of reheated steam; T_{FG}: temperature of flue gas; $p_{HP,IP,LP}$: pressure of streams extracted from the HP, IP and LP steam turbines; p_{CON}: condenser pressure.

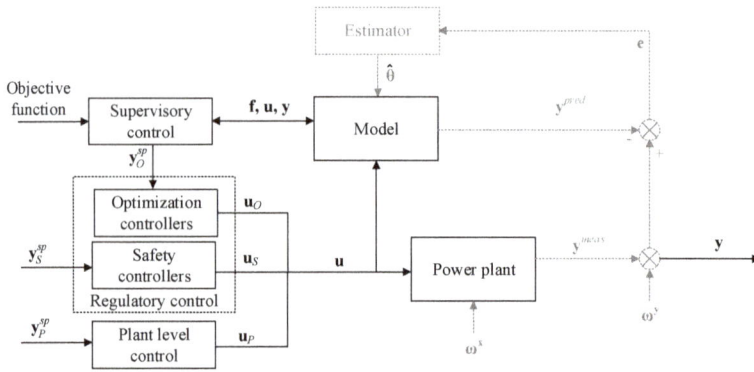

Figure 3. Multilayer control scheme for the reheat regenerative cycle of a 605 MW subcritical-pressure coal-fired power plant. $\mathbf{y}_O^{sp}$: set points of the optimization controllers; $\mathbf{y}_S^{sp}$: set points of the safety controllers; $\mathbf{y}_P^{sp}$: set points of the plant level controllers; $\mathbf{u}_O$: control input of the optimization controllers; $\mathbf{u}_S$: control input of the safety controllers; $\mathbf{u}_P$: control input of the plant level controllers; $\mathbf{u}$: control input; ω^x: disturbance; ω^y: measurement noise; $\mathbf{y}^{pred}$: predicted outputs; $\mathbf{y}^{meas}$: measured outputs; $\mathbf{y}$: system outputs; $\mathbf{e}$: error; $\hat{\theta}$: estimated model parameters; $\mathbf{f}$: system model.

4.3. Optimization Formulation

As described previously, the set points of the optimization controllers are manipulated by the supervisory layer as first-level variables to improve the plant's efficiency (η), Equation (1). This efficiency optimization also translates to coal consumption reduction and decrease of the plant carbon footprint. The intent of this work was to compare the steady-state and dynamic optimization results of the plant shown in Figure 2 to those of the plant operating under nominal conditions. This comparison also included an exploration of the added benefits of dynamic optimization, compared to those from steady-state optimal operation. First, steady-state optimization of the power plant operating at a full load of 605 MW was performed by calculating the optimal set points (constant with time) for the optimization controllers ($\mathbf{y}_O^{sp}$) and specifically, the set points of the superheat steam temperature controller, preheated air temperature controller and mass flow controllers of steam extracted from the steam turbines. The steady-state optimization problem formulation is shown in Equation (2):

$$\max_{\mathbf{y}_O^{sp}} \eta(\mathbf{y}, \mathbf{u}_P)$$

subject to:

$$\mathbf{f}(\mathbf{x}, \mathbf{u}, \theta) = 0,$$
$$\mathbf{u} = \mathbf{F}(\mathbf{y}_O^{sp}, \mathbf{y})$$
$$\mathbf{y} = \mathbf{h}(\mathbf{x}, \mathbf{u}, \theta), \tag{2}$$
$$\mathbf{x}^{min} \leq \mathbf{x} \leq \mathbf{x}^{max},$$
$$\mathbf{u}^{min} \leq \mathbf{u} \leq \mathbf{u}^{max},$$
$$\mathbf{y}^{sp,min} \leq \mathbf{y}^{sp} \leq \mathbf{y}^{sp,max},$$
$$\mathbf{y}^{min} \leq \mathbf{y} \leq \mathbf{y}^{max},$$

where the plant efficiency η is a function of power plant outputs ($\mathbf{y}$) and admissible variable values ($\mathbf{u}_P$) determined by the updated optimization controller set points $\mathbf{y}_O^{sp}$; $\mathbf{f}(\cdot)$ is the vector of steady-state equations describing the system in terms of states, $\mathbf{x}$, admissible inputs, $\mathbf{u}$, and parameters, θ; and

F describes the controller functions at steady-state to account for controller offset with **y** being the measured system outputs, mapped to **x**, **u**, and θ through **h**(·). All variables including unmeasured variables are constrained so that they do not violate plant safety and operability constraints.

Dynamic optimization was performed for the reference power plant operating under time-varying power demand normalized from the New England area data [30]. The objective was to maximize the integral of plant efficiency over a predetermined time horizon, τ. This was accomplished by calculating time-varying optimal set points for the optimization controllers. The generic formulation of the dynamic optimization problem solved for the power plant of Figure 1 is presented in Equation (3), where **f** is the system of differential algebraic equations describing the conservation of mass and energy; **x** is the vector of temporal state variables; $\mathbf{x}^0$ is the vector of initial state variables; $\mathbf{y}_O^{sp}$ is the temporal set points of the optimization controllers; **y** are the temporal system's outputs; $\mathbf{t}_n$ is the vector of control action time points with a constant interval, τ_n; τ is the optimization horizon; and *t* is the time.

$$\max_{\mathbf{y}_O^{sp}(\mathbf{t}_n)} \int_0^\tau \eta(\mathbf{y}_O^{sp}(t), \mathbf{u}_P(t))dt$$

subject to:

$$\mathbf{f}(\dot{\mathbf{x}}, \mathbf{x}, \mathbf{u}, \theta, t) = 0,$$
$$\mathbf{u} = \mathbf{F}(\mathbf{y}_O^{sp}(\mathbf{t}_n), \mathbf{y}(t), t),$$
$$\mathbf{y} = \mathbf{h}(\mathbf{x}, \mathbf{u}, \theta, t),$$
$$\mathbf{x}(t=0) = \mathbf{x}^0, \qquad (3)$$
$$\mathbf{x}^{min} \le \mathbf{x} \le \mathbf{x}^{max},$$
$$\mathbf{u}^{min} \le \mathbf{u} \le \mathbf{u}^{max},$$
$$\mathbf{y}^{sp,min} \le \mathbf{y}^{sp} \le \mathbf{y}^{sp,max},$$
$$\mathbf{y}^{min} \le \mathbf{y} \le \mathbf{y}^{max},$$
$$t \in [0, \tau], \ \mathbf{t}_n \in [0, \tau].$$

Table 3 shows the optimization variables' bounds and time interval constraints for the problems of Equations (2) and (3) for the two cases studied. The set points of the controllers regulating T_{SH}, T_{Air} and $\dot{m}_{ST}$ (including $\dot{m}_{IP1}$, $\dot{m}_{IP2}$, $\dot{m}_{LP1}$ and $\dot{m}_{LP2}$) were manipulated by the supervisory control layer as degrees of freedom to seek an optimal input. In Case Study I, only the set points of T_{SH} and T_{Air} were manipulated. Although not shown in Figure 2, preheating of the air fed to the combustor to T_{Air} was accomplished by manipulating the mass flow of the economizer exhaust gas sent to the air preheater (with the balance being waste heat). In Case Study II, plant optimization was performed by manipulating the set points of $\dot{m}_{ST}$, i.e., the set points of the mass flow rates of steam streams extracted from the first IP turbine (IP1), the second IP turbine (IP2), the first LP turbine (LP1), and the second LP turbine (LP2) ($\dot{m}_{IP1}$, $\dot{m}_{IP2}$, $\dot{m}_{LP1}$, and $\dot{m}_{LP2}$, respectively). The ranges of the admissible inputs, shown in Table 3, were based on common practice and previous work [13–15,27,36,38,41–43,45]. The optimization horizon, τ, was set to 24 h in Case Study I and 4 h in Case Study II, and the control action interval, τ_n, was set to 1 h. Large control actions were not penalized in the solved optimization problems, as the plant load profiles matched during the real-time plant optimization were relatively smooth. For instance, the temperature of the superheated steam feeding the steam turbine was seen to change gradually over time in response to load changes which is adequate for the protection of the steam turbines by thermal stress [14,36].

Table 3. Inputs for the studied optimization problems.

Admissible Inputs (y_O^{sp})	T_{SH}^{sp}(°C)		T_{Air}^{sp}(°C)	
		Case I		
Min	520		150	
Max	610		250	
		Case II		
Admissible Inputs [a] (y_O^{sp})	$\dot{m}_{IP1}^{sp}$ (kg/s)	$\dot{m}_{IP2}^{sp}$ (kg/s)	$\dot{m}_{LP1}^{sp}$ (kg/s)	$\dot{m}_{LP2}^{sp}$ (kg/s)
Min	16	10	10	28
Max	28	28	28	47
		Temporal Inputs [b]		
$\tau_n(h)$	1			
$\tau(h)$	24 for case I (4 for case II)			

[a] $\dot{m}_{IP1}$: mass flow rate of steam stream extracted from the IP1 turbine; $\dot{m}_{IP2}$: mass flow rate of steam stream extracted from the IP2 turbine; $\dot{m}_{LP1}$: mass flow rate of steam stream extracted from the LP1 turbine; $\dot{m}_{LP2}$: mass flow rate of steam stream extracted from the LP2 turbine; sp: set point; [b] if the plant is operating under a time-varying power load.

5. Results

For each case study, the static optimization of the power plant operating at full load with the optimization formulation of Equation (2) is discussed first, followed by the dynamic optimization of the power plant operating under a time-varying power load with the optimization formulation of Equation (3). For the results discussed in the following text, the power plant was formulated with the object-oriented language Modelica [20], in the commercial software Dymola [16], and set point optimization was performed in Matlab [47] using an interior-point algorithm. The model developed in Dymola was flattened (from its object-oriented structure) and translated to a Functional Mockup Unit (FMU) file which includes all the variables and equations of the original plant model. Model exchange between the software packages of Dymola and Matlab was accomplished with use of the Functional Mockup Interface, a tool-independent standard for seamlessly integrating models in various simulation environments [48]. The Functional Mock-up Interface (FMI) enables model exchange of dynamic models in the form of xml-files and compiled C-code.

5.1. Case Study I: Optimization Variables T_{SH} and T_{Air}

Table 4 presents the steady-state optimization results at full load using the superheated steam temperature set point, T_{SH}^{sp}, and that of the preheat air temperature, T_{Air}^{sp}, as the optimization variables. The manipulation of T_{SH}^{sp} and T_{Air}^{sp} led to a power plant efficiency improvement from 38.3% to 40.23%. This efficiency improvement translates to a fuel saving of 3.78%, with the fuel flow rate decreasing from 56.38 kg/s to 54.25 kg/s. The carbon footprint of the plant also decreased from 0.8 kg/kWh to 0.77 kg/kWh. This efficiency optimization was accomplished by increasing T_{Air} from 200 °C to 248 °C, and increasing T_{SH} from 538 °C to 560 °C. This is consistent with earlier reports [13,14,27,38,41–43], showing that increasing T_{SH} and T_{Air} translates to efficiency improvements. The higher T_{SH} enabled the HP turbine to produce the same mechanical torque at a lower rate of coal consumption, while increasing T_{Air} recovered more waste heat from the boiler exhaust gas. It should be noted that the nominal steady-state data used as baseline in Table 4 were as reported by Singer [29] for the reference power plant and corresponded to the design point of this plant. In principle, the set points for T_{SH} and T_{Air} reported by Singer refer to an optimal plant configuration. The further improvement presented here could relate to better integrated plant-level optimization, model-plant differences and relaxation of plant constraints compared to the study reported in [29]. For an off-design operating point, the efficiency benefits of solving Equation (2) would, of course, have been much higher.

Table 4. Steady-state optimization results of Case Study I.

System Output	Nominal	Optimal
T_{SH}^{sp} (°C)	538	560
T_{Air}^{sp} (°C)	200	248
η (%)	38.3	40.23
$\dot{m}_{Coal}$ (kg/s)	56.38	54.25
carbon footprint (kg/kWh)	0.8	0.77

The results of dynamic optimization for a horizon of 24 h of plant operation are presented in Figure 4. The data and plant performance results represent the response to the time-varying power demand normalized from the New England area data [30] shown in Figure 4a. In the absence of disturbances and noise, the solution to Equation (3) in the period $t = 0 - \tau (= 24$ h) is equivalent to an off-line optimal control problem and is valid for the entirety of the time horizon considered. The optimization variables were T_{SH}^{sp} and T_{Air}^{sp}, but, in this case, they were updated in time intervals, $\tau_n = 1$ h. Figure 4 presents the dynamic power plant performance under nominal and optimal operation conditions. The nominal dynamic operation is the result of constant T_{SH}^{sp} at 538 °C and T_{SH}^{sp} at 200 °C. Figure 4d,e shows that the controlled variables, T_{SH} and T_{Air}, were robustly controlled at their optimal set points by the regulatory controller. The values of T_{SH} and T_{Air} from the dynamic optimization solution were always higher than their respective nominal values. In particular, Figure 4d shows that the optimal T_{SH}^{sp} trajectory was inversely proportional to that of the plant load. The optimal temporal, T_{SH}^{sp}, for a plant load higher than maximum, was higher than the 560 °C of the optimal steady-state at full load. This enhanced heat transfer from the flue gas side to the steam side in the superheater under low plant loads. Figure 4e shows that the optimal temporal profile of the temperature of air preheated by the flue gas exiting the boiler varyied proportionally to the plant load. The temperature of the exhaust gas was also proportional to the power load due to the time-varying mass flow rates of the feedwater, air and coal load. As Figure 4c shows, the improvement in time-averaged efficiency was 1.8% points. This efficiency improvement translates to a coal saving of 184.8 tons/day (Figure 4d) and time-averaged carbon footprint decrease of 0.0351 kg/kWh (Figure 4e). In summary, the optimized power plant can operate at a higher $T_{Air}(t)$ and $T_{SH}(t)$, and this is consistent with the results of steady-state optimization.

Figure 5 presents the dynamic performance of the power plant operating with constant nominal set points for T_{SH} and T_{Air}, constant optimal T_{SH}^{sp} and T_{Air}^{sp} (from the steady-state optimization solution), and with time-varying optimal T_{Air}^{sp} and T_{SH}^{sp} (set by the dynamic optimization solution). The coal consumption and carbon footprint of the power plant operating with set points calculated by the static and dynamic optimization problem formulations were both lower those for the power plant under nominal operation conditions. The power plant operating with set points determined by dynamic optimization was the most efficient with the lowest coal consumption and the smallest carbon footprint. As shown in Table 5, the fuel saving accomplished by the power plant with steady-state optimization was 160.9 tons/day, whereas the fuel saving accomplished with dynamic set point optimization was 184.8 tons/day. The reduction in coal load and decrease in the carbon footprint of the dynamically optimal operation were pronounced when the power plant was operating at lower loads. At different loads, the plant had slightly different optimal regulatory control points compared to those of the steady-state optimization at full load, which are exploited by the formulation of Equation (3). As shown in Figure 4d, the values of T_{SH}^{sp} calculated from Equation (3) at low loads were higher than the constant T_{SH}^{sp} calculated from Equation (2) at full load. Dynamically optimizing T_{SH}^{sp} improved the heat transfer in the superheater at low loads and converted more heat from the superheated steam to mechanical torque. This increase in mechanical torque led to improved power generation and efficiency. Moreover, the temperature profile of the preheated air in Figure 4e shows that the values of T_{Air}^{sp} calculated from Equation (3) at low loads were lower than the constant T_{Air}^{sp} calculated from Equation (2) at full load. At low loads, heat transfer between the water side and flue gas side in the boiler is enhanced, leading to a lower flue gas temperature, which, in turn, is used to preheat the air. Thus, the supervisor drives T_{Air}^{sp} down to satisfy system constraints. Depending on the dynamic response times of the plant and the selection of the interval between control actions, τ_n, a multistep steady-state optimization problem could result in similar performance benefits to those of Equation (3). Nonetheless, Equation (3) is more generic and robust for use with a dynamic system. It should be noted that one could execute the same analysis but with an objective function that maximizes the profit for varying electricity prices. This would result in different plant load profiles, but the optimization procedure (not the objective function) and results would be similar.

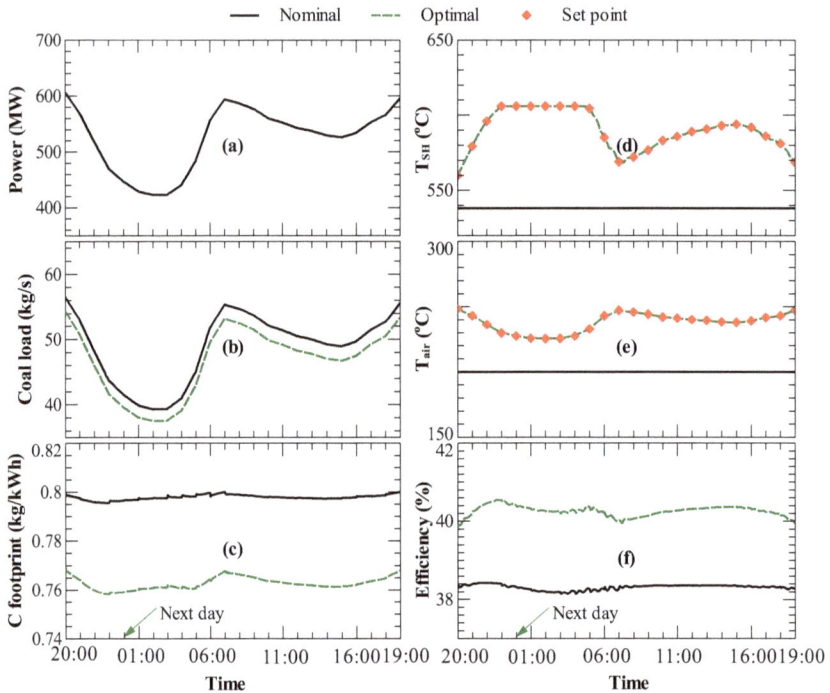

Figure 4. Dynamic optimization results of Case Study I: (**a**) time-varying power load; (**b**) coal load; (**c**) carbon footprint; (**d**) temperature of preheated air; (**e**) temperature of superheated steam; (**f**) power plant efficiency.

Table 5. Comparison of static and dynamic optimization of the power plant for Case Study I *.

Output	Static Optimization	Dynamic Optimization
$\Delta \dot{m}_{coal}$ (tons/day)	160.9	184.8
$\Delta \bar{c}_f$ (kg/kWh)	0.0303	0.0351
$\Delta \dot{m}_{CO_2}$ (tons/day)	440.2	511.9

* $\Delta \dot{m}_{coal}$: coal savings; $\Delta \bar{c}_f$: decrease in the time-averaged carbon footprint; $\Delta \dot{m}_{CO_2}$: reduction of CO_2 emissions.

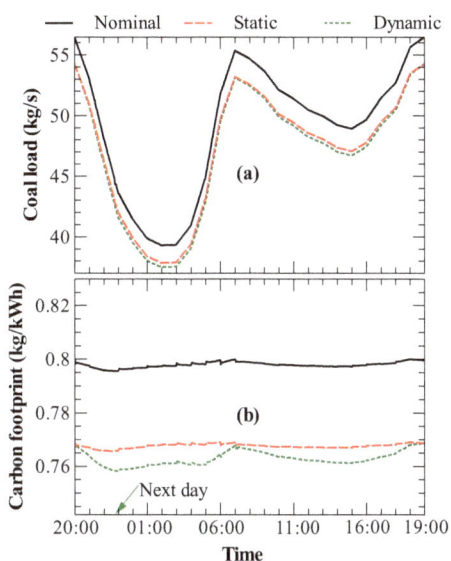

Figure 5. Comparison of the dynamic performance of the power plant with nominal operation set points, steady-state optimal set points, and dynamic optimal set points: (**a**) coal load; (**b**) carbon footprint.

5.2. Case Study II: Optimization Variables $\dot{m}_{IP1}$, $\dot{m}_{IP2}$, $\dot{m}_{LP1}$ and $\dot{m}_{LP2}$

As shown in Figure 1, four proportional–integral (PI) controllers were used to regulate the mass flow rates of steam extracted from the turbines. The parameters of these controllers are presented in Table 6. These controllers manipulate the respective valves to regulate the mass flow rates of streams extracted from the IP1, IP2, LP1 and LP2 turbines. In this case study, the supervisory control variables were the set points of the mass flow controllers of steam extracted from turbines, namely the set points of $\dot{m}_{IP1}$, $\dot{m}_{IP2}$, $\dot{m}_{LP1}$, and $\dot{m}_{LP2}$.

Table 6. Proportional–integral (PI) controllers regulating the mass flow rates of steam extracted from turbines.

Controllers	IP1	IP2	LP1	LP2
Controlled variables	$\dot{m}_{IP1}$	$\dot{m}_{IP2}$	$\dot{m}_{LP1}$	$\dot{m}_{LP2}$
Manipulated variables	Valve opening	Valve opening	Valve opening	Valve opening
K_p	0.1	0.1	0.1	0.1
K_i	0.0001	0.0001	0.0001	0.0001

As before, steady-state optimization was first performed for the plant operating at full load. The set points of $\dot{m}_{IP1}$, $\dot{m}_{IP2}$, $\dot{m}_{LP1}$ and $\dot{m}_{LP2}$ were manipulated by the supervisory control layer to maximize the plant's efficiency, as shown by Equation (2). The bounds of admissible inputs are shown in Table 3, with the optimal values presented in Table 7. The power plant efficiency was improved from 38.3% to 38.78%. The corresponding coal load decreased from 56.38 kg/s to 55.68 kg/s, and the carbon footprint decreased from 0.8 kg/kWh to 0.79 kg/kWh. Compared with the nominal case, the optimal case had a lower $\dot{m}_{IP1}^{sp}$ and higher $\dot{m}_{IP2}^{sp}$, $\dot{m}_{LP1}^{sp}$ and $\dot{m}_{LP2}^{sp}$, as shown in Table 7. The mass flow rate of steam extracted from the IP1 turbine was less than that of other steam turbine extractions. The IP1 turbine extraction had the highest pressure and temperature of all steam extractions. Thus, it would be better utilized for electricity production than water preheating. Meanwhile, the steam extracted from the IP2, LP1 and LP2 turbines would be better utilized for preheating the condensed feedwater to reach a higher temperature before entering the boiler. These results are consistent with the findings of the study by Chaibakhsh and Ghaffari [49] who proposed reducing (or removing) the high pressure and temperature steam extraction stream and increasing the steam extracted from the remaining IP and LP turbine stages.

Table 7. Steady-state optimization results for Case Study II.

System Output	Nominal	Optimal
$\dot{m}_{IP1}^{sp}$ (kg/s)	27.4	16.8
$\dot{m}_{IP2}^{sp}$ (kg/s)	14	23.1
$\dot{m}_{LP1}^{sp}$ (kg/s)	16.5	23.7
$\dot{m}_{LP2}^{sp}$ (kg/s)	30	43.8
η (%)	38.3	38.78
$\dot{m}_{Coal}$ (kg/s)	56.38	55.68
Carbon footprint (kg/kWh)	0.8	0.79

Dynamic optimization was performed for an optimization horizon of 4 h. The time period from 5 a.m. to 9 a.m. was used for the New England power demand data, as shown in Figure 6a. In this interval, the power plant iwa operating in response to a abrupt increase in power demand, with a power load change from 79.9% to 98.1%, followed by a decrease from 98.1% to 95.2%. This time interval includes the most abrupt change in power demand of the New England ISO data used as well as a change in the sign of change in power demand. To solve this problem, the power plant model was first initialized to steady-state at a load of 79.9% ($t = 0$ in Figure 6). As shown in Figure 6a, the power generated by the plant model matched the time-varying power demand which was accomplished by the plant load controllers shown in Figure 1.

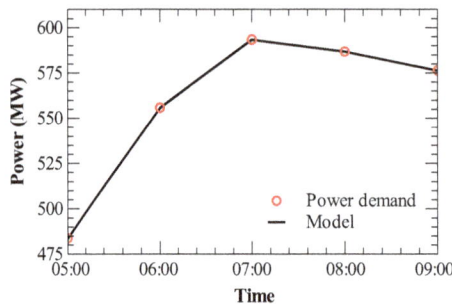

Figure 6. Time-varying power demand and plant load for Case Study II.

Figure 7 presents the transient operation of the virtual power plant in response to nominal inputs and to those calculated with dynamic optimization for the power plant load of Figure 6. The supervisor updated the set points of the controllers regulating $\dot{m}_{IP1}(t)$, $\dot{m}_{IP2}(t)$, $\dot{m}_{LP1}(t)$ and $\dot{m}_{LP2}(t)$ to seek the maximum integral of efficiency over a time horizon of 4 h. The nominal operation of the power plant corresponds to constant set points for the mass flow rate of turbine extraction streams, shown in Table 7. For the optimal dynamic operation, these set points were treated as dynamic optimization variables that were updated every hour by the supervisory controller. Figure 7a shows that the mass flow rate of the steam streams extracted from the turbines was robustly maintained at the respective temporal set points (updated in 1 h intervals), set according to the dynamic optimization solution of the supervisor. Dynamic optimization requires the mass flow rate of IP1 steam extraction to be lower than that of the other steam extractions, similarly to the results for steady-state optimization. The optimal mass flow rates of all the steam extraction streams followed the load profile. This is because the total mass flow rate of water circulating in the steam cycle is proportional to the power load. The improvement in the time-averaged efficiency was 0.43% points, as shown in Figure 7b–d, which shows that the coal saving for four hours and the decrease in the time-averaged carbon footprint were 7.72 tons and 0.00859 kg/kWh, respectively. These benefits became more profound at higher plant loads which is in accordance with the relative contribution of the steam side of the plant to the overall power production.

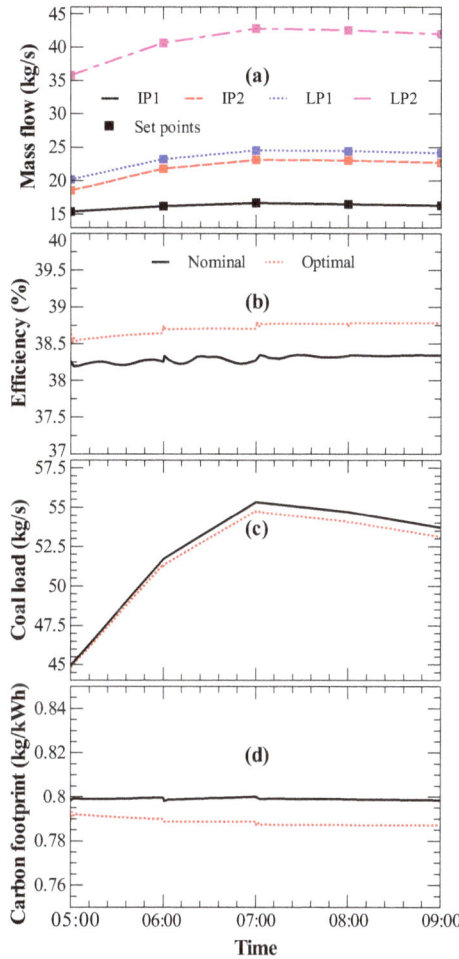

Figure 7. Dynamic optimization results of Case Study II: (**a**) dynamic measurements and set points of mass flow rates of steam extracted from turbines; (**b**) coal load; (**c**) efficiency; (**d**) carbon footprint.

6. Conclusions

A dynamic power plant model was used as a test-bed for dynamic simulation and optimization in response to a variable plant load. Plant-level controllers were added to the plant model to meet the transient market power demand. Thereafter, optimization problems were formulated and solved with the objective of optimizing power plant efficiency at steady-state and dynamically. A supervisory control architecture was designed to manipulate the set points of regulatory controllers according to the solution of the optimization problems explored. The optimization variables T_{SH}^{sp} and T_{Air}^{sp}, and $\dot{m}_{ST}^{sp}$, chosen in this work after a comprehensive literature review, enabled an improvement in time-averaged efficiency of up to 1.95% points with corresponding savings in coal consumption of 184.7 tons/day and a carbon footprint decrease of 0.0352 kg/kWh. A comparison of the static and dynamic optimization formulations serving the supervisory controller showed that dynamic optimization offers higher time-averaged efficiency, fuel savings and CO_2 reduction. Although the power plant model and regulatory control architecture have been validated in previous work [28], validation against transient power plant data would benefit this work in terms of the validity and accuracy of the estimated efficiency benefits.

Author Contributions: Conceptualization, G.M.B. and C.C.; Methodology, G.M.B.; Software, C.C.; Validation, C.C.; Investigation, C.C.; Resources, G.M.B.; Writing—Original Draft Preparation, C.C.; Writing—Review & Editing, G.M.B.; Visualization, C.C.; Supervision, G.M.B.; Project Administration, G.M.B; Funding Acquisition, G.M.B.

Funding: This material is based upon work supported by the National Science Foundation under Grant No. 1054718.

Acknowledgments: Chen Chen gratefully acknowledges support by the GE Graduate Fellowship for Innovation and helpful advice and guidance from Alstom Power. This work was partially sponsored by the United Technologies Corporation Institute for Advanced Systems Engineering (UTC-IASE) of the University of Connecticut. Any opinions expressed herein are those of the authors and do not represent those of the sponsor.

Conflicts of Interest: The authors declare no conflict of interest.

References

1. Han, L.; Bollas, G.M. Chemical-looping combustion in a reverse-flow fixed bed reactor. *Energy* **2016**, *102*, 669–681. [CrossRef]
2. Han, L.; Bollas, G.M. Dynamic optimization of fixed bed chemical-looping combustion processes. *Energy* **2016**, *112*, 1107–1119. [CrossRef]
3. Ibrahim, T.K.; Mohammed, M.K.; Awad, O.I.; Rahman, M.; Najafi, G.; Basrawi, F.; Abd Alla, A.N.; Mamat, R. The optimum performance of the combined cycle power plant: A comprehensive review. *Renew. Sustain. Energy Rev.* **2017**, *79*, 459–474. [CrossRef]
4. Viswanathan, R.; Henry, J.; Tanzosh, J.; Stanko, G.; Shingledecker, J.; Vitalis, B.; Purgert, R. U.S. Program on Materials Technology for Ultra-Supercritical Coal Power Plants. *J. Mater. Eng. Perform.* **2005**, *14*, 281–292. [CrossRef]
5. Hasler, D.; Rosenquist, W.; Gaikwad, R. New coal-fired power plant performance and cost estimates. In *Sargent and Lundy Project*; Environmental Protection Agency: Washington, DC, USA, 2009; pp. 1–82.
6. Cziesla, F. Lünen—State-of-theArt Ultra Supercritical Steam Power Plant Under Construction Andreas Senzel. In Proceedings of the Power Generation Europe 2009, Cologne, Germany, 26–29 May 2009; pp. 1–21.
7. U.S. Energy Information Administration. *Energy Information Administration: Annual Energy Outlook 2018*; U.S. Energy Information Administration: Washington, DC, USA, 2018.
8. Masters, G.M. *Renewable and Efficient Electric Power Systems*; John Wiley & Sons, Inc.: Hoboken, NJ, USA, 2004; p. 712. [CrossRef]
9. Shah, R.; Mithulananthan, N.; Bansal, R. Oscillatory stability analysis with high penetrations of large-scale photovoltaic generation. *Energy Convers. Manag.* **2013**, *65*, 420–429. [CrossRef]
10. Edmunds, R.; Davies, L.; Deane, P.; Pourkashanian, M. Thermal power plant operating regimes in future British power systems with increasing variable renewable penetration. *Energy Convers. Manag.* **2015**, *105*, 977–985. [CrossRef]
11. Critz, D.K.; Busche, S.; Connors, S. Power systems balancing with high penetration renewables: The potential of demand response in Hawaii. *Energy Convers. Manag.* **2013**, *76*, 609–619. [CrossRef]
12. Eser, P.; Singh, A.; Chokani, N.; Abhari, R.S. Effect of increased renewables generation on operation of thermal power plants. *Appl. Energy* **2016**, *164*, 723–732. [CrossRef]
13. Fu, C.; Anantharaman, R.; Jordal, K.; Gundersen, T. Thermal efficiency of coal-fired power plants: From theoretical to practical assessments. *Energy Convers. Manag.* **2015**, *105*, 530–544. [CrossRef]
14. Sanpasertparnich, T.; Aroonwilas, A. Simulation and optimization of coal-fired power plants. *Energy Procedia* **2009**, *1*, 3851–3858. [CrossRef]

15. Tzolakis, G.; Papanikolaou, P.; Kolokotronis, D.; Samaras, N.; Tourlidakis, A.; Tomboulides, A. Simulation of a coal-fired power plant using mathematical programming algorithms in order to optimize its efficiency. *Appl. Therm. Eng.* **2012**, *48*, 256–267. [CrossRef]

16. Elmqvist, H.; Brück, D.; Otter, M. *Dymola User's Manual*; Dynasim AB: Lund, Sweden, 1996

17. Process Systems Enterprise. *gPROMS Advanced User Guide*; Process Systems Enterprise Limited: London, UK, 2004.

18. Chen, C.; Han, L.; Bollas, G.M. Dynamic Simulation of Fixed-Bed Chemical-Looping Combustion Reactors Integrated in Combined Cycle Power Plants. *Energy Technol.* **2016**, *4*, 1209–1220. [CrossRef]

19. Franke, R.; Babji, B.; Antoine, M.; Isaksson, A. Model-based online applications in the ABB Dynamic Optimization framework. In Proceedings of the 6th International Modelica Conference, Bielefeld, Germany, 3–4 March 2008; pp. 279–285.

20. Modelica Association. Modelica—A Unified Object-Oriented Language for Physical Systems Modeling. In Proceedings of the 12th European Conference on Object-Oriented Programming, Brussels, Belgium, 20–24 July 2010.

21. Lind, A.; Sällberga, E.; Velutb, S. Start-up Optimization of a Combined Cycle Power Plant. In Proceedings of the 9th International Modelica Conference, Munich, Germany, 3–5 September 2012.

22. Skogestad, S. Control structure design for complete chemical plants. *Comput. Chem. Eng.* **2004**, *28*, 219–234. [CrossRef]

23. Lestage, R.; Pomerleau, A.; Hodouin, D. Constrained real-time optimization of a grinding circuit using steady-state linear programming supervisory control. *Powder Technol.* **2002**, *124*, 254–263. [CrossRef]

24. Baillie, B.P.; Bollas, G.M. Development, Validation, and Assessment of a High Fidelity Chilled Water Plant Model. *Appl. Therm. Eng.* **2017**, *111*, 477–488. [CrossRef]

25. Mittal, K.; Wilson, J.; Baillie, B.; Gupta, S.; Bollas, G.; Luh, P.B. Supervisory Control for Resilient Chiller Plants under Condenser Fouling. *IEEE Access* **2017**, *5*, 14028–14046. [CrossRef]

26. Sáez, D.; Ordys, A.; Grimble, M. Design of a supervisory predictive controller and its application to thermal power plants. *Optim. Control Appl. Methods* **2005**, *26*, 169–198. [CrossRef]

27. Ponce, C.V.; Sáez, D.; Bordons, C.; Núñez, A. Dynamic simulator and model predictive control of an integrated solar combined cycle plant. *Energy* **2016**, *109*, 974–986. [CrossRef]

28. Chen, C.; Zhou, Z.; Bollas, G.M. Dynamic modeling, simulation and optimization of a subcritical steam power plant. Part I: Plant model and regulatory control. *Energy Convers. Manag.* **2017**, *145*, 324–334. [CrossRef]

29. Joseph, G.; Singer, P. (Eds.) *Combustion Fossil Power: A Reference Book on Fuel Burning and Steam Generation*, 4th ed.; Combustion Engineering: New York, NY, USA, 1991.

30. ISO New England. *Real-Time Maps and Charts*; ISO New England: Holyoke, MA, USA, 2016.

31. Casella, F.; Leva, A. Modelica open library for power plant simulation: Design and experimental validation. In Proceeding of the 2003 Modelica Conference, Linkoping, Sweden, 3–4 November 2003.

32. Hsu, C.; Chen, C. Applications of improved grey prediction model for power demand forecasting. *Energy Convers. Manag.* **2003**, *44*, 2241–2249. [CrossRef]

33. Akay, D.; Atak, M. Grey prediction with rolling mechanism for electricity demand forecasting of Turkey. *Energy* **2007**, *32*, 1670–1675. [CrossRef]

34. Starkloff, R.; Alobaid, F.; Karner, K.; Epple, B.; Schmitz, M.; Boehm, F. Development and validation of a dynamic simulation model for a large coal-fired power plant. *Appl. Therm. Eng.* **2015**, *91*, 496–506. [CrossRef]

35. ABB and Rocky Mountain Institute. *ABB Energy Efficiency Handbook: Power Generation—Energy Efficient Design of Auxiliary Systems in Fossil-Fuel Power Plants*; Technical Report; ABB and Rocky Mountain Institute: Basalt, CO, USA, 2009.

36. Sanpasertparnich, T.; Aroonwilas, A.; Veawab, A. Improved Thermal Efficiency of Coal-Fired Power Station: Monte Carlo Simulation. In Proceeding of the 2006 IEEE EIC Climate Change Conference, Ottawa, ON, Canada, 10–12 May 2006; pp. 1–9. [CrossRef]

37. Lizon-A-Lugrin, L.; Teyssedou, A.; Pioro, I. Appropriate thermodynamic cycles to be used in future pressure-channel supercritical water-cooled nuclear power plants. *Nucl. Eng. Des.* **2012**, *246*, 2–11. [CrossRef]

38. Suresh, M.V.J.J.; Reddy, K.S.; Kolar, A.K. Thermodynamic Optimization of Advanced Steam Power Plants Retrofitted for Oxy-Coal Combustion. *J. Eng. Gas Turbines Power* **2011**, *133*, 063001. [CrossRef]

39. Suresh, M.; Reddy, K.; Kolar, A.K. ANN-GA based optimization of a high ash coal-fired supercritical power plant. *Appl. Energy* **2011**, *88*, 4867–4873. [CrossRef]
40. Wang, L.; Yang, Y.; Dong, C.; Morosuk, T.; Tsatsaronis, G. Multi-objective optimization of coal-fired power plants using differential evolution. *Appl. Energy* **2014**, *115*, 254–264. [CrossRef]
41. Xiong, J.; Zhao, H.; Zhang, C.; Zheng, C.; Luh, P.B. Thermoeconomic operation optimization of a coal-fired power plant. *Energy* **2012**, *42*, 486–496. [CrossRef]
42. Koch, C.; Cziesla, F.; Tsatsaronis, G. Optimization of combined cycle power plants using evolutionary algorithms. *Chem. Eng. Process* **2007**, *46*, 1151–1159. [CrossRef]
43. Espatolero, S.; Cortés, C.; Romeo, L.M. Optimization of boiler cold-end and integration with the steam cycle in supercritical units. *Appl. Energy* **2010**, *87*, 1651–1660. [CrossRef]
44. Regulagadda, P.; Dincer, I.; Naterer, G.F. Exergy analysis of a thermal power plant with measured boiler and turbine losses. *Appl. Therm. Eng.* **2010**, *30*, 970–976. [CrossRef]
45. Kakaras, E.; Ahladas, P.; Syrmopoulos, S. Computer simulation studies for the integration of an external dryer into a Greek lignite-fired power plant. *Fuel* **2002**, *81*, 583–593. [CrossRef]
46. Tatjewski, P. *Advanced Control of Industrial Processes*; Advances in Industrial Control Series; Springer: London, UK, 2007. [CrossRef]
47. The MathWorks Inc. *MATLAB*; MathWorks: Natick, MA, USA, 2013.
48. MODELISAR Consortium. *Functional Mock-up Interface for Model Exchange*, Version 1.0; MODELISAR Consortium: Linköping, Sweden, 2010.
49. Chaibakhsh, A.; Ghaffari, A. Steam turbine model. *Simul. Model. Pract. Theory* **2008**, *16*, 1145–1162. [CrossRef]

Article

Effect of Tariff Policy and Battery Degradation on Optimal Energy Storage

Mariana Corengia and Ana I. Torres *

Instituto de Ingeniería Química, Facultad de Ingeniería, Universidad de la República, Montevideo 11300, Uruguay; corengia@fing.edu.uy
* Correspondence: aitorres@fing.edu.uy; Tel.: +598-2714-2714 (ext. 18102)

Received: 15 September 2018; Accepted: 19 October 2018; Published: 22 October 2018

Abstract: In the context of an increasing participation of renewable energy in the electricity market, demand response is a strategy promoted by electricity companies to balance the non-programmable supply of electricity with its usage. Through the use of differential electricity prices, a switch in energy consumption patterns is stimulated. In recent years, energy self-storage in batteries has been proposed as a way to take advantage of differential prices without a major disruption in daily routines. Although a promising solution, charge and discharge cycles also degrade batteries, thus expected savings in the energy bill may actually be non-existent if these savings are counterbalanced by the capacity lost by the battery. In this work a convex optimization problem that finds the operating schedule for a battery and includes the effects of current-induced degradation is presented. The goal is to have a tool that facilitates for a consumer the evaluation of the convenience of installing a battery-based energy storage system under different but given assumptions of electricity and battery prices. The problem is solved assuming operation of a commercial Li-ion under two very different yet representative electricity pricing policies.

Keywords: demand response; energy management; energy storage; optimal battery operation; battery degradation

---

1. Introduction

Connecting renewable energy sources to the electric grid changes the classical paradigm: as these sources are non-programmable an imbalance between electricity generation and electricity consumption is created. As electricity as such cannot be stored, two main strategies have been proposed to overcome the imbalance: conversion of electricity to a type of energy that can be stored and demand response [1,2]. In the former, energy storage, the amount of electricity that is produced in excess at a certain time is centrally converted to for example potential energy (through mechanical pumping) or chemical-electrochemical energy (e.g., production of H_2 or large-scale batteries). Then, when needed at a later time, this energy is reconverted to electricity and discharged to the grid. In the latter, demand response, a change in electricity consumption habits to match the production rate is sought. This change in habits is promoted by adjusting the price of electricity at different times: when there is an expected large demand or low production of electricity, its price increases, and the opposite when demand is low and supply is large. The term tariff policy or pricing policy refers to the electricity price schedule that companies develop to stimulate demand response. In this context, many energy markets have developed different tariff policies [3,4]. Broadly speaking, tariff policies can be classified as pre-established and on-spot. On on-spot tariffs, electricity prices and even allowed amounts of energy that might be purchased, are announced publicly shortly before the expected consumption (from minutes, up to one day ahead). On pre-established tariffs, electricity prices are known to consumers, but might be different for weekdays and weekends/holidays, and, within each day, they might change hourly. These types of tariffs are also known as time of use pricing (TOU).

Implementation of these policies have given the consumers the opportunity of reducing the total cost of the electricity they use, by increasing their energy consumption when electricity price is low and reducing the consumption when the price is high. Still, not all uses can be managed this way. In particular, in the residential sector, major savings imply a large reorganization of daily routines, which is only worth if there is a strong penalization in the hours that energy consumption is to be reduced. For residents or small-scale businesses, self-storage represents a more suitable non-routine disruptive way to achieve savings in electricity costs. For these purposes, electrochemical, thermal, and even an increase in inventory when the price is low, have been pointed out as valid self-storage alternatives [5].

Along these lines, how to optimally store energy to maximize savings is a problem currently being addressed in the literature, a complete review can be found in Weitzel and Glock [6]. With the foreseen decline in the costs of batteries [7,8] incorporation of this kind of energy storage solution may soon be economically viable for all sectors, including small enterprises and households. This economic viability will depend on both the savings due to TOU and battery lifespan and replacement cost.

Battery lifespan and replacement cost depend on its degradation rate, which in turn depend on how much and how the battery is used. The relevance of including these factors when formulating problems to find an optimal battery operation policy has been recognized in Reniers et al. [9]. Battery degradation mechanisms depend on many factors, in particular, the rate of charge/discharge has a large effect. As in all electrochemical devices, charge/discharge due to electrochemical reaction, and current density are related by Faraday's Law. Larger currents result in shorter charge/discharge periods; however, as electric current increases, so does the overpotentials, which may favor secondary effects. These effects contribute to capacity loss in batteries during operation [10,11].

The goal of this paper is to formulate an optimization problem for finding the optimal operation schedule for a battery that also includes current-induced battery degradation. Other contributions have also considered battery degradation in the formulation of their optimization problems: Sarker et al. [12] showed that not taking into account battery degradation results in a solution that leads to an operating schedule that actually depletes the battery faster than the solution that is found when considering these effects directly in the optimization problem; Yan et al. [13] showed a similar effect but with a slightly different use: that of the battery being used as a frequency regulator; Hu et al. [14] considered the cost of battery degradation in energy management of plug-in hybrid electric vehicles.

Unlike a previous mixed integer linear problem (MILP) [12], in this paper a convex formulation is presented. The advantage of this formulation lies in the fact that global optimality is guaranteed, and also that with the same computing capabilities, problems considering longer time spans (not only a single day) can be solved. This in turn allows the estimation of the leftover capacity of the battery after a certain number of cycles. This is precisely the main novelty of this paper: the capability of predicting the effects that different pricing policies have in long-term optimal operation scheduling, when battery degradation is taken into account.

In all case studies Li-ion batteries, an energy storage device that presents an outstanding specific energy and power, long calendar and cycle lives, high efficiency and reliability [15], were considered. The paper is organized as follows: Section 2 presents the basic optimization problem and its convex form; using this model, the optimal daily operation for two different pricing policies (simple and complex) are presented in Section 3. An extension to the model in Section 2 that captures the cumulative effect of battery usage is presented in Section 4.1. Sections 4.2 and 4.3 present the long-term optimal operation schedule for the same tariffs considered in Section 3. Finally, the key points of the paper are summarized in Section 5.

2. Problem Statement and Model Formulation

2.1. Objective Function

Throughout the paper we will consider a consumer who wants to install a Li-ion battery as a way to take advantage of a given pre-established TOU tariff policy. We will assume that the consumer has already chosen the capacity that needs to be installed and the general problem is how to optimally operate the battery so that savings in the electricity bill are larger than the costs induced by its use. Then, the objective function from the user point of view can be expressed as:

$$\max(electricity savings - capacity loss) \tag{1}$$

As shown in Figure 1, when being charged, at a moment t, the battery takes energy from the grid at charging power p_t^C; when being discharged, at a later time t', the battery feeds a load (for example a house) at discharging power $p_{t'}^D$. Besides, the energy that the battery receives from the grid and the energy the load receives from the battery are affected by efficiency factors denoted as η^+ and η^- respectively. At times, it might be more convenient to by-pass the battery and feed the load directly from the grid. This is shown in Figure 1 as the direct power flow p_t^{direct}.

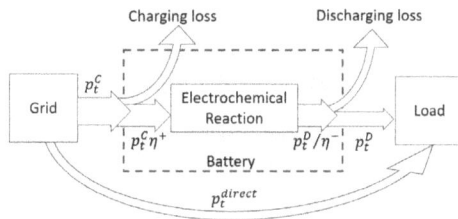

Figure 1. Schematic representation of energy fluxes from the grid to the load. The load can consume energy directly from the grid or from a previously charged battery. p_t represents the power at time t (superscripts C and D denote charge and discharge processes); η^+, η^- represent battery efficiencies.

Using these variables the savings in electricity can be expressed as:

$$electricity\ savings = cost\ of\ energy\ discharged\ at\ time\ t'$$
$$- cost\ of\ energy\ charged\ at\ time\ t \tag{2}$$
$$cost\ of\ energy\ charged\ at\ time\ t = \$_t p_t^C \Delta t \tag{3}$$
$$cost\ of\ energy\ discharged\ at\ time\ t' = \$_{t'} p_{t'}^D \Delta t' \tag{4}$$

where $\$_t$ and $\$_{t'}$ denote the price of the electricity at different times as given in the tariff policy.

On the other hand, the cost of capacity loss due to operation can be written as:

$$capacity\ loss = C^{ES} BC^{ES} x_{t''}^{CL} \tag{5}$$

where C^{ES} denotes the cost of the technology in USD/kWh; BC^{ES} is the installed capacity, a fixed parameter with units kWh; and $x_{t''}^{CL}$ is the fraction of capacity lost during operation time t''. Note that t'' is a dummy variable: $t'' = t$ during charge and $t'' = t'$ during discharge.

Then, for a given period of time τ divided into smaller time steps Δt, with Δt equal to the minimum electricity pricing step, the objective function can be stated in canonical form as:

$$\min \Delta t \sum_{t=1}^{t=\tau} \$_t (p_t^C - p_t^D) + C^{ES} BC^{ES} \sum_{t=1}^{t=\tau} x_t^{CL} \tag{6}$$

2.2. Constraints

In Equation (6) the terms $\$_t$, C^{ES}, BC^{ES} are parameters, whereas the terms p_t^C, p_t^D and x_t^{CL} are decision variables of the problem. These variables are related to each other and to other decision variables through the following constraints:

- The battery cannot charge and discharge at the same time
  This restriction can be mathematically posted as

$$p_t^D p_t^C = 0 \tag{7}$$

- For safety reasons, there is a maximum power that should not be exceeded neither during charge (p^{C-MAX}) nor during discharge (p^{D-MAX}). Additionally, p_t^C and p_t^D will always be assumed as positive variables. Then:

$$0 \le p_t^C \le p^{C-MAX} \tag{8}$$
$$0 \le p_t^D \le p^{D-MAX} \tag{9}$$

- Energy balance: the state of charge at a certain time, soc_t, depends on the state of charge at the immediately previous period of time soc_{t-1} and the amount of energy that is effectively used in electrochemical reactions during charge or discharge processes:

$$soc_t = soc_{t-1} + \Delta t p_t^C \eta^+ - \Delta t p_t^D / \eta^- \tag{10}$$

where the efficiencies η are as defined previously (see Figure 1).
- In order to preserve the battery life, the state of charge should neither be lower than a minimum fixed value (SOC^{MIN}) nor larger than a maximum (SOC^{MAX}). Both these parameters correspond to different fractions of the total available capacity.

$$SOC^{MIN} \le soc_t \le SOC^{MAX} \tag{11}$$

- Equations for battery degradation:

Battery degradation kinetics are studied in the electrochemistry field by using a parameter known as the C_{rate} [16,17]. The C_{rate} is the inverse of the characteristic time that relates electric current during a time step, I_t in units of A or mA, and the electrical charge capacity of the battery, BC_Q^{ES} in units of Ah or mAh.

$$C_{rate,t} = \frac{I_t}{BC_Q^{ES}} \tag{12}$$

This coefficient allows a size-independent study of the kinetics of any reaction happening in batteries. The fraction of capacity lost by the battery x_t^{CL} can then be expressed as a function of the C_{rate}.

$$x_t^{CL} = f(C_{rate,t}) \tag{13}$$

where the functionality f is obtained experimentally by running charge/discharge cycles at different currents, and is usually a non-linear expression. Taking as a basis the experimental data reproduced in Figure S1, it can be seen that $f(C_{rate,t})$ can be modeled as a convex second order polynomial.

$$x_t^{CL} = \alpha_1 C_{rate,t}^2 + \alpha_2 C_{rate,t} \tag{14}$$

with α_1 and α_2 parameters that are adjusted to the experimental data.

To relate the C_{rate} from its definition (Equation (12)) with the variables used in this problem, a nearly constant working potential of the cell needs to be assumed. In practice, this assumption implies disregarding the effects of the overpotentials in the potential vs. current (*E-I*) curve. Then, $E_t = E_t^0$ and:

$$C_{rate,t} = \frac{I_t E_t}{BC_Q^{ES} E_t^0}$$

$$C_{rate,t} = \frac{P_t}{BC^{ES}} \tag{15}$$

Notice that as charge and discharge do not happen simultaneously (see Equations (7) and (15)) can be rewritten to combine both processes in a single equation:

$$C_{rate,t} = \frac{p_t^C + p_t^D}{BC^{ES}} \tag{16}$$

At this point, it is worth commenting that natural aging phenomena (just a function of time) has not been considered here as calendar aging is not dependent on the rate of battery use.

Therefore, given the objective function (Equation (6)) and the restrictions (7)–(11), (14), (16) the problem is a non-linear programming (NLP) problem where the decision variables are $p_t^C, p_t^D, soc_t, x_t^{CL}$ and $C_{rate,t}$. As stated, the problem has two non-linear equality constraints (Equations (7) and (14)) thus it is non-convex. Relaxations, that result in an equivalent convex problem are discussed in the following section.

2.3. Derivation of an Equivalent Convex Problem

2.3.1. Non-Simultaneous Charge and Discharge

Equation (7) is the equation that states that simultaneous charge and discharge processes are not possible. Besides the product-based formulation used here, a formulation employing a binary variable that multiplies either p_t^D or p_t^C has been reported in Jabr et al. [18]. However, this approach leads to a mixed integer problem (MILP or MINLP) which does not solve the fundamental convexity problem.

In this work we will make use of a result in Castillo and Gayme [19].

"For energy storage capacity at bus $n \in \mathcal{N}$ where (the energy storage capacity) $C_n > 0$, if (the locational marginal price) $\lambda_n(t)$ is strictly positive then (...) simultaneous charging and discharging will not occur".

This result was obtained in the context of optimal allocation and operation of a distributed energy storage system, in a network of generators, storage devices and consumers, where the price of the energy was not included as a parameter of the model, instead it was estimated through the dual variable $\lambda_n(t)$ which is referred in the paper as a market-based price signal.

Despite our overall problem formulation is different, from the energy storage device, i.e., the battery, the situation is similar as in both cases the energy storage device consumes from an upstream energy provider and supplies to a downstream load. In our case the electricity prices are given and correspond to the TOU tariffs which are always positive. Then, it is reasonable to assume that the result from Castillo and Gayme [19] will also be valid here, and p_t^D and p_t^C will not be non-zero at the same time. At a more fundamental level, the explanation is that as charge/discharge efficiencies are always lower than 1, simultaneous charge and discharge just dissipates energy. Hence a portion of the energy consumed does not result in an effective change in the state of charge of the energy storage device, an effect that is usually not sought.

Therefore, based on the above discussion, Equation (7) will be excluded from the formulation of the problem; upon finding the optimal solution in different case studies a check will be performed in order to assess the validity of the relaxation.

2.3.2. Battery Degradation

Equation (13) is the equation relating the fraction of capacity lost by the battery for different C_{rates}. As discussed previously, this equation is derived by fitting experimental data and results in a convex expression as the parameter α_1 is positive. As in Equation (6) the variable x_t^{CL} is multiplied by positive values, an increase in the value of x_t^{CL} will directly imply an increase in the objective function. Hence, if the equality constraint in Equation (13) is relaxed to the convex inequality constraint in Equation (17):

$$x_t^{CL} \geq \alpha_1 C_{rate,t}^2 + \alpha_2 C_{rate,t} \tag{17}$$

minimization of the objective function will always result in an active constraint, thus the same optimal solution.

At this point it should be noted that although we started from a particular case (experimental data for Samsung INR 18650 reported in Sarker et al. [12]) it has been experimentally observed that other types of batteries also present a convex functionality between the capacity leftover in the battery after a certain number of charge/discharge cycles (*BC*) and the C_{rate} [20]. Thus, the relaxation presented here can be generalized if other batteries, with other convex functionalities, are considered. Despite the details are not yet clearly understood, the convex functionality is to be expected as increasing the current increases the overpotentials. Thus both, the number of undesired events (physical or chemical) and the rate at which these events happen increase creating a snowball effect in terms of battery degradation [11,21,22].

2.3.3. Problem Statement in Convex Form

Following the discussion in the previous section, we present here the canonical form of the convex NLP problem which is equivalent to the one discussed in Section 2.2.

$$\min_{p_t^C, p_t^D, soc_t, x_t^{CL}, C_{rate,t}} \Delta t \sum_{t \in \tau} \$_t (p_t^C - p_t^D) + C^{ES} BC^{ES} \sum_{t \in \tau} x_t^{CL} \tag{18}$$

$$\text{s.t.} \quad soc_t = soc_{t-1} + \Delta t p_t^C \eta^+ - \Delta t p_t^D / \eta^- \tag{19}$$

$$C_{rate,t} = \frac{p_t^C + p_t^D}{BC^{ES}} \tag{20}$$

$$x_t^{CL} \geq \alpha_1 (C_{rate,t})^2 + \alpha_2 C_{rate,t} \tag{21}$$

$$SOC^{MIN} \leq soc_t \leq SOC^{MAX} \tag{22}$$

$$0 \leq p_t^C \leq p^{C-MAX} \tag{23}$$

$$0 \leq p_t^D \leq p^{D-MAX} \tag{24}$$

The problem is implemented in GAMS 24.8.5 and solved with IPOPT [23].

3. Optimal Scheduling for a 24 h Period

In this section, the optimal charge/discharge schedules for different tariff policies are presented. The simulations were run considering an example of each of the two most extreme TOU pricing policies:

- The "simple tariff" refers to a pricing policy that only has two price steps: cheap (off-peak) and expensive (on-peak), and the daily price pattern is repeated throughout the year. Therefore, there are always some consecutive hours with exactly the same price.

- The "complex tariff" refers to a pricing policy where there are several price steps during a single day (the price is still constant for each hour of the day), different days present different prices (weekdays are priced differently than weekends and holidays and there is also seasonal variation), and prices are dependent on the expected weather.

An example of a "simple tariff" is given by the Uruguayan residential "UTE-Tarifa Inteligente" (smart rate) which consists of an 18 h period at a cheap rate, and a 6 h period at an expensive rate [24]. An example of a "complex tariff" is given by the Californian Southern California Edison tariff [25]. For medium consumers, this tariff classifies days into nine groups according to: High and Low cost Weekend; Extremely Hot, Very Hot, Hot, Moderate and Mild Summer Weekday, High and Low cost

Winter Weekday. Both for winter and weekends classes, "High cost" and "Low cost" depend on the expected temperature.

Other TOU tariffs are in between the previous two extremes. For example the Brazilian "Tarifa Branca" [26], is a three-step simple tariff with the: cheap-medium priced-expensive-medium priced-cheap pattern on weekdays and cheap pattern on weekends. Another example is Iowa's residential hourly pricing option [27] which is a two-step tariff which also differentiates between winter and summer.

In all cases, the maximum C_{rate} is assumed to be 3; then maximum power and capacity ratio are related as $p^{C-MAX} = p^{D-MAX} = 3BC^{ES}$. Upper and lower bounds for the state of charge (SOC^{MAX}, SOC^{MIN}) are assumed to be 80% and 20% of the installed capacity respectively. Both η^+ and η^- are assumed to be 0.95 and installed capacity BC^{ES} is set at 10 kWh. Furthermore, we will assume that the battery is initially discharged (i.e., $soc_0 = SOC^{MIN}$). The coefficients for the battery degradation equation are $\alpha_1 = 1.06 \times 10^{-5}$ and $\alpha_2 = 1.44 \times 10^{-4}$ (see S1 in Supplementary Information). Estimated 2018 battery prices range from 580–750 USD/kWh for commercial and residential use [28]. However, as Li-ion battery systems still have margin to lower their production costs [29], battery prices used in the simulations were selected case by case to show key features of each case study.

3.1. Results and Discussion: Simple tariff

Figure 2a shows the TOU tariff used in Uruguay: electricity is cheap between 0:00 a.m.–5:00 p.m., significantly more expensive between 5:00–11:00 p.m., and cheap again during the last hour of the day (11:00 p.m.–0:00 a.m.). Note that in Figure 2a we have redefined the time vector so that all low prices are together. In practice, this change means that a "storage day" starts at 11:00 p.m., and allows the one-day simulation to exploit the whole off-peak price period.

The optimal results for p_t^C, p_t^D, soc_t and x_t^{CL} for three different battery prices (C^{ES}) are shown in Figure 2b–d.

Notice that for a battery price of 500 USD/kWh the optimal solution neither takes power from the network nor supplies power to the load, and the battery SOC remains constant at the initial level. The reason is that in this case, the economical penalty for degrading the battery is larger than the savings for taking advantage of the TOU tariff, thus the battery should not be installed.

For battery prices of 300 and 400 USD/kWh, the optimal strategy is to slowly charge the battery during the whole low-price period at a steady rate, and then discharge it as slowly as possible during the high price period. This strategy results in low C_{rates} for both charge and discharge thus low degradation factors x_t^{CL}. The strategy itself is the same for both battery prices, then total savings in the energy bill are the same (0.86 USD/d). However, as the penalty due to loss of battery capacity depends on initial battery price, overall savings are different and lower than bill savings: 0.34 USD/d and 0.17 USD/d for the 300 USD/kWh and 400 USD/kWh respectively.

p_t^C and p_t^D are never simultaneously non-zero and, as shown in Supplementary Materials, the value of the Kuhn-Tucker multipliers verifies that Equation (21) is always active. Thus, the assumptions made for convexification of the problem are checked.

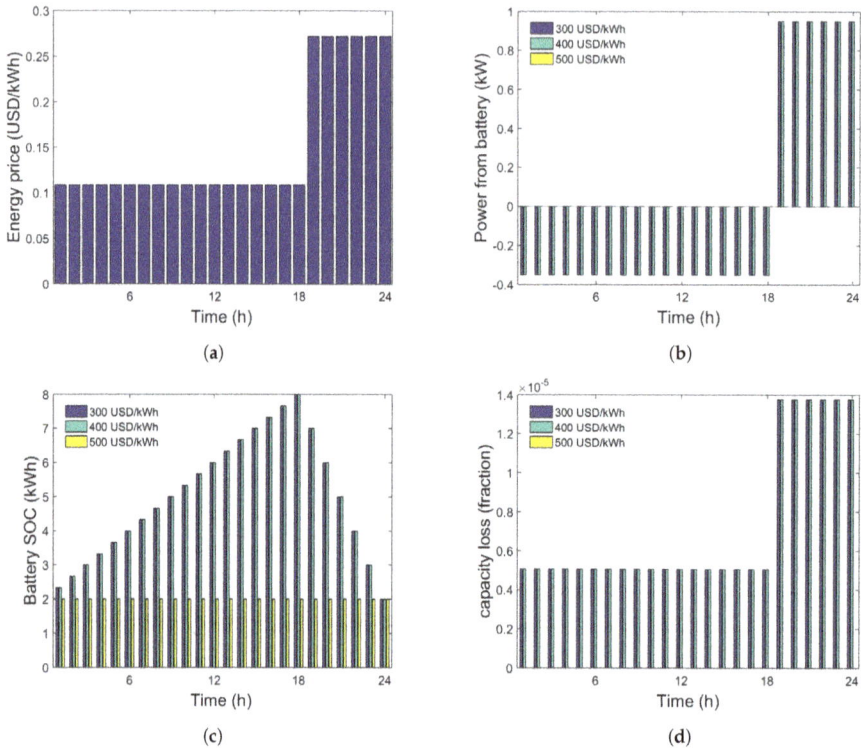

Figure 2. Optimal operation for a "simple tariff" example during a 24 h period. (**a**) Electricity TOU tariff for Uruguay. (**b**) Optimal power consumed (p_t^C, represented as negative for visual purposes) and supplied (p_t^D) by the battery. (**c**) Battery State of Charge at each time (soc_t). (**d**) Fraction of capacity lost by the battery at each time (x_t^{CL}). Accumulated fraction loss in one day is 1.7×10^{-4} for 300 and 400 USD/kWh.

3.2. Results and Discussion: Complex Tariff

Figure 3 shows the TOU tariff used by Southern California Edison Company (TOU-GS-2-RTP); days presenting a large variation in the tariff are shown in Figure 3a, whereas days with moderate variations are shown in Figure 3b.

In this case there are in principle 9 different optimal strategies each corresponding to a different day-type as discussed at the beginning of the section. However, from these nine price schedules, only four result in a non-trivial operation for batteries that cost 300 USD/kWh. The four price schedules that result in non-trivial operation coincide with the ones that have the largest variation between the maximum and the minimum electricity prices (i.e., Figure 3a). Figure 4 shows the optimal charge/discharge schedule for battery prices of 250, 300 and 400 USD/kWh. As the battery price decreases, moderate-priced days start also to show non-trivial operation, see as an example Figure S4.

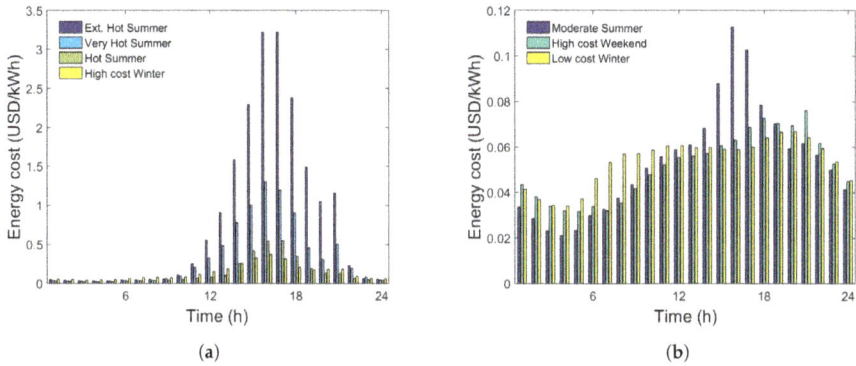

Figure 3. Example of "complex" tariff with different day types and hourly changes. (**a**) Days presenting a large price variation. (**b**) Days presenting a moderate variation.

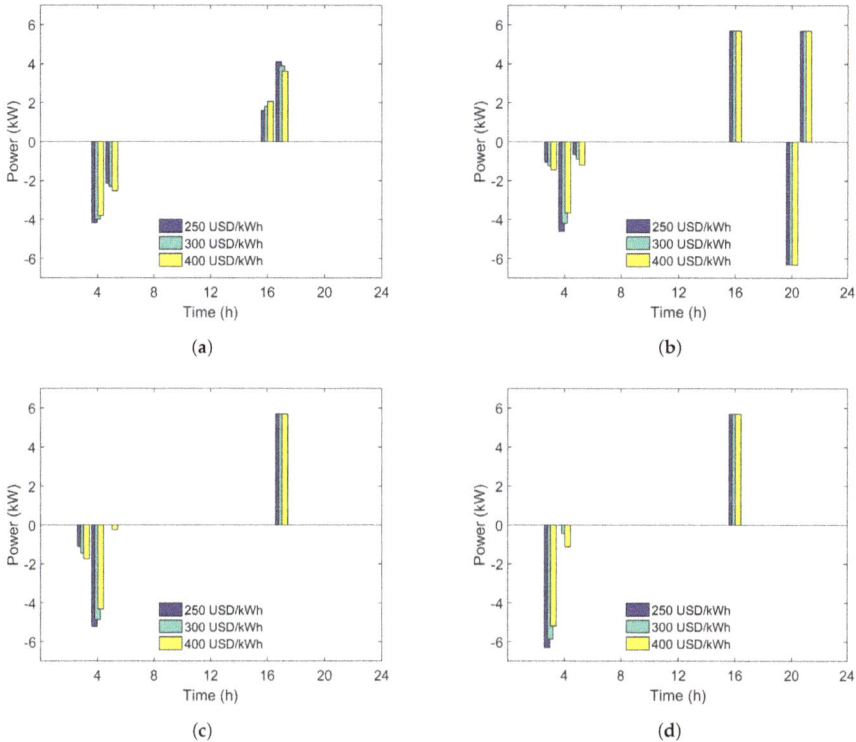

Figure 4. Optimal operation for a "complex tariff" example during a 24 h period, for the days that present a large price variation. p_t^C: optimal power consumed from the grid (represented as negative for visual purposes) and p_t^D: optimal power supplied by the battery (positive). (**a**) Extremely Hot Summer Weekday. (**b**) Very Hot Summer Weekday. (**c**) Hot Summer Weekday. (**d**) High cost Winter Weekday.

Compared to the first case study the most noticeable difference is that charge and discharge are now concentrated and not distributed during the day. For example, for Extremely Hot Summer Days

(EHSD) charge is concentrated in two hours (3:00–5:00 a.m.) and discharge is concentrated between 3:00–5:00 p.m. (see Figure 4a). Notice also that unlike the "simple" tariff the rate of charge/discharge varies. This effect is seen in Figure 4a in the EHSD case: between 3:00–4:00 a.m. the battery consumes more power than between 4:00–5:00 a.m. Logically this is related to the price of the energy during those hours. In absolute terms, for the days that operation is allowed, this "complex" tariff results in a faster degradation of the battery than the "simple" tariff considered before. Figures with degradation rates are in the Supplementary Information (S3).

The difference in pricing also allows for some days to have a double charge/discharge cycle. For example, for a Very Hot Summer Day (VHSD) the first charge is between 2:00–5:00 a.m., the first discharge is between 3:00–4:00 p.m., the second charge is between 7:00–8:00 p.m. and the second discharge is between 8:00–9:00 p.m. This double cycle is allowed by the relatively large difference between the prices corresponding to the 7:00–8:00 p.m. and 8:00–9:00 p.m. interval.

Another salient feature of the complex tariff is that a difference in the cost of the battery affects the optimal operation: although the time of the day at which charge/discharge occurs is the same, the amount charged within each interval is different.

As in the previous case, all assumptions for problem relaxation were satisfied.

4. Optimal Scheduling for Long-Term Periods

4.1. Modifications to the Original Problem

4.1.1. Variable Capacity

In the previous sections we showed how the optimal operation schedule was affected by the TOU tariff and the cost of the storage system itself. In this section, we expand those results to a more realistic setting in which the battery is going to be used during several years. As longer optimization periods are considered, charge and discharge cycles may result in a significant loss of capacity, and the problem presented in Section 2.3.3 needs to be modified in order to allow for a daily change in the capacity of the battery. In other words, capacity is now a variable which will be represented by BC_t.

Without loss of generality we will consider daily changes in BC_t (i.e., $BC_t \equiv BC_d$) and not hourly changes, a consideration that is supported by the small hourly values obtained for x_t^{CL} in Section 3. However, as the tariffs still vary on an hourly basis, two time scales need to be considered: day and hour. As a result the variables in the previous sections will now be represented using both time scales $(p_{d,h}^C, p_{d,h}^D, soc_{d,h}, x_{d,h}^{CL})$.

As BC_d changes daily so do the allowed upper and lower bounds of the state of charge. Then, $SOC^{MIN} \equiv SOC_d^{MIN}$ and $SOC^{MAX} \equiv SOC_d^{MAX}$. If we still keep the same criteria (state of charge between 20% and 80% of the remaining capacity), then Equation (22) should be replaced by Equation (25):

$$0.2BC_d \leq soc_{d,h} \leq 0.8BC_d \tag{25}$$

Also, as BC_d changes, the maximum power at which the battery can be charged or discharged safely, varies. Keeping the same C_{rate} limit as in Section 3 ($C_{rate}^{limit} = 3$) then Equations (26) and (27), substitute Equations (23) and (24).

$$0 \leq p_{d,h}^C \leq 3\frac{BC_d}{\Delta t} \tag{26}$$

$$0 \leq p_{d,h}^D \leq 3\frac{BC_d}{\Delta t} \tag{27}$$

4.1.2. Restricted Time Span between Charge and Discharge Periods

Complex tariff systems that present large differences between the highest and the lowest prices may provide an optimal schedule in which the battery charges months before it discharges. This strategy, even when acceptable, it is not practical as:

- If energy is stored in batteries for a long period of time, self-discharge processes occur. For the sake of simplicity our model has not included these types of processes under the assumption that the charge and discharge cycle of the battery happens in a reasonably short period, most probably during the same day.
- Complex tariffs depend on weather conditions and thus a reliable forecast. In this work we have assumed that reasonable forecasts are given for periods no longer than a week.

Then, to restrict the length between charge and discharge periods, Equation (28) was included as an extra restriction to the long-term optimization problem:

$$\eta^{+}p_{d,h}^{C} \leq \frac{1}{\eta^{-}}\left(\sum_{h'=h}^{24} p_{d,h'}^{D} + \sum_{d'=d+1}^{d+d_{limit}} \sum_{h'=1}^{24} p_{d',h'}^{D}\right) \tag{28}$$

As written, Equation (28) indicates that the net power consumed by the battery at a certain time must be lower than the sum of the power supplied from the battery to the load during the following week ($d_{limit} = 7$). Depending on the technology used, this restriction could be tightened by using a smaller d_{limit}.

4.2. Results and Discussion: Simple Tariff

Figure 5a shows the charge/discharge strategy ($p_{d,h}^{C}$ and $p_{d,h}^{D}$) for the battery for the first and last days of use. As the tariff has two steps and remains constant for all days, the operation policy is the same as in the one-day case: slowly charge and discharge. The difference between the first and last day is the rate at which power is consumed/supplied, which diminishes with time due to battery degradation.

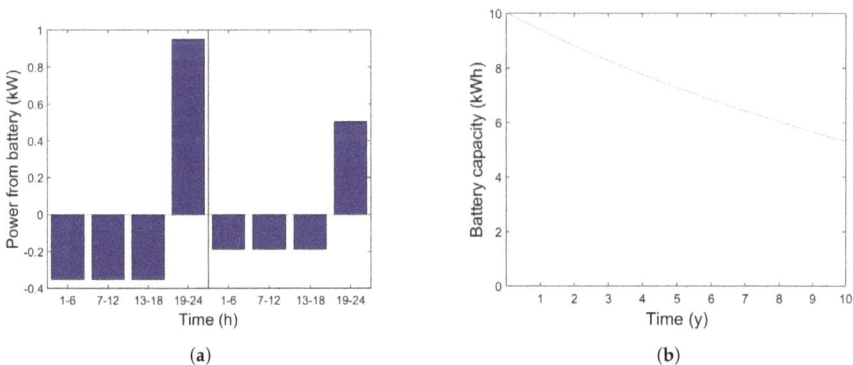

(a)   (b)

Figure 5. Optimal operation for a "simple tariff" example during a 10 years period. (**a**) p_t^{C} (negative) and p_t^{D} (positive). Left: First day. Right: Last day. (**b**) Remaining capacity BC_d.

The capacity remaining in the battery at each time is shown in Figure 5b. As seen after 10 years the battery still retains 53% of its original capacity.

As in the case discussed in Section 3.1, for a non-trivial optimal, the price of the battery affects the value of the objective function, but not the value of the variables $p_{d,h}^{C}$, $p_{d,h}^{D}$, $soc_{d,h}$, $x_{d,h}^{CL}$. Table 1 shows the yearly savings in electricity bills for capital costs below 400 USD/kWh (i.e., a price that guarantees a non-trivial solution). The net savings considering battery degradation over the operation period are 922 USD and 453 USD, for the 300 USD/kWh battery and 400 USD/kWh battery, respectively. The net present values calculated as:

$$NPV = -C^{ES} \cdot BC^{ES} + \sum_{i=1}^{n} \frac{\sum_{d \in n} \sum_{h=1}^{24} \$_{d,h} (p_{d,h}^{C} - p_{d,h}^{D})}{(1+r)^i} \tag{29}$$

are presented in Table 2 for different battery prices and rates (*r*). Roughly, prices below 140 USD/kWh are required for batteries to be economically viable. This value is in agreement with the 125 USD/kWh target established by the DOE for battery prices for electric vehicles [7].

Table 1. Electricity annual savings for each operation year with example simple tariff.

Operation Year	1	2	3	4	5	6	7	8	9	10
Annual savings (USD)	305	286	269	252	237	222	208	196	184	172

Table 2. Simple tariff net present values for different battery prices and rates.

C^{ES} (USD/kWh)	$NPV_{r=8\%}$ (USD)	$NPV_{r=10\%}$ (USD)	$NPV_{r=12\%}$ (USD)
400	−2374	−2497	−2606
300	−1374	−1497	−1606
200	−374	−497	−606
150	126	3	−106
100	626	503	394

4.3. Results and Discussion: Complex Tariff

Figure 6 shows the maximum temperatures recorded in Los Angeles, CA, during each day of 2017 [30]. As mentioned earlier, maximum temperature is the key parameter to establish the electricity price in the example of complex tariff used in this paper. The temperatures themselves are represented with bullets, whereas the colored blocks represent the tariff that would be applied for each day according to season and temperature effects (for clarity values for weekends were omitted in the graph). The full line in the graph represents the historical average for maximum temperature. Note that using averages is not suitable for running the simulations as no day would be classified in the extreme temperatures, which, as discussed in Section 3.2, are precisely the days that provide a non-trivial optimal solution. For this reason, all simulations in this part of the paper where run assuming the temperatures recorded in 2017.

The optimal operation for the battery under the previous assumptions and an assumed battery price of 300 USD/kWh, is shown in Figure 7. As expected, the strategy is to use the energy stored in the battery during the days where the temperature falls in the blue, yellow, orange and red zones indicated in Figure 6, and these are not weekends. This strategy is appreciated in the "Daily discharge energy" subplot, where the total amount of power supplied by the battery to the load during a day ($\sum_{h=1}^{h=24} p_{d,h}^{D}$) is shown. To allow this operation, the battery initiates its charge in the previous days, in some cases distributing it within several days. This effect is clearly seen around day 175 and the stair-like progression in the SOC subplot, and is affected by the value assumed for d_{limit} in Equation (28). The extreme case where d_{limit} is unbounded is discussed in the Supplementary Information (S4.1), and justifies including Equation (28) in the set of restrictions.

An interesting feature is that the battery is not used in every viable day. Instead, battery usage is concentrated towards the end of the simulation time-frame as shown in Figure 8. We interpret this result as a strategy to preserve battery life, a behavior consistent with the fact that the profit from using the battery at each time depends on its leftover capacity. The rationale is as follows: each day that the battery is used implies a loss in battery capacity that diminishes future profits. Days that provide a high profit are clearly always active. However, days that provide a low profit, become active only when the effect of capacity loss over the profit of future active days is counterbalanced by the (low) profit obtained by using the battery at the present day. This condition is more difficult to achieve at the beginning of the simulation, thus low profit days become active only at the end of the simulation time-frame when there are no more possible high profit future days. This effect is verified by running simulations with different simulation time-frames and costs. As expected, as the batteries become more economical, more moderate-priced days become active at earlier times in the simulation, and as the simulation time-frame diminishes, the active days plots become denser. For the interested reader, these plots are shown in Supplementary Materials. At this point, it has to be noticed that despite being logical, a formal proof of strict convexity of the problem is required to guarantee the uniqueness of the strategy discussed in here.

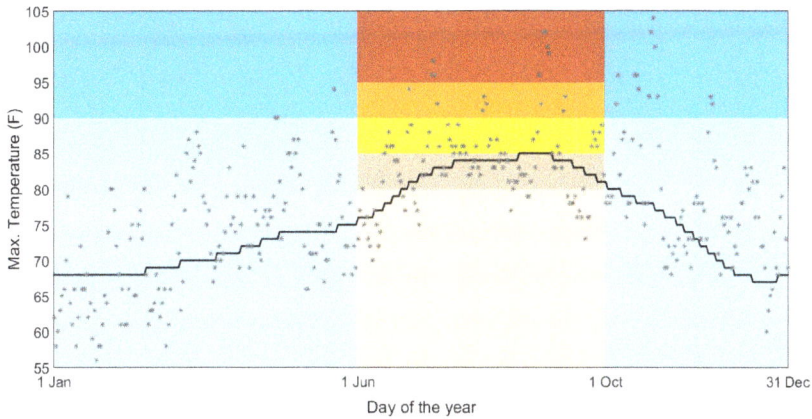

Figure 6. 2017 temperature (*) vs. Average temperature (line) in Los Angeles, CA [30]. Each background color implies a different daily tariff. Only seasonal and temperature effects on tariff are considered.

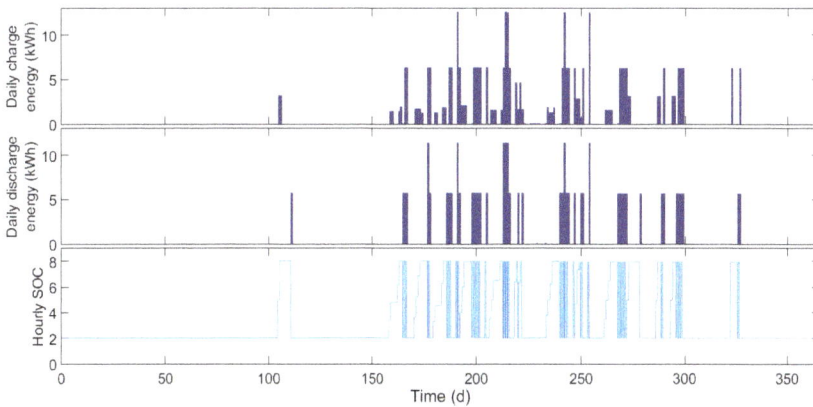

Figure 7. Optimal operation for a "complex tariff" example during a year period and a battery price of 300 USD/kWh.

The net savings considering battery degradation over the operation period are 1006, 977 and 953 USD for 200, 250 and 300 USD/kWh battery costs respectively. Compared to Section 4.2, the difference is not only due to the difference in C^{ES}, but also in the operation scheduling ($p_{d,h}^C$ and $p_{d,h}^D$) and so in battery degradation ($x_{d,h}^{CL}$). Table 3 shows the yearly savings in electricity bills and the annual loss of capacity for different capital costs. Note that electricity savings are similar for each battery cost, because the extra days that the battery is active add little savings.

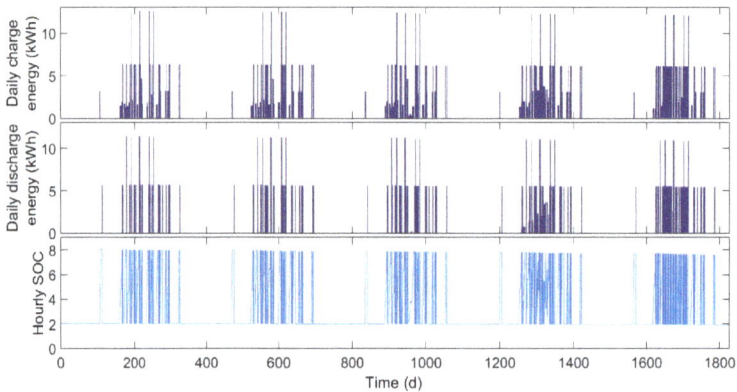

Figure 8. Optimal operation for a "complex tariff" example during a five years period and a battery price of 250 USD/kWh. Notice how battery operation is concentrated towards the end of the simulation period.

Table 3. Electricity annual savings and annual capacity loss for each operation year with complex tariff.

C^{ES} (USD/kWh)	Year 1	Year 2	Year 3	Year 4	Year 5
	Electricity annual savings (USD/year)				
200	223	230	230	226	223
250	223	220	218	221	226
300	223	220	218	216	214
	Annual capacity loss (%)				
200	0.96	1.29	1.37	1.35	1.34
250	0.94	0.93	0.93	1.09	1.34
300	0.94	0.93	0.92	0.92	0.91

5. Conclusions

With the advent of renewable energies, inclusion of batteries as self-storage devices provides an intermediate solution between demand response and grid-level energy storage. In order for this strategy to be widely adopted by small and medium size consumers, a balance between the savings in electricity bills and the cost of replacing the battery must be achieved. In this paper, a new model for optimal battery operation, that includes induced capacity loss due to charge/discharge processes, and its accumulation over time, was developed.

The original problem was relaxed to a convex problem that guarantees global optimality and allows for running the simulations over long time-frames. All the assumptions made in order to relax the original to a convex problem were checked, and are valid as long as battery efficiencies can be considered constant.

The model developed in here allows studying the convenience or not of the installation of a battery-based energy storage system in a residential or small-scale business setting. The model was used to predict the behavior of a Li-ion battery under the assumption of a simple two-price-step day-invariant TOU pricing policy, and a complex temperature-day-season dependent one.

The results of our simulations suggest that simple tariffs, where the price is constant for a large range of hours, operate so that service life is preserved. This preservation is accomplished by charging and discharging every day at very low rates. Our simulations also show that there is a turning point in the cost of the battery that makes the operation not worth. This is, the savings in electricity bill do not compensate the penalization due to capacity loss. This effect is seen even when running the one-day problem, and may be regarded as a first indicator of whether or not it is reasonable to install a battery. For the prices considered in this paper, this turning point is in between 400–500 USD/kWh. Nevertheless, having a battery cost below the turning point does not imply that installing a battery is profitable: positive NPV values for a 10-year lifespan, usual discount rates, and the assumptions made in this case study, are obtained only if the battery price falls below 140 USD/kWh.

For the case of complex tariffs our simulations show that for the same battery price, savings achieved in some individual days could be higher than those of the simple case. However, these savings are counterbalanced with the fact that the battery is not used on a regular (daily) basis. Overall, the battery for the complex tariff considered here is used at most 90 days of the year. Another feature of relevance when considering complex tariffs is that limiting the desired time span between charge and discharge becomes an important input parameter of the optimization problem.

Overall the conclusion is that current Li-ion battery prices (in the order of 500 USD/kWh) and TOU pricing policies are still not attractive enough for acquisition by residential/small-scale consumers. However, as Li-ion batteries systems still have margin to lower their production costs and battery prices are expected to decrease in the near future, small variations in tariff policies may make a difference to favor the installation of this energy storage solutions.

Supplementary Materials: Available online at http://www.mdpi.com/2227-9717/6/10/204/s1.

Author Contributions: A.I.T. is the PhD Advisor of M.C. Problem conceptualization, methodology, and formal analysis were developed by discussions between the two authors. M.C. developed the code and run the simulations under the supervision of A.I.T. Both authors contributed equally to the analysis of the results, writing and editing the manuscript.

Funding: This research received funding from Universidad de la República (UdelaR)–Montevideo, Uruguay.

Acknowledgments: Mariana Corengia and Ana I. Torres gratefully acknowledge the *Comisión Sectorial de Investigación Científica*–UdelaR for the incentive "Dedicación Total" and several travel grants. Ana I. Torres gratefully acknowledges the *Sistema Nacional de Investigadores* for the economic incentive. Both authors would like to express their appreciation to Juan Andrés Bazerque, Instituto de Ingeniería Eléctrica, UdelaR, for bringing to their attention publications that enriched this work.

Conflicts of Interest: The authors declare no conflict of interest.

Abbreviations

The following abbreviations are used in this manuscript:

τ	optimization total period of time
η^+	battery charging efficiency
η^-	battery discharging efficiency
Δt	time step
BC^{ES}	energy storage installed capacity (energy units)
BC_Q^{ES}	initial energy storage installed capacity (electrical charge units)
BC_t	energy storage capacity at t (energy units)
C_{rate}	dimensionless charge/discharge rate
C^{ES}	market cost of battery capacity
E_t	terminal potential difference at time t
E_t^0	open circuit potential difference at time t
I_t	net current at time t
p^{C-MAX}	maximum safe charging power

p_t^C	charge power at time t
p^{D-MAX}	maximum safe discharging power
p_t^D	discharge power at time t
r	discount rate
soc_t	state of charge at time t
SOC^{MIN}	Minimum required state of charge
SOC^{MAX}	Maximum allowed state of charge
x_t^{CL}	fraction of capacity loss during the time step at t
$\$_t$	Energy grid price at time t

References

1. Castillo, A.; Gayme, D.F. Grid-scale energy storage applications in renewable energy integration: A survey. *Energy Convers. Manag.* **2014**, *87*, 885–894. [CrossRef]
2. Morales, J.M.; Conejo, A.J.; Madsen, H.; Pinson, P.; Zugno, M. Facilitating Renewable Integration by Demand Response. In *Integrating Renewables in Electricity Markets: Operational Problems*; Morales, J.M., Conejo, A.J., Madsen, H., Pinson, P., Zugno, M., Eds.; Springer: Boston, MA, USA, 2014; pp. 289–329. [CrossRef]
3. Conejo, A.J.; Sioshansi, R. Rethinking restructured electricity market design: Lessons learned and future needs. *Int. J. Electr. Power Energy Syst.* **2018**, *98*, 520–530. [CrossRef]
4. Mayer, K.; Trück, S. Electricity markets around the world. *J. Commod. Mark.* **2018**, *9*, 77–100. [CrossRef]
5. Samad, T.; Kiliccote, S. Smart grid technologies and applications for the industrial sector. *Comput. Chem. Eng.* **2012**, *47*, 76–84. [CrossRef]
6. Weitzel, T.; Glock, C.H. Energy management for stationary electric energy storage systems: A systematic literature review. *Eur. J. Oper. Res.* **2018**, *264*, 582–606. [CrossRef]
7. United States Department of Energy–Office of Energy Efficiency and Renewable Energy. 2016–2020 Strategic Plan and Implementation Framework. Available online: https://www.energy.gov/ (accessed on 15 August 2018).
8. International Energy Agency. Global EV Outlook 2017. Available online: https://www.iea.org/ (accessed on 10 August 2018).
9. Reniers, J.M.; Mulder, G.; Ober-Blöbaum, S.; Howey, D.A. Improving optimal control of grid-connected lithium-ion batteries through more accurate battery and degradation modelling. *J. Power Sources* **2018**, *379*, 91–102. [CrossRef]
10. Zhao, Y.; Choe, S.Y.; Kee, J. Modeling of degradation effects and its integration into electrochemical reduced order model for Li(MnNiCo)O2/Graphite polymer battery for real time applications. *Electrochim. Acta* **2018**, *270*, 440–452. [CrossRef]
11. Zhao, X.; Bi, Y.; Choe, S.Y.; Kim, S.Y. An integrated reduced order model considering degradation effects for LiFePO4/graphite cells. *Electrochim. Acta* **2018**, *280*, 41–54. [CrossRef]
12. Sarker, M.R.; Murbach, M.D.; Schwartz, D.T.; Ortega-Vazquez, M.A. Optimal operation of a battery energy storage system: Trade-off between grid economics and storage health. *Electr. Power Syst. Res.* **2017**, *152*, 342–349. [CrossRef]
13. Yan, G.; Liu, D.; Li, J.; Mu, G. A cost accounting method of the Li-ion battery energy storage system for frequency regulation considering the effect of life degradation. *Prot. Control Mod. Power Syst.* **2018**, *3*, 4. [CrossRef]
14. Hu, X.; Martinez, C.M.; Yang, Y. Charging, power management, and battery degradation mitigation in plug-in hybrid electric vehicles: A unified cost-optimal approach. *Mech. Syst. Signal Process.* **2017**, *87*, 4–16. [CrossRef]
15. Zubi, G.; Dufo-López, R.; Carvalho, M.; Pasaoglu, G. The lithium-ion battery: State of the art and future perspectives. *Renew. Sustain. Energy Rev.* **2018**, *89*, 292–308. [CrossRef]
16. Crompton, T.R. 1–Introduction to battery technology. In *Battery Reference Book*, 3rd ed.; Crompton, T., Ed.; Newnes: Oxford, UK, 2000; pp. 1–64. [CrossRef]
17. Delacourt, C.; Safari, M. Mathematical Modeling of Aging of Li-Ion Batteries. In *Physical Multiscale Modeling and Numerical Simulation of Electrochemical Devices for Energy Conversion and Storage: From Theory to Engineering to Practice*; Franco, A.A., Doublet, M.L., Bessler, W.G., Eds.; Springer: London, UK, 2016; pp. 151–190. [CrossRef]

18. Jabr, R.A.; Karaki, S.; Korbane, J.A. Robust Multi-Period OPF With Storage and Renewables. *IEEE Trans. Power Syst.* **2015**, *30*, 2790–2799. [CrossRef]

19. Castillo, A.; Gayme, D.F. Profit maximizing storage allocation in power grids. In Proceedings of the 52nd IEEE Conference on Decision and Control, Florence, Italy, 10–13 December 2013; pp. 429–435. [CrossRef]

20. Fortenbacher, P.; Andersson, G. Battery degradation maps for power system optimization and as a benchmark reference. In Proceedings of the 2017 IEEE Manchester PowerTech, Manchester, UK, 18–22 June 2017; pp. 1–6. [CrossRef]

21. Ruan, Y.; Song, X.; Fu, Y.; Song, C.; Battaglia, V. Structural evolution and capacity degradation mechanism of $LiNi_{0.6}Mn_{0.2}Co_{0.2}O_2$ cathode materials. *J. Power Sources* **2018**, *400*, 539–548. [CrossRef]

22. Barré, A.; Deguilhem, B.; Grolleau, S.; Gérard, M.; Suard, F.; Riu, D. A review on lithium-ion battery ageing mechanisms and estimations for automotive applications. *J. Power Sources* **2013**, *241*, 680–689. [CrossRef]

23. GAMS Documentation of IPOPT. Available online: https://www.gams.com/ (accessed on 26 September 2018).

24. UTE. Pliego Tarifario 2018. Available online: https://portal.ute.com.uy (accessed on 7 May 2018).

25. Southern California Edison. TOU-GS-2-RTP. Available online: https://www.sce.com/ (accessed on 20 Febraury 2018).

26. ANEEL. Tarifa Branca. Available online: http://www.aneel.gov.br/tarifa-branca (accessed on 20 August 2018).

27. MidAmerican Energy. Electric Tariffs Iowa. Available online: https://www.midamericanenergy.com (accessed on 20 August 2018).

28. LAZARD. LAZARD's Levelized Cost of Storage Analysis—Version 3.0. Technical Report, 2017. Available online: https://www.lazard.com/perspective/levelized-cost-of-storage-2017/ (accessed on 9 July 2018).

29. Asif, A.; Singh, R. Further Cost Reduction of Battery Manufacturing. *Batteries* **2017**, *3*, 17. [CrossRef]

30. AccuWeather. Available online: https://www.accuweather.com (accessed on 15 August 2018).

![processes logo] *processes*

MDPI

Article

Approximating Nonlinear Relationships for Optimal Operation of Natural Gas Transport Networks

Kody Kazda and Xiang Li *

Department of Chemical Engineering, Queen's University, 19 Division Street, Kington, ON K7L 3N6, Canada; 17kk18@queensu.ca
* Correspondence: xiang.li@queensu.ca; Tel.: +1-613-533-6582

Received: 4 September 2018; Accepted: 15 October 2018; Published: 18 October 2018

Abstract: The compressor fuel cost minimization problem (FCMP) for natural gas pipelines is a relevant problem because of the substantial energy consumption of compressor stations transporting the large global demand for natural gas. The common method for modeling the FCMP is to assume key modeling parameters such as the friction factor, compressibility factor, isentropic exponent, and compressor efficiency to be constants, and their nonlinear relationships to the system operating conditions are ignored. Previous work has avoided the complexity associated with the nonlinear relationships inherent in the FCMP to avoid unreasonably long solution times for practical transportation systems. In this paper, a mixed-integer linear programming (MILP) based method is introduced to generate piecewise-linear functions that approximate the previously ignored nonlinear relationships. The MILP determines the optimal break-points and orientation of the linear segments so that approximation error is minimized. A novel FCMP model that includes the piecewise-linear approximations is applied in a case study on three simple gas networks. The case study shows that the novel FCMP model captures the nonlinear relationships with a high degree of accuracy and only marginally increases solution time compared to the common simplified FCMP model. The common simplified model is found to produce solutions with high error and infeasibility when applied on a rigorous simulation.

Keywords: fuel cost minimization problem; FCMP; piecewise-linear function generation; linearization; natural gas transportation; compressor modeling; compressibility factor; isentropic exponent; friction factor

1. Introduction

Natural gas accounts for an estimated 21% of the current global energy demand, with this share expected to grow to 24% by 2040. The projected 1.5% annual growth in absolute natural gas demand makes it the fastest growing among the fossils fuels [1]. The large and growing natural gas demand puts a heightened importance on natural gas transportation systems, given the enormous quantities of gas being distributed globally. The magnitude of the issue means even marginal improvements in the efficiency of gas transportation can result in substantial energy and cost savings.

This paper addresses a particular area of natural gas transportation research which has been studied for decades, the fuel cost minimization problem (FCMP). The problem exists because pressure drives gas flow in pipelines, but friction between the gas and pipeline wall results in a drop in gas pressure along the length of the pipeline. To maintain gas flow, the gas pressure must be increased at points throughout the pipeline network, which is achieved by compressor stations. The FCMP is concerned with determining the natural gas pressure change at each compressor station in the pipeline network, and the amount of gas to flow through each pipeline segment of the network, to achieve the minimum compressor fuel costs. The system operation must obey gas supply and demand constraints and physical equipment limitations. With an estimated 3–5% of all natural gas transported by transmission pipelines consumed by the compressor stations responsible for maintaining gas flow, substantial energy and cost savings can be realized through optimizing the FCMP. It has been estimated that at least a 20% reduction in compressor station fuel consumption can be obtained through global optimization of pipeline operation [2].

The FCMP is a difficult optimization problem because real pipeline systems have complex network topographies, integer variables are required to represent the pipeline binary operating states (bi-directional flow, ON/OFF compressor units), and gas and compressor physics are governed by highly nonlinear equations. The FCMP is most appropriately modeled as a nonlinear program (NLP) or mixed-integer nonlinear program (MINLP), depending on whether bi-directional flow or ON/OFF compressor units are considered [3]. Given the complexity of the FCMP model, current algorithms can only solve simplified versions for networks large enough to have practical importance. With this, there are numerous simplifications of model parameters that are broadly accepted across FCMP literature because of the computational savings they provide. However, upon inspection of the error such simplifications introduce to the model, the ability for the model to provide a useful representation of the real system is brought into question.

There has been virtually no previous work done on methods of approximating the nonlinear relationships inherent in the FCMP model that balances model accuracy and solution time. This paper introduces a systematic approach to approximating the nonlinear relationships of the parameters that are commonly simplified as constants. The approach is shown to introduce insignificant model error compared to a rigorous FCMP model formulation, and only marginally increases model solution time compared to a common simplified FCMP model. The approach can be summarized as identifying practically convex/concave relationships between model variables that can be approximated by simpler functions, such as piecewise-linear or quadratic functions. This allows the complex nonlinear equations required for rigorous calculation to be eliminated from the model. Here, a practically convex/concave function means a function that does not need be convex/concave according to the mathematical definition per se, but can be approximated well by a function that is mathematically convex/concave. The piecewise-linear approximations are not constructed manually by the modeler, but instead are generated by inputting the nonlinear relationship data into an mixed-integer linear programming (MILP) that optimally determines the piecewise-linear break-points and linear segment orientation so that approximation error is minimized. This MILP relies on the convexity/concavity of the relationship being approximated for its simple formulation. As stated above, the value of this systematic modeling approach is that a model is produced with insignificant error when compared with the most rigorous model. Further, the model is only marginally more difficult to solve by current global MINLP solvers when compared to the highly simplified model. The reason is that the produced model only differs from the simplified model in the additional piecewise-linear functions, which are easily handled by modern global MINLP solvers such as BARON [4,5] and SCIP [6].

Although we have applied this modeling approach to the FCMP, the techniques are transferable to many engineering problems that are difficult to optimize because of complex nonlinear relationships.

This introduction is followed by six main sections. Section 2 introduces the FCMP model that is most commonly used throughout gas transportation literature, and areas of the model that are simplified are discussed. Next, Section 3 presents the method of optimal piecewise-linear function generation for approximating practically convex/concave curves. Section 4 applies this technique to approximate the previously ignored nonlinear relationships. The novel FCMP model that incorporates the piecewise-linear approximations is then developed in Section 5. In Section 6, a case study is presented that compares the optimization results of the simplified, newly developed, and most rigorous FCMP optimization models to the results from a rigorous simulation. Section 7 reiterates the importance of these results and provides a discussion of future work.

2. Common Natural Gas Transportation Model

The model developed in this section is for application specifically to the FCMP, although many of the model components are used across all types of natural gas transportation models. A summary of the model parameters and variables introduced in the following sections can be seen in Table 1.

Table 1. Parameters and variables describing gas transportation.

Symbol	Description	Unit
q	Mass flow rate	kg s^{-1}
P	Pressure	MPa
T	Temperature	K
L	Length	m
D	Diameter	m
A	Cross-sectional area	m^2
λ	Friction coefficient	-
Z	Compressibility factor	-
R	Gas constant	J mol^{-1} K^{-1}
M_w	Molecular weight	kg mol^{-1}
B	Second virial coefficient	m^3 kmol^{-1}
ρ_m	Molar density	kmol m^{-3}
K	Size parameter	m kmol$^{-1/3}$
C_n^*	Compressibility factor coefficient	-
b_n, c_n, k_n	Compressibility factor constant	-
T_c	Critical temperature	K
P_c	Critical pressure	MPa
ϵ	Absolute roughness	m
Re	Reynolds Number	-
μ	Gas dynamic viscosity	Pa s
H_{ad}	Specific change in adiabatic enthalpy	J kg^{-1}
κ	Isentropic exponent	-
Q	Volumetric flow rate	m^{3}s^{-1}
S	Compressor speed	rpm
S_{min}, S_{max}	Compressor speed limit	rpm
$surge, stonewall$	Compressor throughput/speed limit	m^{3}s^{-1}rpm^{-1}
η_{ad}	Adiabatic efficiency	-

2.1. Network Topography

The gas transmission network is most commonly modeled as a directed graph $G = (N, A)$ with nodes N and arcs A [7–10]. The set of nodes is made up of entry nodes N_+, exit nodes N_-, and junctions N_0, representing pipeline supply points, delivery points, and junctions,

$$N = N_+ \cup N_- \cup N_0. \tag{1}$$

Each node in the network is arbitrarily assigned a unique number from 1 to n, with n being the total number of nodes in the network. Nodes are denoted by $i, j \in N$.

The set of arcs is made up of passive arcs A_p, and active arcs A_c, representing pipeline sections that do not contain compressor stations, and sections that contain compressor stations,

$$A = A_p \cup A_c. \tag{2}$$

Differing from nodes, arcs are not arbitrarily assigned numbers but are instead classified according to the nodes they connect. Arcs are denoted by $(i, j) \in A$, corresponding to a pipeline section running from node i to node j.

2.2. Gas Flow Model

The three non-isothermal Euler equations for compressible fluids govern gas dynamics in a pipeline [7], but in their complete form are too complex to be used in an optimization model that is solvable by modern NLP solvers. By making the common approximations of isothermality in pipeline segments and negligible slope in pipeline segments, the equations can be reduced to two manageable equations. These equations are the gas conservation equation (Equation (3)) and pipeline resistance equation (Equation (4)):

$$\sum_{j:(j,i)\in A} q_{ji} - \sum_{j:(i,j)\in A} q_{ij} = -q_{ext,i} \qquad \forall i \in 1,\dots,n, \tag{3}$$

$$(P_i^2 - P_j^2) = \frac{L_{ij}R}{A_{ij}^2 D_{ij}M_w} Z_{ij}T\lambda_{ij}q_{ij}^2 \qquad \forall(i,j)\in A_p, \tag{4}$$

where $q_{ext,i}$ is a model input that represents the mass flow rate of natural gas being supplied to node i (positive for a supply node, negative for a demand node), and Z_{ij} and λ_{ij} are the gas compressibility factor and friction factor for arc (i,j) [7]. It is most common in natural gas transportation literature to assume both parameters to be constants [9,11–15]. In reality, these parameters have complex relationships with the pipeline operating conditions that cannot be ignored without introducing significant model error.

2.2.1. Compressibility Factor

The compressibility factor has a complex dependence on temperature, pressure, and gas composition, which is most accurately calculated implicitly through:

$$Z = 1 + B\rho_m - K^3 \tfrac{P}{ZRT} \sum_{n=13}^{18} C_n^* + \sum_{n=13}^{58} C_n^* \left(b_n - c_n k_n (K^3 \tfrac{P}{ZRT})^{k_n}\right)(K^3 \tfrac{P}{ZRT})^{b_n} exp\left(-c_n(K^3 \tfrac{P}{ZRT})^{k_n}\right), \tag{5}$$

where b_n, c_n, k_n are constants, and B, K, C_n^* are functions of temperature and gas composition [16]. The complexity of Equation (5) is clear, and its avoidance by FCMP researchers is justifiable. Although it is most common to remove this source of complexity by assuming the compressibility factor is constant with both temperature and pressure, on occasion the empirical formulas of the American Gas Association (Equation (6)) or Papay (Equation (7)) are used [7].

$$Z(P,T) = 1 + 0.257\frac{P}{P_c} - 0.533\frac{PT_c}{P_cT}, \tag{6}$$

$$Z(P,T) = 1 - 3.52\frac{P}{P_c}e^{-2.26\frac{T}{T_c}} + 0.274\left(\frac{P}{P_c}\right)^2 e^{-1.878\frac{T}{T_c}} \tag{7}$$

Despite these formulas partially capturing the relationship between the compressibility factor and temperature–pressure, Figure 1 shows that over typical pipeline operating conditions they deviate significantly from the rigorous compressibility factor as calculated through Equation (5). The 0.1–10 MPa pressure range used for comparison encompasses the typical operating range of both natural gas distribution and transmission pipelines [3,17].

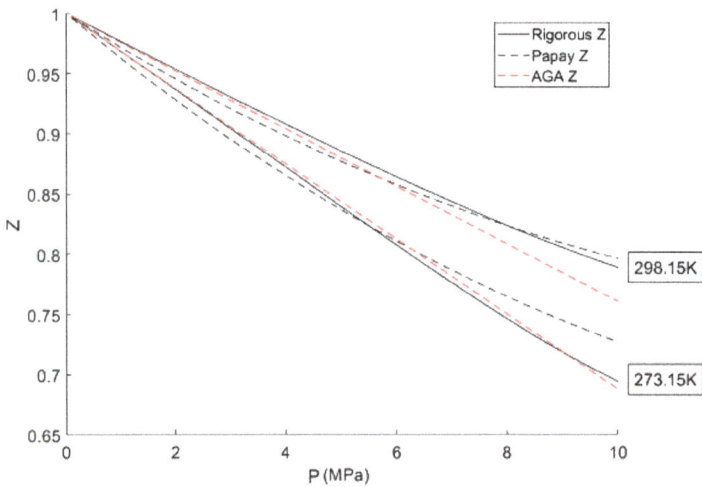

Figure 1. Comparison of the rigorously calculated gas compressibility factor with the American Gas Association and Papay empirical models for natural gas with a composition of: 85% methane, 14% ethane, and 1% nitrogen.

2.2.2. Friction Factor

It is most common in FCMP literature to assume the friction factor to be a constant that depends only on the length, diameter, and roughness of the pipe. The Nikuradse equation is most commonly used to calculate a constant friction factor that has no dependance on the operating condition of the pipeline [18]:

$$\frac{1}{\sqrt{\lambda^N}} = -2\log_{10}\left(\frac{\epsilon/D}{3.71}\right) \tag{8}$$

However, it is widely accepted that in natural gas transmission pipelines the Colebrook–White friction factor (Equation (9)) is more accurate as it captures the dependance of the friction factor on the natural gas mass flow rate [7,18]. Despite this, it is often omitted in favor of the Nikuradse equation because of the complexity required to capture the dependence on mass flow rate through the Reynolds number in Equation (10) [7].

$$\frac{1}{\sqrt{\lambda^{CW}(q)}} = -2\log_{10}\left(\frac{\epsilon/D}{3.71} + \frac{2.51}{Re(q)\sqrt{\lambda^{CW}(q)}}\right) \tag{9}$$

$$Re(q) = \frac{D}{A\mu}|q| \tag{10}$$

Figure 2 shows how the Nikurasde and Colebrook–White friction factors compare over the typical range of transmission pipeline mass flow rates. The range of typical transmission flow rates was determined from data publicly available on all U.S. interstate natural gas pipeline capacities through the U.S. Energy Information Administration [19].

It can be seen in Figure 2 that for mass flow rates on the low range of those typically seen in transmission networks there is a substantial difference between the Nikurasde and Colebrook–White friction factors. The Colebrook–White friction factor approaches the Nikurasde friction factor from above, corresponding to the pressure drop calculated using the Colebrook–White friction factor always being greater than that of the Nikurasde friction factor. As a result, solutions to the FCMP that are deemed feasible when using the Nikurasde friction factor could be infeasible in practice from underestimating the pressure drop throughout the transmission network.

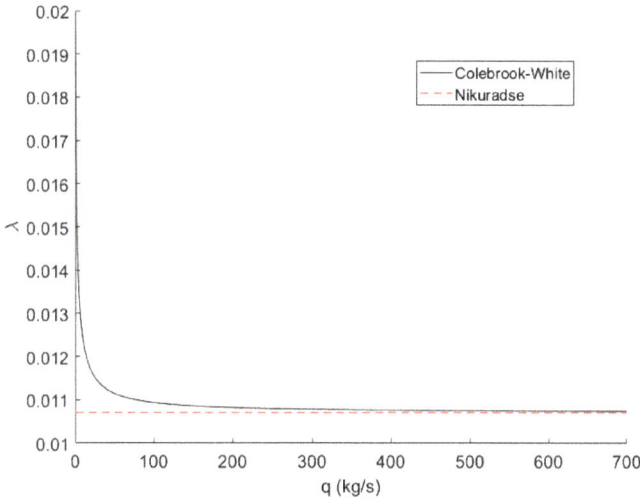

Figure 2. Comparison of the Colebrook–White and Nikuradse friction factor on a pipeline with a diameter of 0.9 m, and roughness of 0.05 mm.

2.3. Compressor Equations

The compressor adiabatic head equation is standard across virtually all FCMP literature, as it is derived from thermodynamic first principles with reasonable assumptions and is not overly complex [20]:

$$H_{ad,ij} = \frac{Z_i TR}{M_w} \frac{\kappa_{ij}}{\kappa_{ij} - 1} \left(\left(\frac{P_j}{P_i} \right)^{\frac{\kappa_{ij}-1}{\kappa_{ij}}} - 1 \right) \quad \forall (i,j) \in A_c. \tag{11}$$

2.3.1. Isentropic Exponent

The isentropic exponent (κ) is the ratio of specific heat at constant pressure to specific heat at constant volume, and is needed to calculate the energy required for compressing a gas [20]. The rigorous calculation of the isentropic exponent is extremely computationally difficult, so in virtually all natural gas transportation literature it is assumed to be constant. At best, the following simplified relationship with temperature is used:

$$\kappa(T) = 1.29 - 5.8824 \times 10^{-4} \left(T - 273.15 \right), \tag{12}$$

where T is defined to be the average of the compressor inlet and outlet temperatures [7].

Similar to the compressibility factor, the isentropic exponent has a complex dependence on temperature, pressure, and gas composition. The rigorous calculation of the isentropic exponent requires over 50 equations, most of which are nonlinear and several are implicitly defined [21]. Figure 3 shows that over typical natural gas pipeline operating conditions significant model error is introduced using the simplifiied Equation (12).

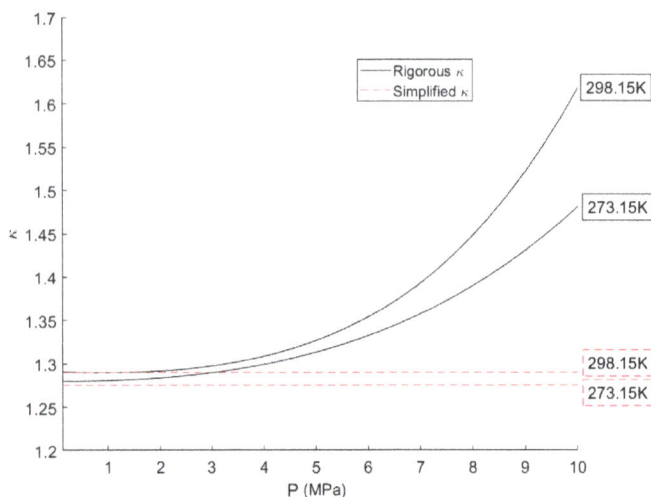

Figure 3. Comparison of the rigorously calculated isentropic exponent with a common simplification for natural gas with a composition of: 85%methane, 14% ethane, and 1% nitrogen.

2.3.2. Compressor Operating Envelop

A restriction on the feasible operating region of the compressor station is necessary to reflect physical equipment limitations. FCMP models typically consider compressor stations made up of centrifugal compressors given their prevalence in natural gas transmission systems. The operating region of a centrifugal compressor is limited by the maximum allowable speed, minimum allowable speed, maximum allowable gas flow rate (surge limit), and minimum allowable gas flow rate (stonewall limit) [22]. The combination of the four limits gives a non-convex operating envelop such as that in Figure 4. The operating envelop in Figure 4 was simulated using the compressor parameters found in Wu et al. [23].

There is some variation in how the compressor operating envelop is modeled across FCMP literature. The method that is widely accepted as the most rigorous can be seen in Equations (13)–(15), which capture the highly nonlinear relationship among compressor head, compressor speed, and volumetric gas flow rate [7,9,15,23]. The coefficients $A_H, B_H, C_H,$ and D_H are constants that are typically estimated by fitting Equation (13) to compressor data of the quantities $Q, S, H,$ and η_{ad} [23]. The complexity that is introduced to the FCMP model by using the rigorous compressor model has given rise to numerous simplified operating envelops. These simplifications involve replacing Equations (13)–(15) with simpler bounds that ignore compressor speed altogether. Some simplifications include reducing the operating window to just an upper limit on the compressor exit gas pressure [12], putting an upper and lower limit on compressor head or power [24,25], or approximating the operating window by a linearly bounded convex region [23].

$$\frac{H_{ad,ij}}{S_{ij}^2} = A_H + B_H \frac{Q_{ij}}{S_{ij}} + C_H \left(\frac{Q_{ij}}{S_{ij}}\right)^2 + D_H \left(\frac{Q_{ij}}{S_{ij}}\right)^3 \qquad \forall (i,j) \in A_c \qquad (13)$$

$$S_{min} \leq S_{ij} \leq S_{max} \qquad \forall (i,j) \in A_c \qquad (14)$$

$$surge \leq \frac{Q_{ij}}{S_{ij}} \leq stonewall \qquad \forall (i,j) \in A_c \qquad (15)$$

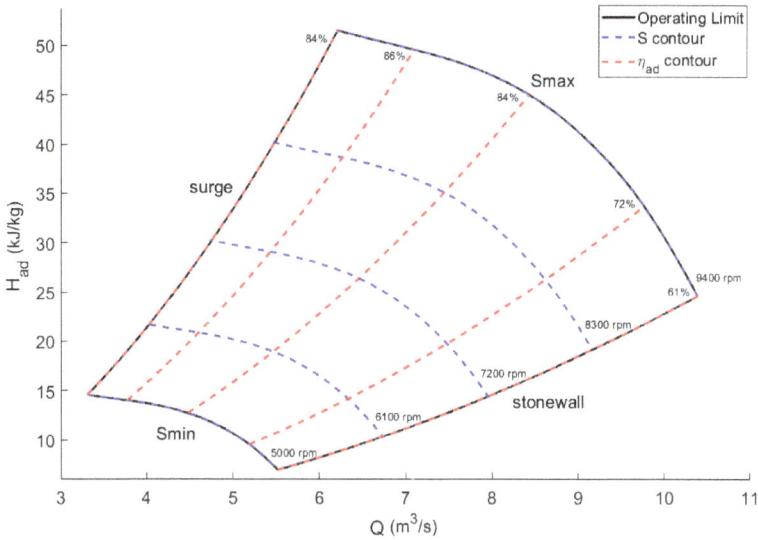

Figure 4. Typical operating envelop of a centrifugal compressor unit, with compressor speed and efficiency contours shown.

2.3.3. Compressor Efficiency

A compressor efficiency parameter (η_{ad}) is required to relate the ideal adiabatic quantity of energy required to achieve a given pressure change, and the actual quantity required by real compressors that are subject to deviations from ideal conditions [20]. The rigorous method of calculating compressor efficiency is through Equation (16), which requires the use of Equation (13) to calculate the compressor speed at a given compressor head and volumetric flow rate [7,23]. As with the coefficients in Equation (13), A_E, B_E, C_E, and D_E are constants that are estimated by fitting Equation (16) to compressor data.

$$\eta_{ad,ij} = A_E + B_E \frac{Q_{ij}}{S_{ij}} + C_E \left(\frac{Q_{ij}}{S_{ij}}\right)^2 + D_E \left(\frac{Q_{ij}}{S_{ij}}\right)^3 \quad \forall (i,j) \in A_c \tag{16}$$

Given that many FCMP models use a simplification of the compressor operating envelop that ignores compressor speed, the method of calculating compressor efficiency often must be simplified as well. Such simplifications include capturing the compressor efficiency dynamics using a simpler polynomial function of compressor mass flow rate, inlet pressure, and outlet pressure [23], or assuming the compressor efficiency to be constant [8,25–27]. The compressor efficiency contour lines in Figure 4 show that there is a large variation in efficiency across the operating envelop which is lost by assuming a constant efficiency. Any error introduced through these simplifications contributes directly to error in the model's objective function (see Problem (17)) because of the dependance of fuel cost on compressor efficiency.

2.4. Common FCMP Model

Combining the above formulas and simplification gives the complete FCMP model that is most common across FCMP literature:

$$\min_{q_{ij},P_i} \sum_{(i,j)\in A_c} \frac{q_{ij}H_{ad,ij}}{\eta_{ad}}$$

$$\text{s.t.} \quad \sum_{j:(j,i)\in A} q_{ji} - \sum_{j:(i,j)\in A} q_{ij} = -q_{ext,i} \qquad \forall i \in 1,\dots,n,$$

$$(P_i^2 - P_j^2) = \frac{L_{ij}R}{A_{ij}^2 D_{ij}M_w} ZT\lambda q_{ij}^2 \qquad \forall (i,j) \in A_p, \tag{17}$$

$$H_{ad,ij} = \frac{ZTR}{M} \frac{\kappa}{\kappa-1} \left(\left(\frac{P_j}{P_i}\right)^{\frac{\kappa-1}{\kappa}} - 1 \right) \qquad \forall (i,j) \in A_c,$$

$$(q_{ij}, P_i, P_j) \in \mathbb{D}_{ij} \qquad \forall (i,j) \in A_c,$$

where $\mathbb{D}_{ij}$ represents the compressor operating envelop as defined through one of the previously discussed methods. The compressibility factor (Z), friction factor (λ), isentropic exponent (κ), and compressor efficiency (η) are all assumed to be constant in this model.

3. Optimal Piecewise-Linear Function Generation

Linearization of nonlinear relationships is often an effective method for reducing the complexity of an optimization model to increase solution efficiency [28–30]. Many of the common linearization techniques manually determine the break-points of the linear segments and the orientation of the linear segments follow, or they manually determine the orientation of the linear segments and the position of the break-points follow. The drawback of these methods is that they offer no guarantee of producing an optimal approximation. When the nonlinear relationships are complex, and especially when they are more than two-dimensional, the manual linearization approach is likely to produce significantly suboptimal approximations.

The linearization technique suggested here is Piecewise-Linear Function Generation (PLFG) that offers the guarantee that for a given number of line segments the approximation is optimal, and can be applied to data for which no analytical function exists. The technique simultaneously determines the break-points and positioning of the linear segments such that they fit the curve with the minimal error. The approximation produced from this method is a convex/concave piecewise-linear function, so must be applied to data that can be approximated closely by a convex/concave curve. This does not restrict the technique to relationships that are mathematically convex/concave.

The technique is a mixed-integer linear program (MILP) that requires data points of the relationship being approximated, and the number of linear segments to approximate the relationship with, as model inputs. The problem is to determine $\mathbb{F}$, which defines a set of continuous piecewise-linear functions, for a finite set of n-dimensional data points (x^i, y_i) [31], that solves:

$$\min \ t$$
$$\text{s.t.} \quad t|y_i| \geq |f_i - y_i|, \qquad \forall i = 1,\dots,m,$$
$$f_i = f(x^i), \qquad \forall i = 1,\dots,m, \tag{18}$$
$$f \in \mathbb{F}.$$

A solution to this problem produces the piecewise-linear function that has the minimum maximum relative difference between the generated function and the data points being approximated.

In MILP PLFG, it is customary to introduce a binary variable z_i^k for each linear segment $k \in 1,\dots,p$, and each data point $i = 1,\dots,m$. The binary variable is used to represent if the linear segment is "active" at the data point (x^i, y_i), where active means the line segment is used for approximation at that part of the domain. The following MILP model is used for PLFG of convex curves:

$$\min_{c^k, d_k, z_i^k} \quad t$$

$$\text{s.t.} \quad |y_i|t \geq f_i - y_i, \qquad \forall i = 1, \ldots, m,$$

$$|y_i|t \geq y_i - f_i, \qquad \forall i = 1, \ldots, m,$$

$$f_i \geq c^k x^i + d_k, \qquad \forall k = 1, \ldots, p, \forall i = 1, \ldots, m, \qquad (19)$$

$$f_i \leq c^k x^i + d_k + M_{Big}(1 - z_i^k), \qquad \forall k = 1, \ldots, p, \forall i = 1, \ldots, m,$$

$$\sum_{k=1}^{p} z_i^k = 1, \qquad \forall i = 1, \ldots, m,$$

where c^k and d^k are the coefficients being determined for the k n-dimensional linear segments, and M_{Big} is some large constant. This model is a variation of the well-known model seen in Toriello and Vielma [31].

Figure 5 helps to demonstrate the importance of each constraint in Problem (19). The $f_i \geq c^k x^i + d_k$ constraint ensures that the approximation of point i is somewhere in the shaded gray region of Figure 5a, that is, above or equal to all of the line segments. Without any additional constraints, any number of line segments could yield an objective function equal to zero by having the line segments below all of the data points, allowing every approximation point to equal every data point exactly. The constraint $f_i \leq c^k x^i + d_k + M_{Big}(1 - z_i^k)$ has no effect unless the binary variable z_i^k is equal to one, in which case the approximation point must be below or equal to the line k at point x_i. Combining these two constants with $\sum_{k=1}^{p} z_i^k = 1$ yields the approximation shown in Figure 5b, as it ensures the approximation point f_i is equal to the greatest line segment at point x_i. An optimal solution that is subject to all of these constraints is one that minimizes the maximum relative error between the points along the line segments and the data points.

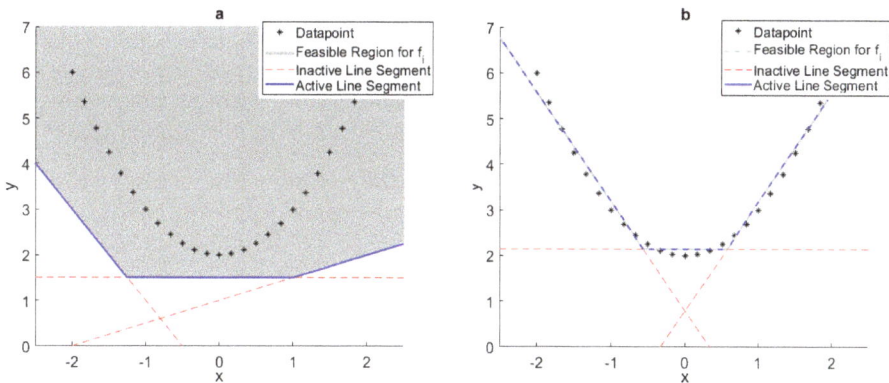

Figure 5. (**a**) Example of potential approximation produced when only the $f_i \geq c^k x^i + d_k$ constraint is used in the PLFG MILP; (**b**) Example of potential approximation produced when all of the constraints are used in the PLFG MILP (PLFG: Piecewise-Linear Function Generation; MILP: mixed-integer linear program).

The MILP in Equation (19) can be easily adapted for fitting concave curves:

$$\min_{c^k, d_k, z_i^k} t$$

$$\text{s.t. } |y_i|t \geq f_i - y_i, \qquad \forall i = 1, \ldots, m,$$

$$|y_i|t \geq y_i - f_i, \qquad \forall i = 1, \ldots, m,$$

$$f_i \geq c^k x^i + d_k - M_{Big}(1 - z_i^k), \qquad \forall k = 1, \ldots, p, \forall i = 1, \ldots, m, \qquad (20)$$

$$f_i \leq c^k x^i + d_k, \qquad \forall k = 1, \ldots, p, \forall i = 1, \ldots, m,$$

$$\sum_{k=1}^{p} z_i^k = 1, \qquad \forall i = 1, \ldots, m.$$

These versions of the MILP produce a piecewise-linear approximation that has the freedom to cross-over the curve being approximated, as shown in Figure 6a. However, in certain applications, it may be desirable to produce an approximation that is entirely above the curve being approximated (Figure 6b), and in other applications for the approximation to be entirely below the curve being approximated (Figure 6c). The MILP can be modified to generate such approximations by adding the constraint in Equation (21) for the approximation always above, and Equation (22) for the approximation always below.

$$f_i \geq y_i, \qquad \forall i = 1, \ldots, m \qquad (21)$$

$$f_i \leq y_i, \qquad \forall i = 1, \ldots, m \qquad (22)$$

The number of data points and linear segments are user inputs to the PLFG MILP, so it is necessary to be deliberate about how these inputs are decided by the modeler. A higher number of data points and linear segments will always yield a more accurate approximation. However, the tractability of the PLFG MILP is worsened as these inputs increase because of the additional constraints and variables in the model. Additionally, the more linear segments used for approximating the relationship the worse the tractability of the optimization model where the approximation is being implemented. A procedure is required for determining the number of data points and linear segments that gives a desirable balance between model tractability and approximation accuracy.

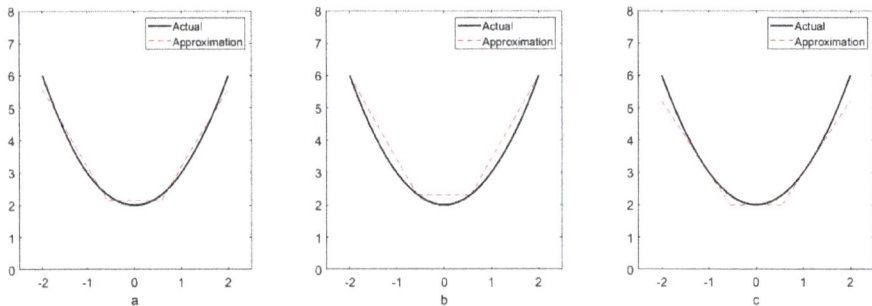

Figure 6. (a) Example of approximation produced when curve cross-over is allowed; (b) Example of approximation when restricted to entirely above the curve; (c) Example of approximation when restricted to entirely below the curve.

The suggested procedure is as follows:

1. Decide what is an acceptable amount of maximum relative error between the piecewise-linear function being generated and the relationship being approximated, this is the stopping criteria E^{tol}. Typically, 1% is appropriate.

2. Start with $m = M$ and $p = 1$, where M is some relatively small number of data points equally dispersed over the approximation domain. Solve the MILP for the piecewise-linear function, $\mathbb{F}_m^p$.
3. Calculate the maximum error between $\mathbb{F}_m^p$ and the entire dataset of the relationship being approximated, $E_m^{max,p}$.
4. Increase m to $m + dM$, and solve the MILP for $\mathbb{F}_m^p$.
5. Calculate $E_m^{max,p}$.
6. Iterate Steps 4 and 5 until the change in $E_m^{max,p}$ is less than 1% for two consecutive iterations, then move to Step 7. If the MILP becomes intractable before this, move to Step 7.
7. If the smallest $E_m^{max,p}$ is less than E^{tol}, or has changed less than 1% for two consecutive increases in p, terminate, the corresponding $\mathbb{F}_m^p$ is either an acceptable approximation or as good as the PLFG method is likely to yield. Else, increase p to $p + 1$, reinitialize $m = M$, and return to Step 4. If the MILP becomes intractable before meeting the stopping criteria, terminate and consider breaking up the domain of approximation.

This procedure can be summarized as an inner loop that increases the number of data points used for PLFG until the maximum approximating error plateaus, and an outer loop that increases the number of line-segments in the piecewise function until the approximation error is smaller than an acceptable tolerance or plateaus.

4. Relationship Approximation

Section 2 shows how key parameters in the FCMP model are commonly simplified to reduce model complexity. In this section, we develop accurate approximations for these commonly ignored nonlinear relationships. The approximations are developed based on the PLFG method introduced in Section 3. The relationships approximated are the compressibility factor, friction factor, isentropic exponent, compressor operating envelop, and compressor efficiency. Two approaches are used to apply the PLFG method to approximate nonlinear parameter relationships. The first method is to approximate the relationship between the parameter and the pipeline operating condition directly. This method is valid whenever the relationship between the parameter and pipeline operating condition is practically convex/concave, as the PLFG MILP can only produce accurate approximation for such relationships. When the rigorous relationship between the parameter and pipeline operating condition does not fit this criteria, there is still the possibility that the PLFG method can be used to approximate the relationship, but a different approach is needed. The alternative approach is to identify a relationship that is practically convex/concave between a particular lumped term of variables and the pipeline operating condition, of which the parameter is one of the variables in the lumped term, and it is the only instance where the parameter is required in the formulation. The lumped term is then approximated by the piecewise-linear approximation, and the term along with the parameter is removed from the formulation entirely. This approach can help reduce the complexity of the formulation by both eliminating the complexity of the lumped term and capturing the rigorous parameter relationship by the piecewise-linear function. The additional complexity that this approach has the potential to remove makes it desirable to be applied even in cases where it is not strictly needed. However, such a lumped term is not always possible to identify so this method is not always valid. The first method is only applied to the compressibility factor, all other parameters are approximated using the alternative method.

4.1. Compressibility Factor

Figure 7a shows the three-dimensional relationship between the compressibility factor and temperature–pressure calculated using Equation (5). It can be seen that the relationship exhibits regions of convexity and concavity, and therefore the PLFG method would not be able to produce a good approximation. However, Figure 7b shows that two-dimensional compressibility factor isotherms are all practically convex and so are well suited to be approximated by the PLFG method. Further, because the isothermal assumption is accepted as necessary across virtually all natural gas transportation literature, the two-dimensional isothermal contours can be used as the basis of approximation without introducing additional model error. The suggested method is then to discritize the compressibility factor along the temperature domain and to approximate the compressibility factor relationship as a function of only pressure. Practitioners implementing the model can then simply choose the isothermal approximation that is closest to their pipeline operating temperature. The 12 isotherms in Figure 7b were chosen so that there is no more than a ±1% error range between each contour. By choosing the isotherm closest to a given operating temperature at most 1% error plus any approximation error is introduced to the model, compared to the rigorous calculation.

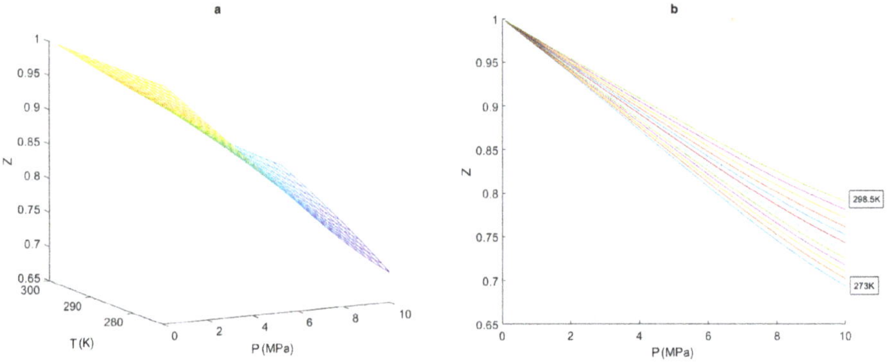

Figure 7. (**a**) Three-dimensional compressibility factor relationship with temperature–pressure; and (**b**) two-dimensional compressibility factor isothermal relationship with pressure for natural gas with a composition of: 85% methane, 14% ethane, and 1% nitrogen.

Figure 8 shows the approximation output from the PLFG method for the 289.5 K compressibility factor isotherm, where approximation cross-over is allowed. The procedure discussed in Section 3 produced an optimal piecewise-linear function comprised of two line segments. The maximum approximation error from the generated approximation is 0.06%.

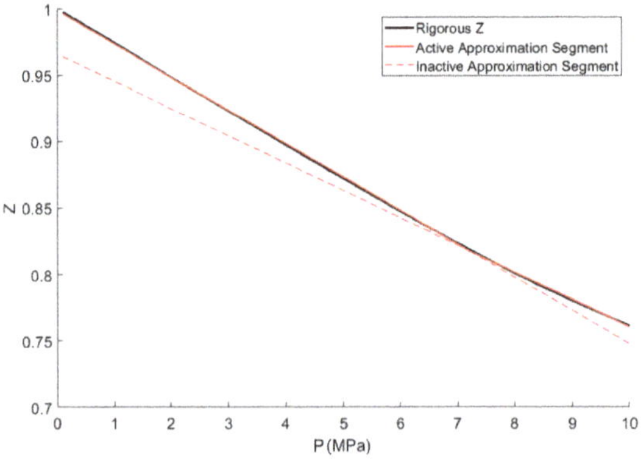

Figure 8. Piecewise-linear function generated for approximating 289.5 K compressibility factor isotherm, consisting of two line segments.

4.2. Friction Factor

The friction factor appears only once in the FCMP model through the pipeline resistance equation (Equation (4)). Although the relationship between the Colebrook–White friction factor and mass flow rate in Figure 2 could be approximated directly, an approximation that removes further model complexity can be made. The friction factor is multiplied by the squared mass flow rate in the pipeline resistance equation. By lumping these terms together, a single function of mass flow rate can be produced:

$$\zeta(q) = \lambda(q)q^2. \tag{23}$$

The plot of $\zeta(q)$ in Figure 9a, where the friction factor is calculated using the Colebrook–White equation (Equation (9)), shows that it has a quadratic relationship with mass flow rate. This relationship is approximated well by performing a nonlinear regression that fits a quadratic function to the curve. No y-intercept coefficient is used in the regression function because at zero mass flow rate anything other than zero for $\zeta(q)$ is not physically realistic. The resulting quadratic regression function can be seen to approximate $\zeta(q)$ closely in Figure 9a. Figure 9b compares the error of the quadratic regression function with a $\zeta(q)$ that uses the constant Nikurasde friction factor. The quadratic regression function contains less error over the entire range of typical natural gas transmission mass flow rates. It is important to note that the regression is specific to pipeline diameter and absolute roughness.

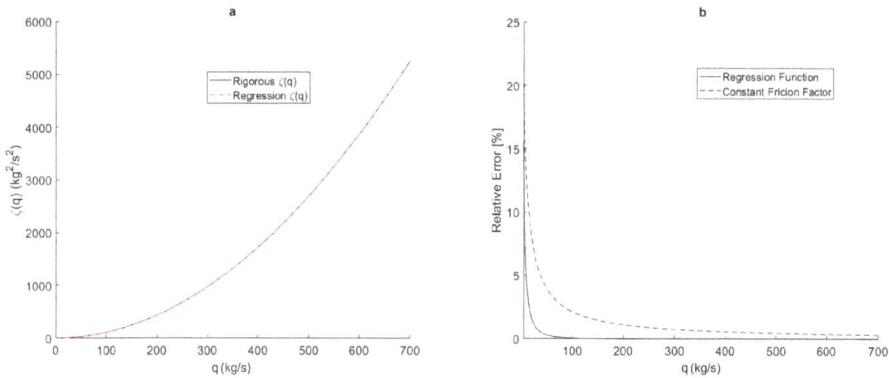

Figure 9. (**a**) Relationship between $\zeta(q)$ and mass flow rate, and quadratic regression approximation; (**b**) Comparison of error for $\zeta(q)$ from the quadratic regression and constant friction factor. All for a pipeline with a diameter of 0.9144 m and absolute roughness of 0.05 mm.

4.3. Isentropic Exponent

The isentropic exponent only appears in the compressor head equation (Equation (11)) in the FCMP model. Additionally, by making the substitution $m(\kappa(T,P)) = \frac{\kappa(T,P)-1}{\kappa(T,P)}$ the isentropic exponent is removed from the formulation entirely:

$$H_{ad,ij} = \frac{Z_i TR}{M_w m_{ij}} \left(\left(\frac{P_j}{P_i}\right)^{m_{ij}} - 1 \right) \quad \forall (i,j) \in A_c. \tag{24}$$

By approximating the relationship of m with temperature and pressure, the complexity of the compressor head equation is reduced. More importantly, if the relationship between the isentropic exponent and temperature–pressure were to be used then any approximation error would be magnified in the formulation through the ratio $\frac{\kappa(T,P)-1}{\kappa(T,P)}$. By approximating the ratio directly, the magnification of error is avoided.

Figure 10a shows that the rigorously calculated three-dimensional relationship between m and temperature–pressure is practically convex, and so the PLFG method can produce a good approximation. A valid approach would be to discritize m in the temperature domain like the compressibility factor, but because the three-dimensional relationship is practically convex it is unnecessary. Figure 10b shows the approximation produced using the procedure described in Section 3, where approximation cross-over is allowed. The optimal piecewise-linear function generated consists of four planes, and has a maximum error from the rigorous relationship of 0.92%.

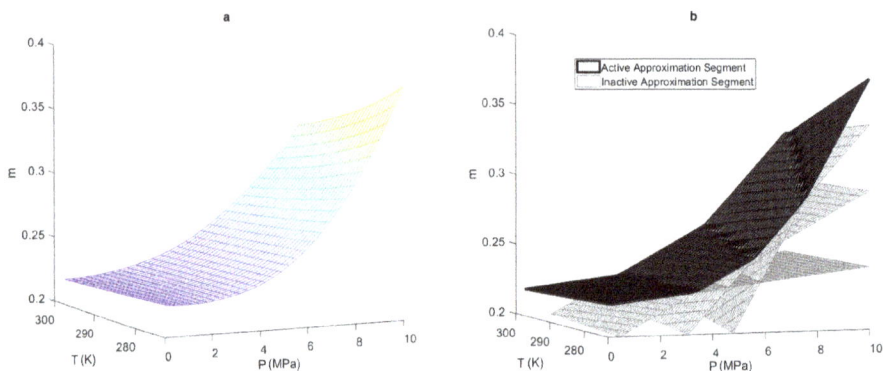

Figure 10. (**a**) Rigorous three-dimensional relationship between m and temperature–pressure; and (**b**) piecewise- linear approximation of m using four planes for natural gas with a composition of: 85% methane, 14% ethane, and 1% nitrogen.

4.4. Compressor Model

4.4.1. Compressor Operating Envelop

Equations (13)–(15) combine to form the compressor operating envelop, which introduce considerable complexity to the FCMP model. It is possible to remove these equations from the formulation by using four piecewise-linear approximations that are functions of volumetric flow rate. This corresponds to one approximation for each of the four curves that bound the operating envelop. By removing these equation from the formulation, the relationship of compressor speed with compressor head and volumetric flow rate is removed, which is important for calculating the compressor efficiency through Equation (16). This issue will be dealt with in the next section.

Figure 4 shows that the surge and stonewall bounds are convex curves, whereas the Smin and Smax bounds are concave curves. Given that the operating envelop bounds the feasible operation of a compressor unit, it is desirable to produce an inner approximation of the envelop so that approximation error does not allow regions of operation to be feasible in the FCMP model that are not feasible in reality. To achieve this, Equation (21) or (22) is used in the PLFG MILP so that approximation cross-over is not allowed. The surge and Smax bounds are approximated by piecewise-linear functions that must be below the actual bounds, and the stonewall and Smin bounds are approximated by piecewise-linear functions that must be above the actual bounds.

Figure 11 shows the inner approximation of the compressor envelop after applying the procedure discussed in Section 3 to each of the four operating bounds, where a stopping criteria of 1% maximum error was used for each bound. The resulting piecewise-linear approximations has three line segments for the surge and stonewall bounds, and four line segments for the Smin and Smax bounds.

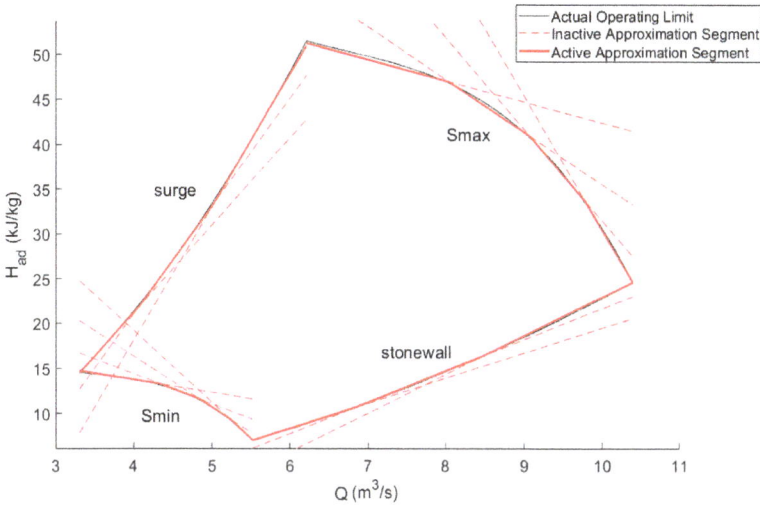

Figure 11. Approximation of the compressor operating envelop shown in Figure 4.

4.4.2. Compressor Efficiency

By removing Equations (13)–(15) from the FCMP model, the compressor speed is meaningless in the compressor efficiency equation (Equation (16)). This is fine because it is desirable to remove Equation (16) from the formulation because of its complexity. Equations (11) and (16) determine that the compressor efficiency is specified for any combination of compressor head and volumetric flow rate, which can be seen in Figure 12. Although the compressor speed is necessary for defining this relationship through an analytical expression, the relationship in Figure 12 can be approximated using a piecewise-linear function of compressor head and volumetric flow rate. This allows compressor speed to be removed from the formulation entirely, as the analytical expression is no longer necessary.

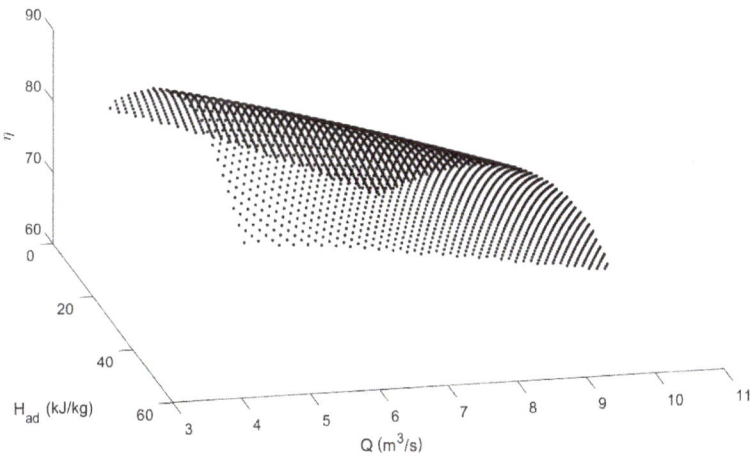

Figure 12. Rigorous relationship of compressor efficiency with compressor head and volumetric flow rate.

While the relationship in Figure 12 is well suited to be approximated using the PLFG method given its concavity, an approximation that removes further model complexity can be made. Aside from in the equation used to calculate compressor efficiency, the compressor efficiency only appears in the FCMP model through the objective function (see Problem (17)). Therefore the compressor efficiency can be removed from the formulation with no effect if its influence on the objective function is still captured. The term $\frac{H_{ad}}{\eta_{ad}}$ appears in the FCMP model objective function, and the relationship of this term with compressor head and volumetric flow rate can be seen in Figure 13a. The relationship is a simple convex curve that can be approximated well by the PLFG method, and by doing so removes compressor efficiency from the formulation entirely. Using the procedure discussed in Section 3, the approximation shown in Figure 13 is generated, which consists of four planes and has a maximum approximation error of 1.12%.

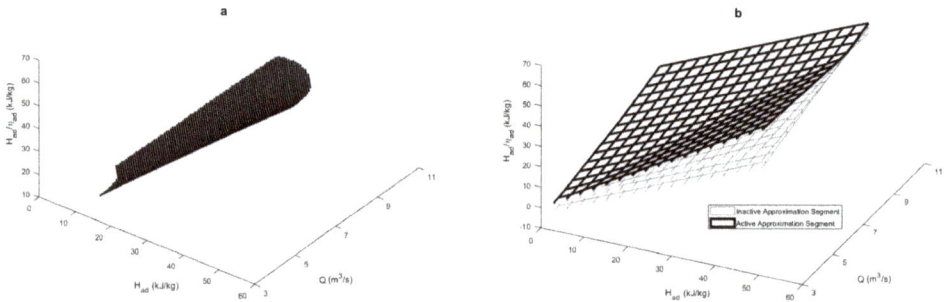

Figure 13. (**a**) Rigorous relationship of $\frac{H_{ad}}{\eta_{ad}}$ with compressor head and volumetric flow rate. (**b**) Approximation of rigorous relationship of $\frac{H_{ad}}{\eta_{ad}}(H_{ad}, Q)$ with a maximum relative error of 1.12%.

5. FCMP Model with Novel Relationship Approximations

There are two distinct methods for implementing the piecewise-linear function approximations developed in Section 4 into the FCMP model. The first method is to simply allow the parameter or term being approximated to be the convex region bounded by the piecewise-linear function. Equation (25) is the expression used by this method for convex piecewise-linear approximations, and Equation (26) is the expression used for concave piecewise-linear approximations.

$$f(x) \geq c^k x + d_k, \qquad \forall k = 1, \ldots, p \tag{25}$$

$$f(x) \leq c^k x + d_k, \qquad \forall k = 1, \ldots, p \tag{26}$$

Equation (25) restricts the parameter or term to be above the maximum of the set of convex line segments, and it corresponds to the epigraph of $f(x)$, as illustrated in Figure 14a. Equation (26) restricts the parameter or term to be below the minimum of the set of concave line segments, and is the hypograph of $f(x)$, as illustrated in Figure 14b. The benefit of this method is that the piecewise-linear function can be implemented without using integer variables. This method is valid only when replacing the function curve with its epigraph or hypograph does not change the optimal solution. For example, if the function $f(x)$ in Figure 14a is the objective function to be minimized, then minimizing its upper bound is same to minimizing itself, and therefore Equation (25) is sufficient for implementing the piecewise-linear approximation.

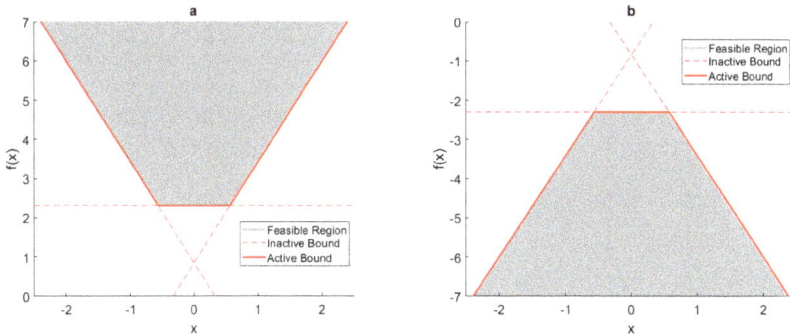

Figure 14. (**a**) Convex feasible region produced by bounding a parameter to be greater than a set of convex line segments. (**b**) Concave feasible region produced by bounding a parameter to be less than a set of concave line segments.

When it is not valid to represent the function curve by its epigraph or hypograph, the use of integer variables is necessary to select which linear segment is active for a given section of the approximation domain. The method for implementing this form of piecewise-linear function follows directly from the PLFG MILP:

$$f(x) \geq c^k x + d_k, \qquad \forall k = 1, \ldots, p,$$
$$f(x) \leq c^k x + d_k + M_{Big}(1 - b_k), \qquad \forall k = 1, \ldots, p,$$
$$\sum_{k=1}^{p} b_k = 1.$$

Table 2 describes the parameters and variables used to implement the approximations into the novel FCMP model.

Table 2. Parameters and variables for implementing the approximations into the Fuel Cost Minimization Problem (FCMP) model.

Symbol	Description	Unit
b_k^Z	Compressibility factor binary variable	-
c_k^Z, d_k^Z	Compressibility factor line coefficients	MPa^{-1}, -
M_{Big}^Z	Compressibility factor sufficiently large number	-
α, β	Friction factor regression coefficients	-, kg s^{-1}
b_k^m	m binary variable	-
$c_{1,k}^m, c_{2,k}^m, d_k^m$	m plane coefficients	K^{-1}, MPa^{-1}, -
c_k^{stw}, d_k^{stw}	Stonewall limit line coefficients	kJ-s kg^{-1}-m^{-3}, kJ kg^{-1}
c_k^{Smax}, d_k^{Smax}	Smax limit line coefficients	kJ-s kg^{-1}-m^{-3}, kJ kg^{-1}
b_k^{Smin}	Smin limit binary variable	-
c_k^{Smin}, d_k^{Smin}	Smin limit line coefficients	kJ-s kg^{-1}-m^{-3}, kJ kg^{-1}
M_{Big}^{Smin}	Smin limit sufficiently large number	kJ kg^{-1}
L_{ij}^{Smin}	Smin limit bound variable	kJ kg^{-1}
b_k^{srg}	Surge limit binary variable	-
c_k^{srg}, d_k^{srg}	Surge limit line coefficients	kJ-s kg^{-1}-m^{-3}, kJ kg^{-1}
M_{Big}^{srg}	Surge limit sufficiently large number	kJ kg^{-1}
L_{ij}^{srg}	Surge limit bound variable	kJ kg^{-1}
$c_{1,k}^g, c_{2,k}^g, d_k^g$	g_c plane coefficients	kJ-s kg^{-1}-m^{-3}, -, kJ kg^{-1}

5.1. Compressibility Factor

The two-dimensional piecewise-linear approximation for the compressibility factor must be implemented in the FCMP model using integer variables. This is because the compressibility factor must be equal to a specific value for each temperature–pressure combination. The following constraints are used in the FCMP model to implement the compressibility factor approximation:

$$Z_{m,ij} \geq c_k^Z P_{m,ij} + d_k^Z, \qquad \forall (i,j) \in A_p, \forall k = 1, \ldots, 2,$$

$$Z_{m,ij} \leq c_k^Z P_{m,ij} + d_k^Z + M_{Big}^Z (1 - b_{ij,k}^Z), \qquad \forall (i,j) \in A_p, \forall k = 1, \ldots, 2,$$

$$\sum_{k=1}^{2} b_{ij,k}^Z = 1, \qquad \forall (i,j) \in A_p,$$

where $P_{m,ij}$ is the average pressure between nodes i and j.

5.2. Friction Factor

The friction factor is approximated through the fitting of a quadratic equation to the lumped friction factor-squared mass flow term in the pipeline resistance equation. Thus, a straightforward substitution of this term for the regressed quadratic function with no y-intercept term is made in the pipeline resistance equation:

$$(P_i^2 - P_j^2) = \frac{L_{ij} R}{M_w A_{ij}^2 D_{ij}} Z_{m,ij} T (\alpha q_{ij}^2 + \beta q_{ij}), \qquad \forall (i,j) \in A_p. \tag{27}$$

5.3. Isentropic Exponent

The three-dimensional piecewise-linear approximation for the function of the isentropic exponent $m(\kappa(T, P))$ must be implemented using the integer based method. Similar to the compressibility factor, m must be equal to a specific value for each temperature–pressure combination and so cannot be defined by a convex region. The following constraints are used in the FCMP model to implement the isentropic exponent approximation:

$$m_{ij} \geq c_{1,k}^m T_i + c_{2,k}^m P_i + d_k^m, \qquad \forall (i,j) \in A_c, \forall k = 1, \ldots, 4,$$

$$m_{ij} \leq c_{1,k}^m T_i + c_{2,k}^m P_i + d_k^m + M_{Big}^m (1 - b_{ij,k}^m), \qquad \forall (i,j) \in A_c, \forall k = 1, \ldots, 4,$$

$$\sum_{k=1}^{4} b_{ij,k}^m = 1, \qquad \forall (i,j) \in A_c.$$

5.4. Compressor Operating Envelop

The two-dimensional piecewise-linear approximations of the compressor operating envelop bounds are implemented using both the convex region method and integer-based method. The stonewall and Smax bounds are able to be approximated using the convex region method, as the stonewall limit is a convex curve that bounds the envelop from below, and the Smax limit is a concave curve that bounds the envelop from above. The surge and Smin limits do not bound a convex region, and thereby need to be implemented using integer variables. Accordingly, only seven binary variables are added to the FCMP model for the 14 lines that approximate the operating envelop. The complete operating envelop approximation is implemented in the FCMP model through the following constraints:

$$H_{ad,ij} \geq c_k^{stw} Q_{ij} + d_k^{stw} \qquad \forall (i,j) \in A_c, k \in 1,\ldots,3,$$

$$H_{ad,ij} \leq c_k^{Smax} Q_{ij} + d_k^{Smax} \qquad \forall (i,j) \in A_c, k \in 1,\ldots,4,$$

$$L_{ij}^{Smin} \geq c_k^{Smin} Q_{ij} + d_k^{Smin} - M_{Big}^{Smin}(1 - b_{k,ij}^{Smin}) \qquad \forall (i,j) \in A_c, k \in 1,\ldots,4,$$

$$L_{ij}^{Smin} \leq c_k^{Smin} Q_{ij} + d_k^{Smin} \qquad \forall (i,j) \in A_c, k \in 1,\ldots,4,$$

$$\sum_{k=1}^{4} b_{k,ij}^{Smin} = 1 \qquad \forall (i,j) \in A_c,$$

$$H_{ad,ij} \geq L_{ij}^{Smin} \qquad \forall (i,j) \in A_c,$$

$$L_{ij}^{srg} \geq c_k^{srg} Q_{ij} + d_k^{srg} \qquad \forall (i,j) \in A_c, k \in 1,\ldots,3,$$

$$L_{ij}^{srg} \leq c_k^{srg} Q_{ij} + d_k^{srg} + M_{Big}^{srg}(1 - b_{k,ij}^{srg}) \qquad \forall (i,j) \in A_c, k \in 1,\ldots,3,$$

$$\sum_{k=1}^{3} b_{k,ij}^{srg} = 1 \qquad \forall (i,j) \in A_c,$$

$$H_{ad,ij} \leq L_{ij}^{srg} \qquad \forall (i,j) \in A_c.$$

5.5. Compressor Efficiency

The compressor efficiency parameter is eliminated from the FCMP formulation by approximating the term $H_{ad}/\eta_{ad} = g_c(H_{ad}, Q)$, which appears in the objective function. For any combination of compressor head and volumetric flow rate, a specific g_c must be calculated for the approximation to be accurate. Implementing the piecewise-linear approximation of g_c would require an integer method if it were not for the approximation being a convex set of planes, and appearing in the objective function that is being minimized. That is to say, the convex region method for implementing the approximation can be used because the minimum of a convex region is equivalent to the boundary of the convex region. The approximation of g_c is implemented into the FCMP as follows:

$$\min \sum_{(i,j) \in A_c} g_{c,ij} q_{ij} \tag{28}$$

$$\text{s.t. } g_{c,ij} \geq c_{1,k}^g Q_{ij} + c_{2,k}^g H_{ad,ij} + d_k^g \qquad \forall (i,j) \in A_c, k \in 1,\ldots,4.$$

The full FCMP model that includes all of the novel approximations can be found in Appendix A.

6. Case Study

Three simple gas networks are used in the case study to investigate the performance of the novel FCMP model. The first two networks investigated are identical to the networks studied in Wu et al. [23]. The Example 3 network is an extension of Example 2 intended to increase complexity. In each network, the compressor stations are comprised of five centrifugal compressor units operated in parallel that can each be switched ON or OFF, and the fitted coefficients of Equations (13) and (16) are the same as in Wu et al. The integer variable n_{ij}^s represents the number of compressor units ON in station (i,j). For each network, the operating temperature is 288.7 K, however the 289.5 K compressibility factor contour (Figure 8) was used for the piecewise-linear approximation to investigate the error introduced by discretizing the temperature domain into sufficiently small intervals.

The three case study gas networks are made up of at most 17 arcs and so are much simpler versions of typical gas networks encountered in practice. In practice, it is common to encounter networks that contain hundreds of arcs, such as those provided in the publicly available gas network test instances library GasLib, which are comparable to real-world networks of Open Grid Europe GmbH [32]. It is important to be able to solve the FCMP on these networks within minutes as the FCMP is an operational problem that needs to be solved in real-time to respond to changing network conditions. When the

FCMP is being solved for long-term planning purposes it is often solved numerous times for a large batch of scenarios so solution times within minutes are still needed to be useful for practitioners.

The optimization software used for the case study was GAMS version 25.1.1 [33], and the MINLP solver used was SCIP version 5.0 [6]. GAMS was ran on the Ubuntu 16.04 operating system, with a CPU frequency of 3592 MHz and 4 GB of RAM. The simulation software used to investigate the optimization results was Matlab version 2018a [34].

6.1. FCMP Models

Three different FCMP models were applied to each of the simple case study gas networks to investigate how the novel FCMP model balances accuracy and solution time. Figure 15 describes the components of the FCMP model with common simplifications ($FCMP_S$), the novel FCMP model which includes the approximations developed here ($FCMP_N$), and a partially rigorous FCMP model ($FCMP_{PR}$). The $FCMP_S$ model found in Appendix B includes all of the common simplifications and so is included to benchmark the lowest reasonable solution time for an FCMP model. Given that the $FCMP_S$ model's components are commonly found throughout FCMP literature, it is also useful for making comparisons to what is currently accepted as a standard FCMP model. The $FCMP_{PR}$ model found in Appendix C is included to benchmark model accuracy, as it is made up of the most rigorous calculations for its components, except for the compressibility factor and isentropic exponent. The $FCMP_{PR}$ model uses the novel approximations for the compressibility factor and isentropic exponent as their rigorous calculations cannot be feasibly implemented in an optimization model, and the novel approximations are the most accurate alternative.

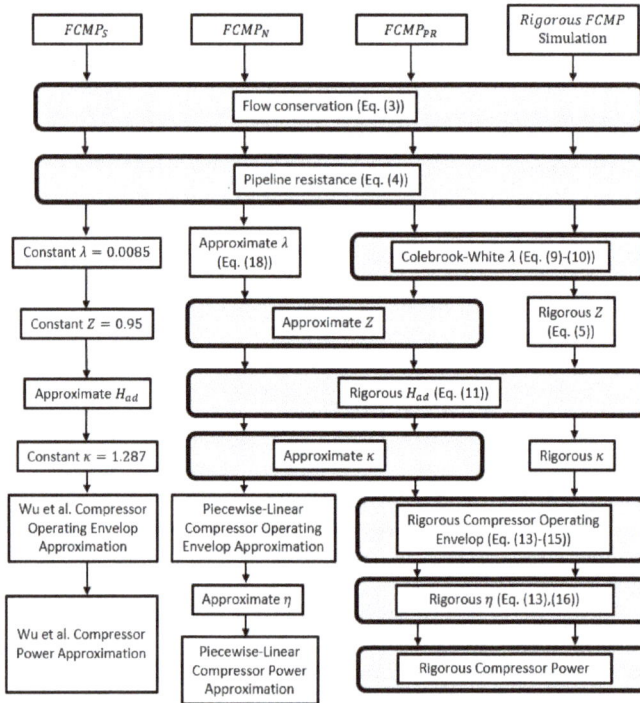

Figure 15. Flow chart detailing the components of each FCMP model applied in the case study, and the rigorous simulation model.

The optimal solutions from the three optimization models are applied on a rigorous simulation model that is composed of the most rigorous modeling components known. The rigorous simulation is ran to assess the accuracy and feasibility of a given optimal solution as follows:

1. The pressure of the first node is set as the first node pressure from the optimal solution.
2. If the next arc is a pipe section, then the pressure of the next node is calculated using the pipeline resistance equation, where Z and λ are calculated using their rigorous relations. Else, move to Step 3.
3. The next arc is a compressor station, so the outlet pressure is set as the outlet pressure from the optimal solution. The simulated compressor fuel cost is then calculated using the rigorous calculations for Z, κ, H_{ad}, and η_{ad} so that a theoretically exact fuel cost is obtained.
4. Return to Step 2 and iterate until all node pressures and compressor station fuel costs are calculated.

The optimal solutions obtained from the three FCMP optimization models are all feasible with respect to their individual simplifications and constraints, but when applied on a rigorous simulation that calculates the pressure drop and compressor physics rigorously certain constraints may be violated. It is important to note that the simulation will complete its calculation of node pressures and compressor station fuel costs even if constraints are violated. In the case of the simulation finding physical constraint violations, the optimal solution is noted as infeasible, but the simulated fuel cost is still computed so that the error in the optimal solution can be quantified.

6.2. Example 1

The Example 1 gas network in Figure 16 is a simple gun-and-barrel network with two compressor stations. The optimization and simulation results in Table 3 show that the $FCMP_N$ and $FCMP_{PR}$ models are highly accurate, with their optimal fuel cost differing by less than 1% compared to the fuel cost obtained from the simulation. The $FCMP_S$ model had the quickest solution time, but the optimal solution differed by 11.41% compared to the fuel cost obtained from the simulation. Further, the optimal solution obtained from the $FCMP_S$ model was proven to be infeasible by the rigorous simulation, whereas both the $FCMP_N$ and $FCMP_{PR}$ models produced optimal solutions that were feasible in the simulation. The reason for the infeasibility obtained in the $FCMP_S$ simulation is that the model underestimated the pressure drop in pipe sections due to error introduced in both Z and λ. This resulted in the simulation pressure at the end of the pipeline being below the lower pressure limit, and the simulation compressor stations operating outside the operating envelop due to the larger pressure increase required. The 0.09% difference between the simulated fuel costs of the $FCMP_N$ and $FCMP_{PR}$ models demonstrates that the compressor model and friction factor approximations developed introduce an insignificant amount of model error compared to their rigorous calculations.

For this gas network, the $FCMP_N$ model produced an optimal fuel cost that deviates from its solution's simulated fuel cost less than the optimal fuel cost produced by the $FCMP_{PR}$ model deviates from its solution's simulated fuel cost. This gives the false impression that the $FCMP_N$ model is more accurate than the $FCMP_{PR}$ model, and, given that the $FCMP_{PR}$ model contains fewer approximations than the $FCMP_N$ model, this is an unexpected result. The reason for this unexpected result is that in this particular case the error introduced by approximating some parameters in the $FCMP_N$ model cancels out with the error introduced by approximating some other parameters. Both $FCMP_N$ and $FCMP_{PR}$ models contain approximation error that produces an overestimation of pressure drop compared to the simulated pressure drop, so this results in an overestimation of the compressor fuel cost compared to the simulated fuel cost. However, the $FCMP_N$ model contains approximation error in the objective function while the $FCMP_{PR}$ model does not, and, depending on where the optimal solution of the $FCMP_N$ places the operation of the compressor stations, the objective function approximation error can result in an underestimation of the compressor fuel costs. For the $FCMP_S$ model, the overestimation of compressor fuel cost through overestimating pressure drop, and the underestimation of compressor fuel costs introduced by approximating the objective function can result in a cancellation of model error that results in the optimal fuel cost being closer to the simulated fuel cost compared to that of the $FCMP_{PR}$ model. It is important to note that the $FCMP_S$ model will not always benefit from the cancellation of model errors. There will be cases where the optimal solution is on a part of the compressor operating domain where the objective function approximation error results in fuel cost being overestimated, and therefore the pressure drop error is compounded with the objective function error.

Figure 16. Example 1 gas network consisting of three pipes and two compressor stations. $P_i^L = 4.14$ MPa, and $P_i^H = 5.52$ MPa for all $i \in N$. $D_{ij} = 0.9144$ m and $L_{ij} = 80.47$ km for all $(i, j) \in A_p$.

Table 3. Example 1 case study results. Both compressor stations only have one unit active in all three model solutions. Lines in bold represent outlet pressures from compressor stations.

Variable	$FCMP_S$ Optimization	$FCMP_S$ Simulation	$FCMP_N$ Optimization	$FCMP_N$ Simulation	$FCMP_{PR}$ Optimization	$FCMP_{PR}$ Simulation
P_1 (MPa)	4.99	4.99	5.01	5.01	5.01	5.01
P_2	**4.47**	**4.37**	**4.39**	**4.39**	**4.39**	**4.39**
P_3	4.87	4.87 **	4.92	4.92	4.92	4.92
P_4	4.33	4.22	4.28	4.29	4.28	4.29
P_5	**4.70**	**4.70 ****	**4.80**	**4.80**	**4.80**	**4.80**
P_6	4.14	4.02 *	4.14	4.14	4.14	4.14
Fuel Cost (kJ s^{-1})	39.76	44.88 ***	47.31	47.33	47.58	47.28
% Diff	11.41%		0.05%		0.61%	
Solution Time (s)	0.32 s		1.26 s		5.97 s	
# of Constraints	31		123		86	
# of Continuous Variables	23		71		68	
# of Discrete Variables	2		34		20	

* Pressure is outside of pressure bound by more than 1%. ** Compressor operating point is outside the operating envelop by more than 1%. *** Fuel cost calculated on the basis of a solution that violates physical constraints.

6.3. Example 2

The Example 2 gas network shown in Figure 17 is a tree network with three compressor stations. Table 4 shows that the optimal solutions obtained from the $FCMP_N$ and $FCMP_{PR}$ models are both highly accurate, differing from their simulated fuel costs by 1.02% and 0.78%, respectively. Both optimal solutions produced by these models were found to be feasible in the rigorous simulation. The $FCMP_S$ model was found to be infeasible in the simulation due to the compressor stations on arcs $(3, 4)$ and $(3, 8)$ operating outside the feasible operating envelop. Further, the optimal fuel cost of the $FCMP_S$ model was found to be 28.87% different than the fuel cost obtained from the simulation. As in Example 1, this difference can largely be attributed to underestimating the pressure drop in pipeline segments. The solution times of the $FCMP_S$ and $FCMP_N$ models are comparable, both being under 2 s. The $FCMP_{PR}$ solution time is much higher, at 54.26 s. Given that the $FMCP_{PR}$ simulated fuel cost is only 0.28% lower than that of the $FCMP_N$ model, the drastically higher solution time of the $FCMP_{PR}$ model shows the value of the $FCMP_N$ model.

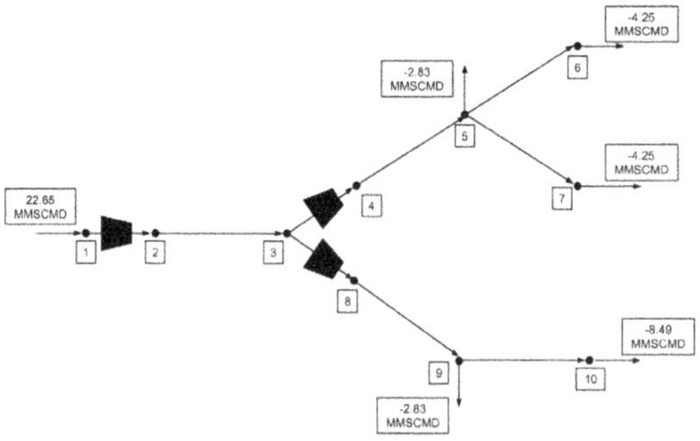

Figure 17. Example 2 gas network consisting of six pipes and three compressor stations. $P_1^L = P_2^L = 4.14$ MPa, $P_3^L = P_5^L = P_6^L = P_7^L = P_9^L = 3.10$ MPa, $P_4^L = 3.45$ MPa, $P_8^L = 3.79$ MPa, $P_{10}^L = 2.76$ MPa. $P_1^H = 4.83$ MPa, $P_i^H = 5.52$ MPa for all $i > 1$. $D_{ij} = 0.9144$ m and $L_{ij} = 80.47$ km for all $(i,j) \in A_p$.

Table 4. Example 2 case study results. All three compressor stations only have one unit active in all three model solutions. Lines in bold represent outlet pressures from compressor stations.

Variable	$FCMP_S$ Optimization	$FCMP_S$ Simulation	$FCMP_N$ Optimization	$FCMP_N$ Simulation	$FCMP_{PR}$ Optimization	$FCMP_{PR}$ Simulation
P_1 (MPa)	4.26	4.26	4.38	4.38	4.38	4.38
P_2	**4.53**	**4.53**	**4.69**	**4.69**	**4.69**	**4.69**
P_3	3.42	3.15	3.37	3.38	3.38	3.38
P_4	**3.79**	**3.79 ****	**3.79**	**3.79**	**3.79**	**3.79**
P_5	3.49	3.42	3.42	3.42	3.42	3.42
P_6	3.45	3.36	3.36	3.36	3.36	3.36
P_7	3.45	3.36	3.36	3.36	3.36	3.36
P_8	**3.79**	**3.79 ****	**3.79**	**3.79**	**3.79**	**3.79**
P_9	3.49	3.42	3.42	3.42	3.42	3.42
P_{10}	3.31	3.18	3.18	3.18	3.18	3.18
Fuel Cost (kJ s^{-1})	56.55	79.51 ***	58.47	57.88	58.17	57.72
% Diff	28.87%		1.02%		0.78%	
Solution Time (s)	1.03 s		1.88 s		54.26 s	
# of Constraints	50		197		146	
# of Continuous Variables	30		115		115	
# of Discrete Variables	3		54		33	

* Pressure is outside of pressure bound by more than 1%. ** Compressor operating point is outside the operating envelop by more than 1%. *** Fuel cost calculated on the basis of a solution that violates physical constraints.

6.4. Example 3

The Example 3 network shown in Figure 18 is an extension of the Example 2 network, with two additional compressor stations and six additional pipes. Table 5 shows that the accuracy of the $FCMP_N$ and $FCMP_{PR}$ models was unaffected by the additional complexity of the Example 3 network, with the error between the optimal and simulated fuel costs being less than 1% for both models. The accuracy of the $FCMP_S$ model did suffer from the extension of the network, with the fuel cost error increasing to 47.50% compared to the simulation. Again, the simulation showed that the $FCMP_S$ model solution is infeasible and the $FCMP_N$ and $FCMP_R$ model solutions are feasible. The solution time of the $FCMP_{PR}$ model increased drastically, going from 54.26 s in Example 2 to 994.98 s in Example 3. The $FCMP_N$ model solution time also increased, but only to 8.21 s from 1.88 s in Example 2. The increased solution efficiency of the $FCMP_N$ model over the $FCMP_{PR}$ model does not come at a loss of model accuracy, with the simulated fuel costs differing by only 0.41% between the models. As in Example 1, the $FCMP_N$ optimal solution is observed to have slightly less model error than the $FCMP_{PR}$ model. The explanation for this unexpected result is the same as was given for Example 1.

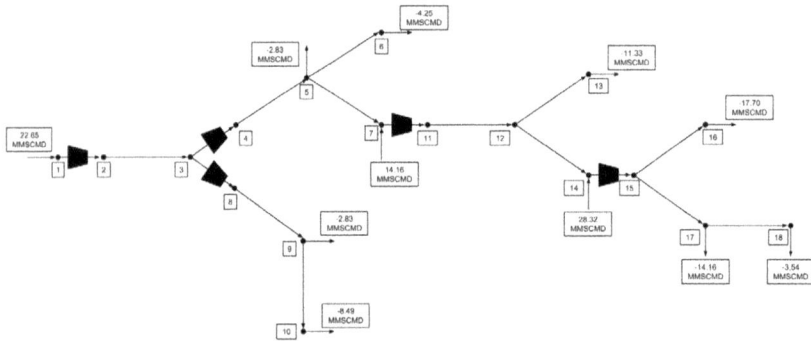

Figure 18. Example 3 gas network consisting of 12 pipes and five compressor stations. $P_1^L = P_2^L = 4.14$ MPa, $P_3^L = P_5^L = P_6^L = P_7^L = P_9^L = 3.10$ MPa, $P_4^L = 3.45$ MPa, $P_8^L = 3.79$ MPa, $P_{10}^L = 2.76$ MPa, $P_{11}^L = P_{12}^L = P_{13}^L = P_{14}^L = P_{15}^L = P_{16}^L = P_{17}^L = P_{18}^L = 0.69$ MPa. $P_1^H = 4.83$ MPa, $P_i^H = 5.52$ MPa for all $i > 1$. $D_{ij} = 0.9144$ m and $L_{ij} = 80.47$ km for all $(i, j) \in A_p$.

Table 5. Example 3 case study results. Lines in bold represent outlet pressures from compressor stations.

Variable	$FCMP_S$		$FCMP_N$		$FCMP_{PR}$	
	Optimization	Simulation	Optimization	Simulation	Optimization	Simulation
P_1 (MPa)	4.26	4.26	4.46	4.46	4.45	4.45
P₂	**4.53**	**4.53 ****	**4.81**	**4.81**	**4.80**	**4.80**
P_3	3.42	3.15	3.56	3.56	3.54	3.54
P₄	**3.79**	**3.79 ****	**4.01**	**4.01**	**4.01**	**4.01**
P_5	3.49	3.42	3.66	3.66	3.66	3.66
P_6	3.45	3.36	3.60	3.60	3.60	3.60
P_7	3.45	3.36	3.60	3.60	3.60	3.60
P₈	**3.79**	**3.79 ****	**4.01**	**4.01**	**3.98**	**3.98**
P_9	3.49	3.42	3.66	3.66	3.63	3.63
P_{10}	3.31	3.18	3.44	3.44	3.41	3.41
P₁₁	**3.66**	**3.66 ****	**3.84**	**3.84**	**3.84**	**3.84**
P_{12}	2.76	2.49	2.75	2.76	2.75	2.76
P_{13}	2.32	1.85	2.19	2.19	2.19	2.19
P_{14}	2.59	2.26	2.55	2.55	2.55	2.55
P₁₅	**2.76**	**2.76 ****	**2.77**	**2.77**	**2.77**	**2.77**
P_{16}	1.49	0.85	0.88	0.88	0.88	0.88
P_{17}	1.49	0.85	0.88	0.88	0.88	0.88
P_{18}	1.42	0.65 *	0.69	0.69	0.69	0.69
$n_{1,2}^s$	2		1		1	
$n_{3,4}^s$	1		1		1	
$n_{3,8}^s$	1		1		1	
$n_{7,11}^s$	2		1		1	
$n_{14,15}^s$	4		3		3	
Fuel Cost (kJ s⁻¹)	98.21	187.08 ***	123.02	122.26	123.74	122.77
% Diff	47.50%		0.62%		0.79%	
Solution Time (s)	3.99 s		8.21 s		994.98 s	
# of Constraints	88		345		266	
# of Continuous Variables	68		205		211	
# of Discrete Variables	5		94		59	

* Pressure is outside of pressure bound by more than 1%. ** Compressor operating point is outside the operating envelop by more than 1%. *** Fuel cost calculated on the basis of a solution that violates physical constraints.

A final note on the three examples is that the the solution time for the $FCMP_{PR}$ model grows much more quickly than that for the $FCMP_N$ model. It can be seen that, between each of the case study examples, the problem size is roughly doubled while the solution time for $FCMP_{PR}$ is increased an order of magnitude. Following the observed exponential time complexity, one could imagine that, for a real industrial gas network (10–100 times larger than Example 3), the solution time for $FCMP_{PR}$ would be several hours to days which is too slow for real-time operation or even off-line analysis. On the other hand, the solution time for the $FCMP_N$ model is much less for each example and it does not increase with problem size dramatically. The computational advantage of the proposed MILP modeling approach makes it more appealing for real-world problems.

7. Conclusions and Future Work

The common way of modeling the FCMP has been to assume key modeling parameters as constant to avoid the complexity involved with their rigorous calculation. We have shown that, even for three simple gas networks, the model error that is introduced by making these simplifications are significant and often cause the optimal solution obtained from these models to be infeasible in reality. A method for approximating the complex relationships that have typically been avoided has been demonstrated. The method is to apply an optimal piecewise-linear function generating MILP to the rigorously simulated data points of the relationship being approximated. An optimal solution to the MILP determines the orientation and break-points of a specified number of linear segments so that the maximum relative error between the approximation and the actual data points is minimized. For the MILP to be applicable, the relationship being approximated must be practically convex or concave, meaning that it is a relationship that can be approximated by a convex or concave function without significant error.

The piecewise-linear approximations developed have been implemented in a novel FCMP model and applied to three simple gas networks. The novel FCMP model was shown to marginally increase the solution time compared to the common simplified model, but significantly reduce model error. When compared to a theoretically exact simulation, the optimal solutions produced by the novel FCMP model had a maximum error of 1.02%. A comparison was also made to a rigorous FCMP model and the novel FCMP model was found to have a much great solution efficiency, and a similar model accuracy.

The proposed piecewise linearization approach is believed to be more advantageous than classical piecewise linearization approaches in that it determines the break-points in an optimal way. Since there has been no other work that linearizes the nonlinear relationships considered in this paper, it will be an interesting future work to apply the classical piecewise linearization approaches to FCMP and compare their performance to that of the proposed piecewise linearization approach. Given that the proposed method cannot be applied to clearly non-convex functions, future work will also seek to formulate a piecewise-linear function generating MILP that can be applied to non-convex relationships. This MILP will also aim to have the property of determining the optimal orientation and break-points of linear segments so that approximation error is minimized. This will allow the piecewise-linear method of linearization to be applied to any type of relationship and to benefit a wide variety of NLPs and MINLPs which are difficult to solve.

Author Contributions: Data curation, K.K.; Formal analysis, K.K.; Funding acquisition, X.L.; Investigation, K.K.; Methodology, K.K. and X.L.; Project administration, X.L.; Resources, X.L.; Software, K.K.; Supervision, X.L.; Validation, X.L.; Visualization, K.K.; Writing—original draft, K.K.; and Writing—review and editing, X.L.

Funding: This research was funded by Natural Sciences and Engineering Research Council of Canada through the the Discovery Grant (RGPIN 418411-13) and the Collaborative Research and Development Grant (CRDPJ 485798-15).

Conflicts of Interest: The authors declare no conflict of interest.

Abbreviations

The following abbreviations are used in this manuscript:

FCMP   Fuel Cost Minimization Problem
NLP   Nonlinear Program
MINLP   Mixed-Integer Nonlinear Program
PLFG   Piecewise-Linear Function Generation
MILP   Mixed-Integer Linear Program

Appendix A. Novel FCMP Model ($FCMP_N$)

$$\min \sum_{(i,j) \in A_c} g_{c,ij} q_{ij}$$

$$\text{s.t. } g_{c,ij} \geq c_{1,k}^g Q_{ij} + c_{2,k}^g H_{ad,ij} + d_k^g \qquad \forall (i,j) \in A_c, k \in 1, \ldots, 4,$$

$$\sum_{j:(j,i) \in A} q_{ji} - \sum_{j:(i,j) \in A} q_{ij} = -q_{ext,i} \qquad \forall i \in 1, \ldots, n,$$

$$(P_i^2 - P_j^2) = \frac{L_{ij} R}{M_w A_{ij}^2 D_{ij}} Z_{m,ij} T (\alpha q_{ij}^2 + \beta q_{ij}) \qquad \forall (i,j) \in A_p,$$

$$P_{m,ij} = \frac{P_i + P_j}{2} \qquad \forall (i,j) \in A_p,$$

$$Z_{m,ij} \geq c_k^Z P_{m,ij} + d_k^Z, \qquad \forall (i,j) \in A_p, \forall k = 1, \ldots, 2,$$

$$Z_{m,ij} \leq c_k^Z P_{m,ij} + d_k^Z + M_{Big}^Z (1 - b_{ij,k}^Z), \qquad \forall (i,j) \in A_p, \forall k = 1, \ldots, 2,$$

$$\sum_{k=1}^2 b_{ij,k}^Z = 1, \qquad \forall (i,j) \in A_p,$$

$$H_{ad,ij} = \frac{Z_i T R}{M m_{ij}} \left(\left(\frac{P_j}{P_i} \right)^{m_{ij}} - 1 \right) \qquad \forall (i,j) \in A_c,$$

$$m_{ij} \geq c_{1,k}^m T + c_{2,k}^m P_i + d_k^m, \qquad \forall (i,j) \in A_c, \forall k = 1, \ldots, 4,$$

$$m_{ij} \leq c_{1,k}^m T + c_{2,k}^m P_i + d_k^m + BigM^m (1 - b_{ij,k}^m), \qquad \forall (i,j) \in A_c, \forall k = 1, \ldots, 4,$$

$$\sum_{k=1}^4 b_{ij,k}^m = 1, \qquad \forall (i,j) \in A_c,$$

$$Z_i \geq c_k^Z P_i + d_k^Z, \qquad \forall (i,j) \in A_c, \forall k = 1, \ldots, 2,$$

$$Z_i \leq c_k^Z P_i + d_k^Z + M_{Big}^Z (1 - b_{i,k}^Z), \qquad \forall (i,j) \in A_c, \forall k = 1, \ldots, 2,$$

$$\sum_{k=1}^2 b_{i,k}^Z = 1, \qquad \forall (i,j) \in A_c,$$

$$Q_{ij} = \frac{Z_i q_{ij} R T}{P_i n_{ij}^s M_w} \qquad \forall (i,j) \in A_c,$$

$$H_{ad,ij} \geq c_k^{stw} Q_{ij} + d_k^{stw} \qquad \forall (i,j) \in A_c, k \in 1, \ldots, 3,$$

$$H_{ad,ij} \leq c_k^{Smax} Q_{ij} + d_k^{Smax} \qquad \forall (i,j) \in A_c, k \in 1, \ldots, 4,$$

$$L_{ij}^{Smin} \geq c_k^{Smin} Q_{ij} + d_k^{Smin} - M_{Big}^{Smin} (1 - b_{k,ij}^{Smin}) \qquad \forall (i,j) \in A_c, k \in 1, \ldots, 4,$$

$$L_{ij}^{Smin} \leq c_k^{Smin} Q_{ij} + d_k^{Smin} \qquad \forall (i,j) \in A_c, k \in 1, \ldots, 4,$$

$$\sum_{k=1}^4 b_{k,ij}^{Smin} = 1 \qquad \forall (i,j) \in A_c,$$

$$H_{ad,ij} \geq L_{ij}^{Smin} \qquad \forall (i,j) \in A_c,$$

$$L_{ij}^{srg} \geq c_k^{srg} Q_{,ij} + d_k^{srg} \qquad \forall (i,j) \in A_c, k \in 1, \ldots, 3,$$

$$L_{ij}^{srg} \leq c_k^{srg} Q_{,ij} + d_k^{srg} + M_{Big}^{srg} (1 - b_{k,ij}^{srg}) \qquad \forall (i,j) \in A_c, k \in 1, \ldots, 3,$$

$$\sum_{k=1}^3 b_{k,ij}^{srg} = 1 \qquad \forall (i,j) \in A_c,$$

$$H_{ad,ij} \leq L_{ij}^{srg} \qquad \forall (i,j) \in A_c,$$

$$P_i^L \leq P_i \leq P_i^U \qquad \forall i \in N,$$

$$q_{ij} \geq 0 \qquad \forall (i,j) \in A.$$

Appendix B. Simplified FCMP Model ($FCMP_S$)

$$\min \sum_{(i,j) \in A_c} g_{c,ij} q_{ij}$$

$$\text{s.t.} \; g_{c,ij} = A_c \left(\frac{q_{ij}}{P_i n_{ij}^s}\right)^2 + B_c \left(\frac{P_j}{P_i}\right)^2 + C_c \frac{q_{ij} P_j}{P_i^2 n_{ij}^s} + D_c \frac{q_{ij}}{P_i n_{ij}^s}$$

$$+ E_c \frac{P_j}{P_i} + F_c \qquad \forall (i,j) \in A_c,$$

$$\sum_{j:(j,i) \in A} q_{ji} - \sum_{j:(i,j) \in A} q_{ij} = -q_{ext,i} \qquad \forall i \in 1, \ldots, n,$$

$$(P_i^2 - P_j^2) = \frac{L_{ij} R}{M_w A_{ij}^2 D_{ij}} ZT\lambda q_{ij}^2 \qquad \forall (i,j) \in A_p,$$

$$P_j \leq a_1 q_{ij} + b_1 P_i \qquad \forall (i,j) \in A_c,$$

$$P_j \leq b_2 P_i \qquad \forall (i,j) \in A_c,$$

$$P_j \leq \frac{a_3 q_{ij}}{n_{ij}^s} + b_3 P_i \qquad \forall (i,j) \in A_c,$$

$$P_j \geq a_4 q_{ij} + b_4 P_i \qquad \forall (i,j) \in A_c,$$

$$P_j \geq b_5 P_i \qquad \forall (i,j) \in A_c,$$

$$P_j \geq \frac{a_6 q_{ij}}{n_{ij}^s} + b_6 P_i \qquad \forall (i,j) \in A_c,$$

$$P_i^L \leq P_i \leq P_i^U \qquad \forall i \in N,$$

$$q_{ij} \geq 0 \qquad \forall (i,j) \in A.$$

Appendix C. Rigorous FCMP Model ($FCMP_R$)

$$\min \sum_{(i,j)\in A_c} \frac{H_{ad,ij}q_{ij}}{\eta_{ad,ij}}$$

$$\text{s.t.} \quad \sum_{j:(j,i)\in A} q_{ji} - \sum_{j:(i,j)\in A} q_{ij} = -q_{ext,i} \quad \forall i \in 1,\ldots,n,$$

$$(P_i^2 - P_j^2) = \frac{L_{ij}R}{M_w A_{ij}^2 D_{ij}} Z_{m,ij} T \lambda_{ij} q_{ij}^2 \quad \forall(i,j) \in A_p,$$

$$\frac{1}{\sqrt{\lambda_{ij}}} = -2log_{10}\left(\frac{\epsilon/D_{ij}}{3.71} + \frac{2.51}{Re_{ij}\sqrt{\lambda_{ij}}}\right) \quad \forall(i,j) \in A_p,$$

$$Re_{ij} = \frac{D_{ij}}{A_{ij}\mu}q_{ij} \quad \forall(i,j) \in A_p,$$

$$P_{m,ij} = \frac{P_i + P_j}{2} \quad \forall(i,j) \in A_p,$$

$$Z_{m,ij} \geq c_k^Z P_{m,ij} + d_k^Z, \quad \forall(i,j) \in A_p, \forall k = 1,\ldots,2,$$

$$Z_{m,ij} \leq c_k^Z P_{m,ij} + d_k^Z + M_{Big}^Z(1 - b_{ij,k}^Z), \quad \forall(i,j) \in A_p, \forall k = 1,\ldots,2,$$

$$\sum_{k=1}^{2} b_{ij,k}^Z = 1, \quad \forall(i,j) \in A_p,$$

$$H_{ad,ij} = \frac{Z_i T R}{M_w m_{ij}}\left(\left(\frac{P_j}{P_i}\right)^{m_{ij}} - 1\right) \quad \forall(i,j) \in A_c,$$

$$m_{ij} \geq c_{1,k}^m T + c_{2,k}^m P_i + d_k^m, \quad \forall(i,j) \in A_c, \forall k = 1,\ldots,4,$$

$$m_{ij} \leq c_{1,k}^m T + c_{2,k}^m P_i + d_k^m + BigM^m(1 - b_{ij,k}^m), \quad \forall(i,j) \in A_c, \forall k = 1,\ldots,4,$$

$$\sum_{k=1}^{4} b_{ij,k}^m = 1, \quad \forall(i,j) \in A_c,$$

$$Z_i \geq c_k^Z P_i + d_k^Z, \quad \forall(i,j) \in A_c, \forall k = 1,\ldots,2,$$

$$Z_i \leq c_k^Z P_i + d_k^Z + M_{Big}^Z(1 - b_{i,k}^Z), \quad \forall(i,j) \in A_c, \forall k = 1,\ldots,2,$$

$$\sum_{k=1}^{2} b_{i,k}^Z = 1, \quad \forall(i,j) \in A_c,$$

$$Q_{ij} = \frac{Z_i q_{ij} RT}{P_i n_{ij}^s M_w} \quad \forall(i,j) \in A_c,$$

$$\frac{H_{ad,ij}}{S_{ij}^2} = A_H + B_H\frac{Q_{ij}}{S_{ij}} + C_H\left(\frac{Q_{ij}}{S_{ij}}\right)^2 + D_H\left(\frac{Q_{ij}}{S_{ij}}\right)^3 \quad \forall(i,j) \in A_c,$$

$$\eta_{ad,ij} = A_E + B_E\frac{Q_{ij}}{S_{ij}} + C_E\left(\frac{Q_{ij}}{S_{ij}}\right)^2 + D_E\left(\frac{Q_{ij}}{S_{ij}}\right)^3 \quad \forall(i,j) \in A_c,$$

$$S_{min} \leq S_{ij} \leq S_{max} \quad \forall(i,j) \in A_c,$$

$$surge \leq \frac{Q_{ij}}{S_{ij}} \leq stonewall \quad \forall(i,j) \in A_c,$$

$$P_i^L \leq P_i \leq P_i^U \quad \forall i \in N,$$

$$q_{ij} \geq 0 \quad \forall(i,j) \in A.$$

References

1. IEA. *World Energy Outlook 2016*; International Energy Agency: Paris, France, 2016; p. 684.
2. Schroeder, D. Hydraulic analysis in the natural gas industry. In *Advances in Industrial Engineering Applications and Practice I*; International Journal of Industrial Engineering: Houston, TX, USA, 1996; pp. 960–965.
3. Ríos-Mercado, R.Z.; Borraz-Sánchez, C. Optimization problems in natural gas transportation systems: A state-of-the-art review. *Appl. Energy* **2015**, *147*, 536–555. [CrossRef]
4. Tawarmalani, M.; Sahinidis, N.V. A polyhedral branch-and-cut approach to global optimization. *Math. Program.* **2005**, *103*, 225–249. [CrossRef]
5. Sahinidis, N.V. *BARON 17.8.9: Global Optimization of Mixed-Integer Nonlinear Programs, User's Manual*; BARON Software: Pittsburgh, PA, USA, 2017.
6. Gleixner, A.; Eifler, L.; Gally, T.; Gamrath, G.; Gemander, P.; Gottwald, R.L.; Hendel, G.; Hojny, C.; Koch, T.; Miltenberger, M.; et al. *The SCIP Optimization Suite 5.0*; ZIB-Report 17-61; Zuse Institute: Berlin, Germany, 2017.
7. Schmidt, M.; Steinbach, M.C.; Willert, B.M. High detail stationary optimization models for gas networks. *Optim. Eng.* **2015**, *16*, 131–164. [CrossRef]
8. Misra, S.; Fisher, M.W.; Backhaus, S.; Bent, R.; Chertkov, M. Optimal Compression in Natural Gas Networks: A Geometric Programming Approach. *IEEE Trans. Control Netw. Syst.* **2015**, *2*, 47–56. [CrossRef]
9. Borraz-Sánchez, C.; Haugland, D. Minimizing fuel cost in gas transmission networks by dynamic programming and adaptive discretization. *Comput. Ind. Eng.* **2011**, *61*, 364–372. [CrossRef]
10. Martin, A.; Moller, M.; Moritz, S. Mixed Integer Models for the Stationary Case of Gas Network Optimization. *Math. Program.* **2006**, *105*, 563–582. [CrossRef]
11. Wong, P.; Larson, R. Optimization of Natural-Gas Pipeline Systems Via Dynamic Programming. *IEEE Trans. Autom. Control* **1968**, *13*, 475–481. [CrossRef]
12. Wolf, D.D.; Smeers, Y. The Gas Transmission Problem Solved by an Extension of the Simplex Algorithm. *Manag. Sci.* **2000**, *46*, 1454–1465. [CrossRef]
13. Borraz-Sanchez, C.; Rios-Mercado, R.Z. Improving the Operation of Pipeline Systems on Cyclic Structures by Tabu Search. *Comput. Chem. Eng.* **2009**, *33*, 58–64. [CrossRef]
14. Flores-Villarreal, H.; Rios-Mercado, R. Computational Experience With a GRG Method for Minimizing Fuel Consumption on Cyclic Natural Gas Networks. In *Computational Methods in Circuits System Applications*; World Scientific and Engineering Academy and Society Press: Cambridge, UK, 2003; pp. 90–94.
15. Chebouha, A.; Yalaoui, F.; Smati, A.; Younsi, K.; Tairi, A. Optimization of Natural Gas Pipeline Transportation Using Ant Colony Optimization. *Comput. Oper. Res.* **2009**, *36*, 1916–1923. [CrossRef]
16. ISO-12213-2. *Natural Gas—Calculation of Compression Factor—Part 2: Calculation Using Molar-Composition Analysis*; Technical Report; International Organization for Standardization: Geneva, Switzerland, 2006.
17. Westhoff, M.A. *Using Operating Data at Natural Gas Pipelines*; Technical Report; Colorado Interstate Gas Company: Houston, TX, USA, 2004.
18. Coelho, P.; Pinho, C. Considerations About Equations for Steady State Flow in Natural Gas Pipelines. *J. Braz. Soc. Mech. Sci. Eng.* **2007**, *29*, 262–273. [CrossRef]
19. About U.S. Natural Gas Pipelines—Transporting Natural Gas. Available online: https://www.eia.gov/naturalgas/archive/analysis_publications/ngpipeline/interstate.html (accessed on 27 March 2018).
20. Kelkar, M. *Natural Gas Production Engineering*; PennWell Books: Tulsa, OK, USA, 2008; Chapter Gas Compression, pp. 445–464.
21. Maric, I.; Galovic, A.; Smuc, T. Calculation of natural gas isentropic exponent. *Flow Measur. Instrum.* **2005**, *16*, 13–20. [CrossRef]
22. Mokhatab, S.; Poe, W.A.; Speight, J.G.; (Eds.) *Handbook of Natural Gas Transmission and Processing*; Elsevier: Amsterdam, The Netherlands, 2006; Chapter Natural Gas Compression, pp. 295–322.
23. Wu, S.; Rios-Mercado, R.; Boyd, E.; Scott, L. Model Relaxations for the Fuel Cost Minimization of Steady-State Gas Pipeline Networks. *Math. Comput. Model.* **2000**, *31*, 197–220. [CrossRef]
24. Wu, X.; Li, C.; Jia, W.; He, Y. Optimal Operation of Trunk Natural Gas Pipelines Via an Inertia-Adaptive Particle Swarm Optimization Algorithm. *J. Nat. Gas Sci. Eng.* **2014**, *21*, 10–18. [CrossRef]
25. Nørstebø, V.S.; Rømo, F.; Hellemo, L. Using operations research to optimise operation of the Norwegian natural gas system. *J. Nat. Gas Sci. Eng.* **2010**, *2*, 153–162. [CrossRef]

26. Tabkhi, F.; Pibouleau, L.; Hernandez-Rodriguez, G.; Azzaro-Pantel, C.; Domenech, S. *Improving the Performance of Natural Gas Pipeline Networks Fuel Consumption Minimization Problems*; Wiley Interscience: Hoboken, NJ, USA, 2009.

27. Rios-Mercado, R.Z.; Kim, S.; Boyd, E.A. Efficient operation of natural gas transmission systems: A network-based heuristic for cyclic structures. *Comput. Oper. Res.* **2006**, *33*, 2323–2351. [CrossRef]

28. Camponogara, E.; Nazari, L. Models and Algorithms for Optimal Piecewise-Linear Function Approximation. *Math. Probl. Eng.* **2015**, *2015*. [CrossRef]

29. Codas, A.; Camponogara, E. Mixed-integer linear optimization for optimal lift-gas allocation with well-separator routing. *Eur. J. Oper. Res.* **2011**, *217*, 222–231. [CrossRef]

30. Carrión, M.; Arroyo, J.M. A Computationally Efficient Mixed-Integer Linear Formulation for the Thermal Unit Commitment Problem. *IEEE Trans. Power Syst.* **2006**, *21*, 1371–1378. [CrossRef]

31. Toriello, A.; Vielma, J. Fitting piecewise linear continuous functions. *Eur. J. Oper. Res.* **2012**, *219*, 86–95. [CrossRef]

32. GasLib. Available online: http://gaslib.zib.de/ (accessed on 25 September 2018).

33. GAMS Development Corporation. *General Algebraic Modeling System (GAMS) Release 25.1.1*; GAMS Development Corporation: Washington, DC, USA, 2018.

34. The MathWorks Inc. *MATLAB and Statistics Toolbox Release 2018a*; The MathWorks Inc.: Natick, MA, USA, 2018.

Article

Modelling of a Naphtha Recovery Unit (NRU) with Implications for Process Optimization

Jiawei Du and William R. Cluett *

Department of Chemical Engineering & Applied Chemistry, University of Toronto,
Toronto, ON M5S 3E5, Canada; gracedu0213@gmail.com
* Correspondence: will.cluett@utoronto.ca; Tel.: 1-416-275-4220

Received: 30 May 2018; Accepted: 18 June 2018; Published: 22 June 2018

Abstract: The naphtha recovery unit (NRU) is an integral part of the processes used in the oil sands industry for bitumen extraction. The principle role of the NRU is to recover naphtha from the tailings for reuse in this process. This process is energy-intensive, and environmental guidelines for naphtha recovery must be met. Steady-state models for the NRU system are developed in this paper using two different approaches. The first approach is a statistical, data-based modelling approach where linear regression models have been developed using Minitab® from plant data collected during a performance test. The second approach involves the development of a first-principles model in Aspen Plus® based on the NRU process flow diagram. A novel refinement to this latter model, called "withdraw and remix", is proposed based on comparing actual plant data to model predictions around the two units used to separate water and naphtha. The models developed in this paper suggest some interesting ideas for the further optimization of the process, in that it may be possible to achieve the required naphtha recovery using less energy. More plant tests are required to validate these ideas.

Keywords: naphtha recovery unit; statistical model; simulation; optimization

---

Note to reader regarding figures: The units of some of the variables in the figures have been normalized or expressed as a percentage of the working range for proprietary reasons.

1. Introduction

The naphtha recovery unit (NRU) is an important process used to recover the hydrocarbon diluent (naphtha) that is used to assist with bitumen extraction from oil sands [1]. Oil extraction from oil sands has drawn considerable public attention due to its energy consumption and ecological footprint. The NRU plays a key role in both respects in that it requires energy (steam) to recover the naphtha for reuse in the process and it discharges unrecovered naphtha to the environment (tailings pond).

Figure 1 gives a simplified process flow diagram to help put the NRU in the context of the bitumen extraction process. The froth that is produced from a hot water extraction process is a highly viscous fluid containing approximately 60% bitumen, 30% water and 10% solids on a weight basis. This froth is diluted with naphtha and is sent to a centrifuge system where the bitumen is separated and sent downstream for upgrading. The tailings, consisting of mostly water, sand and trace amounts of diluent and bitumen, are sent to the NRU. The NRU consists of a vacuum stripping tower with steam injected at the bottom of the deck. As its name implies, this unit recovers naphtha from the tailings for reuse and the remnants are discharged to the tailings pond.

Figure 2 offers a closer look at the NRU. The feed to the NRU column consists of the tailings from the upstream centrifuge system. Because the column is operated under vacuum, the feed flashes with a vapour mixture of water and naphtha rising up the column. A demister pad at the top of the column knocks out liquid droplets and solid particles entrained in the vapour. The vapour stream then enters the overhead system, where cooling water and a heat exchanger is used to condense the vapour into a liquid stream. The condensed liquid is then separated into naphtha and water by decantation in two separators that are connected in series. The recovered naphtha is sent to storage tanks for reuse and the water is recycled back to the column. The feed to the column that does not leave in the overhead vapour stream settles down in the liquid pool at the bottom of the column and is a mixture of unrecovered naphtha, solids, bitumen and water. Steam is injected directly into the liquid pool to provide heat that causes the naphtha in the pool to vaporize and travel up the column. Water from upstream upgrading facilities is also added to the column feed as an additional heating source.

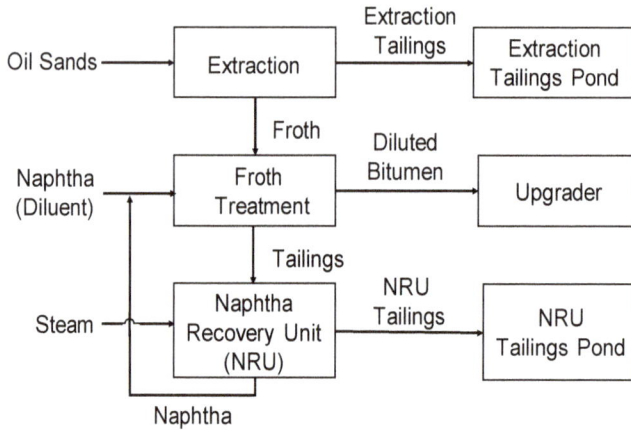

Figure 1. Simplified process flow diagram of bitumen extraction from oil sands.

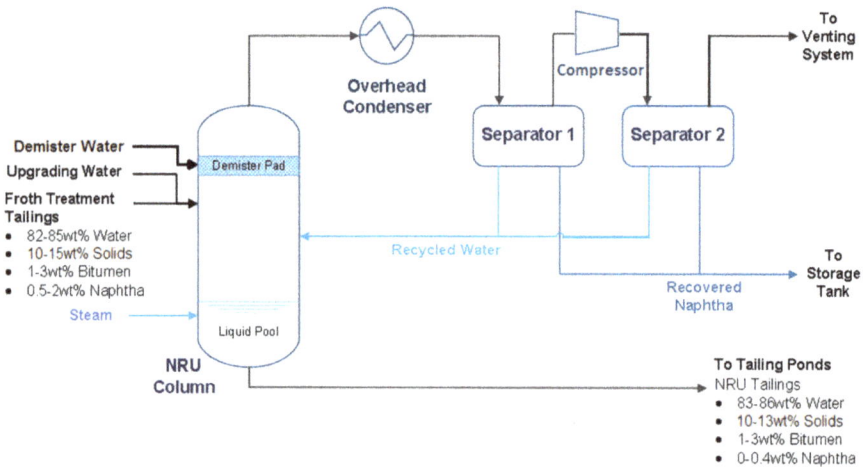

Figure 2. Process flow diagram of naphtha recovery unit.

In this paper, we are interested in exploring two different modelling approaches, with the goal being to optimize the NRU process, i.e., to maximize naphtha recovery while minimizing energy consumption. The first approach is based on a statistical model of the NRU process that was built using steady-state data collected during a plant trial when the steam flow to the column was varied over a wide operating range. The second approach is based on a first-principles model of the NRU process using knowledge of the process and measurements taken of key process variables.

It turns out that there is a dearth of papers in the literature that deal with the optimization of the NRU. Special issues of the Canadian Journal of Chemical Engineering were published in both 2004 and 2013 with a focus on the oil sands industry. The importance of models and soft sensors for monitoring, control and optimization of oil sands processes is discussed in [2]. In the case of the NRU, the paper's focus is on the development of a soft sensor for measuring the composition of the product coming off the top of the NRU column. Process control and optimization are also mentioned in [3]. Here, the focus is on the use of NIR spectroscopy to analyze bitumen and solvent-diluted bitumen samples from the froth stream produced by the hot water extraction process and the development of a soft sensor based on the NIR measurements for fast routine analysis with the potential for on-line applications. These two papers, along with the overview in [4], highlight the importance of accurate measurements and the need for novel techniques to obtain such measurements in the oil sands industry, including the work presented here.

2. Performance Test on the NRU

Based on years of experience, operators at the oil sands plant where this study was performed believe that naphtha recovery is improved by injecting more steam into the NRU. Therefore, the current optimization strategy consists of injecting the maximum flowrate of steam available. Recently, however, on-site process engineers have suspected that too much steam is being injected, causing overload issues in the overhead heat exchange system, and the cost of generating additional steam is a major economic concern.

To examine this issue more carefully, a performance test was conducted in 2015 over a three-week period with the goal being to determine if the steam injection rate could be reduced without negatively affecting the naphtha recovery. During the test, the steam injection rate was reduced from its maximum value to approximately 50% of that value through a series of step changes. The steam rate was held constant for at least one day following each step. Once the steam rate reached its lower limit, it was returned to its original value in two large steps. Process data consisting of the key process variables were collected using automatic sensors every 10 min based on the process engineers' knowledge that the time for the NRU to reach steady-state is on the order of 5 min. The steam injection rate over the test period is shown in Figure 3a.

The definition for naphtha recovery, a dimensionless quantity, used in this study is a standard in the industry and is given by Equation (1), where *naphtha* on the right-hand side of the equation refers to its mass flowrate:

$$naphtha\ recovery = \frac{naphtha\ in\ Feed\ to\ NRU - naphtha\ in\ Tailings\ of\ NRU}{naphtha\ in\ Feed\ to\ NRU} \qquad (1)$$

The mass flow rate of naphtha in the feed and tailings is calculated using Equation (2):

$$mass\ flowrate\ of\ naphtha = total\ mass\ flowrate \times mass\ fraction\ of\ naphtha \qquad (2)$$

Ultrasonic flowmeters are used to measure the total volumetric flowrate (v) of the feed and tailings streams. These volumetric flowrates (expressed in L/h) can be further converted into total mass flowrates using Equation (3):

$$total\ mass\ flowrate = v \times [100 / \sum_i (\frac{wt.\%\ of\ component\ i}{specific\ gravity\ of\ component\ i})] \qquad (3)$$

where the feed and tailings streams consist of four major components, namely water, naphtha, bitumen and solids. Near-infrared (NIR) sensors are used to measure the mass fractions (wt.%) of each of these four major components in the feed and tailings streams.

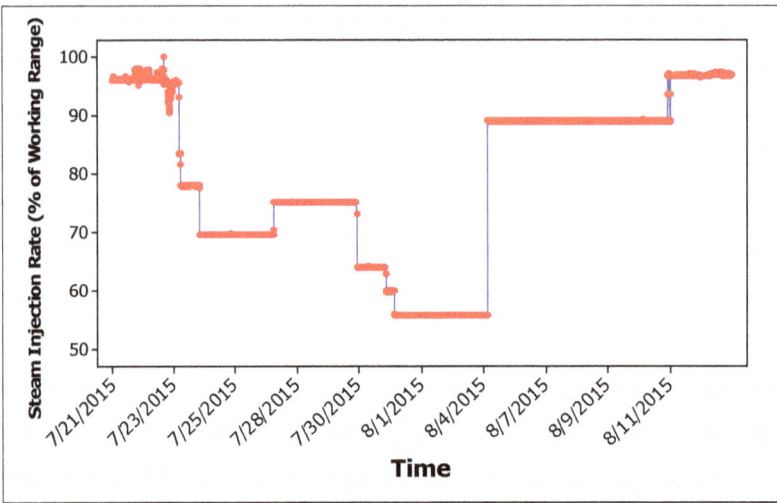

(a)

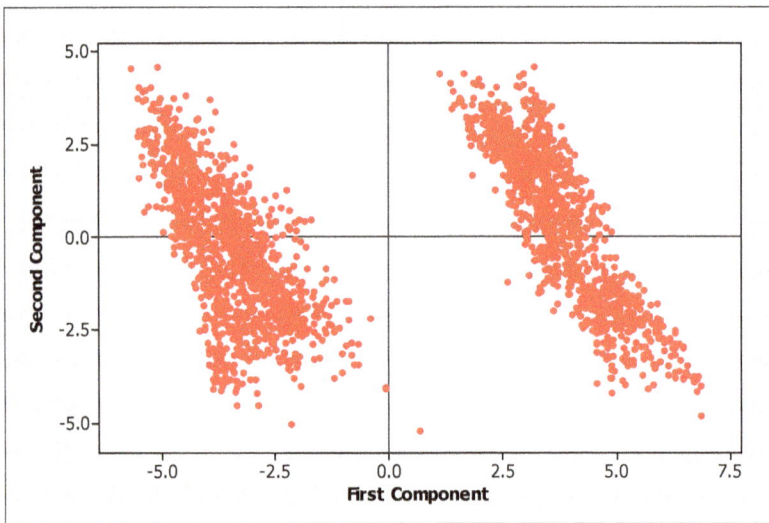

(b)

Figure 3. Data and preliminary results from the performance test: (**a**) steam injection rate used during the performance test, and (**b**) principal component analysis (PCA) of data collected from the performance test.

3. Statistical Modelling of the NRU Using Minitab® (V16)

3.1. Principal Component Analysis (PCA)

The process variables collected during the performance test and included in the PCA are summarized here in two tables, with the first being the NRU stream variables (Table 1) and the second being the NRU operating variables (Table 2). The units in these two tables reflect the units associated with the automatic sensors. An overall mass balance was initially performed around the NRU to verify that the sensors being used were behaving consistently throughout the test period. Results showed that variations in total input and total output mass flows were being tracked consistently.

Table 1. NRU stream variables.

Measured Variables	Description	Units	Measured Variables	Description	Units
F_{feed}	Feed flow rate	bbl/h	$C_{tailing, H2O}$	Water composition in tailings	wt.%
T_{feed}	Feed temperature	°F	$C_{tailing, min}$	Solid composition in tailings	wt.%
$C_{feed, dil}$	Naphtha composition in feed	wt.%	F_{RW}	Recycled water flowrate	USGPM
$C_{feed, bit}$	Bitumen composition in feed	wt.%	D_{RW}	Recycled water density	SG
$C_{feed, H2O}$	Water composition in feed	wt.%	$C_{RW, H2O}$	Water composition in recycled water	wt.%
$C_{feed, min}$	Solid composition in feed	wt.%	$C_{RW, HC}$	Hydrocarbon composition in recycled water	wt.%
F_{steam}	Steam injection rate	MLB/h	F_{PN}	Recovered naphtha flowrate	USGPM
$F_{upgrading}$	Upgrading water flow rate	bbl/h	D_{PN}	Recovered naphtha density	SG
$F_{demister}$	Demister water flow rate	USGPM	$C_{PN, H2O}$	Water composition in recovered naphtha	wt.%
$F_{tailing}$	Tailings flow rate	USGPM	$C_{PN, HC}$	Hydrocarbon composition in recovered naphtha	wt.%
$T_{tailing}$	Tailings temperature	°F	F_{CW}	Cooling water flowrate	USGPM
$C_{tailing, dil}$	Naphtha composition in tailings	wt.%	$T_{CW, in}$	Cooling water inlet temperature	°F
$C_{tailing, bit}$	Bitumen composition in tailings	wt.%	$T_{CW,out}$	Cooling water outlet temperature	°F

Table 2. NRU operating variables.

Measured Variables	Description	Units	Measured Variables	Description	Units
$T_{C1,bottom}$	NRU column bottom temperature	°F	$P_{C1,top}$	NRU column top pressure	psi
$P_{C1,bottom}$	NRU column bottom pressure	psi	T_{S2}	Second separator temperature	°F
$T_{C1,top}$	NRU column top temperature	°F	P_{Comp1}	Overhead compressor suction pressure	psi

PCA was used to help determine if the data set was representative of one or more operating modes. This is similar to the classical discrimination problem proposed in [5] in their tutorial and the analysis of historical process data sets suggested in [6] in their tutorial. Minitab was used and the correlation matrix was selected where each variable was mean-centred and scaled by its standard deviation. Data points with missing values were removed from the data set manually. Outliers were removed through a combination of visual inspection and with the help of the PCA analysis itself [5]. Missing data and outliers represented less than 10% of the original data set.

Figure 3b shows a plot of the first two principal components. The data set in this reduced dimension clearly clusters into two distinct regions. By comparing the data points in each cluster to the time series plot of the steam injection schedule during the performance test (Figure 3a), it was determined that all the data points in the left-hand cluster came from data collected during the performance test up to August 4th (referred to as OP1) and all the data points found in the right-hand cluster came from the data collected after August 4th (referred to as OP2). This would seem to indicate that a distinct shift in process behaviour occurred on August 4th, coinciding with the large increase in steam flowrate that occurred at that time. Therefore, if statistical models were going to be built for prediction purposes, it would probably be best to identify separate models for the process in OP1 and OP2 [7].

3.2. Linear Regression Models

In this part of our analysis, our goal was to generate an equation that describes the statistical relationship (model) between one or more predictors (regressors) and the primary response variable of interest in this study, naphtha recovery (NR), as defined in Equation (1). Separate models were developed using Minitab for each operating mode (OP1 and OP2) based on our PCA. Each data set was divided randomly in half, and one half of the data was used for model development (training) while the other half was used for model validation (testing). Given that we have assumed the data is representative of the system at steady-state, partitioning the data in this way is valid.

Several dimensionless variables were examined as possible predictors based on our engineering knowledge of the NRU. A sequence of models was constructed iteratively by starting with a large number of predictors and then reducing the model size based on the p-value of each predictor, one predictor at a time. In the end, only the predictors that made a significant and therefore meaningful contribution to the model were retained.

The predictors that were determined to make a significant contribution in both OP1 and OP2 to NR are given here:

Composition of naphtha in the NRU feed (NF):

$$NF = C_{feed,\ dil}\ (\boldsymbol{wt.\%}) \tag{4}$$

Normalized steam injection rate in the NRU feed (SF):

$$SF = \{(F_{steam}/2.2)/(v_{feed} \times [100/\sum_i(\frac{wt.\%\ of\ component\ i}{specific\ gravity\ of\ component\ i})]\} \times 100 \tag{5}$$

Here, we have made use of Equation (3) to normalize the steam injection rate by the feed mass flowrate in Equation (5). The 2.2 factor is needed for unit conversion and the multiplier of 100 at the right end of Equation (5) is used to bring *SF* to the same order of magnitude as *NF* and *NR*.

The fact that these variables have appeared in these models makes physical sense because, from an input-output point of view, *NR* is the primary output variable of interest, *SF* is a key manipulated input variable and *NF* is an important disturbance input variable. The modelling results are summarized in Tables 3 and 4 and plots related to these models may be found in Figures 4 and 5.

Table 3. Statistical model results for operating mode 1 (OP1).

LS Model	$NR_{prediction} = 0.347 + 0.735\ NF - 0.264\ (NF)^2 - 0.077\ SF$	
R square	Training Set	Testing Set
	0.927	0.912
Validation [1]	$NR_{measured} = -0.0095 + 1.011\ NR_{prediction}$	

[1] Validation corresponds to fit obtained using testing data. A perfect fit between predicted and measured values of *NR* would correspond to slope = 1 and intercept = 0.

Table 4. Statistical model results for OP2.

LS Model	$NR_{prediction} = 0.605 + 0.475\,NF - 0.129\,(NF)^2 - 0.375\,SF$	
R square	Training Set	Testing Set
	0.810	0.817
Validation [1]	$NR_{measured} = 0.0264 + 0.963\,NR_{prediction}$	

[1] Validation corresponds to fit obtained using testing data. A perfect fit between predicted and measured values of *NR* would correspond to slope = 1 and intercept = 0.

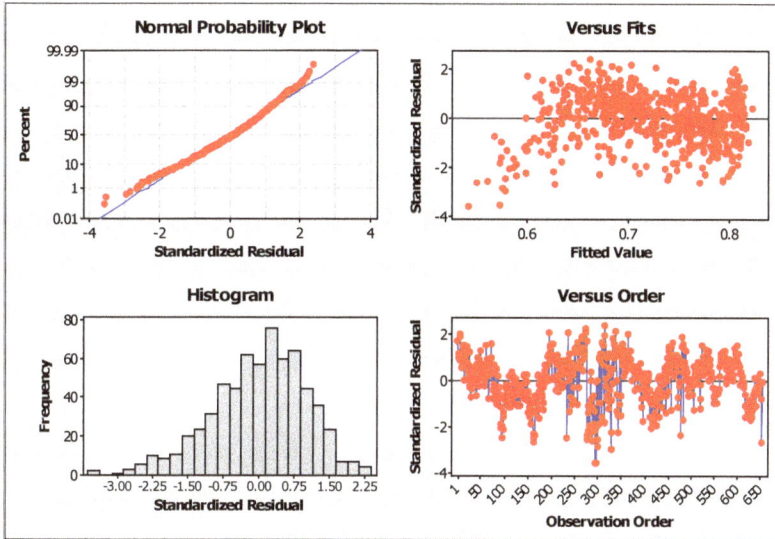

(a)

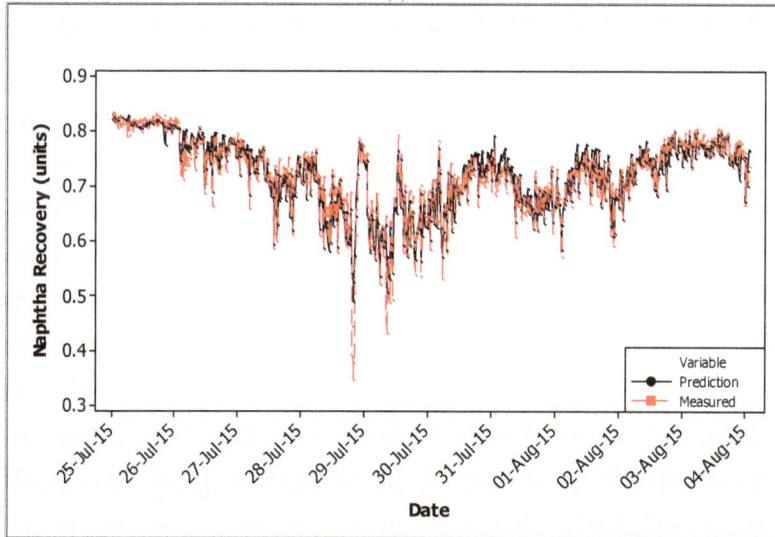

(b)

Figure 4. (**a**) Statistical modelling results for OP1 (residuals developed using training data); (**b**) Statistical modelling results for OP1 comparing measured versus predicted *NR* using testing data.

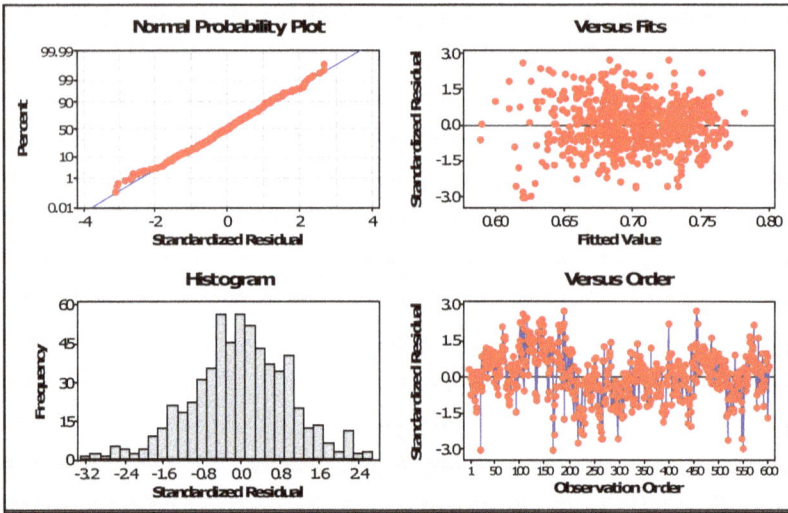

(a)

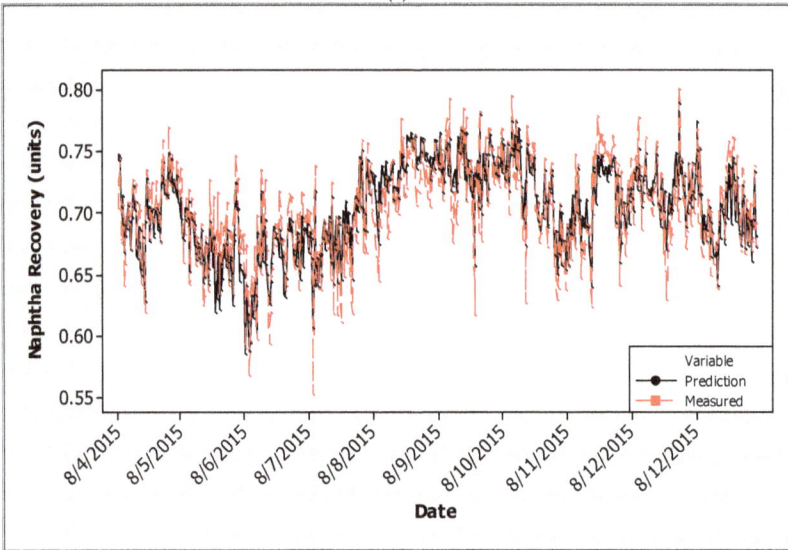

(b)

Figure 5. (**a**) Statistical modelling results for OP2 (residuals developed using training data); (**b**) Statistical modelling results for OP2 comparing measured versus predicted *NR* using testing data.

Based on these linear regression models, Figures 6 and 7 were generated to look at the prediction of the individual effects of *NF* and *SF* on *NR*, respectively. The positive slope in Figure 6 associated with the *NF* to *NR* relationship may at first appear counterintuitive, i.e., one might expect based on mass and energy balance considerations that an increase in the mass fraction of naphtha in the feed would cause a higher naphtha loss and therefore a drop in naphtha recovery. However, given the way that naphtha recovery is defined in Equation 1, this is not necessarily the case. For example, say there are 100 units of naphtha coming in initially. If the naphtha recovery is 0.8, the naphtha loss would be 20 units. Now, let us assume the naphtha inflow increases from 100 to 130 units. The expected naphtha loss would increase as well when using the same steam flowrate; let us say it increases from 20 to 24 units. In this case, the recovery actually increases from 0.8 to 0.82 (106/130).

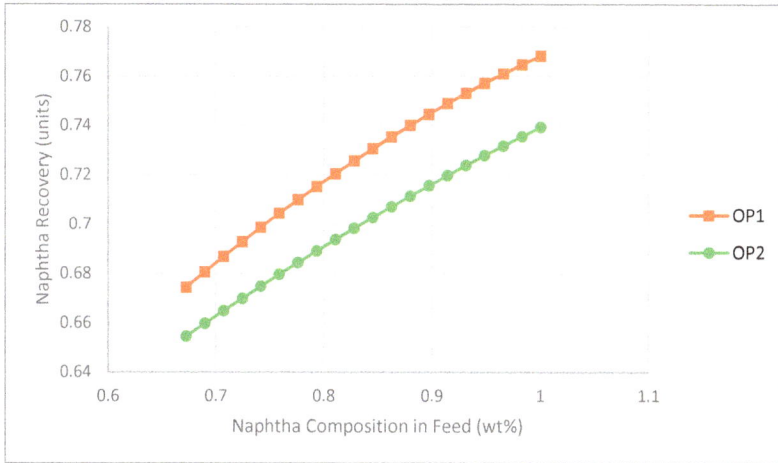

Figure 6. Impact of naphtha mass fraction in the feed on naphtha recovery (NR) as predicted by statistical models ($SF = 0.562$).

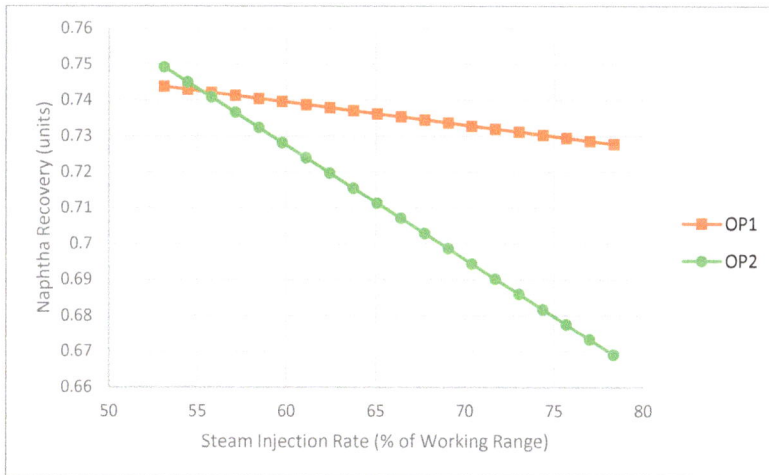

Figure 7. Impact of steam injection rate on naphtha recovery (NR) as predicted by statistical models ($NF = 0.863$ wt.%, $F_{feed} = 4247$ bbl/h).

The negative slope in Figure 7 associated with the SF to NR relationship is also counterintuitive based on mass and energy balance considerations and is contrary to the general belief of the operators. However, it is important to point out that the NRU does not consist of only a vacuum stripping column but also includes an overhead heat exchanger system and two separators connected in series with material recycled from both separators flowing back to the column. In addition, these modelling results seem to reinforce the suspicions of the process engineers that more steam does not necessarily improve recovery because of the interaction between the column and the overhead heat exchange system. Altogether, these interesting findings are what encouraged us to dig more deeply into the problem and turn our attention away from a data-based modelling approach towards a first-principles modelling approach, as described in the following section.

Before moving to the next section, we would like to close this section by illustrating one possible application of these linear regression models. A measurement of the mass flowrate of naphtha in the tailings is given by Equations (2) and (3) written for the tailings:

$$
\begin{aligned}
\textit{measured mass } \quad & \textit{flowrate of naphtha in tailings} \\
&= \textit{total mass flowrate of tailings} \\
&\times \textit{mass fraction of naphtha in tailings}
\end{aligned}
\tag{6}
$$

$$
\textit{total mass flowrate of tailings} = v_{tailings} \times [100 / \sum_i (\frac{\textit{wt.\% of component i in tailings}}{\textit{specific gravity of component i}})]
\tag{7}
$$

Note that this measurement of naphtha in the tailings requires a measurement of the volumetric flowrate of the tailings, $v_{tailings}$, and composition analysis of the tailings (*wt.% of each component i in the tailings*).

$$
\begin{aligned}
\textit{predicted mass } \quad & \textit{flowrate of naphtha in tailings} \\
&= \textit{total mass flowrate of naphtha in feed} \times (1 - NR)
\end{aligned}
\tag{8}
$$

Figures 8 and 9 show comparisons of the measured and predicted mass flowrates of naphtha in the tailings for OP1 and OP2, respectively. In these plots, the training and testing data have been included to illustrate the overall fit obtained using these regression models to generate a prediction for the mass flowrate of naphtha in the tailings. The correlation between the measured mass flowrate and the predicted value is approximately 0.8 for both OP1 and OP2.

This represents a simple soft-sensor application for these linear regression models in that they could be used to predict naphtha in the tailings based solely on the measured mass flowrate and composition analysis of the feed and a predicted value for naphtha recovery without requiring measurements of the flowrate and composition of the tailings. This application highlights the need for multiple models and the ability to detect when a system has shifted from one operating mode to another. PCA could be used in an on-line manner for this purpose [7] as could a Bayesian approach [8].

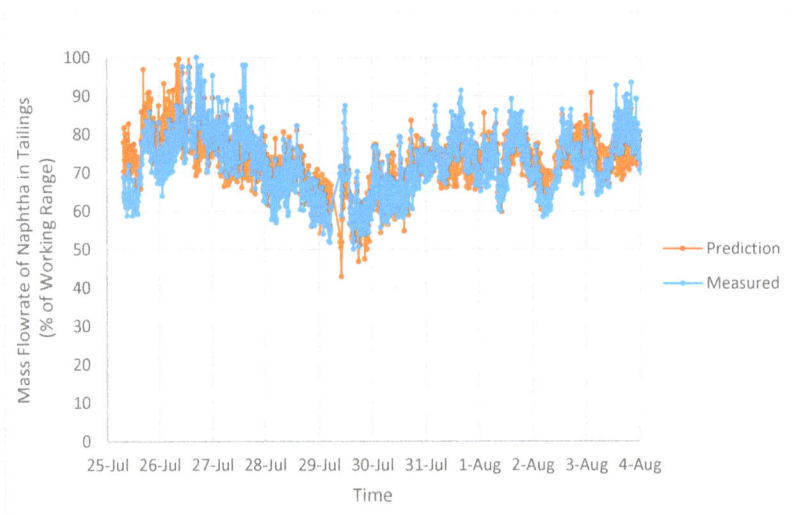

Figure 8. Measured and predicted mass flowrates of naphtha in the tailings for OP1.

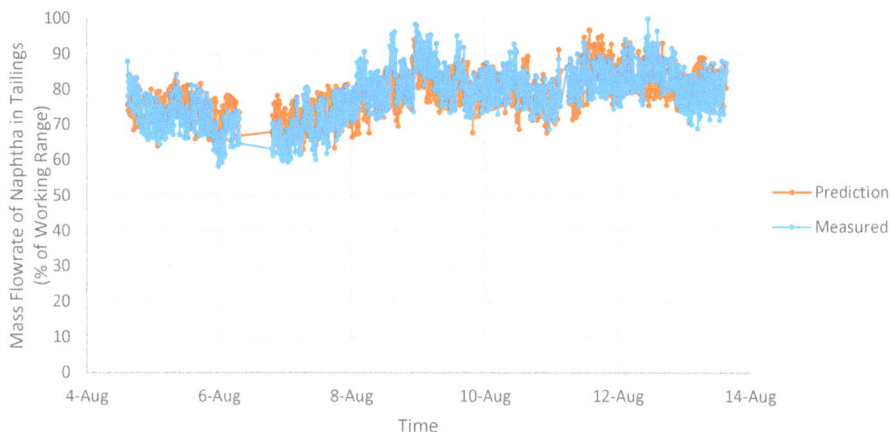

Figure 9. Measured and predicted mass flowrates of naphtha in the tailings for OP2.

4. First-Principles Modelling of the NRU in Aspen Plus® (V8.6)

Aspen Plus is a chemical process simulation and optimization software package widely used in the petroleum industry. It was used here to model the actual plant operation of the NRU system at steady-state. The composition of naphtha taken from laboratory analysis was used to define the naphtha stream in the simulation model. The analysis revealed approximately 475 components in naphtha, among which 353 components had a mass fraction less than 0.1%. Only components more than 1% were retained in the list. The rest of the components with mass fractions between 0.1% and 1% were grouped based on their chemical composition. Within each group, the component with the highest mass fraction was chosen to represent all the components in the group, and the mass fraction of this particular component was replaced with the sum of the mass fractions of all components in the corresponding group. Additionally, attention was also given to components with the same chemical composition but different properties due to their structural differences, i.e., aromatic structure verse chain structure. Under such circumstances, subgroups were created based on the similarity of their chemical properties. After grouping was completed, a new list with 25 components was created and scaled so that the total added up to 100% (Table 5).

Table 5. Composition of naphtha stream used in the Aspen Plus model.

Component	Mass Fraction (Total = 100)
n-Pentane	0.25
Benzene	0.27
Methylcyclopentane	5.25
n-Hexane	9.23
Toluene	2.38
Methylcyclohexane	9.32
n-Heptane	9.33
2-Methylhexane	8.43
m-Xylene	3.21
Ethylbenzene	0.79
1-Methyl-1-ethycyclopentane	5.90
n-Octane	6.42

Table 5. *Cont.*

Component	Mass Fraction (Total = 100)
2-Methylheptane	8.60
2,3-Dihydroindene	0.50
1-Ethyl-3-methylbenzene	3.66
tert-Butylcyclopentane	1.83
n-Nonane	4.33
2,2,5-Trimethylhexane	11.71
n-Decane	4.13
C10-Naphthene	1.18
1,3-Dimethyl-4-ethylbenzene	0.46
Isobutylcyclohexane	0.30
3,6-Dimethyloctane	1.82
n-Undecane	0.70

Other major components, namely solids and bitumen, were defined as follows: solids were assumed to be 50% SiO_2 and 50% Al_2O_3 (on a mass basis) and bitumen was assumed to be $C_{40}H_{80}$: 1-Tetracontene.

NRTL (non-random-two-liquid) was chosen as the physical method in Aspen Plus because it has a large number of built-in binary parameters. This method is also excellent for handling highly non-ideal chemical mixtures as well as vapor-liquid equilibrium and liquid-liquid equilibrium calculations.

4.1. Conceptual Design of the NRU Column

The RadFrac unit in Aspen Plus was the first of two designs considered for the NRU column (Figure 10). The RadFrac unit was specified to have two equilibrium stages but no reboiler or overhead drum. It was set up to operate at a vacuum pressure of 2.7 psi at the top stage of the column with a 0.3 psi pressure drop across the column. The top equilibrium stage (stage 1) represents the separation occurring in the flashing zone. The bottom equilibrium stage (stage 2) represents the separation taking place in the liquid pool. The feed to the column is set at 948 t/h and is made up of naphtha (DIL), bitumen (BIT), water (H2O) and solids (SOLID) coming in as four separate streams and then combined to create a single feed stream (FEED). The weight percentage of each stream in terms of the total feed is 1.03%, 1.87%, 85.1% and 12%, respectively. This feed stream enters the NRU column at 163.5 °F and atmospheric pressure (14.7 psi). The demister water and upgrading water streams are combined (UPGRADIN) and enter the column at 211 °F and 14.7 psi and a flowrate or 36 t/h. Finally, 11.8 t/h of steam enters the column at 291.5 °F and 43 psi. The two liquid streams, FEED and UPGRADIN, enter the column at the top stage (stage 1) and STEAM enters at the second stage (stage 2) to create the liquid–vapor contact that is necessary for the naphtha separation. On the output side, the vapor (OVERHEAD) is sent to the overhead system whereas the remaining liquid (TAILINGS) is discharged through the NRU tailings.

The second design considered for the NRU column is shown in Figure 11 and consists of two interconnected Flash2 units. Each Flash2 unit has one equilibrium stage. The first Flash2 unit (FLASHING) is where the feed (FEED) and the upgrading water (UPGRADIN) enter. It is used to represent the flashing zone in which the vapor–liquid equilibrium is established at 2.7 psi and with a vapor fraction of 0.95. The vapor (OVERHEAD) generated from the first Flash2 unit is sent to the overhead system, whereas the liquid bottom (INTERLIQ) becomes the feed to the second Flash2 unit (LIQPOOL). LIQPOOL, with pressure and vapor fraction 3 psi and 0.15, respectively, is used to represent the liquid pool where the steam is directly injected. The overhead vapor (INTERVAP) from LIQPOOL is sent to FLASHING while the liquid product (TAILINGS) leaves as the NRU tailings. The specification of column temperature is avoided in both Flash2 units because the temperature is subject to the amount of heat provided by the steam. The conditions of all input streams are the same as those used in the RadFrac design.

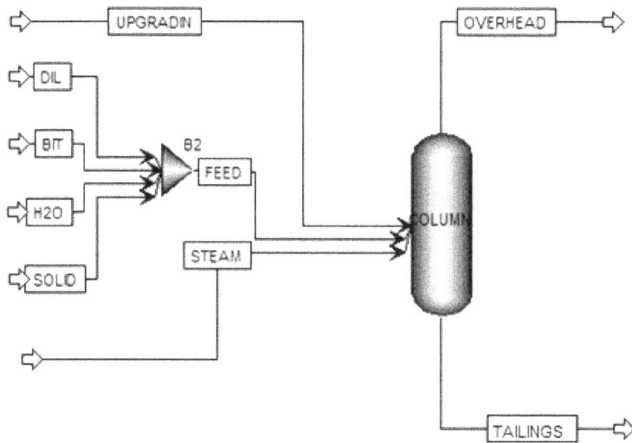

Figure 10. Configuration of the RadFrac model of the NRU column.

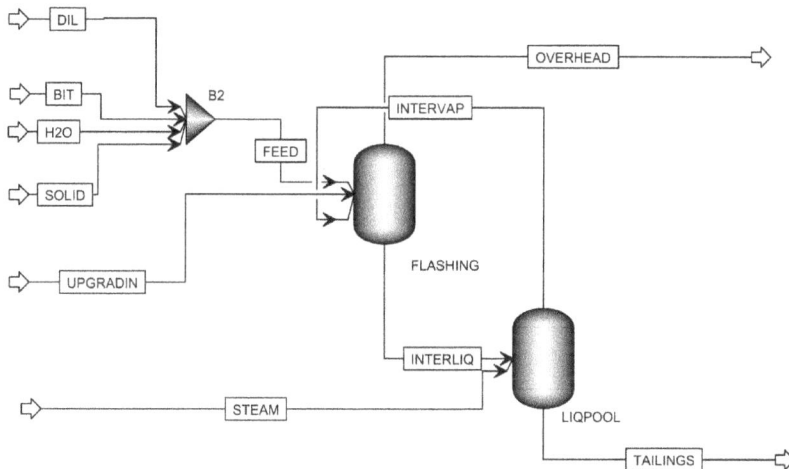

Figure 11. Configuration of the interconnected Flash2 model of the NRU column.

These two conceptual designs for the NRU column were evaluated based on how well they estimated the column temperature and the steam impact on the naphtha recovery as compared to the process data.

With the same specifications for the input streams, the simulations of both designs overestimated the column temperature by 5 °F as compared to what was recorded in the plant (135 °F). This difference between the predicted and actual temperatures can be explained by the mal-distribution of steam in the actual column. The poor distribution of steam limits the liquid–vapour contact, which further limits the heat transfer. Thus, the NRU column may very well experience lower temperatures.

With the steam injection rate set at a mid-range value (referred to now as the nominal value), it was found that both designs provide a similar estimate for naphtha recovery consistent with the actual process. Then, the steam injection rate was varied to see the effect on naphtha recovery in both designs. What was discovered is that with an increasing steam rate, an increasing naphtha recovery is predicted with the RadFrac design, whereas no change in naphtha recovery is predicted by the Flash2 design. Because the actual process does exhibit a change in the naphtha recovery when the steam injection rate is changed, the RadFrac design was chosen to simulate the NRU column in Aspen Plus.

4.2. Conceptual Design of the Overhead System

Moving now to the overhead system, we begin by examining what we will refer to as the base case design, consisting of one overhead condenser, two separators and one compressor, that mimics the actual process design (see Figure 12). The vapor (OVERHEAD) coming from the NRU column first enters the overhead condenser (HX1). HX1 partially condenses the vapor (CONDENSA) and then sends it to the first separator. The first separator is modeled in Aspen Plus by one Flash2 unit (SEP11) and one decanter (SEP12). SEP11 separates the vapor from the condensate and passes the condensate to the decanter (SEP12) where the organic liquid phase that is rich in naphtha is separated from the non-organic phase that is rich in water. The decanter is chosen because it allows separation efficiencies to be specified for each component. The second separator (SEP2), modeled by a Flash3 unit, is used to recover the remaining naphtha from the non-condensed overhead vapor after being pressurized in a compressor (COMP). The Flash3 model is selected for the second separator because the inputs (temperature and pressure) required to define this model are available from the process data. The water streams from both separators are recycled back to the NRU column whereas the recovered naphtha is sent to storage. Table 6 summarizes the model selection and specifications for the key units.

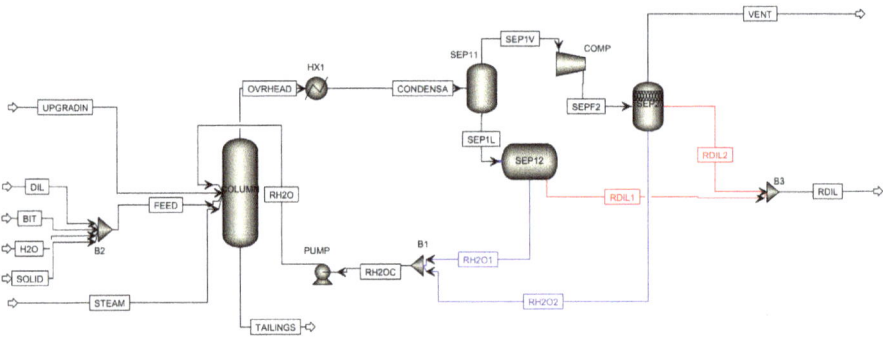

Figure 12. Configuration of the base case design for the overhead system.

Table 6. Aspen Plus unit specifications for both base case and "withdraw and remix" designs.

Industrial Unit	Symbol	Simulation Model	Specification		
			Variable	Value	Unit
NRU Column	COLUMN	RadFrac	Number of Stage	2	stage
			Pressure @ 1st stage	2.7	psi
			Pressure Drop	0.3	psi
Overhead Condenser	HX1	HEATER	Pressure	2.5	psi
			Duty	−75	MMBTU/h
Compressor	COMP	Compr(Isentropic)	Discharge Pressure	3.8	psi
Pump	PUMP	Pump	Discharge Pressure	2.7	psi
First Separator	SEP11	Flash 2	Pressure	2.5	psi
			Duty	0	MMBTU/h
	SEP12	Decanter	Pressure	2.5	psi
			Temperature	82	°F
Second Separator	SEP2	Flash 3	Temperature	100	°F
			Pressure	3.8	psi

The major limitation of the base case design is that the separator models available in Aspen Plus normally perform equilibrium calculations with the assumption of ideal conditions, e.g., the decanter model assumes sufficient residence time to establish phase equilibrium such that a complete separation between the two liquid phases can be achieved. Thus, the two product streams—recovered naphtha and recycled water streams—have extremely high purity (almost 99%). However, this does not match data collected from the actual process which indicates, under normal operation, that the produced naphtha contains at least 20 wt.% water and the recycled water stream contains approximately 3–7 wt.% naphtha. The operating data also indicates that a significant amount (10 wt.%) of the total naphtha enters the column via the recycled water stream.

To simulate this non-ideal separation that is occurring in both overhead separators, the separation efficiency of water in the decanter associated with the first separator was varied from 0 to 0.5. However, little change was observed in the composition and flowrate of the recycled water. Therefore, a novel design in Aspen Plus was needed to give a more accurate representation of the actual process.

At this point, an innovative "withdraw and remix" concept was proposed. The idea is to take a small portion from both the water and naphtha streams and mix them with the other stream. As illustrated in Figure 13, the naphtha and water streams coming from the same separator are each divided into two sub-streams on a 10/90 split. The sub-stream which contains 10 wt.% of the original naphtha flow is remixed with the sub-stream containing 90 wt.% of the original water flow to represent the formulated recycled water stream. The other two sub-streams, one with 90 wt.% of the original naphtha flow and one with 10 wt.% of the original water flow, are mixed to form the formulated recovered naphtha stream. This concept artificially adds impurity to both two product streams coming from the overhead system to mimic the non-ideal separation. The actual split ratio can be adjusted iteratively to match the composition of the formulated streams with the actual produced streams.

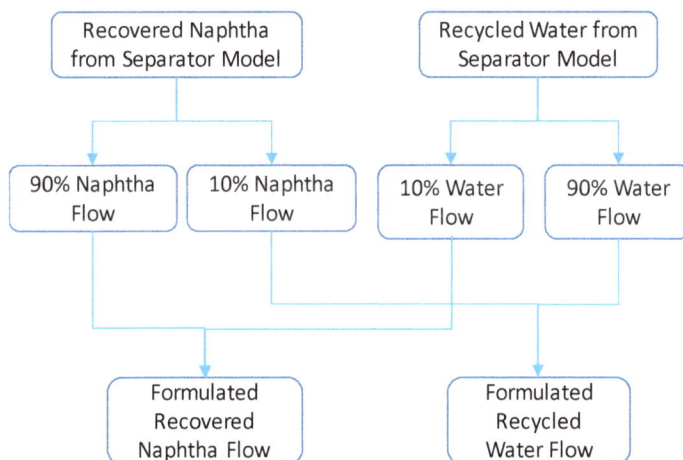

Figure 13. Illustration of "withdraw and remix" strategy to mimic the non-ideal NRU separators.

This new design is shown in Figure 14, where we have taken the base case design as our starting point and have then incorporated the "withdraw and remix" strategy. All Aspen Plus models used in this design have the same specifications summarized in Table 6. The Aspen files for both the base case design and the "withdraw and remix" design are provided in the Supplementary Materials.

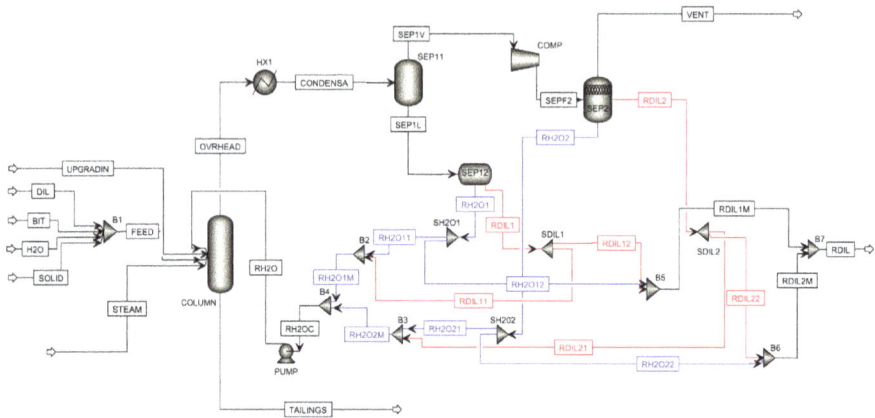

Figure 14. Configuration of the "withdraw and remix" design for the overhead system.

5. Implications for Process Optimization

In this section of the paper, we are going to examine the effects of cooling duty and steam feed rate or injection rate on naphtha recovery as predicted by our "withdraw and remix" Aspen Plus model of the NRU (Figure 14). These two independent variables have been chosen with a view to gaining some insight into process optimization. The cooling duty associated with the overhead condenser under nominal operating conditions is calculated as 75 MMBTU/h. For this sensitivity analysis, the cooling duty is varied from 65 to 75 MMBTU/h and the steam feed rate is varied from 68% to 80% of its working range.

Figure 15 shows the relationship between steam injection rate and naphtha recovery that is predicted by the model. With the condenser duty fixed at 75 MMBtu/h, the naphtha recovery initially goes down and then up as the steam rate increases. Over this range, there is a minimum value for naphtha recovery that occurs. Beyond a certain steam rate, the predicted recovery levels off.

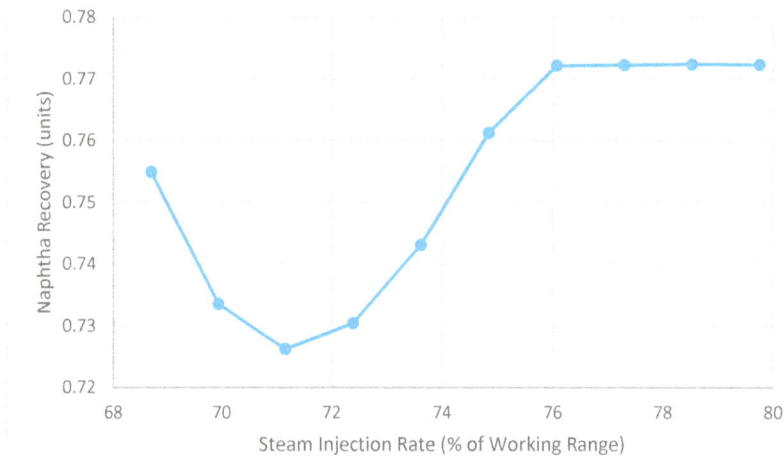

Figure 15. Impact of steam on naphtha recovery as predicted by Aspen Plus model.

Insight into the reason for the shape of this relationship in Figure 15 can be gleaned from Figure 16, which shows the predicted naphtha being recovered from the first separator and the second separator separately. One can see at the lower steam rates that most of the "work", in terms of recovering the naphtha, is being done by the first separator. As the steam rate increases, the load on the overhead condenser system increases, resulting in a higher percentage of uncondensed vapor at a higher temperature going over to the second separator. Thus, more "work", again in terms of recovering the naphtha, ends up being done by the second separator while less naphtha is recovered by the first separator. This trend continues with increasing steam rates. However, when taken together, the combined recovery of naphtha initially decreases, goes through a minimum, and then rises back up before leveling off.

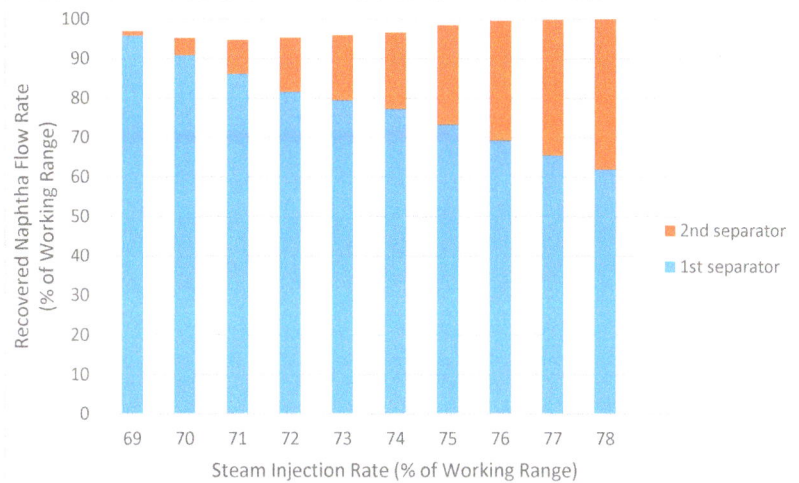

Figure 16. Impact of steam on recovered naphtha from separators 1 and 2 as predicted by Aspen Plus model.

Returning to Figure 15, the parabolic shape of this relationship between steam rate and naphtha recovery has interesting implications for the NRU system. It indicates that at lower steam rates, the relationship between steam rate and naphtha recovery has a negative slope, but at higher steam rates this relationship takes on a positive slope. The negative relationship is contrary to what operators believe but is consistent with the findings of the linear regression models summarized in Figure 7. On the other hand, the positive relationship is consistent with what the operators believe.

These findings also have interesting implications for process optimization. The parabolic shape in Figure 15 indicates that it might be possible to achieve the same naphtha recovery by operating the NRU system at two different steam rates, a higher value and a lower value. Therefore, to minimize cost, a lower steam rate could be chosen while still ensuring that the required naphtha recovery is achieved. This phenomenon is referred to as "input multiplicity" and has been observed in a heat-integrated double-column air separation unit [9] as well as both isothermal and adiabatic chemical reactors [10].

The flat section of the relationship in Figure 15 at high steam rates is another interesting result. Here, the model predicts that increasing the steam rate beyond a certain point has no effect on the naphtha recovery and therefore represents a waste of energy. This result is consistent with the beliefs of the on-site process engineers.

The impact of the cooling duty of the overhead condenser on naphtha recovery has also been examined and the results as predicted by the model are shown in Figure 17. The cooling duty is set to four different values (65, 67.5, 70 and 75 MMBTU/h) and a series of four parabolic curves like the one found in Figure 15 shows the impact of steam on naphtha recovery at these different duties. As the condenser duty decreases, the curves shift left towards lower steam injection rates. This is because the cooling capacity limits the amount of steam that the overhead system can handle. The minimum naphtha recovery decreases slightly with lower cooling duty, but the change is negligible.

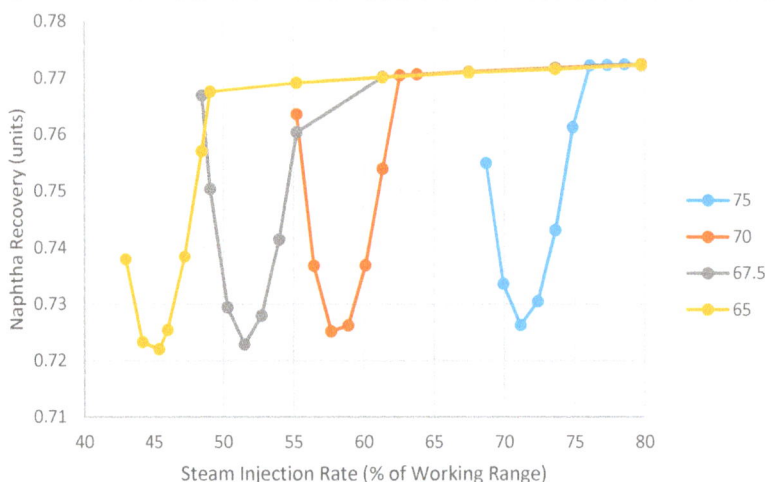

Figure 17. Impact of steam and cooling water flowrate on naphtha recovery as predicted by Aspen Plus model (4 different cooling water duties shown with units of MMBTU/h).

Figure 17 indicates that when more steam is injected into the NRU system, a higher cooling duty is required to achieve the same recovery. This is another finding that has interesting implications for the optimization of the NRU process. In terms of minimizing the cost of both steam production and cooling water, it appears that it might be possible to achieve the target naphtha recovery using a combined strategy of less steam and less cooling water. For example, if the target recovery was 0.75, a steam injection rate of approximately 47–48% of the working range and a cooling duty in the range 65–67.5 MMBTU/h would be optimal.

6. Conclusions

The goal in this paper was to develop steady-state models for a naphtha recovery unit (NRU) found in the oil sands industry. Two approaches have been examined with one being a statistical, data-based modelling approach and the other being a first-principles modelling approach. Some of the findings in this paper are consistent with the process understanding of operators and engineers at the plant. However, some of the results are counterintuitive and provide interesting suggestions and implications for optimization of the NRU system. A second performance test has been suggested to plant personnel to explore and validate the ideas presented here.

Our paper suggests that it may be possible to reduce steam consumption without compromising naphtha recovery. This is counterintuitive because the general belief is that a higher steam injection rate will translate into more naphtha being vaporized in the stripping column, which in turn should lead to a higher naphtha recovery. Instead, our first-principles modelling work predicts that increasing the steam injection rate beyond a certain threshold does not have a significant effect on naphtha recovery because the overhead condenser system cannot handle the increased volume of overhead vapour. Our results suggest that the overhead system plays an important role and that it is the interconnections between the NRU column and the overhead system that produce this counterintuitive behaviour. Hence, it is recommended that both parts of the NRU system be studied in an integrated way to further improve naphtha recovery.

Supplementary Materials: The Aspen files for the base case design in Figure 12 and the "withdraw and remix" design in Figure 14 are available online at http://www.mdpi.com/2227-9717/6/7/74/s1.

Author Contributions: Conceptualization, J.D. and W.R.C.; Methodology, J.D. and W.R.C.; Software, J.D.; Validation, J.D. and W.R.C.; Formal Analysis, J.D. and W.R.C.; Investigation, J.D.; Resources, J.D. and W.R.C.; Data Curation, J.D.; Writing-Original Draft Preparation, J.D.; Writing-Review & Editing, J.D. and W.R.C.; Visualization, J.D. and W.R.C.; Supervision, W.R.C.; Project Administration, J.D. and W.R.C.

Funding: This research received no external funding.

Conflicts of Interest: The authors declare no conflict of interest.

References

1. Speight, J.G. *Oil Sand Production Processes*; Elsevier: New York, NY, USA, 2013.
2. Khatibisepehr, S.; Huang, B.; Domlan, E.; Naghoosi, E.; Zhao, Y.; Miao, Y.; Shao, X.; Khare, S.; Keshavarz, M.; Feng, E.; et al. Soft sensor solutions for control of oil sands processes. *Can. J. Chem. Eng.* **2013**, *91*, 1416–1426. [CrossRef]
3. Long, Y.; Dabros, T.; Hamza, H. Analysis of solvent-diluted bitumen from oil sands froth treatment using NIR spectroscopy. *Can. J. Chem. Eng.* **2004**, *82*, 776–781. [CrossRef]
4. Feng, E.; Domlan, E.; Kadali, R. Spectroscopic measurements in oil sands industry—From laboratories to real-time applications. In Proceedings of the 9th International Symposium on Advanced Control of Chemical Processes, The International Federation of Automatic Control, Whistler, BC, Canada, 7–10 June 2015; pp. 199–204.
5. Wold, S.; Esbensen, K.; Geladi, P. Principle component analysis. *Chemom. Intell. Lab. Syst.* **1987**, *2*, 37–52. [CrossRef]
6. Kourti, T.; MacGregor, J.F. Process analysis, monitoring and diagnosis, using multivariate projection methods. *Chemom. Intell. Lab. Syst.* **1995**, *28*, 3–21. [CrossRef]
7. Zhao, S.J.; Zhang, J.; Xu, Y.M. Monitoring of processes with multiple operating modes through multiple principle component analysis models. *Ind. Eng. Chem. Res.* **2004**, *43*, 7025–7035. [CrossRef]
8. Khatibisepehr, S.; Huang, B.; Xu, F.; Espejo, A. A Bayesian approach to design of adaptive multi-model inferential sensors with application in the oil sand industry. *J. Process Control* **2012**, *22*, 1913–1929. [CrossRef]
9. Mahapatra, P.; Bequette, B.W. Design and control of an elevated-pressure air separations unit for IGCC power plants in a process simulator environment. *Ind. Eng. Chem. Res.* **2013**, *52*, 3178–3191. [CrossRef]
10. Sistu, P.B.; Bequette, B.W. Model predictive control of processes with input multiplicities. *Chem. Eng. Sci.* **1995**, *50*, 921–936. [CrossRef]

![processes logo] **processes**

MDPI

Article

A General Model for Estimating Emissions from Integrated Power Generation and Energy Storage. Case Study: Integration of Solar Photovoltaic Power and Wind Power with Batteries

Ian Miller [1,2], Emre Gençer [1] and Francis M. O'Sullivan [1,*]

[1]   MIT Energy Initiative, Massachusetts Institute of Technology, 77 Massachusetts Avenue, Cambridge, MA 02139, USA; igm@mit.edu (I.M.); egencer@mit.edu (E.G.)
[2]   Department of Chemical Engineering, Massachusetts Institute of Technology, 77 Massachusetts Avenue, Cambridge, MA 02139, USA
*   Correspondence: frankie@mit.edu; Tel.: +1-617-715-5433

Received: 17 October 2018; Accepted: 8 December 2018; Published: 18 December 2018

Abstract: The penetration of renewable power generation is increasing at an unprecedented pace. While the operating greenhouse gas (GHG) emissions of photovoltaic (PV) and wind power are negligible, their upstream emissions are not. The great challenge with the deployment of renewable power generators is their intermittent and variable nature. Current electric power systems balance these fluctuations primarily using natural gas fired power plants. Alternatively, these dynamics could be handled by the integration of energy storage technologies to store energy during renewable energy availability and discharge when needed. In this paper, we present a model for estimating emissions from integrated power generation and energy storage. The model applies to emissions of all pollutants, including greenhouse gases (GHGs), and to all storage technologies, including pumped hydroelectric and electrochemical storage. As a case study, the model is used to estimate the GHG emissions of electricity from systems that couple photovoltaic and wind generation with lithium-ion batteries (LBs) and vanadium redox flow batteries (VFBs). To facilitate the case study, we conducted a life cycle assessment (LCA) of photovoltaic (PV) power, as well as a synthesis of existing wind power LCAs. The PV LCA is also used to estimate the emissions impact of a common PV practice that has not been comprehensively analyzed by LCA—solar tracking. The case study of renewables and battery storage indicates that PV and wind power remain much less carbon intensive than fossil-based generation, even when coupled with large amounts of LBs or VFBs. Even the most carbon intensive renewable power analyzed still emits only ~25% of the GHGs of the least carbon intensive mainstream fossil power. Lastly, we find that the pathway to minimize the GHG emissions of power from a coupled system depends upon the generator. Given low-emission generation (<50 gCO$_2$e/kWh), the minimizing pathway is the storage technology with lowest production emissions (VFBs over LBs for our case study). Given high-emission generation (>200 gCO$_2$e/kWh), the minimizing pathway is the storage technology with highest round-trip efficiency (LBs over VFBs).

Keywords: solar PV; wind power; life cycle analysis; energy storage

---

1. Introduction

1.1. Background

The electric power sector produces 40% of global greenhouse gas (GHG) emissions [1]. To limit human-caused global warming to 2 °C, the International Energy Agency (IEA) projects that power sector emissions must decline by 90% by 2050, from 13 to 1.4 gigatons of carbon dioxide per year [1].

To help achieve this decline, many governments have increased their support for low-carbon power sources. Because of this support and large cost reductions [2], photovoltaic (PV) and wind power have grown from ~1% of global electricity in 2007, to ~6% in 2017 [3], and are projected to reach 15–33% by 2040 [4,5]. The most widely deployed PV cell types are multi-crystalline silicon (mc-Si), single-crystalline silicon (sc-Si), and cadmium telluride (CdTe), with market shares of 70%, 24%, and 4%, respectively, of new PV capacity installed in 2016 [6]. The most widely deployed wind turbines are 1.6–3 MW models installed at onshore wind farms, with generation from larger offshore turbines comprising a small but growing share of the total (5% in 2017) [7,8].

While the operating GHG emissions of PV and wind power are negligible, their upstream emissions are not. To be concise, this paper refers to the total life cycle GHG emissions of AC electricity as "carbon intensity", and the units of grams-CO_2-equivalent/kilowatthour as "gCO_2e/kWh". The carbon intensity of PV power has been estimated at 76, 53, and 27 gCO_2e/kWh for sc-Si, mc-Si, and CdTe PV, respectively, installed in northern Europe circa 2015, and at 33, 22, and 13 gCO_2e/kWh for PV installed in the United States Southwest circa 2015 [9]. The carbon intensity of wind power has been estimated at approximately 12, 10, and 9 gCO_2e/kWh, respectively, for onshore wind farms in low, medium, and high wind-speed areas, respectively, and at 14 gCO_2e/kWh for offshore wind farms (which are typically located in high wind-speed areas) [10–18]. For context, representative GHG emissions from combined cycle natural gas (CCNG) and supercritical pulverized coal power (SCPC) are approximately 488 and 965 gCO_2e/kWh, respectively [19].

The leading method for estimating carbon intensity is the life cycle assessment (LCA). As outlined by ISO standards 14040 and 14044 [ISO], LCA quantifies a product's environmental impacts through the input–output accounting of cradle-to-grave or cradle-to-gate processes. Since 2007, over 50 studies have conducted LCA of PV power, and have found that the top drivers of PV carbon intensity are upstream electricity source, cell type, module efficiency, irradiance at installation site, and system lifetime [20,21]. Peng et al. [20] provided a detailed literature review of PV LCA and underlined gaps in the field, including the treatment of cell temperature and module orientation. IEA reports from 2008, 2011, and 2016 [22] provided methodology guidelines for PV LCA.

Important differences between this work and prior LCAs include the following: (1) This work addresses a major development in PV that prior LCAs have not addressed in depth or with explanatory equations (solar tracking); and (2) this work provides a model and equations that show how to estimate emissions from coupled generation and storage, whereas prior battery LCAs have focused on modeling emissions from battery production.

1.2. Motivation for PV LCA

One major development in PV power production, the growth of solar tracking, has not been comprehensively analyzed by LCA. In the United States from 2008 to 2014, only 19% of new utility-scale (AC capacity >5 MW) CdTe projects had tracking (16 of 86 projects); in 2015 and 2016, the number was 56% (44 of 79 projects), including locations outside the exceptionally sunny United States Southwest, such as Colorado, Tennessee, and Georgia [8]. For all cell types, tracking was used on 53% of cumulative panel capacity and 70% of new capacity at utility-scale sites in the United States in 2016 [8]. Several LCAs have analyzed the emissions impact of solar tracking, but with a limited geographic scope and with tracking set-ups that are not (and do not claim to be) representative of industry practice [23–25]. Bayod-Rújula [23] analyzed a two-axis tracking system in Spain. Beylot et al. [25] and Desideri et al. [24] analyzed hypothetical one-axis tracking systems with a 30° tilt in Italy. In contrast, the industry norm for PV tracking is horizontal one-axis tracking; in the United States in 2016, 97% of utility-scale tracking PV projects used horizontal one-axis tracking (255 of 263 projects) [26].

Two PV LCAs did analyze industry-representative tracking set-ups. Leccisi et al. [9] found that horizontal one-axis tracking reduced carbon intensity by 11% and 1% for mc-Si and CdTe PV, respectively, given installation in the United States Southwest. Sinha et al. [27] estimated that tracking reduced the carbon intensity of CdTe PV by 3% in the United States Southwest. However, neither study calculated tracking's impact outside the United States Southwest. This paper aims to build on these studies by calculating tracking's impact on PV carbon intensity over a range of locations.

1.3. Motivation for Modeling Emissions from Coupled Generation and Energy Storage

A barrier to the continued growth of both PV and wind power, and one reason that projections span a large range, is intermittency and uncertainties over how intermittency will be managed as renewables grow. As PV rises above 20% of daily generation in a region, this can lead to curtailment of existing PV, less favorable economics for future PV, and inefficient "peaker" use of other power systems, including gas-fired turbines [28,29]. "Peaker" power plants are only used to meet peak demand and are otherwise are idle. Similar dynamics apply to wind at a high penetration [30,31].

Energy storage can contribute to solving the challenges of intermittency. Storage coupled with generation has the potential to (1) flatten power supplied by intermittents to the grid, and (2) reduce peak generation and thus capacity required of non-intermittents. Partly because of this potential, the growth of non-hydro storage capacity in the United States has been accelerating, from ~1000 MWh added between 2013 and 2017, to a projected ~1000 MWh in 2018, and more in each year through 2023 [32]. Lithium-ion batteries (LBs) and vanadium redox flow batteries (VFBs) are two of the most common non-hydro storage types. LBs constitute a large majority of current and projected non-hydro storage deployment [32]. VFBs constitute a significant share of utility-scale storage installations with durations over 3 h [33]. Storage duration is the ratio of rated storage energy capacity to rated storage power capacity, and represents the time required to discharge the fully charged storage unit at its rated power output. This paper focuses on storage applications that require durations of 3 to 7 h, such as smoothing intermittent generation to better match demand.

While coupling storage with generation can mitigate intermittency, the impact on the GHG emissions of electricity supply is less well understood. Spanos et al. [34] conducted cradle-to-gate LCAs of lead–acid, nickel–zinc, and manganese dioxide–zince batteries, and thus estimated their GHG emissions per battery mass produced. The authors did not analyze the operating life or coupling of the batteries with generation. Rydh and Sandén [35] evaluated the energy required to produce eight common battery types, including LBs, VFBs, lead–acid batteries (LABs), and sodium sulfur batteries (SSBs), and found that, on a mass basis, LABs require significantly more energy to be produced than LBs or VRBs. Denholm et al. [36] analyzed the coupling of generation with VRBs, pumped hydro storage (PHS), and compressed air energy storage (CAES). The authors found that electricity from nuclear or renewables coupled with PHS had a lower carbon intensity than electricity from nuclear or renewables coupled with VRBs or CAES. On the other hand, with fossil generation, systems employing CAES had significantly lower GHG emissions than those with VRBs or PHS. Denholm et al. and Mitavachan et al. [37] both emphasized that the carbon intensity of electricity from a coupled system depended on both generation type and energy storage type.

These prior battery LCAs have comprehensively modeled production of the most common battery types for power plant applications. This paper complements these prior works by presenting an easy to implement model to estimate emissions from a combined generation and storage system. The model is general and applies to emissions of all pollutants, not just GHGs, and to all storage methods, not only electrochemical. As a case study, we used the model to estimate the impact of adding LBs and VFBs to typical mainstream PV, wind, gas, and coal-based generation.

2. Methodology

2.1. Model of Combined Generation and Storage

2.1.1. Model of Electricity Supply to Grid

Figure 1 shows an example of a generation curve (yellow) and a supply curve (blue). The generation curve shows how much electric power the generator is producing, while the supply curve shows how much power is supplied to the grid. When supply exceeds generation, the difference comes from storage, and when generation exceeds supply, the difference goes to storage. In Figure 1, typical variable PV generation is converted to constant supply. This is a simple example to illustrate the model. Infinite shapes are possible for the two curves, as long as the area under the supply curve does not exceed the area under the generation curve. The area under the supply curve will always be less than the area under the generation curve, because of round-trip efficiency losses in storage.

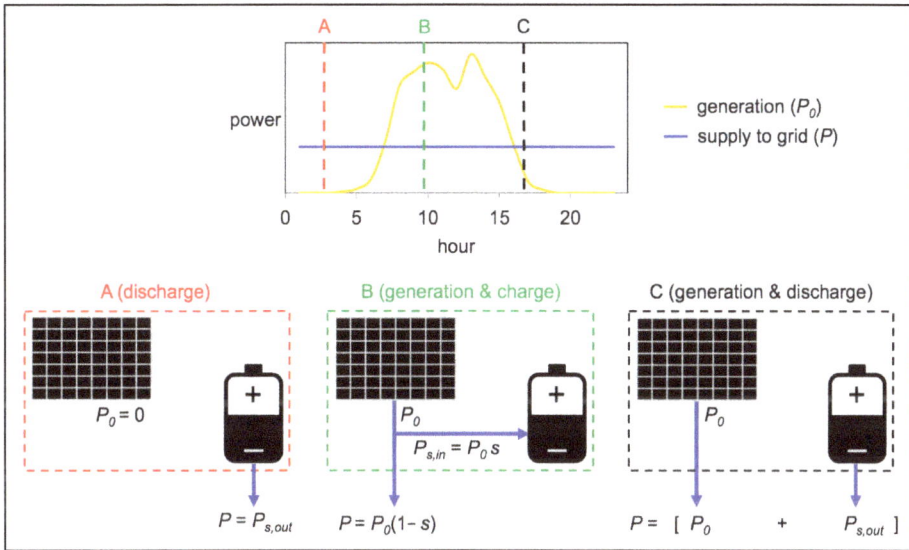

Figure 1. Three scenarios corresponding to three times of day for a combined photovoltaic (PV) and storage system. P_0 is power generation (kW), P is power supply to grid (kW), $P_{s,in}$ is the amount of power generation sent to storage (kW), s is the instantaneous fraction of power generation going to storage, and $P_{s,out}$ is the power supply to grid from storage.

Averaging the power flows in Figure 1 produces Figure 2, which shows how the average power to grid can be calculated, as follows:

$$\overline{P} = \overline{P}_0[1 - f(1 - \eta)]$$

where η is the round-trip efficiency of the storage, and f is the fraction of the energy generation sent to storage (vs. directly to grid). f is the generation-weighted average of s ($<sP_0>/<P_0>$), not the time-average ($<s>$). The lifetime electricity to grid can then be calculated as follows:

$$E = t\overline{P} = t\overline{P}_0[1 - f(1 - \eta)] = E_0[1 - f(1 - \eta)] \tag{1}$$

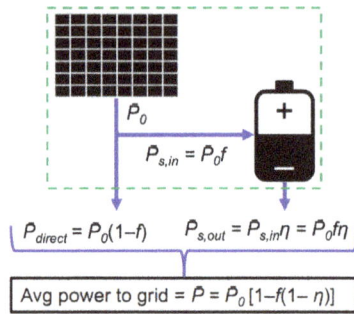

Figure 2. Average power flows of combined generation and storage. Over-bars indicate time averages, η is the round-trip efficiency of the storage, and f is the fraction of energy generation sent to storage (vs. directly to grid).

f depends on the generation and supply curves. Figure 3 illustrates this dependence, as well as how f changes with the degree of peak-smoothing.

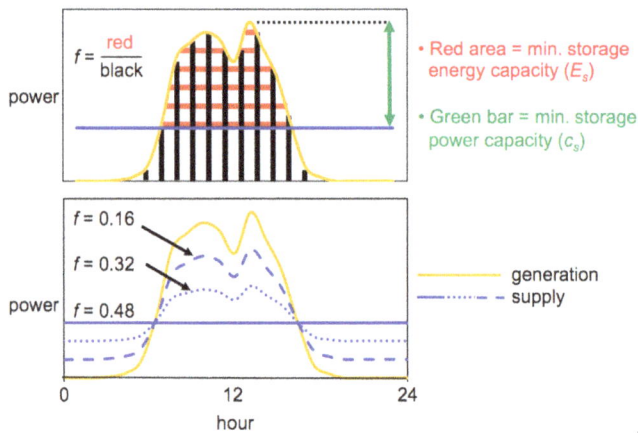

Figure 3. Dependence of f (fraction of generation sent to storage) on power generation and supply curves. Assumes no curtailment.

Figure 3 shows the minimum storage energy capacity and power capacity needed to convert the given generation to the desired supply. This paper does not attempt to prescribe the optimum amounts of storage energy capacity, storage power capacity, or generation curtailment for a given desired f. For example, developers of a combined PV and storage plant that aims for constant supply must decide whether to install enough storage energy capacity to maintain a supply to the grid for three or four or more consecutive cloudy days. These decisions involve risk-analyses and design choices that are beyond the scope of this paper. This paper provides tools to estimate how those different design choices impact emissions.

2.1.2. Model Equations and Parameters

Without storage, the emissions intensity I (g/kWh) of the power generated and delivered to the grid can be calculated as follows:

$$I_0 = e_g / E_0$$

where e_g is the life emissions of the generator (g), and E_0 is the life electricity generated (kWh). The subscripts 0 in I_0 and E_0 indicate the no-storage base case. In this paper, the generator is assumed to operate nearly identically in the no-storage and storage-added cases, with the only difference being that, in the latter case, some of the generation is sent through storage on its way to the grid. In reality, the addition of storage to a power plant might change the way its generator is operated. For example, adding storage to a wind or gas power plant might enable reduced curtailment or increased fuel-to-power efficiency, respectively.

With storage in general and battery storage in particular, the emissions intensity of the power delivered to the grid can be calculated as follows, respectively:

$$I = [I_0 + f\eta(e_s/E_{sT})/(\overline{D}C)]/[1 - f(1 - \eta)] \tag{2}$$

and

$$I = [I_0 + f\eta(e_s/m_s)/(\rho\overline{D}C)]/[1 - f(1 - \eta)] \tag{3}$$

where f is the fraction of electricity generated that is stored (vs. sent directly to the grid), η is the roundtrip efficiency of the storage, e_s is the life emissions of the storage (g), $\overline{D}$ is the average depth of discharge of the storage (DoD), E_{sT} is the total rated energy capacity of the storage used over the power plant life (kWh), C is the cycle life of the storage (the number of charge–discharge cycles before a storage unit is retired), m_s is the mass of batteries used over the power plant life (kg), and ρ is the specific energy capacity of the batteries (kWh/kg). E_{sT} is the total rated energy capacity of all of the storage used and retired over the power plant life. For example, if the power plant storage system is rated at 1 MWh for energy capacity (E_s = 1 MWh), and four storage units are used and retired over the plant lifetime, then E_{sT} is 4 MWh.

The parameters in Equations (2) and (3) depend on both the design and technology, as shown in Table 1. The values and ranges used in our case study for the combined generation and battery storage come from Mitavachan [37] and include the following: LB average roundtrip efficiency of 90% (low 85% and high 98%), VFB average roundtrip efficiency of 75% (low 60% and high 80%), LB average cycle life of 10,250 cycles (low 5000 and high 15,000), VFB average cycle life of 13,000 cycles (low 10,000 and high 15,000), LB average DoD of 80%, VFB average DoD of 90%; LB emissions per mass of 22 gCO$_2$e/g, VFB emissions per mass of 2.7 gCO$_2$e/kWh, LB specific energy capacity of 140 Wh/kg, and VFB specific energy capacity of 20 Wh/kg. These ranges for efficiency and cycle life reflect uncertainties and variety in battery technologies. The purpose of this paper, with regard to storage, is to provide an exploitable model for modeling emissions from combined generation and storage, and to demonstrate the use of that model, given realistic literature-supported parameter ranges. Narrowing these ranges further would improve the model and its specificity, potentially via the incorporation of correlations between cycle life and DoD, temperature, and charge–discharge rate. Such correlations can be found in the literature [38,39]. This is one possible focus of future work. The battery emissions per mass are based on previous detailed LCAs of LB and VFB battery production, which are described in the Introduction. Battery production can be elaborated as shown in Figure 4.

Equations (2) and (3) apply not only to emissions (g) and emissions intensity (g/kWh), but also to any quantitative characteristic of both generation and storage. For example, the emissions values of I_0 and e_s can be replaced with cost values to estimate the "cost intensity" (i.e., the cost of electricity delivered to the grid from the combined generation and storage system ($/kWh)).

Table 1. Parameters in model equations and examples of dependencies. No example does not mean no dependence. GHG—greenhouse gas; LBs—lithium-ion batteries; LABs—lead–acid batteries; VFBs—vanadium redox flow batteries.

Symbol	Description	Units	Example of Dependence on Technology	Example of Dependence on Design
I_0	emissions intensity of electricity w/o storage	g/kWh	Coal gen. has higher GHG emissions than wind generation	
f	fraction of electricity generation that is stored			Greater smoothing of generation peaks requires greater f (see Figure 3)
η	round-trip efficiency of storage		LBs usually have higher η than VFBs	
e_s / E_{sT}	emissions per unit of rated storage energy capacity produced	g/kWh	Compared to LABs, LBs generally have higher GHG emissions per rated energy	Two VFBs w/same rated energy and different rated power will have different emissions
e_s / m_s	emissions per unit of storage mass produced	g/kg		
ρ	specific energy capacity	kWh/kg		Two VFBs w/same rated energy but different power will have different ρ
$\overline{D}$	average depth of discharge (DoD)		LBs are usually run at higher DoD than LABs	More "excess" storage energy capacity means lower D for same f
C	cycle life	cycles	LBs have higher C than LABs	

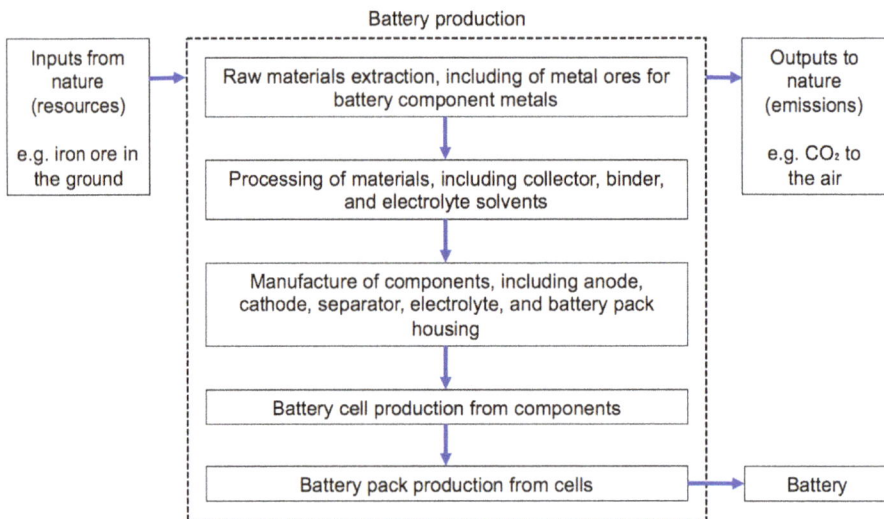

Figure 4. Life cycle stages of battery production.

2.1.3. Model Equations Derivations

The total number of storage charge–discharge cycles over a power plant's life can be calculated as follows:

$$\#_{cyc} = E_{s,out}/E_{1cyc}$$

where $E_{s,out}$ is the total energy discharged from storage over the power plant's life (kWh), and E_{1cyc} is the average energy discharged per cycle (kWh). This equation can be expanded as follows:

$$\#_{cyc} = \eta E_{s,in}/(\overline{D}E_s)$$

$$\#_{cyc} = \eta f E_0/(\overline{D}E_s) \tag{4}$$

where $E_{s,in}$ is the total energy sent to storage over the power plant's life (kWh), E_0 is the life electricity produced by the generator before storage (kWh), $\overline{D}$ is the average depth of discharge (DoD), and E_s is the rated energy capacity of the storage (kWh).

Let a "unit" refer to the rated storage system in place and operating at any given time. The number of units that "die" over the power plant's life can then be calculated as follows, if the storage unit reaches its cycle life before its calendar life:

$$\#_{deaths} = \#_{cyc}/C \;\rightarrow\; \text{plug in (4)}$$

$$\#_{deaths} = f\eta E_0/(\overline{D}CE_s) \tag{5}$$

where C is the cycle life and the other variables are defined above. The total storage energy capacity used and retired over the power plant's life can then be calculated as follows:

$$E_{sT} = \#_{deaths}E_s \;\rightarrow\; \text{plug in (5)}$$

$$E_{sT} = f\eta E_0/(\overline{D}C) \tag{6}$$

The life emissions of the storage can then be expanded as follows:

$$e_s = e_s/E_{sT}\; E_{sT} \;\rightarrow\; \text{plug in Equation (6)}$$

$$e_s = e_s/E_{sT}\; f\eta E_0/(\overline{D}C) \tag{7}$$

$$e_s = e_s/m_s\; m_s/E_{sT}\; f\eta E_0/(\overline{D}C) \;\rightarrow\; \text{if batteries, definition of specific energy capacity}$$

$$e_s = e_s/m_s\; 1/\rho\; f\eta E_0/(\overline{D}C) \tag{8}$$

Finally, Equation (7) or (8) can be plugged in to calculate the emissions intensity (g/kWh) of power delivered to the grid from a combined generation and storage system

$$I = (e_g + e_s)/E \;\rightarrow\; \text{account for losses from storage}$$

$$= (e_g + e_s)/\{E_0[1 - f(1 - \eta)]\}$$

$$= (I_0 + e_s/E_0)/[1 - f(1 - \eta)] \;\rightarrow\; \text{plug in (7) or (8)}$$

by plugging in (7), Equation (2) is derived and by plugging in (8) Equation (3) is derived.

2.1.4. Model Assumptions and Approximations Include the Following:

(1) Storage efficiency is treated as constant.
(2) The number of storage units retired over the system's lifetime is treated as continuous in the model (see Equation (5)).

(3) In Table 2, the battery production emissions (e_s) do not include emissions from producing the battery inverter and control system, because of a lack of data. An alternative is estimating battery inverter emissions by approximating that the battery inverters are equivalent to PV inverters, and then using the PV inventories discussed in Section 2.2.1. [40,41]. These inventories report inverter production GHGs of ~27.4 gCO_2e per watt of AC inverter capacity, and inverter lifetime of 15 years. For the cases analyzed in this paper, this would add approximately 1.5 gCO_2e/kWh to the total carbon intensity, and increase the storage production emissions by ~15% for LBs and ~30% for VFBs. Unlike the battery inverter case, we do not identify a reasonable approximation for the emissions from producing the battery control system.

(4) Charging from the grid is not analyzed. Such an analysis would include the expansion of the generator definition to encompass all generators connected to a grid.

(5) Cycle life is considered as controlling, not calendar life. Our analysis is focused on storage applications that require regular cycling at high DoD, and thus cause cycling-induced degradation to be the primary degradation mode. The storage unit is assumed to reach its cycle life before its calendar life. For applications that do not involve a high-use of battery capacity, such as frequency regulation, this might not be a valid approximation [37].

(6) Temperature-induced degradation is not considered. More information on battery degradation modes can be found in the literature [38,39].

(7) DC–DC coupling is possible for PV and battery storage, but is not analyzed here. It is unclear whether DC–DC coupling of PV and batteries actually increases the round-trip efficiency, given the particular equipment and system architecture required. For more discussions, see the literature [42].

(8) Possible storage impacts on curtailment and generator efficiency are not analyzed. For example, storage can be used to store off-peak wind generation that would otherwise be curtailed, for later supply to the grid. Thus, in some cases, storage might actually increase the generator capacity factor F by offsetting storage efficiency losses with curtailment reductions. As another example, storage can be used to increase the fuel-to-power efficiency of the simple cycle gas turbines (SCGT). A preliminary analysis suggests that, for an SCGT with an original generation profile typical of California peaker plants, adding enough storage to allow for flat generation might increase generation-averaged fuel-to-power efficiency by up to ~3%, thus partly offsetting efficiency losses in storage.

Table 2. Carbon intensities of wind generation, from wind life cycle assessment (LCA) synthesis [42–51]. Assumptions include onshore turbine size ~2 MW, offshore turbine size ~5 MW, onshore wind farm 20 km from grid with 3% transmission losses (transmission just to grid), offshore wind farm 70 km from grid with 8% transmission losses, wind farm size >50 MW, hub height 95 m, wake losses 6%, availability 97%, and offshore foundation monopile.

	Carbon Intensity (gCO_2e/kWh)		
	Avg	Low	High
Onshore high wind speed (~9 m/s)	8.9	5.5	12
Onshore med wind speed (~8 m/s)	9.7	6.4	12
Onshore low wind speed (~7 m/s)	12	7.3	15.8
Onshore high wind speed (~9 m/s)	14	10.8	18.2

2.2. Life Cycle Assessments of Generation

As shown in Equations (2) and (3), calculating the emissions intensity requires knowing both the emissions from producing a unit mass of storage, and the emissions of generation without storage. The GHG emissions per storage mass are given in Section 2.1.2. GHG emissions from combined-cycle gas and coal-fired generation have been estimated at 488 and 965 gCO_2e/kWh, respectively [19].

For our case studies on solar tracking and on combined renewables and batteries, we conducted an LCA of solar PV generation, and synthesized the prior LCAs of wind generation.

2.2.1. PV LCA Methodology

We developed a solar life cycle assessment tool (SoLCAT) [43] following the ISO 14040 and 14044 standards [44] and the IEA PV LCA guidelines [22]. Miller et al. [43] summarizes how SoLCAT corresponds to these guidelines. To estimate the GHG emissions from PV power, SoLCAT integrates the following four main elements: published PV life cycle inventories (LCIs); background emission factors from the Ecoinvent database [45]; known physical correlations; and capacity factors from PVWatts Version 5, a software tool from the U.S. National Renewable Energy Laboratory (NREL).

PV LCA-Goal, System Boundary, and Functional Unit

The goal is estimating the carbon intensity of PV power. The system is electricity production by PV. Electricity production can be considered a black box that consumes resources and produces electricity and emissions. The system's primary function is producing electricity, and thus the functional unit is a kilowatthour of AC electricity supplied to the grid. In addition to electricity, the other system output we analyzed in the case study is GHG emissions. These two system outputs are combined into our central metric—GHGs emitted per AC electricity generated (gCO_2e/kWh), or carbon intensity.

PV electricity production can be elaborated as shown in Figures 1 and 2. Commercial thin film module production is relatively less complex and more vertically integrated [40,47], and is not further segmented beyond Figure 5. Crystalline silicon module production is further elaborated here, as shown in Figure 6.

LCA refers to the explicit stages in Figures 5 and 6 as the "foreground" and their implied component processes as the "background". Although not explicitly drawn, background processes are inside the system border and include the following: production of chemicals used in silicon processing; construction and operation of module-manufacturing infrastructure; and raw material extraction processes (e.g., iron ore acquisition) that "pull" resources from nature, across the system boundary, and into the electricity production system. In this analysis, transport is treated as a background process that contributes to multiple foreground stages. For example, chemicals are transported to the site of silicon processing. The processes not included in our system include the following: electricity transmission from generation site to end-use; displacement of competing power sources; and growth of a carbon-absorbing biomass underneath PV modules, which growth can be reduced by up to ~75% by module shading in temperate climates [48].

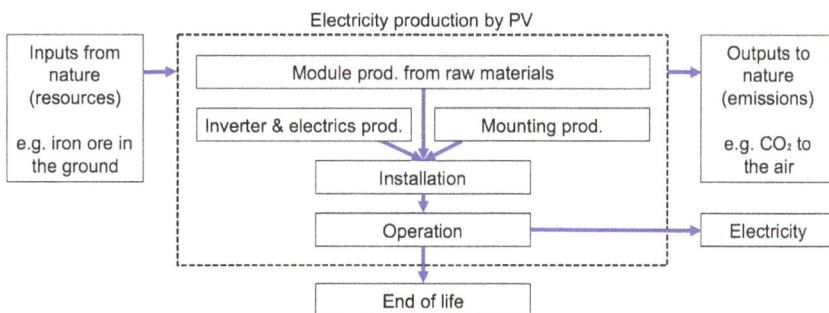

Figure 5. Life cycle stages of PV electricity production.

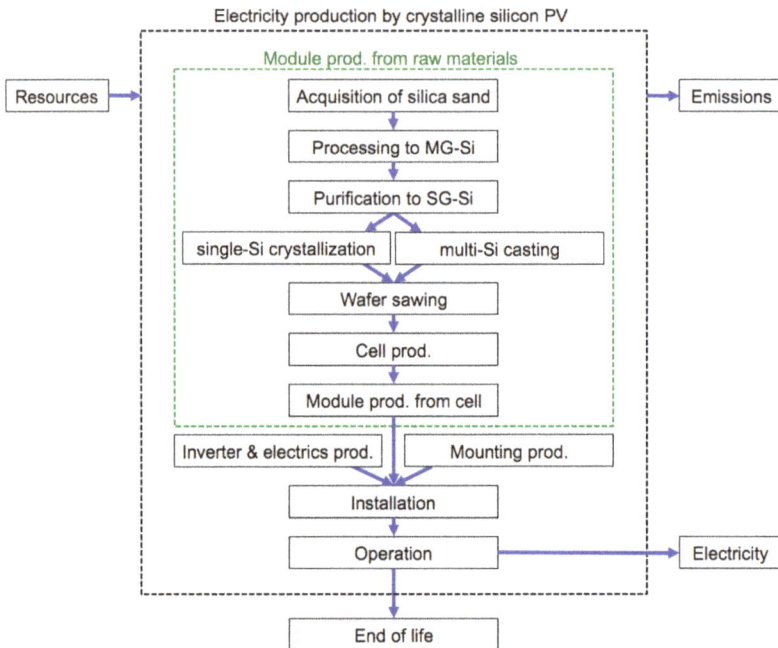

Figure 6. Life cycle stages of electricity production by crystalline silicon PV (MG-Si: metallurgical grade silicon; SG-Si: solar grade silicon).

PV LCA-Data Sources

A list quantifying inputs and outputs of a stage is called a life cycle inventory (LCI). Our primary sources for PV foreground LCIs are the IEA's 2015 Report "Life Cycle Inventories and Life Cycle Assessments of PV Systems" [41] and the 2012 Report "Life Cycle Inventories of Photovoltaics" by ESU-services [40]. These sources aggregate large numbers of LCIs and explicitly aim to represent production typical of the PV industry. The IEA report provides "mainstream" and "best technology" LCIs for Chinese mc-Si module production. We use the "mainstream" LCIs. Other sources for foreground LCIs are the Ecoinvent Database V3 [45] for an LCI of silica sand acquisition, and Sinha et al. [27] for an LCI of a horizontal one-axis tracking system. For emission factors of background processes, our primary data source is the Ecoinvent V3 database [45]. The impact assessment method used by Ecoinvent V3, and thus by our model, to calculate emission factors in units of CO_2-equivalent is the Intergovernmental Panel on Climate Change's 2013 GWP 100a method [49].

The referenced LCIs do not provide data on end of life (EOL), partly because the large majority of PVs installed have not reached their end of life. In the absence of data, our model does not account for emissions from PV EOL processes. Wind power LCAs face a similar lack of EOL data, but benefit from more established recycling processes for wind's primary materials (steel from towers, fiber glass from blades, etc.).

PV LCA-Model Structure

SoLCAT uses foreground LCIs and background emissions factors in several ways to estimate the carbon intensity of PV power. Figure 7 gives an overview of SoLCAT's operation and utilization of data sources. Capacity refers to rated DC power capacity, unless otherwise indicated.

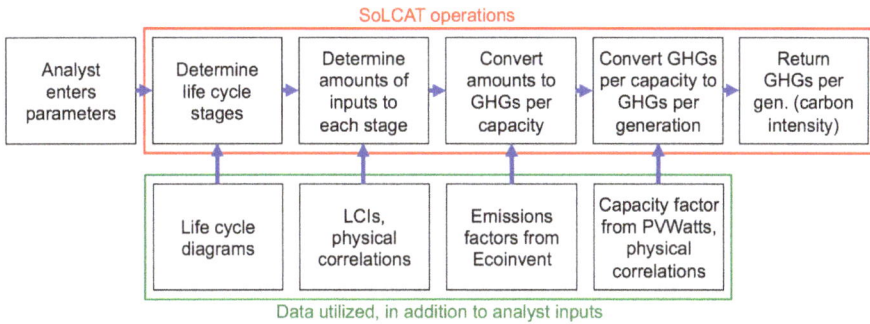

Figure 7. Flowchart of SoLCAT operations and utilization of data sources. Capacity is rated DC capacity. Generation is AC electricity generation (SoLCAT: solar life cycle assessment tool; GHGs: greenhouse gas; LCIs: life cycle inventories).

SoLCAT converts amounts (a) to GHG emissions (e_{total}) using three general equations, shown below with indented examples after each equation, as follows:

$$e_{total} = \sum_i e_{stage\ i} \tag{9}$$

$$e_{total,mc-Si} = e_{Si-sand\ acquistion} + e_{MG-Si\ processing} \cdots + e_{cell\ prod.} \cdots + e_{installation} + e_{operation}$$

$$e_{stage\ i} = \sum_j e_{input\ j\ to\ stage\ i} - \sum_k^{stages\ before\ i} e_{stage\ k} \tag{10}$$

$$e_{MG-Si\ process.} = \left(e_{charcoal\ input\ to\ MG-Si\ process.} + e_{Si-sand\ input\ to\ MG-Si\ process.} + \cdots \right) - e_{Si-sand\ acq.}$$

$$e_{input\ j\ to\ stage\ i} = a_{input\ j\ to\ stage\ i}\ EF_j \tag{11}$$

$$e_{chromium\ steel\ input\ to\ mounting\ prod.} = a_{chromium\ steel\ input\ to\ mounting\ prod.}\ EF_{chromium\ steel}$$

where e is emissions (gCO_2e), a is amount (e.g., kg-iron, m^3-acetone, etc.), and EF is emission factor (e.g., gCO_2e/kg-iron, gCO_2e/m^3-acetone, etc.). Calculating the life cycle emissions for one scenario involves >100 variations of Equation (11) corresponding to different stages and inputs. Amounts ($a_{input\ j\ to\ stage\ i}$) are provided by the PV LCIs [27,40,41,45] or are determined by input variables to SoLCAT. Emission factors (EF_j) are provided by Ecoinvent [45] or SoLCAT inputs. Miller et al. [43] provide more detail on how SoLCAT utilizes the input variables impact amounts and emission factors.

PV LCA-Converting GHGs per Capacity to GHGs per Generation (Carbon Intensity)

SoLCAT's last operation requires a capacity factor, or equivalently an energy yield. Our model utilizes capacity factor estimates from PVWatts [51]. PVWatts outputs a year-one DC capacity factor based on inputs of location, installation type, orientation, cell type, inverter loading ratio, rated inverter efficiency, ground coverage ratio, and system losses.

Our model adjusts capacity factors from PVWatts to account for shading, snow, light-induced degradation (LID), non-LID degradation (commonly called "degradation"), and tracker energy consumption, in order to calculate a lifetime average capacity factor (F). Finally, carbon intensity is calculated as follows:

$$I = e_{total}/(Fc_gt_{hr}) \tag{12}$$

where I is carbon intensity (gCO_2e/kWh), e_{total} is life cycle GHG emissions (gCO_2e), c_g is rated power capacity, and t_{hr} is PV system lifetime (h).

PV LCA-Tracking Energy Gain Methodology

An analysis of solar tracking's impact on carbon intensity requires the calculation of tracking energy gain (TEG). TEG is the percentage increase in PV power output that results from tracking the sun, relative to a fixed-position system, and can be estimated using PVWatts as follows:

$$TEG = \left(\overline{P}_{AC,track} - \overline{P}_{AC,fixed} \right) / \overline{P}_{AC,fixed} \times 100\% \tag{13}$$

or equivalently

$$TEG = \left(F_{track} - F_{fixed} \right) / F_{fixed} \times 100\% \tag{14}$$

where F_{track} is the capacity factor of a PV system with tracking, and F_{fixed} is the capacity factor of a PV system with fixed orientation but otherwise identical features (location, modules, etc.). When discussing TEG, the details of both the tracking system (whether one-axis or two-axis, axis of rotation, rotation limits, etc.) and the fixed base case (tilt and azimuth) should be specified. In this paper, the fixed base case orientation is always irradiance-maximizing, with equator-facing azimuth (south in the Northern Hemisphere anfd north in the Southern Hemisphere) and near-latitude tilt.

2.2.2. Wind LCA Synthesis

Wind electricity production can be elaborated as shown in Figure 8. To be consistent with our PV LCA, our wind system likewise excludes EOL and transmission from generation site to end-use.

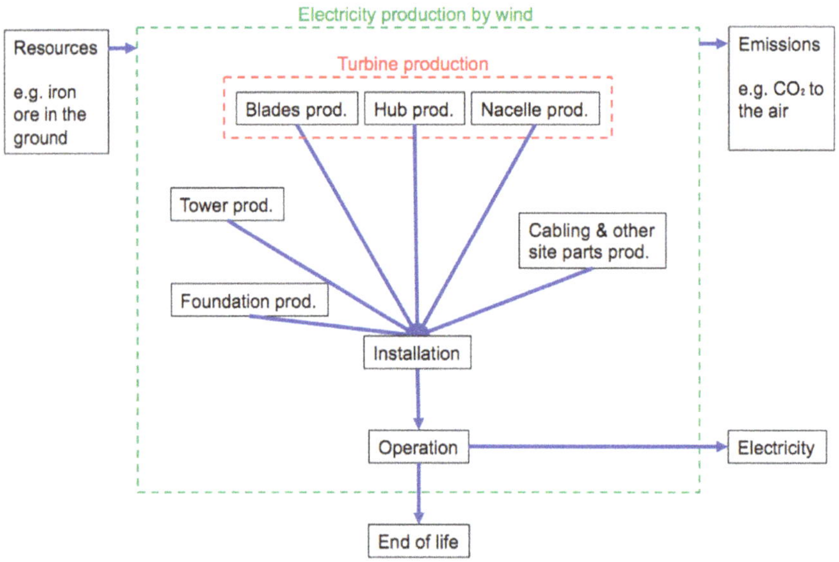

Figure 8. Life cycle stages of wind electricity production.

Our main data sources are 10 wind power LCAs conducted by the turbine manufacturers Vestas and Siemens-Gamesa (also called SWP) in 2011–2017 [10–17,52,53]. Vestas and SWP together accounted for ~25% and ~50% of the new turbines installed in the world and in the United States, respectively, in 2016 [7,54]. SWP accounted for over half of both cumulative and new offshore installations in 2016 [7]. These LCAs have been reviewed for compliance with ISO 14040 LCA standards, by external reviewers in Vestas' case and internal reviewers in SWP's case. The 10 turbine types modeled are widely deployed in conditions matching our four categories of interest, namely: onshore low wind speed (~7 m/s), onshore medium (~8 m/s), onshore high (~9 m/s), and offshore high (~9 m/s). In the United States, four of the ten models are deployed in onshore low wind speed conditions, seven in onshore medium speed conditions, and five in onshore high speed conditions [8]. In Europe, three of ten are deployed in offshore high speed conditions [16,17,55].

Each LCA estimates the carbon intensity of generation from a "typical" wind farm of turbines of the modeled type. The definition of "typical" varies slightly across the 10 LCAs. Thus, before averaging the emissions figures in each of the four categories, we first harmonize several parameters. For example, we assume a distance to grid of 20 km for onshore wind farms [11]. If one LCA assumes 15 km to grid and emissions from grid cable production of 1 gCO_2e/kWh, we multiply those emissions by 20 km/15 km = 1.33. We also harmonize lifetime (20 years), hub height (95 m), offshore distance to grid (70 km [7]), and hourly wind speed data (four data sets for four categories). The wind speed data are input into NREL's System Advisor Model program, along with the turbine model, to estimate the capacity factors. Table 2 shows the results.

3. Results and Discussion

3.1. Base Cases for PV Generation (No Tracking, No Storage)

Figure 9 shows how PV carbon intensity varies by location, irradiance, and cell type. The parameter values used in Figure 9 are assumed throughout the paper, unless otherwise indicated. Figure 10 shows the sensitivity of PV carbon intensity to lifetime and degradation rate.

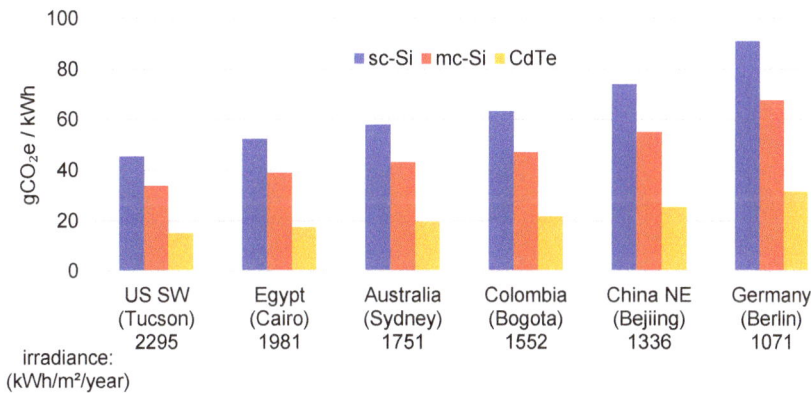

Figure 9. Carbon intensities of power from PV power plants installed in different locations around 2015. Installation type is large-scale (AC capacity >1 MW), open-ground, and fixed. Irradiance is lifetime-average irradiance incident on modules. Lifetime is 30 years. Rated module efficiencies are 17%, 16%, and 15.6% for sc-Si, mc-Si, and CdTe, respectively [9]. Degradation is 0.7%/year [56]. GHG emissions of upstream electricity are 660 gCO_2e/kWh for module production and 510 for balance of system (BOS) production, corresponding to Chinese and global averages in 2015 [57]. These parameter values are assumed throughout the paper, unless otherwise indicated.

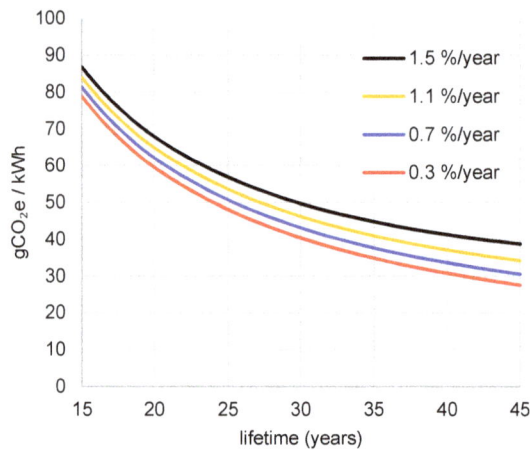

Figure 10. Carbon intensity of mc-Si PV power vs. lifetime at different degradation rates. Location is Sydney, Australia. Rated module efficiency is 16%. Other parameters match values in Figure 9.

3.2. The Impact of Solar Tracking

We find that in the United States Southwest, for mc-Si PV, horizontal one-axis tracking reduces carbon intensity by 12% relative to the fixed-tilt base case (from 34 to 30 gC/kWh), consistent with previously published results [9]. Tracking produces this reduction despite requiring ~50% more structural metal (iron and aluminum) and ~30% more copper cable per module, compared to fixed-tilt mounting [27]. Emissions from producing extra tracker materials are offset by increased generation from tracking, such that the overall carbon intensity decreases.

To determine whether this ~10% emissions reduction holds for other regions and cell types, we applied analogous calculations to hypothetical mc-Si and CdTe PV systems at 10 other locations (the base cases from Figure 9), and have presented them in Figure 11, which underlines several related findings. (1) Location influences the emissions impact of tracking, via TEG; (2) tracking decreases the carbon intensity of mc-Si PV in most locations; (3) consistent with Sinha et al. [27], tracking reduces the carbon intensity of CdTe PV in the United States Southwest by ~3%; and (4) the United States Southwest is the exception to the rule—for most locations tested, tracking actually increases the carbon intensity of CdTe PV power. This includes many places with favorable economics for tracking and TEG above 13%. Near Sydney, Australia, for example, horizontal single-axis tracking increases both electricity output (by 14%) and carbon intensity (by 5%, from 19 to 20 gCO$_2$e/kWh). In Germany, where tracking is less common, tracking increases power output by 6% and emissions by 13% (from 31 to 35 gCO$_2$e/kWh).

The dependence on location is mainly driven by latitude and cloud cover. The greater the latitude, the greater the module tilt that maximizes incident irradiance, and the more irradiance is lost by "reclining" to a horizontal tilt for one-axis tracking (as described in Section 1, horizontal one-axis tracking is the norm for large-scale tracking; tilted one-axis trackers do exist, but are much less common for reasons including self-shading and wind concerns). Greater latitude also means more atmosphere for sunlight to travel through. This increases light scattering, as does greater cloud cover. The greater the fraction of ambient light that is scattered (i.e., diffuse), the less energy there is to be gained from tracking the sun's non-diffuse direct beam irradiance. Lower tracking energy gain (TEG) means less extra electricity over which to amortize extra emissions from tracker-production. For both module types, this explains why, as TEG decreases left to right in Figure 11, tracking's emissions impact increases in relative terms (the bars).

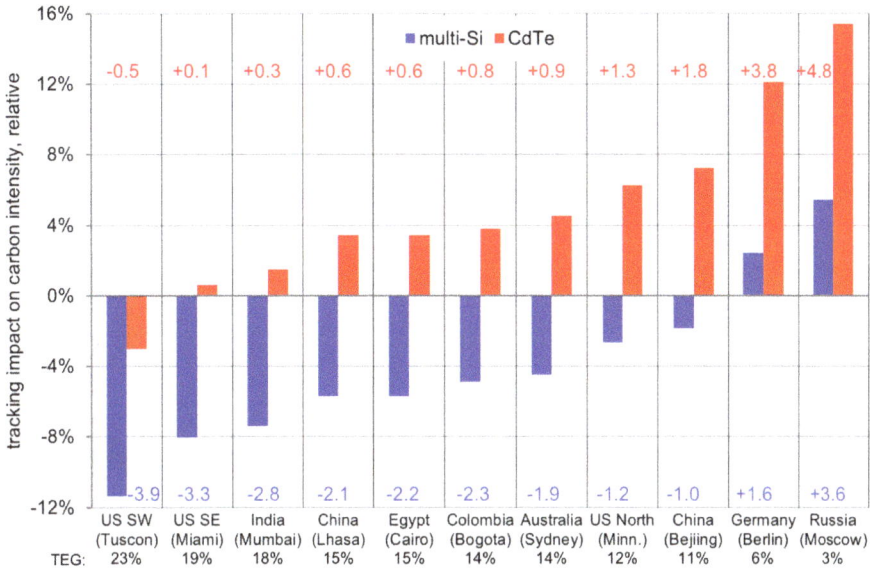

Figure 11. The impact of horizontal one-axis solar tracking on PV carbon intensity in different locations, for mc-Si PV (blue), and CdTe PV (red). The y-axis and bars indicate relative impact, that is, percentage change in carbon intensity compared to fixed PV at the same location: Percentage change = $(CI_{tracking\ PV} - CI_{fixed\ PV})/CI_{fixed\ PV} \times 100\%$. The red and blue numbers indicate the absolute impact (gCO2e/kWh): absolute change = $(CI_{tracking\ PV} - CI_{fixed\ PV})$. Tracking energy gains (TEGs) are at the bottom.

The varying impact by module type can be explained with the following equations:
Let:

$T \equiv$ the factor by which tracking increases electricity generation. E.g., If TEG = 20%, $T = 1.2$.

$e_{f,i} \equiv$ emissions of fixed PV system (gCO$_2$e). i = mc-Si or CdTe

$e_t \equiv$ emissions from adding tracking (gCO$_2$e)

$E_f \equiv$ generation from fixed PV system (kWh)

$E_t \equiv$ generation from tracking PV system (kWh)

$I_f \equiv$ emissions per generation (carbon intensity) of fixed PV system (gCO$_2$e/kWh)

$I_t \equiv$ emissions per generation (carbon intensity) of tracking PV system (gCO$_2$e/kWh)

$M \equiv$ factor by which tracking changes carbon intensity.

$(M - 1) \times 100\%$, the percentage by which tracking changes the carbon intensity.

Using these definitions,

$$M = I_t/I_f$$
$$= \left[\left(e_{f,i} + e_t \right) / E_t \right] / \left[e_{f,i} / E_f \right]$$
$$= \left[\left(e_{f,i} + e_t \right) / \left(T \times E_f \right) \right] / \left[e_{f,i} / E_f \right]$$

$$M = \left(e_{f,i} + e_t\right) / \left(T \times e_{f,i}\right) \tag{15}$$

Consider Equation (15) when $e_{f,i} \gg e_t$, that is, when emissions from module production are much larger than emissions from tracker production, as follows:

$$M_{f\ large} = e_{f,i} / \left(T \times e_{f,i}\right)$$

$$M_{f\ large} = 1/T \tag{16}$$

$M_{f\ large}$ will always be less than 1, because T is always greater than 1. In other words, for a module type with large production emissions, adding tracking will reduce carbon intensity. This explains why adding tracking reduces the carbon intensity of mc-Si PV in most locations (blue bars in Figure 11). Multi-Si module production is significantly more carbon intensive than CdTe module production, as seen in Figure 6 and previously reported [41]. $e_{f,multi-Si}$ is approximately $11 \times e_t$, whereas $e_{f,CdTe}$ is approximately $5 \times e_t$. Equation (15) thus also explains why adding tracking increases CdTe PV's carbon intensity in most locations (red bars in Figure 11), as follows:

$$M_{CdTe} = \left(e_{f,CdTe} + e_t\right) / \left(T \times e_{f,CdTe}\right)$$

$$\approx \left(e_{f,CdTe} + e_{f,CdTe}/5\right) / \left(T \times e_{f,CdTe}\right)$$

$$M_{CdTe} \approx 1.2/T \tag{17}$$

For M_{CdTe} to be less than 1, T must be greater than 1.2. In other words, CdTe PV requires a TEG above 20% for tracking to reduce its carbon intensity, a TEG only possible in exceptionally sunny regions like the United States Southwest.

Figure 12 illustrates the general principle behind both scenarios (high- and low-module emissions). In the future, if the ratio of module production emissions to tracker production emissions increases, independent of absolute emission values, tracking will more commonly decrease PV carbon intensity.

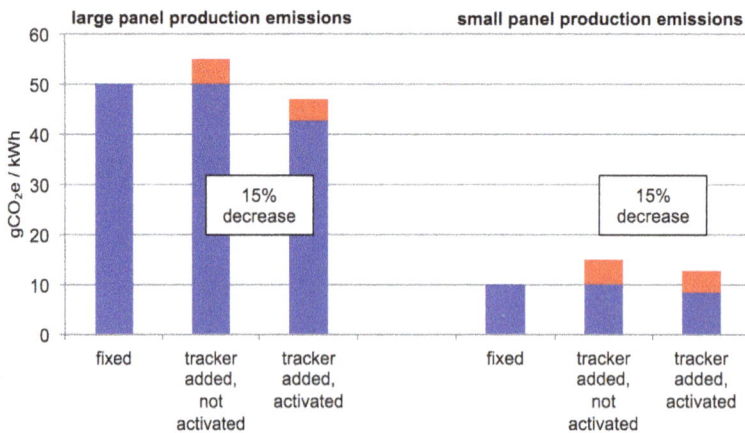

Figure 12. Illustration of how panel production emissions influence whether tracking decreases (left) or increases (right) the carbon intensity, relative to the fixed PV. Left and right systems have the same location and power output per panel. Activating tracking on both boosts power by 17%, a typical TEG for a sunny location. Red represents the tracker contribution to emissions. In terms of Equation (15) definitions, red represents e_t/g_f in the "not activated" bars, and e_t/g_t in the "activated" bars. Emissions values in this figure are arbitrary and chosen to illustrate the concept. A TEG of 17% is assumed, giving a 15% reduction in carbon intensity upon activation.

3.3. Hybrid Power Generation and Storage

Figures 13–15 show the carbon intensity of electricity from coupled generation and batteries for PV, wind, and fossil-based generators, respectively. These carbon intensities are estimated using the model and LCAs described in Section 2. Figure 16 shows the linear correlation between coupled system carbon intensity and generator carbon intensity, and how storage technology affects that correlation.

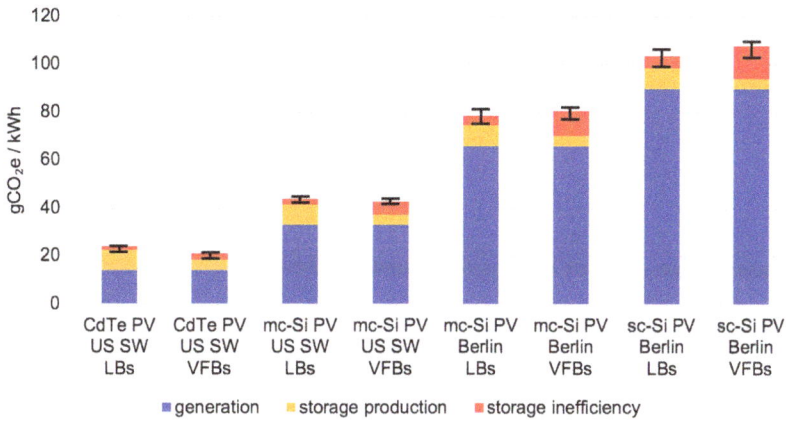

Figure 13. GHG emissions of power from different PV systems coupled with LBs or VFBs. PV systems differ by cell type and installation location. Other parameters match base case values in Figure 9. Bars indicate the difference between "best" and "worst" case scenarios for each storage technology. Best assumes the high values of efficiency and cycle life in Section 2.1.2, while worst assumes the low values. Fraction of generation sent to storage *f* is 50%, to correspond to the approximate upper bound of *f* needed to flatten the intermittent generation, or conversely needed to convert constant fossil-based generation into peak supply (CdTe: cadmium telluride; mc-Si: multi-crystalline silicon; sc-Si: single-cyrstalline silicon; LBs: lithium-ion batteries; VFBs: vanadium redox flow batteries).

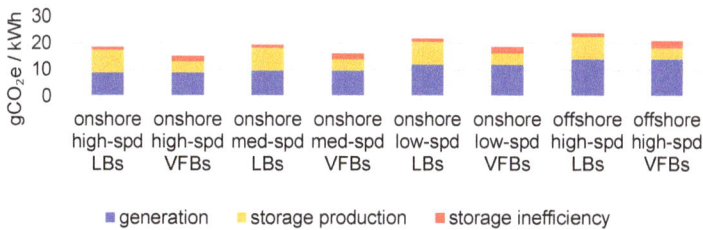

Figure 14. GHG emissions of power from different wind power systems coupled with LBs or VFBs. Wind systems differ by average local wind-speed and installation onshore or offshore. Other parameters match assumptions in Table 2. Bars indicate the difference between "best" and "worst" case scenarios for each storage technology. Best assumes the high values of efficiency and cycle life in Table 2, while worst assumes the low values. The fraction of generation sent to storage *f* is 50%, to correspond to the approximate upper bound of *f* needed to flatten the intermittent generation, or conversely needed to convert constant fossil-based generation into peak supply.

The most important result is that, across all scenarios displayed, renewable electricity has much lower GHG emissions than fossil electricity. Figures 13–16 show this to hold true, independent of renewable energy source, renewable power generation technology, location of generator, and integration of energy storage. As seen in Figure 13, the most carbon-intensive mainstream PV power (sc-Si PV produced in China, installed in Berlin, and coupled with sufficient LBs to store

50% of it's generation) still emits only ~25% of the GHGs emitted by the least carbon intensive mainstream fossil power (CCNG with no storage). Therefore, to understand the GHG emissions impacts of renewables, future work should not focus on more accurately accounting for GHGs from the production of battery control systems or other system components. Whether wind power coupled with LBs emits 20 or 21 or even 40 gCO_2e/kWh, it is still an order of magnitude less carbon intensive than fossil generation without carbon capture. Instead, future work on renewables' emissions impacts should focus on consequential LCA, including questions of which generation types are most likely to be displaced by renewables deployment, dependent on location and economics, among other factors.

The percentage impact panel in Figure 15 illustrates another result—the storage parameter that controls emissions impact depends upon the generator to which the storage is coupled. Given low-emission generation, such as wind or CdTe PV, the storage feature that dominates emissions impact is storage production emissions (see orange sections in the left two bars of Figure 13). Given high-emission generation, such as gas or coal, the storage feature that dominates emissions impact is storage efficiency (see red sections in bottom four bars of Figure 15). One implication is that, if storage is being added to micro-grids or regional grids with low-emission generation ($I_{avg} < 50$ gCO_2e/kWh), GHG emissions can be minimized by deploying storage products with the lowest production emissions. Conversely, if storage is being added to regions with high-emission generation ($I_{avg} > 200$ gCO_2e/kWh), GHG emissions can be minimized by deploying storage products with the highest efficiency.

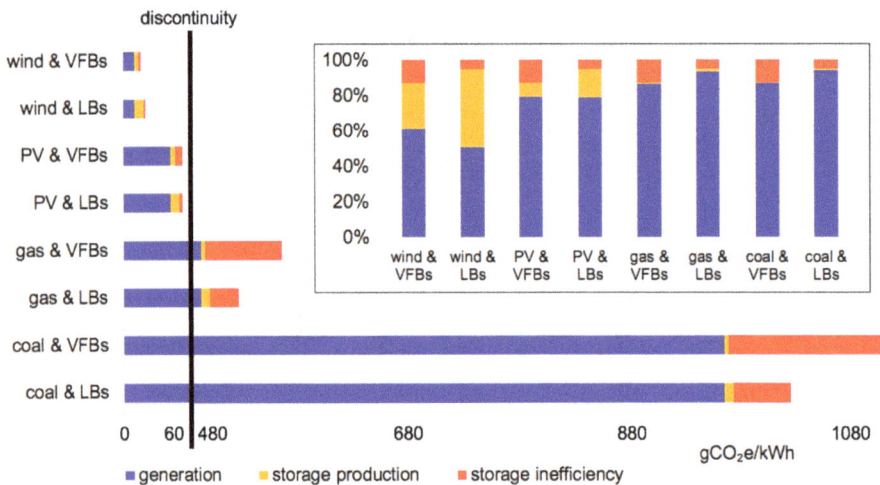

Figure 15. GHG emissions of power from different generators coupled with LBs and VFBs. PV case assumes installation in Sydney, Australia, and mc-Si cell type. Other PV parameters match the base case values in Figure 8. Wind case assumes onshore installation in a medium wind-speed (~8 m/s) area. Other parameters match the values in Section 2.1.2. Gas and coal cases assume typical combined cycle natural gas (CCNG) and supercritical pulverized coal power (SCPC) emissions values of 488 and 965 gCO_2e/kWh, respectively [19]. The fraction of generation sent to storage f is 50%, to correspond to the approximate upper bound of f needed to flatten intermittent generation, or conversely needed to convert constant fossil-based generation into peak supply.

Figure 16 illustrates the linear relationship between the coupled system carbon intensity and generator carbon intensity. The parameters of the storage technology influence both the y-axis intersection and the slope of the line, according to the model equations in Section 2.1, as follows:

$$y - axis\ intersection = \eta(e_s/E_{sT})/(DC)/(1/f - 1 + \eta)$$

$$slope = [1 - f(1 - \eta)]$$

These values are not particular to GHGs. Analogous qualitative behavior will thus hold for emissions of other pollutants, as well as "emissions" of dollars (costs/kWh). That is, emission and cost intensities of coupled systems, as a function of generator intensity, will have different slopes dependent on storage technology. Choosing a specific storage technology for a coupled system can minimize emissions (or cost) intensity for one generator, but not for another.

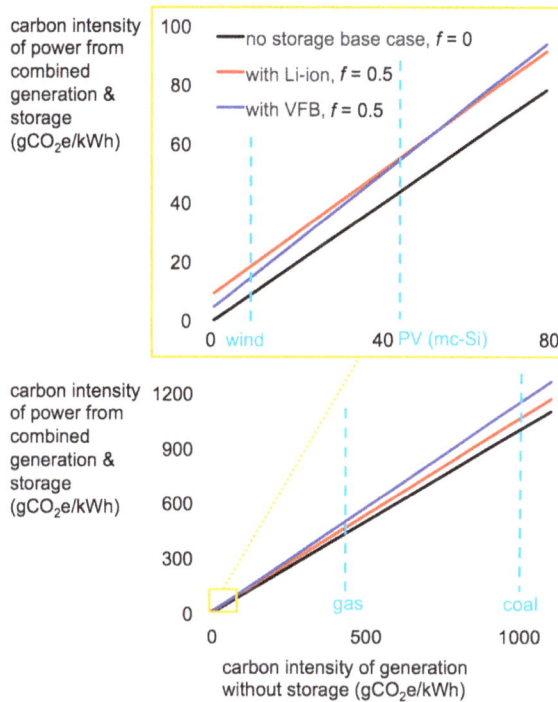

Figure 16. Carbon intensity of coupled generation and storage vs. carbon intensity of generation without storage. The wind, PV, gas, and coal numbers shown are respresetnative of onshore wind power in high wind-speed areas, mc-Si PV in Australia, combined cycle natural gas power, and supercritical pulverized coal, respectively.

4. Conclusions

A model for estimating emissions from integrated power generation and energy storage is presented. The model applies to the emissions of all pollutants, including greenhouse gases (GHGs), and to all storage technologies, including pumped hydroelectric and electrochemical storage. As a case study, the model is used to estimate the GHG emissions of electricity from systems that couple photovoltaic and wind generation with lithium-ion batteries (LBs) and vanadium redox flow batteries (VFBs). To facilitate the case study, we conduct a life cycle assessment (LCA) of photovoltaic (PV) power, and a synthesis of existing wind power LCAs. The PV LCA is also used to estimate the emissions impact of a common PV practice that has not been comprehensively analyzed by LCA—solar tracking. Relative to stationary mounting, solar tracking is found to decrease the GHG emissions of power from multi-crystalline silicon PV in most of the regions analyzed (by 0 to ~12%, or 0 to ~4 gCO$_2$e/kWh), and to increase the emissions of power from cadmium telluride PV in most regions analyzed (by 0 to ~12%, or 0 to ~4 gCO$_2$e/kWh). For either PV cell type, if the ratio of module production emissions to tracker production emissions increases in the future, independent of absolute emission values, tracking will more commonly decrease PV carbon intensity. The case study of renewables and battery storage indicates that PV and wind power remain much less carbon intensive than fossil-based generation, even when coupled with large amounts of LBs or VFBs. The most carbon intensive renewable power analyzed (single-crystalline silicon PV produced in China, installed in Berlin, and coupled with sufficient VFBs to store 50% of it's generation) still emits only ~25% of the GHGs of the least carbon intensive mainstream fossil power (combined cycle natural gas turbine with no storage). Lastly, we find that the pathway to minimize GHG emissions of power from a coupled system depends upon the generator—given low-emission generation (<50 gCO$_2$e/kWh), the minimizing pathway is the storage technology with lowest production emissions (VFBs over LBs for our case study), and given high-emission generation (>200 gCO$_2$e/kWh), the minimizing pathway is the storage technology with highest round-trip efficiency (LBs over VFBs). The latter case applies to a majority of the world's power generation today. Directions for future work include the application of the coupled storage and generation model to other non-battery technologies, including pumped hydroelectric storage.

Author Contributions: I.M., E.G., and F.M.O. designed the research; I.M and E.G. developed the models and derived equations; I.M. performed all systems calculations; I.M., E.G., and F.M.O. analyzed data; I.M. and E.G. drafted the manuscript; all authors reviewed, edited, and approved the final version of the manuscript. F.M.O. directed the overall research.

Funding: This research was supported by The Low Carbon Energy Centers of the MIT Energy Initiative.

Conflicts of Interest: The authors declare no conflict of interest.

Nomenclature (In Order of Appearance)

GHG	greenhouse gas
IEA	International Energy Agency
PV	photovoltaic
mc-Si	multi-crystalline silicon
sc-Si	single-crystalline silicon
CdTe	cadmium telluride
gCO$_2$e	grams CO$_2$ equivalent
CCNG	combined cycle natural gas
SCPC	supercritical pulverized coal
cradle-to-grave	covering all parts of a product's life cycle from raw material acquistion (cradle) through disposal and/or recycling (grave)
cradle-to-gate	covering all parts of a product's life cycle from raw material acquistion (cradle) through production (factory gate)
LCA	life cycle assessment
utility-scale	having AC capacity >5 MW, used to describe power plants
LB	lithium-ion battery
VFB	vanadium redox flow battery
LAB	lead acid battery
SSB	sodium sulfur battery
PHS	pumped hydroelectric storage
CAES	compressed air energy storage
P_0	power generation (kW)
P	power supply to grid (kW)
$P_{s,in}$	is the amount of power generation sent to storage (kW)
s	is the instantaneous fraction of power generation going to storage
$P_{s,out}$	is the power supply to grid from storage
η	round-trip efficiency of the storage
f	the fraction of energy generation sent to storage (vs. directly to grid)
E	lifetime electricity supply to grid from power plant (kWh)

E_0	lifetime electricity supply to grid from power plant without storage (kWh)
I	emissions intensity (g/kWh)
e_g	the life emissions of the generator (g)
e_s	the life emissions of the storage (g)
$\overline{D}$ or DoD	the average depth of discharge of the storage (DoD)
E_{sT}	the total rated energy capacity of storage used over the power plant life (kWh)
C	the cycle life of the storage (the number of charge–discharge cycles before a storage unit is retired)
m_s	the mass of batteries used over the power plant life (kg)
ρ	the specific energy capacity of the batteries (kWh/kg)
E_{sT}	the total rated energy capacity of all storage used and retired over the power plant life (kWh)
c_g	rated power capacity
F	capacity factor
TEG	tracking energy gain
EOL	end of life

References

1. IEA. *Energy, Climate Change & Environment*; IEA: Paris, France, 2016.
2. The Energy Initiative MIT. *The Future of Solar Energy: An Interdisciplinary MIT Stud*; MIT: Cambridge, MA, USA, 2015; pp. 3–6. [CrossRef]
3. GlobalData. *Solar PV Generation Statistics, Wind Generation Statistics*; GlobalData: London, UK, 2018.
4. ExxonMobil. *2017 Outlook for Energy: A View to 2040*; ExxonMobil: Irving, TX, USA, 2017.
5. BNEF. *Bloomberg New Energy Finance Report 2017*; BNEF: New York, NY, USA, 2017.
6. Fraunhofer Institute for Solar Energy Systems. *Photovoltaics Report*; Fraunhofer Institute for Solar Energy Systems: Freiburg, Germany, 2017.
7. WindEurope. *The European Offshore Wind Industry—Key Trends And Statistics 2017*; WindEurope: Brussels, Belgium, 2018.
8. EIA. *Form EIA-860: Annual Electric Generator Report*; EIA: Paris, France, 2016.
9. Leccisi, E.; Raugei, M.; Fthenakis, V. The energy and environmental performance of ground-mounted photovoltaic systems—A timely update. *Energies* **2016**, *9*, 622. [CrossRef]
10. Garrett, P.; Rønde, K. *Life Cycle Assessment of Electricity Production from an Onshore V90-3.0 MW Wind Plant September 2012*; Vestas Wind Systems A/S: Aarhus, Denmark, 2012; pp. 1–106.
11. Garrett, P.; Rønde, K.; Finkbeiner, M. *Life Cycle Assessment of Electricity Production from an Onshore V110-2.0 MW Wind Plant*; Vestas Wind Systems A/S: Aarhus, Denmark, 2012; pp. 1–106.
12. Garrett, P.; Klaus, R. *Life Cycle Assessment of Electricity Production from a V100-1.8 MW Gridstreamer Wind Plant*; Vestas Wind Systems A/S: Aarhus, Denmark, 2011; p. 105.
13. Razdan, A.P.; Garrett, P. *Life Cycle Assessment of Electricity Production from an Onshore V100-2.0 MW Wind Plant*; Vestas Wind Systems A/S: Aarhus, Denmark, 2015.
14. Siemens. Onshore Wind Power Plant Employing SWT-2.3-108. In *A Clean Energy Solut—From Cradle to Grave*; Siemens: Munich, Germany, 2014; 16p.
15. Siemens. Onshore Wind Power Plant Employing SWT-3.2-113. In *A Clean Energy Solut—From Cradle to Grave*; Siemens: Munich, Germany, 2014; 16p.
16. Siemens. Offshore Wind Power Plant Employing SWT-4.0-130. In *A Clean Energy Solut—From Cradle to Grave*; Siemens: Munich, Germany, 2014; 16p.
17. Siemens. Offshore Wind Power Plant Employing SWT-7.0-154. In *A Clean Energy Solut—From Cradle to Grave*; Siemens: Munich, Germany, 2014; 16p.
18. Weinzettel, J.; Reenaas, M.; Solli, C.; Hertwich, E.G. Life cycle assessment of a floating offshore wind turbine. *Renew. Energy* **2009**, *34*, 742–747. [CrossRef]
19. Skone, T.J.; Littefield, J.; Cooney, G.; Marriott, J. *Power Generation Technology Comparison from a Life Cycle Perspective*; National Energy Technology Laboratory: Pittsburgh, PA, USA, 2013.
20. Peng, J.; Lu, L.; Yang, H. Review on life cycle assessment of energy payback and greenhouse gas emission of solar photovoltaic systems. *Renew. Sustain. Energy Rev.* **2013**, *19*, 255–274. [CrossRef]

21. Hsu, D.D.; O'Donoughue, P.; Fthenakis, V.; Heath, G.A.; Kim, H.C.; Sawyer, P.; Choi, J.K.; Turney, D.E. Life Cycle Greenhouse Gas Emissions of Crystalline Silicon Photovoltaic Electricity Generation: Systematic Review and Harmonization. *J. Ind. Ecol.* **2012**, *16*, S122–S135. [CrossRef]

22. Frischknecht, R.; Heath, G.; Raugei, M.; Sinha, P.; de Wild-Scholten, M.; Fthenakis, V. *Methodology Guidelines on Life Cycle Assessment of Photovoltaic Electricity*, 3rd ed.; IEA PVPS Task 12; National Renewable Energy Lab. (NREL): Golden, CO, USA, 2016.

23. Bayod-Rújula, Á.A.; Lorente-Lafuente, A.M.; Cirez-Oto, F. Environmental assessment of grid connected photovoltaic plants with 2-axis tracking versus fixed modules systems. *Energy* **2011**, *36*, 3148–3158. [CrossRef]

24. Desideri, U.; Zepparelli, F.; Morettini, V.; Garroni, E. Comparative analysis of concentrating solar power and photovoltaic technologies: Technical and environmental evaluations. *Appl. Energy* **2013**, *102*, 765–784. [CrossRef]

25. Beylot, A.; Payet, J.Ô.; Puech, C.; Adra, N.; Jacquin, P.; Blanc, I.; Beloin-Saint-Pierre, D. Environmental impacts of large-scale grid-connected ground-mounted PV installations. *Renew. Energy* **2014**, *61*, 2–6. [CrossRef]

26. Bolinger, M.; Seel, J.; La Commare, K.H. *Utility-Scale Solar 2016*; LBNL-2001055; Lawrence Berkeley National Laboratory: Foshan, Guangdong, China, September 2017.

27. Sinha, P.; Schneider, M.; Dailey, S.; Jepson, C.; De Wild-Scholten, M. Eco-efficiency of CdTe photovoltaics with tracking systems. In Proceedings of the 39th IEEE Photovoltaic Specialists Conference (PVSC), Tampa, FL, USA, 16–21 June 2013.

28. Sivaram, V.; Kann, S. Solar power needs a more ambitious cost target. *Nat. Energy* **2016**, *1*, 16036. [CrossRef]

29. Zhuk, A.; Zeigarnik, Y.; Buzoverov, E.; Sheindlin, A. Managing peak loads in energy grids: Comparative economic analysis. *Energy Policy* **2016**, *88*, 39–44. [CrossRef]

30. Krieger, E.M.; Casey, J.A.; Shonkoff, S.B.C. A framework for siting and dispatch of emerging energy resources to realize environmental and health benefits: Case study on peaker power plant displacement. *Energy Policy* **2016**, *96*, 302–313. [CrossRef]

31. Denholm, P.; Hand, M. Grid flexibility and storage required to achieve very high penetration of variable renewable electricity. *Energy Policy* **2011**, *39*, 1817–1830. [CrossRef]

32. Simon, B.; Finn-Foley, D.; Gupta, D. *GTM Research/ESA U.S. Energy Storage Monitor*; GTM: Boston, MA, USA, 2018.

33. Yang, G. Is this the ultimate grid battery? *IEEE Spectrum* **2017**, *54*, 36–41. [CrossRef]

34. Spanos, C.; Turney, D.E.; Fthenakis, V. Life-cycle analysis of flow-assisted nickel zinc-, manganese dioxide-, and valve-regulated lead-acid batteries designed for demand-charge reduction. *Renew. Sustain. Energy Rev.* **2015**, *43*, 478–494. [CrossRef]

35. Rydh, C.J.; Sandén, B.A. Energy analysis of batteries in photovoltaic systems. Part I: Performance and energy requirements. *Energy Convers. Manag.* **2005**, *46*, 1957–1979. [CrossRef]

36. Denholm, P.; Kulcinski, G.L. Life cycle energy requirements and greenhouse gas emissions from large scale energy storage systems. *Energy Convers. Manag.* **2004**, *45*, 2153–2172. [CrossRef]

37. Mitavachan, H.; Derendorf, K.; Vogt, T. Comparative life cycle assessment of battery storage systems for stationary applications. *Environ. Sci. Technol.* **2015**. [CrossRef]

38. Cui, Y.; Du, C.; Yin, G.; Gao, Y.; Zhang, L.; Guan, T.; Yang, L.; Wang, F. Multi-stress factor model for cycle lifetime prediction of lithium ion batteries with shallow-depth discharge. *J. Power Source* **2015**, *279*, 123–132. [CrossRef]

39. Alsaidan, I.; Khodaei, A.; Gao, W. A Comprehensive Battery Energy Storage Optimal Sizing Model for Microgrid Applications. *IEEE Trans. Power Syst.* **2018**, *33*, 3968–3980. [CrossRef]

40. Jungbluth, N.; Stucki, M.; Flury, K.; Frischknecht, R.; Büsser, S. *Life Cycle Inventories of Photovoltaics*; ESU-services Ltd.: Ulster, Switzerland, 2012.

41. Frischknecht, R.; Itten, R.; Sinha, P.; de Wild-Scholten, M.; Zhang, J.; Fthenakis, V. *Life Cycle Inventories and Life Cycle Assessment of Photovoltaic Systems*; Report IEA-PVPS T12-04:2015; International Energy Agency (IEA): Paris, France, 2015.

42. Eaton, C.T. *AC vs. DC Coupling in Utility-Scale Solar Plus Storage Projects*; Eaton: Cleveland, OH, USA, 2016.

43. Miller, I.; Gençer, E.; Vogelbaum, H.S.; Brown, P.R.; Torkamani, S.; O'Sullivan, F.M. Parametric modeling of life cycle greenhouse gas emissions from photovol-taic power. *Appl. Energy* **2018**. Under Review.

44. ISO 14040. *The International Standards Organisation. Environmental Management—Life cycle Assessment—Principles and Framework*; ISO 14040 2006; ISO: Geneva, Switzerland, 2006; pp. 1–28. [CrossRef]

45. Steubing, B.; Wernet, G.; Reinhard, J.; Bauer, C.; Moreno-Ruiz, E. The ecoinvent database version 3 (part II): Analyzing LCA results and comparison to version 2. *Int. J. Life Cycle Assess.* **2016**, *21*, 1269–1281. [CrossRef]
46. Gagnon, P.; Margolis, R.; Melius, J.; Phillips, C.; Elmore, R. *Rooftop Solar Photovolatic Technical Potential in the United States: A Detailed Assessment*; National Renewable Energy Lab.(NREL): Golden, CO, USA, 2016; p. 82. [CrossRef]
47. Jean, J.; Brown, P.R.; Jaffe, R.L.; Buonassisi, T.; Bulović, V. Pathways for solar photovoltaics. *Energy Environ. Sci.* **2015**, *8*, 1200–1219. [CrossRef]
48. Gençer, E.; Miskin, C.; Sun, X.; Khan, M.R.; Bermel, P.; Alam, M.A.; Agrawal, R. Directing solar photons to sustainably meet food, energy, and water needs. *Sci. Rep.* **2017**, *7*, 3133. [CrossRef] [PubMed]
49. Pachauri, R.K.; Allen, M.R.; Barros, V.R.; Broome, J.; Cramer, W.; Christ, R.; Church, J.A.; Clarke, L.; Dahe, Q.; Dasgupta, P. *Climate Change 2014: Synthesis Report*; Contribution of Working Groups, I II and III to the Fifth Assessment Report of the Intergovernmental Panel on Climate Change; IPCC: Geneva, Switzerland, 2014. [CrossRef]
50. Gilman, P.; Dobos, A.; DiOrio, N.; Freeman, J.; Janzou, S.; Ryberg, D. *SAM Photovoltaic Model Technical Reference Update*; NREL: Golden, CO, USA, 2018.
51. Dobos, A.P. *PVWatts Version 5 Manual*; NREL/TP-6A20-62641; National Renewable Energy Lab. (NREL): Golden, CO, USA, 2014.
52. Garrett, P.; Ronde, K. *Life Cycle Assessment of Electricity Production from an Onshore V90-2.6 MW Wind Plant*; Vestas Wind Systems A/S: Aarhus, Denmark, 2013; p. 107.
53. Razdan, A.P.; Garrett, P. *December 2015 Life Cycle Assessment of Electricity Production from an Onshore V110-2.0 MW Wind Plant*; Vestas Wind Systems A/S: Aarhus, Denmark, 2015.
54. GWEC. *Global Wind Statistics 2016*; GWEC: Brussels, Beligum, 2017.
55. Siemens. Offshore wind power plant employing SWT-6.0-154. In *A Clean Energy Solut—From Cradle to Grave*; Siemens: Munich, Germany, 2014; 16p.
56. Jordan, D.C.; Kurtz, S.R.; VanSant, K.; Newmiller, J. Compendium of photovoltaic degradation rates. *Prog. Photovolt. Res. Appl.* **2016**, *24*, 978–989. [CrossRef]
57. IEA. *CO_2 Emissions from Fuel Combustion 2017*; International Energy Agency: Paris, France, 2017. [CrossRef]

![processes logo]

processes

MDPI

Article

Energy and Exergy Analysis of the S-CO$_2$ Brayton Cycle Coupled with Bottoming Cycles

Muhammad Ehtisham Siddiqui [1,*]**, Aqeel Ahmad Taimoor** [2] **and Khalid H. Almitani** [1]

[1] Mechanical Engineering Department, King Abdulaziz University, Jeddah 21589, Saudi Arabia;
    kalmettani@kau.edu.sa

[2] Department of Materials and Chemical Engineering, Ghulam Ishaq Khan Institute of Engineering and
    Technology, Topi 23640, Pakistan; taimooruet@gmail.com

*   Correspondence: mesiddiqui@kau.edu.sa or ehtisham.siddiqui@gmail.com; Tel.: +966-55-218-4681

Received: 11 July 2018; Accepted: 28 August 2018; Published: 1 September 2018

Abstract: Supercritical carbon dioxide (S-CO$_2$) Brayton cycles (BC) are soon to be a competitive and environment friendly power generation technology. Progressive technological developments in turbo-machineries and heat exchangers have boosted the idea of using S-CO$_2$ in a closed-loop BC. This paper describes and discusses energy and exergy analysis of S-CO$_2$ BC in cascade arrangement with a secondary cycle using CO$_2$, R134a, ammonia, or argon as working fluids. Pressure drop in the cycle is considered, and its effect on the overall performance is investigated. No specific heat source is considered, thus any heat source capable of providing temperature in the range from 500 °C to 850 °C can be utilized, such as solar energy, gas turbine exhaust, nuclear waste heat, etc. The commercial software 'Aspen HYSYS version 9' (Aspen Technology, Inc., Bedford, MA, USA) is used for simulations. Comparisons with the literature and simulation results are discussed first for the standalone S-CO$_2$ BC. Energy analysis is done for the combined cycle to inspect the parameters affecting the cycle performance. The second law efficiency is calculated, and exergy losses incurred in different components of the cycle are discussed.

Keywords: supercritical carbon dioxide; recompression cycle; combined cycle; efficiency; organic Rankine cycle; exergy loss; second law efficiency

---

1. Introduction

Gas turbines (GT) are inevitable in modern power generation. The simple GT cycle has poor efficiency due to the elevated temperature of flue gases. To improve the fuel efficiency, the simple GT cycle is generally coupled with a bottoming cycle in stationary variants of GT, like the conventional steam Rankine cycles (RC) and the organic Rankine cycles (ORC) [1,2]. The rapid development of industries around the world has resulted in an increasing demand of energy. The shortage of fossil energy prompts people to investigate more efficient gas turbine combined cycles to meet the energy requirements. Exploiting low-grade waste heat for energy production is an attractive option for its potential to reduce fossil fuel consumption. When exploiting medium-temperature heat sources, supercritical carbon dioxide (S-CO$_2$) is advantageous because of its high efficiency, compactness, and cost [3,4]. Many pilot-scale facilities have been developed in the last few years to investigate the performance of S-CO$_2$ BC (Brayton cycles) [5–8]. When utilizing low-grade waste heat, the traditional steam Rankine cycle does not give satisfying results because of poor thermal efficiency, thus the organic Rankine cycles and transcritical CO$_2$ (t-CO$_2$) cycles are proposed [9–12].

The supercritical state of CO$_2$ as working fluid in BC has various advantages. Its critical temperature is low (31.1 °C), which allows to use the natural resources of water as a cooling medium in the condenser. The density of CO$_2$ close to the critical point is similar to that of a liquid and allows to decrease the compressor work significantly. CO$_2$ in its supercritical state is almost twice as dense as steam. This results in very high power density, which allows to drastically reduce the compressor and turbine size. S-CO$_2$ BC is compatible with a variety of renewable heat sources, thus it has little impact on the environment with low to no ozone depletion potential. Atif and Al-Sulaiman [13,14] presented the energy and exergy analysis of solar-driven S-CO$_2$ and its applicability in desert climates, like in Saudi Arabia. Kun Wang [15] studied S-CO$_2$ recompression Brayton cycle (RBC) with molten salt solar power. He developed a model to investigate the effects of salt temperature, compressor inlet conditions, and heliostat orientation on the overall cycle efficiency. Hou et al. recently examined S-CO$_2$ recompression and regenerative cycle utilizing waste heat energy from a marine gas turbine [16]. They found a 13% overall thermal efficiency improvement with the combined cycle. Wang et al. [17] worked on the cascaded S-CO$_2$ cycle integrating solar and biomass.

S-CO$_2$ working with a medium-temperature source may be coupled with a bottoming cycle, which utilizes low-grade waste heat of the primary cycle and rejects in a low-temperature sink [18]. Wang et al. investigated the t-CO$_2$ cycle to exploit low-grade geothermal sources for electricity production [19]. They used liquefied natural gas (LNG) as a low-temperature heat sink to allow low back pressure of the CO$_2$ turbine, thus greatly improving the overall performance of the cycle. Ahmadi et al. established a similar energy conversion system and showed a significant contribution of t-CO$_2$ in geothermal energy utilization [20]. Amini et al. exploited the low-grade energy of exhaust gas (150 °C) from a combined cycle power plant to run the t-CO$_2$ cycle. The results indicated a significant improvement of power output and efficiency [21]. Walnum et al. concluded that a dual-stage t-CO$_2$ system performs better in offshore gas turbines [22]. Wu et al. reported that t-CO$_2$ has the potential to recover medium-grade heat and suggested that more stages should be designed with the increase of waste heat temperature [23].

Much research has been dedicated to the thermodynamic analysis of supercritical CO$_2$ BC utilizing medium-temperature heat either from renewable sources or from gas turbine exhausts. Some studies have been carried out on the utilization of low-grade heat of S-CO$_2$ BC using the t-CO$_2$ cycle as a bottoming cycle. However, little attention has been paid to the effect of pressure drop in the system. S-CO$_2$ RBC is efficient for medium to high-temperature sources, which offer low-grade heat energy (with a temperature of about 100 °C to 120 °C). In the present study, a simulation investigation is done to seek parameters that could possibly improve the overall cycle's efficiency. We consider various working fluids, including CO$_2$, in the bottoming cycle, utilizing low-grade heat energy from S-CO$_2$ RBC as a primary cycle. Implementing the bottoming cycle with a low-temperature heat source requires to maintain the sink medium at a very low temperature. Liquefied natural gas (LNG) contains a large amount of cold energy, naturally making it a suitable candidate for providing a low-temperature sink medium [24–27].

The selection of appropriate working fluids depends on many factors and properties, such as critical temperature and pressure, chemical stability at the operating temperature, environment friendliness, economic convenience, and allows a high utilization of the energy available from the heat source. Considering the maximum temperature available for the bottoming cycle, i.e., about 120 °C, CO$_2$, R134a, ammonia, and argon were chosen as potential candidates for the present study. R134a has zero ozone depletion potential (ODP). It has already been used commercially, and the necessary equipment, such as heat exchangers and turbo-machines, is readily available. It has a critical temperature of 101.1 °C, allowing it to be used in the temperature range of interest. Ammonia, despite being toxic and flammable, is one of the most environment friendly working fluid with zero ODP and zero global warming potential (GWP). It is being used in industries and is considered a highly efficient refrigerant. It has a critical temperature of 132.4 °C, which allows to adopt it as a working fluid for the available heat source. The bottoming cycle utilizing CO$_2$, R134a, and ammonia is similar to the Rankine cycle, thus it requires a pump to maintain the cycle pressure ratio. Moreover, the performance of argon was also studied in the bottoming cycle, which is similar to the Brayton cycle. Argon was selected because of its high density, which allows to employ compact turbo-machines with fewer stages.

The first part of the article discusses the configuration of the simulation environment, the adjustment of the operational parameters, and the mathematical model for energy and exergy analysis. The model is validated by comparing the results for the supercritical CO_2 recompression Brayton cycle with previous findings. The second part of the paper presents the results of the energy analysis. The potential improvements in the overall efficiency with the bottoming cycle utilizing different working fluids (CO_2, argon, ammonia, R134a) are outlined. The last part of the paper addresses the results of the exergy analysis. The exergy loss in various components of the cycle is calculated. This analysis could help in selecting the working fluids suitable for the bottoming cycle.

2. System Configuration, Modeling, and Simulation Environment

2.1. Primary Cycle (Base Model) Layout for the S-CO$_2$ Cycle

The base model is a recompression Brayton cycle (RBC) with partial condensation. This layout was originally proposed by Feher [28] and Angelio [29], then refined by Dostal [30]. They observed that large internal irreversibility losses occurred in the case of a fully condensing cycle due to heat recuperation between the low heat capacity turbine exhaust and the high heat capacity flow from the pump. Thus, partial condensation was proposed to reduce these losses. This was done by splitting the stream from the turbine exhaust. The term flow ratio 'x' is generally used to indicate the magnitude of the divided mass flow rate. The layout of the cycle is shown in Figure 1. The stream leaving the low-temperature recuperator (LTR) is divided into two streams. The first stream (State 3a) goes to Compressor 1 and the second stream (State 3b) goes to Compressor 2, which is a recompressor operating at the exit temperature and pressure of the LTR. The stream leaving Compressor 1 (State 5) passes through the LTR where it receives energy from the hotter stream (State 2) and is then mixed with the stream (State 7) leaving Compressor 2. The mixed stream (State 8) is further heated in the high-temperature recuperator (HTR) before it receives heat from the heat source. Finally, it expands in Turbine 1. Part of the work produced in Turbine 1 is used to drive Compressor 1 and Compressor 2.

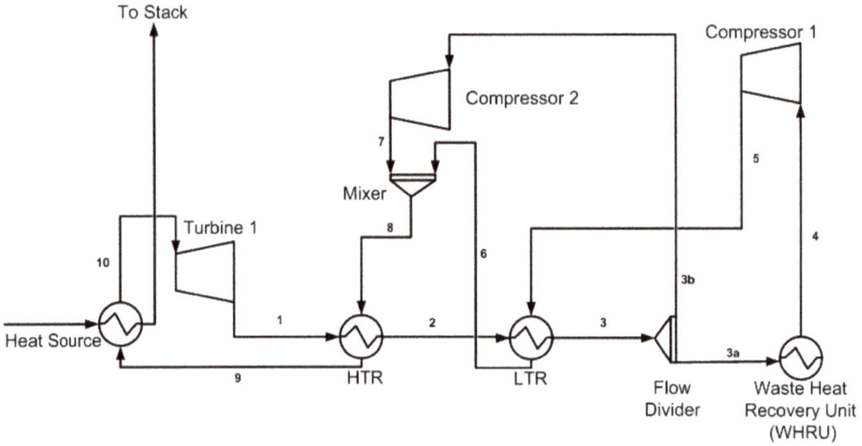

Figure 1. Schematic diagram of the supercritical carbon dioxide (S-CO$_2$) recompression Brayton cycle. HTR: high-temperature recuperator; LTR: low-temperature recuperator.

2.2. Cycle Cascade of the S-CO$_2$ Combined Cycle

Figure 2 illustrates the schematic diagram of the combined cycle based on the cold energy utilization of LNG. The secondary cycle runs on the energy recovered from the primary cycle in the waste heat recovery unit (WHRU). The working fluid in the secondary cycle receives heat energy from the WHRU and then expands in Turbine 2. After expansion, it rejects heat to the low-temperature sink, which is a liquefied natural gas heat exchanger (LNG HEX). The stream then goes to a pump or a compressor, depending on the state of the working fluid. Part of the work produced by Turbine 2 is consumed to drive the pump or compressor of the secondary cycle.

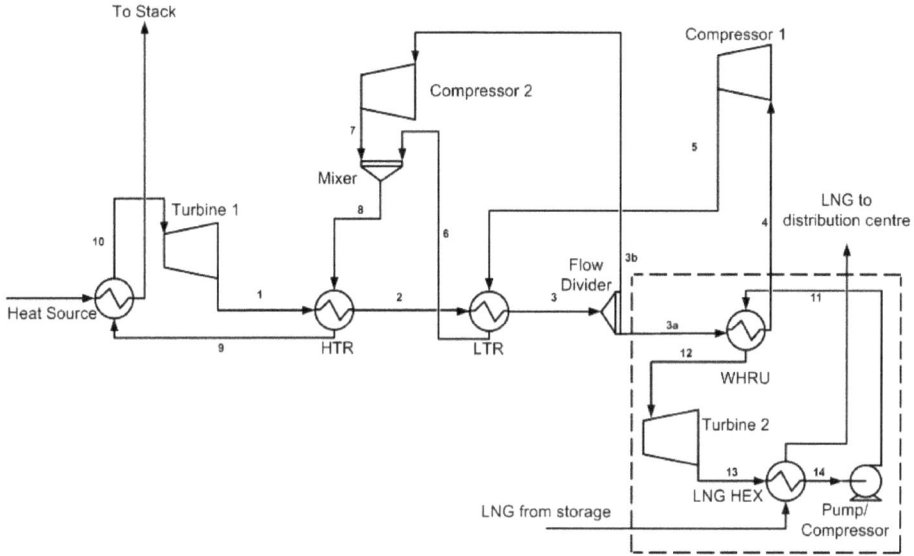

Figure 2. Schematic diagram of the S-CO$_2$ cycle coupled with a bottoming cycle. The dashed lines encircle the secondary or bottoming cycle. LNG: liquefied natural gas; HEX: heat exchanger.

3. Mathematical Model

3.1. Energy Analysis—Governing Equations

The primary cycle efficiency (η_{pc}) and combined cycle efficiency (η_{cc}) are calculated by:

$$\eta_{pc} = (W_{T1} - W_{C1} - W_{C2})/Q_{IN} \tag{1}$$

$$\eta_{cc} = (W_{T1} + W_{T2} - W_{C1} - W_{C2} - W_P)/Q_{IN} \tag{2}$$

$$W_{T1} = m_{PC} \, (h_{10} - h_1) \tag{3}$$

$$W_{T2} = m_{SC} \, (h_{12} - h_{13}) \tag{4}$$

$$W_{C1} = m_{C1} \, (h_4 - h_5) \tag{5}$$

$$W_{C2} = m_{C2} \, (h_8 - h_3) \tag{6}$$

$$W_{Pump} = m_{SC} \, (h_{11} - h_{14}) \tag{7}$$

$$Q_{IN} = m_{PC} \, (h_{10} - h_9) \tag{8}$$

where m_{PC}, m_{SC}, m_{C1}, m_{C2} are the mass flow rates through the primary cycle, secondary cycle, Compressor 1, and Compressor 2, respectively. 'Q_{IN}' represents the total heat input per unit time given to the cycle; 'h' represents enthalpy, and the subscript numbers are associated with the state points shown in Figure 2. The ratio of total mass flow rate in the primary cycle returning to Compressor 1 is defined by the variable 'x' and is equal to:

$$x = m_{C1}/m_{PC} \tag{9}$$

The heat exchange between primary and secondary cycle via WHRU is governed by the following energy balance:

$$m_{C1} (h_4 - h_3) = m_{SC} (h_{12} - h_{11}) \tag{10}$$

A minimum temperature approach is enforced and set to 10 °C. Heat transfer in the HTR and LTR obeys the following governing equations, which are also set to exchange heat using a minimum temperature approach of 10 °C

$$(h_1 - h_2) = (h_9 - h_8) \tag{11}$$

$$(h_2 - h_3) = x (h_8 - h_5) \tag{12}$$

3.2. Exergy Analysis—Governing Equations

The exergy destroyed ($X_{destroyed}$) in each of the components of the cycle is calculated according to the following equations. The kinetic and potential energy change is neglected.

$$X_{destroyed, C1} = m_{C1} T_{surr} (s_5 - s_4) \tag{13}$$

$$X_{destroyed, C2} = m_{C2} T_{surr} (s_7 - s_{3b}) \tag{14}$$

$$X_{destroyed, T1} = m_{PC} T_{surr} (s_1 - s_{10}) \tag{15}$$

$$X_{destroyed, T2} = m_{SC} T_{surr} (s_{13} - s_{12}) \tag{16}$$

$$X_{destroyed, HTR} = m_{PC} T_{surr} [(s_2 - s_1) + (s_9 - s_8)] \tag{17}$$

$$X_{destroyed, LTR} = m_{PC} T_{surr} [x(s_6 - s_5) + (s_3 - s_2)] \tag{18}$$

$$X_{destroyed, WHRU} = T_{surr} [m_{SC} (s_{12} - s_{11}) + m_{PC} x(s_4 - s_{3a})] \tag{19}$$

$$X_{destroyed, LNG HEX} = T_{surr} [m_{SC} (s_{14} - s_{13}) + m_{LNG} (\Delta s_{LNG})] \tag{20}$$

$$X_{destroyed, Pump} = m_{SC} T_{surr} (s_{11} - s_{14}) \tag{21}$$

where T_{surr} is the surrounding temperature, and 's' represents entropy.

The second law efficiency of the cycle is calculated as

$$\eta_{II} = (W_{T1} + W_{T2} - W_{C1} - W_{C2} - W_P)/(\psi_{T1} + \psi_{T2} - \psi_{C1} - \psi_{C2} - \psi_{Pump}) \tag{22}$$

where ψ represents reversible work and is defined as

$$\psi_{T1} = m_{PC} [(h_{10} - h_1) - T_{surr} (s_{10} - s_1)] \tag{23}$$

$$\psi_{T2} = m_{SC} [(h_{12} - h_{13}) - T_{surr} (s_{12} - s_{13})] \tag{24}$$

$$\psi_{C1} = m_{PC} (x) [(h_5 - h_4) - T_{surr} (s_5 - s_4)] \tag{25}$$

$$\psi_{C2} = m_{PC} (1 - x) [(h_7 - h_{3b}) - T_{surr} (s_7 - s_{3b})] \tag{26}$$

$$\psi_{Pump} = m_{SC} [(h_{11} - h_{14}) - T_{surr} (s_{11} - s_{14})] \tag{27}$$

4. Simulation Environment and Procedure

The commercial software Aspen HYSYS V9 (Aspen Technology, Inc., Bedford, MA, USA)was used to simulate the cycle. The Peng-Robinson model was considered for state properties calculation. The analysis was done with the following restrictions imposed:

1. The cycle operates under steady-state conditions.
2. Surrounding temperature is 25 °C.
3. Primary cycle mass flow rate is 100 kg/s.

4. LNG in the storage tank is maintained at $-162\ °C$.
5. Energy losses in the pipelines are neglected.
6. Compression and expansion processes are adiabatic.
7. Compressor and turbine adiabatic efficiencies are 85% and 90%, respectively.
8. Adiabatic efficiency of a centrifugal pump in the secondary cycle is 80%.
9. Minimum temperature approach is set to 10 degrees in HTR, LTR, WHRU, and LNG HEX.
10. The state of CO_2 is kept close to the critical point at the inlet of Compressor 1, ($P = 7.2$ MPa and $T = 30\ °C$).

5. Primary Cycle Parametric Adjustments

Thermodynamically, the cycle performance greatly depends on a number of parameters, such as operating pressure ratio, flow ratio, operating temperatures [28,31]. This section discusses the steps undertaken in the selection of the parameters for the cycle. Three cases were considered with different assumed pressure drop in the system. For each case, the pressure drop was considered uniform in the cycle: for example, a 2% pressure drop indicated the reduction of 2% of the inlet pressure across each heat exchanger in the cycle for both sides (hot and cold sides).

The case study was set in Aspen HYSYS to calculate the cycle efficiency for a range of flow ratios 'x' and cycle's pressure ratios. The results are presented as contour plots in Figure 3a–c. It is worth noting that the efficiency for each colored patch can be read with a degree of error of ± 0.5. A parametric simulation study was set up to seek operational parameters (pressure ratio and flow ratio) that approximate a near-optimal operation of the cycle with respect to thermal efficiency. The best possible combination found for each case is listed in Table 1.

Table 1. Optimal values of flow ratio and pressure ratio for maximum efficiency of the primary cycle.

Pressure Drop	Flow Ratio x	Compressor 1 Compression Ratio	Compressor 2 Compression Ratio	Turbine Inlet Pressure (P_{10}) MPa
No Drop	0.66	2.40	2.40	16.60
2%	0.69	3.00	3.14	20.40
4%	0.71	3.60	3.85	24.45

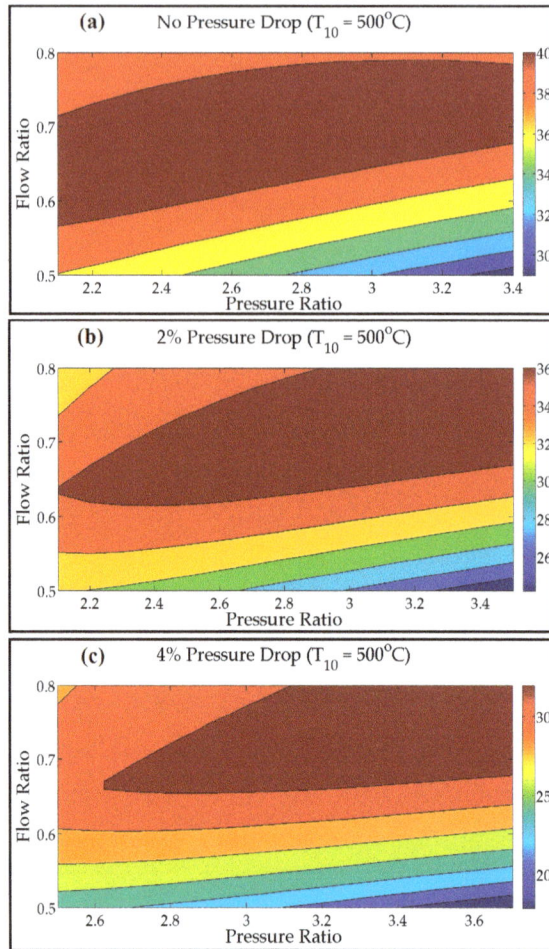

Figure 3. (**a**) Contour plot of the thermal efficiency of the primary cycle for the no-pressure-drop condition plotted as a function of the flow ratio and pressure ratio for a turbine inlet temperature of 500 °C. (**b**) Contour plot of the thermal efficiency of the primary cycle for the 2% pressure drop condition plotted as a function of the flow ratio and pressure ratio for a turbine inlet temperature of 500 °C. (**c**) Contour plot of the thermal efficiency of the primary cycle for the 4% pressure drop condition plotted as a function of flow ratio and pressure ratio for a turbine inlet temperature of 500 °C.

Figure 4 represents the temperature–entropy T–S diagram of the primary cycle with a turbine inlet temperature of 500 °C. It can be observed that the LTR and HTR units recovered and recycled more than 60% of the heat.

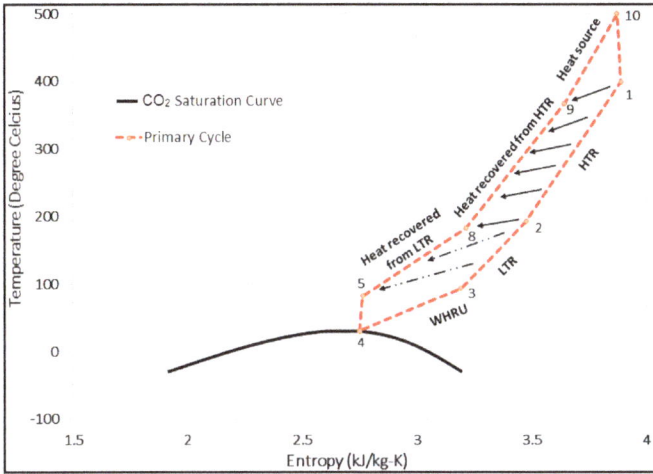

Figure 4. Temperature–entropy (T–S) diagram of the primary cycle (S-CO$_2$ recompression Brayton cycle (RBC)) plotted for a turbine inlet temperature of 500 °C. All state points correspond to the numbers shown in Figure 1.

6. Primary Cycle (Base Model) Validation and Performance

The primary cycle (S-CO$_2$ RBC) serves as a base model, thus it is imperative to validate the results with previously published data. Figure 5 represents the thermal efficiency plotted against the turbine inlet temperature. The simulation results are plotted along with the results published by Kun Wang [32] and Turchi et al. [33] (this study did not consider the pressure drop in the system). It is evident from the plot that the base model, developed in Aspen HYSYS V9, produced results in agreement with the previously published data. It is worth mentioning that Kim et al. [34] investigated the amount of pressure loss that could occur in a printed circuit heat exchanger (PCHE) for S-CO$_2$. Their study indicated a pressure loss of about 2%. Thus, if we assume a similar pressure drop, then the efficiency of 40% or above is achievable with a turbine inlet temperature of 550 °C or higher.

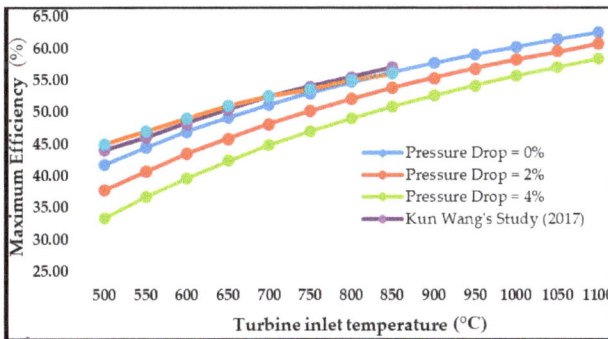

Figure 5. Thermal efficiency of the primary cycle plotted against turbine inlet temperature for the no-pressure-drop, the 2% pressure drop, and the 4% pressure drop conditions. Data points taken from references [32,35] are also plotted for comparison.

Figure 6 represents the back work ratio (BWR) plotted as a function of the turbine inlet temperature. The BWR is the fraction of work produced in the turbine that is consumed by the compressors in the cycle. Considering the curve for no pressure drop in Figure 6, the BWR is below 40%, which is uncommon for a standard Air Brayton cycle. The smaller BWR of the S-CO$_2$ RBC is due to the fact that the CO$_2$ is brought to the critical state before the inlet to Compressor 1, which results in reduced compressor work and improvement in the overall efficiency of the cycle.

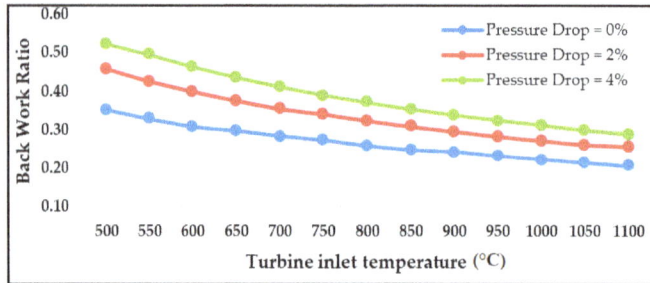

Figure 6. Back work ratio of the primary cycle plotted as a function of turbine inlet temperature for the no-pressure-drop, the 2% pressure drop, and the 4% pressure drop conditions. For all cases, the back work ratio decreases with the increase of the turbine inlet temperature.

Figure 7 represents the magnitude of efficiency improvement with respect to turbine inlet temperature. It is worth noting that the S-CO$_2$ recompression cycle is best suited for medium temperature range (approx. up to 850 °C). Further increase to turbine inlet temperature would increase the cost of high-temperature resistance material with no significant increase in the efficiency improvements.

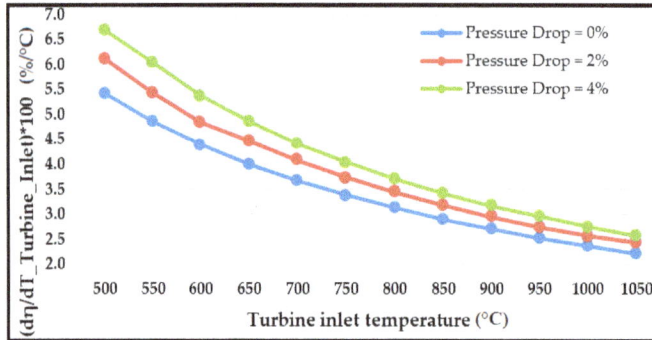

Figure 7. Change in the cycle's thermal efficiency per unit change in turbine inlet temperature plotted as a function of turbine inlet temperature for the no-pressure-drop, the 2% pressure drop, and the 4% pressure drop conditions. For all cases, the improvement in the cycle's thermal efficiency declines with the increase of then turbine inlet temperature.

7. Combined Cycle Parametric Adjustments

The secondary cycle in cascade with the primary S-CO$_2$ cycle was investigated using carbon dioxide, ammonia, R134a, and argon as working fluids. A minimum temperature approach of 10 degrees was used for the heat exchangers (WHRU and LNG HEX). The secondary fluid inlet temperature to the WHRU was fixed at −25 °C for all cases. The combined cycle efficiency was maximized by searching the best possible combination of primary and secondary cycle pressure ratios for all temperatures and pressure drops considered in the study. Figure 8 displays the contour plot of efficiency against a range of pressure ratios of primary and secondary cycles. It reveals that the

Processes **2018**, *6*, 153

optimum value of primary cycle pressure ratio remained at nearly the same value as when there was no bottoming cycle (see Table 1). These plots are only shown for two temperatures at different pressure drops, with CO_2 as a working fluid for the secondary or bottoming cycle; however, optimum values of primary and secondary cycle pressure ratios were obtained in the same manner for all cases discussed in this study. After expansion, the working fluid exchanged heat in the LNG HEX, where the LNG inlet temperature was fixed at -162 °C (refer to Figure 2). Afterwards, the pump or compressor raised the pressure of the working fluid according to cycle's operating pressure ratio.

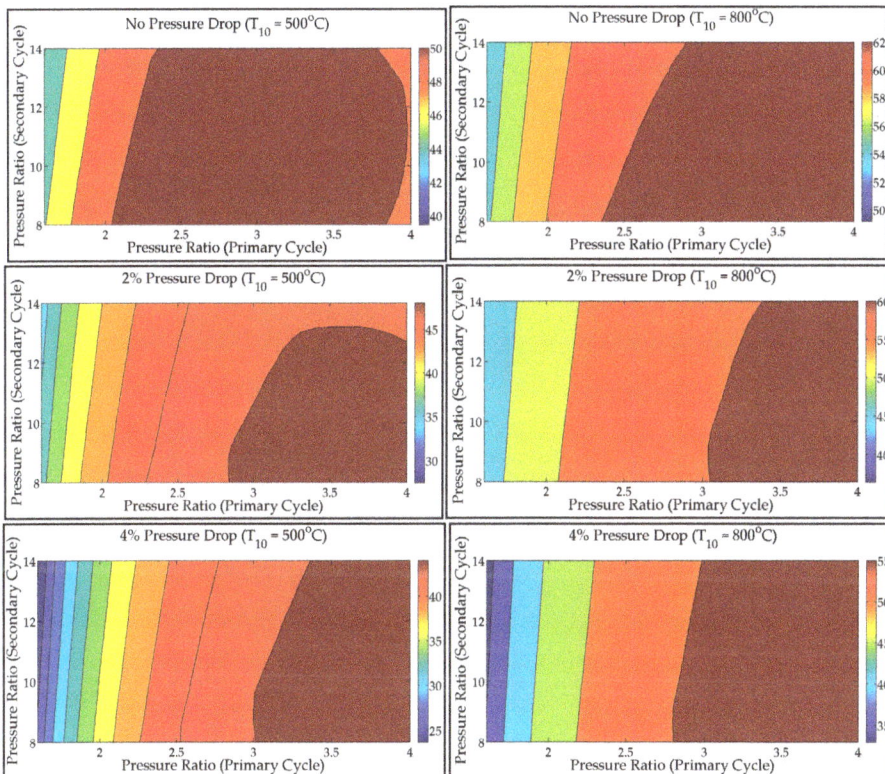

Figure 8. Contour plots of the thermal efficiency of the combined cycle as a function of pressure ratios of the primary and secondary cycles for turbine inlet temperatures of (**left**) 500 °C and (**right**) 800 °C in (**top**) the no-pressure-drop condition, (**center**) the 2% pressure drop condition, and (**bottom**) the 4% pressure drop condition. CO_2 is considered as the working fluid in the secondary cycle.

8. Combined Cycle Performance and Overall Improvement

8.1. Combined Cycle Energy Analysis

This section discusses the essential outcomes of the energy analysis for the combined cycle governed by the set of equations from 1 to 12. The maximum efficiency of the cascade S-CO_2 combined cycle was plotted against the turbine inlet temperature (T_{10}) and is shown in Figure 9. A general monotonic behavior was observed for the efficiency of the combined cycle, which increased with the turbine inlet temperature. A similar behavior of the cycle's efficiency was observed for the standalone primary cycle (see Figure 5). The effect of pressure drop in the system was reflected in the reduction of the overall cycle's efficiency. Ammonia and R134a were found to be the least efficient. Argon appeared to be more efficient than the other candidates. This is due to the fact that the argon cycle is similar to the Brayton cycle and incurs much smaller losses (due to no phase change). Thus, argon can expand to lower temperatures than the other candidates that run on cycles similar to the Rankine cycle.

Figure 10 illustrates the overall efficiency improvement coming from the secondary cycle with argon, CO_2, ammonia, and R134a. It is observed that the role of the secondary cycle in the improvement of the cycle's overall efficiency rises with the increase of the pressure drop in the system and it is more pronounced for medium turbine inlet temperatures (approximately up to 850 °C). This is because the primary cycle's efficiency drastically decreases with the increase of the pressure drop (refer to Figures 5 and 7). The efficiency improvement due to the secondary cycle declined rapidly with the rise of the turbine inlet temperature (T_{10}). This was due to a higher efficiency of the primary cycle at elevated temperatures (refer to Figures 5 and 6).

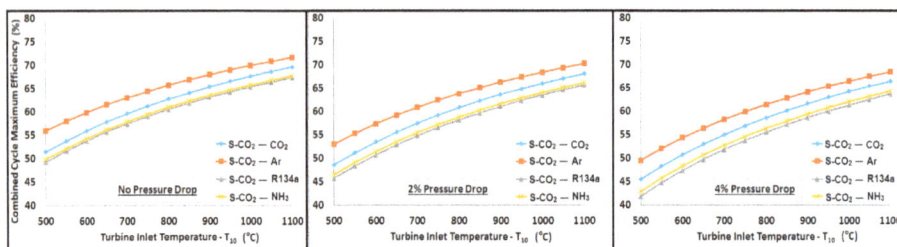

Figure 9. Thermal efficiency of the combined cycle plotted against the turbine inlet temperature (T_{10}) for (**left**) the no-pressure-drop condition, (**center**) the 2% pressure drop condition, and (**right**) the 4% pressure drop condition. For all cases, the cycle's thermal efficiency monotonically increases with the turbine inlet temperature; the S-CO_2–Argon cycle appears the most efficient.

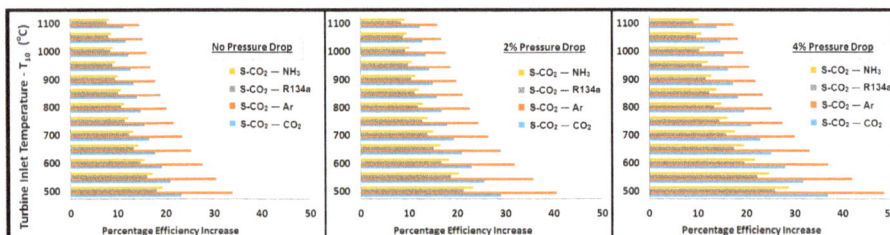

Figure 10. Improvement in the thermal efficiency of the combined cycle (in percentage) in comparison to the standalone S-CO_2 cycle plotted against the turbine inlet temperature (T_{10})) for (**left**) the no-pressure-drop condition, (**center**) the 2% pressure drop condition, and (**right**) the 4% pressure drop condition.

The process of heat recovery from the primary cycle to the secondary cycle is illustrated graphically using the T–S plots in Figure 11. This plot was constructed for a turbine inlet temperature (T_{10}) of 500 °C (note: a similar qualitative behavior was observed for higher values of T_{10}). There was a turbine between the state points 10 and 1, and two compressors between the state points 4 and 5 and the state points 3 and 8. The compressor ratios for both compressors were the same, as there was no pressure drop considered in the system, (the values are listed in Table 1). The bottoming cycle received waste heat from the primary cycle through the WHRU between the state points shown in Figure 11. The inlet of hotter fluid from the primary cycle was at state 3, whereas state 12 represented the condition of the secondary working fluid after heat gain in the WHRU. The bottoming cycle with CO_2, ammonia, and R134a was similar to the Rankine cycle, with the liquid state after the heat rejection process between state points 13 and 14, thus requiring a pump to increase the pressure. However, the argon cycle was similar to the standard Brayton cycle and required a compressor. Since no phase change was involved in the case of argon, the working fluid expanded at a temperature much lower than that required for the other candidates (see Figure 11). Moreover, the bottoming cycle of CO_2, R134a, and ammonia was similar to the Rankine cycle, which is inherently inefficient, as most of the heat addition and heat rejection are done isothermally. It is worth noting that the minimum temperature of the bottoming cycle (T_{14}) was nearly -25 °C in all cases except for argon, for which the value dropped to -150 °C. This makes CO_2 a better candidate in the bottoming cycle for cold regions of the world where the environment temperature favors heat rejection. Argon could be a better choice if LNG's cold energy is readily available for heat exchange.

Figure 12 represents the T–S plots related to different pressure drop conditions for a turbine inlet temperature (T_{10}) of 500 °C. The effect of the pressure drop in the cycle resulted in the increase of the temperature T_3. Thus, the temperature at the turbine inlet of the bottoming cycle increased. The increased availability of energy for the bottoming cycle led to an increased contribution of the secondary cycle to the performance with the increase of the pressure drop in the cycle. However, as mentioned earlier, this improvement was more pronounced if the turbine inlet temperature (T_{10}) was less than 850 °C (refer to Figure 10). Beyond that value, the efficiency improvement due to the secondary cycle became less prominent.

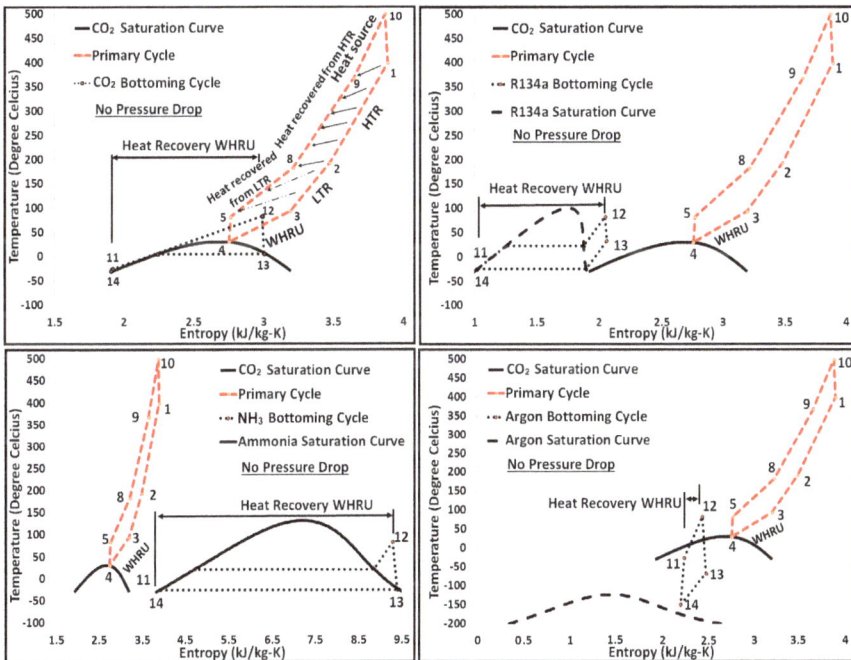

Figure 11. T–S plots of the S-CO_2 cycle coupled with the secondary (**top left**) CO_2 cycle, (**top right**) R134a cycle, (**bottom left**) ammonia cycle, and (**bottom right**) argon cycle. No pressure drop is considered.

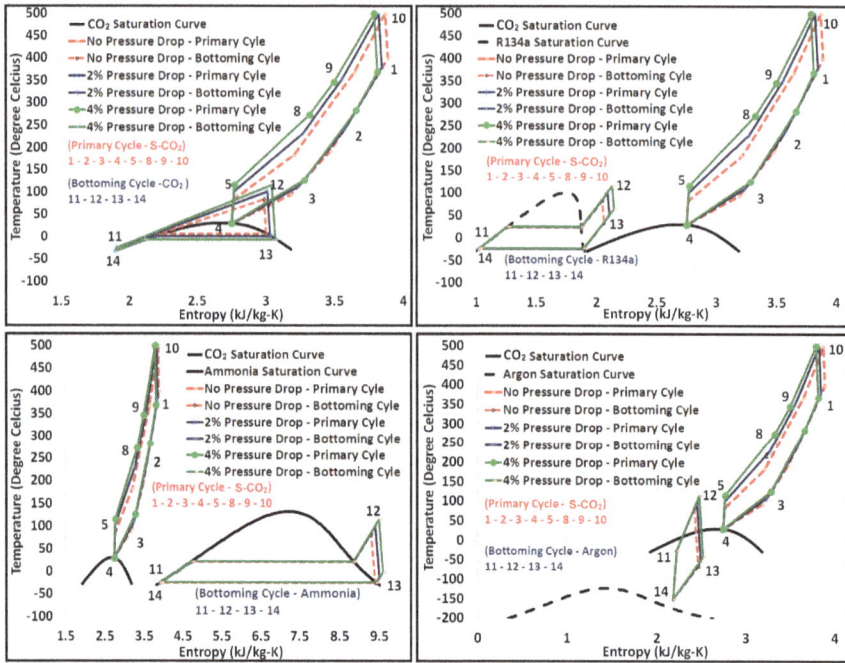

Figure 12. Effect of pressure drop on the T–S plots of the S-CO$_2$ cycle coupled with the secondary (**top left**) CO$_2$ cycle, (**top right**) R134a cycle, (**bottom left**), ammonia cycle and (**bottom right**), argon cycle.

8.2. Combined Cycle Exergy Analysis

This section encapsulates the results of exergy analysis done for a combined cycle. Exergy analysis is governed by the set of equations from 13 to 27. Figure 13 shows the exergy destruction taking place in each of the components of the cycle. Maximum exergy was lost in the LNG HEX, which was expected, as it is the sink for the cycle where waste heat was rejected. Ammonia offered maximum exergy destruction due to its highest value of specific heat and latent heat of condensation. Argon, on the other hand, had minimum exergy loss due to no phase change. CO$_2$ offered minimum exergy loss during heat exchange from primary to bottom cycle (in WHRU), which resulted in maximum exergy available for the secondary cycle. It is interesting to note that the mass flow rate required for energy balance between the primary and secondary cycle was the highest for argon and the lowest for ammonia; the ratio of the mass flow rate for secondary cycle to primary cycle is shown in Table 2. The higher mass flow rate for argon is the result of its low specific heat value, because of which it experienced maximum exergy loss in the turbine and compressor of the secondary cycle. Figure 14 represents the total exergy loss in the cycle plotted against increasing turbine inlet temperature. It comes as no surprise that the total exergy loss decreased with the turbine inlet temperature (T_{10}), as the overall cycle efficiency increased with T_{10}. The exergy loss with R134a and CO$_2$ was almost the same when no pressure drop was considered in the cycle. However, with the increasing pressure drop in the system, R134a performed better in the energy conversion, while ammonia was the worst candidate.

Figure 15 shows the cycle's second law efficiency plotted against the turbine inlet temperature. Similar to the first law efficiency, the second law efficiency showed to increase monotonically with temperature. However, the rate of efficiency rise appeared to saturate near the turbine inlet temperature of 800 °C. Argon seemed to be the worst candidate in terms of second law efficiency, which is a result of a large exergy loss in the turbine and the compressor. R134a and ammonia behaved identically, offering the highest second law efficiency. CO_2 performed marginally less well than R134a and ammonia.

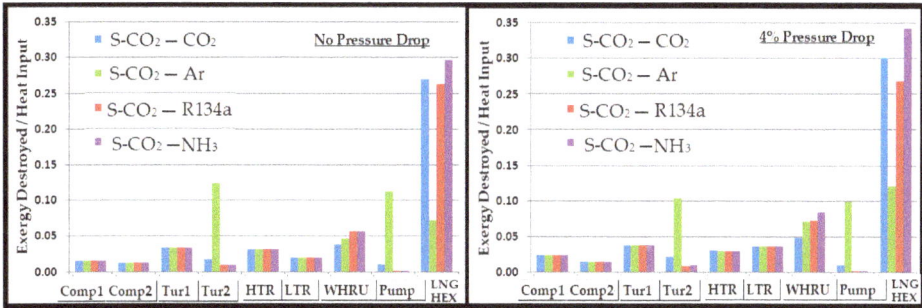

Figure 13. Exergy destroyed (normalized by heat input) in various components of the combined cycle plotted for (**left**) the no-pressure-drop condition and (**right**) the 4% pressure drop condition with a turbine inlet temperature of $T_{10} = 500\ °C$.

Table 2. Ratio of mass flow rates of the secondary cycle to primary cycle (m_{SC}/m_{PC}).

Pressure Drop	S-CO$_2$—CO$_2$	S-CO$_2$—Ar	S-CO$_2$—R134a	S-CO$_2$—NH$_3$
No Drop	33%	156%	36.5%	6.5%
2%	34.4%	164%	37%	7.7%
4%	36%	168%	38.6%	8.7%

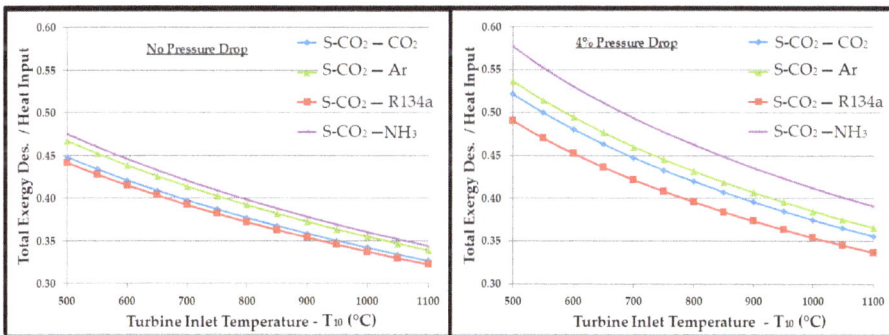

Figure 14. Total exergy destruction (normalized by heat input) plotted as a function of turbine inlet temperature for each combined cycle for (**left**) the no-pressure-drop condition and (**right**) the 4% pressure drop condition. For all cases, total exergy destruction exhibits a monotonically decreasing behavior with the turbine inlet temperature.

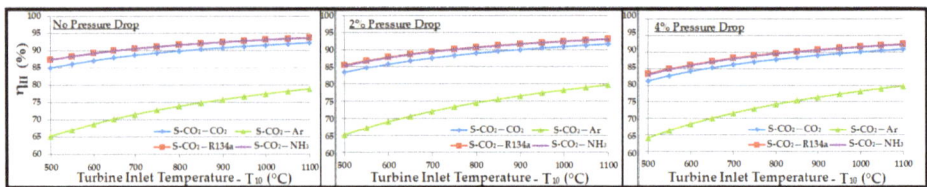

Figure 15. Second law efficiency of the combined cycle plotted as a function of turbine inlet temperature for (**left**) the no-pressure-drop condition, (**center**) the 2% pressure drop condition, and (**right**) the 4% pressure drop condition. For all cases, argon stands out as the least efficient candidate.

9. Conclusions

Energy and exergy analyses of a supercritical CO_2 recompression Brayton cycle, with or without bottoming cycle, were performed. The following key results were obtained:

- The supercritical CO_2 recompression Brayton cycle is efficient, with a potential to provide high-efficiency values for medium-range source temperatures.
- As expected, the S-CO_2 RBC thermal efficiency declined with the pressure drop. The pressure drop in the heat exchangers resulted in increased compressor work required to maintain the optimum cycle pressure. The optimum cycle pressure was 16.60 MPa, which raised to 24.45 MPa for a 4% pressure drop.
- Implementing the bottoming cycle is an attractive option with a sink temperature as low as $-50\ ^{\circ}C$.
- The pressure drop in the primary cycle reduced the efficiency but, in turn, offered higher temperature and exergy available to the WHRU, which appeared as an increased contribution from the secondary cycle.
- The combined cycle efficiency increased monotonically with the turbine inlet temperature (T_{10}); the rise was more evident for temperatures up to 850 $^{\circ}C$, beyond which the curve appeared to saturate.
- Regardless of the pressure drop in the system, the bottoming cycle with argon as a working fluid gave the highest thermal efficiency. On the other side, it required a mass flow rate approximately 5 times higher than R134a and CO_2. A high mass flow rate would require a large equipment (turbine, compressor, heat exchangers, etc.), which ultimately increases the capital and operational costs. Significantly high exergy losses in compressor, turbine, and WHRU were also associated with argon, which was manifested by the smaller second law efficiency.
- Ammonia required a mass flow rate approximately 4 to 5 times lower than those of R134a and CO_2. Exergy analysis revealed a higher second law efficiency associated with ammonia. However, a substantial amount of exergy was lost in the WHRU as a result of the high latent heat of vaporization for ammonia. Thus, a significantly smaller exergy was available for the secondary cycle, which resulted in a smaller contribution in the overall thermal efficiency in comparison to other working fluids.
- R134a can be a good candidate for the bottoming cycle. It offered an overall thermal efficiency improvement between 20% and 25%, as shown in Figure 10. It showed minimum exergy loss and high second law efficiency.
- CO_2 provided a significantly higher contribution than ammonia and R134a in the overall cycle efficiency improvement, with a sink temperature of about $-25\ ^{\circ}C$. It provided a thermal efficiency improvement of 30% to 35%.

Considering the above energy and exergy analyses, CO_2 could be a good option for a combined cycle with S-CO_2 being the primary cycle. R134a could be the second viable option as a working fluid for the bottoming cycle.

Processes **2018**, *6*, 153

Author Contributions: M.E.S. and A.A.T. conceived and set up the simulation in Aspen HYSYS V9; M.E.S., A.A.T. and K.H.A. analyzed the simulation results; M.E.S. wrote the paper. K.H.A. managed research financials, wherever required. The final submission was proofread by all authors.

Funding: This research received no external funding.

Conflicts of Interest: The authors declare no conflict of interest.

References

1. Chacartegui, R.; Sánchez, D.; Muñoz, J.M.; Sánchez, T. Alternative ORC bottoming cycles for combined cycle power plants. *Appl. Energy* **2009**, *86*, 2162–2170. [CrossRef]
2. Cao, Y.; Dai, Y. Comparative analysis on off-design performance of a gas turbine and ORC combined cycle under different operation approaches. *Energy Convers. Manag.* **2017**, *135*, 84–100. [CrossRef]
3. Wang, X.; Dai, Y. Exergoeconomic analysis of utilizing the transcritical CO_2 cycle and the ORC for a recompression supercritical CO_2 cycle waste heat recovery: A comparative study. *Appl. Energy* **2016**, *170*, 193–207. [CrossRef]
4. Santini, L.; Accornero, C.; Cioncolini, A. On the adoption of carbon dioxide thermodynamic cycles for nuclear power conversion: A case study applied to Mochovce 3 Nuclear Power Plant. *Appl. Energy* **2016**, *181*, 446–463. [CrossRef]
5. Conboy, T.; Pasch, J.; Fleming, D. Control of a supercritical CO_2 recompression Brayton cycle demonstration loop. *J. Eng. Gas Turbines Power* **2013**, *135*, 111701. [CrossRef]
6. Conboy, T.; Wright, S.; Pasch, J.; Fleming, D.; Rochau, G.; Fuller, R. Performance characteristics of an operating supercritical CO_2 Brayton cycle. *J. Eng. Gas Turbines Power* **2012**, *134*, 111703. [CrossRef]
7. Wang, J.; Huang, Y.; Zang, J.; Liu, G. Research Activities on Supercritical Carbon Dioxide Power Conversion Technology in China. In Proceedings of the ASEM Turbo Expo 2014: Turbine Technical Conference and Exposition, Dusseldorf, Germany, 16–20 June 2014.
8. Vesely, L.; Dostal, V.; Hajek, P. Design of Experimental Loop With Supercritical Carbon Dioxide. In Proceedings of the 2014 22nd International Conference on Nuclear Engineering, Prague, Czech Republic, 7–11 July 2014.
9. Angelino, G.; Colonna Di Paliano, P. Multicomponent working fluids for organic Rankine cycles (ORCs). *Energy* **1998**, *23*, 449–463. [CrossRef]
10. Astolfi, M.; Romano, M.C.; Bombarda, P.; Macchi, E. Binary ORC (Organic Rankine Cycles) power plants for the exploitation of medium-low temperature geothermal sources—Part B: Techno-economic optimization. *Energy* **2014**, *66*, 435–446. [CrossRef]
11. Chen, Y.; Lundqvist, P.; Johansson, A.; Platell, P. A comparative study of the carbon dioxide transcritical power cycle compared with an organic Rankine cycle with R123 as working fluid in waste heat recovery. *Appl. Therm. Eng.* **2006**, *26*, 2142–2147. [CrossRef]
12. Chen, H.; Goswami, D.Y.; Stefanakos, E.K. A review of thermodynamic cycles and working fluids for the conversion of low-grade heat. *Renew. Sustain. Energy Rev.* **2010**, *14*, 3059–3067. [CrossRef]
13. Al-Sulaiman, F.A.; Atif, M. Performance comparison of different supercritical carbon dioxide Brayton cycles integrated with a solar power tower. *Energy* **2015**, *82*, 61–71. [CrossRef]
14. Atif, M.; Al-Sulaiman, F.A. Energy and exergy analyses of solar tower power plant driven supercritical carbon dioxide recompression cycles for six different locations. *Renew. Sustain. Energy Rev.* **2017**, *68*, 153–167. [CrossRef]
15. Wang, K.; He, Y.L.; Zhu, H.H. Integration between supercritical CO_2 Brayton cycles and molten salt solar power towers: A review and a comprehensive comparison of different cycle layouts. *Appl. Energy* **2017**, *195*, 819–836. [CrossRef]
16. Hou, S.; Wu, Y.; Zhou, Y. Performance analysis of the combined supercritical CO_2 recompression and regenerative cycle used in waste heat recovery of marine gas turbine. *Energy Convers. Manag.* **2017**, *151*, 73–85. [CrossRef]
17. Wang, X.; Liu, Q.; Bai, Z.; Lei, J.; Jin, H. Thermodynamic analysis of the cascaded supercritical CO_2 cycle integrated with solar and biomass energy. *Energy Procedia* **2017**, *105*, 445–452. [CrossRef]
18. Cao, Y.; Ren, J.; Sang, Y.; Dai, Y. Thermodynamic analysis and optimization of a gas turbine and cascade CO_2 combined cycle. *Energy Convers. Manag.* **2017**, *144*, 193–204. [CrossRef]

19. Wang, J.; Wang, J.; Dai, Y.; Zhao, P. Thermodynamic analysis and optimization of a transcritical CO_2 geothermal power generation system based on the cold energy utilization of LNG. *Appl. Therm. Eng.* **2014**, *70*, 531–540. [CrossRef]

20. Ahmadi, M.H.; Mehrpooya, M.; Pourfayaz, F. Thermodynamic and exergy analysis and optimization of a transcritical CO_2 power cycle driven by geothermal energy with liquefied natural gas as its heat sink. *Appl. Therm. Eng.* **2016**, *109*, 640–652. [CrossRef]

21. Amini, A.; Mirkhani, N.; Pakjesm Pourfard, P.; Ashjaee, M.; Khodkar, M.A. Thermo-economic optimization of low-grade waste heat recovery in Yazd combined-cycle power plant (Iran) by a CO_2 transcritical Rankine cycle. *Energy* **2015**, *86*, 74–84. [CrossRef]

22. Walnum, H.T.; Nekså, P.; Nord, L.O.; Andresen, T. Modelling and simulation of CO_2(carbon dioxide) bottoming cycles for offshore oil and gas installations at design and off-design conditions. *Energy* **2013**, *59*, 513–520. [CrossRef]

23. Wu, C.; Yan, X.J.; Wang, S.S.; Bai, K.L.; Di, J.; Cheng, S.F.; Li, J. System optimisation and performance analysis of CO_2 transcritical power cycle for waste heat recovery. *Energy* **2016**, *100*, 391–400. [CrossRef]

24. Qiang, W.; Yanzhong, L.; Jiang, W. Analysis of power cycle based on cold energy of liquefied natural gas and low-grade heat source. *Appl. Therm. Eng.* **2004**, *24*, 539–548. [CrossRef]

25. Kim, C.W.; Chang, S.D.; Ro, S.T. Analysis of the power cycle utilizing the cold energy of LNG. *Int. J. Energy Res.* **1995**, *19*, 741–749. [CrossRef]

26. Zhang, N.; Lior, N. A novel near-zero CO_2 emission thermal cycle with LNG cryogenic exergy utilization. *Energy* **2006**, *31*, 1666–1679. [CrossRef]

27. Deng, S.; Jin, H.; Cai, R.; Lin, R. Novel cogeneration power system with liquefied natural gas (LNG) cryogenic exergy utilization. *Energy* **2004**, *29*, 497–512. [CrossRef]

28. Feher, E.G. The Supercritical thermodynamic power cycle. *Energy Convers.* **1968**, *8*, 85–90. [CrossRef]

29. Angelino, G. Carbon dioxide condensation cycles for power production. *J. Eng. Gas Turbines Power* **1968**, *90*, 287. [CrossRef]

30. Dostal, V.; Driscoll, M.J.; Hejzlar, P. A Supercritical Carbon Dioxide Cycle for Next Generation Nuclear Reactors. *Tech. Rep. MIT-ANP-TR-100* **2004**, 1–317. Available online: http://web.mit.edu/22.33/www/dostal.pdf (accessed on 28 August 2018).

31. Ahn, Y.; Lee, J.; Kim, S.G.; Lee, J.I.; Cha, J.E.; Lee, S.W. Design consideration of supercritical CO_2 power cycle integral experiment loop. *Energy* **2015**, *86*, 115–127. [CrossRef]

32. Wang, K.; He, Y.L. Thermodynamic analysis and optimization of a molten salt solar power tower integrated with a recompression supercritical CO_2 Brayton cycle based on integrated modeling. *Energy Convers. Manag.* **2017**, *135*, 336–350. [CrossRef]

33. Turchi, C.S.; Ma, Z.; Neises, T.W.; Wagner, M.J. Thermodynamic study of advanced supercritical carbon dioxide power cycles for concentrating solar power systems. *J. Sol. Energy Eng.* **2013**, *135*, 041007. [CrossRef]

34. Kim, I.H.; No, H.C. Physical model development and optimal design of PCHE for intermediate heat exchangers in HTGRs. *Nucl. Eng. Des.* **2012**, *243*, 243–250. [CrossRef]

35. Neises, T.; Turchi, C. A comparison of supercritical carbon dioxide power cycle configurations with an emphasis on CSP applications. *Energy Procedia* **2013**, *49*, 1187–1196. [CrossRef]